Lecture Notes in Computer Science 11754

More information about this series at http://www.springer.com/series/7407

Dimitrios Tzovaras · Dimitrios Giakoumis ·
Markus Vincze · Antonis Argyros (Eds.)

Computer Vision Systems

12th International Conference, ICVS 2019
Thessaloniki, Greece, September 23–25, 2019
Proceedings

 Springer

Editors
Dimitrios Tzovaras
Centre for Research and Technology
Hellas (CERTH-ITI)
Thessaloniki, Greece

Dimitrios Giakoumis
Centre for Research and Technology
Hellas (CERTH-ITI)
Thessaloniki, Greece

Markus Vincze
Vienna University of Technology
Vienna, Austria

Antonis Argyros
Foundation for Research and Technology
Hellas (FORTH)
Heraklion, Greece

ISSN 0302-9743   ISSN 1611-3349   (electronic)
Lecture Notes in Computer Science
ISBN 978-3-030-34994-3   ISBN 978-3-030-34995-0   (eBook)
https://doi.org/10.1007/978-3-030-34995-0

LNCS Sublibrary: SL1 – Theoretical Computer Science and General Issues

This Springer imprint is published by the registered company Springer Nature Switzerland AG
The registered company address is: Gewerbestrasse 11, 6330 Cham, Switzerland

Preface

Vision constitutes one of the key components of perception. In order to enable future intelligent machines to understand and interpret the environment in a similar manner as that of humans, they should be endowed with the ability to "see". During the last decades, laborious research endeavors have brought to light important scientific findings for the Computer Vision community, however the contemporary vision systems still lack the ability to operate in unconstrained, real-life environments. Tackling this challenge constitutes the main mission of the International Conference on Computer Vision Systems (ICVS), which with its 12th edition in Greece, 2019, celebrated 20 years since its debut.

ICVS 2019 received 114 submissions, out of which 29 papers were accepted as oral presentations in the main conference. The conference also hosted three workshops and a poster session. Each paper was reviewed by at least two members of the Program Committee and the authors of accepted submissions were then asked to submit their final versions taking into consideration the comments and suggestions of the reviewers. The best paper of the conference was selected by the general and program chairs, after suggestions from the Program Committee. Accepted papers cover a broad spectrum of issues falling under the wider scope of computer vision in real-life applications, including among others, integrated vision systems for robotics, industrial and social collaboration, as well as energy efficiency. Among them, the two most represented topics were Robot Vision and Vision Systems Applications, proving the dedication of the community in developing and deploying vision systems in real environments.

The frame of the technical program of ICVS 2019 was outlined by two invited speakers. First, Prof. Kostas Daniilidis (University of Pennsylvania, Philadelphia, USA) through his speech entitled: "Learning Geometry-Aware Representations: 3D Object and Human Pose Inference" described the essence of intelligence with advanced learning tools in computer vision. The second invited speaker, Prof. Antonis Argyros (University of Crete, Greece), with his speech entitled "Computer Vision Methods for Capturing and Interpreting Human Motion" covered the analysis of humans' motion in image and video sequences in contemporary vision systems.

Three workshops were selected and organized in conjunction with the main conference. The "Workshop on Cognitive and Computer Vision Assisted Systems for Energy Awareness and Behavior Analysis" was organized by Dr. Stelios Krinidis and Mr. Konstantinos Arvanitis. The "Workshop on Movement Analytics and Gesture Recognition for Human-Machine Collaboration in Industry 4.0" was organized by Dr. Sotiris Manitsaris, Dr. Alina Glushkova, and Dr. Dimitrios Menychtas. The "Workshop on Vision-Enabled UAV and Counter-UAV Technologies for Surveillance and Security of Critical Infrastructures" was organized by Dr. Dimitrios Zarpalas and Dr. Konstantinos Votis.

One tutorial was selected and organized in conjunction with the main conference. The "Adaptive Vision for Human Robot Collaboration" tutorial was organized by Prof. Dimitri Ognibene, Prof. Manuela Chessa, Prof. Giovanni Maria Farinella, Prof. Fabio Solari, Prof. David Rudrauf, Dr. Tom Foulsham, Prof. Fiora Pirri, Prof. Guido De Croon, and Prof. Lucas Paletta.

We wish to thank the Information Technologies Institute of Centre for Research and Technology Hellas and, mostly, we feel the need to thank the people who made ICVS 2019 happen, our workshop and tutorial chairs: Prof. Zoe Doulgeri and Prof. Georgios Triantafyllidis, our publication chairs: Prof. Konstantinos Moustakas and Dr. Ioannis Kostavelis, our local chair: Dr. Sotiris Malasiotis, the 48 members of our Program Committee, as well as all the authors who submitted their work to ICVS 2019.

October 2019

Dimitrios Tzovaras
Dimitrios Giakoumis
Markus Vincze
Antonis Argyros

Organization

Conference General Chairs

Dimitrios Tzovaras   Centre for Research and Technology Hellas, Greece
Dimitrios Giakoumis   Centre for Research and Technology Hellas, Greece

Program Co-chairs

Markus Vincze   Vienna University of Technology, Austria
Cornelia Fermüller   University of Maryland Institute for Advanced
                                 Computer Studies, USA
Ming Liu   Hong Kong University of Science and Technology,
                                 Hong Kong, China
Antonis Argyros   Foundation for Research and Technology Hellas,
                                 Greece

Workshop and Tutorial Chairs

Zoe Doulgeri   Aristotle University of Thessaloniki, Greece
Georgios Triantafyllidis   Aalborg University, Denmark

Publicity Chairs

Anastasios Drosou   Centre for Research and Technology Hellas, Greece
Emmanouil Eleftheroglou   National and Kapodistrian University of Athens,
                                 Greece

Publication Chairs

Konstantinos Moustakas   University of Patras, Greece
Ioannis Kostavelis   Centre for Research and Technology Hellas, Greece

Local Chair

Sotiris Malasiotis   Centre for Research and Technology Hellas, Greece

Program Committee

Helder Araujo   University of Coimbra, Portugal
Loukas Bampis   Democritus University of Thrace, Greece
Jorge Batista   University of Coimbra, Portugal

Jeannette Bohg	Stanford University, USA
Richard Bormann	Fraunhofer IPA, Germany
Elisavet Chatzilari	Centre for Research and Technology Hellas, Greece
Dimitrios Chrysostomou	Aalborg University, Denmark
Carlo Colombo	University of Florence, Italy
Dmitrij Csetverikov	Eötvös Loránd University, Budapest
Nikolaos Dimitriou	Centre for Research and Technology Hellas, Greece
Anastasios Drosou	Centre for Research and Technology Hellas, Greece
Azade Farshad Technical	University of Munich, Germany
Robert Fisher	University of Edinburgh, UK
Antonios Gasteratos	Democritus University of Thrace, Greece
Stübl Gernot	University of Trier, Germany
Dimitrios Kanoulas	Istituto Italiano di Tecnologia, Italy
Andreas Kargakos	Centre for Research and Technology Hellas, Greece
Petros Karvelis	Luleå University of Technology, Sweden
Hannes Kisner	Technische Universität Chemnitz, Germany
Ioannis Kostavelis	Centre for Research and Technology Hellas, Greece
Stelios Krinidis	Centre for Research and Technology Hellas, Greece
Baoquan Li	Tianjin Polytechnic University, China
Timm Linder	University of Freiburg, Germany
Dimitrios Makris	Kingston University, UK
Sotiris Malassiotis	Centre for Research and Technology Hellas, Greece
Ioannis Mariolis	Centre for Research and Technology Hellas, Greece
Carlos Hernandez Matas	Foundation of Research and Technology, Hellas, Greece
Lazaros Nalpantidis	Aalborg University, Denmark
George Nikolakopoulos	Luleå University of Technology, Sweden
Christophoros Nikou	University of Ioannina, Greece
Lucas Paletta Joanneum	Research Forschungsgesellschaft mbH, Austria
Costas Panagiotakis	Foundation of Research and Technology, Hellas, Greece
Paschalis Panteleris	Foundation of Research and Technology, Hellas, Greece
Timothy Patten	TU Wien, Austria
Georgios Pavlakos	University of Pennsylvania, USA
Giannis Pavlidis	University of Houston, USA
Georgia Peleka	Centre for Research and Technology Hellas, Greece
Fiora Pirri	University of Rome, Italy
Christian Potthast	University of Southern California, USA
Ioannis Pratikakis	Democritus University of Thrace, Greece
Paolo Remagnino	Kingston University, UK
Evangelos Skartados	Centre for Research and Technology Hellas, Greece
Georgios Stavropoulos	Centre for Research and Technology Hellas, Greece

Angeliki Topalidou-Kyniazopoulou	Centre for Research and Technology Hellas, Greece
Panos Trahanias	Foundation of Research and Technology, Hellas, Greece
Manolis Vasileiadis	Centre for Research and Technology Hellas, Greece
George Vogiatzis	Aston University, UK
Zichao Zhang	University of Zurich, Switzerland

Contents

Vision Systems Applications

High-Level and Learning Vision Systems

Cognitive Vision Systems

**Workshop on: Vision-Enabled UAV and Counter-UAV Technologies
for Surveillance and Security of Critical Infrastructures**

Hardware Accelerated and Real Time Vision Systems

Hardware Accelerated Image Processing on an FPGA-SoC Based Vision System for Closed Loop Monitoring and Additive Manufacturing Process Control

Dietmar Scharf[1]([✉]), Bach Le Viet[1], Thi Bich Hoa Le[1], Janine Rechenberg[1], Stefan Tschierschke[1], Ernst Vogl[1], Ambra Vandone[2], and Mattia Giardini[2]

[1] Ramteid GmbH, Amberg, Germany
dietmar.scharf@ramteid.gmbh
[2] SUPSI, University of Applied Sciences, Manno, Switzerland

Abstract. In many industrial sectors such as aeronautics, power generation, oil & gas, complex metal parts especially the critical ones are constructed and manufactured for a very long lifespan (more than 10 years). 4D Hybrid, an EU research project develops a new concept of hybrid additive manufacturing (AM) modules to ensure first time right production. To achieve that, the in-line process monitoring activity can be persistently realized by a complex sensing and vision system mounted on the equipment to ensure that the process responds to nominal operating conditions. This paper presents concepts and design of the vision system for process monitoring activities with hardware accelerated image processing by using camera hardware based on SoC FPGA devices, a hybrid of FPGA and ARM-based Cortex-A9 dual core CPU.

Keywords: Image processing · FPGA · SoC (System on Chip) · Machine vision · Additive manufacturing · Algorithmic optimization

1 Introduction

In many industrial sectors additive manufacturing as well as numerous maintenances and repairing activities of complex and large-scale metal parts embrace major challenges across their lifecycles. 4D Hybrid is an EU research project with the main goal of developing a new concept of hybrid additive manufacturing, maintenance, and repair operations based on the modular integration of compact modules including laser source, deposition head, sensors and control [1].

To ensure the first-time right production requirement, a complex sensing and vision system is developed to monitor the manufacturing process. This paper specifically focuses on the vision system designed for Direct Energy Deposition (DED) additive manufacturing processes in which a laser source is moved along a specific path melting the metal powder blown out through a nozzle and fused them together to a solid part.

© Springer Nature Switzerland AG 2019
D. Tzovaras et al. (Eds.): ICVS 2019, LNCS 11754, pp. 3–12, 2019.
https://doi.org/10.1007/978-3-030-34995-0_1

The vision system described in this paper is conceptually designed based on the requirements of fast data processing and data reduction. It is shown in [19–22] that on the field of real-time/hardware-accelerated image processing, high performance parallel image processing could be achieved by using hardware processing power of multiple GPUs or in [23, 24] with Field Programmable Gate Array (FPGA) in combination with Digital Signal Processor (DSP). Following another approach, cameras within the introduced vision system were equipped with processing capabilities provided by a SoC-FPGA (System on Chip Field Programmable Gate Array) for processing most of the captured image data directly on the camera itself and interconnecting with other system sensors by efficiently transferring processing data [1]. SoC-FPGA devices integrate both processor and FPGA architectures into a single device, ideal for timing critical and high-performance embedded vision systems.

In this paper we introduced our design and implementation concept of split image processing and the underlying algorithms within cameras in the vision system. Split processing combines the advantages of the massive parallel computing capabilities of the FPGA with the effectiveness and flexibility of the ARM processor. By this procedure the vast amount of raw image data could be processed on the FPGA in massive parallel mode and simultaneously outsource mathematical operations within the library functions of the ARM processor.

The programmable logic part of the SoC-FPGA device is best suited for massive parallel image operations. Section 2.1 described algorithms such as run-length based melt pool detection as well as several object feature extraction methods which are implemented in the programmable logic FPGA part of the SoC-FPGA device. Section 2.2 discussed frame-based algorithms which are implemented in the processor part of the SoC-FPGA hybrid device.

2   Design and Implementation

The vision system is designed primarily for analyzing the melt pool along with its thermal characteristic, inspecting the metal surface as well as monitoring the powder ejection process within the DED (direct energy deposition) additive manufacturing process. Figure 1 shows the vision system hardware setup which consists of several cameras for various vision tasks: a high-speed camera for monitoring the power ejection process, a camera in line with laser head for analyzing the melt pool, an infrared camera for thermal data acquisition and a stereo camera system using projected light patterns for surface reconstruction.

The melt pool monitoring camera system and its underlying hardware and processing algorithms address the main subject of this paper – hardware accelerated on-camera split processing. The melting pool monitoring consists of melting pool region detection and extraction of two sets of parameters including shape and intensity parameters, center points, major and minor axis length, the area and perimeter, orientation (rotation angle, direction), central moments (m11, m20, m02), extent, eccentricity, diameter, average, maximum and minimum, median and majority intensity.

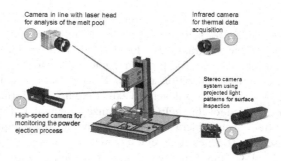

Fig. 1. Vision system hardware

Figure 2 illustrates the implementation goal of segmenting two overlapping regions R1 and R2 of the melt pool area, extracting the geometrical shape features and properties of the melting pool temperature and its distribution. The customized melt pool monitoring camera is positioned in the machine head to capture images directly within the optical axis of the laser beam, which enables monitoring and measuring the melt pool from a best view possible.

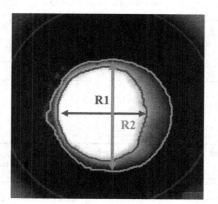

R1 is the melt pool core region (red curve)
R2 is the melt pool tail region (green curve)

Fig. 2. Melt pool region detection. (Color figure online)

The traditional approach based on cameras without on-camera processing capability and data to be transferred to an external device like a PC for computational purposes could not fulfill the requirements of real-time processing behavior, because of the latency of several frames introduced through the chain of image buffering, image transmission to external device for computing.

SoC FPGA devices integrate both processor and FPGA architectures into a single device. Consequently, they provide higher integration, lower power, smaller board size, and higher bandwidth communication between the processor and FPGA [4].

The camera processing pipeline is shown in Fig. 3. The hybrid architecture of FPGA and Dual ARM CPU is ideal for split processing. Several image processing algorithms are implemented within the programmable logic FPGA part, as the image data line by line streaming into the device. The programmable logic part handles time critical operations like sensor timing generation as well. The periodical acquisition of image data, filtering the data direct at sensor level and extracting image information of interest using run-length encoding algorithm helps to minimize the amount of transmitted data.

In the vision system each camera for melt pool monitoring is equipped with a 5 Megapixel CMOS sensor. The melt pool region represents the area of interest and has a size of 100×100 pixel. On this region the image processing pipeline implemented in the FPGA can process up to 500 frames per second but is limited to 100 Hz due to the limitation of the process control system behind the vision system.

The computation results will be sent over Gigabit Ethernet to the control hardware via a project specific protocol based up on the GigE Vision protocol [26]. The standard is based on UDP/IP instead of TCP/IP to fulfill the data throughput and latency requirements.

Traditional HDL programming language on programmable logic side in FPGA is used for implementing the image processing algorithms. On the processor side, post processing is done by user-implemented applications in conjunction with the OpenCV library (Open Source Computer Vision) with a customized embedded Linux distribution. The processor part will also take over the communication handling with control hardware outside the vision system for closed loop control.

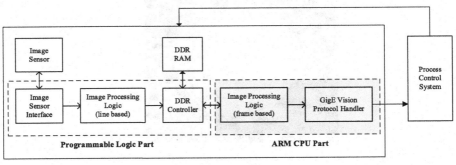

Fig. 3. Closed loop and image processing pipeline on cameras in the vision system

2.1 Processing on Programmable Logic Part

One of the major tasks in melt pool process monitoring is the detection of the melt pool itself. The melt pool region detection is based upon the image intensity distribution, which consists of many groups of connected pixels that have similar values for intensity but different from the ones surrounding it. It has been shown that detection algorithms of those regions could be implemented directly on the programmable logic within the SoC FPGA [2, 5].

Figure 4 shows the implemented processing pipeline on the camera for melt pool region detection and parameters extraction. The processing pipeline is instantiated two times with different parameter settings, one for the melt pool core region detection (R1) and one for the melt pool tail region detection (R2).

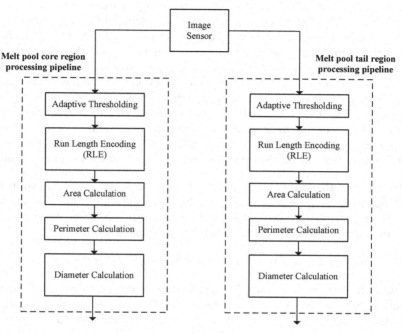

Fig. 4. Melt pool region processing pipeline

Adaptive Thresholding. Thresholding is a very fundamental first step for detecting the melt pool region of interest within the full image. The result of this operation is a binary image, consisting of only ones and zeros. In contrast to conventional global threshold for all pixels in the image, adaptive thresholding implemented in our imaging pipeline changes the threshold dynamically over the image [10, 11, 25], depending on the neighboring pixel intensities of each pixel. The local threshold is determined by subtracting a constant parameter from the mean intensity values of the local 5×5 neighborhood of each pixel [28]. Any pixel intensity that is greater than the calculated threshold value can be set to 1 and pixel intensities that are less or equal to the threshold value can be set to 0. This adaptive threshold allows automatic adaption to varying illumination conditions induced from rapidly modified laser intensity changes without the need to have a fast responding auto exposure algorithm at the beginning of the imaging pipeline.

Melt Pool Region Detection with Run Length Encoding. All the following processing steps of the melt pool parameters calculation are based on the Run Length Encoding (RLE) method presented in [2]. This RLE is applied line by line as the image data from the image sensor streams into the camera. On every single line of the frame a run starts with the first detected pixel with intensity above the threshold determined on the adaptive thresholding step. The number of all following pixels adjacent to the first detected pixel, which has the binary value 1 as well, determines the length of a run. The number of detected pixels or run lengths and the coordinates of the first detected pixel in a run are stored efficiently in a single register and summed up at the end of each line with the result of the previous line. Prior to area and centroid calculation, minimum and maximum pixel coordinates are determined in this step as well. The coordinates of the first and the last detected pixel in a line are stored for comparison with the coordinates of the next pixel pair of the next line. The RLE method for melt pool detection reduces the number data to be stored, thus reducing memory usage within the FPGA. This method is applied to detect the melt pool core region, as well as the tail region of the melt pool shown in Fig. 2. The RLE data is the basis for the next processing steps within the imaging pipeline.

Melt Pool Area Calculation. The area calculation of the melt pool core and its tail is the next step in the image processing chain. The area is the number of pixels the region consists of. It is calculated from the results of previous RLE processing step by summing up the run-length determined after each run for every image line in frame.

$$\text{Melt pool area} = \text{Total number of detected pixels for the region} \quad (1)$$

Centroid Calculations. The centroid or center of mass computation logic for the melt pool core region and its tail region is done by finding the smallest rectangular box which contains the detected pixels [10]. It is defined by the minimum and maximum X coordinate values as well as the minimum and maximum Y coordinate values.

Several algorithms for effective centroid calculation in FPGA were introduced in [2]. One algorithm measures the center of the detected melt pool region by checking for the minimum and maximum X and Y coordinates of the region which was precalculated from the previous processing step. The center of gravity of the captured area can be calculated by averaging the sum of the X and Y coordinates separately. The calculation is described by the following Eqs. (2) and (3).

$$\text{Melt pool center X} = \frac{X_{max} + X_{min}}{2} \quad (2)$$

$$\text{Melt pool center Y} = \frac{Y_{max} + Y_{min}}{2} \quad (3)$$

The second algorithm calculates the center point coordinates of the detected melt pool region as a weighted sum, described in the Eqs. (4) and (5). The disadvantage of this method lies on division operations which require additional logic resources in FPGA and minor delay in its computation. Despite of this minor disadvantage this algorithm was implemented in our imaging pipeline. It delivers a better approximation of the center of mass in comparison with the first algorithm.

$$\text{Melt pool center X} = \frac{\sum X \text{ of all detected pixels}}{\text{number of detected pixels}} \tag{4}$$

$$\text{Melt pool center Y} = \frac{\sum Y \text{ of all detected pixels}}{\text{number of detected pixels}} \tag{5}$$

Perimeter Calculation. Perimeter of each area of the detected melt pool region is the length of the contour of the region. The perimeter is determined by total number of pixels detected on the outer boundary of the melt pool region, which can be obtained from the results of the previous RLE processing step.

Diameter Calculation. Under the assumption that the melt pool region is circular, the diameter of the detected melt pool region can be approximated by the largest side of the region to its bounding box area.

$$\text{Diameter} = \begin{cases} X_{max} - X_{min} & \text{if } X_{max} - X_{min} > Y_{max} - Y_{min} \\ Y_{max} - Y_{min} & \text{otherwise} \end{cases} \tag{6}$$

2.2 Processing on the ARM CPU Part

Extracting the orientation of the detected melt pool region, calculating several parameters including major and minor axis length, eccentricity was done on the ARM CPU.

Parallel to the image processing in the programable logic part, image sensor data is stored pixel by pixel in the external shared DDR RAM. As soon as a frame is finished written into the memory, embedded image processing tasks which run on the CPU starts their operation. The image processing application running on the CPU embedded the OpenCV library.

Tracking Melt Pool Region Orientation, Major and Minor Axis Length. One of the favorite methods for tracking object's orientation is based on image moments, where the second-order moments M20, M11 and M02 can be used to extract the orientation of the major and/or minor axes [9, 16–18]. The following equations show the mathematical operations behind the computation of the major axis orientation θ and the major and minor axis lengths l and ω:

$$\theta = \frac{1}{2} \cdot \tan^1 \left(\frac{2m_{11}}{m_{20} - m_{02}} \right) \tag{7}$$

$$l = \sqrt{8 \left(m_{20} + m_{02} + \sqrt{4m_{11}^2 + (m_{20} - m_{02})^2} \right)} \tag{8}$$

$$\omega = \sqrt{8\left(m_{20} + m_{02} - \sqrt{4m_{11}^2 + (m_{20} - m_{02})^2}\right)} \tag{9}$$

The second-order central moments m_{20}, m_{11} and m_{02} are derived from the centroid coordinates and the central moments M_{20}, M_{11} and M_{02}.

$$m_{20} = \frac{M_{20}}{M_{00}} - x_{centroid}^2 \tag{10}$$

$$m_{11} = \frac{M_{11}}{M_{00}} - x_{centroid} \cdot y_{centroid} \tag{11}$$

$$m_{02} = \frac{M_{02}}{M_{00}} - y_{centroid}^2 \tag{12}$$

Interfacing. Besides the image processing tasks, the dual core processor part takes over the communication handling with control hardware outside the vision system for closed loop control. The communication to other processing parts is handled through the Gigabit Ethernet of the camera. The communication protocol is project-specific defined based on the standard GigE Vision protocol, an interface standard on UDP protocol for transmitting image and control data over Gigabit Ethernet [27].

3   Conclusion

The paper presented the concepts and design of a vision system with hardware accelerated image processing by using small form factor cameras with hardware based on SoC FPGA devices. It is shown that fast data processing could be achieved with pixel-based implementation of algorithms and data reduction with run-length based method on the programmable logic part. Frame-based image processing, computation of object central moments and orientation as well as the communication handling to other control hardware was implemented on the processor part of the SoC-FPGA device.

Final Remark. Due to commercial and license restrictions this paper did not reveal any form of implementation details and performance measurements of the image processing pipeline algorithms of the introduced on-camera vision-based melt pool monitoring and measuring system.

References

1. European Commission: 4D-Hybrid: Novel ALL-IN-ONE machines, robots and systems for affordable, worldwide and lifetime Distributed 3D hybrid manufacturing and repair operations. https://4dhybrid.eu/. Accessed 28 Apr 2019
2. Bochem, A., Kent, K.B., Herpers, R.: Hardware acceleration of blob detection for image processing. In: International Conference on Advances in Circuits, Electronics and Microelectronics, pp. 28–33 (2010). https://doi.org/10.1109/CENICS.2010.12

3. MicroZed Chronicles: Creating a Zynq or FPGA-based, image processing platform. https://blog.hackster.io/microzed-chronicles-creating-a-zynq-or-fpga-based-image-processing-platform-bd30d70bb928. Accessed 28 Apr 2019
4. Altera SoC architecture brief. https://www.intel.com/content/dam/www/programmable/us/en/pdfs/literature/ab/ab1_soc_fpga.pdf. Accessed 28 Apr 2019
5. Trein, J., Schwarzbacher, A., Hoppe, B., Noffz, K.-H., Trenschel, T.: Development of a FPGA based real-time blob analysis circuit. In: Irish Systems and Signals Conference, Derry, Northern Ireland, pp. 121–126 (2007)
6. Thesholding. https://www.cse.unr.edu/~bebis/CS791E/Notes/Thresholding.pdf. In: Jain et al., Sections 3.2.1, 3.2.2, Petrou et al., Chap. 7
7. BLOB Analysis (Introduction to Video and Image Processing). http://what-when-how.com/introduction-to-video-and-image-processing/blob-analysis-introduction-to-video-and-image-processing-part-1/. Accessed 28 Apr 2019
8. Regionprops. https://de.mathworks.com/help/images/ref/regionprops.html. Accessed 28 Apr 2019
9. Rocha, L., Velho, L., Carvalho, P.C.P.: Image moments-based structuring and tracking of objects (2002)
10. Bailey, D.G.: Design for Embedded Processing on FPGAs, 1st edn. Wiley, Hoboken (2011)
11. Davies, E.R.: Computer and Machine Vision: Theory, Algorithms, Practicalities, 4th edn. Academic Press, Cambridge (2012)
12. Pratt, W.K.: Digital Image Processing, 3rd edn. Wiley, Hoboken (2001)
13. Petrou, M., Petrou, C.: Image Processing: The Fundamentals. Wiley, Hoboken (2010)
14. Gonzalez, R.C., Woods, R.E.: Digital Image Processing, 2nd edn. Prentice Hall, Upper Saddle River (2002)
15. Nixon, M.: Feature Extraction and Image Processing for Computer Vision, 3rd edn. Academic Press, Cambridge (2012)
16. Flusser, J., Suk, T., Zitova, B.: 2D and 3D Image Analysis by Moments. Wiley, Hoboken (2016)
17. Flusser, J., Suk, T., Zitova, B.: Moments and Moment Invariants in Pattern Recognition, 1st edn. Wiley, Hoboken (2009)
18. Flusser, J.: Moment invariants in image analysis. In: Proceedings of World Academy of Science, Engineering and Technology, vol. 11, pp. 196–201 (2006)
19. Fung, J., Mann, S.: OpenVIDIA: parallel GPU computer vision. In: Proceedings of the 13th Annual ACM International Conference on Multimedia. ACM (2005)
20. Bampis, L., et al.: Real-time indexing for large image databases: color and edge directivity descriptor on GPU. J. Supercomput. 71(3), 909–937 (2015)
21. Bampis, L., et al.: A LoCATe-based visual place recognition system for mobile robotics and GPGPUs. Concurr. Comput. Pract. Exp. 30(7), e4146 (2018)
22. Bampis, L., Karakasis, E.G., Amanatiadis, A., Gasteratos, A.: Can speedup assist accuracy? An on-board GPU-accelerated image georeference method for UAVs. In: Nalpantidis, L., Krüger, V., Eklundh, J.-O., Gasteratos, A. (eds.) ICVS 2015. LNCS, vol. 9163, pp. 104–114. Springer, Cham (2015). https://doi.org/10.1007/978-3-319-20904-3_10
23. Schnell, M.: Development of an FPGA-based data reduction system for the Belle II DEPFET pixel detector (2015)

24. Xiao, J., Li, S., Sun, B.: A real-time system for lane detection based on FPGA and DSP. Sens. Imaging **17**(1), 6 (2016)
25. Adaptive Thresholding. http://homepages.inf.ed.ac.uk/rbf/HIPR2/adpthrsh.htm. Accessed 28 Apr 2019
26. GigE Vision. https://en.wikipedia.org/wiki/GigE_Vision. Accessed 28 Apr 2019
27. AIA Vision Online, GigE Vision – True Plug and Play Connectivity. https://www.visiononline.org/vision-standards-details.cfm?type=5. Accessed 28 Apr 2019
28. Adaptive Thresholding. http://hanzratech.in/2015/01/21/adaptive-thresholding.html. Accessed 28 Apr 2019

Real-Time Binocular Vision Implementation on an SoC TMS320C6678 DSP

Rui Fan[1], Sicheng Duanmu[2], Yanan Liu[3]([✉]), Yilong Zhu[4], Jianhao Jiao[1], Mohammud Junaid Bocus[3], Yang Yu[1], Lujia Wang[5], and Ming Liu[1]

[1] The Hong Kong University of Science and Technology, Kowloon, Hong Kong
[2] China Unionpay Data Services Co., Ltd., Shanghai, China
[3] University of Bristol, Bristol, UK
yanan.liu@bristol.ac.uk
[4] Unity Drive, Shenzhen, China
[5] Shenzhen Institutes of Advanced Technology, Chinese Academy of Sciences, Shenzhen, China

Abstract. In recent years, computer binocular vision has been commonly utilized to provide depth information for autonomous vehicles. This paper presents an efficient binocular vision system implemented on an SoC TMS320C6678 DSP for real-time depth information extrapolation, where the search range propagates from the bottom of an image to its top. To further improve the stereo matching efficiency, the cost function is factorized into five independent parts. The value of each part is pre-calculated and stored in the DSP memory for direct data indexing. The experimental results illustrate that the proposed algorithm performs in real time, when processing the KITTI stereo datasets with eight cores in parallel.

Keywords: Computer binocular vision · Autonomous vehicles · DSP

1 Introduction

Since 2012, Google has been conducting various prototype vehicle road tests in the US. Its subsidiary X plans to commercialize their self-driving vehicles from the next year [1]. Volvo also carried out a series of self-driving experiments involving about 100 cars in 2016 [2]. The race to commercialize autonomous vehicles by industry giants, e.g., Waymo, Tesla and GM, has been fiercer than

This work is supported by grants from the Shenzhen Science, Technology and Innovation Commission (JCYJ20170818153518789), National Natural Science Foundation of China (No. 61603376) and Guangdong Innovation and Technology Fund (No. 2018B050502009) awarded to Dr. Lujia Wang. This work is also supported by grants from the Research Grants Council of the Hong Kong SAR Government, China (No. 11210017, No. 16212815, No. 21202816, NSFC U1713211) awarded to Prof. Ming Liu.

© Springer Nature Switzerland AG 2019
D. Tzovaras et al. (Eds.): ICVS 2019, LNCS 11754, pp. 13–23, 2019.
https://doi.org/10.1007/978-3-030-34995-0_2

ever [3,4]. Advanced Driver Assistance Systems (ADAS) provide a variety of techniques, such as binocular vision (BV), to enhance the driving safety [2]. The 3D depth map used in the ADAS is supplied by either active sensors or passive sensors [5]. Although the active sensors can acquire depth image with a better accuracy, passive sensors, e.g., cameras, are more commonly used in practice, due to advantages such as high resolution, high depth range and low cost [2].

Compared with other subsystems in the ADAS, e.g., lane detection [6,7] and vehicle state estimation [8], BV takes up the biggest portion of the execution time. Therefore, it should be implemented in such a way to provide a faster processing and consequently accommodate the other subsystems, such that the overall system performs in real time [9]. The state-of-the-art BV algorithms can be classified as local and global ones [2]. Local BV algorithms simply match a series of blocks and select the shifting distance corresponding to the lowest cost. This optimization technique is also referred to as winner-take-all (WTA) [2]. On the other hand, global algorithms realize BV using some more sophisticated optimization techniques, e.g., graph cuts (GC) [10], belief propagation (BP) [11], etc. These techniques are based on the Markov Random Fields (MRF) [12], where disparity estimation is formulated as a probability maximization problem. However, the occluded areas always make it difficult to find the optimum smoothness term (ST) values, as over-penalizing the ST can help minimize the errors around discontinuous areas, but on the other hand, can produce errors in continuous areas [13]. Hence, many researchers have proposed to break down the global BV problems into multiple local BV problems, each of which can be solved with local optimization techniques [14]. For example, an image is divided into a series of slanted planes in [14,15]. The disparities in each plane are then estimated with local constraints. Nevertheless, these algorithms are highly computationally complex, rendering the real-time performance on resource-limited hardware, e.g., Digital Signal Processor (DSP), very challenging.

In recent years, many researchers have utilized seed-and-grow (SAG) local BV algorithms to further minimize the trade-off between speed and accuracy. In these algorithms, the disparity map is grown from a collection of reliable seeds. This not only minimizes expensive computations but also reduces mismatches caused by ambiguities. For example, growing correspondence seeds (GCS), an efficient quasi-dense BV algorithm, was proposed in [16–18] to iteratively estimate disparities, with the search range propagated from a group of reliable seeds. Similarly, various Delaunay triangulation-based binocular vision (DT-BV) algorithms have also been presented in [19–21] to generate semi-dense disparity maps with the support of piece-wise planar meshes. Furthermore, Fan et al. [2] proposed an efficient BV algorithm named SRP, whereby the search range at row v propagates from three estimated neighboring disparities on the row $v + 1$. SRP outperforms both GCS and DT-BV in terms of dense road disparity estimation. Therefore, SRP is more feasible to acquire depth information for autonomous vehicles. This paper mainly focuses on the implementation of [2] using a state-of-the-art System-on-Chip (SoC) DSP. To improve the performance of the implementation, the algorithm is programmed using linear assembly language and its processing speed is accelerated using Open Multiple-Processing (OpenMP).

The remainder of this paper is structured as follows: Sect. 2 presents the proposed BV algorithm. Section 3 provides details on the DSP implementation. Section 4 discusses the experimental results and evaluates the implementation performance. Finally, Sect. 5 concludes the paper and provides recommendations for future work.

2   Algorithm Description

2.1   Cost Computation

The stereo image pairs processed in this paper are assumed to be perfectly-rectified. The cost function is as follows [22]:

$$c(u,v,d) = \frac{n\sigma_l\sigma_r - \sum\limits_{(u,v)\in W}(i_l(u,v)-\mu_l)(i_r(u-d,v)-\mu_r)}{n\sigma_l\sigma_r}, \tag{1}$$

where $c(u,v,d)$ denotes the matching cost. W represents the left block. n denotes the number of pixels within W. (u,v) is the center of the left block. μ_l and μ_r are the intensity means of the left and right blocks, respectively [23,24]. σ_l and σ_r are the intensity standard deviations of the left and right blocks [25]:

$$\sigma_l = \sqrt{\sum_{(u,v)\in W}\frac{i_l{}^2(u,v)}{n}-\mu_l{}^2}, \quad \sigma_r = \sqrt{\sum_{(u,v)\in W}\frac{i_r{}^2(u-d,v)}{n}-\mu_r{}^2}, \tag{2}$$

where $\sum i_l^2(u,v)$ and $\sum i_r^2(u-d,v)$ are two dot products [26]. The denominator of Eq. 1 can therefore be pre-calculated. Similarly, we rearrange the numerator of Eq. 1 to obtain the following expression [27]:

$$c(u,v,d) = \frac{n\sigma_l\sigma_r - \sum\limits_{(u,v)\in W}i_l(u,v)i_r(u-d,v)+n\mu_l\mu_r}{n\sigma_l\sigma_r}. \tag{3}$$

According to [28], the micro-operations of subtraction, addition, division, multiplication, and square root are almost the same. Therefore, the process of computing the matching costs can be more efficient by pre-calculating μ and σ and fetching their values directly from the DSP memory. More details on cost computation implementation will be discussed in Sect. 4.

2.2   Disparity Estimation

The disparity estimation performs iteratively from the image bottom to its top. The search range for the bottom row is $[d_{\min}, d_{\max}]$, where $d_{\min}$ and $d_{\max}$ are set to 0 and 80 in this paper, respectively. Then, the search range at the position of

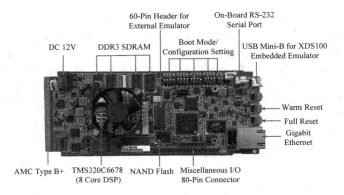

Fig. 1. An illustration of TMS320C6678 DSP.

(u, v) propagates from three adjacent neighbors on row $v + 1$. Due to the fact that the disparities at $(u, v+1)$, $(u-1, v+1)$ and $(u+1, v+1)$ have already been obtained in the previous iteration, the search range for estimating the disparity at the position of (u, v) is controlled using the following equation [29]:

$$sr = \bigcup_{k=u-1}^{u+1} \{d \mid d \in [\ell(k, v + 1) - \tau, \ell(k, v + 1) + \tau]\}, \tag{4}$$

where ℓ represents the disparity map, sr denotes the search range, and τ is the search range propagation tolerance. In this paper, τ is set to 1. More details on the implementation of the proposed BV algorithm are given in Sect. 3.

3 Notes on Implementation

3.1 Hardware Overview

This subsection gives some details on the hardware we use. An example of TMS320C6678 DSP is shown in Fig. 1. The DSP is designed based on Texas Instruments (TI) Keystone I multi-core architecture. The TMS320C6678 DSP has different levels of memories: 32 KB level 1 P, D cache, 512 KB level 2 cache per core, 4 MB multi-core shared memory and 512 MB DDR3 [2]. It is integrated with eight C66X CorePac DSPs, each of which runs at 1.25 GHz. Benefiting from the multi-core processing of the Keystone I architecture, the TMS320C6678 DSP supports a variety of high-performance autonomous applications such as depth perception, lane detection, etc. The next subsection will discuss the multi-core processing using OpenMP.

3.2 Multi-core Processing

Keystone I platform supports OpenMP 3.0 APIs, which can break a serial code into several independent chunks. Therefore, the implementation on the DSP is

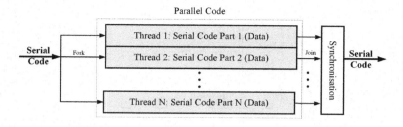

Fig. 2. Fork-join model.

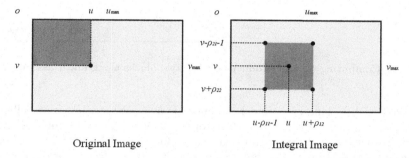

Original Image   Integral Image

Fig. 3. Integral image initialization and value indexing.

also accelerated by processing the algorithm with eight cores in parallel. OpenMP mainly consists of data-sharing, work-sharing and synchronization. A general structure of OpenMP is shown in Fig. 2 which is also known as a fork-join model. Data-sharing specifies an appropriate scheduling model, work-sharing specifies a part of serial code to be parallelized, and synchronization determines how the data is shared [2]. In the implementation, four lookup tables are firstly created to store the values of μ_l, μ_r, σ_l and σ_r. The *for* loops in this stage are divided among eight cores using *omp for* clause. In addition, *dynamic* is selected as the scheduling model, because it performs better when the subtasks distributed to each core are unequal. In the meantime, the values of μ_l, μ_r, σ_l and σ_r are declared as *private* variables, and thus, each core has a copy of these values. For synchronization, *nowait* clause is utilized to ignore the implicit barrier of *for* pragma. The rest of the algorithm is parallelized using *omp sections*, where the serial code is equally divided into eight sub-blocks, which are executed in parallel. Finally, the disparities estimated by each core are transferred into a Dynamic Random Access Memory (DRAM) and a dense disparity map is obtained. In the next subsection, some details on the implementation of cost computation are provided.

3.3   Implementation of Cost Computation

In our implementation, we programme the proposed algorithm in linear assembly language to exploit the single instruction multiple data (SIMD) features.

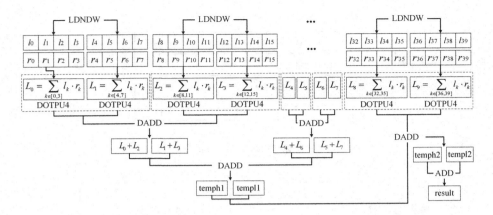

Fig. 4. Computing the dot product between two blocks using linear assembly.

Compared with C language, linear assembly is a lower-level but more efficient programming language which takes care of the pipeline structure and generates highly parallel assembly code automatically [30]. The data type of a pixel in gray-scale images is unsigned char (8 bits), and moving data between memory and registers is a time-consuming process. Typically, loading 1-byte data to a register costs 5 cycles [30], which is very inefficient. In our experiments, we utilize Load Non-Aligned Double Word (LDNDW) instruction to load 8-byte data to a pair of registers (32-bit for each). This process takes the same number of cycles as Load Double Word (LDW). Therefore, the block size is set as 5×8 in this paper and only five LDNDWs are required to load the data of each block (Fig. 4).

Furthermore, we also utilize four integral images I_{μ_l}, I_{μ_r}, I_{σ_l} and I_{σ_l} to accelerate the computations of μ_l, μ_r, σ_l and σ_r, respectively. An example of integral image initialization and value indexing is shown in Fig. 3, and the expressions of the integral images are as follows:

$$
\begin{aligned}
I_{\mu_l}(u, v) = \sum_{x \le u, \; y \le v} i_l(x, y), \quad I_{\mu_r}(u, v) = \sum_{x \le u, \; y \le v} i_r(x, y), \\
I_{\sigma_l}(u, v) = \sum_{x \le u, \; y \le v} i_l{}^2(x, y), \quad I_{\sigma_r}(u, v) = \sum_{x \le u, \; y \le v} i_r{}^2(x, y).
\end{aligned}
\tag{5}
$$

The values of μ_l and μ_r can now be computed using the following equations:

$$
\begin{aligned}
\mu_l = \frac{1}{n}\big(& I_{\mu_l}(u + \rho_{12}, v + \rho_{22}) - I_{\mu_l}(u - \rho_{11} - 1, v + \rho_{22}) \\
& - I_{\mu_l}(u + \rho_{12}, v - \rho_{21} - 1)) + I_{\mu_l}(u - \rho_{11} - 1, v - \rho_{21} - 1)\big),
\end{aligned}
\tag{6}
$$

$$
\begin{aligned}
\mu_r = \frac{1}{n}\big(& I_{\mu_r}(u + \rho_{12}, v + \rho_{22}) - I_{\mu_r}(u - \rho_{11} - 1, v + \rho_{22}) \\
& - I_{\mu_r}(u + \rho_{12}, v - \rho_{21} - 1)) + I_{\mu_r}(u - \rho_{11} - 1, v - \rho_{21} - 1)\big).
\end{aligned}
\tag{7}
$$

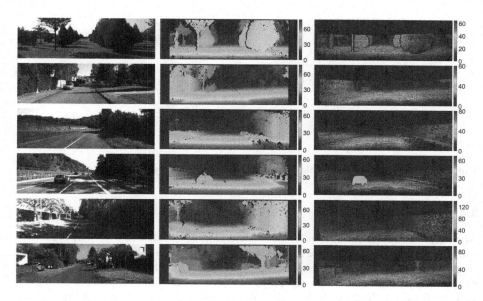

Fig. 5. Experimental results of the KITTI datasets. The first column illustrates the left images. The second column shows the estimated disparity maps. The third column gives the ground truth.

Similarly, the values of σ_l and σ_r can be computed using the following equations:

$$\sigma_l = \sqrt{t_l - \mu_l^2}, \qquad \sigma_r = \sqrt{t_r - \mu_r^2} \tag{8}$$

where

$$t_l = \frac{1}{n}\big(I_{\sigma_l}(u + \rho_{12}, v + \rho_{22}) - I_{\sigma_l}(u - \rho_{11} - 1, v + \rho_{22})$$
$$- I_{\sigma_l}(u + \rho_{12}, v - \rho_{21} - 1)\big) + I_{\sigma_l}(u - \rho_{11} - 1, v - \rho_{21} - 1)\big), \tag{9}$$

$$t_r = \frac{1}{n}\big(I_{\sigma_r}(u + \rho_{12}, v + \rho_{22}) - I_{\sigma_r}(u - \rho_{11} - 1, v + \rho_{22})$$
$$- I_{\sigma_r}(u + \rho_{12}, v - \rho_{21} - 1)\big) + I_{\sigma_r}(u - \rho_{11} - 1, v - \rho_{21} - 1)\big). \tag{10}$$

To further minimize the number of load instructions, we save the values of μ_l, μ_r, σ_l and σ_r into four lookup tables. Also, Double Addition (DADD) is utilized instead of Addition (ADD) to perform two 32-bit additions and return as two 32-bit results. DOTPU4 performs four 8-bit wide multiplies and adds the results together. The compiler will optimize the code by rearranging the instructions and the pipeline. An example of computing dot product using linear assembly code is shown in Fig. 5. Moreover, left-right consistency (LRC) check is also performed to remove the half occluded areas from the disparity maps [2].

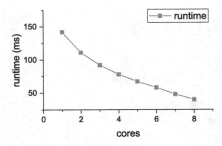

Fig. 6. The runtime of the implementation with respect to different number of cores.

3.4   Post-processing

Although the LRC check doubles the computational complexity by re-projecting the computed disparity values from one image to the other one, most of the incorrectly matched pixels can be removed and an outline can be found [2]. For the estimation of the right disparity map, the process of memorization is unnecessary because the lookup tables have already been set up. The performance of the proposed binocular vision system is discussed in Sect. 4.

4   Experimental Results

The proposed algorithm is evaluated using the KITTI stereo 2012 and 2015 datasets [31–33]. Some examples of the experimental results are shown in Fig. 5, where the first column illustrates the left images; the second column shows the estimated dense disparity maps; and the third column illustrates the ground truth disparity maps. In [2], the accuracy of the proposed algorithm has already been evaluated by comparing the estimated disparity maps with their ground truth, where the percentage of the incorrectly matched disparities (the threshold is 2 pixels) account for around 6.82% when the block size is 7×7. Hence in this paper, we will mainly discuss the performance of the implementation in terms of the processing speed.

The runtime with respect to different number of cores is shown in Fig. 6, where it can be observed that the runtime of the implementation using a single C6678 core is approximately 150 ms. The processing speed is boosted by around 3.5 times when we use eight cores to process the proposed algorithm. The runtime with respect to eight cores is only 40 ms, thereby allowing for a higher processing time for other subsystems in the ADAS.

5   Conclusion

In this paper, an efficient BV algorithm was successfully implemented on a TMS320C6678 DSP. To improve the computational complexity, the cost function was divided into five independent parts, each of which was pre-calculated before performing binocular matching. In addition, the process of computing the mean and standard deviation values of a given block was also optimized using two integral images. Furthermore, we fully exploited the benefits of linear assembly and OpenMP when implementing the proposed algorithm. The image data was loaded more efficiently from the memory of the DSP using LDNDW instruction. The dot product between a pair of chosen blocks was computed using the DOTPU4 instruction. To further accelerate the processing speed of the implementation, OpenMP broke the serial code into eight independent chunks which were then processed using eight cores in parallel. The implementation achieves a high processing speed of 25 fps when estimating the disparity maps for the KITTI stereo datasets. The real-time performance of the proposed algorithm allows other subsystems in the ADAS, such as obstacle detection and lane detection, to have a higher processing time.

References

1. Brink, J.A., Arenson, R.L., Grist, T.M., Lewin, J.S., Enzmann, D.: Bits and bytes: the future of radiology lies in informatics and information technology. Eur. Radiol. **27**(9), 3647–3651 (2017)
2. Fan, R., Ai, X., Dahnoun, N.: Road surface 3D reconstruction based on dense subpixel disparity map estimation. IEEE Trans. Image Process. **27**(6), 3025–3035 (2018)
3. Nordhoff, S.: Mobility 4.0: are consumers ready to adopt google's self-driving car? Master's thesis, University of Twente (2014)
4. Fan, R., Jiao, J., Ye, H., Yu, Y., Pitas, I., Liu, M.: Key ingredients of self-driving cars. arXiv preprint arXiv:1906.02939
5. Ozgunalp, U., Fan, R., Ai, X., Dahnoun, N.: Multiple lane detection algorithm based on novel dense vanishing point estimation. IEEE Trans. Intell. Transp. Syst. **18**(3), 621–632 (2017)
6. Fan, R., Dahnoun, N.: Real-time stereo vision-based lane detection system. Meas. Sci. Technol. **29**(7), 074005 (2018)
7. Bertozzi, M., Broggi, A.: Gold: a parallel real-time stereo vision system for generic obstacle and lane detection. IEEE Trans. Image Process. **7**(1), 62–81 (1998)
8. Fan, R., Wang, L., Liu, M., Pitas, I.: A robust roll angle estimation algorithm based on gradient descent. arXiv preprint arXiv:1906.01894
9. Fan, R., Prokhorov, V., Dahnoun, N.: Faster-than-real-time linear lane detection implementation using SoC DSP TMS320C6678. In: Proceedings IEEE International Conference Imaging Systems and Techniques (IST), pp. 306–311, October 2016
10. Boykov, Y., Veksler, O., Zabih, R.: Fast approximate energy minimization via graph cuts. IEEE Trans. Pattern Anal. Mach. Intell. **23**(11), 1222–1239 (2001)
11. Ihler, A.T., John III, W.F., Willsky, A.S.: Loopy belief propagation: convergence and effects of message errors. J. Mach. Learn. Res. **6**, 905–936 (2005)
12. Tappen, M.F., Freeman, W.T.: Comparison of graph cuts with belief propagation for stereo, using identical MRF parameters. In: Proceedings Ninth IEEE International Conference on Computer Vision, p. 900. IEEE (2003)
13. Mozerov, M.G., van de Weijer, J.: Accurate stereo matching by two-step energy minimization. IEEE Trans. Image Process. **24**(3), 1153–1163 (2015)

14. Sinha, S.N., Scharstein, D., Szeliski, R.: Efficient high-resolution stereo matching using local plane sweeps. In: Proceedings of the IEEE Conference on Computer Vision and Pattern Recognition, pp. 1582–1589 (2014)
15. Bleyer, M., Rhemann, C., Rother, C.: Extracting 3D scene-consistent object proposals and depth from stereo images. In: Fitzgibbon, A., Lazebnik, S., Perona, P., Sato, Y., Schmid, C. (eds.) ECCV 2012. LNCS, vol. 7576, pp. 467–481. Springer, Heidelberg (2012). https://doi.org/10.1007/978-3-642-33715-4_34
16. Šára, R.: Finding the largest unambiguous component of stereo matching. In: Heyden, A., Sparr, G., Nielsen, M., Johansen, P. (eds.) ECCV 2002. LNCS, vol. 2352, pp. 900–914. Springer, Heidelberg (2002). https://doi.org/10.1007/3-540-47977-5_59
17. Sara, R.: Robust correspondence recognition for computer vision. In: Rizzi, A., Vichi, M. (eds.) COMPSTAT 2006-Proceedings in Computational Statistics, pp. 119–131. Springer, Heidelberg (2006). https://doi.org/10.1007/978-3-7908-1709-6_10
18. Cech, J., Sara, R.: Efficient sampling of disparity space for fast and accurate matching. In: IEEE Conference on Computer Vision and Pattern Recognition, CVPR 2007, pp. 1–8. IEEE (2007)
19. Spangenberg, R., Langner, T., Rojas, R.: Weighted semi-global matching and center-symmetric census transform for robust driver assistance. In: Wilson, R., Hancock, E., Bors, A., Smith, W. (eds.) CAIP 2013. LNCS, vol. 8048, pp. 34–41. Springer, Heidelberg (2013). https://doi.org/10.1007/978-3-642-40246-3_5
20. Miksik, O., Amar, Y., Vineet, V., Pérez, P., Torr, P.H.: Incremental dense multimodal 3D scene reconstruction. In: 2015 IEEE/RSJ International Conference on Intelligent Robots and Systems (IROS), pp. 908–915. IEEE (2015)
21. Pillai, S., Ramalingam, S., Leonard, J.J.: High-performance and tunable stereo reconstruction. In: 2016 IEEE International Conference on Robotics and Automation (ICRA), pp. 3188–3195. IEEE (2016)
22. Fan, R., Liu, Y., Bocus, M.J., Wang, L., Liu, M.: Real-time subpixel fast bilateral stereo. In: 2018 IEEE International Conference on Information and Automation (ICIA), pp. 1058–1065. IEEE, August 2018
23. Fan, R., Jiao, J., Pan, J., Huang, H., Shen, S., Liu, M.: Real-time dense stereo embedded in a UAV for road inspection. In: Proceedings of the IEEE Conference on Computer Vision and Pattern Recognition Workshops (2019)
24. Zhang, Z.: Advanced stereo vision disparity calculation and obstacle analysis for intelligent vehicles. Ph.D. dissertation, University of Bristol (2013)
25. Fan, R., Liu, Y., Yang, X., Bocus, M.J., Dahnoun, N., Tancock, S.: Real-time stereo vision for road surface 3-D reconstruction. In: Proceedings of IEEE International Conference Imaging Systems and Techniques (IST), pp. 1–6, October 2018
26. Ai, X.: Active based range measurement systems and applications. Ph.D. dissertation, University of Bristol (2014)
27. Fan, R., Dahnoun, N.: Real-time implementation of stereo vision based on optimised normalised cross-correlation and propagated search range on a GPU. In: Proceedings of IEEE International Conference Imaging Systems and Techniques (IST), pp. 1–6, October 2017
28. Mano, M.M.: Computer System Architecture (2003)
29. Fan, R.: Real-time computer stereo vision for automotive applications. Ph.D. dissertation, University of Bristol, July 2018
30. Texas Instruments: Multicore fixed and floating-point digital signal processor. Literature Number SPRS691E (2014)

31. Geiger, A., Lenz, P., Urtasun, R.: Are we ready for autonomous driving? The KITTI vision benchmark suite. In: Proceedings of IEEE Conference on Computer Vision and Pattern Recognition, pp. 3354–3361, June 2012
32. Menze, M., Heipke, C., Geiger, A.: Joint 3D estimation of vehicles and scene flow. In: ISPRS Workshop on Image Sequence Analysis (ISA), vol. 8 (2015)
33. Menze, M., Geiger, A., Heipke, C.: Object scene flow. ISPRS J. Photogram. Remote Sens. (JPRS) **140**, 60–76 (2018)

Real-Time Lightweight CNN in Robots with Very Limited Computational Resources: Detecting Ball in NAO

Qingqing Yan[(✉)], Shu Li, Chengju Liu, and Qijun Chen

Tongji University, Shanghai 201804, China
qyan_0131@163.com

Abstract. This paper proposed a lightweight CNN architecture called *Binary-8* for ball detection on NAO robots together with a labelled dataset of 1000+ images containing balls in various scenarios to address the most basic and key issue in robot soccer games: detecting the ball. In contrast to the existing ball detection methods base on traditional machine learning and image processing, this paper presents a lightweight CNN object detection approach for CPU. In order to deal with the problems of tiny size, blurred image, occlusion and many other similar objects during detection, the paper designed a network structure with strong enough feature extraction ability. In order to achieve real time performance, the paper uses the ideas of depthwise separable convolution and binary weights. Besides, we also use SIMD (Single Instruction Multiple Data) to accelerate the operations. Full procedure and net structure have been given in this paper. Experimental results show that the proposed CNN architecture can run at full frame rate (140 Fps on CPU) with an accurate percentage of 97.13%.

Keywords: RoboCup · Ball detection · Depthwise separable convolution · Weight binarization · SIMD acceleration

1 Introduction

In the RoboCup Soccer Standard Platform League (SPL), ball detection is a fundamental and crucial ability for robotics, which is used to provide target distance and specific location for robots in various light environment. In addition, as the only standard device used in the SPL, the Softbank Robotics NAO has very limited resources, such as constrained computational abilities, and limited camera resolution. So, designing a real-time and efficient ball detection system has been a challenging task to address in the games: tiny size, blurred image, uneven illumination, occlusion and many similar objects. Traditional machine learning and image processing methods for ball recognition usually lead to a lot of and false positives and missed recognition.

State-of-the-art CNN shows excellent abilities of classification and object detection, but existing CNN-based detectors suffer from massive computational cost with server-class GPUs. When it comes to the application of CNN to mobile devices, there are several progresses in lightweight object detectors based on CNN, like YOLO-LITE, tiny-Yolo,

© Springer Nature Switzerland AG 2019
D. Tzovaras et al. (Eds.): ICVS 2019, LNCS 11754, pp. 24–34, 2019.
https://doi.org/10.1007/978-3-030-34995-0_3

Xception, MobileNet, XNOR-Net [1, 11–16], to improve the less computational cost on GPUs. However, the real-time operation performances are not able to meet the competition requirements when the nets are transferred to NAO robots vision application. In order to ensure the competitiveness of the game, we have to consider the balance between the real-time performance and detection accuracy. Furthermore, because the input resolutions are related to the model operation performance, we use the NAO's camera to capture a lot of images in various light conditions in our lab and competition field. In total, we captured 1008 unique images. While guaranteeing the detection performance, we pay more attention to the detection efficiency.

In this paper, we firstly provide a dataset, and then we investigate the effectiveness of depthwise convolution with binary weight in achieving real-time operational ability and desired detection accuracy on NAO robot. In the network design part, we described the process of building the network structure step by step in detail. The experimental results show that the computing time of the network structure designed by us decreases gradually without significant performance degradation. Satisfactory results have also been achieved in practical application.

The remainder of this letter is organized as follows. Section 2 introduces related works and analyses the existing shortcomings. Section 3 introduces our dataset in detail. Section 4 describes the procedure of developing the network structure step by step. Followed by Sect. 5 experiments and Sect. 6 conclusions.

2 Related Works

Lightweight CNN. As state-of-the-art one-stage object detection algorithms, YOLO [12–14] and SSD [18] enable to run in real time on GPUs with high accuracy. And YOLOv3-tiny [14] further improve efficiency of detection with acceptable accuracy on GPUs. But all of them suffer from massive computational cost. Recently, there have been several progresses in developing object detection algorithms to attribute to mobile and embedded vision applications, like MobileNet [15], and ShuffleNet [17]. However, these architecture designs are inspired by depthwise separable convolution which lacks efficient implementation. And other Pelee [3] enables to be executed on mobile devices at low frame rates. Compared with SSD MobileNet V1, YOLO-LITE [1] achieves the progress of computational speed improvement, but at the cost of losing the detection accuracy. So considering the balance between the real-time performance and detection accuracy on NAO robot vision application, we propose a real time lightweight CNN based on depthwise convolution with binary weight in NAO Robots for ball detection with better performance. Our design is mainly focused on efficiency.

Compression of CNNs. Generally, compression of CNNs enables to reduce the parameters and storage space of the model by means of related methods, such as pruning, quantization and approximation. Different methods have been proposed for pruning a network in [4–7]. Besides, quantization techniques were shown in [8, 9] for weights and representation of layers quantized in CNNs. With respect to approximation method, the authors proposed using FFT to compute the required convolutions in [10]. [11] Proposed a novel CNN which introduced two efficient approximations to CNNs by weight binary: Binary-Weight-Networks and XNOR Networks. In Binary-Weight-Networks, the weight values are approximated with the closest binary values, resulting in a 32x size smaller. Furthermore, XNOR Networks in which both the weights and the input of convolutional layers are binary values offers 58x speed up on a CPU, but 12.4% accuracy dropping in top-1 measure, by utilizing mostly XNOR and bit-counting operations. Inspired from the idea, our work utilizes the binary weight to compress our model.

Detection Algorithms on NAO. In response to the ball detection task, different recognition algorithms are proposed by the teams participating in the competition from all over the world. UChile, the Chilean team, has proposed a classification algorithm based on pentagonal recognition, which can better classify the positive and negative samples of the ball, but when the image is blurred, there are many missing recognition. German team HULK proposed a classification algorithm using Haar features. Although it enables to improve the accuracy of recognition to some extent, it takes a long time to compute and operation, resulting in the slow reaction of the robotics. Nao-Team HTWK and UT Austin Villa utilize a shallow CNN classifier, but they have to add other traditional image processing methods to generate Hypotheses first, and then use the CNN classifier to determine whether each hypothesis is a ball. However, in this way, the process of generating hypotheses with traditional methods will lead to leak recognition in all likelihood. Additionally, good features of the ball will be lost in the resize process. Except those, the shallow network with only 1–2 convolutional layers, generally consisting of convolution, Batch Normalization, ReLU activation and Max pooling layers, results in the weak feature extraction ability and poor generalization of the classifier.

3 Data Set

The proposed dataset was collected in our lab and real RoboCup competition fields, consisting of 1008 unique images with ball. Generally, the original images captured from the NAO's cameras is YUV format and the size of the images are lowered to 640 × 480 pixels and 320 × 240 pixels from the upper and lower camera, respectively. In order to speed up the operation process and improve the robustness under various scenarios, only the luminance (Y) channel of each image was extracted from NAO in action with various light conditions. Only when the original dataset obtained, were the ball pixels manually labelled. For the purpose of acceleration while ensuring detection accuracy, we resized the input images with label to a middle size of 416 × 416 pixels for later training and testing. An example of the proposed dataset is shown as Fig. 1. And the specific Dataset is online at https://github.com/qyan0131/Binary-8-DataSet.git.

Fig. 1. Dataset for training and testing: It consists of 1008 Y channel images with ball captured from the NAO's cameras. The size of the images is lowered to 640×480 pixels and 320×240 pixels from the upper and lower camera, respectively

4 Network Design

In this section, we demonstrate the network design procedure in detail. The proposed network mainly focuses on the effectiveness and still maintain acceptable performances when transferred to NAO. And the design guidelines are composed of four parts. In the backbone part, we first design a small standard CNN as the fundamental network, which is enable to extract enough features for NAO vision detection. And under the premise of maintaining the accuracy for detection, we compress the standard CNN model as the backbone by reducing the number of layers and filters as much as possible. Then inspired from MobileNet ideas, we use the depthwise separable convolution based on the backbone to greatly reduce the number of parameters. Besides, a binary-weights approach is utilized in the point-wise convolution operation to further speed up the computational performance, because point-wise convolution operation takes more than 80% float type computation cost in MobileNet architecture while binary-weights use mostly XNOR and bit-counting operation. Finally, in view of the Intel CPU used by the NAO robot, we rewrite the convolution operation, batch normalization operation and ReLU none-linear activation operation of CNN network with SIMD instructions, which once again increases the speed by many times.

4.1 Backbone

In this paper, we no longer only apply CNN to classifier, we hope to use CNN to achieve end-to-end object detection. As consequence, we first build a backbone with sufficient feature extraction and generalization capabilities. Then, in order to deal with tiny size problem, we use anchor mechanism and design three anchors for different size objects. Finally, the output data structure of the network is given.

In the backbone design procedure, we adopt the sequential iteration method. We weigh the running time, accuracy against the number of layers and channels of the network. Because the number of network layers and layer filters will affect the network parameters and computation costs. The more network layers, the stronger the non-linear ability of the whole network, the stronger the ability to extract features, the stronger the robustness and generalization ability. The more layer filters, the more information flows between adjacent two layers of network, the richer and more accurate the extracted features are. However, increasing the number of network layers or increasing the number of layer filters will result in real-time performance degradation.

Darknet Reference Model is a small but efficient network proposed by [20]. Inspired by it, we prune Darknet Reference network layer by layer and keep training and testing. When the accuracy drops dramatically, we stop pruning the network layer. Then we start reducing layer filters. Similarly, when the accuracy on training and test sets declines significantly, we stop reducing the number of filters. In this way, the backbone containing 8 convolutional layers has been built, as shown in Table 1. We call it *Backbone-8*.

Table 1. *Backbone-8* architecture

Type/stride	Filter shape	Input size
Conv/s2	$3 \times 3 \times 3 \times 16$	$416 \times 416 \times 3$
Conv/s1	$3 \times 3 \times 16 \times 32$	$208 \times 208 \times 16$
Conv/s2	$3 \times 3 \times 32 \times 64$	$104 \times 104 \times 32$
Conv/s1	$3 \times 3 \times 64 \times 64$	$104 \times 104 \times 64$
Conv/s2	$3 \times 3 \times 64 \times 128$	$52 \times 52 \times 64$
Conv/s2	$3 \times 3 \times 128 \times 256$	$26 \times 26 \times 128$
Conv/s1	$3 \times 3 \times 256 \times 256$	$26 \times 26 \times 256$
Conv/s2	$3 \times 3 \times 256 \times 512$	$13 \times 13 \times 256$
Conv/s1	$1 \times 1 \times 512 \times 18$	$13 \times 13 \times 512$
Yolo		

4.2 Using Depthwise Convolution

MobileNet [15] uses depthwise separable convolutions, as opposed to YOLO's method, to lighten a model for real-time object detection. The idea of depthwise separable convolutions combines depthwise convolution and point-wise convolution. Depthwise convolution applies one filter on each channel then pointwise convolution applies a 1×1 convolution [15] to expand channels.

Based on [15], related to standard convolutions, using depthwise separable convolutions can get a reduction in computational cost of:

$$\frac{D_K \cdot D_K \cdot M \cdot D_F \cdot D_F + M \cdot N \cdot D_F \cdot D_F}{D_K \cdot D_K \cdot M \cdot N \cdot D_F \cdot D_F} = \frac{1}{N} + \frac{1}{D_K^2} \quad (1)$$

where $D_K \cdot D_K \cdot M \cdot N$ is the size of a parameterized convolution kernel K, and $D_F \times D_F \times M$ is the size of the feature map taken as input.

In order to make the network mentioned in Sect. 4.1 more real-time, we rewrite the backbone in MobileNet structure called *Depthwise-8*. In the model, we also use 3×3 depthwise separable convolutions to achieve 8 to 9 times less computation than *Backbone-8* (Table 2).

Table 2. *Depthwise-8* architecture

Type/stride	Filter shape	Input size
Conv/s2	$3 \times 3 \times 3 \times 16$	$416 \times 416 \times 3$
Conv dw/s1	$3 \times 3 \times 16$ dw	$208 \times 208 \times 16$
Conv/s1	$1 \times 1 \times 16 \times 32$	$208 \times 208 \times 32$
Conv dw/s2	$3 \times 3 \times 32$ dw	$104 \times 104 \times 32$
Conv/s1	$1 \times 1 \times 32 \times 64$	$104 \times 104 \times 64$
Conv dw/s1	$3 \times 3 \times 64$ dw	$104 \times 104 \times 64$
Conv/s1	$1 \times 1 \times 64 \times 64$	$104 \times 104 \times 64$
Conv dw/s2	$3 \times 3 \times 64$ dw	$52 \times 52 \times 64$
Conv/s1	$1 \times 1 \times 64 \times 128$	$52 \times 52 \times 128$
Conv dw/s2	$3 \times 3 \times 128$ dw	$26 \times 26 \times 128$
Conv/s1	$1 \times 1 \times 128 \times 256$	$26 \times 26 \times 128$
Conv dw/s1	$3 \times 3 \times 256$ dw	$26 \times 26 \times 256$
Conv/s1	$1 \times 1 \times 256 \times 256$	$26 \times 26 \times 256$
Conv dw/s2	$3 \times 3 \times 256$ dw	$13 \times 13 \times 128$
Conv/s1	$1 \times 1 \times 256 \times 512$	$13 \times 13 \times 128$
Conv/s1	$1 \times 1 \times 512 \times 18$	$13 \times 13 \times 512$
Yolo		

4.3 Using Weight Binarization

Floating-point operation is time-consuming for device CPU, which is one of the most important factors restricting CNN running on CPU. Weights binarization can convert complex floating-point operations into simple XOR operations to accelerate the computation procedure. In the experiment we found that in Mobilenet architecture, point-wise convolution has limited feature extraction ability but it takes more than 80% of the total computation cost, while depthwise convolution can extract features effectively. Based on the above findings, we apply binary-weight operation to point-wise convolution to further accelerate the whole computation process. According to [11], the convolutional weight can be approximated by:

$$\mathcal{A}_{lk} = \frac{1}{n} \| \mathcal{W}_{lk}^t \|_{l1} \tag{2}$$

$$\mathcal{B}_{lk} = sign\left(\mathcal{W}_{lk}^t \right) \tag{3}$$

$$\widetilde{\mathcal{W}}_{lk} = \mathcal{A}_{lk} \mathcal{B}_{lk} \tag{4}$$

Where $\mathcal{W} \in \mathbb{R}^n$, ℓ, k represent k^{th} filter in l^{th} layer.

Since the number of network layers proposed in this paper is small and the input is 8-bit unsigned integer, if we binaries the network input i.e. the image or the output of each layer, it will take even more time to traverse the whole image. Therefore, this paper only binaries the network weight. The structure of binary-weighted network structure is shown in Table 3. We call it *Binary-8*.

Table 3. *Binary-8* architecture

Type/stride	Filter shape	Input size	Binary or not
Conv/s2	$3 \times 3 \times 3 \times 16$	$416 \times 416 \times 3$	0
Conv dw/s1	$3 \times 3 \times 16$ dw	$208 \times 208 \times 16$	0
Conv/s1	$1 \times 1 \times 16 \times 32$	$208 \times 208 \times 32$	1
Conv dw/s2	$3 \times 3 \times 32$ dw	$104 \times 104 \times 32$	0
Conv/s1	$1 \times 1 \times 32 \times 64$	$104 \times 104 \times 64$	1
Conv dw/s1	$3 \times 3 \times 64$ dw	$104 \times 104 \times 64$	0
Conv/s1	$1 \times 1 \times 64 \times 64$	$104 \times 104 \times 64$	1
Conv dw/s2	$3 \times 3 \times 64$ dw	$52 \times 52 \times 64$	0
Conv/s1	$1 \times 1 \times 64 \times 128$	$52 \times 52 \times 128$	1
Conv dw/s2	$3 \times 3 \times 128$ dw	$26 \times 26 \times 128$	0
Conv/s1	$1 \times 1 \times 128 \times 256$	$26 \times 26 \times 128$	1
Conv dw/s1	$3 \times 3 \times 256$ dw	$26 \times 26 \times 256$	0
Conv/s1	$1 \times 1 \times 256 \times 256$	$26 \times 26 \times 256$	1
Conv dw/s2	$3 \times 3 \times 256$ dw	$13 \times 13 \times 128$	0
Conv/s1	$1 \times 1 \times 256 \times 512$	$13 \times 13 \times 128$	1
Conv/s1	$1 \times 1 \times 512 \times 18$	$13 \times 13 \times 512$	0
Yolo			

4.4 Boost Real Time Performance

The network in Sect. 4.3 already has strong real-time performance, but we can still use the SIMD instructions provided by Intel CPU to accelerate the operation on NAO robots to further enhance real-time performance. SIMD stands for Single Instruction Multiple Data. It can copy multiple operands and package them in a set of instructions in a single register. SSE is one of the instructions sets of SIMD which is supported by NAO's CPU. NAO uses 32-bit Intel CPU with 128-bit register length and 8-bit unsigned integer for CNN image input. Therefore, the operation of 32 pixel values can be processed at one time with SSE, leading to several times faster CNN calculation.

We rewrite the convolution operation, batch normalization operation and ReLU nonelinear activation operation of CNN network with SIMD instructions.

5   Experimental Results

5.1   Comparison Among Proposed Networks

We evaluate the performance of our proposed approach on the task of NAO camera image. After successfully training models for our dataset, the network architectures accompany with their respective weights were test on the customized test set. The number of parameters, inference time and accuracy of the three proposed network is shown in Table 4. Figure 2 shows the APs during training phase.

Table 4. Comparison of three proposed network (Intel Atom 1.9 Hz CPU @ 320 * 240 pix)

Model	Million parameters	Inference time	Test set accuracy
Backbone-8	2.208	94 ms	98.17%
Depthwise-8	0.2612	17.5 ms	98.02%
Binary-8	0.2612	14.9 ms	97.13%
Binary-8-SSE	–	7.1 ms	–

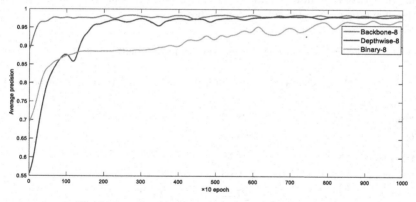

Fig. 2. Average precision of the three propose network

As shown in the table, every step of the network design improves the speed of computation while keep the correct rate is similar. And the final designed network with SSE optimization only takes 7.1 ms to process an image at the resolution of 320 × 240, which is fast enough to run on NAO robot.

Figure 3 shows the IoU and Loss during training phase of our proposed network. From the figure we can discover that the order of IoU rising speed is Backbone-8 > Depthwise-8 > Binary-8, however, with the increase of training epochs, the IoU of the three networks tends to be stable and the values are very similar. Loss in the figure is similar: although Backbone-8 declined the fastest, Binary-8 declined the slowest, they eventually tend to be stable. The difference is that, the Backbone-8 network's stable loss is the smallest, while the Binary-8 network is the highest. But considering the trade-off between computation time and performance, Binary-8 is the most efficient network.

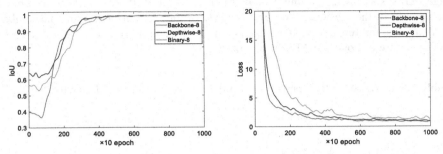

Fig. 3. IoU and Loss among proposed networks

5.2 Comparison Among Typical CNN Models

We also compare our proposed network with some of the famous or state-of-the-art lightweight network as shown in Table 5. Weights size, accurate percentage, BFLOPS and inference time on NAO robot are considered.

Table 5. Comparison of typical CNN models on Ball Dataset

Model	Weights size (MB)	AP@IOU0.5	BFLOPS	Inference time on NAO [ms]
AlexNet	238	98.78	2.862	1029
VGG	528	98.83	46.978	17310
Tiny-yolov3	35.19	98.86	1.773	566
MobilenetV1	12.22	98.65	3.798	113
Binary-8-SSE	1.05 M	97.22	0.189	7.1

According to the result in Table 5, compared with other typical lightweight models, the network we designed shows superior performance. The network structure proposed in this paper improves the computation speed greatly when accuracy is similar.

Experiments show that the proposed network has strong real-time performance (about 140 Fps on NAO robot CPU), and the accuracy (above 97%) can meet the recognition requirements.

6   Conclusion and Future Work

We propose a simple, efficient, and accurate CNN for object detection (ball) on NAO robots. We train a neural network that learns to find binary values for weights with depthwise convolution. In order to speed up execution on NAO CPU, we present a method of rewriting convolution layer, batch normalization layer and ReLU activation function using SSE. We also present a RoboCup Standard Platform image dataset with annotations, allowing other RoboCup researchers to train new models. The proposed network can detect balls accurately and run on NAO CPU in real-time.

In the future, we may continue focus on investigating more real-time CNNs on NAO robots with new tricks or new network architecture. We may research on the group point-wise convolution as proposed in ShuffleNet [17], as the point-wise convolution takes up a lot of computation in our network. We many also research on concatenating different layers to combine more feature information and further reduce the parameters. As for the RoboCup competition, we may use this network to detect all objects in games (i.e. ball, robot, obstacle, goalpost etc.). We may also use the backbone and similar tricks to build a real-time semantic segmentation algorithm on NAO robots to segment different objects/regions on the field.

Acknowledgements. The authors would like to acknowledge Team TJArk in RoboCup Standard Platform League for providing NAO robots and robot soccer field to capture dataset images. They also offer a lot of technical supports on operating and programming on NAO.

References

1. Pedoeem, J., Huang, R.: YOLO-LITE: a real-time object detection algorithm optimized for non-GPU computers. arXiv preprint arXiv:1811.05588 (2018)
2. Chollet, F.: Xception: deep learning with depthwise separable convolutions. In: Proceedings of the IEEE Conference on Computer Vision and Pattern Recognition, pp. 1251–1258 (2017)
3. Wang, R.J., Li, X., Ling, C.X.: Pelee: a real-time object detection system on mobile devices. In: Advances in Neural Information Processing Systems, pp. 1963–1972 (2018)
4. Van Nguyen, H., Zhou, K., Vemulapalli, R.: Cross-domain synthesis of medical images using efficient location-sensitive deep network. In: Navab, N., Hornegger, J., Wells, W., Frangi, A. (eds.) MICCAI 2015. LNCS, vol. 9349, pp. 677–684. Springer, Cham (2015). https://doi.org/10.1007/978-3-319-24553-9_83
5. Han, S., Mao, H., Dall, W.J.: Deep compression: compressing deep neural networks with pruning, trained quantization and Huffman coding. arXiv preprint arXiv:1510.00149 (2015)
6. Chen, W., Wilson, J.T., Tyree, S., Weinberger, K.Q., Chen, Y.: Compressing neural networks with the hashing trick. arXiv preprint arXiv:1504.04788 (2015)
7. Jaderberg, M., Vedaldi, A., Zisserman, A.: Speeding up convolutional neural networks with low rank expansions. arXiv preprint arXiv:1405.3866 (2014)
8. Gong, Y., Liu, L., Yang, M., Bourdev, L.: Compressing deep convolutional networks using vector quantization. arXiv preprint arXiv:1412.6115 (2014)
9. Lin, Z., Courbariaux, M., Memisevic, R., Bengio, Y.: Neural networks with few multiplications. arXiv preprint arXiv:1510.03009 (2015)
10. Jaderberg, M., Vedaldi, A., Zisserman, A.: Speeding up convolutional neural networks with low rank expansions. arXiv:1405.3866 [cs.CV]

11. Rastegari, M., Ordonez, V., Redmon, J., Farhadi, A.: XNOR-Net: ImageNet classification using binary convolutional neural networks. arXiv preprint arXiv:1603.05279 (2016)
12. Redmon, J., Divvala, S., Girshick, R., Farhadi, A.: You only look once: unified, real-time object detection. In: Proceedings of the IEEE Conference on Computer Vision and Pattern Recognition, pp. 779–788 (2016)
13. Redmon, J., Farhadi, A.: Yolo9000: better, faster, stronger. In: Proceedings of the IEEE Conference on Computer Vision and Pattern Recognition, pp. 7263–7271 (2017)
14. Redmon, J., Farhadi, A.: Yolov3: an incremental improvement. arXiv preprint arXiv:1804.02767 (2018)
15. Howard, A.G., et al.: Mobilenets: efficient convolutional neural networks for mobile vision applications. CoRR, abs/1704.04861 (2017)
16. Sandler, M., Howard, A., Zhu, M., Zhmoginov, A., Chen, L.-C.: MobileNetV2: inverted residuals and linear bottlenecks. In: Proceedings of the IEEE Conference on Computer Vision and Pattern Recognition, pp. 4510–4520 (2018)
17. Zhang, X., Zhou, X., Lin, M., Sun, J.: Shufflenet: an extremely efficient convolutional neural network for mobile devices. In: Proceedings of the IEEE Conference on Computer Vision and Pattern Recognition, pp. 6848–6856 (2018)
18. Liu, W., et al.: SSD: single shot multibox detector. In: Leibe, B., Matas, J., Sebe, N., Welling, M. (eds.) ECCV 2016. LNCS, vol. 9905, pp. 21–37. Springer, Cham (2016). https://doi.org/10.1007/978-3-319-46448-0_2
19. Iandola, F.N., Han, S., Moskewicz, M.W., Ashraf, K., Dally, W.J., Keutzer, K.: Squeezenet: alexnet-level accuracy with 50x fewer parameters and <0.5 MB model size. arXiv preprint arXiv:1602.07360 (2016)
20. Redmon, J., Lu, Y., Agirbau, L.: Huang ImageNet Classification [EB/OL], 3 November 2013. https://pjreddie.com/darknet/imagenet/

Reference-Free Adaptive Attitude Determination Method Using Low-Cost MARG Sensors

Jian Ding[1], Jin Wu[2], Mingsen Deng[1(✉)], and Ming Liu[2]

[1] International Joint Research Center for Data Science and High-Performance Computing, School of Information, Guizhou University of Finance and Economics, Guizhou, China
msdeng@mail.gufe.edu.cn
[2] Department of Electronic and Computer Engineering, Hong Kong University of Science and Technology, Hong Kong, China

Abstract. In this paper, an improved iterative method for attitude determination using microelectromechanical-system (MEMS) Magnetic, Angular Rate, and Gravity (MARG) sensors is proposed. The proposed complementary filter is motivated by several existing algorithms and it decreases the amount of variables for iteration which consequently lowers the convergence time. To enhance the adaptive ability i.e. the performance under external acceleration, of the proposed method, a novel scheme is designed, where the gravity estimation residual is utilized for adaptive tuning of the complementary gain. Experiments are carried out to demonstrate the advantages of the proposed method. The comparisons with representative methods show that the proposed method is more effective, not only in convergence speed, but in dynamic performance under harsh conditions as well.

Keywords: MEMS sensors · Attitude determination · MARG sensors · Complementary filter · Gradient Descent Algorithm

1 Introduction

Measurement technology is boosting the development of industrial and consumer electronics [1]. Inertial sensor fusion is a key technology in many applications including the pedestrian navigation, smart wearables, unmanned aerial vehicles, and etc. [2–4]. Among all related sensors, the Magnetic, Angular Rate and Gravity (MARG, [5]) sensors i.e. the highly compact 3-axis magnetometer, gyroscope and accelerometer, are most widely used [6]. Such sensor combination was successfully integrated in a single chip just a few years ago [7]. This led to its mass production in many areas. In this way, many fusion algorithms are designed in order to give accurate and reliable estimation of attitude, gyro-stabilized gravity, external acceleration, and etc. [8–10]. However, due to the differences in the

© Springer Nature Switzerland AG 2019
D. Tzovaras et al. (Eds.): ICVS 2019, LNCS 11754, pp. 35–48, 2019.
https://doi.org/10.1007/978-3-030-34995-0_4

internals of various algorithms, their time consumption, accuracy and convergence speed are quite different. Moreover, for industrial applications, the above factors would have much correlation on power consumption, which is a highly limited specification in product design [11–14]. Hence it is possible to develop an algorithm with high accuracy, low time consumption and high convergence speed [15].

As the most important method in state estimation, Kalman filter (KF) has been combined into the MARG sensor fusion [16]. However, KF takes too much time consumption which limits its use on low-cost platforms [9,17–20]. The complementary filter, on the other hand, can perform attitude estimation with relatively low time consumption [21]. Many efficient algorithms have been proposed to solve this problem. Madgwick et al. used the Gradient Descent Algorithm (GDA) to form a fixed-gain complementary filter [22]. This method was then improved by Tian et al. who adopted the Improved Gauss Newton Algorithm (IGNA) [2,23]. Moreover, the Levenberg-Maquardt Algorithm (LMA) for attitude estimation is investigated by Fourati et al. as well [24]. However, these optimization-based methods are iterative i.e. they may be time-costly and hard-to-converge in real applications. Would it be possible to lower the target execution time? Another question is that since representative methods mainly concern the fixed-gain complementary filter, their attitude-tracking performances are limited. In this case, motivated by the two problem raised above, this paper has the following contributions:

1. The GDA-based complementary filter is reviewed. Improvement is proposed regarding the determination of the magnetometer's reference vector, which enhances the convergence speed and the time consumption.
2. An adaptive scheme is designed in order to enhance the real-time motion-tracking ability of the filter in the presence of large external acceleration.

The developed system using MARG sensors can also provide aided accurate attitude estimation for robotic applications e.g. hand-eye calibration [25].

This paper is briefly structured as follows: Sect. 2 contains the sensor brief and modeling. Section 3 involves the theory of the complementary filter, proposed improvement to optimization and adaptive scheme. Experiments, results and comparisons are presented in Sect. 4. Section 5 contains the concluding remarks.

2   Sensor Brief and Modeling

The MARG sensors mainly include the 3-axis accelerometer, gyroscope and magnetometer. The accelerometer measures a certain object's acceleration while the gyroscope gives the angular rate information. Magnetometer provides us with the magnetic sensing of the local earth-magnetic field. The sensor observations can be modeled by the following vectors

$$\begin{cases} \boldsymbol{A}^S = (a_x, a_y, a_z)^T = \boldsymbol{A}_{true}^S + \boldsymbol{o}_{\boldsymbol{A}^S}(T) + \zeta_{\boldsymbol{A}^s} \\ \boldsymbol{\omega}^S = (\omega_x, \omega_y, \omega_z)^T = \boldsymbol{\omega}_{true}^S + \boldsymbol{o}_{\omega^s}(T) + \zeta_{\omega^b} \\ \boldsymbol{M}^S = (m_x, m_y, m_z)^T = \boldsymbol{M}_{true}^S + \boldsymbol{o}_{\boldsymbol{M}^s}(T) + \zeta_{\boldsymbol{M}^s} \end{cases} \quad (1)$$

where the vector $\boldsymbol{A}, \boldsymbol{\omega}, \boldsymbol{M}$ denote the observations of accelerometer, gyroscope and magnetometer respectively. The superscript S stands for the sensor frame. $\boldsymbol{o}$ is the sensor bias which may differ with temperature T and ζ stands for the white Gaussian noise with satisfies

$$\zeta \sim N(\boldsymbol{0}, \boldsymbol{\Sigma}_\zeta) \tag{2}$$

where $\boldsymbol{\Sigma}_\zeta$ denotes the autocovariance matrix of the noise. All we need are the true outputs of the sensors and the sensor biases can be determined by the mean values of the output residuals

$$\begin{cases} \bar{\boldsymbol{o}}_{A^S} = \frac{1}{n} \sum\limits_{i=1}^{n} \left[\boldsymbol{A}_i^S - \boldsymbol{A}_{true,i}^S \right] \\ \bar{\boldsymbol{o}}_{\omega^S} = \frac{1}{n} \sum\limits_{i=1}^{n} \left[\boldsymbol{\omega}_i^S - \boldsymbol{\omega}_{true,i}^S \right] \\ \bar{\boldsymbol{o}}_{M^S} = \frac{1}{n} \sum\limits_{i=1}^{n} \left[\boldsymbol{M}_i^S - \boldsymbol{M}_{true,i}^S \right] \end{cases} \tag{3}$$

Since the sensor biases especially for the gyroscope are related to the operating temperature, it is possible to form the temperature-bias diagram by altering the environment temperature. With the diagram, using techniques like curve fitting, we can easily model the bias. The geographic frame adopted in this paper is the in North-East-Down (NED) frame.

3   Sensor Fusion

3.1   Gyroscope Update Equation

Gyroscope provides the information of angular rate that can be used for calculating rotation. The quaternion differential equation can be given by the following quaternion kinematic equation [5]

$$\dot{\boldsymbol{q}}_{\omega,t} = \frac{1}{2} [\boldsymbol{\Omega} \times]_t \boldsymbol{q}_{\omega,t-1} \tag{4}$$

where $\dot{\boldsymbol{q}}$ denotes the derivative of the quaternion with respect to the time t. $[\boldsymbol{\Omega} \times]_t$ can be given by angular rate at time t [26]. After integration, the quaternion can be calculated by

$$\boldsymbol{q}_{\omega,t} = \boldsymbol{q}_{\omega,t-1} + \dot{\boldsymbol{q}}_{\omega,t} \Delta t \tag{5}$$

where Δt denotes the time interval. Note that the bias exists in the sensor output thus Eq. (5) would diverge when t approaches $+\infty$.

The Fusion of Accelerometer and Magnetometer. As the gyroscopic integral may diverge, there must be some other sources to compensate for the error. In this case the accelerometer and magnetometer are fused together and the produced quaternion is used for gyroscope error compensation.

The accelerometer-magnetometer fusion can be represented by direction cosine matrix (DCM), such that [27,28]

$$\begin{cases} \boldsymbol{A}^S = \boldsymbol{C}_R^S \boldsymbol{A}^R \\ \boldsymbol{M}^S = \boldsymbol{C}_R^S \boldsymbol{M}^R \end{cases}, \begin{cases} \boldsymbol{A}^R = (0,0,1)^T \\ \boldsymbol{M}^R = (m_N, 0, m_D)^T \end{cases} \tag{6}$$

where $\boldsymbol{C}$ is the DCM satisfying $\boldsymbol{C}\boldsymbol{C}^T = \boldsymbol{I}$. R stands for the reference frame (here is NED) while $\boldsymbol{C}_R^S$ represents the rotation from reference frame to the sensor frame. Using the quaternion representation of the DCM, Eq. (6) can be further calculated by

$$\begin{cases} 2(q_1 q_3 - q_0 q_2) - a_x = 0 \\ 2(q_0 q_1 + q_2 q_3) - a_y = 0 \\ 1 - 2q_1^2 - 2q_2^2 - a_z = 0 \\ m_N(1 - 2q_2{}^2 - 2q_3{}^2) + 2m_D(q_1 q_3 - q_0 q_2) - m_x = 0 \\ 2m_N(q_1 q_2 - q_0 q_3) + 2m_D(q_0 q_1 + q_2 q_3) - m_y = 0 \\ 2m_N(q_0 q_2 + q_1 q_3) + m_D(1 - 2q_1{}^2 - 2q_2{}^2) - m_z = 0 \end{cases} \tag{7}$$

which is a set of nonlinear equations that can be solved with numerical iterative methods like Gauss Newton Algorithm (GNA), Gradient Descent Algorithm (GDA), and etc. The formula of GDA is given as follows:

$$x_{n+1} = x_n - \gamma \nabla F(x_n) \tag{8}$$

where $\nabla F(x_n)$ represents the stepest gradient of the function $F(x_n)$ and γ represents the step length. Define the error function of the accelerometer estimation as

$$\boldsymbol{f}_a(\boldsymbol{q}, \boldsymbol{A}^S) = \begin{bmatrix} 2(q_1 q_3 - q_0 q_2) - a_x \\ 2(q_0 q_1 + q_2 q_3) - a_y \\ 1 - 2q_1^2 - 2q_2^2 - a_z \end{bmatrix} \tag{9}$$

Our work is to minimize this function hence its gradient can be calculated by

$$\nabla \boldsymbol{f}_a(\boldsymbol{q}, \boldsymbol{A}^S) = \boldsymbol{J}_a^T(\boldsymbol{q}) \boldsymbol{f}_a(\boldsymbol{q}, \boldsymbol{A}^S) \tag{10}$$

where the Jacobian matrix $\boldsymbol{J}_a^T(\boldsymbol{q})$ can be given by

$$\boldsymbol{J}_a(\boldsymbol{q}) = \frac{d\boldsymbol{f}_a}{d\boldsymbol{q}} = \begin{bmatrix} \frac{\partial f_{a_x}}{dq_0} & \frac{\partial f_{a_x}}{dq_1} & \frac{\partial f_{a_x}}{dq_2} & \frac{\partial f_{a_x}}{dq_3} \\ \frac{\partial f_{a_y}}{dq_0} & \frac{\partial f_{a_y}}{dq_1} & \frac{\partial f_{a_y}}{dq_2} & \frac{\partial f_{a_y}}{dq_3} \\ \frac{\partial f_{a_z}}{dq_0} & \frac{\partial f_{a_z}}{dq_1} & \frac{\partial f_{a_z}}{dq_2} & \frac{\partial f_{a_z}}{dq_3} \end{bmatrix} = \begin{bmatrix} -2q_2 & 2q_3 & -2q_0 & 2q_1 \\ 2q_1 & 2q_0 & 2q_3 & 2q_2 \\ 0 & -4q_1 & -4q_2 & 0 \end{bmatrix} \tag{11}$$

Note that the convexity of this optimization is detailed in [29]. In this case the quaternion can be optimized recursively via

$$\boldsymbol{q}_{\nabla,t} = \boldsymbol{q}_{\nabla,t-1} - \mu_t \nabla \boldsymbol{f}_a(\boldsymbol{q}_{\nabla,t-1}, \boldsymbol{A}^S) \tag{12}$$

where μ_t indicates the descent step length of the algorithm and is determined by object's motion.

With the same methodology, the error function of the magnetometer estimation can be defined as

$$\boldsymbol{f}_m(\boldsymbol{q}, \boldsymbol{M}^R, \boldsymbol{M}^S) = \begin{bmatrix} m_N(1 - 2q_2{}^2 - 2q_3{}^2) + 2m_D(q_1q_3 - q_0q_2) - m_x \\ 2m_N(q_1q_2 - q_0q_3) + 2m_D(q_0q_1 + q_2q_3) - m_y \\ 2m_N(q_0q_2 + q_1q_3) + m_D(1 - 2q_1{}^2 - 2q_2{}^2) - m_z \end{bmatrix} \quad (13)$$

The magnetometer-wise quaternion optimization can be given by

$$\boldsymbol{q}_{\nabla,t} = \boldsymbol{q}_{\nabla,t-1} - \mu_t \nabla \boldsymbol{f}_m(\boldsymbol{q}_{\nabla,t-1}, \boldsymbol{M}^R, \boldsymbol{M}^S) \quad (14)$$

where

$$\nabla \boldsymbol{f}_m(\boldsymbol{q}, \boldsymbol{M}^R, \boldsymbol{M}^S) = \boldsymbol{J}_m^T(\boldsymbol{q}, \boldsymbol{M}^R) \boldsymbol{f}_m(\boldsymbol{q}, \boldsymbol{M}^R, \boldsymbol{M}^S) \quad (15)$$

Augmenting the $\boldsymbol{f}_a$ and $\boldsymbol{f}_m$, we have

$$\boldsymbol{f}_{a,m}\left(\boldsymbol{q}, \boldsymbol{A}^S, \boldsymbol{M}^R, \boldsymbol{M}^S\right) = \begin{bmatrix} \boldsymbol{f}_a(\boldsymbol{q}, \boldsymbol{A}^S) \\ \boldsymbol{f}_m(\boldsymbol{q}, \boldsymbol{M}^R, \boldsymbol{M}^S) \end{bmatrix} \quad (16)$$

Yet we can obtain the augmented Jacobian matrix by

$$\boldsymbol{J}_{a,m}^T\left(\boldsymbol{q}, \boldsymbol{M}^R\right) = \begin{bmatrix} \boldsymbol{J}_a^T(\boldsymbol{q}) \\ \boldsymbol{J}_m^T(\boldsymbol{q}, \boldsymbol{M}^R) \end{bmatrix} \quad (17)$$

Finally, the optimization can be given by

$$\boldsymbol{q}_{\nabla,t} = \boldsymbol{q}_{\nabla,t-1} - \mu_t \nabla \boldsymbol{f} \quad (18)$$

where

$$\nabla \boldsymbol{f} = \begin{cases} \boldsymbol{J}_a^T(\boldsymbol{q}_{\nabla,t-1}) \boldsymbol{f}_a(\boldsymbol{q}_{\nabla,t-1}, \boldsymbol{A}^S), & \text{acc. only} \\ \boldsymbol{J}_{a,m}^T(\boldsymbol{q}_{\nabla,t-1}, \boldsymbol{M}^R) \boldsymbol{f}_{a,m}(\boldsymbol{q}_{\nabla,t-1}, \boldsymbol{M}^R, \boldsymbol{M}^S), & \text{acc. \& mag.} \end{cases} \quad (19)$$

which depends on the availability of the adopted sensors.

Proposed Adaptive Scheme. The complementary filter can be designed as

$$\begin{aligned} \boldsymbol{q}_{est,t} &= \beta_t \boldsymbol{q}_{\omega,t} + (1 - \beta_t) \boldsymbol{q}_{\nabla,t} \\ &= \beta_t \left[\boldsymbol{q}_{est,t-1} + \tfrac{1}{2} [\Omega\times] \boldsymbol{q}_{est,t-1} \right] + (1 - \beta_t) \left[\boldsymbol{q}_{est,t-1} - \mu_t \nabla \boldsymbol{f} \right] \\ &= \left\{ \boldsymbol{I} + \tfrac{\beta_t}{2} [\Omega\times] \right\} \boldsymbol{q}_{est,t-1} + (\beta_{t-1} - 1)\mu_t \nabla \boldsymbol{f} \end{aligned} \quad (20)$$

The gravity estimation residual can be given by

$$\boldsymbol{r}_g = \boldsymbol{G}^S - \boldsymbol{A}^S \quad (21)$$

where $\boldsymbol{G}^S$ denotes the estimated gravity which can be given by

$$\boldsymbol{G}^S = \begin{bmatrix} 2(q_1q_3 - q_0q_2) \\ 2(q_0q_1 + q_2q_3) \\ 1 - 2q_1^2 - 2q_2^2 \end{bmatrix} \quad (22)$$

In fact the norm of the residual reflects the intensity of the motion and it can be adopted for adaptive compensation for the $(\beta_{t-1} - 1)\mu_t$. Here the regulation is set empirically by

$$(\beta_{t-1} - 1)\mu_t = \begin{cases} c, \|\boldsymbol{r}_g\| \leq 0.1 \\ 10c, 0.1 < \|\boldsymbol{r}_g\| \leq 2.0 \\ 100c, \|\boldsymbol{r}_g\| > 2.0 \end{cases} \tag{23}$$

where c is a pre-determined parameter which makes the motion tracking ability real-time. Three different choices defines the motion in normal, slightly dynamic and highly dynamic states.

Improved Optimization. The above optimization for accelerometer and magnetometer involves 6 variables: $q_0, q_1, q_2, q_3, m_N, m_D$. This leads to the iterative computation for m_N and m_D. Actually, their determination can be done with rotation theory. Using Eq. (6), we obtain

$$(\boldsymbol{A}^S)^T \boldsymbol{M}^S = (\boldsymbol{C}_R^S \boldsymbol{A}^R)^T \boldsymbol{C}_R^S \boldsymbol{M}^R = (\boldsymbol{A}^R)^T (\boldsymbol{C}_R^S)^T \boldsymbol{C}_R^S \boldsymbol{M}^R \tag{24}$$

Since the DCM is an orthogonal matrix, it satisfies

$$(\boldsymbol{C}_R^S)^T \boldsymbol{C}_R^S = \boldsymbol{C}_R^S (\boldsymbol{C}_R^S)^T = \boldsymbol{I} \tag{25}$$

Hence Eq. (24) becomes

$$(\boldsymbol{A}^S)^T \boldsymbol{M}^S = (\boldsymbol{A}^R)^T \boldsymbol{M}^R \tag{26}$$

which can be expanded by

$$m_D = a_x m_x + a_y m_y + a_z m_z \tag{27}$$

In fact m_N, m_D are determined by the local magnetic dip angle α and can be given with

$$\begin{cases} m_N = \cos\alpha \\ m_D = -\sin\alpha \end{cases} \tag{28}$$

As $\alpha \in \left[-\frac{\pi}{2}, \frac{\pi}{2}\right]$, m_N should always be positive. Consequently, it is computed by

$$m_N = \sqrt{1 - m_D^2} \tag{29}$$

This provides us with an important information that lowers the amount of variables to be iteratively determined. Hence this improvement significantly facilitates the convergence of the complementary filter.

$$f_m(q, M^R, M^S) = \begin{bmatrix} 2\sqrt{1-\alpha^2}(\frac{1}{2} - q_2{}^2 - q_3{}^2) + 2\alpha(q_1 q_3 - q_0 q_2) - m_x \\ 2\sqrt{1-\alpha^2}(q_1 q_2 - q_0 q_3) + 2\alpha(q_0 q_1 + q_2 q_3) - m_y \\ 2\sqrt{1-\alpha^2}(q_0 q_2 + q_1 q_3) + 2\alpha(\frac{1}{2} - q_1{}^2 - q_2{}^2) - m_z \end{bmatrix}$$

$$J_m(q, M^R) = \begin{bmatrix} -2\alpha q_2 & 2\alpha q_3 & -4\sqrt{1-\alpha^2}q_2 - 2\alpha q_0 & -4\sqrt{1-\alpha^2}q_3 + 2\alpha q_1 \\ -2\sqrt{1-\alpha^2}q_3 + 2\alpha q_1 & 2\sqrt{1-\alpha^2}q_2 + 2\alpha q_0 & 2\sqrt{1-\alpha^2}q_1 + 2\alpha q_3 & -2\sqrt{1-\alpha^2}q_0 + 2\alpha q_2 \\ 2\sqrt{1-\alpha^2}q_2 & 2\sqrt{1-\alpha^2}q_3 - 4\alpha q_1 & 2\sqrt{1-\alpha^2}q_0 - 4\alpha q_2 & 2\sqrt{1-\alpha^2}q_1 \end{bmatrix} \tag{30}$$

Summary of the Proposed Algorithm. The overall fusion process is given in Algorithm 1.

Algorithm 1. Calculation procedure of the proposed method.

Requirements:

$t = 0$, $\boldsymbol{q}_{t=0}$ is the initialized quaternion, $c = c_0$.

while no stop commands received **do**

1. Get one set of measurement $\boldsymbol{A}^S$, $\omega = (\omega_x, \omega_y, \omega_z)^T$ and $\boldsymbol{M}^S$;
2. $t = t + 1$; Get the time interval Δt.
3. **Magnetometer is available:**
   Conduct the accelerometer-magnetometer fusion with: $\boldsymbol{q}_{\nabla,t} = \boldsymbol{q}_{\nabla,t-1} - \mu_t \nabla \boldsymbol{f}$.
   The improved formula is in Equation (30).
   Else: Use the first equation in Equation (19).
4. Calculate the acceleration residual: $\boldsymbol{r}_g = \boldsymbol{G}^S - \boldsymbol{A}^S$.
5. With the empirical regulation in Equation (23), the adaptive gain of the comple-
   mentary filter can be computed: $(\beta_{t-1} - 1)\mu_t = \begin{cases} c, \|\boldsymbol{r}_g\| \le 0.1 \\ 10c, 0.1 < \|\boldsymbol{r}_g\| \le 2.0 \\ 100c, \|\boldsymbol{r}_g\| > 2.0 \end{cases}$
6. Perform the time update: $\boldsymbol{q}_{est,t} = \left\{\boldsymbol{I} + \frac{\beta_t}{2}[\Omega\times]\right\}\boldsymbol{q}_{est,t-1} + (\beta_{t-1} - 1)\mu_t \nabla \boldsymbol{f}$
7. Normalize the estimated quaternion: $\boldsymbol{q}_{est,t} = \frac{\boldsymbol{q}_{est,t}}{\|\boldsymbol{q}_{est,t}\|}$
8. If possible, calculate the Euler angles.

end

In fact, there are some parameters to be determined before the overall fusion process. The c_0 is a factor that influences the adaptive ability of the algorithm. It can be adjusted using the response of the attitude estimation i.e. the users can tune this parameter by sensing the delay of the estimated attitude angles. The initialized quaternion is usually computed using the accelerometer and magnetometer combination forming two triads of sensor axes for attitude determination from two vector observations. Such problem can be solved via the famous Wahba's problem [30]. And using one algorithm for Wahba's solution from two vector observations [31], the problem is solved. The only one we should notice is that the conventional Wahba's problem uses the magnetometer's reference information while this can be extracted from Eq. (30).

4   Experiments and Results

4.1   Experimental Platform Configuration

Attitude estimation requires real-time sampling of inertial sensors due to the discrete strapdown integral process. Besides, the price of the sensors should not be too high that matches the consumer electronics applications.

In this way the experimental platform is designed with a STM32F4 micro controller, a MPU6500 MEMS Inertial Measurement Unit (IMU) and an HMC5983

MEMS magnetometer. The sensor are mounted properly in the direction of NED frame. An commercial-level Attitude and Heading Reference System (AHRS) is firmly attached to the experimental platform to produce reference angles. All the sensors are calibrated before experiments to cancel the sensor biases according to the sensor modelling presented in Sect. 2. The data sampling frequency is set to 1 KHz for accelerometer and gyroscope and 200 Hz for magnetometer ensuring real-time estimation. The AHRS is connected with the microcontroller via serial port interrupt with the output frequency of 300 Hz. One wireless transmitter is applied to monitor the attitude estimates more easily. In the following subsections, the MATLAB r2016 software is used for data presentation and analysis (Fig. 1).

Fig. 1. Experimental platform which contains a STM32F4 micro controller with onboard MEMS gyroscopes, accelerometers and magnetometers. Wireless transmission and TF-card data logging are supported. Apart from these, a commercial-level attitude and heading reference system (AHRS) is firmly attached to the platform giving reference angle outputs. Robust and accurate data acquisition is ensured under such architecture.

4.2 Convergence Performance

First, an experiment is performed, which indicates a period of common arbitrary motion with relatively small external acceleration. Corresponding reference attitude angles from the AHRS are given in the third figure in Fig. 3. In the experiment, the initial quaternions are both set to $(0,0,0,1)^T$ which models a very harsh initial attitude state for yaw. For the two methods, the complementary gains are fixed at 0.1 and the adaptive scheme is not taken into effect. The evolution of m_N and m_D from the proposed method and Madgwick's method is plotted in Fig. 2. Attitude estimation from the two methods are plotted in Fig. 3.

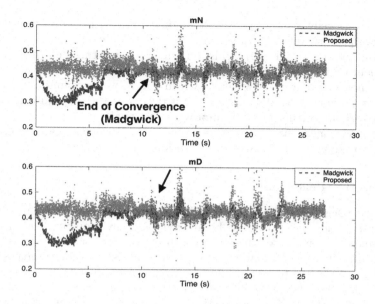

Fig. 2. The evolution of m_N and m_D. The x-axis represents the time.

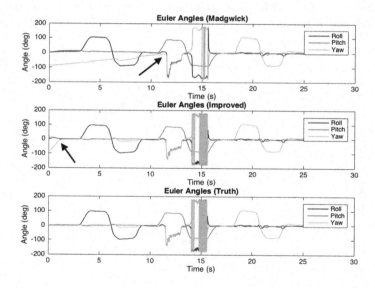

Fig. 3. Euler angles from different sources. The black arrows denote the end of the convergence period (initial alignment).

As can be easily seen from Fig. 2, Madgwick's method undergoes a long convergence process until the values of m_N, m_D basically becomes equal to the findings in Eqs. (27) and (29). This verifies the correctness of the proposed improvement on the magnetometer's reference vector.

From another point of view, as shown in Fig. 3, we may find out that the proposed improvement is obvious because it decreases the original convergence time. Throughout calculation, the improved time is about 60% to 75% of that of the previous one. This proves the proposed improvement can be used for efficiently speeding up the related optimization.

4.3   Adaptive Performance

The adaptive scheme is designed in order to enhance the performance of the attitude estimation system under highly dynamic environments. In this case, an experiment is conducted where the vibration intensity is quite big. The fast complementary filter (FCF) proposed by Wu et al. is used for comparison [5]. The FCF's gain is set to 0.05 while the proposed improved filter utilizes the proposed adaptive scheme where the empirical constant c is set to 0.08. As is shown in Fig. 4, the attitude estimation performances are quite different. For FCF, the attitude estimation is largely affected by the vibration. Although this can be improved by simply tuning the gain to a smaller one, the convergence performance of the tuned filter would then be very terrible. In other words, it could not find a balance between the convergence speed and anti-vibration performance.

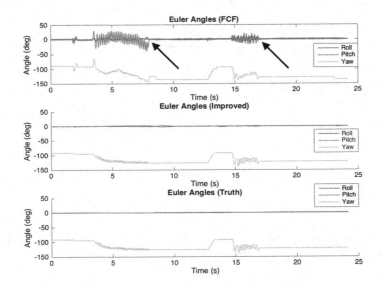

Fig. 4. Attitude estimation results in the presence of large external acceleration. The black arrows stand for the large attitude errors of the FCF.

The overall mean attitude error of the two algorithms are calculated as: 3.44° (FCF); 0.72° (Proposed Filter). The proposed adaptive scheme, however, successfully solved this problem. As can be seen from the figure, the attitude estimation from the proposed improved filter is basically not influenced by external vibration signals. The estimates are comparable with the reference angles from the AHRS. Moreover, as the AHRS adopts the KF-based algorithms, it may be much more time-costly than the proposed filter.

4.4   Magnetic Distortion

The magnetic distortion is a crucial factor considering the performance of an attitude estimator. In this sub-section, we conduct one experiment where the magnetic field is significantly distorted with a smart phone in the stand-still mode. The conventional Madgwick's GDA filter and our proposed filter are applied to solve the problem. The estimation results are shown in Fig. 5. The yaw results of our improved method are basically identical with that produced by the attitude truth output. The results of the roll and pitch angles from conventional Madgwick GDA filter and our proposed one fit with each other showing that they are jointly slightly affected by the magnetic distortion. However, as depicted in Fig. 5, the yaw distortion of Madgwick's GDA filter is much larger than that generated by the improved one. That is to say, with the same magnetic distortion, the conventional method is not able to compute reliable enough yaw estimates. In this circumstance, the improved one can also replace the Madgwick's GDA filter for better performances.

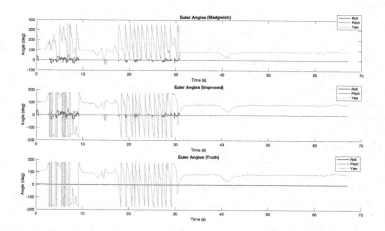

Fig. 5. The attitude estimation results in the presence of irregular magnetic distortion

5 Conclusions

The sensor fusion problem of MARG sensors can be solved via complementary filter techniques. The design of the complementary filter involves the attitude determination from accelerometer-magnetometer combination. This problem can be solved with numerical optimization algorithms. However, related optimization-based methods may too brute-force i.e they adopts the attitude quaternions along with other related reference vectors as unknown variables. In this case, this paper proposed an improvement to existing methods which enhances the original filter's convergence speed. Yet, related algorithms mainly includes the fixed-gain complementary filter that may have quite limited performance for various environments. In order to deal with this problem, we successfully established an adaptive scheme for dynamically tuning the filter gain. Experiments and comparisons are given that show the advantages of the proposed filter. We think we successfully find a balance between the time consumption and accuracy.

However, it is shown in the last experiment that the roll and pitch angles are still influenced by outer magnetic disturbances slightly. How to avoid such condition from happening would be our next research task in the future.

Acknowledgment. This work was financially supported by Joint Foundation, Ministry of Commerce and GUFE (No. 2016SWBZD04); One Hundred Person Project of the Guizhou Province (No. QKH-PTRC[2016]5675); Plan Project for Guizhou Provincial Science and Technology (No. QKH-PTRC[2018]5803).

References

1. Cheng, Y., Tian, L., Yin, C., Huang, X., Bai, L.: A magnetic domain spots filtering method with self-adapting threshold value selecting for crack detection based on the MOI. Nonlinear Dyn. **86**(2), 741–750 (2016)
2. Tian, Y., Hamel, W.R., Tan, J.: Accurate human navigation using wearable monocular visual and inertial sensors. IEEE Trans. Instrum. Meas. **63**(1), 203–213 (2014)
3. Yang, Z.L., Wu, A., Min, H.Q.: Deployment of wireless sensor networks for oilfield monitoring by multiobjective discrete binary particle swarm optimization. J. Sens. **2016**, 1–15 (2016)
4. Yun, X., Calusdian, J., Bachmann, E.R., McGhee, R.B.: Estimation of human foot motion during normal walking using inertial and magnetic sensor measurements. IEEE Trans. Instrum. Meas. **61**(7), 2059–2072 (2012)
5. Wu, J., Zhou, Z., Chen, J., Fourati, H., Li, R.: Fast complementary filter for attitude estimation using low-cost MARG sensors. IEEE Sens. J. **16**(18), 6997–7007 (2016)
6. Yun, X., Lizarraga, M., Bachmann, E., McGhee, R.: An improved quaternion-based Kalman filter for real-time tracking of rigid body orientation. In: IEEE IROS 2003, vol. 2, October 2003
7. Leclerc, J.: MEMS for aerospace navigation. IEEE Aerosp. Elect. Syst. Mag. **22**(10), 31–36 (2007)
8. Li, W., Wang, J.: Effective adaptive Kalman filter for MEMS-IMU/magnetometers integrated attitude and heading reference systems. J. Navig. **66**(01), 99–113 (2012)

9. Marantos, P., Koveos, Y., Kyriakopoulos, K.J.: UAV state estimation using adaptive complementary filters. IEEE Trans. Control Syst. Technol. **50**(7), 1573–1582 (2015)
10. Markley, F.L.: Attitude error representations for Kalman filtering. AIAA J. Guid. Control Dyn. **26**(2), 311–317 (2003)
11. Fourati, H., Manamanni, N., Afilal, L., Handrich, Y.: Posture and body acceleration tracking by inertial and magnetic sensing: application in behavioral analysis of free-ranging animals. Biomed. Signal Process. Control **6**(1), 94–104 (2011)
12. Fourati, H., Manamanni, N., Afilal, L., Handrich, Y.: A nonlinear filtering approach for the attitude and dynamic body acceleration estimation based on inertial and magnetic sensors: bio-logging application. IEEE Sens. J. **11**(1), 233–244 (2011)
13. Fourati, H.: Heterogeneous data fusion algorithm for pedestrian navigation via foot-mounted inertial measurement unit and complementary filter. IEEE Trans. Instrum. Meas. **64**(1), 221–229 (2015)
14. Makni, A., Fourati, H., Kibangou, A.: Energy-aware adaptive attitude estimation under external acceleration for pedestrian navigation. IEEE/ASME Trans. Mechatron. **21**(3), 1366–1375 (2016)
15. Wu, J., Zhou, Z., Gao, B., Li, R., Cheng, Y., Fourati, H.: Fast linear quaternion attitude estimator using vector observations. IEEE Trans. Auto. Sci. Eng. **15**(1), 307–319 (2018)
16. Kannan, R.: Orientation estimation based on LKF using differential state equation. IEEE Sens. J. **15**(11), 6156–6163 (2015)
17. Euston, M., Coote, P., Mahony, R., Kim, J., Hamel, T.: A complementary filter for attitude estimation of a fixed-wing UAV. In: IEEE IROS 2008, pp. 340–345 (2008)
18. Sabatini, A.M.: Quaternion-based extended Kalman filter for determining orientation by inertial and magnetic sensing. IEEE Trans. Biomed. Eng. **53**(7), 1346–1356 (2006)
19. Vasconcelos, J.F., Cardeira, B., Silvestre, C., Oliveira, P., Batista, P.: Discrete-time complementary filters for attitude and position estimation: design, analysis and experimental validation. IEEE Trans. Control Syst. Technol. **19**(1), 181–198 (2011)
20. Yun, X., Bachmann, E.: Design, implementation, and experimental results of a quaternion-based Kalman filter for human body motion tracking. IEEE Trans. Robot. **22**(6), 1216–1227 (2006)
21. Higgins, W.: A comparison of complementary and Kalman filtering. IEEE Trans. Aerosp. Elect. Syst. **11**(3), 321–325 (1975)
22. Madgwick, S.O.H., Harrison, A.J.L., Vaidyanathan, R.: Estimation of IMU and MARG orientation using a gradient descent algorithm. In: 2011 IEEE ICRR, pp. 1–7 (2011)
23. Tian, Y., Wei, H., Tan, J.: An adaptive-gain complementary filter for real-time human motion tracking with MARG sensors in free-living environments. IEEE Trans. Neural Syst. Rehabil. Eng. **21**(2), 254–264 (2013)
24. Fourati, H., Manamanni, N., Afilal, L., Handrich, Y.: Complementary observer for body segments motion capturing by inertial and magnetic sensors. IEEE/ASME Trans. Mechatron. **19**(1), 149–157 (2014)
25. Wu, J., Sun, Y., Wang, M., Liu, M.: Hand-eye calibration: 4D procrustes analysis approach. IEEE Trans. Instrum. Meas. (2019)
26. Wu, J., Zhou, Z., Fourati, H., Li, R., Liu, M.: Generalized linear quaternion complementary filter for attitude estimation from multi-sensor observations: an optimization approach. IEEE Trans. Auto. Sci. Eng. **16**(3), 1–14 (2019)

27. Yun, X., Bachmann, E.R., McGhee, R.B.: A simplified quaternion-based algorithm for orientation estimation from earth gravity and magnetic field measurements. IEEE Trans. Instrum. Meas. **57**(3), 638–650 (2008)
28. Wu, J., Zhou, Z., Fourati, H., Cheng, Y.: A super fast attitude determination algorithm for consumer-level accelerometer and magnetometer. IEEE Trans. Consum. Elect. **64**(3), 375–381 (2018)
29. Wu, J., Zhou, Z., Song, M., Fourati, H., Liu, M.: Convexity analysis of optimization framework of attitude determination from vector observations. In: 2019 IEEE CODIT, pp. 440–445 (2019)
30. Wahba, G.: A least squares estimate of satellite attitude. SIAM Rev. **7**(3), 409 (1965)
31. Markley, F.L., Mortari, D.: How to estimate attitude from vector observations. Adv. Astronaut. Sci. **103**(PART III), 1979–1996 (2000)

Feature-Agnostic Low-Cost Place Recognition for Appearance-Based Mapping

S. M. Ali Musa Kazmi[✉][iD], Mahmoud A. Mohamed[iD], and Bärbel Mertsching[iD]

GET Lab, University of Paderborn, Pohlweg 47–49, 33098 Paderborn, Germany
{kazmi,mohamed,mertsching}@get.upb.de

Abstract. The agent's ability to locate itself in an unfamiliar environment is essential for a reliable navigation. To address this challenge, place recognition methods are widely adopted. A common trend among most of these methods is that they are either tailored to work in specific environments or need prior training overhead [11]. Whereas, others demand extreme computational resources, such as CNN [8]. In this paper, we study the existing GSOM-based place recognition framework [12] and investigate the question of translating the system to other feature spaces, such as HOG, for low-cost place recognition. The experiments performed on four challenging sequences demonstrate the algorithm's ability to learn the representation of the new feature space without parameter tuning, provided the scaling factor along each dimension of the descriptor is taken into account. This highlights the feature-agnostic characteristic of the algorithm. We further observed that despite the low dimensionality of the HOG descriptor, the algorithm shows comparable place recognition results to the gist features, while offering threefold speed-ups in execution time.

Keywords: Place recognition · Appearance-based mapping · Global descriptors · Gist of the scene · Growing self-organizing networks

1 Introduction

Place recognition has been enormously addressed in the recent literature of robotic mapping, referred to as appearance-based mapping. As the motion sensors of the robot are noisy, a precise position estimate is not possible. In order to reduce the errors, a robust solution to place recognition is desirable. One drawback of the existing place recognition solutions is that they rely on prior knowledge of the environment or extensive training procedures before the system is functional. Consequently, when these systems are applied to previously unseen environments, they fail to deliver high precision performance. As a workaround, the parameters are usually tuned to tailor the system according to environment. This raises the question on applicability of these systems to unknown scenarios.

© Springer Nature Switzerland AG 2019
D. Tzovaras et al. (Eds.): ICVS 2019, LNCS 11754, pp. 49–59, 2019.
https://doi.org/10.1007/978-3-030-34995-0_5

To overcome this shortcoming, several state-of-the-art works have addressed the challenges of online place recognition (cf. [1–4]).

One of the key factors in any recognition system is the uniqueness of the features exploited. Distinct features facilitate robust place recognition. In this regard, researchers are divided between crafting hand-engineered features [5,6] or using the state-of-the-art deep learning algorithms to automatically learn the features from the training data [7,8]. Although convolutional neural networks (CNN) have shown outstanding performance on object recognition tasks [9], their usefulness for place recognition is still under evolution [7,10]. Besides robust feature extraction, place recognition performance also relies on the machine learning model. For instance, a robust model takes into account the visual similarities as well as temporal dependencies between the observations [11].

In this work, we study the existing place learning and recognition framework [12] and investigate the extensibility of the algorithm to other feature spaces, such as histogram of oriented gradients (HOG) [13], without the need of prior parameter tuning. To this end, we obtained the visual description of a place from the weighted average of the feature descriptor of the nearby locations. The new descriptor is then fed to the growing self-organizing map (GSOM). Given the query image and learned representation, the expected network activity is estimated and the query location is recognized as a familiar or a novel location. We tested the algorithm on four benchmark datasets, holding the settings fixed throughout the experiments. The obtained results show the feature-agnositc strength of the algorithm on the selected datasets. Moreover, despite the low dimensionality of the HOG features, the algorithm has outperformed the baseline approaches almost on all the datasets, although its place recognition performance remains quite close to gist-based variant.

Further discussion in the paper is organized as follows: Sect. 2 gives overview of the place learning and recognition frameworks. In Sect. 3, we describe the procedure to obtain gist and HOG features. Following the discussion on global descriptors, Sect. 4 presents the evaluation and analysis of the results, including comparison with the baseline methods. Finally, Sect. 5 concludes the present work with the future directions.

2   Methodology -

The process of remembering and recalling places is the core of navigation in biophysical systems (e.g. humans) [14]. These skills develop progressively and subconsciously in a mammalian's brain. However, teaching a technical system these skills need ample efforts. For instance, a place has different interpretations subject to the context of navigation [15]. At a country level, different cities are regarded as places, while within a city famous landmarks could be the individual places. This interpretation further diverges at a building level and so on. As our approach purely relies on the visual input, we consider an input image as a potential location (or place). In order to learn the visual representation of the physical space, a modified version of GSOM is used [17], which preserves the topological distances between places in the feature space. The motivations for such a choice are already discussed in [16]. Finally, given the trained network, the expected network activity criterion is leveraged to distinguish between the familiar and novel places.

The following discussion presents a brief overview of the learning dynamics of the network, while for a deeper insight the interested reader is referred to [12,17]. Let $\mathbf{x}^{(t)}$ be the feature descriptor of an input image $I^{(t)}$ at a present time t. At first, we exploit the nearby context of a place to obtain the visual description of the current location $\hat{\mathbf{x}}^{(t)}$:

$$\hat{\mathbf{x}}^{(t)} = (1 - \gamma)\hat{\mathbf{x}}^{(t-1)} + \gamma\mathbf{x}^{(t)} \tag{1}$$

where γ is a smoothness factor. It assigns weights to the current descriptor in relation to the previous observations. Because the weights assigned to the previous observations decay exponentially, the descriptors of the nearby places contribute more to the visual description of the place. We initialize the network with a single neuron and a suitable γ value (see Sect. 4), whereas the initial condition in the above equation is set to a null vector. Given the visual description of the place $\hat{\mathbf{x}}^{(t)}$ at the present time, the winner neuron c is determined according to [17, Eq. (4)] and the weights vectors of the winner and its topological neighbors $\mathbf{w}_j^{(t)} \; \forall j \in \{c, N_c\}$ are adapted following a soft-competitive learning scheme [17, Eq. (6)], aka winner-takes-most mechanism. Note that the topological neighbors of the winner N_c are all the neurons for which $\mathrm{k}(c, j) > 0$ satisfies, as given by Eq. (7) in [17]. Here, rather than using a fixed learning rate, we start with the initial learning rate $\alpha_0 = 0.3$ and reduced it in relation to the activation frequency[1] of the winner neuron $\alpha = \alpha_0 \left(1/f_c^{(t)}\right)$.

For each query input, we account for the error $\varepsilon_c^{(t)}$ in the winner, which if exceeds a certain threshold value τ, results in the creation of a new neuron in the network. The weight vector of the new neuron $\mathbf{w}_n^{(t)}$ is calculated from the learned network statistics, as follows:

$$\mathbf{w}_n^{(t)} = \frac{1}{\sigma + 1}\hat{\mathbf{x}}^{(t)} + \frac{\sigma}{\sigma + 1}\mathbf{w}_c^{(t)} \tag{2}$$

where the parameter σ, defined as $\sigma = P(I^{(t)} \,|\, C = c)$, is calculated using Eq. (7) in [17]. Finally, the error in the network is slightly smoothed and the pruning is performed to remove the unused neurons from the network, as explained in [17, Sect. IIIB-4 and -5]. Given the learned network weights at any time instant t, we find the neuron that maximizes aposteriori estimate (MAP) for the query image. Nonetheless, finding a single best neuron for the place recognition decision may yield false detections. Therefore, similar to [12], we observe the activation strength of the sequence of maximally active neurons in the network and average

---

[1] The activation frequency $f_c^{(t)}$ is the number of times a neuron is selected to represent an arbitrary input vector $\hat{\mathbf{x}}^{(k)}$, where $k = 1, \ldots, t$.

their activity, called as *expected network activity*:

$$\mathbb{E}[a(L)] = \frac{1}{t - T} \sum_{k=t-T}^{t} a(l_k) \tag{3}$$

where L is the random variable that instantiates to the maximally active neuron. A particular instance of it is denoted as $L = l_k$. Hence, the function $a(l_k)$ determines the activation strength of the maximally active neuron l_k at an arbitrary time instance k:

$$a(l_k) = \max_{i \in [1, N_p^{(k)}]} P(C = i \mid I^{(k)}) \tag{4}$$

In the above equation, $N_p^{(k)}$ is the number of neurons in the network at a time instant k, discarding the p recently created neurons. A curious reader might have noticed that the above equation is similar to Eq. (9) in [17], except that here we determine the activation strength of the maximally active neuron rather than finding the maximizer neuron itself. So, the right-hand side of Eq. (4) can be expanded according to [17, Eq. (10)]. Finally, the decision of a loop-closure detection at a certain time is taken, if the expected network activity exceeds a pre-determined threshold, i.e., $\mathbb{E}[a(L)] >$ threshold. Hence, rather than tackling the place recognition problem in a conventional way, we formulated it in hierarchical steps. For instance, finding a highly active neuron in the network is a multi-class classification problem. Given expected network's activity, the loop-closure decision is a binary classification task. Once the expected activity and highly active neuron are known, it is then straight forward to associate the places based on direct descriptors matching or geometric validation [18].

3   Image Descriptors

In this section, we briefly introduce the gist and the HOG features to compute the global description of images for the place recognition task.

3.1   Gist Features

Gist features capture the spatial layout of the scenes without the need of explicit segmentation or image analysis. To compute them, the input image is patch normalized, followed by a downsampling and cropping process, which yields a pre-processed image of resolution 256×256 pixels. The pre-processed image is then transformed to the Fourier domain and its power spectrum is convoluted with a bank of 32 Gabor filters at different frequencies and orientations. The extracted features responses are transformed back to the spatial domain and the results are averaged in 4×4 blocks, resulting in a $D = 512$ dimensional vector of gist features. For further details, we refer to the original work [19].

3.2   HOG Features

The histogram of oriented gradients (HOG) [13] characterizes the appearance and shape of the objects in a scene by the local distribution of intensity gradients. This makes the representation relatively invariant to the shadows and illumination changes. To compute HOG, the image gradients along the horizontal and vertical directions are computed within a local window using a centered derivative mask, see for instance [20]. From the resulting derivatives, the magnitudes and orientations of gradients are computed for whole image. For constructing the global descriptor, we divide the image into $N = 4 \times 4$ blocks. For each block, we computed the localized histogram of $M = 12$ bins, which describes the edge orientation in the range 0 to 2π radians. The value of each bin is obtained by summing the magnitudes of the gradients whose orientations are mapped to that bin. Typically, the obtained histogram is normalized using different schemes, e.g., L1- and, L2-norms, etc. However, in the present work, we normalized the localized histogram bins by their sum.

4   Experiments

The performance of the algorithm is evaluated on four challenging datasets recorded under extreme lighting conditions and in the presence of dynamic objects. The CityCenter (CC) dataset [21] comprises 1237 left-right camera image pairs, having the resolution of 640×480 pixels, of which we picked the right camera images for evaluation. Among the KITTI sequences [22], we selected the sequence 00 (K00), which consists of 4541 images of resolution 1241×376 pixels. The Malaga Parking 6L (M6) imagery [23] has 3474 examples recorded at 1024×768 pixels resolution. Amongst all the datasets, the St. Lucia (SL) sequence 190809 [24] is the largest route traveled ~17 km and consists of 21,815 images. The image resolution of SL dataset is same as CC. To evaluate the algorithm, the GPS error of 10 m and 4 m was taken into account for SL and M6, respectively. For the K00 dataset, we obtained the ground truth matrices from [25], while for CC the ground truth matrix provided with the dataset is used.

In Eq. (1), we exploit the nearby context to obtain visual description of a location. As the datasets have different frame rates, we obtain the smoothness level as follows:

$$\gamma = 1 - \exp\left(\frac{-\rho}{\max(\rho, \texttt{frame\_rate})}\right)$$

All the parameter configurations, including ρ, were set in accordance with [17, Tab. I]. The above equation collects, at a high frame rate, more frames to compute the description of the places, while low frame rate picks up few images. We refer to the respective papers to select the value of the frame_rate parameter.

4.1   Precision–Recall Performance: Gist vs. HOG

Following discussion evaluates the performance of our algorithm on gist and HOG descriptors using precision–recall metric:

$$\text{Precision}\,(\%) = \frac{\text{Correct detections}}{\text{All detections}} \times 100$$

$$\text{Recall}\,(\%) = \frac{\text{Correct detections}}{\text{Actual correct detections}} \times 100$$

The correct detections are the cases where algorithm successfully recognizes the query image as a familiar location, while all detections are the images which are flagged as a familiar place by the algorithm (i.e. all positive cases). Actual correct detections are the locations which are marked familiar according to ground truth. Hence, precision score describes the robustness of the algorithm to recognize a place without committing a mistake, whereas recall determines the strength of the algorithm in detecting the familiar places without missing any. In order to test the performance of the algorithm on the selected feature descriptors, the precision–recall curves are generated by varying the `threshold` parameter in the range $(0, 1]$, see Fig. 1. The plotted curves in the figure demonstrate that the high precision remains persistent across the datasets, even when the HOG feature descriptor is utilized. This highlights the robustness of the algorithm for place recognition. Moreover, the high precision–recall scores in different feature spaces shows that the algorithm is feature-agnostic. It is not just the choice of the feature space, rather the strength of the algorithm to discriminate between the places.

For an autonomous vision-based navigation system, the high precision scores are important. This implies that the false positive detections are intolerable for a coherent map building and reliable navigation. Hence, we record the recall rate

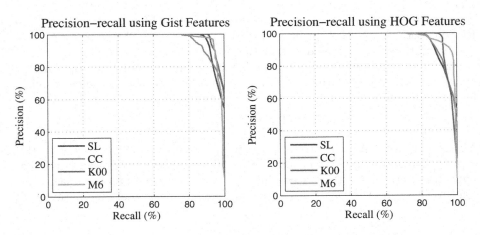

Fig. 1. Gist vs. HOG: Precision–recall rate of the algorithm on the benchmark datasets (for details see text).

of the algorithm at 100% precision and compared the gist and HOG features on the selected sequences. Figure 2 shows the maximum recalls achieved by our algorithm (at 100% precision) for the gist and HOG features. Clearly, gist features show better performance compared to the HOG descriptors. However, the differences in recalls at the pre-set precision criterion are quite comparable on all the datasets. It is worth noticing here that the HOG descriptor is very low dimensional ($D = 192$) compared to the gist descriptor, follow details in Sect. 3. In experiments, we observed that increasing the dimensionality of the HOG descriptor does not improve the recognition performance any further. Indeed, increasing the dimensions of the HOG descriptor means that we are keen at finding the specific gradient details in the image. As a result, the descriptor becomes sensitive to even small changes due to variable lighting or dynamic objects. By contrast, a low dimensional descriptor will capture the coarse gradient information in the image. One should, however, be careful as a very low dimensionality will degenerate place recognition performance. In the present setup, we have utilized the default parameter configurations to evaluate the performance of the HOG descriptor. Hence, it could also be the reason why HOG did not show superior performance. Optimizing the parameter configurations (in particular, error tolerance and spatial resolution [17, Tab. I]) for the HOG-based feature space fall beyond the scope of this work and we consider it as a future direction.

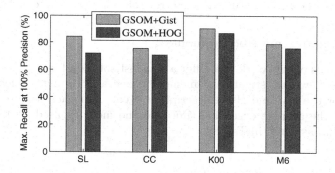

Fig. 2. Maximum recall achieved at 100% precision for the gist and HOG features.

Table 1. Highest recall rate achieved by various methods at 100% precision

Dataset	FAB-MAP	SeqSLAM+G	iBoW-LCD	GSOM+G	GSOM+H
SL	53.98	63.29	41.32	**84.22**	71.93
CC	40.11	75.12	75.4	**75.58**	70.77
K00	61.22	78.33	76.50	**90.39**	86.95
M6	21.83	35.58	46.36	**79.15**	76.06

Next, we compare the gist- and HOG-based variants of the GSOM, referred as GSOM+G and GSOM+H, respectively, with the well-known place recognition methods, namely FAB-MAP 2.0 [11], SeqSLAM [26] with gist (SeqSLAM+G), and iBoW-LCD [1]. Amongst the selected approaches, the FAB-MAP method needs offline vocabulary training. Whereas, SeqSLAM and iBoW-LCD are online approaches that learn the representation of the environment on-the-fly. For evaluation of the baseline methods, the parameter settings described in the respective works are utilized. For the FAB-MAP, we adjusted the "false positive probability" in the interval [0.01, 0.07] with regard to the dataset; otherwise, the algorithm never showed 100% precision. In the same vein, the sequence length parameter of SeqSLAM was set to 20 to obtain high recall rate at 100% precision. Table 1 enlists the maximum recalls achieved by the selected place recognition methods in comparison to current work (at 100% precision). It can clearly be observed that the variants of the GSOM approach outperform the baseline approaches on all the datasets, except for the CC, in which case the HOG-based variant of GSOM (GSOM+H) showed comparable performance to other approaches. These results demonstrate that our algorithm offers feature-agnostic solution to place recognition and thus can be extended to other feature spaces, without parameter tuning. However, one must take care of the scale of each dimension of the feature space so that the numerical values along each dimension do not exceed unity. Otherwise, parameters have to be optimized, as for a different scale, the current value of the error tolerance parameter is not useful.

4.2   Space and Time Requirements

One of the goals of this work is to offer a low-cost solution for place recognition. In this respect, we investigate the memory usage and timing efficiency of the algorithm for the gist and HOG descriptors. Hence, for each feature modality, we recorded the average execution time and the memory used by the learned representation, as listed in Table 2.

Table 2. Gist vs. HOG: Memory requirements and timing efficiency of the algorithm

Dataset	Network size[‡]		Mean exec. time[†]	
	Gist	HOG	Gist	HOG
CC	0.85	0.71	95.4	29.91
K00	1.96	1.74	96.9	30.15
M6	2.73	1.4	95.6	30.0
SL	5.44	6.01	99.1	33.45

[‡]Network size in physical memory in megabytes (MB).
[†]Execution time is measured in milliseconds (ms).

Note that on small datasets, the HOG-based GSOM variant occupies less memory compared to gist. However, for SL (longest route), HOG-based representation learned by the network reserves more physical memory. We observed that this behavior is the consequence of creation of more neurons in case of HOG descriptor. On the other hand, one can observe that the mean execution time for the HOG (including the pre-processing and feature extraction) is only 33.45 ms on the SL sequence, i.e., 3 times faster than gist. Here, it is worth mentioning that most of the computation overhead belong to the feature extraction time in either cases. However, for longer sequences, these differences in algorithm's execution time tend to be indifferent, as the network is inclined to produce more neurons for HOG compared to gist features. Due to the use of default parameter settings, we believe that parameters can be further optimized to overcome this challenge.

5   Conclusion

In this paper, we studied the existing growing self-organizing network and evaluated its performance for various feature modalities, i.e., gist and HOG features. To this end, the descriptor of a location is obtained as a weighted average of the description of the current location and its neighbors. Given the trained network and the query image, the decision of place recognition is taken based-on expected network activity. The experiments demonstrate the strength of our algorithm to learn other feature spaces without the need of parameter tuning. However, to obtain the best performance, it is preferable to learn different settings for spatial resolution and error tolerance parameters (see [12]). Hence, in future we look for the parameter optimization methods to automatically learn the optimal settings for the given feature space.

References

1. Garcia-Fidalgo, E., Ortiz, A.: iBoW-LCD: an appearance-based loop-closure detection approach using incremental bags of binary words. IEEE Robot. Autom. Lett. **3**(4), 3051–3057 (2018)
2. Zhang, G., Lilly, M.J., Vela, P.A.: Learning binary features online from motion dynamics for incremental loop-closure detection and place recognition. In: IEEE International Conference on Robotics and Automation, pp. 765–772 (2016)
3. Nicosevici, T., Garcia, R.: Automatic visual bag-of-words for online robot navigation and mapping. IEEE Trans. Robot. **28**(4), 886–898 (2012)
4. Kawewong, A., Tongprasit, N., Tangruamsub, S., Hasegawa, O.: Online and incremental appearance-based SLAM in highly dynamic environments. Int. J. Robot. Res. **30**(1), 33–55 (2011)
5. McManus, C., Upcroft, B., Newman, P.: Learning place-dependant features for long-term vision-based localisation. Auton. Rob. **39**(3), 363–387 (2015)

6. Lategahn, H., Beck, J., Kitt, B., Stiller, C.: How to learn an illumination robust image feature for place recognition. In: IEEE Intelligent Vehicles Symposium (IV), pp. 285–291 (2013)
7. Zhou, B., Lapedriza, A., Xiao, J., Torralba, A., Oliva, A.: Learning deep features for scene recognition using places database. In: Advances in Neural Information Processing Systems, pp. 487–495 (2014)
8. Chen, Z., Lam, O., Jacobson, A., Milford, M.: Convolutional neural network-based place recognition. arXiv preprint arXiv:1411.1509 (2014)
9. Krizhevsky, A., Sutskever, I., Hinton, G.E.: ImageNet classification with deep convolutional neural networks. In: Advances in Neural Information Processing Systems, pp. 1097–1105 (2012)
10. Sünderhauf, N., Dayoub, F., Shirazi, S., Upcroft, B., Milford, M.: On the performance of convnet features for place recognition. arXiv preprint arXiv:1501.04158 (2015)
11. Cummins, M., Newman, P.: Appearance-only SLAM at large scale with FAB-MAP 2.0. Int. J. Robot. Res. **30**(9), 1100–1123 (2011)
12. Kazmi, S.A.M., Mertsching, B.: Detecting the expectancy of a place using nearby context for appearance-based mapping. IEEE Trans. Robot. (2019, accepted)
13. Dalal, N., Triggs, B.: Histograms of oriented gradients for human detection. In: IEEE International Conference on Computer Vision & Pattern Recognition, vol. 1, pp. 886–893 (2005)
14. Wyeth, G., Milford, M.: Spatial cognition for robots. IEEE Robot. Autom. Mag. **16**(3), 24–32 (2009)
15. Lowry, S., et al.: Visual place recognition: a survey. IEEE Trans. Robot. **32**(1), 1–19 (2015)
16. Kazmi, S.M., Mertsching, B.: Gist+RatSLAM: an incremental bio-inspired place recognition front-end for RatSLAM. In: Proceedings of the 9th EAI International Conference on Bio-inspired Information and Communications Technologies (formerly BIONETICS), pp. 27–34 (2016)
17. Kazmi, S.A.M., Mertsching, B.: Simultaneous place learning and recognition for real-time appearance-based mapping. In: IEEE/RSJ International Conference on Intelligent Robots and Systems, pp. 4898–4903 (2016)
18. Turcot, P., Lowe, D.G.: Better matching with fewer features: the selection of useful features in large database recognition problems. In: ICCV Workshop on Emergent Issues in Large Amounts of Visual Data, vol. 4 (2009)
19. Oliva, A., Torralba, A.: Modeling the shape of the scene: a holistic representation of the spatial envelope. Int. J. Comput. Vis. **42**(3), 145–175 (2001)
20. Rashwan, H.A., Mohamed, M.A., García, M.A., Mertsching, B., Puig, D.: Illumination robust optical flow model based on histogram of oriented gradients. In: Weickert, J., Hein, M., Schiele, B. (eds.) GCPR 2013. LNCS, vol. 8142, pp. 354–363. Springer, Heidelberg (2013). https://doi.org/10.1007/978-3-642-40602-7_38
21. Cummins, M., Newman, P.: FAB-MAP: probabilistic localization and mapping in the space of appearance. Int. J. Robot. Res. **27**(6), 647–665 (2008)
22. Geiger, A., Lenz, P., Urtasun, R.: Are we ready for autonomous driving? The KITTI vision benchmark suite. In: IEEE International Conference on Computer Vision and Pattern Recognition, pp. 3354–3361 (2012)
23. Blanco, J.L., Moreno, F.A., Gonzalez, J.: A collection of outdoor robotic datasets with centimeter-accuracy ground truth. Auton. Robots **27**(4), 327 (2009)

24. Glover, A.J., Maddern, W.P., Milford, M.J., Wyeth, G.F.: FAB-MAP+ RatSLAM: appearance-based SLAM for multiple times of day. In: IEEE International Conference on Robotics and Automation, pp. 3507–3512 (2010)
25. Arroyo, R., Alcantarilla, P.F., Bergasa, L.M., Yebes, J.J., Bronte, S.: Fast and effective visual place recognition using binary codes and disparity information. In: IEEE/RSJ International Conference on Intelligent Robots and Systems, pp. 3089–3094 (2014)
26. Milford, M.J., Wyeth, G.F.: SeqSLAM: visual route-based navigation for sunny summer days and stormy winter nights. In: IEEE International Conference on Robotics and Automation, pp. 1643–1649 (2012)

Robotic Vision

Semi-semantic Line-Cluster Assisted Monocular SLAM for Indoor Environments

Ting Sun[1]([✉]), Dezhen Song[2], Dit-Yan Yeung[3], and Ming Liu[1]

[1] Department of Electronic and Computer Engineering,
Hong Kong University of Science and Technology, Hong Kong, China
{tsun,eelium}@ust.hk

[2] Department of Computer Science and Engineering,
Texas A&M University, College Station, TX, USA
dzsong@cse.tamu.edu

[3] Department of Computer Science,
Hong Kong University of Science and Technology, Hong Kong, China
dyyeung@cse.ust.hk

Abstract. This paper presents a novel method to reduce the scale drift for indoor monocular simultaneous localization and mapping (SLAM). We leverage the prior knowledge that in the indoor environment, the line segments form tight clusters, e.g. many door frames in a straight corridor are of the same shape, size and orientation, so the same edges of these door frames form a tight line segment cluster. We implement our method in the popular ORB-SLAM2, which also serves as our baseline. In the front end we detect the line segments in each frame and incrementally cluster them in the 3D space. In the back end, we optimize the map imposing the constraint that the line segments of the same cluster should be the same. Experimental results show that our proposed method successfully reduces the scale drift for indoor monocular SLAM.

Keywords: SLAM · Monocular · Indoor · Line-segment · Clustering

1 Introduction

Visual simultaneous localization and mapping (vSLAM) is widely adopted by robots to explore indoor environments where a GPS signal is unavailable. There are three dominant types of vSLAM, which are classified according to the sensors used: monocular, stereo and RGB-D. Compared with stereo and RGB-D, monocular vSLAM is the most hardware-economical and is free of the trouble of calibration and synchronization. However, monocular vSLAM suffers from scale drift [21]. Without any extra sensor, currently this drift is corrected by loop closure with each camera pose represented by 7 DoF (degree of freedom) rather than 6 DoF in the graph optimization [6,21]. There are three major limitations of purely relying on loop closure: (1) the robot may never return to the same

© Springer Nature Switzerland AG 2019
D. Tzovaras et al. (Eds.): ICVS 2019, LNCS 11754, pp. 63–74, 2019.
https://doi.org/10.1007/978-3-030-34995-0_6

place; (2) the same place may have a different appearance due to different lighting conditions, viewing angles, occlusion caused by moving objects etc., and (3) some places simply cannot be distinguished merely by appearance, e.g. all the segments of a long corridor could have an identical appearance.

In this paper we present a novel method to reduce the scale drift for indoor monocular SLAM by leveraging two key observations about the indoor environment and the property of scale drift:

1. Buildings are carefully crafted by humans and are full of standardized elements, e.g. all the door frames in a straight corridor are of the same shape, size and orientation; all the bricks tessellating the floor are also the same; etc. Most of these elements appear very frequently and their edges form tight line segment clusters.
2. Scale drift happens gradually, i.e. within the distance of adjacent doors, a robot can still correctly identify the edges of the two doors are of the same length, and this gives the robot a chance to correct the small drift before it accumulates.

Our method identifies and builds line segment clusters incrementally in the front end, and optimizes the map in the back end by imposing the constraints that the segments of the same cluster should be the same. We implement our method in ORB-SLAM2 [16,17], and the experimental results show that our proposed method successfully reduces the scale drift for indoor monocular SLAM and the whole system runs in real time. We describe our method as 'semi-semantic' since it does NOT fall into the type of using line feature for matching in odometry as many others did, but leverages the observation of 'how the line segments appear indoors' and impose global constraints in a SLAM system, yet this prior knowledge is not as 'high level' and 'semantic' as commonly mentioned 'reappear object' and 'scene understanding'.

There are 4 appealing properties of our proposed method:

1. Without requiring extra sensors, the scale regulation is explored from frequently appearing standardized indoor elements, i.e. line segments of the same length and orientation.
2. It does not need pre-training nor access to a model database during running.
3. It tightly integrates to the ORB-SLAM2 [16,17] system to achieve the most benefit with the least computation, i.e. it takes advantage of the key points detected by ORB-SLAM2 [16,17] to filter the line segments' end points, then directly uses the 3D map points to calculate the 3D representation of the line segments.
4. Compared with approaches using object recognition etc., our method makes a relatively weaker assumption and thus is very generalizable in indoor environments.

Notice that our method is not a replacement of loop closure for the following reasons: (1) they are designed for different application scenarios, i.e. loop closure is for when the robot revisits the same place, while our method is for indoor environments; (2) though both impose regulation in the optimization graph, loop closure only takes effect during revisiting, while our method keeps regulating the SLAM system on its way; (3) their implementations have no conflict nor dependency. Loop closure is a standard and mature module in SLAM system, it can eliminate the drift once successful. Our method is a novel approach that leverages the regularity of indoor line segments to reduce drift, and its current implementation can not completely remove the scale drift by itself.

The remainder of this paper is organized as follows. Sect. 2 reviews previous related works. Our proposed method is presented in Sect. 3, which is then followed by experimental results in Sect. 4. Section 5 concludes this paper.

2   Related Work

The scale drift in monocular vSLAM system is mainly handled by two approaches: fusion of other sensors and revisit known places. In [18], two methods, i.e. spline fitting and multi rate Extended Kalman Filter(EKF) are proposed to estimate the scale from an Inertial Measurement Unit (IMU). A more common approach relies on loop closure with each camera pose represented by 7 DoF rather than 6 DoF in the graph optimization [6, 21]. Our method leverages the regularity of indoor environment to reduce scale drift without using extra sensors nor relying on revisiting known places.

Some attempts have been made to integrate object recognition results into the SLAM system [2, 7, 8, 24], so the scale information can be obtained from known objects. These methods require training a detector or classifier ahead, and assume that particular objects can be encountered by the robot, which does not usually occur in a real application scenario. Our method does not need pre-training, nor access to a model database during running. The only assumption made in our algorithm is that the line segments in an indoor environment are quantized and form tight clusters, which is usually true in a modern building.

The specialty of indoor environments has been explored in SLAM related research for over a decade [1, 19]. At 'low feature level', the abundant line features in buildings are commonly used [12–14, 20, 26–29] because they are more robust and contain more structure information than key points. In contrast to using lines for feature matching and structure representation, we observe that not only the appearance of line segments, but the regularity of how they appear can be used to assist SLAM. Specifically, the frequent existence of line segments of the same length and orientation gives the robot a hint to regulate its map built.

Highly semantic knowledge of the indoor environment are mainly studied for two purposes: using semantic knowledge to assist SLAM, or vice-versa. A common related research topic is semantic mapping [4, 9, 10, 22], which targets identifying and recording the meaningful signs in human-inhabited areas. These approaches take in the map built by the SLAM system together with other inputs, then generate a semantic map that is an enhanced representation of the environment with labels understandable by humans [3, 10]. The highly semantic information, like door numbers and place types (i.e. living room, corridor etc.), are convenient for human-instructed navigation, but it is not easy to use them to benefit SLAM. It is difficult to integrate information like 'scene type' into the formulation of a SLAM system, and practical issues like camera's view point, resolution, stability can hardly guarantee reliable recognition of door number etc. Our method is based on the assumption that standardized line segments of the same length and orientation appear frequently indoor, so that the line segments observed by the robot form tight clusters. This knowledge is more abstract and impose more global constraints than local line features, yet it is not too highly semantic to formulate in the graph optimization. The performance of proposed method depends on how frequent the line segments within the same cluster appear, whether there is occlusion etc. One cluster may not be able to regulate the whole trajectory, but as long as its member line segments are detected, our method can help to reduce the scale drift to some extent.

3   Proposed Method

As mentioned previously, the key observation and the assumption made in our method is that in the indoor environment there exists many quantized line segments that form tight clusters. By leveraging these line segment clusters the scale drift in monocular SLAM can be reduced. Our implementation is based on ORB-SLAM2 [16,17]. In the tracking thread, our method detects line segments in each frame, and builds clusters agglomeratively. In the local mapping thread, after local bundle adjustment, we construct another optimization graph, which contains the map points corresponding to the detected line segments' ends, and imposes the constraint that the segments within one cluster have to be the same. The updated ORB-SLAM2 [16,17] system is shown in Fig. 1 (better viewed in color), where the modules added by our method are shown in light orange, and the details of the 'loop closing' thread and place recognition are omitted. The original ORB-SLAM2 [16,17] maintains two things: the place recognition database, and the map which includes map points and key frames. With our method equipped, we also need to maintain the line segment clusters. (The information stored for a cluster is shown in the left part of Fig. 3.)

The proposed method consists of three main steps : line segment detection, building clusters, and graph optimization to update the map. The details of each step are given in the following three subsections.

3.1   Line Segment Detection

We adopt the LSD (Line Segment Detector) [25] in our algorithm. The outputs of the LSD are filtered in the image domain. The line segments that are either too short (i.e. shorter than 1000 in the image coordinate system) or are truncated by the frame boundary are omitted first. A problem for the LSD is that it cannot detect the end points of a line segment precisely, as shown in the first row of images in Fig. 2. By taking advantage of ORB-SLAM2 [16,17], we select the line segments whose end points coincide with the detected key points. The second

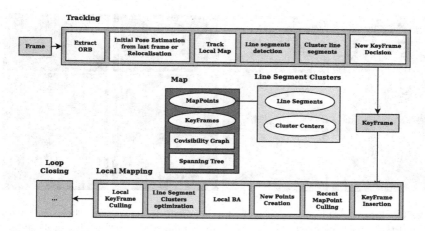

Fig. 1. The updated ORB-SLAM2 [16,17] system with the proposed method integrated. The'loop closing' thread and place recognition are omitted. The modules added by our method are shown in light orange. In the front end, we detect and cluster the line segments in each frame incrementally. In the back end, we optimize the map imposing the constraint of clustering information. The information maintained by the system is shown in the center.

row in Fig. 2 shows the corresponding frames with key points marked, and the last row shows the final line segment detection results in image space, which are then identified and clustered in 3D space (detailed in the next subsection). Notice that our method does not rely on all the line segments being detected in a frame, nor one line segment being detected in all the frames it appears. In order for a line segment to contribute in the constraints, it only needs to be detected once in the video sequence it appears.

3.2 Building Line Segment Clusters

Building line segment clusters is conducted in the 3D space that is constructed by ORB-SLAM2 [16,17] when the map is initialized. Since we select the line segments whose end points coincide with the detected key points, and ORB-SLAM2 [16,17] already handles the projection from 2D to 3D for all the key points, given the ith line segment in image space, we obtain its 3D end points $\mathbf{p}_1^i$, $\mathbf{p}_2^i$ for free. In our method, a line segment is represented by a 3D vector $\mathbf{v}_{seg}^i$, which is the difference between its end points: $\mathbf{v}_{seg}^i = \mathbf{p}_2^i - \mathbf{p}_1^i$. The center of jth cluster $\mathcal{C}^j$ is the mean vector $\mathbf{v}_c^j$ of its elements:

$$\mathbf{v}_c^j = \frac{1}{|\mathcal{C}^j|} \sum_{\mathbf{v}_{seg}^i \in \mathcal{C}^j} \mathbf{v}_{seg}^i \qquad (1)$$

where $|\mathcal{C}^j|$ is the cardinality or number of elements of cluster $\mathcal{C}^j$. $\mathbf{v}_c^j$ is incrementally updated when a new element is added. When we are to decide if

Fig. 2. The first row of images shows the line segment detection with those that are too short and truncated removed. The second row of images shows the key points detected by ORB-SLAM. The third row of images shows our final line segment detection results.

a newly detected line segment $\mathbf{v}^i_{seg}$ belong to some existing cluster $\mathcal{C}^j$, the criterion is whether the Euclidean norm of the difference between this vector and the cluster center is within 0.5% of the length of the cluster center: $||\mathbf{v}^i_{seg} - \mathbf{v}^j_c|| < 0.005 * ||\mathbf{v}^j_c||$. Since there is no guarantee that the angle between $\mathbf{v}^i_{seg}$ and $\mathbf{v}^j_c$ is smaller than 90°, both $||\mathbf{v}^i_{seg} - \mathbf{v}^j_c||$ and $||\mathbf{v}^i_{seg} + \mathbf{v}^j_c||$ are calculated and the minimum distance is considered. If $\mathbf{v}^i_{seg}$ is decided to belong to $\mathcal{C}^j$, its end points $\mathbf{p}^i_1, \mathbf{p}^i_2$ are stored in $\mathcal{C}^j$, and $\mathbf{v}^j_c$ is updated. If no existing cluster is for $\mathbf{v}^i_{seg}$, a new cluster is created for it with itself being the cluster center. The Euclidean distance is directly used to determine the clustering without countermeasure for the drift problem, because the assumption is that the drift happens gradually, while the standardized line segments appear frequently. Hopefully the elements of the same cluster are identified frequently enough to correct slight drift before a large error accumulates.

3.3   Graph Optimization of Line Segment Clusters

The clustering information stored is our semi-semantic knowledge, which propagates to the SLAM system by updating the map points involved. ORB-SLAM2 [17] extracts a part of g2o [11] to conduct bundle adjustment. We replace it with the full version of g2o [11] and define a new type of edge connecting two end points (map points) of a line segment and its cluster center. The error vector is the difference between the line segment and its cluster center. Figure 3 shows the information stored for cluster i and the optimization graph constructed for it. The error vector of edge E^{ij} is

$$\mathbf{e}^{ij} = \mathbf{v}^i_c - (\mathbf{p}^{ij}_2 - \mathbf{p}^{ij}_1) \tag{2}$$

and the overall objective function is

$$\underset{\mathbf{p}_1^{ij},\mathbf{p}_2^{ij}}{\operatorname{argmin}} \sum_i \sum_j (\mathbf{e}^{ij})^T \mathbf{e}^{ij}. \tag{3}$$

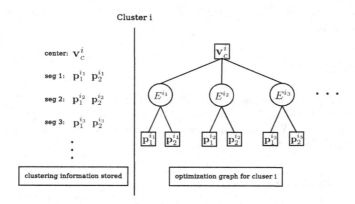

Fig. 3. The left part of this figure shows the information stored for cluster i, and the right part shows the graph constructed for this cluster.

We claim that the above objective function helps to reduce the scale drift, impose limited constraint on orientation and have no effect on translational error. The analysis is as follows.

Suppose $\hat{\mathbf{p}}_1^{ij}$, $\hat{\mathbf{p}}_2^{ij}$ are the map points constructed by the SLAM, and relate to their ground truth position $\mathbf{p}_1^{ij}$, $\mathbf{p}_2^{ij}$ by

$$\hat{\mathbf{p}}_1^{ij} = s * R * \mathbf{p}_1^{ij} + T \tag{4}$$

$$\hat{\mathbf{p}}_2^{ij} = s * R * \mathbf{p}_2^{ij} + T. \tag{5}$$

The same rotation R, translation t and scaler s are used in both (4) and (5) since they are observed in the same frame, i.e. related to one camera pose. From (2) it can be seen that the line segment clustering constraint applies to the difference between two map points:

$$\begin{aligned} \mathbf{v}_c^i &= \hat{\mathbf{p}}_2^{ij} - \hat{\mathbf{p}}_1^{ij} \\ &= s * R * (\mathbf{p}_2^{ij} - \mathbf{p}_1^{ij}). \end{aligned} \tag{6}$$

The translation T is eliminated. The rotation R remains but a vector is invariant by rotation with respect to it. Unfortunately this is a common case in a corridor, e.g. many door frame edges are vertical line segments which are parallel to the axis of rotation of the corridor turning. This explains why our proposed method is particularly helpful in reducing scale drift.

After the optimization, all the involved map points (i.e. all the **p**s) are updated according to the clustering constraint. We conduct our clustering graph optimization after local bundle adjustment, so the map is optimized alternatively to minimize re-projection error and the clusters' variance. We separate these two graph optimizations for the sake of implementation simplicity, since the map points involved in the clustering graph may be out of the concern of local bundle adjustment, and adding them could ruin the sparse structure of the graph for local bundle adjustment.

4   Experiments

The effectiveness of our method is shown by comparing of the trajectories built by a monocular SLAM system with and without the assist of the proposed line-cluster optimization. We pick the test sequences that are taken in a typical corridor without loop closure.

4.1   Implementation

Our implementation is based on ORB-SLAM2 [17], which also serves as our baseline. With the proposed method integrated, the augmented SLAM system still runs in real time. The results of the proposed method are presented with 2 different configurations named *Seg* and *SegGlobal*. In *Seg*, local bundle adjustment and clustering optimization are conducted alternatively, while in *SegGlobal*, additional global bundle adjustment is conducted after each clustering optimization, helping to propagate the semi-semantic knowledge to the whole map built.

We adopt the widely used absolute trajectory error (ATE) [23] and relative pose error (RPE) [23] as our numeric measurement for location only. The obtained camera pose trajectories are aligned to the ground truth by finding the transformation that minimizes their sum of squared error.

4.2   Datasets

Finding the proper test sequences turns out to be challenging, and there are two main problems: (1) missing ground truth and (2) the quality of the video content. For the trajectories of very small scale, i.e. within one room, the ground truth can be collected by optical tracking system or markers deployed in the room; for the outdoor environments, GPS is available; while the corridors are neither easy to deploy a system nor have GPS, but that is where vSLAM is most needed. As for the content, we want to extract a long sequence without loop closure, and the monocular mode ORB-SLAM2 [17] will not lose tracking during it. Some datasets contain sharp turnings in front of blank walls, causing the SLAM system losing tracking frequently. We show our experimental results on two datasets: HRBB4 [15] and IT3F [26].

HRBB4 dataset [15] contains an image sequence of 12,000 frames of 640×360 pixels captured from the 4th floor of the H.R. Bright Building, Texas A&M

University using a Nikon 5100 camera. Some sample images are shown in Fig. 4a. The ground truth camera positions (no orientation) of a subset of frames are offered, and we interpolate both ground truth and the SLAM results by spline interpolation [5].

IT3F dataset [26] is collected from the 3rd floor of the IT building at Hanyang University, with the dimensions 24×11.5 m. A calibrated Bumblebee BB2-08S2C-38 is used as the vision sensor, and all the images are undistorted and rectified. Some sample images are shown in Fig. 4b. This dataset is proposed for stereo SLAM, and it contains multiple loop closures but no ground truth is offered. The stereo results from ORB-SLAM2 [16,17] are used as ground truth in our experiments, and we conduct monocular SLAM using an extracted sequence captured by the left camera.

      (a) HRBB4 dataset [15]   (b) IT3F dataset [26]

Fig. 4. The sample images from HRBB4 dataset [15] and IT3F dataset [26].

4.3 Results

Results on HRBB4 Dataset. We use the whole sequence of the HRBB4 dataset [15] which does not contain loop closure and the monocular ORB-SLAM2 [17] can run through the whole sequence without losing tracking. The aligned trajectories are shown in Fig. 5a, and the corresponding ATE and RPE results are shown in Table 1a. It can be seen that our proposed method with configuration *SegGlobal* achieves the best performance, far exceeding our baseline, and with configuration *Seg*, our method still improves the baseline.

Table 1. The ATE and PRE results. The best results achieved are highlighted in boldface.

	ORB-SLAM2	Seg	SegGlobal		ORB-SLAM2	Seg	SegGlobal
ATE	1.9879	1.8248	**1.0062**	ATE	3.5841	**2.5380**	2.9918
PRE	3.8383	3.5086	**1.9801**	PRE	6.0966	**4.4054**	5.1996

      (a) HRBB4 dataset [15]   (b) IT3F dataset [26]

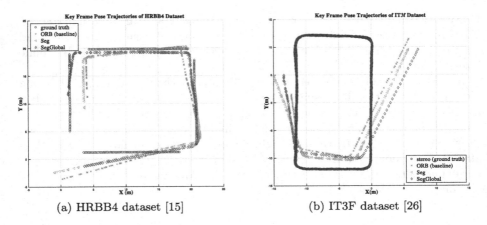

(a) HRBB4 dataset [15]   (b) IT3F dataset [26]

Fig. 5. The aligned trajectories of HRBB4 dataset [15] and IT3F dataset [26].

Results on IT3F Dataset. The aligned trajectories are shown in Fig. 5b, and the corresponding ATE and RPE results are shown in Table 1b. We found that monocular SLAM is incapable of keeping tracking of a complete circle (but stereo SLAM can), which is why there is a missing part in all the results. (Tracking is lost after turning in front of a blank wall.) We extract the longest sequence (about 4600 frames) that can be followed by monocular ORB-SLAM2 [17] to show our results. It can be seen from Fig. 5b that with or without our method equipped, the orientation errors accumulate with each turning alike, but Table 1b shows that the results with our method are closer to the ground truth. According to the analysis in Subsect. 3.3, this improvement comes from the reduced scale drift.

Computational Time. We run the monocular ORB-SLAM2 [17] with the proposed method integrated on a desktop with Intel® Core™ i5-4570 CPU @ 3.20 GHz x 4 processors. For input image of size 480 × 640, our method costs about 0.031 s per frame in the front end, i.e. the two orange modules in tracking thread in Fig. 1, and about 0.012 s per round optimization (10 iterations) in the back end, i.e. the orange module in local mapping thread in Fig. 1. For your information, the local bundle adjustment of ORB-SLAM2 [17] cost about 0.21 s per round (15 iterations), while the time cost by global bundle adjustment increases as the map grows.

5   Conclusions

In this paper we proposed a novel method to reduce the scale drift for indoor monocular SLAM. Our method assists the SLAM system by leveraging the regularity of the abundant standardized line segments in the indoor environment. We observed that the indoor line segments are quantized and form tight clusters, which can be used to regulate the map built by SLAM. Our method makes a very weak assumption and does not need pre-training nor to access model database while running, and thus is very generalizable. We implemented the proposed method in the popular ORB-SLAM2 [17] and took advantage of its results to save computation. In the front end we detect the line segments in each frame and incrementally cluster them in the 3D space. In the back end, we optimize the map imposing the constraint by our newly defined edges. The performance of our method depends on how frequent the line segments within the same cluster appear, whether there is occlusion etc. Experimental results showed that our proposed method successfully reduces the scale drift for indoor monocular SLAM.

References

1. Beevers, K.R., Huang, W.H.: Inferring and enforcing relative constraints in SLAM. In: Akella, S., Amato, N.M., Huang, W.H., Mishra, B. (eds.) Algorithmic Foundation of Robotics VII. STAR, vol. 47, pp. 139–154. Springer, Heidelberg (2008). https://doi.org/10.1007/978-3-540-68405-3_9
2. Botterill, T., Mills, S., Green, R.: Correcting scale drift by object recognition in single-camera SLAM. IEEE Trans. Cybern. **43**(6), 1767–1780 (2013)
3. Case, C., Suresh, B., Coates, A., Ng, A.Y.: Autonomous sign reading for semantic mapping. In: ICRA, pp. 3297–3303. IEEE (2011)
4. Civera, J., Gálvez-López, D., Riazuelo, L., Tardós, J.D., Montiel, J.: Towards semantic SLAM using a monocular camera. In: IROS, pp. 1277–1284. IEEE (2011)
5. De Boor, C., De Boor, C., Mathématicien, E.U., De Boor, C., De Boor, C.: A Practical Guide to Splines, vol. 27. Springer, New York (1978)
6. Engel, J., Schöps, T., Cremers, D.: LSD-SLAM: large-scale direct monocular SLAM. In: Fleet, D., Pajdla, T., Schiele, B., Tuytelaars, T. (eds.) ECCV 2014. LNCS, vol. 8690, pp. 834–849. Springer, Cham (2014). https://doi.org/10.1007/978-3-319-10605-2_54
7. Fioraio, N., Di Stefano, L.: Joint detection, tracking and mapping by semantic bundle adjustment. In: CVPR, pp. 1538–1545. IEEE (2013)
8. Gálvez-López, D., Salas, M., Tardós, J.D., Montiel, J.: Real-time monocular object SLAM. Robot. Auton. Syst. **75**, 435–449 (2016)
9. Kostavelis, I., Charalampous, K., Gasteratos, A., Tsotsos, J.K.: Robot navigation via spatial and temporal coherent semantic maps. Eng. Appl. Artif. Intell. **48**, 173–187 (2016)
10. Kostavelis, I., Gasteratos, A.: Semantic mapping for mobile robotics tasks: a survey. Robot. Auton. Syst. **66**, 86–103 (2015)
11. Kümmerle, R., Grisetti, G., Strasdat, H., Konolige, K., Burgard, W.: G2O: a general framework for graph optimization. In: ICRA, pp. 3607–3613. IEEE (2011)
12. Lemaire, T., Lacroix, S.: Monocular-vision based SLAM using line segments. In: ICRA, pp. 2791–2796. IEEE (2007)
13. Lu, Y., Song, D.: Robust RGB-D odometry using point and line features. In: ICCV, pp. 3934–3942 (2015)
14. Lu, Y., Song, D.: Visual navigation using heterogeneous landmarks and unsupervised geometric constraints. IEEE Trans. Robot. **31**(3), 736–749 (2015)

15. Lu, Y., Song, D., Yi, J.: High level landmark-based visual navigation using unsupervised geometric constraints in local bundle adjustment. In: ICRA, pp. 1540–1545. IEEE (2014)
16. Mur-Artal, R., Montiel, J.M.M., Tardos, J.D.: ORB-SLAM: a versatile and accurate monocular SLAM system. IEEE Trans. Robot. **31**(5), 1147–1163 (2015)
17. Mur-Artal, R., Tardós, J.D.: ORB-SLAM2: an open-source SLAM system for monocular, stereo, and RGB-D cameras. IEEE Trans. Robot. **33**(5), 1255–1262 (2017)
18. Nützi, G., Weiss, S., Scaramuzza, D., Siegwart, R.: Fusion of IMU and vision for absolute scale estimation in monocular SLAM. J. Intell. Robot. Syst. **61**(1–4), 287–299 (2011)
19. Parsley, M.P., Julier, S.J.: Towards the exploitation of prior information in SLAM. In: IROS, pp. 2991–2996. IEEE (2010)
20. Pumarola, A., Vakhitov, A., Agudo, A., Sanfeliu, A., Moreno-Noguer, F.: PL-SLAM: real-time monocular visual SLAM with points and lines. In: ICRA, pp. 4503–4508. IEEE (2017)
21. Strasdat, H., Montiel, J., Davison, A.J.: Scale drift-aware large scale monocular SLAM. Robot.: Sci. Syst. VI **2** (2010)
22. Stückler, J., Biresev, N., Behnke, S.: Semantic mapping using object-class segmentation of RGB-D images. In: IROS, pp. 3005–3010. IEEE (2012)
23. Sturm, J., Engelhard, N., Endres, F., Burgard, W., Cremers, D.: A benchmark for the evaluation of RGB-D SLAM systems. In: IROS, pp. 573–580. IEEE (2012)
24. Tomono, M., Yuta, S.: Mobile robot navigation in indoor environments using object and character recognition. In: ICRA, vol. 1, pp. 313–320. IEEE (2000)
25. Von Gioi, R.G., Jakubowicz, J., Morel, J.M., Randall, G.: LSD: a fast line segment detector with a false detection control. TPAMI **32**(4), 722–732 (2010)
26. Zhang, G., Lee, J.H., Lim, J., Suh, I.H.: Building a 3-D line-based map using stereo SLAM. IEEE Trans. Robot. **31**(6), 1364–1377 (2015)
27. Zhang, G., Suh, I.H.: A vertical and floor line-based monocular SLAM system for corridor environments. IJCAS **10**(3), 547–557 (2012)
28. Zhang, J., Song, D.: On the error analysis of vertical line pair-based monocular visual odometry in urban area. In: IROS, pp. 3486–3491. IEEE (2009)
29. Zhang, J., Song, D.: Error aware monocular visual odometry using vertical line pairs for small robots in urban areas. In: AAAI (2010)

Appearance-Based Loop Closure Detection with Scale-Restrictive Visual Features

Konstantinos A. Tsintotas$^{(\boxtimes)}$ (ID), Panagiotis Giannis (ID), Loukas Bampis (ID), and Antonios Gasteratos (ID)

Laboratory of Robotics and Automation, School of Engineering,
Department of Production and Management Engineering,
Democritus University of Thrace, 67132 Xanthi, Greece
{ktsintot,panagian8,lbampis,agaster}@pme.duth.gr
https://robotics.pme.duth.gr/robotics/

Abstract. In this paper, an appearance-based loop closure detection pipeline for autonomous robots is presented. Our method uses scale-restrictive visual features for image representation with a view to reduce the computational cost. In order to achieve this, a training process is performed, where a feature matching technique indicates the features' repeatability with respect to scale. Votes are distributed into the database through a nearest neighbor method, while a binomial probability function is responsible for the selection of the most suitable loop closing pair. Subsequently, a geometrical consistency check on the chosen pair follows. The method is subjected into an extensive evaluation via a variety of outdoor, publicly-available datasets revealing high recall rates for 100% precision, as compared against its baseline version, as well as, other state-of-the-art approaches.

Keywords: Localization · Mapping · Visual-based navigation · Mobile robots

1 Introduction and Literature Review

Nowadays, a major research focus is devoted in robots' map formulation techniques via the utilization of several exteroceptive sensors, such as laser scanners, cameras, odometry and inertial measurement units [4,13,20,27]. An autonomous system in an unknown environment needs to construct a map of the working area in order to perform various tasks, such as path planning, exploration, and collision avoidance. Simultaneous Localization And Mapping (SLAM) is the process

This research has been co-financed by the European Union and Greek national funds through the Operational Program Competitiveness, Entrepreneurship and Innovation, under the call RESEARCH – CREATE – INNOVATE (project code: T1EDK-00737). The paper was partially supported by project ETAA, DUTH Research Committee 81328.

© Springer Nature Switzerland AG 2019
D. Tzovaras et al. (Eds.): ICVS 2019, LNCS 11754, pp. 75–87, 2019.
https://doi.org/10.1007/978-3-030-34995-0_7

where the robot concurrently constructs a model of the environment (the map), while at the same time is able to estimate its position as moving within it [12,32]. Thus, SLAM is a sine qua non procedure in any modern autonomous system. One of the essential components for any SLAM architecture is place recognition, a process which allows an intelligent mechanism to realize if a location has been already visited, widely known as loop closure detection [1,33,34].

Owed to the increased availability of computational power during the last years, cameras overcame range type sensors, e.g., laser, ultrasonic, radar, due to the rich textural information provided by visual data. Appearance-based place recognition is the ability to trigger a loop closure in the environment using vision as the main sensory modality. Traditionally, place recognition is casted as an image retrieval task since the query instance (the current robot view) seeks for the most visually similar one, by searching into the database. Each database visual information is represented using invariant local features, such as SURF [7], SIFT [22], or binary equivalents like BRISK [21] and ORB [29]. Comparisons are performed via voting techniques [17,23], in order to highlight the proper candidate. Gehrig et al. [17] propose a loop closure framework depending on the votes' aggregation which each database instance pools through a k Nearest Neighbor (k-NN) scheme, while a probabilistic score highlights the proper pair. Accompanied with a geometrical verification step, outliers are avoided. The authors in [23] built a k-d tree from projected BRISK descriptors and via a similar NN method previsited locations are identified. Despite the fact that the outcome of a voting procedure is more robust, the searching process is computational costly.

To address the challenge of complexity, recent studies [1,6,11,15,26,36], adopted the Bag-of-Words (BoW) model [30] originally proposed for text retrieval tasks [3]. BoW approaches make use of a visual vocabulary, generated off-line through a training procedure, in order to represent images with visual word histograms. Owing to histogram comparisons the proper instance is selected. In [1], a BoW algorithm for scene recognition is proposed. Along the navigation, two parallel visual vocabularies (one for image-descriptors and one for color histograms) are constructed and combined. Candidate pairs are highlighted via a Bayesian filter and validated via a epipolar geometry constraint. In our previous work we propose the representation of a group of instances by a common visual word histogram [5,6]. Matches are indicated through these histogram comparisons and a quantitative interpretation of temporal consistency enhances the results. A probabilistic appearance-based pipeline based on a pretrained vocabulary of SIFT descriptors is proposed in [11]. In addition, this approach includes a Chow and Liu tree to learn the co-occurrence probabilities among visual words [9]. Similarly, a binary vocabulary accompanied with geometrical and temporal checks prevents the system from fault detections [15,26]. Providing the well known sequence-based place recognition algorithm SeqSLAM [25] with the BoW model a significant performance improvement is presented in [36].

Although, BoW has shown efficient performance in loop closure detection frameworks, it has a key drawback which related to the training procedure. The visual vocabulary is generated offline via a set of descriptors extracted from a generic environment. This practically means that the system may not be able to represent the incoming images appropriately and false detections may appear due to perceptual aliasing (high similarity between different locations). In order to avoid such situations, incremental dictionaries which are build on-line appeared in the robotics community [10,16,19,33,35,37]. Cieslewski et al. [10] proposed a voting scheme where a search into the database is performed by the usage of an incremental vocabulary tree, in order to retrieve the appropriate match. In [16], an incremental visual dictionary is built on a hierarchical structure of visual words. Similarly, in [19,35] visual words are generated on-line through a local feature-tracking, while voting techniques are responsible for the detection of loop closures. The authors in [37] propose a binary codebook with perspective invariance to the camera's motion, while unique visual words are generated on-line and assigned to dynamic places of the traversed map in [33].

Convolutional Neural Networks-based approaches were recently introduced with a view to solve the place recognition task [2,28,31]. Convolutional or fully connected layers are used as image descriptors and comparisons are performed among them. Despite their highly efficient performance, these frameworks are known for their excess demand in computational resources [24].

This paper presents a straightforward appearance-based loop closure pipeline, which relies on the images' description by a selective subset of raw SURF visual features, with the aim to reduce the computational complexity. The selected keypoints are chosen with respect to the scale which are extracted, similarly to BoRF [38]. In order to define the most informative visual features, we examine their repeatability among consecutive images though a features' matching technique. This decision is based on the observation that a feature's extracted scale, whose repeatability is strong, is not affected by variations in velocity or view point. Following this pipeline a vast database reduction is achieved, minimizing the computational complexity of a large feature multitude. Subsequently, a voting technique is performed to the descriptors' space via a k-NN technique, and a binomial probability function determines the proper candidates. Finally, a geometrical verification check between the chosen pair suppresses the system's false detections. Due to the careful features' selection, our method is capable of achieving high recall rates with 100% precision, as evaluated on three outdoor, community datasets.

The rest of the paper is organized as follows: Sect. 2 contains the formulation of the method in detail. The experimental results demonstrating the feasibility of the proposed pipeline are in Sect. 3, while in Sect. 4, the conclusion and future work are discussed.

2   Methodology

In this section an extended description of the proposed pipeline is presented in detail. The core algorithm of the system is based on the usage of scale-restrictive visual features in order to represent the incoming visual sensory information for

Online Stage

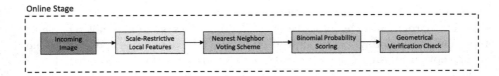

Fig. 1. Method's overview.

loop closure detection. A k-NN scheme follows, aiming to implement the votes' pooling, while a binomial distribution function is adopted with a view to classify pre-visited and non-visited locations, as proposed in [35]. An overview of the method is illustrated in Fig. 1.

2.1 Scale-Restrictive Visual Features Projection

Evaluation Stage

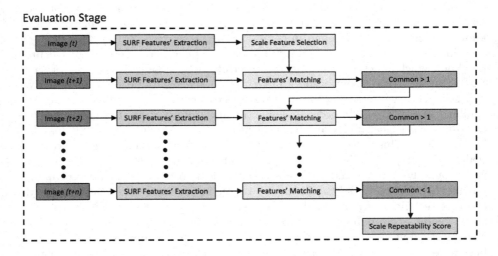

Fig. 2. Proposed scheme for measuring the features' scaling repeatability.

The features extraction procedure finds points of interest or keypoints in the incoming visual data that are distinct in a certain way. Beyond the achieved compression in the amount of data, these features have various invariance properties which define them, viz. scale, rotation, intensity, giving them the capability for images' comparison. Although robust results can be provided, the searching process is computational costly, especially in cases where the robot encounters a highly informative texture. Aiming to a system with improved matching success and operational frequency, we examine the SURF visual features' repeatability with respect to scale along consecutive images. In this off-line process, we are able to selectively detect these scale-restrictive keypoints which are meant to be used during the course of the main procedure. More specifically, at time t, a low quantity of scale-restrictive features $D_{S(t)}$ are extracted from the incoming visual sensory information. These features are matched with the total of extracted descriptors $D_{(t+1)}$ in the following image $I_{(t+1)}$:

$$\left| D_{S(t)} \bigcap D_{(t+1)} \right|, \tag{1}$$

where $|X|$ denotes the cardinality of set X. Common features are maintained and subsequently matched to the next ones $D_{(t+2)}$ until the correlation between the images' descriptors cease to exit:

$$\left| D_{S(t)} \bigcap_{i=t+1}^{i=t+n} D_{(i)} \right| \leq 1, \tag{2}$$

whilst a counter representing the passed images is retained. In this process, we are not interested about the number of frames each feature is presented along the incorporated instances, but only for the duration of feature matching. Our scaling repeatability algorithm is shown in Fig. 2.

2.2 Nearest Neighbor Vote Assignment

When a query image I_Q is captured, we retain only the local features detected in scales with the highest repeatability, as measured from the procedure presented in Sect. 2.1. Then, a searching scheme is performed with all the pre-visited locations in the database, seeking for the most suitable loop closing candidates. The proposed framework utilizes a voting mechanism between the query's generated scale-restrictive features and the database's ones. A k-NN ($k = 1$) search defines the most similar descriptors in the traversed path and votes are distributed into the corresponding locations. The vote density $x_l(t)$ of each visited location l constitutes the primary factor for the binomial probability function. Subsequently, in order to avoid erroneous detections in cases where the robot' velocity decreases or the platform remains still, the proposed pipeline seeks into the database frames which are recorded prior to a temporal constant of 40 s [33]. This way, the system maintains the certainty that I_Q does not share common features with early visited locations.

2.3 Database Probabilistic Assignment

After votes' aggregation, each location in the database receive a matching score, obtained via a binomial probability function [17], which evaluates the similarity with the query image. Using the probabilistic score, the naïve approach of selecting a heuristic threshold over the aggregated votes is avoided. The proposed score examines the rareness of an event and is based on the assumption that if a robot visits a new location, which has never encountered before, votes should be distributed randomly over the total of the traversed map. Accordingly, in case where a location has been seen in the past, the corresponding votes' density should be high indicating the existence of a loop closure candidate:

$$X_l(t) \sim Bin(n, p), n = N(t), p = \frac{\lambda_i}{\Lambda(t)}, \tag{3}$$

where $X_l(t)$ represents the random variable for the number of aggregated votes of the pre-visited location l at time t, N denotes the multitude of query's projected scale-restrictive features, λ is the total of features in l, and Λ corresponds to the size of database searching area $(\Lambda = \sum_{i=1}^{i=t-40s} D_{(i)})$.

The probabilistic score is calculated for each traversed location which gathers more than one votes, in order for a time saving to be performed, while two conditions have to be satisfied before an instance is indicated as loop closure candidate. First, a binomial probability threshold θ needs to be met:

$$Pr(X_l(t) = x_l(t)) < \theta < 1, \tag{4}$$

and additionally the number of accumulated votes has to be greater than the distribution's expected value:

$$x_l(t) > E[Xl(t)], \tag{5}$$

such that the system being able to discard cases with low votes' pooling.

2.4 Candidate Selection and Geometrical Verification Approvement

Up to this point, the proposed pipeline is capable of highlighting a set of pre-visited locations in the navigated map as candidates for loop closure events. Since the decision threshold may provide more than one detections, the algorithm selects as proper the one with the highest votes' accumulation (I_L). The chosen image is then processed for further validation. Aiming to a robust system without any false detection, the matched pair (I_Q, I_L) is subjected to a geometrical check through a RANSAC scheme [14] for the estimation of a representative fundamental matrix. If the computation of this matrix fails or the number of inlier points to the corresponding transformation is lower than a factor ($\varphi <$ 9), the candidate loop closure detection is ignored. The parameterization of the applied RANSAC method is based on [35].

3 Experimental Validation

In this section, the experimental protocol followed by this work is described in detail. A total of three outdoor image-sequences is selected and used for the method's assessment. The system's evaluation is presented and comparisons against its baseline version and one state-of-the-art method are also reported and discussed. Experiments were performed on an Intel i7-6700HQ 2.6 GHz processor with 8 GB of RAM.

Table 1. Datasets' synopsis

Name	Description	Images	Frequency	Resolution
KITTI 05 [18]	Dynamic, urban area observing mostly buildings and cars	2761	10 Hz	1241 × 371
Lip 6 Outdoor [1]	Highly dynamic, urban dataset of crowed street recorded from a hand-held camera	301	1 Hz	240 × 192
Malaga 2009 6L [8]	University parking containing cars and trees	3474	7.5 Hz	1024 × 768

3.1 Datasets

Publicly-available outdoor datasets are chosen with a view to achieve a variety of different data characteristics e.g., vehicle's velocity, camera's frame-rate. Table 1 provides a summary of each image-sequence utilized. In the case of KITTI 05 dataset [18], the incoming visual stream is obtained through a stereo camera system mounted on a car, with high resolution instances depicting houses, cars and trees. Several loop closure events are performed under different platform velocity. The second data-sequence belongs to the Lip 6 Outdoor environment [1]. The visual information is provided through a hand-held camera with low acquisition frequency and resolution, recording an urban environment with many buildings. A high amount of loop closure events are encountered along the navigated path. In Malaga 2009 6L [8], the recorded data offers high resolution images with accurate odometry information from a University parking area. Plenty loop closing examples are accounted from a stereo vision system mounted on an electric buggy-typed vehicle. Since the proposed approach aims to an appearance-based pipeline, only the right visual information was utilized for the system's evaluation.

3.2 Visual Features Selection

The scale-restrictive visual features has to be sufficiently reliable so as to accurately describe the incoming camera measurements. Towards this aim, KITTI 05 is selected as the evaluation dataset since it includes high resolution images with rich texture and considerable frame-rate along the traversed path. For our experiments two of the longest routes in the dataset are used with the SURF [7] detector and descriptor being adopted for keypoint extraction. To evaluate the features' repeatability with respect to keypoint scale, a total of four value ranges cases wherein the scaling value ranges cases was assessed ($\sigma = \{(0,3], (3,6], (6,9], (6,9)\}$). The specific scales are selected experimentally through a quantitative estimation of the extracted visual features. Throughout

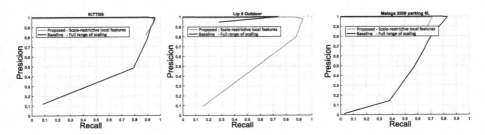

Fig. 3. Precision and recall curves evaluating the utilized scale-restrictive local features. The proposed approach (red lines) performs comparably, reaching high recall rates for perfect precision, as compared to the baseline version (black lines) where extracted keypoints are used arbitrarily. (Color figure online)

the experiments, features which belong to $\sigma = (0, 3]$ showed a dominance in repeatability against the higher ranges, and thus they are selected for the main pipeline.

3.3   Loop Closure Performance Evaluation Protocol

Precision-Recall Metric: In order to evaluate the system's performance against the chosen datasets, the precision-recall metrics are illustrated. Precision is defined as the ratio of the system's correct detected loop closure matches over the total method's identifications:

$$\text{Precision} = \frac{\text{True Positives}}{\text{True Positives} + \text{False Positives}}. \tag{6}$$

Recall is the ratio between the detected true positive events and the actual loop closures declared through the ground truth:

$$\text{Recall} = \frac{\text{True Positives}}{\text{True Positives} + \text{False Negatives}}. \tag{7}$$

A true positive match is considered to be any database association occurring within a small distance radius of 10 locations from the query image, while a false positive corresponds to any match lies outside this area. On the contrary, a false-negative is considered the incoming image that ought to be matched but the system was unable to achieve any detection. Ground Truth (GT) information is defined as the binary matrix (GT) whose elements correspond to the absence ($\text{GT}_{ij} = \text{false}$) or existence ($\text{GT}_{ij} = \text{true}$) of a loop closure event. In the cases of KITTI 05 and Malaga 2009 6L datasets, the used GTs were constructed manually within the scope of our previous work [33]. The evaluation of Lip 6 Outdoor is established through the GT data offered by the authors [1].

Method's Evaluation: In order to monitor the system's performance through precision-recall metrics, a variety of binomial probability thresholds θ were

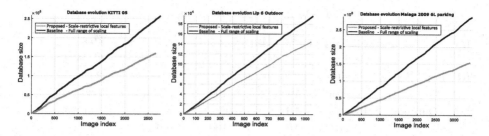

Fig. 4. Database evolution of the proposed pipeline for the scale-restrictive (red line) version, as well as for the unscaled one (black line). The generated visual database is about half in the proposed approach, resulting into a significant data reduction for each assessed dataset. (Color figure online)

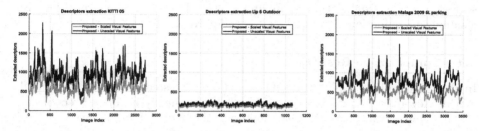

Fig. 5. Number of extracted SURF [7] features per incoming camera measurement. Red lines depict the scale-restrictive approach which achieves great reduction of the utilized keypoints. This effect is mostly highlighted in the case of Malaga 2009 6L parking. (Color figure online)

tested. In Fig. 3, the illustrated scores prove the ability of the proposed scale-restrictive pipeline to perform comparably as compared to the baseline method. In the KITTI 05 dataset, the system reaches a recall rate over 90% for perfect precision in both cases. In case of Lip 6 Outdoor environment the proposed version shows a superiority over the baseline reaching nearly 90% recall, while the precision remains at 100%. This is owed to the fact that the images' acquisition frequency is low and the camera's orientation changes along the navigated path, resulting into a robust database construction of scale-restrictive with distinct sets of votes during the query process. In the Malaga 2009 6L data-sequence, the performance in the baseline version is increased due to the high perception aliasing of the dataset, which allows the system to identify more loop closing events through the more information available.

Descriptors and Database Evolution: Aiming to analyze the computational complexity of the proposed pipeline, in Fig. 4, the system's overall database evolution is illustrated, while Fig. 5 shows the number of features extracted per image. It is noteworthy that for each tested dataset, the number of

Table 2. Comparative results

Dataset	Metric (%)	Proposed restrictive	Baseline	Tsintotas et al. [35]	Gehring et al. [17]
KITTI 05 [18]	Recall	92	91	92.6	**94**
	Precision	100	100	100	**100**
Lip 6 Outdoor [1]	Recall	**73**	67	50	Not available
	Precision	**100**	100	100	
Malaga 6L [8]	Recall	70	75	**85**	Not available
	Precision	100	100	**100**	

scale-restrictive visual features is about half of the baseline version, resulting into a significant reduction of computational complexity.

3.4   Comparative Results

In Table 2, the algorithm's obtained results for each tested dataset are presented. With a view to carry out a fair comparison between the two evaluated versions, as well as against other state-of-the-art methods, the binomial probability threshold was selected as the single value which performs the highest recall for 100% precision. The chosen decision values, $\theta_{restrictive} = 1e^{-11}$ and $\theta_{baseline} = 1e^{-16}$, remain the same for every image-sequence. As it can be seen, the proposed scale-restrictive pipeline can achieve remarkable recall scores for perfect precision in every tested dataset. When comparing the KITTI 05 image-sequence, the proposed version exhibits over 90% recall score performing comparably to the rest of the approaches. In Lip 6 Outdoor sequence, high recall rates are achieved, while in Malaga 6L the proposed framework performs unfavorable against the other methods. This in mainly due to low-texture environment that the robot encounters, as well as the perceptual aliasing occurring, which prevents the scale-restrictive features to reach high recall scores.

4   Conclusion

The paper in hand presented a scale-restrictive visual loop closure detection framework. Through an evaluation procedure, local features are assessed for their repeatability with respect to their scale. The mechanism represents each visited location through the scale-specific local features extracted via SURF detector in order to reduce the computational cost. At query time, a probability score is generated for every instance in the database, based on the votes' aggregation which are collected via a k-NN technique. The system's loop closure belief generator is based on a probabilistic framework, while a geometrical check is also adopted in order to reduce possible false detections. The proposed method is tested on several environments demonstrating a substantial performance as compared to

KITTI 05   Lip 6 Outdoor   Malaga 2009 parking 6L

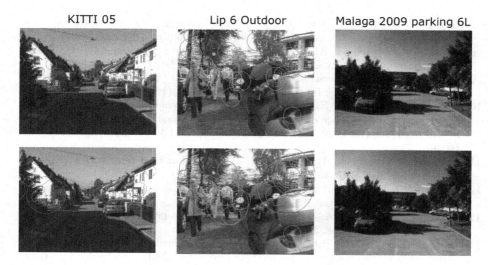

Fig. 6. Different local features extracted for the proposed method (top) and its baseline version (bottom) on KITTI 05 [18] (left), Lip 6 Outdoor [1] (center) and Malaga 2009 Parking 6L [8] (right) datasets.

its baseline version, as well as to other state-of-the-art approaches, offering high recall rates for perfect precision. Future work will focus on a more extensive evaluation and the utilization of the same informative scales for the generation of an on-line incremental vocabulary. Examples of extracted scale-restrictive visual features are illustrated in Fig. 6.

References

1. Angeli, A., Filliat, D., Doncieux, S., Meyer, J.A.: A fast and incremental method for loop-closure detection using bags of visual words. IEEE Trans. Robot. 1027–1037 (2008)
2. Arandjelovic, R., Gronat, P., Torii, A., Pajdla, T., Sivic, J.: NetVLAD: CNN architecture for weakly supervised place recognition. In: Proceedings of IEEE International Conference on Computer Vision and Pattern Recognition, pp. 5297–5307 (2016)
3. Baeza-Yates, R., Ribeiro-Neto, B., et al.: Modern Information Retrieval, vol. 463. ACM Press, New York (1999)
4. Balaska, V., Bampis, L., Gasteratos, A.: Graph-based semantic segmentation. In: Proceedings of International Conference on Robotics in Alpe-Adria Danube Region, pp. 572–579 (2018)
5. Bampis, L., Amanatiadis, A., Gasteratos, A.: Encoding the description of image sequences: a two-layered pipeline for loop closure detection. In: Proceedings of IEEE/RSJ International Conference on Intelligent Robots and Systems, pp. 4530–4536 (2016)
6. Bampis, L., Amanatiadis, A., Gasteratos, A.: Fast loop-closure detection using visual-word-vectors from image sequences. Int. J. Robot. Res. **37**(1), 62–82 (2018)

7. Bay, H., Tuytelaars, T., Van Gool, L.: Surf: speeded-up robust features. In: Proceedings of European Conference on Computer Vision, pp. 404–417 (2006)
8. Blanco, J.L., Moreno, F.A., Gonzalez, J.: A collection of outdoor robotic datasets with centimeter-accuracy ground truth. Auton. Robots **27**(4), 327 (2009)
9. Chow, C., Liu, C.: Approximating discrete probability distributions with dependence trees. IEEE Trans. Inf. Theory **14**(3), 462–467 (1968)
10. Cieslewski, T., Stumm, E., Gawel, A., Bosse, M., Lynen, S., Siegwart, R.: Point cloud descriptors for place recognition using sparse visual information. In: Proceedings of IEEE International Conference on Robotics and Automation, pp. 4830–4836 (2016)
11. Cummins, M., Newman, P.: Appearance-only SLAM at large scale with FAB-MAP 2.0. Int. J. Robot. Res. **30**(9), 1100–1123 (2011)
12. Durrant-Whyte, H., Bailey, T.: Simultaneous localization and mapping: Part I. IEEE Robot. Autom. Mag. **13**(2), 99–110 (2006)
13. Erkent, Ö., Bozma, H.I.: Bubble space and place representation in topological maps. Int. J. Robot. Res. **32**(6), 672–689 (2013)
14. Fischler, M.A., Bolles, R.C.: Random sample consensus: a paradigm for model fitting with applications to image analysis and automated cartography. Commun. ACM **24**(6), 381–395 (1981)
15. Gálvez-López, D., Tardos, J.D.: Bags of binary words for fast place recognition in image sequences. IEEE Trans. Robot. **28**(5), 1188–1197 (2012)
16. Garcia-Fidalgo, E., Ortiz, A.: iBoW-LCD: an appearance-based loop-closure detection approach using incremental bags of binary words. IEEE Robot. Autom. Lett. **3**(4), 3051–3057 (2018)
17. Gehrig, M., Stumm, E., Hinzmann, T., Siegwart, R.: Visual place recognition with probabilistic voting. In: Proceedings of IEEE International Conference on Robotics and Automation, Singapore, pp. 3192–3199, May 2017
18. Geiger, A., Lenz, P., Urtasun, R.: Are we ready for autonomous driving? The KITTI vision benchmark suite. In: Proceedings of Conference on Computer Vision and Pattern Recognition (2012)
19. Khan, S., Wollherr, D.: IBuILD: incremental bag of binary words for appearance based loop closure detection. In: Proceedings of IEEE International Conference on Robotics and Automation, pp. 5441–5447 (2015)
20. Kostavelis, I., Gasteratos, A.: Semantic mapping for mobile robotics tasks: a survey. Robot. Auton. Syst. **66**, 86–103 (2015)
21. Leutenegger, S., Chli, M., Siegwart, R.: BRISK: binary robust invariant scalable keypoints. In: Proceedings of IEEE International Conference on Computer Vision, pp. 2548–2555 (2011)
22. Lowe, D.G.: Distinctive image features from scale-invariant keypoints. Int. J. Comput. Vis. **60**(2), 91–110 (2004)
23. Lynen, S., Bosse, M., Furgale, P.T., Siegwart, R.: Placeless place-recognition. In: Proceedings of IEEE International Conference on 3D Vision, pp. 303–310 (2014)
24. Maffra, F., Chen, Z., Chli, M.: Tolerant place recognition combining 2D and 3D information for UAV navigation. In: Proceedings of IEEE International Conference on Robotics and Automation, pp. 2542–2549 (2018)
25. Milford, M.J., Wyeth, G.F.: SeqSLAM: visual route-based navigation for sunny summer days and stormy winter nights. In: Proceedings of IEEE International Conference on Robotics and Automation, pp. 1643–1649 (2012)
26. Mur-Artal, R., Tardós, J.D.: Fast relocalisation and loop closing in keyframe-based SLAM. In: Proceedings of IEEE International Conference on Robotics and Automation, pp. 846–853 (2014)

27. Newman, P., Cole, D., Ho, K.: Outdoor SLAM using visual appearance and laser ranging. In: Proceedings of IEEE International Conference on Robotics and Automation, pp. 1180–1187 (2006)
28. Radenović, F., Tolias, G., Chum, O.: CNN image retrieval learns from BoW: unsupervised fine-tuning with hard examples. In: Proceedings of European Conference on Computer Vision, pp. 3–20 (2016)
29. Rublee, E., Rabaud, V., Konolige, K., Bradski, G.: ORB: an efficient alternative to SIFT or SURF. In: Proceedings of IEEE International Conference on Computer Vision, pp. 2564–2571, November 2011
30. Sivic, J., Zisserman, A.: Video Google: a text retrieval approach to object matching in videos, p. 1470 (2003)
31. Sünderhauf, N., Dayoub, F., Shirazi, S., Upcroft, B., Milford, M.: On the performance of convnet features for place recognition. arXiv preprint arXiv:1501.04158 (2015)
32. Thrun, S., Leonard, J.J.: Simultaneous localization and mapping. In: Siciliano, B., Khatib, O. (eds.) Springer Handbook of Robotics, pp. 871–889. Springer, Heidelberg (2008)
33. Tsintotas, K.A., Bampis, L., Gasteratos, A.: Assigning visual words to places for loop closure detection. In: Proceedings of IEEE International Conference on Robotics and Automation, pp. 1–7 (2018)
34. Tsintotas, K.A., Bampis, L., Gasteratos, A.: DOSeqSLAM: dynamic on-line sequence based loop closure detection algorithm for SLAM. In: Proceedings of IEEE International Conference on Imaging Systems and Techniques, pp. 1–6 (2018)
35. Tsintotas, K.A., Bampis, L., Gasteratos, A.: Probabilistic appearance-based place recognition through bag of tracked words. IEEE Robot. Autom. Lett. 4(2), 1737–1744 (2019)
36. Tsintotas, K.A., Bampis, L., Rallis, S., Gasteratos, A.: SeqSLAM with bag of visual words for appearance based loop closure detection. In: Proceedings of International Conference on Robotics in Alpe-Adria Danube Region, pp. 580–587 (2018)
37. Zhang, G., Lilly, M.J., Vela, P.A.: Learning binary features online from motion dynamics for incremental loop-closure detection and place recognition. In: Proceedings of IEEE International Conference on Robotics and Automation, pp. 765–772. IEEE (2016)
38. Zhang, H.: BoRF: loop-closure detection with scale invariant visual features. In: Proceedings of IEEE International Conference on Robotics and Automation, pp. 3125–3130 (2011)

Grasping Unknown Objects by Exploiting Complementarity with Robot Hand Geometry

Marios Kiatos[1,2]([⊠]) and Sotiris Malassiotis[2]

[1] Department of Electrical and Computer Engineering,
Aristotle University of Thessaloniki, 54124 Thessaloniki, Greece
`mkiatos@auth.gr`
[2] Information Technologies Institute (ITI), Center for Research and Technology
Hellas (CERTH), 57001 Thessaloniki, Greece
{`kiatosm,malasiot`}`@iti.gr`

Abstract. Grasping unknown objects with multi-fingered hands is challenging due to incomplete information regarding scene geometry and the complicated control and planning of robot hands. We propose a method for grasping unknown objects with multi-fingered hands based on shape complementarity between the robot hand and the object. Taking as input a point cloud of the scene we locally perform shape completion and then we search for hand poses and finger configurations that optimize a local shape complementarity metric. We validate the proposed approach in MuJoCo physics engine. Our experiments show that the explicit consideration of shape complementarity of the hand leads to robust grasping of unknown objects.

Keywords: Grasping · Perception for grasping and manipulation

1 Introduction

General purpose robots need the capability of robustly grasping previously unseen objects in unstructured environments. A robot with multi-fingered hand should be able to grasp different objects with various surfaces and sizes, from simple boxes to tools with complicated shapes. Recent grasp planning methods employ a data driven approach for the detection of plausible grasps but assume top grasps and parallel grippers. Exploiting the geometry of multi-finger hands to achieve highly stable grasps remains a challenging goal, considering the high dimensionality of the configuration space of the robot hand in which a valid robot grasp is searched. In addition, in most real world scenarios the objects are unknown and thus the robot is unable to plan robust grasps due to incomplete information of the scene geometry.

Previous works in grasping with multi-fingered hands focus on planning precision grasps for known objects. However, these approaches require full knowledge

© Springer Nature Switzerland AG 2019
D. Tzovaras et al. (Eds.): ICVS 2019, LNCS 11754, pp. 88–97, 2019.
https://doi.org/10.1007/978-3-030-34995-0_8

of the object properties such as its shape, pose, material properties and mass in order to plan a successful grasp. This reliance on object specific modeling makes the grasp-planning infeasible for real world scenarios, where the objects are unknown. On the contrary, a lot of recent research focused on grasping unknown objects with parallel grippers due to simpler gripper structures and fewer constraints. Most of these approaches take as input a noisy and partially occluded point cloud and produce as output pose estimates of viable grasps. The underlying idea is to infer a mapping between the hand pose and the local geometry of graspable surfaces by labeling poses of successful grasps on a large dataset.

In this paper we adapt a geometric approach to efficiently plan grasps with multi-finger hands by considering the effect of shape complementarity between the hand and the object. Intuitively, grasp success is greatly affected by the ability of the hand to complement the shape of the object. This adaptation increases the contact area and thereby the robustness of the grasp. This is in contrast with past methods that seek for precision grasps and thus try to achieve force closure with only a few contact points.

Our approach involves three contributions. The first is an objective function that evaluates the shape complementarity between the hand and the object. Experimental results demonstrate a positive correlation between this metric and grasp success. Since this metric works locally on a point cloud we also propose a local 3D shape completion method that predicts the points of the occluded parts of the scene around the current grasp candidate. Finally, an approach for efficient search for optimal hand poses and grasp configurations that maximize the above metric is proposed.

2 Related Work

Robotic grasping has evolved a lot over the last two decades. Bohg et al. [2] divide the various methods of robotic grasping to analytic and data-driven. Analytic methods assume full knowledge about geometry and physics of the object in order to calculate stable grasps [20], but tend to not transfer well to the real world due to the difficulty of modelling physical interactions between a manipulator and an object [1,18]. For this reason data-driven approaches have risen in popularity in recent years. Instead of relying exclusively on analytic understanding of the physics and geometry of the object, data driven methods learn to predict either human-inspired grasps or empirically estimated grasp outcomes. Some approaches precompute good grasp points from a database of object models [6,16]. Grasp simulators are used to process 3D meshes of objects and compute the stability of a grasp based upon the grasp wrench space. However, this limits the approach to the range of objects previously seen.

The recent trend in grasping of unknown objects is to detect object grasp poses directly from images relying on local features rather than the whole object geometry. The idea of searching an image for grasp targets independently of object identity was probably first explored in Saxena's [19] early work that used a sliding window classifier to localize good grasps based on broad collection of local visual features. Instead of learning a single grasp region, in [12] multiple contact points are learned from the image to implement a good grasp. Jiang et al. [8] represented a grasp as a 2D oriented rectangle in image, with two edges corresponding to the gripper plates, using surface normals to determine the grasp approach direction. Lenz et al. [14] train a deep network on Cornell Grasp Dataset [9] to predict the probability that a particular pose will be graspable, by passing the corresponding image patch through the network. Using the same dataset, Redmon et al. [17] pose grasp detection task as a regression problem and solve it using a convolutional neural network by passing the entire image through the network rather than individual patches. Mahler et al. [15] train a convolutional neural network to predict the robustness of grasp candidates directly from depth images.

All the above approaches consider the robotic grasping of a single object in an uncluttered environment. Ten Pas et al. [7] proposed a grasp pose detection method which samples a number of grasp candidates from a 3D point cloud and assigns quality scores to each candidate through a convolutional neural network. This approach produces good grasp candidates under the parallel gripper assumption. Our approach also works for cluttered scenes but it is not restricted to parallel grasps thus generating more stable candidates.

Relatively little work has been done for multi-finger grasp planning. Eppner et al. [4] fit geometric shape primitives to depth measurements. Hand geometry is abstracted into a set of preshape-configurations and grasps are chosen according to those shape primitives. Kappler et al. [10] learn to predict if a given palm pose will be successful for multi-finger grasps using a fixed preshape and perform planning by evaluating a number of sampled grasp poses in a physics simulator. Fan et al. [5] plan precision grasps by performing a dual stage-optimization based on known object geometry and properties.

Unlike previous approaches, we perform grasp planning without the need of any external planner or physics simulator. Furthermore, we take into consideration a reconstructed part of the scene, leading to more robust grasps.

3   Problem Formulation

The input to our algorithm is a 3D point cloud $\mathcal{C}$ of the scene, where each point is represented by its coordinates $\boldsymbol{p}_i \in \mathbb{R}^3$, $i = 1, \cdots, N$ and the associate normal $\boldsymbol{n}_i \in \mathbb{R}^3$. The algorithm also takes as input the geometry of the robot hand to be used and its kinematics. The output of the algorithm is a set of grasp candidates, where each candidate is represented by the preshape $\boldsymbol{s} \in \mathcal{S}$ and is selected so that it optimizes a suitable grasp quality metric. A preshape is defined as:

$$\mathcal{S} = \{(\boldsymbol{q}, H), \boldsymbol{q} \in \mathbb{R}^d, H \in SE(3)\} \tag{1}$$

where $\boldsymbol{q}$ is the hand configuration parameters and H the wrist pose. The hand configuration is defined as the vector $\boldsymbol{q} = [q_1, \cdots, q_d]$ where q_i is the value of i-th degree of freedom and the pose as the rotation $R \in SO(3)$ and position $\boldsymbol{t} \in \mathbb{R}^3$ of the hand. The parameters q_i may correspond to finger joints or to some other reduced parameterization such as eigengrasps [3].

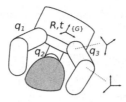

Fig. 1. Illustration of grasping using a three-fingered hand. A local frame is attached at each finger link and one in the palm.

4   Shape Complementarity in Grasping

The general underlying idea of our method is to characterize the complementarity between the shape of an arbitrary unknown object and the shape of a known robot hand under a preshape $s \in \mathcal{S}$. For efficiency, we approximate the contact surface of the hand with a set of M contacts $c = [c_1^T, \cdots, c_M^T]$ spread over the robot hand surface. Each contact $c_i \in \mathbb{R}^6$ is composed of a 3D point $c_{p,i} \in \mathbb{R}^3$ and its associate normal $c_{n,i} \in \mathbb{R}^3$. The contacts are placed at the interior of the palm and finger links and are expressed in the associate local frames. Note that local frames are attached to the palm and finger links as shown in Fig. 1. In our case, we uniformly spread contacts over the robot hand surface as shown in Fig. 2. The forward kinematics of the hand f are used in order to compute the contacts $\hat{c}_i = f(s, c_i)$ in a new preshape s. The updated contacts $\hat{c}_i$ are all expressed in the palm frame $\{G\}$.

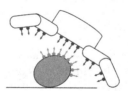

Fig. 2. Illustration of the shape complementarity metric. The contact points $c_{p,i}$ are the red dots while the red arrows represent the associate normals $c_{n,i}$. The green dots are the points p_i and the green arrows are the associate normals n_i. (Color figure online)

We propose a shape complementarity metric that penalizes preshapes where contact points are far from nearest object points and whose normals do not align well with the associate object normals. We define the distance penalty d_p and the alignment penalty d_n as:

$$d_p(\hat{c}_i) = \exp(w\|\hat{c}_{p,i} - \pi_{\mathcal{C}}(\hat{c}_i)\|)$$
$$d_n(\hat{c}_i) = 1 - \langle -\hat{c}_{n,i}, n_{\mathcal{C}}(\hat{c}_i) \rangle \tag{2}$$

where $\pi_C(\hat{c}_i)$ is closest point to contact $\hat{c}_i$ in the point cloud C defined as:

$$\pi_C(\hat{c}_i) = \arg\min_{p \in C} \|\hat{c}_{p,i} - p\|, \tag{3}$$

$n_C(\hat{c}_i)$ is the associate normal of the closest point and w is a scalar weight. We can now define the shape complementarity metric as:

$$Q(s) = \frac{1}{M} \sum_{i=1}^{M} d_n(\hat{c}_i) d_p(\hat{c}_i) \tag{4}$$

Computing this metric over all points and configurations is expensive. However, the majority of candidate preshapes lead to collision between the hand and the point cloud. Therefore, we first perform a fast collision detection step using an approximation of the robot hand geometry. Similar to Lei et al. [13], we approximate the robot hand as a C-shaped cylinder (Fig. 3b), and we test for collision of this simple geometry against the point cloud. We use a massively parallel CPU implementation to obtain fast collision free candidates.

(a)   (b)

Fig. 3. (a) A three-finger hand that is approximated by the C-shape cylinder (b) The C-shape cylinder

Our grasp detection algorithm then works as follows. At first, we sample points randomly from C. For each sample, we compute an axis of principle curvature of the surface in the neighborhood of that point similar to [7]. Combining the associate normal of the point and the principal curvature axis we generate candidate poses. We align the C-shape with each candidate pose and check for collisions as described above. After this step only a few candidate poses survive. For a collision-free pose, we 'push' the hand forward along the normal until a collision is detected. Then, we extract the points $C_G \in C$ that are contained inside a sphere with center the hand's position and radius defined according to the size of the robot hand. This subset of points is used for the computation of the shape complementarity metric as explained in the following sections.

5   Local Shape Completion

The metric above would not be a good measure of shape complementarity if only the visible surface part is used in the computation. Therefore, we deal

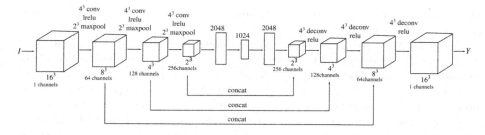

Fig. 4. Autoencoder architecture

with the unseen part via shape completion. We propose a method to predict the hidden surface points from a partial 3D point cloud C_G using a variation of an autoencoder similar to [21]. Given the predicted surface points we can compute the associate surface normals. To achieve this task, we represent both the input and the output of the autoencoder with a 3D voxel grid. Specifically, we use occupancy grids aligned with the candidate pose so as to enclose the sphere points C_G, where 1 represents an occupied voxel and 0 an empty one. Specifically, both the input denoted as I and the output denoted as Y are 16^3 occupancy grids. The task is to learn a mapping $I \rightarrow Y$ that is able to generalize to previously unseen observations.

5.1   Training

The network is a variation of [21] and is based on an autoencoder with skip connections between encoder and decoder (Fig. 4). Specifically, the encoder has three 3D convolutional layers, each of which has a bank of $4 \times 4 \times 4$ filters with strides of $1 \times 1 \times 1$. Each one is followed by a leaky ReLU activation function and a max pooling layer which has $2 \times 2 \times 2$ filters and strides of $2 \times 2 \times 2$. The number of output channels of max pooling layers starts with 16 and doubles at each subsequent layer and ends up with 256. The encoder is lastly followed by two fully-connected layers. To make the autoencoder more descriptive we add a separate input layer for the grasp pose. The decoder is composed by three symmetric up-convolutional layers which are followed by ReLU activation functions. The skip connections between the encoder and decoder guarantee propagation of local structures. The last layer is composed by an up-convolutional layers, corresponding to the occupancy of the predicted shape. The objective function of our network $\mathcal{L}$ is a modified cross-entropy loss function used by [21]

$$\mathcal{L}(y, y') = -\alpha y \log(y') - (1 - \alpha)(1 - y)\log(1 - y') \tag{5}$$

where y is the target value in $\{0, 1\}$, y' is the estimated value for each occupancy voxel from the autoencoder and α is a hyperparameter which weights the relative importance of false positives against false negatives due to the fact that most of the occupancy grid voxels tends to be empty and the network gets easily a false positive estimation. We set α to 0.85.

To generate training and test data we use the KIT object database [11]. Specifically, a subset of 70 CAD models is selected for training and 10 CAD models for testing. For each CAD model, we create depth maps by rendering each object from 25 different angles and then obtain the corresponding synthetic point clouds. For each point cloud, we search for collision-free grasp poses and we keep the points that are contained in the enclosing region as described in Sect. 4. We transform the extracted point cloud w.r.t. candidate poses. Ground truth point clouds are extracted by transforming the full point cloud w.r.t. candidate poses and keeping the points inside the enclosing sphere. For each pair, we derive 16^3 resolution grids. Overall, around 40K training pairs and 7k testing pairs are generated.

The model is trained using an Adam optimizer with batch size of 32 for 15 epochs. The other three Adam parameters are set as $\beta_1 = 0.9$, $\beta_2 = 0.999$ and $\epsilon = 1e - 8$ and the learning rate is set to 0.0001.

6   Grasp Optimization

The initial grasp proposals using the approach of Sect. 4 are guaranteed to be collision free and enclose the target object. We proceed to find a stable grasp around this initial preshape by optimizing the complementarity metric by using the predicted complete surface.

We define a finite set of preshapes around the initial preshape and perform an exhaustive search to find the one with the best complementarity metric. In particular, we generate 30 hand poses in the neighborhood of the initial pose by translating and rotating the robot hand around each axis. At each pose, we further generate 10 different finger configurations resulting in 300 candidate preshapes. We use a parallel GPU implementation to compute the metric at each preshape since the metric computation is local.

7   Experimental Results

We evaluate grasp success by simulating grasp using the MuJoCo physics engine as shown in Fig. 6. The Barret Hand is used in all experiments and a subset of 10 objects of various shapes from the KIT object database as test objects. To execute a grasp, the robot hand is positioned in a pose under a specific configuration and a fixed constant force is applied to the finger joints for a fixed time. Subsequently, the hand is raised upwards by 20 cm. We counted a grasp to be successful if the object is fully lifted and pertained to the hand for 5 s.

At first, we ran a series of experiments without the local shape completion and optimization to correlate the grasp success with the proposed metric. Then, we evaluate the shape complementarity metric with local shape completion and finally we integrated the optimization stage.

7.1 Shape Complementarity and Grasp Success

To measure the correlation of the proposed metric with grasp success we conducted a simplified experiment. We created 5 synthetic scenes for each object by placing it in different orientations on the top of a planar surface. A point cloud of the synthetic scene is obtained and used as input to the grasp planner. For each object 10 collision free grasps were generated and executed. For each trial the final value of the complementarity metric and the result of the simulated grasp were recorded. Then the correlation between successful grasps and low metric values was calculated (Fig. 5). Whenever a high quality grasp e.g. low value of shape complementarity metric, was predicted, the likelihood that the corresponding grasp in simulation also succeeded was high.

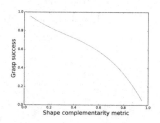

Fig. 5. Correlation between the shape complementarity metric and grasp success

7.2 Performance with Local Shape Completion

We evaluated the impact of the local shape completion in the performance of the proposed grasping method. Following the previous setup, the grasp planner took as input the point cloud of the scene, and outputted 100 collision free grasps for each object. Then, we computed the shape complementarity metric for each collision free grasp with and without the local shape completion. From these sets of grasps we executed only the top five. The results showed that the integration of the local shape completion led to higher grasp success (Table 1). The grasp planner produced intuitively better grasps since the unseen part was taken into consideration.

7.3 Performance with Grasp Optimization

We continued analyzing the impact of the optimization stage on the planned grasps. In the above sections we evaluated the performance of the proposed planner by taking into consideration only the initial collision free grasps. Similar to previous experiments, we placed each object in 5 random orientations on the top of a surface and obtained a point cloud of the scene as input for the grasp planner. By integrating the optimization stage and executing the resulting grasps in simulation, the performance of grasp planner was boosted around 10% (Table 1).

Table 1. Grasp success

Approaches	Grasp success
Partial view	55%
Local shape completion	72%
Local shape completion + grasp optimization	80%

Fig. 6. Experiments of the proposed grasp planning method in MuJoCo

8   Conclusion

In this paper we addressed the problem of grasping unknown objects with multi-fingered hand. We proposed a way to evaluate the grasp success based on the shape complementarity between the robot hand and the object. Furthermore, the integration of local shape completion and an optimization stage led to more robust grasps. Future works include experiments on cluttered environments.

References

1. Bicchi, A., Kumar, V.: Robotic grasping and contact: a review. In: Proceedings 2000 ICRA. Millennium Conference. IEEE International Conference on Robotics and Automation. Symposia Proceedings (Cat. No. 00CH37065), vol. 1, pp. 348–353. IEEE (2000)
2. Bohg, J., Morales, A., Asfour, T., Kragic, D.: Data-driven grasp synthesis: a survey. IEEE Trans. Robot. **30**(2), 1–21 (2013). https://doi.org/10.1109/TRO.2013.2289018
3. Ciocarlie, M., Goldfeder, C., Allen, P.: Dexterous grasping via eigengrasps: a low-dimensional approach to a high-complexity problem. In: Robotics: Science and Systems Manipulation Workshop-Sensing and Adapting to the Real World. Citeseer (2007)
4. Eppner, C., Brock, O.: Grasping unknown objects by exploiting shape adaptability and environmental constraints. In: IEEE International Conference on Intelligent Robots and Systems, pp. 4000–4006 (2013). https://doi.org/10.1109/IROS.2013.6696928

5. Fan, Y., Tang, T., Lin, H.C., Tomizuka, M.: Real-time grasp planning for multi-fingered hands by finger splitting. In: 2018 IEEE/RSJ International Conference on Intelligent Robots and Systems (IROS), pp. 4045–4052. IEEE (2018)
6. Goldfeder, C., Allen, P.K., Lackner, C., Pelossof, R.: Grasp planning via decomposition trees (2007)
7. Gualtieri, M., Pas, A.T., Saenko, K., Platt, R.: High precision grasp pose detection in dense clutter. In: IEEE International Conference on Intelligent Robots and Systems, November 2016, pp. 598–605 (2016). https://doi.org/10.1109/IROS.2016.7759114
8. Jiang, Y., Moseson, S., Saxena, A.: Efficient grasping from RGBD images: learning using a new rectangle representation. In: Proceedings - IEEE International Conference on Robotics and Automation, pp. 3304–3311 (2011). https://doi.org/10.1109/ICRA.2011.5980145
9. Jiang, Y., Moseson, S., Saxena, A.: Efficient grasping from RGBD images: learning using a new rectangle representation. In: 2011 IEEE International Conference on Robotics and Automation, pp. 3304–3311. IEEE (2011)
10. Kappler, D., Bohg, J., Schaal, S.: Leveraging big data for grasp planning. In: Proceedings - IEEE International Conference on Robotics and Automation, June 2015, pp. 4304–4311 (2015). https://doi.org/10.1109/ICRA.2015.7139793
11. Kasper, A., Xue, Z., Dillmann, R.: The KIT object models database: an object model database for object recognition, localization and manipulation in service robotics (2012). https://doi.org/10.1177/0278364912445831
12. Le, Q.V., Kamm, D., Kara, A.F., Ng, A.Y.: Learning to grasp objects with multiple contact points. In: 2010 IEEE International Conference on Robotics and Automation, pp. 5062–5069. IEEE (2010)
13. Lei, Q., Meijer, J., Wisse, M.: Fast C-shape grasping for unknown objects. In: IEEE/ASME International Conference on Advanced Intelligent Mechatronics, AIM, pp. 509–516 (2017). https://doi.org/10.1109/AIM.2017.8014068
14. Lenz, I., Lee, H., Saxena, A.: Deep learning for detecting robotic grasps (2013). https://doi.org/10.1177/0278364914549607. http://arxiv.org/abs/1301.3592
15. Mahler, J., et al.: Dex-Net 2.0: deep learning to plan robust grasps with synthetic point clouds and analytic grasp metrics (2017). https://doi.org/10.15607/RSS.2017.XIII.058. http://arxiv.org/abs/1703.09312
16. Miller, A.T., Knoop, S., Christensen, H.I., Allen, P.K.: Automatic grasp planning using shape primitives (2003)
17. Redmon, J., Angelova, A.: Real-time grasp detection using convolutional neural networks, pp. 1316–1322 (2015). https://doi.org/10.1109/ICRA.2015.7139361
18. Sahbani, A., El-Khoury, S., Bidaud, P.: An overview of 3D object grasp synthesis algorithms. Robot. Auton. Syst. **60**(3), 326–336 (2012). https://doi.org/10.1016/j.robot.2011.07.016
19. Saxena, A., Driemeyer, J., Ng, A.Y.: Robotic grasping of novel objects using vision. Int. J. Robot. Res. **27**(2), 157–173 (2008)
20. Siciliano, B., Khatib, O.: Springer Handbook of Robotics. Springer, Heidelberg (2016). https://doi.org/10.1007/978-3-540-30301-5
21. Yang, B., Rosa, S., Markham, A., Trigoni, N., Wen, H.: 3D object dense reconstruction from a single depth view. In: ICCV, pp. 679–688 (2017). https://doi.org/10.1109/ICCVW.2017.86. http://arxiv.org/abs/1802.00411

Grapes Visual Segmentation for Harvesting Robots Using Local Texture Descriptors

Eftichia Badeka, Theofanis Kalabokas, Konstantinos Tziridis, Alexander Nicolaou, Eleni Vrochidou, Efthimia Mavridou, George A. Papakostas$^{(\boxtimes)}$, and Theodore Pachidis

HUman-MAchines INteraction Laboratory (HUMAIN-Lab), Department of Computer Science, International Hellenic University, Agios Loukas, 65404 Kavala, Greece
efthimia.mavridou@gmail.com, {evbadek,theokala,kenaaske, alexniko,evrochid,gpapak,pated}@teiemt.gr

Abstract. This paper investigates the performance of Local Binary Patterns variants in grape segmentation for autonomous agricultural robots, namely Agrobots, applied to viniculture and winery. Robust fruit detection is challenging and needs to be accurate to enable the Agrobot to execute demanding tasks of precise farming. Segmentation task is handled by classification with the supervised machine learning model k-Nearest Neighbor (k-NN), including extracted features from Local Binary Patterns (LBP) and their variants in combination of color components. LBP variants are tested for both varieties of red and white grapes, subject to performance measures of accuracy, recall and precision. The results for red grapes indicate an approximate intended accuracy of 94% of detection, while the results relating to white grapes confirm the concerns of complex indiscreet visual cues providing accuracies of 83%.

Keywords: Visual computing · Computer vision · Grapes detection · Image segmentation · Local binary patterns

1 Introduction

In recent years, *precision farming* has come to the fore. With the help of *computer* vision manually performed agricultural tasks have been automated in the context of precision farming. Manual methods, based on human operators, are prone to errors. Moreover, the modern growing concept, involving vast fields, is rather demanding in terms of money and time. Computer vision can provide both extreme accurate and efficient solutions to support agricultural practices and enable the efficient analysis of big data. The main challenge of Agrobots is to work on a plant scale with extreme accuracy. Agrobots can save labor costs, prevent workers from performing risky operations, and provide the farmer with up-to-date and precise information for management decision-making. Thus, developing cost-effective Agrobots is a newly introduced field of science especially in industrialized countries [1, 2]. Agrobots typically use RGB sensors, thermal sensors, infrared sensors, laser and stereo cameras under changing lighting conditions or artificial illumination, in order to recognize and perform tasks in the farm.

© Springer Nature Switzerland AG 2019
D. Tzovaras et al. (Eds.): ICVS 2019, LNCS 11754, pp. 98–109, 2019.
https://doi.org/10.1007/978-3-030-34995-0_9

Computer vision is considered as a significant part for developing intelligent Agrobots. Harvesting robots use visual cues, with RGB sensors, such as spectral reflectance, thermal response, texture and shape information. Detection of fruits, estimation, recognition of variety etc. depend on visual cues in a digital image. The detection of fruit, discrimination and localization of space, are computer vision tasks that allow an Agrobot to perform precision agricultural tasks in a human-like way.

Our interest is focused on autonomous robot harvesters in precise farming applications, in particular grape harvesters in vineyards. The challenge here is the detection of grapes, which can be found in many varieties, forms and stages of harvesting. Tasks in vineyards need specific actions for keeping the high quality of the grapes guiding to high quality winery. Images of grapes under variable illumination contain large variations in color intensities, scale and size of grapes, weather conditions, obstacles occurrences e.g. leaves or branches and shadows that may changes the real shape of grapes. After the successful detection of the grapes, many tasks can be applied; classification of grapes variety, estimation of grapes maturation state, localization of grapes during the leaf thinning task or calculation of cutting point of the peduncle for harvesting the grapes [3].

In this paper, a visual computing method for grapes detection is examined, based on local texture descriptors. The primary goal of this work is to analyze the ability for background subtraction of the conventional LBP texture descriptor and its variants $LBP_{riu}, VARLBP, OCLBP, SILTP, CSLBP, BGLBP$.

The rest of the paper is organized as follows: Sect. 2 highlights the role of visual computing as a tool for harvesting robots, Sect. 3 briefly presents the fundamental theory of the used local texture descriptors, while Sect. 4 introduces the applied grapes segmentation methodology. Section 5 presents the results of the contacted experiments and Sect. 6 summarizes the main conclusions of this work and puts ahead the next research actions that need to be scheduled.

2 Visual Computing for Harvesting Robots

The first task of a visual computing based Agrobot in grapes harvesting is to locate the grapes in a vine area using artificial vision. There are many applications of computer vision in agriculture [4–9].

A robust grapes segmentation method has been proposed in [4] based on the color channels of RGB images, with average detection accuracy 93.74%. The lighting conditions of the depicted bunches of grapes were varied and the vineyard environment was complex. After testing the color channels RGB, HSV, and L^*a^*b, the proposed method chooses a combination of color components as input to linear classifiers under the Adaboost ensemble framework. Color has proved to play an important role for the detection of a bunch of grapes especially in the HSV color space [5, 6]. HSV color space represents hue as color shade, saturation as the combination of light intensity and the quantity of distribution across the spectrum of different wavelengths, and the value component as intensity. Other methods use artificial illumination with high resolution images with HSV color space [7] as harvesting task can also be performed at night with pixelwise segmentation for red grapes. A novel method [5] uses Zernike moments for grape detection in outdoor images with RGB and HSV color spaces based on shape detection. SVM was used for classification with less of 0.5% error for few of the samples.

LBP with Adaboost classifier was tested in [8] for removing the false positives on detection and counting of immature green citrus fruit, with overall succeed accuracy 85.6%. A HOG & LBP combination with $SVM - RBF$ classifier for both white and red varieties by using grayscale normalized images under natural illumination was proposed in [9]. In [9], the proposed method involves a combination of robust techniques. Preprocessing steps include down sampling of the images and searching of circular forms with a predefined range to help the location of interest points of grape clusters. Processing steps include the computation of a combined feature vector with HOG and Uniform LBP used for classification, and DBSCAN for elimination of isolated berries. The reported performance was 95% accuracy, 92.5% recall and 99.41% precision.

Deep learning methods for object detection are also reported in the literature [10, 11]. The reported methods You Only Look Once (YOLO) [10] and fast Region-based Convolutional Neural Network (R-CNN) [11] can process images in real-time, providing both high speed and high detection performance with GPU support. However, the proposed technique is considered adequate to be embedded on an Agrobot, since LBPs can run fast on a CPU, requiring comparatively fewer samples for training.

3 Local Binary Patterns

Texture in real world has some significant statistical properties with similar repeating structures, usually with some degree of randomness. The mean texture is a set of texture elements, or *texels*, which can be found in a repeated pattern. Visually can be formed by shape, reflectance, shadows, absorption and mostly by illumination. Repeating patterns can occur in many scales in real world. As the digital depiction of them can be exploited as forming textures from intensities, they can be effective in case color is not a discrete visual cue.

Ordinary Local Binary Pattern (LBP) is a texture measure that can aid in recognition of textures even under varying illumination. It is a non-parametric method; it is tolerant to monotonic illumination changes and is characterized by computational simplicity. It was introduced by Ojala et al. [12], who implemented the method experimentally. LBP is proven to exhibit high performance and robustness. In its original form, LBP can detect small scale textures due to a binary thresholding method of central pixel with its closest neighborhood, in a 3 × 3 region. The threshold comes from the central pixel value and its expression is described in Fig. 1. Every pixel of an image is calculated to a new 8-bit binary value, which expresses the sum of the passed bit from the central's pixel threshold.

Rotation Invariance
LBP is implemented by changing the radius and the equally spaced pixels of the local neighborhood, forming a circularly symmetric set of neighbors as described by Ojala et al. [13]. The definition of texture is described as T and the values of gray levels as $P + 1$ where $P > 0$. Texture can be described by g_c as a central pixel value and g_p where $p = 0 \ldots P - 1$ as gray values of P pixels that belong to a circle of radius R ($R > 0$).

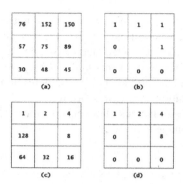

Fig. 1. Local Binary Pattern calculation: (a) is a region 3×3 pixel values and the central value is 75. If the neighbor other pixels are smaller from the central value, then their new value will be 0, whilst other that are bigger from the central value will have 1, (b) based on the clockwise binary sequence, each 1 will be calculated as decimal value from corresponding table (c) and the result will be the sum of the values in (d) as the new value of LBP image.

$LBP_{P,R}$ is calculated by the binomial weight with the transforming difference in a neighborhood as shown in Eqs. (1) and (2):

$$LBP_{P,R} = \sum_{p=0}^{P-1} s(g_p - g_c) 2^p \tag{1}$$

$$s(g_p - g_c) = \begin{cases} 1 & g_p \geq g_c \\ 0 & g_p < g_c \end{cases} \tag{2}$$

In order to achieve rotation invariance, $LBP_{P,R}$ is calculated with the function $ROR(x, y)$ which shifts circularly the P-bit binary number, i times to the right according to Eq. (3)

$$LBP_{P,R}^{ri} = \min\{ROR\{LBP_{P,R}, i\} \mid i = 0, 1 \ldots P - 1\} \tag{3}$$

Uniformity

This mechanism of binary code offers more characteristics and implementations. In is implemented a simple LBP with uniformity (ELBP). Uniformity can be expressed by the changing binary sequence of bits, when the next of 1 is 0, or the opposite; this expression is a binary transition. When the transitions are less or equal to 2, then the condition is called uniform and the operator can be expressed as $LBP_{P,R}^{u2}$ as defined in Eq. (4).

$$LBP_{P,R}^{riu2} = \begin{cases} \sum_{p=0}^{P-1'} s(g_p - g_c) \, If \, U(N(P, R)) < 2 \end{cases} \tag{4}$$

Invariance Measure

Invariant Texture Classification (VARLBP) [14, 15] is an enhanced $LBP_{P,R}^{riu2}$ with a rotation invariance measure $VAR_{P,R}$. The operator can measure the spatial structure of a local image, which lacks one significant ability; the contrast. The $VAR_{P,R}$ measure adds the factor of contrast. This generalized method is computationally attractive because only few operations are needed. For this method $LBP_{P,R}^{riu2}$ is calculated along with the uniformity measure. This calculation also loses the definition of contrast; therefore, it can be measured with a rotation invariant measure of local variance $VAR_{P,R}$, which is invariant against gray scale changes. This enhancement offers a powerful rotation invariant measure of local image texture.

Opponent Color LBP
Texture and color [16] are combined and the LBP is extracted in 6 color components by using the theory of opponent color. Initially, the use of color distribution in texture is tested. Every couple of color components is collected in a way to form a color pattern with center pixel of one-color component with neighborhood from different color components. This method can be efficient in static illumination, and it can be used when color is a visual cue of texture. From OCLBP, a 6-dimension LBP feature is extracted.

Back Ground LBP
For background extraction, Back Ground LBP (BGLBP) [17] can deal efficiently with assigned complexities in term of time and computational memory. BGLBP is based on ULBP with a significantly reduced number of histogram bins compared to original LBP. Its computation has an anti-clockwise ability and is described in (5).

$$BGLBP_{P,R} = \begin{cases} \sum_{i=0}^{(P/2)-1} s\left(g_p, m, g_{\frac{P}{2}+i}\right) \times 2^i & U\left(LB\frac{P}{2}_{P,R}\right) \leq 2 \\ 2^{\frac{P}{2}} & otherwise \end{cases} \quad (5)$$

Center Symmetric LBP
Center Symmetric LBP (CSLBP) is a combined variant with SIFT descriptor [18]. Both feature descriptors inherit the benefits of robustness, tolerance to illumination changes and computational simplicity. The value of pixels is not compared like it is done with LBP, which compares the center with its neighbors, but each pixel of the neighbor of the center can be compared with the opposite side (Fig. 2) as shown in Eq. (6).

Fig. 2. The function of CS-LBP, the pixels are calculating without the central pixel

$$CS - LBP_{R,N,T}(x, y) = \sum_{p=0}^{\left(\frac{N}{2}\right)-1} s\left(g_p - g_{p+(N/2)}\right) 2^p, s(x) = \begin{cases} 1 & x > T \\ 0 & otherwise \end{cases} \quad (6)$$

Scale Invariant Local Ternary Pattern

Scale Invariant Local Ternary Pattern (SILTP) is an extended LTP [19]. I_c is the value of the center pixel in location (x_c, y_c) and $\{I_k\}_{k=0}^{N-1}$ describes the values of the N pixels inside the neighbor, which they are equally spaced on a circle of radius R. The calculation of SILTP is defined in (6).

$$SILTP_{N,R}^{\tau}(x_c, y_c) = \oplus_{k=0}^{N-1} s_\tau(I_c, I_k) \tag{7}$$

It is worth to note that after the first introduction of the standard LBP descriptor, many attempts to enhance the original LBP or to develop different texture descriptors following the main encoding mechanism of the LBP, was presented in the last two decades [20, 21]

4 LBP-Based Grapes Segmentation

Segmentation is a common computer vision process to detect the regions of interest inside an image. For object recognition, segmentation task divides the image into interest regions in the form of a binary mask. Foreground segmentation is a task where the object is separated by the background, providing 2 or more classes depending on the number of objects that need to be detected. The objective of this work is to segment the image into two classes to indicate the interest regions that contain a grape bunch.

The ground truth images are formed manually, each pixel value represent the class of pixel, '0' (black) for 'no grape bunch' class or '1' (white) for 'grape bunch' class. The aim of the ground truth images is to label the data for supervised machine learning to build an accurate classifier that can extract in real time the segmented images.

One big challenge in computer vision is the detection of white varieties of grapes in a green leaf background. The similarity of color between the grapes and the leaves makes the detection quite difficult, something that also happens for the human perception. Moreover, the captured images have varied illumination, orientation and scale. Under these circumstances, LBP can be a robust texture descriptor with a discriminant power able to distinguish the grapes from the background. The used LBP variants and their characteristics are summarized in Table 1.

The proposed methodology is divided into four steps; Pre-processing, Feature extraction, Classification and Post-processing:

Pre-processing

Histogram equalization is applied to enhance the quality of the images. The images are transformed to HSV color space since this color space seemed to be more informative for grapes detection [5–7]. This procedure is applied to all LBP variants except $OCLBP$, which does not require the extraction of color, as it takes for input the RGB normalized downscaled image for extraction of a 6-dimensional LBP feature based on opponent color theory.

Feature Extraction

After the color transformation, LBP and variants [22] are applied on the three-color components and three LBP images were extracted in each case. After experimentation with different values for the radius and number of neighboring pixels, it is decided to use radius 1 and 8 neighboring pixels. The LBP feature vector is constructed as follows: after the computation of the LBP images for each color component, the 8 LBP values of the 3×3 neighbor were selected, excluding the central value. Finally, a 24 size feature vector is formed to describe the central pixel of a 3×3 neighbor, which along with the class information (label: background or foreground) are used to train the classifier.

Classification

The k-NN classifier is used to classify each pixel into background and foreground classes. GridSearch is conducted with 10-fold cross validation and the best parameters are k = 23 and k = 24 for red and white grapes respectively.

Post-processing

In order to improve the final segmentation result of the k-NN classifier, morphological filtering is applied. First, a morphological opening applied for removing small noise areas and after a morphological closing for filling holes of the foreground object. For each morphological operation, elliptical structure elements were used.

Table 1. LBP variants and their characteristics

LBP variants	Characteristics
LBP	Measures the spatial structure of local image
BGLBP	Less sensitivity to noise, uses histogram for each block
CSLBP	Tolerance to illumination changes, computational simplicity
SILTP	Small tolerate range, counteract illumination variations
ELBP	Contrast measures and fewer bins of histogram
VARLBP	Powerful rotation invariant measure of local image texture
OCLBP	Uses opponent color

5   Simulations

A set of experiments was conducted for studying the performance of the proposed methodology and each one of the LBP texture descriptors. The experiments were executed using MATLAB (LBP implementations) and Python (using OpenCV, Scikit-learn and Scikit-image libraries) programming environments.

Two datasets of both white and red grape images were formed to test the proposed algorithm. Each dataset consists of 50 images; 45 images for training and 5 for testing.

Initially, the images were converted from RGB to grayscale and the k-NN model was applied to segment both varieties, with accuracy 60% on average for all LBP descriptors, whereas $OCLBP$, which applied to the RGB images, showed better performance with accuracy over 90% for the case of the red variety. The used dataset consists of close up viewpoints of grapes, however the method could also work reliably for more distant views of high-resolution images.

In a second stage, color channel wise operations were examined by converting the images to, RGB, HSV and L^*a^*b, with the performance measures being improved compared to the grayscale images. Figure 3 shows the color components for the three examined color spaces.

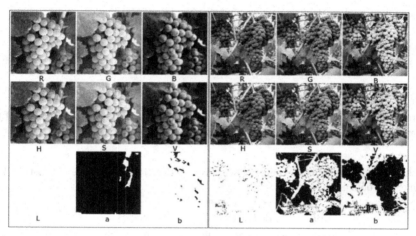

Fig. 3. Color components of white (left) and red (right) varieties of grapes. (Color figure online)

The original LBP descriptor was applied for each color component of the three different color spaces, and its segmentation performance in each case was studied for both grapes varieties. Table 2 summarizes the segmentation accuracy (%) for each color component and for red and white grapes. From Table 2, it is deduced that the H channel of the HSV color model outperforms the other channels for the red as well as for the white grapes, and thus this color representation was selected for the next experiments.

Note that for white varieties of grapes, the conducted experiments had very low accuracy compared to the red variety, so several experiments for investigating the performance of some free parameters were executed. More precisely, different radius and number of neighbor points, e.g. $LBP_{1,8}$, $LBP_{2,16}$, $LBP_{3,24}$, were examined yielding almost the same performance. Moreover, an attempt to combine different LBP variants, e.g. $VARLBP/LBP$, $CSLBP/LBP$ and $SILTP/LBP$, was done without any performance improvement. Finally, several color channels combinations of the color spaces RGB, HSV and L^*a^*b were also tested without having significant increase in segmentation accuracy.

After the above mentioned sensitivity analysis, radius 1 and 8 neighboring pixels were selected as parameters, while the color components of the HSV color space was used to extract grapes texture information. The performance indices of accuracy, recall

Table 2. Accuracy evaluation of the color components.

Color components	Red grapes	White grapes
R	66%	65%
G	85%	61%
B	87%	62%
H	92%	68%
S	50%	63%
V	61%	62%
L	60%	62%
a	61%	57%
b	58%	59%

and precision of the proposed methodology for both grape varieties, for the case of the LBP based descriptors under evaluation are summarized in Table 3.

Table 3. Performance indices Accuracy (Acc.), Recall (Rec.) and Precision (Pre.) of the LBP descriptors for both grapes varieties.

LBP	Red grapes			White grapes		
Descriptor	Acc.	Rec.	Pre.	Acc.	Rec.	Pre.
LBP	94%	93%	96%	83%	78%	77%
ELBP	93%	95%	93%	82%	78%	79%
VARLBP	93%	95%	92%	83%	80%	76%
CSLBP	54%	52%	63%	78%	81%	74%
SILTP	73%	67%	76%	68%	56%	61%
BGLBP	58%	53%	48%	71%	72%	49%
OCLBP	94%	98%	92%	73%	65%	74%

From the above results it is concluded that the segmentation of the white grapes constitutes a difficult task due to the high similarity of the grape bunch with the background including leaves. However, there are some LBP variants e.g. $VARLBP$, $ELBP$ that provide an acceptable segmentation accuracy up to 83%. On the other hand, the segmentation of the red grapes seems to be more reliable for the LBP, $VARLBP$, $ELBP$ and $OCLBP$, achieving high accuracy up to 94%.

An example of the segmentation performance of the k-NN classifier using the $OCLBP$ features for the case of the red variety having accuracy 94% is depicted in Fig. 4. From this figure, one can also see the result of the post-processing morphological filtering, which shapes the final segmentation result.

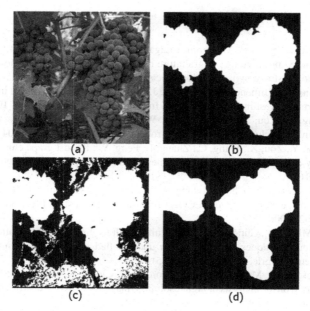

Fig. 4. (a) An image that contains a bunch of grapes from red variety, (b) the corresponding ground truth manual segmented image, (c) the segmented image using OCLBP and k-NN, (d) final result after post-processing stage.

6   Discussions and Conclusions

In the previous section, the conventional LBP texture descriptor and its variants were evaluated in grapes detection under varying illuminations for red and white grapes varieties. The LBP, $VARLBP$, $ELBP$ and $OCLBP$ descriptors, proved to be efficient grapes bunch detectors in red grapes variety with up to 94% accuracy. In white grapes, they did not perform very well, with the highest accuracy being at 83%, due to the high similarity between foreground and background.

This study reveals that the LBP descriptors can be applied for a robust red grape detection, allowing an Agrobot to perform the needful actions of grape estimation, measuring, calculation of finding the peduncle of grape and harvesting, combined data from equipped sensors (Lidar, Thermal camera, NIR, stereoscopy). However, for detecting the white variety the LBP descriptors cannot distinguish the grapes from their background very accurately. Contouring is a simple process that simplifies the shape of grape bunch and could be applied at post-processing step in order to improve the segmentation result. Moreover, spectral imaging can be obtained for a robust calculation, as it gives more visual clues to distinguish the white variety. Finally, we propose testing of the algorithm with alternative classification schemes, with bigger grapes dataset.

Acknowledgment. This research has been co-financed by the European Union and Greek national funds through the Operational Program Competitiveness, Entrepreneurship and Innovation, under the call RESEARCH – CREATE – INNOVATE (project code: T1EDK-00300).

References

1. Rosenberger, C., Arguenon, V., Bergues-Lagarde, A., Bro, P., Rosenberger, C., Smari, W.: Multi-agent based prototyping of agriculture robots using additional soft biometric features to enhance the keystroke dynamics biometric system view project (2006)

2. Ceres, R., Pons, J.L., Jiménez, A.R., Martín, J.M., Calderón, L.: Design and implementation of an aided fruit-harvesting robot (Agribot). Ind. Robot Int. J. **25**, 337–346 (1998)

3. Luo, L., Tang, Y., Zou, X., Ye, M., Feng, W., Li, G.: Vision-based extraction of spatial information in grape clusters for harvesting robots. Biosys. Eng. **151**, 90–104 (2016)

4. Luo, L., Tang, Y., Zou, X., Wang, C., Zhang, P., Feng, W.: Robust grape cluster detection in a vineyard by combining the AdaBoost framework and multiple color components. Sensors **16**, 2098 (2016)

5. Chamelat, R., Rosso, E., Choksuriwong, A., Rosenberger, C., Laurent, H., Bro, P.: Grape detection by image processing. In: IECON Proceedings of Industrial Electronics Conference, pp. 3697–3702 (2006)

6. Aquino, A., Millan, B., Gutiérrez, S., Tardáguila, J.: Grapevine flower estimation by applying artificial vision techniques on images with uncontrolled scene and multi-model analysis. Comput. Electron. Agric. **119**, 92–104 (2015)

7. Font, D., Tresanchez, M., Martínez, D., Moreno, J., Clotet, E., Palacín, J.: Vineyard yield estimation based on the analysis of high resolution images obtained with artificial illumination at night. Sensors (Switzerland) **15**, 8284–8301 (2015)

8. Wang, C., Lee, W.S., Zou, X., Choi, D., Gan, H.: Detection and counting of immature green citrus fruit based on the local binary patterns (LBP) feature using illumination - normalized images. Precis. Agric. **19**, 1062–1083 (2018)

9. Pérez-zavala, R., Torres-torriti, M., Cheein, F.A., Troni, G.: Original papers a pattern recognition strategy for visual grape bunch detection in vineyards. Comput. Electron. Agric. **151**, 136–149 (2018)

10. Redmon, J., Divvala, S., Girshick, R., Farhadi, A.: You only look once: unified, real-time object detection (2016)

11. He, K., Gkioxari, G., Dollár, P., Girshick, R.: Mask R-CNN (2017)

12. Ojala, T., Pietikäinen, M., Harwood, D.: A comparative study of texture measures with classification based on featured distributions. Pattern Recogn. **29**, 51–59 (1996)

13. Ojala, T., Pietikainen, M., Harwood, D.: Performance evaluation of texture measures with classification based on Kullback discrimination of distributions. In: Proceedings of 12th International Conference on Pattern Recognition, vol. 1, pp. 582–585. IEEE Computer Society Press (1999)

14. Ojala, T., Pietikainen, M., Maenpaa, T.: Multiresolution gray-scale and rotation invariant texture classification with local binary patterns. IEEE Trans. Pattern Anal. Mach. Intell. **24**, 971–987 (2002)

15. Pietikäinen, M., Ojala, T., Xu, Z.: Rotation-invariant texture classification using feature distributions. Pattern Recogn. **33**, 43–52 (2000)

16. Mäenpää, T., Pietikäinen, M.: Classification with color and texture: jointly or separately? Pattern Recogn. **37**, 1629–1640 (2004)

17. Davarpanah, S.H., Khalid, F., Nurliyana Abdullah, L., Golchin, M.: A texture descriptor: background local binary pattern (BGLBP). Multimed. Tools Appl. **75**, 6549–6568 (2016)

18. Heikkilä, M., Pietikäinen, M., Schmid, C.: Description of interest regions with center-symmetric local binary patterns. In: Kalra, P.K., Peleg, S. (eds.) ICVGIP 2006. LNCS, vol. 4338, pp. 58–69. Springer, Heidelberg (2006). https://doi.org/10.1007/11949619_6

19. Liao, S., Zhao, G., Kellokumpu, V., Pietikainen, M., Li, S.Z.: Modeling pixel process with scale invariant local patterns for background subtraction in complex scenes. In: Proceedings of the 2010 IEEE Computer Society Conference on Computer Vision and Pattern Recognition, pp. 1301–1306. IEEE (2010)
20. Papakostas, G.A., Koulouriotis, D.E., Karakasis, E.G., Tourassis, V.D.: Moment-based local binary patterns: a novel descriptor for invariant pattern recognition applications. Neurocomputing **99**, 358–371 (2013)
21. Brahnam, S., Jain, L.C., Nanni, L., Lumini, A. (eds.): Local Binary Patterns: New Variants and New Applications. Studies in Computational Intelligence, vol. 506. Springer, Heidelberg (2013). https://doi.org/10.1007/978-3-642-39289-4
22. Silva, C., Bouwmans, T., Frélicot, C.: An extended center-symmetric local binary pattern for background modeling and subtraction in videos. In: Proceedings of the 10th International Conference on Computer Vision Theory and Applications, pp. 395–402 (2015)

Open Space Attraction Based Navigation in Dark Tunnels for MAVs

Christoforos Kanellakis$^{(\boxtimes)}$, Petros Karvelis, and George Nikolakopoulos

Robotics Team, Department of Computer, Electrical and Space Engineering,
Luleå University of Technology, SE-97187 Luleå, Sweden
chrkan@ltu.se

Abstract. This work establishes a novel framework for characterizing the open space of featureless dark tunnel environments for Micro Aerial Vehicles (MAVs) navigation tasks. The proposed method leverages the processing of a single camera to identify the deepest area in the scene in order to provide a collision free heading command for the MAV. In the sequel and inspired by haze removal approaches, the proposed novel idea is structured around a single image depth map estimation scheme, without metric depth measurements. The core contribution of the developed framework stems from the extraction of a 2D centroid in the image plane that characterizes the center of the tunnel's darkest area, which is assumed to represent the open space, while the robustness of the proposed scheme is being examined under varying light/dusty conditions. Simulation and experimental results demonstrate the effectiveness of the proposed method in challenging underground tunnel environments [1].

Keywords: Depth map estimation · Open space attraction · Visual navigation · Micro Aerial Vehicles

1 Introduction

1.1 Motivation and Related Works

Deploying MAVs in harsh subterranean environments for long term and large scale operations challenge their integrity over time. Lately, the concept of aerial scout robots is emerging in a way to support the operation of the aerial platform that carries the expensive inspection and navigation sensorial units for accomplishing subterranean missions. These platforms are defined as low-cost and lightweight aerial robots that are capable to fly fast in subterranean environments. Their main task is to explore unknown areas, collect data and transmit them back to a ground station.

Briefly, harsh underground environments pose obstacles for flying MAVs, like the narrow passages, reduced visibility due to rock falls, dust, wind gusts and

This work has been partially funded by the European Unions Horizon 2020 Research and Innovation Program under the Grant Agreement No. 730302 SIMS.

D. Tzovaras et al. (Eds.): ICVS 2019, LNCS 11754, pp. 110–119, 2019.
https://doi.org/10.1007/978-3-030-34995-0_10

lack of proper illumination, all of which constitute necessary the development of elaborate control, navigation, and perception modules for these vehicles. This work focuses on a method that will allow MAVs to fly into the darkness of the tunnel, characterizing the open space into a 2D centroid in the image plane. More specifically, inspired by the low illumination challenge, the presented work tries to identify alternative ways for generating a proper guidance command (heading) by using low cost equipment, like a single camera. When flying in underground tunnels, that lack natural illumination, can be really challenging for the visual sensors to provide any useful input. Nevertheless, in this article the main drawback is twisted into the main advantage for the visual sensors, by following the intuition that the darkest area in the image consists of the front looking open space of the tunnel. More specifically, the robot is equiped to carry onboard an illumination source that uniformally illuminates the surrounding walls and ground, while the illumination descreases in longer distances.

Several works in the existing literature have addressed the navigation of MAVs in challenging environments using various sensor configurations. In [8] the fields of estimation, control and mapping for the MAV's autonomous navigation along penstocks, have been studied. In this work the major sensors used were a laser range finder and four cameras for the task of state estimation and mapping. In [10], a range based sensor array approach has been developed to navigate along right-rectangular tunnels and cylindrical shafts. The authors proposed a range sensor configuration, to improve the localization in such environments and provide the means for autonomous navigation. In [7], the authors presented a multi-modal sensor unit for mapping applications, a means for aerial robots to navigate in dark underground tunnels. In this work, the unit consists of a stereo camera, a depth sensor, an IMU and led lights syncs with the camera capture for artificial lightning. Furthermore, the unit has been integrated with a volumetric exploration method, demonstrating the capabilities of the overall system. In [5], a method that fuses visual and depth information with IMU data in an Extended Kalman filter framework has been presented. The proposed system initially extracts keypoints from both the visual and depth sensor using a feature extraction and edge detection method, generating a multi modal feature map. Afterwards, a feature descriptor has been employed and the identified features are fused with the IMU data in the EKF framework.

1.2   Contributions

Nowadays perception algorithms have reached high performance levels in localization and mapping with a major assumption from the scientific and engineering communities that these machines have to operate in environments with adequate features, while the robot motion should be smooth to avoid affecting the outcome of the utilized perception algorithms. It becomes a challenge to accomplish a fully autonomous mission in dark subterranean areas. Thus, in this article, the proposed novel architecture, approaches the challenges of the tunnel like environment from another point of view, incorporating specific image processing steps, suitable for the navigation purposes. More specifically, this work takes advantage of the area darkness, designing a method that identifies open space in the tunnel by processing single image frame streams. The core concept of the proposed approach is to provide an estimation of the depth-map of the tunnel based on the light scattering method. The depth-map is then further processed to deduce the tunnel free space horizontally and vertically. The process steps include: (1) depth map morphology filtering, (2) region clustering on the filtered depth map, and (3) image binarization and central moment calculation for extracting the area centroid. Secondly, a set of simulation and experimental efforts are presented to demonstrate the performance of the method in challenging environments under varying conditions e.g. illumination, dust and various sensing modalities. The results include: (a) a proof-of-concept validation test inside a dark tunnel in a simulated environment, and (b) results on dataset from an actual aerial robot, flying in subterranean tunnel using onboard illumination.

1.3   Outline

The rest of the article is structured as it follows. Section 2 describes the overall framework of the proposed centroid extraction method, while Sect. 3 provides an overview of the simulation results, as well as the results from field datasets, collected from real life experiments with the corresponding analysis and discussion on the results. Finally, Sect. 4 presents the concluding remarks of the developed system.

2   Methodology

In the proposed approach the characterization of the open space along the tunnel is expressed in the centroid position in the image plane and is based on: (a) depth map estimation, (b) image binarization, and (c) centroid calculation, while Fig. 1 depicts the proposed approach.

Fig. 1. An overview of the proposed approach.

2.1   Singe Image Depth Estimation

For the proposed depth estimation scheme the starting point is the light scattering [2] method, a well known process where the light is deflected to other directions and the formation of an image can be defined as follows:

$$I(u,v) = O(u,v) \cdot tr(u,v) + a[1 - tr(u,v)] \tag{1}$$

where $I : [0...M-1] \times [0...N-1] \to \mathbb{N}^2$ is the observed image, $O : [0...M-1] \times [0...N-1] \to \mathbb{N}^2$ is the original image, a is the color of the atmospheric light, $tr(u,v)$ is the transmission term, (u,v) are the pixel coordinates where $u = 0, ..., M-1$ and $v = 0, ..., N-1$ with M the width and N the height of the image. The first term $O(u,v) \cdot tr(u,v)$ is called direct attenuation [11] and the second term $a[1 - tr(u)]$ is called airlight. The transmission term describes the portion of the light that is not scattered and reaches the camera and could be defined as:

$$tr(u,v) = e^{-\beta d(u,v)} \tag{2}$$

where β is the scattering coefficient of the atmosphere and $d(u,v)$ is the depth of the scene for pixel coordinates (u,v). Although (2) can be utilized for the estimation of a depth map from the original image, for the estimation of the terms $tr(u,v)$ the a is required. A widely used method to extract the transmission map is the DCP method, proposed by [4], in order to estimate the depth map of an image, which can be defined as:

$$tr(u,v) = 1 - \omega \left[\frac{I^{dark}(u,v)}{a} \right], I^{dark}(u,v) = \min_{C \in R,G,B} [\min_{z \in \Omega(u,v)} I^c(z)] \tag{3}$$

where ω is a number controlling the desired level of restoration with 1 the highest possible value, $I^{dark}(u,v)$ is the dark channel, $\Omega(u,v)$ is a patch of 15×15 pixels centered on (u,v), I^C is the color channel of the image I and z represents the index of the pixel of $\Omega(u,v)$.

2.2   Open Area Centroid Extraction

Morphological Operation and Clustering. As mentioned before, $d(u,v)$ represents the depth estimation of the captured scene, without providing metric measurements, but rather a normalized representation. Since in this work metric depth information is not required, a refinement procedure is performed on the $d(u,v)$ to smooth the image using a grey scale morphological operation [9]. More specifically, the algorithm employs a morphological closing $\gamma(d)$ defined in Eq. 5, including dilation δ (Eq. 4) followed by erosion ϵ (Eq. 4) operations with an elliptical structuring element S. The structuring element is passed over the whole image $d(u,v)$ and at each spatial position (u,v), the relationship between the element and the image is analyzed. For an image I and a structuring element S the erosion and dilation operations are defined as follows:

$$\begin{aligned} \epsilon(I) := I(u,v) \ominus S(u,v) = \min_{\bar{u},\bar{v} \in S} \{ I(u-\bar{u}, v-\bar{v}) - S(u-\bar{u}, v-\bar{v}) \} \\ \delta(I) := I(u,v) \oplus S(u,v) = \max_{\bar{u},\bar{v} \in S} \{ I(u-\bar{u}, v-\bar{v}) + S(u-\bar{u}, v-\bar{v}) \} \end{aligned} \tag{4}$$

where $\bar{u}$, $\bar{v}$ are set by the size of the structuring element. Overall, the morphological closing tries to brighten small, dark areas to match the values of their neighbours in the image, producing the smoothed image $\gamma : [0...M-1] \times [0...N-1] \to \mathbb{N}^2$.

$$\gamma(d) = \epsilon(\delta(d)) \tag{5}$$

Afterwards γ is processed through a segmentation method to divide into a discrete number of regions that include pixels with high similarities. In the proposed approach, the computationally efficient k-means [12] clustering algorithm is employed in order to segment the depth image into a predefined number of clusters $C = \{k_i\}$, where $k_i \in R^2$ denote the cluster centers and $i = 1, ..., 10$. Initially, the algorithm assigns random k_i for each cluster. Then the euclidean distance of each point $\gamma(u_M, v_N)$ in the image with every k_i is computed as shown in Eq. 6.

$$dist\,(k_i, \gamma(u_M, v_N)) = \sqrt{(\gamma(u_M) - k_{i,u})^2 + (\gamma(v_N) - k_{i,v})^2} \tag{6}$$

Afterwards all image points are assigned to the respective cluster based in Eq. 7.

$$\underset{C_i \in C}{\arg \min}\; dist\,(k_i, \gamma(u_M, v_N))^2 \tag{7}$$

Then, the cluster centroids are calculated again, based on average of all points that currently belong to the cluster and the process is repeated until convergence based on termination criteria. The overall process leads to to the clustered image $d_{clustered} : [0...M-1] \times [0...N-1] \to \mathbb{N}^2$.

In the sequel, $d_{clusterd}$ is further processed to provide a binary image d_{binary} that isolates the dark area from the rest of the image surroundings. More specifically, the cluster center k_i with the minimum image intensity is extracted and the pixels intensities are set to 1, while the rest pixels are set to 0.

Centroid. Finally, the centroid (s_x, s_y) of the binarized depth image $d_{binary}(u, v)$ is extracted using the image moments [3]. More specifically, the geometric moments M_{pq} of $d_{binary}(u, v)$ are defined using a discrete sum approximation as shown in Eq. 8, while (s_x, s_y) are calculated according to Eq. 9.

$$M_{pq} = \sum_u \sum_v u^p v^q d_{binary}(u, v) \tag{8}$$

$$s_x = \frac{M_{10}}{M_{00}}, \; s_y = \frac{M_{01}}{M_{00}} \tag{9}$$

where M_{00} represents the area in the binary image, M_{10} is the sum over x and M_{01} is the sum over y. The normalized pixel coordinates s_x and $s_y \in \mathbb{R}$ can be defined as:

$$s_x = (u - o_x)/f_x, \; s_y = (v - o_y)/f_y, \tag{10}$$

where $o_x \in \mathbb{N}$ and $o_y \in \mathbb{N}$ are the principal points in pixels, f_x, $f_y \in \mathbb{R}$ are the camera focal length for the pixel columns and rows respectively.

The calculated centroid is visualized with a red circle in the sequential frames, while the visual processing architecture provides the centroid with update rates of 20 Hz. Figure 2 depicts an example of the extracted centroid overlaid in the onboard captured image as well as the generated single image depth map.

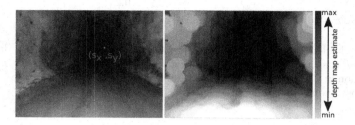

Fig. 2. On the left onboard image frame with the extracted centroid denoted with the red circle, while on the right the estimated depth map from a single image is depicted (Color figure online).

3 Results

This section reports results both from simulation and experimental trials using the proposed method. More specifically, the simulation results present the overall navigation concept depicting the guidance of a MAV along a dark tunnel, while the experimental results focus on the performance analysis of the centroid extraction from collected datasets. Moreover, the code is written in C++ within Robot Operating System[1] (ROS) framework.

3.1 Simulation Results

The proposed framework has been initially evaluated in the Gazebo [6] robot simulation environment interfaced to ROS. The simulated quadrotor is equipped with a front-facing camera, while this simulation demonstrates the envisioned application of the tunnel open area extraction method, described in Sect. 2. The centroid extraction method is part of the overall navigation stack, which also includes state estimation and control that are considered black boxes. In a nutshell, the s_x, s_y are one part of the inputs to the controller, which is assigned to provides the motor commands for the quadrotor to navigate along the tunnel. In this simulation, the gazebo world[2] *"tunnel_practice_2.world"*, developed for the DARPA Subterranean challenge[3], has been selected for deploying the aerial platform.

---

[1] http://www.ros.org/.

[2] https://bitbucket.org/osrf/subt/wiki/Home.

[3] https://subtchallenge.com/.

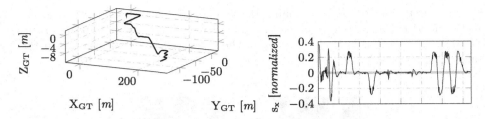

Fig. 3. On the left ground truth 3D performed translation of the MAV during the navigation in the simulated tunnel, on the right centroid s_x evolution over time during the navigation in the simulated tunnel. The following video demonstrates the simulation results https://youtu.be/x5T72ndiMxc.

Figure 3 depicts the total ground truth path that the MAV followed. Additionally, this Figure showcases the evolution of the normalized s_x over time. For the first 20 s, the aerial platform starts from an off-center pose outside the tunnel and therefore the centroid on the x axis is slightly oscillating in a way that guides the MAV inside the tunnel. For the time instances 38 s, 65 s, 140 s, 150 s and 165 s the centroid reaches peak values, depicting that the tunnel is taking a turn and the MAV has to yaw along the axis.

3.2 Experimental Results

Dataset Acquisition and Description. In this section, the proposed novel method is evaluated using datasets collected from real autonomous experiments in an underground mine located in Sweden at 790 m deep, where the underground tunnels did not have strong corrupting magnetic fields, their morphology resembled an S shape environment with small inclination. The dimensions of the area where the MAV navigates autonomously were 6(width) × 4(height) × 20(length)m³. The dataset has been collected from an PlayStation Eye camera, mounted onboard the aerial vehicle, operated at 30 fps and with a resolution of 640 × 480 pixels, a 2D Lidar, while the LED light bars provided 460 lux illumination in 1 m distance. The MAV was flying at 1 m altitude relative to the ground, moving forward with a constant velocity $v_{d,x}$ of 0.2 m/s. Figure 1 provides a visualization of different parts of the underground tunnel.

Performance Analysis. This Section presents the application of the proposed method on the collected datasets, focusing on the performance in real environments, as well as the challenges, such as different illumination conditions and also dust that the algorithm come across. Figure 4 depicts snapshots with the extracted s_x, s_y for different time instances, where on the bottom part provides an overall visualization of the direction commands at various time instances over the traversed course. More specifically, using the data from the onboard 2D lidar, a 2D occupancy grid map is generated, showing the area covered in the tunnel.

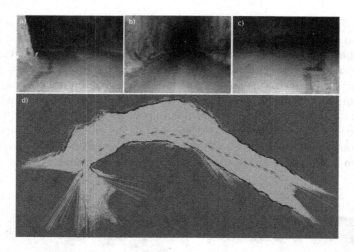

Fig. 4. (a)–(c) depict onboard image frames with the extracted centroid denoted with the red circle, (d) depicts the direction commands with green arrow, the current pose with red arrow both overlaid on a 2D occupancy grid map. The following video demonstrates results https://youtu.be/x5T72ndiMxc from a real subterranean environment (Color figure online).

The map has been augmented with red and green arrows that show the current pose and the instant direction command respectively. The map and the red arrows are part of the collected dataset and cannot be modified, while the green arrows are the result of the proposed centroid extraction method. This figure is representative of the overall performance of the method and can be divided into three parts: (1) the straight tunnel path, (2) the curve, and (3) the open areas. For the straight tunnel path, the majority of the direction commands tend to orient the aerial robot closer to the tunnel axis. For the curve part of the tunnel the centroid is affected by the combination of curve geometry and the illumination. Figure 4(a) demonstrates that there could be cases where the curve is visible long before reaching it and therefore the direction command will lean towards it, which in the sequel will drive the MAV closer to the wall. Overall, this is mostly the case for wide tunnels and the navigation is not majorly affected. The third part of the map shows an open space, where in this situation the proposed algorithm will drive the MAV to the area that is less illuminated.

Applying the proposed method on datasets from real underground tunnels, apart from providing an insight on performance, it also brings up challenges that should be addressed. More specifically, Fig. 5 showcases various occasions where the performance is affected due to harsh conditions. The starting point is the tunnel geometry where the tunnel width forms a critical factor. The algorithm showed to be able to perform equally well in both narrow and wide tunnels and in areas where the tunnels lack any natural illumination (Fig. 5d). Moreover, the developed method is partly able to address the dust floating in the tunnel, while flying (Fig. 5b). The final point of discussion focuses on areas where an additional illumination source exists apart from the onboard, distorting the estimated depth images and overall the guidance command. For wide tunnels the additional illumination source affects slightly the commands when the MAV is flying closely (Fig. 5a), while in narrow tunnels the illumination source affects majorly the guidance command, where is some cases the centroid is pointed in darker areas (e.g. the wall) away from the tunnel open space (Fig. 5c). In a summary, the proposed method has been developed to address the guidance command generation for completely dark areas, assuming that the aerial robot has an onboard illumination source, providing substantial performance characteristics.

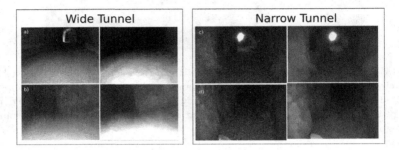

Fig. 5. Challenging test cases including: (a) varying illumination in a wide tunnel, (b) dust in a wide tunnel, (c) varying illumination in a narrow tunnel, and (d) narrow tunnel general case. All cases depict the onboard image with the depth map.

Finally, the centroid method has been tested in image frames collected from an underground tunnel environment with extremely low illumination of 40 lx. The method seems that is able to handle the low illumination levels and provides the estimated depth map from a real dark image, as depicted in Fig. 6. Future directions will focus on a generalized characterization method that will address the problem in cases where dust and varying illumination sources are evident in the scene.

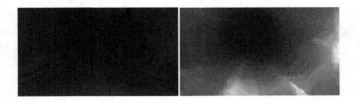

Fig. 6. Centroid extraction method in dark images

4   Conclusions

This work proposed a framework for characterizing open space in featureless dark tunnel-like environments using information from the image stream from a single onboard camera. The proposed method leverages the processing of a single camera to identify the area in the scene with the largest depth value, which is expressed though a 2D centroid in the image plane, without using metric depth measurements. The proposed method has been developed to address the guidance command generation for completely dark areas, providing substantial performance characteristics. Future directions will focus on a generalized characterization method that will address the problem in cases where dust and varying illumination sources are evident in the scene.

References

1. DARPA SubTerranean Challenge. https://www.subtchallenge.com/. Accessed 06 May 2019
2. Cozman, F., Krotkov, E.: Depth from scattering. In: Proceedings of IEEE Computer Society Conference on Computer Vision and Pattern Recognition, pp. 801–806 (June 1997). https://doi.org/10.1109/CVPR.1997.609419
3. Gonzalez, R.C., Woods, R.E.: Digital Image Processing, 3rd edn. Prentice-Hall Inc., Upper Saddle River (2006)
4. He, K., Sun, J., Tang, X.: Single image haze removal using dark channel prior. IEEE Trans. Pattern Anal. Mach. Intell. **33**(12), 2341–2353 (2011). https://doi.org/10.1109/TPAMI.2010.168
5. Khattak, S., Papachristos, C., Alexis, K.: Vision-depth landmarks and inertial fusion for navigation in degraded visual environments. In: Bebis, G., et al. (eds.) ISVC 2018. LNCS, vol. 11241, pp. 529–540. Springer, Cham (2018). https://doi.org/10.1007/978-3-030-03801-4_46
6. Koenig, N., Howard, A.: Design and use paradigms for gazebo, an open-source multi-robot simulator. In: 2004 IEEE/RSJ International Conference on Intelligent Robots and Systems (IROS) (IEEE Cat. No. 04CH37566), vol. 3, pp. 2149–2154. IEEE (2004)
7. Mascarich, F., Khattak, S., Papachristos, C., Alexis, K.: A multi-modal mapping unit for autonomous exploration and mapping of underground tunnels. In: 2018 IEEE Aerospace Conference, pp. 1–7. IEEE (2018)
8. Özaslan, T., et al.: Autonomous navigation and mapping for inspection of penstocks and tunnels with MAVs. IEEE Robot. Autom. Lett. **2**(3), 1740–1747 (2017)
9. Soille, P.: Morphological Image Analysis: Principles and Applications, 2nd edn. Springer, Berlin (2003). https://doi.org/10.1007/978-3-662-05088-0
10. Tan, C.H., Sufiyan, D., Ang, W.J., Win, S.K.H., Foong, S.: Design optimization of sparse sensing array for extended aerial robot navigation in deep hazardous tunnels. IEEE Robot. Autom. Lett. **4**(2), 862–869 (2019)
11. Tan, R.T.: Visibility in bad weather from a single image. In: 2008 IEEE Conference on Computer Vision and Pattern Recognition, pp. 1–8 (June 2008). https://doi.org/10.1109/CVPR.2008.4587643
12. Theodoridis, S., Koutroumbas, K.: Pattern Recognition, 4th edn. Academic Press Inc., Orlando (2008)

6D Gripper Pose Estimation from RGB-D Image

Qirong Tang$^{(\boxtimes)}$, Xue Hu, Zhugang Chu, and Shun Wu

Laboratory of Robotics and Multibody System, School of Mechanical Engineering,
Tongji University, Shanghai 201804, People's Republic of China
qirong.tang@outlook.com

Abstract. This paper proposes an end-to-end system to directly estimate the 6D pose of gripper given RGB and depth images of an object. A dataset containing RGB-D images and 6D poses of 20 kinds, 10 for known objects and 10 for unknown ones, is developed in the first place. With all coordinates information gained from successful grasp, the separation between object properties and grasping strategies could be avoided. To improve the usability and uniformity of raw data, distinctive data preprocessing approach is illustrated immediately after the creation of the dataset. Entire convolutional neural network frame is given subsequently and the training with unique loss function adjusts the model to desired accuracy. Testing on both known and unknown objects verifies our system when it comes to grasping precision.

Keywords: Robotic grasping · 6D pose · RGB-D · Deep learning

1 Introduction

Being the most intelligent creature in our planet, human's invaluable ability to grasp objects, familiar or not, is an impetus for consistent robotic grasping research. The emergence of You Only Look Once (YOLO) [1], GoogLeNet [2], deep residual learning for image recognition (ResNet) [3], Visual Geometry Group net (VGG) [4] and many other deep learning methods makes significant breakthrough in robotic grasp. Datasets of grasping objects are proposed in multiple forms. Since human labeling is not a trivial task and is inevitably biased by semantics, over 700 h of robot grasping is collected in [5]. In the meanwhile, associative learning of multiple robotic tasks including grasping and pushing is studied in [6].

Accurate, real-time robotic grasp based on convolutional neural network (CNN) is gaining the prevalence, however, no one gives direct gripper pose for grasping from RGB-D image and that's what makes our research significant. Technique to seek synergies between pushing and grasping using Reinforcement Learning (RL) is demonstrated

This work is supported by the projects of National Natural Science Foundation of China (No. 61603277, No. 61873192), the Key Pre-Research Project of the 13th-Five-Year-Plan on Common Technology (No. 41412050101), Field Fund (No. 61403120407), the Fundamental Research Funds for the Central Universities and the Youth 1000 program project, the Key Basic Research Project of Shanghai Science and Technology Innovation Plan (No. 15JC1403300), and China Academy of Space Technology and Launch Vehicle Technology.

© Springer Nature Switzerland AG 2019
D. Tzovaras et al. (Eds.): ICVS 2019, LNCS 11754, pp. 120–125, 2019.
https://doi.org/10.1007/978-3-030-34995-0_11

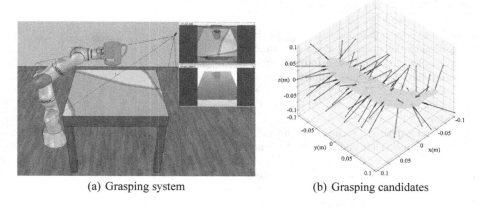

(a) Grasping system   (b) Grasping candidates

Fig. 1. Grasping system and grasping candidates

in [7]. And much more complex situation including clutter, larger variety of objects and the performance in the real world are considered in other related works. Nevertheless, the proliferation of algorithms makes it demanding to distinguish particular approach best suited for rich, diverse grasping task.

Figure 1(a) shows our environment to construct a dataset involving 70, 000 groups of data closely related to the task of grasping itself. Six coordinates of gripper are trained in our distinctive CNN, Anticipation Lies in Vision (ALV), the loss function of which is especially devised to transform the deviation so as to acquire ideal accuracy. Our major contributions are summarized as follows,

(1) 6D gripper pose dataset creation,
(2) individual RGB-D image preprocessing by calculating average images,
(3) special loss function in ALV to conceptualise the deviation.

2   Dataset

Evenly take 5, 000 samples on the trimesh surface of 3D model and calculate their normals. Normalize the length of the vectors to 0.07 m which is exactly among the optimum distance for Barrett Hand to grasp common objects and part of the candidates are shown in Fig. 1(b). Initialize the mass, centroid and inertia of the object, then drop it on the table (0.65 m in height) randomly in the region of

$$D = \{(x,y)| -0.15 \leqslant x, y \leqslant 0.15\}, \tag{1}$$

to make the gripper system effective to an appropriate range of operating distance and orientation instead of a fixed one. Both RGB and depth vision sensors are set in the same rigid position, $(0, 0.45, 1.15, -\frac{5}{6}\pi, 0, 0)$, to maintain consistent perspective.

Algorithm 1. 6D gripper pose and RGB-D from 3D object

---

input: 3D grasp object
output: 6D gripper pose and RGB-D object images
 1: generate $5,000$ points on object and calculate normals
 2: load object and transform coordinates from $\{O\}$ to $\{W\}$
 3: **while** unused candidates number > 0 **do**
 4:     drop object in D
 5:     choose unused candidate
 6:     **for** $i = 1 \to 5$ **do**
 7:         set gripper and rotate wrist randomly
 8:         **if** fingers all in contact with object **and** number of collisions < 70 **then**
 9:             lift object
10:             **if** proximity sensor signal > 0 **then**
11:                 save 6D pose $\boldsymbol{a} = (x_0, y_0, z_0, \alpha_0, \beta_0, \gamma_0)$
12:                 reset object and gripper
13:                 take RGB and depth photos in 5 lights
14:             **end if**
15:         **end if**
16:     **end for**
17: **end while**

Transform the coordinates of candidates from object's body-attached system $\{O\}$ to world coordinate system $\{W\}$, the homogeneous transformation of which is

$$\begin{bmatrix} {}^{W}\boldsymbol{p} \\ 1 \end{bmatrix} = \begin{bmatrix} {}^{W}_{O}\boldsymbol{R} & {}^{W}\boldsymbol{p}_{O_O} \\ 0 & 1 \end{bmatrix} \begin{bmatrix} {}^{O}\boldsymbol{p} \\ 1 \end{bmatrix}, \tag{2}$$

where ${}^{W}_{O}\boldsymbol{R}$ and ${}^{W}\boldsymbol{p}_{O_O}$ represent respectively the orientation and the displacement of $\{O\}$ relative to $\{W\}$. The gripper is positioned in accordance with ${}^{W}\boldsymbol{p}$ for grasping attempt to generate training data.

Algorithm 1 illustrates the whole procedure of grasping. Close the palm until all three fingers are in contact with the object, and the number of collisions between fingers and other entities is under certain threshold. Lift the object to given height and check whether it is still within the reach of proximity sensor fixed at the center of the palm. If the answer is true, the 6D pose of the gripper before lifting is saved and RGB and depth information are taken with the object reset to the original position.

3   CNN

Image preprocessing is of vital importance since it's hard to focus on the key variables in raw data. As is shown in Fig. 2, average images of RGB (b) and depth (e) are calculated by adding all the pixels together and dividing the sum of the images. The object is conformed to Gaussian distribution, and the central area is darker than the border due to the difference of probability. Subsequently, average images are subtracted from the original images (a) and (d) to highlight the characteristic of this particular group of data. Ultimately, depth information is added to the RGB image as the fourth channel and entire RGB-D image is compressed to 128×128.

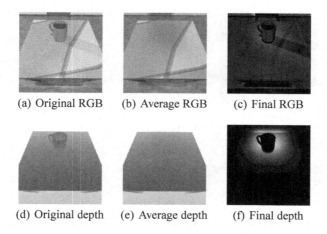

(a) Original RGB   (b) Average RGB   (c) Final RGB

(d) Original depth   (e) Average depth   (f) Final depth

Fig. 2. Data preprocessing

ALV is inspired by YOLO [1] as is shown in Fig. 3. Instead of pretraining the convolutional layers on the ImageNet, we train the whole net in every epoch. RReLu activation function is applied to introduce non-linear features, the mathematical expression of which is

$$\phi(x) = \begin{cases} x, & \text{if } x \geq 0 \\ ax, & \text{if } x < 0 \end{cases}, \tag{3}$$

where $a \in [0.1, 0.3]$ is randomly picked each time. To reflect the gap between model and reality, the loss function is set to be

$$loss = \frac{\sum \left(e^{C \cdot |b-a|} - 1\right)}{B}, \tag{4}$$

where C refers to the coefficient of the 6 coordinates, $b = (x, y, z, \alpha, \beta, \gamma)$ represents the predicted results while a is the ground truth gained through successful grasping. In ideal occasion, the loss should be near to 0, therefore 1 is subtracted from the numerator. Thus the loss can be transformed to the accuracy rate of the prediction, and help adjust the frame to a satisfactory accuracy.

4 Evaluation

10 categories of objects are earnestly chosen to become good representatives of all normal grasping objects, each containing 3 different entities. After training ALV on $10 \times 5,000$ groups (each containing 5 pair of RGB and depth images taken in different lights) of training dataset, $10 \times 1,000$ testing dataset is loaded for evaluation. Table 1 infers that the simpler the model is, the better the performance achieved. Non convex surface calculation is much slower than convex one, and the scene of fingers sticking to object shows quite frequently. Fortunately, this unpleasant simulation problem does not affect the judgement of grasping success. Prolate models are tougher since it is difficult

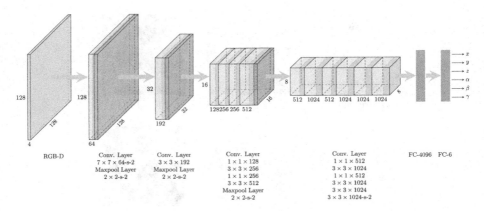

Fig. 3. 6D pose training from RGB-D image

for three fingers to operate elaborately. Delicate protuberance for decoration is another barrier for grasping learning.

As is shown in Table 2, the success rate of unknown objects is far below that of known ones, which is the main problem in current grasping system but, pleasantly, our approach is making progress in that aspect. YOLO has the nearest performance to ALV in both known and unknown objects for ALV is originated from YOLO, and then transformed to be more suitable for our specific task.

Only four most popular nets besides ALV are listed in the tables, and many other polished versions may achieve better results thus deserve more exploration. For networks intended for classification, it is practicable to calculate the maximum and

Table 1. Success rate of known objects grasping

CNN	Avg.	Cup	Mango	Knife	Vase	Cat	Tower	Banana	Car	Egg	Camera
ALV	**87.58**	**91.7**	**96.66**	72.38	**90.34**	**86.12**	**80.2**	89.08	82.00	**80.68**	**92.86**
YOLO	84.56	86.16	91.64	**73.12**	87.04	81.8	78.62	**90.62**	**91.78**	76.82	87.98
ResNet	80.52	81.08	84.4	70.46	86.2	72.82	78.72	85.04	86.4	76.24	83.82
GoogLeNet	77.48	79.64	80.28	68.28	79.9	73.22	77.00	81.06	80.78	75.44	79.16
VGG	71.13	74.62	75.84	62.26	66.22	71.52	63.60	75.58	78.5	69.02	74.14

Table 2. Success rate of unknown objects grasping

CNN	Avg.	Bottle	Baseball	Spoon	Ashtray	Pig	Racket	Spanner	Mouse	Yoyo	Box
ALV	**71.72**	**85.7**	**82.66**	51.38	61.34	**82.08**	61.20	56.08	**79.88**	**84.68**	**85.90**
YOLO	70.90	79.14	77.90	**53.12**	**62.76**	78.10	**63.60**	**56.66**	74.18	80.60	82.98
ResNet	61.48	70.08	66.14	50.60	53.74	64.94	55.72	46.00	65.40	70.28	71.88
GoogLeNet	59.73	68.64	64.26	49.28	55.92	63.22	51.08	45.50	66.78	64.46	68.18
VGG	52.02	61.60	59.84	40.26	45.22	55.46	47.60	41.58	57.50	59.02	52.10

minimum of each coordinate and then divide consecutive values into 50 normalized discrete values to transfer the regression issue into classification. Meanwhile, low success rate does not necessarily negate the net completely since grasping in situation set in this paper is a highly particular one.

5 Conclusions

This paper presents an end-to-end system to predict the 6D pose of a gripper given RGB-D image. First of all, a pose estimation dataset consisting of both known and unknown objects is developed. Data preprocessing is expatiated in the following step, after which ALV is given for training 10 known objects. Unique loss function is derived for estimating the performance of the network and adjusting it to desired accuracy. The performance of ALV is compared with other most popular nets in known and unknown grasping tasks and achieves leading results.

Experiments in real world will be carried out in the near future. The candidate vector and the homogeneous transformation matrix are also saved and may contribute to better performance. Moreover, grasping soft objects with large deformation exerted force has broad application prospects and is an area we eager to set foot in.

References

1. Redmon, J., Divvala, S., Girshick, R., Farhadi, A.: You only look once: unified, real-time object detection. In: IEEE Conference on Computer Vision and Pattern Recognition (CVPR), pp. 779–788 (2016)
2. Szegedy, C., et al.: Going deeper with convolutions. In: IEEE Conference on Computer Vision and Pattern Recognition (CVPR), pp. 1–9 (2015)
3. He, K., Zhang, X., Ren, S., Sun, J.: Deep residual learning for image recognition. In: IEEE Conference on Computer Vision and Pattern Recognition (CVPR), pp. 770–778 (2016)
4. Simonyan, K., Zisserman, A.: Very deep convolutional networks for large scale image recognition. In: International Conference on Learning Representations (ICLR) (2015)
5. Pinto, L., Gupta, A.: Supersizing self-supervision: learning to grasp from 50k tries and 700 robot hours. In: IEEE International Conference on Robotics and Automation (ICRA), pp. 3406–3413 (2016)
6. Pinto, L., Gupta, A.: Learning to push by grasping: using multiple tasks for effective learning. In: IEEE International Conference on Robotics and Automation (ICRA), pp. 2161–2168 (2017)
7. Zeng, A., Song, S., Welker, S., Lee, J., Rodriguez, A., Funkhouser, T.: Learning synergies between pushing and grasping with self-supervised deep reinforcement learning. In: IEEE/RSJ International Conference on Intelligent Robots and Systems (IROS), pp. 4238–4245 (2018)

Robust Rotation Interpolation
Based on $SO(n)$ Geodesic Distance

Jin Wu[1], Ming Liu[1]([✉]), Jian Ding[2], and Mingsen Deng[2]

[1] Department of Electronic and Computer Engineering,
Hong Kong University of Science and Technology, Hong Kong, China
`eelium@ust.hk`
[2] International Joint Research Center for Data Science and High-Performance
Computing, School of Information, Guizhou University of Finance and Economics,
Guizhou, China

Abstract. A novel interpolation algorithm for smoothing of successive rotation matrices based on the geodesic distance of special orthogonal group $SO(n)$ is proposed. The derived theory is capable of achieving optimal interpolation and owns better accuracy and robustness than representatives.

Keywords: Rotation interpolation · Special orthogonal group · Motion analysis

1 Introduction

The n-dimensional rotation matrices take place everywhere requiring motion sensing such as high-precision satellite attitude control [1], scene reconstruction [2], laser-scan matching [3], human motion tracking [4], autonomous assembly [5] and etc. However, since most rotation estimation tasks rely on digital sampling of sensors, the obtained sequences of rotation matrices are always discrete. Many applications have high demands on smoothed rotation estimates so that better industrial engineering quality could be reached. For instance, the laser scanners mounted on a robot typically own the sampling frequency of $1\sim 20$ Hz for indoor obstacle avoidance and robotic egomotion estimation. But with limitations on its original sampling speed, many accurate tasks e.g. real-time visual servoing can not be efficiently implemented [6]. For instance, in a recent study, the hand-eye calibration is solved by a mapping from the special Euclidean group $SE(3)$ to the special orthogonal group $SO(4)$ [7]. Therefore, such hand-eye calibration method will also encounter the data mismatch problems in the presence of unsynchronoused images and robotic readings. For the 3D case, a simple approach dealing with this issue is that we can interpolate the rotation sequence so that smoother motion will be generated. As rotation is usually parameterized with the unit quaternion, the linear interpolation (LERP) and spherical linear interpolation (SLERP) of quaternions are extensively applied [8]. The normalized version of LERP (NLERP) is

© Springer Nature Switzerland AG 2019
D. Tzovaras et al. (Eds.): ICVS 2019, LNCS 11754, pp. 126–132, 2019.
https://doi.org/10.1007/978-3-030-34995-0_12

also designed to maintain the unit norm of quaternion. LERP and SLERP interpolate the rotation using linear weighting and nonlinear weighting on a unit sphere, respectively. They are proposed to cope with the motion discontinuities in computer animation. Many later extended algorithms are based on LERP and SLERP while some include more motion constraints [9–11]. However, LERP and SLERP do not consider the optimality inside the rotation interpolation. As the special orthogonal group $SO(n)$ consists of all n-dimensional rotation matrices, minimizing the geodesic distances to two discrete rotation matrices will produce an optimal interpolation in $SO(n)$. The $SO(n)$-based minimum can better describe the optimal rotational details and will definitely achieve interpolation with less errors. In this paper, we introduce our theory for such a purpose. Main theory and simulation results are presented to show the proposed method along with its efficiency in the following sections.

2   Main Theory

The notations in this paper are analogous to those in [12]. The $n \times n$ rotation matrix $\boldsymbol{R}$ is defined in the special orthogonal group $SO(n)$ such that $SO(n) := \{\boldsymbol{R} | \boldsymbol{R}^T \boldsymbol{R} = \boldsymbol{I}, \det(\boldsymbol{R}) = 1\}$. For any $\boldsymbol{R}$, there is always an associated unit quaternion $\boldsymbol{q}$ representing the same rotation. Given two successive rotation matrices $\boldsymbol{A}, \boldsymbol{B} \in SO(n)$ and a weighting factor $0 < w < 1$, our purpose is to find the optimal interpolated rotation $\boldsymbol{R}$ based on the geodesic of $SO(n)$ to maintain motion smoothness. The error matrix between two arbitrary rotation matrices $\boldsymbol{R}_1$ and $\boldsymbol{R}_2$ can be expressed by

$$\tilde{\boldsymbol{R}} = \boldsymbol{R}_1 \boldsymbol{R}_2^T \tag{1}$$

When $\boldsymbol{R}_1$ and $\boldsymbol{R}_2$ are exactly the same, this error turns into the identity matrix $\boldsymbol{I}$. Hence $\text{tr}\left[(\tilde{\boldsymbol{R}} - \boldsymbol{I})^T (\tilde{\boldsymbol{R}} - \boldsymbol{I})\right]$ represents the error geodesic distance in $SO(n)$. Since all n-D rotations belong to $SO(n)$, using such geodesic distance will be more effective than previous quaternion SLERP on a unit sphere. Figure 1 shows the relationship between $\boldsymbol{A}, \boldsymbol{B}$ and $\boldsymbol{R}$. Therefore, our problem is then cast into finding $\boldsymbol{R}$ via the following optimization:

$$\underset{\boldsymbol{R} \in SO(n)}{\arg \min} \text{tr} \begin{bmatrix} w(\boldsymbol{A}\boldsymbol{R}^T - \boldsymbol{I})^T (\boldsymbol{A}\boldsymbol{R}^T - \boldsymbol{I}) + \\ (1-w)(\boldsymbol{R}\boldsymbol{B}^T - \boldsymbol{I})^T (\boldsymbol{R}\boldsymbol{B}^T - \boldsymbol{I}) \end{bmatrix}$$

$$\Rightarrow \underset{\boldsymbol{R} \in SO(n)}{\arg \min} \text{tr} \begin{bmatrix} w(2\boldsymbol{I} - \boldsymbol{R}\boldsymbol{A}^T - \boldsymbol{A}\boldsymbol{R}^T) + \\ (1-w)(2\boldsymbol{I} - \boldsymbol{R}\boldsymbol{B}^T - \boldsymbol{B}\boldsymbol{R}^T) \end{bmatrix} \tag{2}$$

$$\Rightarrow \underset{\boldsymbol{R} \in SO(n)}{\arg \max} \text{tr} \left\{ \boldsymbol{R} \left[w\boldsymbol{A}^T + (1-w)\boldsymbol{B}^T\right] \right\}$$

Let

$$\boldsymbol{D} = w\boldsymbol{A} + (1-w)\boldsymbol{B} \tag{3}$$

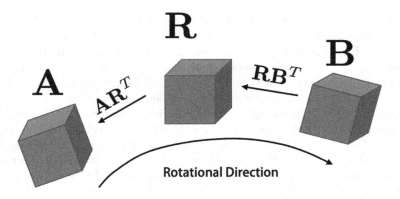

Fig. 1. The rotational relation and differences between $\boldsymbol{A}, \boldsymbol{B}$ and $\boldsymbol{R}$.

Then the problem is solved via a tantamount orthogonal Procrustes solver [13]

$$\boldsymbol{R} = \boldsymbol{U}\mathrm{diag}[1, 1, \det(\boldsymbol{UV})]\boldsymbol{V}^T \tag{4}$$

in which $\boldsymbol{D} = \boldsymbol{USV}^T$ is the singular value decomposition of $\boldsymbol{D}$ with $\boldsymbol{U}^T\boldsymbol{U} = \boldsymbol{V}^T\boldsymbol{V} = \boldsymbol{I}$ and $\boldsymbol{S}$ being the matrix containing singular values of $\boldsymbol{D}$.

When $\boldsymbol{A}$ and $\boldsymbol{B}$ are not strictly rigid i.e. $\boldsymbol{A}^T\boldsymbol{A} \neq \boldsymbol{I}, \boldsymbol{B}^T\boldsymbol{B} \neq \boldsymbol{I}, \det(\boldsymbol{A}) \neq 1, \det(\boldsymbol{B}) \neq 1$, (2) still holds since

$$\begin{aligned} \mathrm{tr}\left(\boldsymbol{RAA}^T\boldsymbol{R}^T\right) = \mathrm{tr}\left(\boldsymbol{AA}^T\right) \\ \mathrm{tr}\left(\boldsymbol{RBB}^T\boldsymbol{R}^T\right) = \mathrm{tr}\left(\boldsymbol{BB}^T\right) \end{aligned} \tag{5}$$

for any $\boldsymbol{R} \in SO(n)$. That is to say even if $\boldsymbol{A}, \boldsymbol{B} \notin SO(n)$, we can still find out such a rotation interpolation $\boldsymbol{R}$ guaranteeing the minimum $SO(n)$ geodesic errors to $\boldsymbol{A}, \boldsymbol{B}$. This attribute allows for the optimal interpolation between two improperly orthonormalized rotation matrices. The non-orthonormality is usually caused by some roundoff errors or number storage limitations, which occurs frequently in engineering practices [14]. For the 3D case, let us recall the LERP and SLERP techniques for quaternion interpolation. The LERP method weights the two successive quaternions $\boldsymbol{q}_1$ and $\boldsymbol{q}_2$ linearly with

$$\boldsymbol{q}_{\mathrm{LERP}} = (1 - w)\boldsymbol{q}_1 + w\boldsymbol{q}_2 \tag{6}$$

The SLERP obtains interpolation by finding the vector part on a unitary sphere in order to preserve the unit norm and maintain the minimum spherical error. Although LERP and SLERP are all feasible in most cases, they have a joint preliminary that $\boldsymbol{A}$ and $\boldsymbol{B}$ must be rigid enough to produce converted quaternions. If $\boldsymbol{A}, \boldsymbol{B}$ are not rigid, pre-orthonormalization is required to be performed on them in advance of interpolation [14]. This in fact will add up multiple computationally expensive operations. However, the proposed algorithm, shown in (4), is free of such preliminary, and would be more robust than both LERP and SLERP.

3 Simulation Results

We simulate cases in this section for 3D motion interpolation. In the following simulations, the weighting factor w is chosen as $w = 0.5$ for equalized interpolation. We simulate the true reference quaternions by

$$q_{\text{true}} = \sin\left(akp + b\right), q = q/\|q\| \tag{7}$$

where $p = 1, 2, 3, \cdots, N$ are indices; $N = 10000$ is the total number of simulated quaternions; $k = 1/500$ denotes the time base factor for a $500\,\text{Hz}$ sampled smooth motion; ε stands for the noise item and here we choose

$$\begin{aligned} a &= (0.21574, -0.43553, -1.7781, 0.49345)^T \\ b &= (1.8575, 0.22406, -1.52913, -0.79564)^T \end{aligned} \tag{8}$$

Using the true quaternions, we construct a true sequence R_{true} of rotation matrices. The sequence R_{true} is then corrupted by white-Gaussian noises ε

$$\begin{aligned} R_{\text{noised}} &= R_{\text{true}} + \varepsilon \\ vec(\varepsilon) &\sim (0, \rho I) \end{aligned} \tag{9}$$

in which $\rho \geqslant 0$ denotes the noise scale. R_{noised} is converted to quaternions for LERP, NLERP and SLERP using the typical method in [15]. The discontinuity problems between $-q$ and $+q$ for all methods are fixed fairly with the algorithm in [16]. The results of the proposed method are compared with that of LERP, NLERP and SLERP. In the first simulation, ρ is picked up from 10^{-5} to 2×10^{-1} while R_{noised} here are orthonormalized using method in [14] for proper rigidity. Defining η as the root mean-squared error (RMSE) between interpolated quaternions and true values, the comparisons are shown in Table 1.

Table 1. Numerical interpolation errors η for strictly orthonormalized rotation sequence.

Algorithms	$\rho = 10^{-5}$	$\rho = 10^{-2}$	$\rho = 2 \times 10^{-1}$
LERP	$3.24178966458071 \times 10^{-5}$	$3.07795249503934613 \times 10^{-3}$	0.062265521206481
NLERP	$3.24178966458036 \times 10^{-5}$	$3.07795249503934687 \times 10^{-3}$	0.062265521206481
SLERP	$3.24178966457941 \times 10^{-5}$	$3.07795249503934730 \times 10^{-3}$	0.062265521206481
Proposed $SO(n)$	$3.24178966457762 \times 10^{-5}$	$3.07795249503934600 \times 10^{-3}$	0.062265521206481

It can be seen from the results that for interpolation with exactly rigid sequence, the interpolated results from $SO(n)$ are slightly better than those from the others. However, the differences between each other can totally be omitted in a modern computer with sufficient processing word-length. Therefore we can say that the four algorithms are almost identical in such case.

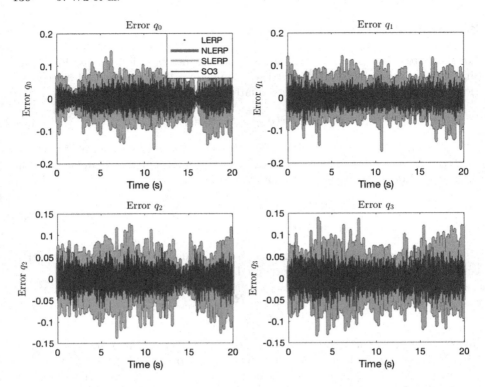

Fig. 2. Quaternion errors of various algorithms with $\rho = 10^{-1}$.

Meanwhile, things will be a little different in the second simulation where R_{noised} is not orthonormalized. The proposed $SO(n)$ method first computes the rotation interpolation and then corresponding quaternion is extracted from the interpolated matrix. For a case when $\rho = 10^{-1}$, the errors between the interpolated and true quaternions are depicted in Fig. 2. We can see that the errors of the proposed $SO(n)$ method are the least. The reason, as mentioned in last section, is that LERP, NLERP and SLERP can not minimize the geodesic error in $SO(n)$ and thus will produce larger errors. As SLERP is better than LERP and NLERP, the quaternion trajectories are also presented in Fig. 3.

The noticeable larger errors of SLERP indicate that it may be less reliable than the proposed $SO(n)$ method for highly accurate rotation-matrix interpolation. We can demonstrate the relationship between η and noise scale ρ in Fig. 4 where $-20\log(\eta)$ is employed to enlarge the differences. The larger value of $-20\log(\eta)$ indicates smaller error. One can see that the proposed method always obtains the best result beyond all algorithms. That is to say the proposed $SO(n)$ method also owns better robustness than other representatives.

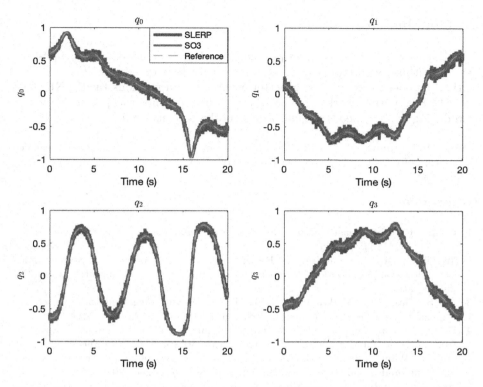

Fig. 3. The computed interpolation results from SLERP and proposed method with $\rho = 10^{-1}$.

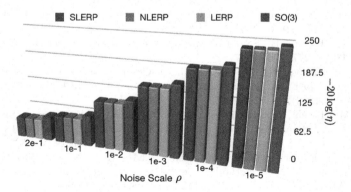

Fig. 4. The relationship between the error magnitude and noise scale.

4   Conclusion

In this paper, the rotation interpolation problem is revisited. We propose a new algorithm for interpolation based on the geodesic distance in $SO(n)$. The method has been proven to own better accuracy than previous LERP, NLERP and SLERP. Future works will be devoted to improving the performance by introducing heterogeneous angular and smoothness constraints.

Acknowledgement. This work has been supported by National Natural Science Foundation of China under the grant of No. 41604025.

References

1. Chang, G.: Total least-squares formulation of Wahba's problem. Electron. Lett. **51**(17), 1334–1335 (2015)
2. Wedel, A., Brox, T., Vaudrey, T., Rabe, C., Franke, U., Cremers, D.: Stereoscopic scene flow computation for 3D motion understanding. Int. J. Comput. Vis. **95**(1), 29–51 (2011)
3. Xu, S., Zhu, J., Li, Y., Wang, J., Lu, H.: Effective scaling registration approach by imposing emphasis on scale factor. Electron. Lett. **54**(7), 422–424 (2018)
4. Sabatini, A.M.: Quaternion based attitude estimation algorithm applied to signals from body-mounted gyroscopes. Electron. Lett. **40**(10), 584–586 (2004)
5. Niu, X., Wang, T.: C2-continuous orientation trajectory planning for robot based on spline quaternion curve. Assemb. Autom. **38**(3), 282–290 (2018)
6. Assa, A., Janabi-Sharifi, F.: Virtual visual servoing for multicamera pose estimation. IEEE/ASME Trans. Mech. **20**(2), 789–798 (2015)
7. Wu, J., Sun, Y., Wang, M., Liu, M.: Hand-eye calibration: 4D procrustes analysis approach. IEEE Trans. Instrum. Meas. (2019, in Press)
8. Shoemake, K.: Animating rotation with quaternion curves. ACM SIGGRAPH Comput. Graph. **19**(3), 245–254 (1985)
9. Park, F.C., Ravani, B.: Smooth invariant interpolation of rotations. ACM Trans. Graph. **16**(3), 277–295 (1997)
10. Leeney, M.: Fast quaternion SLERP. Int. J. Comput. Math. **86**(1), 79–84 (2009)
11. Eberly, D.: A fast and accurate algorithm for computing SLERP. J. Graph. GPU Game Tools **15**(3), 161–176 (2011)
12. Wang, M., Tayebi, A.: Hybrid pose and velocity-bias estimation on SE(3) using inertial and landmark measurements. IEEE Trans. Autom. Control (2018, in Press)
13. Arun, K.S., Huang, T.S., Blostein, S.D.: Least-squares fitting of two 3-D point sets. IEEE Trans. Pattern. Anal. Mach. Intell. **PAMI-9**(5), 698–700 (1987)
14. Bar-Itzhack, I.Y.: New method for extracting the quaternion from a rotation matrix. AIAA J. Guid. Control Dyn. **23**(6), 1085–1087 (2000)
15. Markley, F.L.: Unit quaternion from rotation matrix. AIAA J. Guid. Control Dyn. **31**(2), 440–442 (2008)
16. Wu, J.: Optimal continuous unit quaternions from rotation matrices. AIAA J. Guid. Control Dyn. **42**(2), 919–922 (2019)

Estimation of Wildfire Size and Location Using a Monocular Camera on a Semi-autonomous Quadcopter

Lucas Goncalves de Paula[1], Kristian Hyttel[1], Kenneth Richard Geipel[1],
Jacobo Eduardo de Domingo Gil[1], Iuliu Novac[1],
and Dimitrios Chrysostomou[2]([⊠]) (iD)

[1] Department of Materials and Production, Aalborg University,
Fibigerstraede 16, 9220 Aalborg East, Denmark
[2] Robotics and Automation Group, Department of Materials and Production,
Aalborg University, Fibigerstraede 16, 9220 Aalborg East, Denmark
dimi@mp.aau.dk, http://robotics-automation.aau.dk/

Abstract. This paper addresses the problem of estimating the location and size of a wildfire, within the frame of a semi-autonomous recon and data analytics quadcopter. We approach this problem by developing three different algorithms, in order to accommodate this problem. Two of these taking into the account that the middle of the camera's FOV is horizontal with respect to the drone it is mounted. The third algorithm relates to the bottom point of the FOV, directly under the drone in 3D space. The evaluation shows that having the pixels correlate to ratios in percentages rather than predetermined values, with respect to the edges of the fire, will result in better performance and higher accuracy. Placing the monocular camera horizontally in relation to the drone will provide an accuracy of 68.20%, while mounting the camera with an angle, will deliver an accuracy of 60.76%.

Keywords: Wildfire recognition · Quadcopter analytics · Image processing · Area estimation

1 Introduction

As the frequency of extreme weather increases and become more violent due to climate change [6], so does the initial attack success of wildfires [9]. Emergency response personnel, across the globe, struggle to contain the increase of wildfire size, numbers and severity. In the summer of 2018, California experienced the largest wildfire in the state's history [8], in Sweden emergency personnel were overwhelmed by the numbers of forest fires [12], and in Greece, two violent fires left 250 injured and 105 dead [4]. Considering this, emergency services are investigating how flying robotic technologies can facilitate a faster and more accurate data gathering procedure from wildfires, to increase the efficacy of firefighting operations [3,11].

© Springer Nature Switzerland AG 2019
D. Tzovaras et al. (Eds.): ICVS 2019, LNCS 11754, pp. 133–142, 2019.
https://doi.org/10.1007/978-3-030-34995-0_13

The needs of firefighters combatting wildfires were determined through a collaboration with the Danish Emergency Management Agency, DEMA and interviews with Evan Bek Jensen, the second in command of the Herning drone unit. Discussing a potential new product versus current methods, it became evident that the location, size, intensity and direction of a wildfire were essential for better allocation of their resources. This can be achieved by developing a semi-autonomous quadcopter, that will be controlled remotely via a handheld device. An area of interest is provided as input to the quadcopter by the operator, which later autonomously explores the area. The proposed solution utilizes MobileNet v2 [10], a deep convolutional neural network architecture, as well as a custom built database for recognition and processing of the data of the wildfire, in regards to previous mentioned needs. The contribution of this work lies on the use of a monocular camera for fast calculation and accurate estimation of the location and size of the wildfire area, by comparing the nearest and farthest detected point of the fire with respect to the quadcopter's GPS location.

1.1   Related Works

A number of systems have been developed in order to triangulate the location of a wildfire, most of these include multiple sensors in order to get an accurate location of the fire. One study group recorded the same fire or smoke from different angles. Comparing the position of the sources recording the fire and key terrain features to each other using four sensors, two UAVs and two ground based cameras [2]. Another work used multiple sensors, mounted on three UAVs, and compared the location according to each of the UAVs while recorded the contour of the fire, in order to predict the direction of the fire spread [7]. Recently, Amanatiadis et al. introduced a real-time surveillance detection system for UAVs based on GPGPUs and FPGAs enabling the accurate and fast detection of ambiguous objects [1]. As the presented work is based on a low-cost, commercial drone, the on board computation is kept at minimum due to lack of computational resources and weight restrictions on the drone.

Further Studies in regards to fire recognition have been achieved via a consumer grade monocular camera system. Merino et al. were able to detect wildfire, but also predict its development with regards to vegetation or burned area [5]. Furthermore, Yuan et al. showed that detecting fire can be based not only on the fire palette, but also on the optical flow of the moving flames [13]. Additionally, Zhao et al. presented a novel 15-layered self-learning deep learning architecture to extract fire features and successfully classify them as wildfire [14]. The work presented in this paper will combine elements of these different approaches by detecting the fire using a deep convolutional neural network and later estimate the size and location using a monocular low-budget camera mounted on a quadcopter.

1.2 System Description

The system used for this work relies on an open-source quadcopter, the Intel Aero Ready-To-Fly drone coupled with the Intel Realsense R-200 RGBD camera (Fig. 1). In order to have the computational power needed to process the acquired data, a computer will act as a master, although the system is controlled by an intuitive user interface on the handheld device. Figure 2 illustrates the system's functionalities. Upon arriving at the site of an emergency, the quadcopter's operator in collaboration with the incident commander, identifies in which area the quadcopter must be put into action (Fig. 2a). Utilising the GUI on the tablet, the quadcopter's operator is presented with a topological map, to select an area of interest (Fig. 2b). The quadcopter autonomously flies to this area and begins an area search of a fire (Fig. 2c), flying in a snake pattern when arrives in the area of interest (Fig. 3a). In the incident of the detection of a wildfire, the quadcopter will stream the collected data to the master computer for further processing and calculation of the location, size, intensity and direction of the wildfire. This information will then be overlaid onto the topological map of the user interface, providing a current data interpretation for the firefighters to take into account when allocating resources (Fig. 2d).

2 Proposed Method

Due to the collaboration with DEMA, video footage of real firefighting operations in Denmark, was obtained as Fig. 3b depicts. This is footage from Dokkedal, Denmark, where a fire in a field is shown. DEMA operates with thermal imagery, and scans the area within a range of 80–800 °C, in order to eliminate any noise. In order to apply any calculation in regards to estimate the area of the fire, the algorithm will first determine the edges of the fire. The presented image, of a real wildfire, is required to be in grey scale, while the algorithm detects differences in intensity, and by utilising a dynamic threshold, it sets the edge pixel to 1 while non-edge to 0. Once the edges have been identified, the second part of the algorithm will calculate the distances to the edges, compared to that of the quadcopter's GPS position, in order to calculate the size of the fire.

(a) Intel Aero Ready-To-Fly Quadcopter   (b) Intel RealSense R200

(c) Microsoft Surface Go

Fig. 1. Hardware used for the setup of the project

(a) The firefighter arrives on site and assess the situation

(b) The firefighter selects the area of interest

(c) The quadcopter scans the selected area

(d) The operator receives a visual feedback of the fire's size and location

Fig. 2. Infographic of the stepwise process on the use of the developed framework

Figure 4 illustrates a general approach of this, visualized viewing from the side to highlight how this is calculated. The height of the quadcopter is represented with H which is known from the quadcopter's altitude sensor. Knowing two angles, and the altitude of the quadcopter, $L2 - L1$ represents the fire size and can be calculated by Eq. 1.

$$L2 - L1 = tan(A2 + A1) * H - tan(A1) * H \tag{1}$$

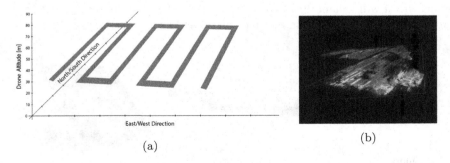

(a)

(b)

Fig. 3. (a) The quadcopter navigates in a snake pattern and (b) a snippet of thermal imagery provided by DEMA.

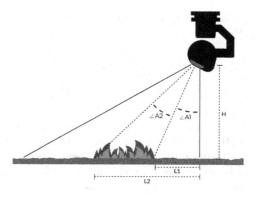

Fig. 4. Side view of the fire size calculation

Based on this, three algorithms were developed, gradually having more generic requirements, in order to calculate the location and the size of the detected wildfire, and determine the optimal way of extinguishing the wildfire. Figure 5 illustrates generic drone viewpoints for both front facing camera (Fig. 5a) and the angled camera (Fig. 5b) approaches with their correspondent total area seen by the drone.

Front-Facing Pixel Value Approach. The first algorithm is based on the premise that the camera faces straightforward in front of the quadcopter, as originally mounted, which means that the upper half of the image does not capture the ground. As a result, only the bottom half of the image can be used, and the input image is cut in half, discarding the top half of the image.

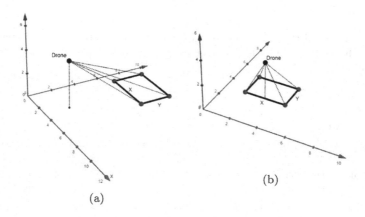

(a)   (b)

Fig. 5. Generic representation of proposed approaches and the respective coordinate frames

The lower half is then divided into 44 equally sized pieces. The lowest visible angle can be seen in Eq. 2.

$$90° - \frac{FOV}{2} = 68.5° \tag{2}$$

The difference from 68.5° to 90° can be divided into 44 steps of 0.5°, hence why the lower half of the image is divided into 44 pieces. Each of these equal parts corresponds to a specific pixel value in the input image. The input images have a resolution of 1920 × 1080 pixels and since only half of each image is used, 540 pixels would be divided equally by 44. Each of these pieces will then be equally spaced by ≈12.53 pixels, see Table 1 for clarification. After finding the edges of the fire in the input image in the form of a pixel value, it will be correlated to the array of pixel values. After finding the pixel value in the table which has the smallest difference to the edge pixel values, the angle can be found in the row above. Using this angle, the distance to the fire can be found. The same approach is taken to find the end of the fire. These two distances are then subtracted to find the location and size of the fire.

Front-Facing Image Ratio Approach. The second algorithm uses the same fundamental principles as the first, but it makes use of a different method to calculate the angle from the centre of the lens to the wildfire's edges. Instead of approximating the pixel value to an array of predetermined pixel values, the second algorithm makes use of the angle of the camera and ratios for increased accuracy. Equation 3 calculates a ratio of where the wildfire's initial position in relation to the height of the image, given as a percentage.

$$FireStart = 100 - \frac{\frac{FireEdgeStart}{ImageHeight}}{2} * 100 \tag{3}$$

Table 1. Angles and their Respective Pixel Values - The specific angle to each point is given by the first row in each line. The second row represents the pixel value of the above angle.

Angle	68.5°	69°	69.5°	70°	70.5°	71°	71.5°	72°	72.5°	73°
Pixel Value	1	12.53	25.06	37.6	50.13	62.67	75.2	87.74	100.27	112.81
Angle	73.5°	74°	74.5°	75°	75.5°	76°	76.5°	77°	77.5°	78°
Pixel Value	125.34	137.88	150.41	162.95	175.48	188.02	200.55	213.09	225.62	138.16
Angle	78.5°	79°	79.5°	80°	80.5°	81°	81.5°	82°	82.5°	83°
Pixel Value	250.69	263.23	275.76	288.3	300.83	313.37	325.9	338.44	350.97	363.51
Angle	83.5°	84°	84.5°	85°	85.5°	86°	86.5°	87°	87.5°	88°
Pixel Value	376.04	388.58	401.11	413.65	426.18	438.72	451.25	463.79	476.32	488.86
Angle	88.5°	89°	89.5°	90°						
Pixel Value	501.39	513.93	526.46	539						

Where, *FireEdgeStart* is the position of the pixel at which the fire starts and *ImageHeight* is the entire height of the input image. This value is subtracted from 100 to find the ratio from the bottom of the image instead of the top. This is necessary because the origin of pixels in an image is in the top-left corner. This equation is calculated twice to find the initial and final position of the fire. Using these ratios, the angles related to the two points can be calculated with the Eq. 4.

$$FireStartDeg = \frac{FireStart * \frac{FOV}{2}}{100} \tag{4}$$

Where, *FireStart* is the value of Eq. 3 and *FOV* is the vertical field of view of the camera. This calculation is also done twice. The results are angles in degrees from the bottom of the image, which represents the point at which the fire starts and ends. After these two calculations have been completed, the distance to each point can be established using the trigonometric functions as shown in Eq. 5.

$$FireDistStart = \tan(FireStartDeg + 68.5°) * Height \tag{5}$$

Where, *FireStartDeg* is the result of Eq. 4 and *Height* is the altitude of the quadcopter when the input image was taken. As it is necessary to take the part below the visible FOV of the camera into consideration, another 68.5° are added.

Angled Camera Image Ratio Approach. The previously described algorithms have been based on the premise that the camera would point directly in front of the quadcopter, this algorithm is based on the camera's lowest FOV point aimed straight down from the quadcopter's position. Therefore, in the *Angle-Improved Image Ratio Approach*, the entire FOV of the camera can be used rather than only half. This is done due to the necessity to address the error that arises near the horizon of the input image. Consequently, the camera is now mounted onto the quadcopter using a 3D printed custom-made mount.

Consequently, the calculations used in *Front-Facing Pixel Value Approach* and *Front-Facing Image Ratio Approach* must be altered to accommodate for the increased FOV and change in angle. Equation 3 will then transform into Eq. 6.

$$FireStart = 100 - \frac{FireEdgeStart}{ImageHeight} * 100 \tag{6}$$

Likewise, Eq. 4 transform into the Eq. 7.

$$FireStartDeg = FireStart * \frac{FOV}{100} \tag{7}$$

Equation 5 transforms in the following way, as the 68.5°, which was added, is now incorporated into the orientation of the camera and, therefore, should no longer be part of the equation, and will have the expression as seen in Eq. 8.

$$FireDistStart = \tan(FireStartDeg) * Height \qquad (8)$$

3   Evaluation

In order to evaluate the robustness and accuracy of the proposed algorithms, additional tests were required to correlate the data from the thermal imagery provided by DEMA, with the quadcopter's altitude, position and FOV data. A white blanket positioned in the middle of a field simulates an area of interest and acts as an input to the edge detection algorithm, Furthermore, the same test setup will be used for all three algorithms to compare the results and reflect the differences in accuracy. This test setup is shown in Fig. 6a and b, as captured from the camera's perspective. As the camera is angled in algorithm 3, so the lowest point of the FOV correlates to the point directly under the quadcopter, the input image will therefore be as seen in Fig. 6b from the camera's perspective.

 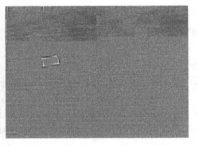

(a) Test setup environments for the   (b) Test setup environment for the
first two algorithms   third algorithm

Fig. 6. Test setup environments for the proposed algorithms

3.1   Results

In order to evaluate the accuracy of the algorithms, the acquired results are shown in comparison to the estimated area and location of the test subject. The distribution of the calculated points of interest generated from the test of the *Front-Facing Pixel Value Approach*, are depicted in Fig. 7a, and resulted in an accuracy of 46.06%. As the *Front-Facing Image Ratio Approach* does not attempt to match an angle to a pixel value, but rather uses a more precise ratio, the results were vastly improved. The resulting calculation provided an accuracy of 68.20%, and the respective distribution of the point of interest is illustrated in Fig. 7b. Lastly, the distribution of the point of interest used for the *Angled Camera Image Ratio Approach* are highlighted in Fig. 7c. Although this test differs from the first two, the resulting calculation of the area within the polygon corresponds to an accuracy of 60.76%.

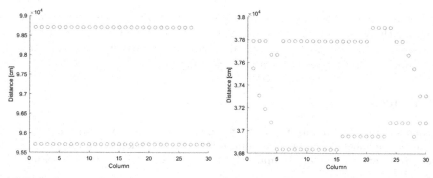

(a) Distribution of points of interest for the test of the *Front-Facing Pixel Value Approach*

(b) Distribution of points of interest for the test of the *Front-Facing Image Ratio Approach*

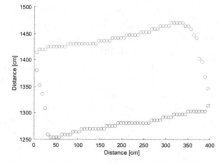

(c) Distribution of points of interest for the test of the *Angle-Improved Image Ratio Approach*

Fig. 7. Test results for all three proposed algorithms

4   Conclusion

This paper proposed three different lightweight algorithms, in order to accommodate the estimation of a specific area of interest for a semi-autonomous, wildfire recognition quadcopter. The main outcome of the performed tests is that in the case of a real scenario, the first two approaches provide an increased safety distance of the quadcopter from the fire. The *Angled Camera Image Ratio Approach* required that the quadcopter is significantly closer to the area of interest, risking the integrity of the quadcopter. As a result, this approach will provide a slightly lower accuracy than that of the *Front-Facing Image Ratio Approach*, with a difference of ≈8%. For the purpose of this study, these tests are designed to simulate the input from a DEMA quadcopter collecting thermal imagery, as previously described. Although, before these algorithms can be implemented into real emergency situations, further testing using images and videos from real wildfires is needed in order to test the accuracy on a multitude of situations.

Additionally, part of the future directions of this project is to develop a way to determine the wildfire's size even if it is not fully included in a single frame and to explore a highly efficient on board algorithm to process the detection and size estimation without the need for a ground computer.

References

1. Amanatiadis, A., Bampis, L., Karakasis, E.G., Gasteratos, A., Sirakoulis, G.: Real-time surveillance detection system for medium-altitude long-endurance unmanned aerial vehicles. Concurr. Comput.: Pract. Exp. **30**(7), e4145 (2018). https://doi.org/10.1002/cpe.4145
2. Martínez-de Dios, J., Merino, L., Caballero, F., Ollero, A.: Automatic forest-fire measuring using ground stations and unmanned aerial systems. Sensors **11**(6), 6328–6353 (2011). https://doi.org/10.3390/s110606328
3. Gates, D.: Drone tracks fire hot spots in successful olympic forest test (2015). https://tinyurl.com/y4egvn9o. Accessed 13 Feb 2019
4. Hansen, S.M., Pedersen, M.S.: 83 dræbt i græske skovbrande (2018). https://tinyurl.com/y2jnvnvo. Accessed 04 Feb 2019
5. Lum, C.W., Summers, A., Carpenter, B., Rodriguez, A., Dunbabin, M.: Automatic wildfire detection and simulation using optical information from unmanned aerial systems. Technical report, SAE Technical Paper (2015)
6. Mackay, A.: Climate change 2007: impacts, adaptation and vulnerability. Contribution of working group II to the fourth assessment report of the intergovernmental panel on climate change. J. Environ. Qual. **37**(6), 2407 (2008). https://doi.org/10.2134/jeq2008.0015br
7. Merino, L., Caballero, F., Martinez-de Dios, J., Ollero, A.: Cooperative fire detection using unmanned aerial vehicles. In: Proceedings of the 2005 IEEE International Conference on Robotics and Automation, pp. 1884–1889. IEEE (2005). https://doi.org/10.1109/ROBOT.2005.1570388
8. Park, M.: California fire explodes in size, is now largest in state history (2018). https://tinyurl.com/y5t7uapf. Accessed 04 Feb 2019
9. Romps, D.M., Seeley, J.T., Vollaro, D., Molinari, J.: Projected increase in lightning strikes in the United States due to global warming. Science **346**(6211), 851–854 (2014). https://doi.org/10.1126/science.1259100
10. Sandler, M., Howard, A., Zhu, M., Zhmoginov, A., Chen, L.C.: MobileNetV 2: inverted residuals and linear bottlenecks. In: Proceedings of the IEEE Conference on Computer Vision and Pattern Recognition, pp. 4510–4520 (2018). https://doi.org/10.1109/CVPR.2018.00474
11. Smith, B.: Are drones the future of firefighting? (2014). https://tinyurl.com/y5yvqv7k. Accessed 13 Feb 2019
12. Toft, E.: 80 skovbrande i sverige: Nu skal folk forlade et helt omrde (2018). https://tinyurl.com/dr-sverige. Accessed 04 Feb 2019
13. Yuan, C., Liu, Z., Zhang, Y.: Vision-based forest fire detection in aerial images for firefighting using UAVs. In: 2016 International Conference on Unmanned Aircraft Systems (ICUAS), pp. 1200–1205. IEEE (2016). https://doi.org/10.1109/10.1109/ICUAS.2016.7502546
14. Zhao, Y., Ma, J., Li, X., Zhang, J.: Saliency detection and deep learning-based wildfire identification in UAV imagery. Sensors **18**(3), 712 (2018). https://doi.org/10.3390/s18030712

V-Disparity Based Obstacle Avoidance for Dynamic Path Planning of a Robot-Trailer

Efthimios Tsiogas[(✉)], Ioannis Kostavelis, Dimitrios Giakoumis,
and Dimitrios Tzovaras

Centre for Research and Technology Hellas - Information Technologies Institute
(CERTH/ITI), 6th Km Charilaou-Thermi Road, 57100 Thessaloniki, Greece
{etsiogas,gkostave,dgiakoum,Dimitrios.Tzovaras}@iti.gr

Abstract. Structured space exploration with mobile robots is imperative for autonomous operation in challenging outdoor applications. To this end, robots should be equipped with global path planners that ensure coverage and full exploration of the operational area as well as dynamic local planners that address local obstacle avoidance. The paper at hand proposes a local obstacle detection algorithm based on a fast stereo vision processing step, integrated with a dynamic path planner to avoid the detected obstacles in real-time, while simultaneously keeping track of the global path. It considers a robot-trailer articulated system, based on which the trailer trace should cover the entire operational space in order to perform a dedicated application. This is achieved by exploiting a model predictive controller to keep track of the trailer path while performing stereo vision-based local obstacle detection. A global path is initially posed that ensures full coverage of the operational space and during robot's motion, the detected obstacles are reported in the robot's occupancy grid map, which is considered from a hybrid global and local planner approach to avoid them locally. The developed algorithm has been evaluated in a simulation environment and proved adequate performance.

Keywords: Stereo vision · V-disparity · Path planning · Obstacle avoidance · Model predictive control · Dubins path

1 Introduction

Scene analysis in computer vision frequently requires great computational effort for the modelling and annotation of the existing elements [11]. Normally, one has to make certain assumptions regarding environmental parameters in order to ensure real-time performance. The paper at hand tackles the problems of obstacle extraction and dynamic path planning in order to enable obstacle avoidance for a robot-trailer system that performs autonomous navigation in a flat surface area. Several computer vision based methods have been proposed for

© Springer Nature Switzerland AG 2019
D. Tzovaras et al. (Eds.): ICVS 2019, LNCS 11754, pp. 143–152, 2019.
https://doi.org/10.1007/978-3-030-34995-0_14

scene understanding and obstacle modelling such as the ones described in [3] and [7], however the closed loop obstacle detection and avoidance in exploration of unknown environments remains an active research topic.

In particular, authors in [9] utilized stereo vision to obtain depth information about the scene followed by feature extraction from V-disparity images to determine whether or not the terrain is traversable by a mobile robot. Another work introduced in [8], a visual odometry algorithm that implements outlier detection and motion estimation based on stereo image pairs resulting in minimal positioning errors has been presented. In [11], an obstacle detection method using dynamic pitching of the vehicle to deal with non-flat terrains has been exhibited. The approach relied on V-disparity images to gain geometric information about the road scene and performed semi-global matching without explicitly extracting coherent structures in the stereo image pair. A combination of a stereo camera with a monocular has been used to perform obstacle detection and state estimation respectively, for outdoor flights of micro air vehicles [13]. The method allows for scalability and generalizations regarding obstacle representation and additional sensor fusion. Work in [19] tackles obstacle extraction and 3D position estimation using disparity and V-disparity images for the purposes of driving support. Gao et al. [5] employ an uncalibrated 3D camera to generate a depth map used to calculate U-disparity and V-disparity images that contain line representations of obstacles and ground surface. Line extraction is realized with an optimized Hough transform [4] after the addition of steerable filters for noise reduction. In [1], an artificial intelligence (AI) vision method for obstacle detection in real-time in unstructured environments is proposed. Again, the V-disparity image is exploited to estimate the camera pitch oscillation caused by the vehicle's motion while stereo matching is used for the extraction and mapping of the obstacles in 3D coordinates. Authors in [16], demonstrated an obstacle detection algorithm based on U-disparity. The method examined the uncertainty of stereo vision by incorporating a probabilistic representation taking into account distance and disparity values. Mandelbaum et al. [12], used a next-generation video processor to develop real-time stereo processing as well as obstacle/terrain detection from stereo cameras installed on a moving vehicle. Stereo vision has been used for road and free space detection by separating them through color segmentation [18]. The proposed approach utilized an extended V-disparity algorithm that also incorporated U-disparity to successfully classify varying road profiles. The work presented in [22] discussed a global correlation method for ground plane extraction from V-disparity images in featureless roads. The method applied for on and off-road navigation by vehicles with vision-based obstacle detection. Authors in [6] presented a road scene analysis algorithm based on stereo vision as a supplement to radar-based analysis. U-V-disparity have been used for 3D road scene classification into surface planes characterizing the road's features. Sappa et al. [17] introduced two road plane approximation techniques, the first utilizing U-disparity and Hough transform and the second using least squares fitting to find the optimal fitting plane for the whole point cloud data. The benefits of stereo vision compared to other sensing technologies have been emphasized in [14], which proposed another usage of U-V-disparity for the purpose of modelling an urban environment. V-disparity has been also used by an unmanned ground vehicle (UGV) to perform obstacle detection, in order to obtain terrain information and enable navigation in more complex surfaces [21]. Authors in [15] proposed a new method for 3D scene representation named theta-disparity which has been calculated from a disparity image as a 2D angular depth histogram and revealed radial distribution of existing objects relative to a point of interest.

In this work, obstacle extraction based on V-disparity images incorporated into a hybrid local/global path planning algorithm has been developed, with the purpose of being integrated in a real robot-trailer system. Regarding the models used in simulation, the mobile robot is a SUMMIT XL robot produced by Robotnik and a two-wheeled trailer that carries Ground Penetrating Radar produced by IDS GeoRadar. The GPR is not exploited in this work, however the embodiment of the GPR antenna is considered in the robot's path planning and re-planning, to address the demand in similar application scenarios, where the sensor/device mounted on the robot's trailer is utilized for a dedicated purpose, such as plowing in agricultural robotic applications. The ZED stereo camera is mounted in front of the robot-trailer's tow hitch. The goal of this paper is twofold; firstly, the presentation of a method that assures obstacle avoidance from both the robot vehicle and the trailer part, assuming a flat ground plane in the vicinity of the vehicles, where obstacle extraction is employed and, secondly, the method guarantees that the GPR-equipped trailer keeps on following the updated path after local obstacle avoidance.

2 Processing of Stereo Image Pairs

2.1 Disparity and V-Disparity Computation

The ZED camera has been mounted on the robot's body with a pitch angle of around $17°$, since most of visual input should be the ground surface along with existing obstacles, to maximize input for robot navigation. Disparity images have been calculated using a standard local block matching stereo vision algorithm as 8-bit grayscale images using the pair of left and right images captured from the stereo camera. The left rectified stereo image has been considered as reference

| (a) | (b) | (c) |

Fig. 1. (a) Left camera image, (b) right camera image, (c) resulting disparity image

for the disparity image calculations. D_{min} and D_{max} are the parameters that quantify the expected minimum and maximum disparity, and in our case they are set equal to 0 and 111 respectively, assuming planar robot operation. A pair of grayscale stereo images captured from the camera as well as the calculated disparity image are illustrated in Fig. 1.

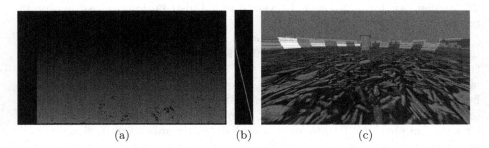

(a)   (b)   (c)

Fig. 2. (a) Disparity image, (b) respective V-disparity image, (c) reference image with highlighted obstacles

After obtaining depth information from the disparity image, the V-disparity image is calculated to extract knowledge about obstacles and the plane of the ground surface. Each row of the V-disparity image is calculated as a histogram of the varying disparities in the corresponding row of the disparity image. The number of columns is equal to the discrete disparity values. Thus, given a disparity image of size (W, H) and the parameters D_{min} and D_{max}, the resulting V-disparity image will have a size equal to $(D_{max} - D_{min} + 1, H)$. As an example, we may consider a disparity image of size $(10, H)$, $D_{min} = 0$ and $D_{max} = 5$ with the first row being equal to [2, 0, 4, 4, 1, 0, 2, 4, 5, 2]. After applying the binning principle, the resulting first row in the V-disparity image will be equal to [2, 1, 3, 0, 3, 1]. After finishing the calculations, the pixel values are mapped to [0, 255] range for visualization purposes. A pair of disparity and V-disparity images along with the reference image with highlighted obstacles are depicted in Fig. 2.

2.2   Obstacle Extraction from V-Disparity

Using the sparse disparity map obtained from the stereo correspondence algorithm, a reliable V-disparity image is computed where each pixel has a positive integer value that denotes the number of pixels in the input image that lie on the same image line (ordinate) and have disparity value equal to its abscissa [10]. The information resulting from the V-disparity image is of great importance because it easily distinguishes the pixels belonging to the ground plane from the ones belonging to obstacles. The ground plane in the V-disparity image can be modeled by a linear equation, the parameters of which can be found using Hough transform, if the geometry of the camera-environment system is unknown. However, if the geometry of the system is constant and known, as in a case of a camera firmly mounted on a robot exploring a planar outdoors environment, then the two parameters can be easily computed beforehand and used in all the image pairs during the exploration. Looking top-down in Fig. 2(a), the increasing pixel intensity expresses the decreasing distance between the camera and the ground plane. Thus, it holds true that pixels corresponding to obstacles present in the camera's field of view (FoV), have a positive intensity and appear above the characteristic line in a V-disparity image. This effect is demonstrated in Fig. 2(b) which illustrates an obstacle that has been placed away from the camera.

Yet, the hypothesis of a perfectly flat floor never stands in practice, either because of the floor actually not being perfectly flat or because of small disparity estimation and discretization errors. As a result, a tolerance region extending on both sides of the linear segment of the terrain has to be defined and any point outside of it can then be safely considered as originating from an obstacle. For each row in the V-disparity image, pixels that satisfy the aforementioned criterion have their column numbers (disparity values) stored in a set. Then, pixels with disparities contained in the corresponding sets are traced back in the disparity image and marked as obstacles in the original reference image, as depicted in Fig. 2(c).

After tracking the pixels (x_p, y_p) that correspond to obstacles in the original left camera image of size (W, H), 3D reconstruction of the pixels that correspond to obstacles in world frame are realized taking into consideration the stereo camera model. Thus, if f is the camera's focal length measured in pixels and b is the camera's baseline, transformed (x, y, z) coordinates with respect to the camera's frame are given by:

$$(x, y, z) = \left(\frac{z \cdot (x_p - x_c)}{f}, \frac{z \cdot (y_p - y_c)}{f}, \frac{f \cdot b}{disp(x_p, y_p)} \right) \tag{1}$$

Where $(x_c, y_c) = (\frac{W}{2}, \frac{H}{2})$ is the image's central pixel and $disp(x_p, y_p)$ is the disparity value of pixel (x_p, y_p).

Figure 3(a) depicts the robot in a stationary position observing an obstacle in the simulation environment and Fig. 3(b) depicts the 3D reconstructed obstacles extracted from the V-disparity process as point cloud in front of the robot. It is apparent that the method manages to extract every obstacle in the camera's FoV without being limited to the robot's vicinity. However in our case obstacles that are greater that a predefined threshold add extra computational cost thus, in order to achieve real-time performance, the camera's maximum depth range is limited to 2.5 m, an adequate distance for the robot to detect and avoid an obstacle in time.

3 Dynamic Path Planning and Obstacle Avoidance

To achieve robot-trailer coverage of the entire surface multiple passes are required. The global path to be followed by the robot-trailer is constructed as a Dubins path that utilizes the boustrophedon motion algorithm [2] although

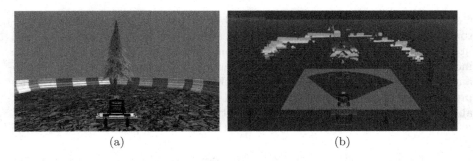

(a)   (b)

Fig. 3. (a) Simulation environment and (b) extracted obstacles from V-disparity as a 3D point cloud

differing from a standard boustrophedon path since it involves unidirectional instead of bidirectional turns (see Fig. 4(b)). During the robot-trailer exploration, existence of obstacles is unknown to the global planner. Thus, obstacles detected by the robot during navigation are reported to the occupancy grid map as occupied cells in order to be avoided (see Fig. 4(c)).

For the obstacle modeling inside the robot navigation environment, point clouds of extracted obstacles (see Fig. 4(d)) are projected to the occupancy grid map (OGM) that expresses occupancy states in the ground plane. The value of each cell of OGM represents the probability of the cell to be occupied. Cell's state is mainly interpreted as free, occupied or unknown according to how the user handles the cell values. The OGM, used as a costmap herein, is initialized around the robot's initial position. As the robot-trailer explores the previously unknown environment, it constantly expands the visible costmap in accordance to the obtained obstacles from the V-disparity procedure, to cover previously unknown area that is currently inside the camera's FoV and under the 2.5 m threshold.

By taking into consideration occupied cells that intersect with planned way-points by the Dubins path planner, a custom made parametric local obstacle avoidance has been developed. The latter considers the intersected way-points of the global planner with the obstacles and infers an arc trajectory that bypasses the reported obstacles. The inferred arc obeys to the robot-trailer kinematic constraints and does not exceed the maximum hitch angle between the robot and the trailer (see Fig. 4(d)). The length of the arc is proportional to the distance required for the trailer to avoid the obstacle. Thus each time an obstacle is presented at the robot's retina, i.e. V-disparity obstacle detection system, the robot re-plans a trajectory around the obstacle that forces the robot-trailer module to avoid the obstacle and keep the co-linear trajectory to the Dubins path. This method allows maximum coverage to the explored environment and deviation from the global plan only for the avoidance of small previously unseen obstacles. In situations where the robot-trailer cannot meet again the Dubins path i.e. the trailer exhibits a predefined threshold of 0.5 m, a re-plan of the Dubins path is requested from the global planner.

In order to ensure accurate robot-trailer exploration that obeys to the drawn paths, robot pose estimation and path tracking modules have also been incorporated into the system. Robot pose estimation is performed through stereo-based visual odometry, which is mostly relied in 3D features tracking [8], exploiting the disparity map computed for the obstacle detection. For the local navigation and in order to closely follow the Dubins path, a model predictive controller (MPC) has been implemented as an optimal approach to solve the path tracking problem for the robot-trailer system [20]. MPC is a control method used to control a process while satisfying a set of constraints and in our approach, a linear model has been implemented in order to enhance real-time performance. The goal of MPC, quite common in robot-trailer systems, is the elimination of the trailer's lateral position error.

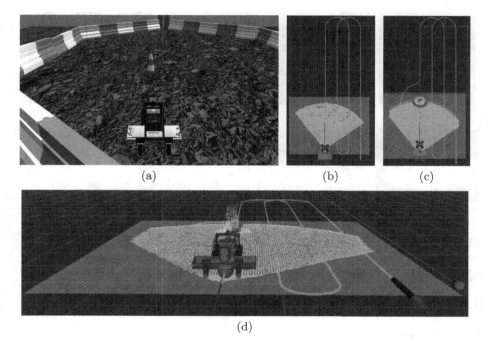

(a)   (b)   (c)

(d)

Fig. 4. (a) Gazebo environment containing obstacle, (b) Boustrophedon path, (c) Replanned path with curved segment, (d) Re-planned path with obstacle's 3D point cloud reconstruction

4   Experimental Evaluation

The performance of the implemented methods has been evaluated in a simulation environment. Figure 4(a) depicts the environment with the robot-trailer in its stationary initial position. The Dubins path to be followed by the trailer, is illustrated in green color in Fig. 4(b). Figure 4(c) shows a top view of the

scene after obstacle extraction through V-disparity and re-planning application
of the re-planning procedure. Specifically, the cone's area marked on the costmap
along with the added path segment that bypasses the obstacle are exhibited in
the same figure whereas the cone's point cloud reconstruction is better perceived
in Fig. 4(d). In a more large-scale experiment, the robot-trailer bypasses several
obstacles during the traversal of the Dubins path as illustrated in Fig. 5. The
paths followed by the robot and trailer are depicted in red and blue color respectively along with the Dubins path. It is revealed that the proposed dynamic
obstacle avoidance and planning algorithm manages to bypass the obstacles and
at the same time closely follows the imposed path with absolute deviation error
in the trajectory less than 0.5 m, measured through the ground-truth observations of the Gazebo simulation environment.

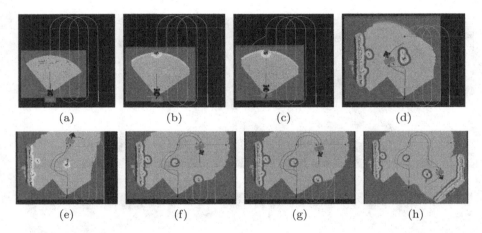

Fig. 5. (a) Initialization with Dubins path, (b) detected and reported obstacle in the
cost map, (c) local replan on the Dubins path and (d)–(h) continuous path tracking
and replanning. Replanned reference Dubins path (green color) with actual robot path
(red color) and actual trailer path (blue color) (Color figure online)

5   Conclusions

In this work, a method for maximum coverage of a surface area with an articulated robot-trailer setup has been presented. During the exploration, obstacle
detection has been performed based on stereo vision, which is then exploited by
a dynamic path planner in a closed loop obstacle avoidance scheme. V-disparity
has been utilized for the detection of obstacles above the surface plane which
are isolated as pixels in the stereo reference image and 3D reconstructed to be
expressed as occupied cells in the robot's occupancy grid map. Robot exploration is performed through structured Dubins path planning which is dynamically updated with smaller local paths that allow real-time obstacle avoidance
and enable co-linearity with the initial global path after the avoidance routine.

The method has been integrated with stereo-based visual motion estimation and MPC, that allows constant path tracking during robot-trailer exploration. The method assures that the robot-trailer is able to avoid small obstacles while the trailer is kept in the added segment and whenever possible, keeps on following the initial path after avoiding the obstacle.

Acknowledgment. This work has been supported by the EU Horizon 2020 funded project "BADGER (RoBot for Autonomous unDerGround trenchless opERations, mapping and navigation)" under the grant agreement with no: 731968.

References

1. Broggi, A., Caraffi, C., Fedriga, R.I., Grisleri, P.: Obstacle detection with stereo vision for off-road vehicle navigation. In: 2005 IEEE Computer Society Conference on Computer Vision and Pattern Recognition (CVPR 2005)-Workshops, pp. 65–65. IEEE (2005)
2. Choset, H.: Coverage of known spaces: the boustrophedon cellular decomposition. Auton. Robots **9**(3), 247–253 (2000)
3. De Cubber, G., Doroftei, D., Nalpantidis, L., Sirakoulis, G.C., Gasteratos, A.: Stereo-based terrain traversability analysis for robot navigation. In: IARP/EURON Workshop on Robotics for Risky Interventions and Environmental Surveillance, Brussels, Belgium (2009)
4. Duda, R.O., Hart, P.E.: Use of the Hough transformation to detect lines and curves in pictures. Tech. rep, Sri International Menlo Park Ca Artificial Intelligence Center (1971)
5. Gao, Y., Ai, X., Wang, Y., Rarity, J., Dahnoun, N.: UV-disparity based obstacle detection with 3D camera and steerable filter. In: 2011 IEEE Intelligent Vehicles Symposium (IV), pp. 957–962. IEEE (2011)
6. Hu, Z., Lamosa, F., Uchimura, K.: A complete UV-disparity study for stereovision based 3D driving environment analysis. In: Fifth International Conference on 3-D Digital Imaging and Modeling (3DIM 2005), pp. 204–211. IEEE (2005)
7. Khan, M., Hassan, S., Ahmed, S.I., Iqbal, J.: Stereovision-based real-time obstacle detection scheme for unmanned ground vehicle with steering wheel drive mechanism. In: 2017 International Conference on Communication, Computing and Digital Systems (C-CODE), pp. 380–385. IEEE (2017)
8. Kostavelis, I., Boukas, E., Nalpantidis, L., Gasteratos, A.: Stereo-based visual odometry for autonomous robot navigation. Int. J. Adv. Robot. Syst. **13**(1), 21 (2016)
9. Kostavelis, I., Nalpantidis, L., Gasteratos, A.: Supervised traversability learning for robot navigation. In: Groß, R., Alboul, L., Melhuish, C., Witkowski, M., Prescott, T.J., Penders, J. (eds.) TAROS 2011. LNCS (LNAI), vol. 6856, pp. 289–298. Springer, Heidelberg (2011). https://doi.org/10.1007/978-3-642-23232-9_26
10. Kostavelis, I., Nalpantidis, L., Gasteratos, A.: Collision risk assessment for autonomous robots by offline traversability learning. Robot. Auton. Syst. **60**(11), 1367–1376 (2012)
11. Labayrade, R., Aubert, D., Tarel, J.P.: Real time obstacle detection in stereovision on non flat road geometry through "v-disparity" representation. In: Intelligent Vehicle Symposium, 2002, vol. 2, pp. 646–651. IEEE (2002)

12. Mandelbaum, R., McDowell, L., Bogoni, L., Reich, B., Hansen, M.: Real-time stereo processing, obstacle detection, and terrain estimation from vehicle-mounted stereo cameras. In: Proceedings Fourth IEEE Workshop on Applications of Computer Vision. WACV 1998 (Cat. No. 98EX201), pp. 288–289. IEEE (1998)
13. Matthies, L., Brockers, R., Kuwata, Y., Weiss, S.: Stereo vision-based obstacle avoidance for micro air vehicles using disparity space. In: 2014 IEEE International Conference on Robotics and Automation (ICRA), pp. 3242–3249. IEEE (2014)
14. Musleh, B., de la Escalera, A., Armingol, J.M.: UV disparity analysis in urban environments. In: Moreno-Díaz, R., Pichler, F., Quesada-Arencibia, A. (eds.) EURO-CAST 2011. LNCS, vol. 6928, pp. 426–432. Springer, Heidelberg (2012). https://doi.org/10.1007/978-3-642-27579-1_55
15. Nalpantidis, L., Kragic, D., Kostavelis, I., Gasteratos, A.: Theta-disparity: an efficient representation of the 3D scene structure. In: Menegatti, E., Michael, N., Berns, K., Yamaguchi, H. (eds.) Intelligent Autonomous Systems 13. AISC, vol. 302, pp. 795–806. Springer, Cham (2016). https://doi.org/10.1007/978-3-319-08338-4_57
16. Perrollaz, M., Spalanzani, A., Aubert, D.: Probabilistic representation of the uncertainty of stereo-vision and application to obstacle detection. In: 2010 IEEE Intelligent Vehicles Symposium, pp. 313–318. IEEE (2010)
17. Sappa, A.D., Herrero, R., Dornaika, F., Gerónimo, D., López, A.: Road approximation in Euclidean and v-disparity space: a comparative study. In: Moreno Díaz, R., Pichler, F., Quesada Arencibia, A. (eds.) EUROCAST 2007. LNCS, vol. 4739, pp. 1105–1112. Springer, Heidelberg (2007). https://doi.org/10.1007/978-3-540-75867-9_138
18. Soquet, N., Aubert, D., Hautiere, N.: Road segmentation supervised by an extended v-disparity algorithm for autonomous navigation. In: 2007 IEEE Intelligent Vehicles Symposium, pp. 160–165. IEEE (2007)
19. Suganuma, N., Fujiwara, N.: An obstacle extraction method using virtual disparity image. In: 2007 IEEE Intelligent Vehicles Symposium, pp. 456–461. IEEE (2007)
20. Tsiogas, E., Kostavelis, I., Kouros, G., Kargakos, A., Giakoumis, D., Tzovaras, D.: Surface exploration with a robot-trailer system for autonomous subsurface scanning. In: IEEE Smart World Congress (2019)
21. Yang, C., Jun-Jian, P., Jing, S., Lin-Lin, Z., Yan-Dong, T.: V-disparity based UGV obstacle detection in rough outdoor terrain. Acta Automatica Sin. **36**(5), 667–673 (2010)
22. Zhao, J., Katupitiya, J., Ward, J.: Global correlation based ground plane estimation using v-disparity image. In: Proceedings 2007 IEEE International Conference on Robotics and Automation, pp. 529–534. IEEE (2007)

Intersection Recognition Using Results of Semantic Segmentation for Visual Navigation

Hiroki Ishida[1], Kouchi Matsutani[1], Miho Adachi[1], Shingo Kobayashi[1], and Ryusuke Miyamoto[2]([✉])

[1] Department of Computer Science, Graduate School of Science and Technology, Meiji University, 1-1-1 Higashimita, Tama-ku, Kawasaki-shi, Japan
{ishida,coach,miho,sin}@cs.meiji.ac.jp
[2] Department of Computer Science, School of Science and Technology, Meiji University, 1-1-1 Higashimita, Tama-ku, Kawasaki-shi, Japan
miya@cs.meiji.ac.jp

Abstract. It is popular to use three-dimensional sensing devices such as LiDAR and RADAR for autonomous navigation of ground vehicles in modern approaches. However, there are significant problems: the price of 3D sensing devices, the cost for 3D map building, the robustness against errors accumulated in long-term moving. Visual navigation based on a topological map using only cheap cameras as external sensors has potential to solve these problems; road-following and intersection recognition can enable robust navigation. This paper proposes a novel scheme for intersection recognition using results of semantic segmentation, which has a high affinity for vision-based road-following strongly depending on semantic segmentation. The proposed scheme mainly composed of mode filtering for a segmented image and similarity computation like the Hamming distance showed that good accuracy for the Tsukuba-Challenge 2018 dataset constructed by the authors: perfect results were obtained for more than half intersections included in the dataset. In addition, a running experiment using the proposed scheme with vision-based road-following showed that the proposed scheme could classify intersections appropriately in actual environments.

Keywords: Intersection recognition · Semantic segmentation · Robot navigation · Simple similarity computation

1 Introduction

The autonomous navigation of ground vehicles has recently become a significant technology, allowing automatic systems to be used in practical applications, and several studies have been conducted on self-driving cars, security services [1], and automated agriculture, among other areas. It is important for ground vehicles to observe their surrounding external environments using external sensors. To obtain sensing results that are sufficiently robust and accurate for localization and the updating of local maps, accurate three-dimensional (3D) sensing devices

© Springer Nature Switzerland AG 2019
D. Tzovaras et al. (Eds.): ICVS 2019, LNCS 11754, pp. 153–163, 2019.
https://doi.org/10.1007/978-3-030-34995-0_15

such as LiDAR [2] and RADAR have generally been adopted. Using 3D maps constructed from 3D sensing results, ground vehicles can move toward a given destination while avoiding obstacles. Recent studies have shown that 3D maps and 3D sensors enable accurate and robust navigation, although schemes based on 3D sensors have some significant problems, namely, the high price of accurate 3D sensors, the costs required for 3D map building, and robustness against errors accumulated during long-term movement.

Vision-based navigation using only cheap visible cameras is a solution to reduce the price of sensors and the cost required for map building. The problem of vision-based navigation, however, is its accuracy and robustness. To improve the reliability of vision-based navigation, novel concepts using only a topological map consisting of positional relationships among landmarks have been shown, and vision-based road-following methods with obstacle avoidance have been proposed [11]. In the proposed scheme, a robot can follow the road region in urban scenes without dense or accurate 3D metric maps, which are indispensable for modern navigation schemes using 3D sensors. Only the results of semantic segmentation are used to extract the runnable area and follow the road region.

Using the concepts proposed in [11], vision-based navigation will become available when an appropriate recognition of intersections, in addition to road-following, is realized. With this concept, a robot moves along a road corresponding to an edge of a given topological map until the robot reaches an intersection where the robot should change course. When the robot reaches the intersection, the robot makes a turn and re-starts the road-following along the next road to be followed. This procedure is similar to a human movement scheme, which does not require an accurate odometry, using the number of steps, accurate 3D maps, and accurate 3D sensing results.

To achieve such visual navigation using a topological map, this paper proposes a novel scheme for intersection recognition based on the concepts proposed in [11]. In the proposed scheme, only images obtained after semantic segmentation that are also used for the vision-based road-following proposed in [11] are applied to reduce the number of computations required for visual navigation. The proposed scheme attempts to estimate the similarity between the obtained image at the current location and the prior registered images corresponding to intersections through a simple similarity computation. To improve the robustness and reduce the number of computations required for computing the similarity, a mode filter that provides the most frequent value from a computation sub-window is applied after semantic segmentation is applied. For the similarity computation itself, a novel approach for further improvement of the robustness is adopted.

To validate the proposed scheme for intersection recognition, a quantitative evaluation using two types of an actual datasets is shown, namely, Tsukuba and Ikuta datasets created for the Tsukuba-Challenge, the most popular robot competition on autonomous navigation in Japan, and for typical experiments at Ikuta Campus, respectively. In addition to a quantitative evaluation, the results of experiments using the proposed scheme and the vision-based road-following approach proposed in [11] at the Ikuta Campus are shown.

The remainder of this paper is organized as follows. Section 2 describes recent advances in semantic segmentation and summarizes the concepts of visual navigation proposed in [11]. Section 3 describes the proposed novel scheme for intersection recognition using a simple computation for feature similarity, which is evaluated in Sect. 4. Finally, Sect. 5 provides some concluding remarks.

2   Related Work

This section describes the recent advances in the area of semantic segmentation and summarizes the concepts of visual navigation proposed in [11].

2.1   Recent Advances in Semantic Segmentation

Semantic segmentation is a traditional problem in the field of image recognition. Although several schemes have been proposed to solve this problem, good results have yet to be obtained; sub-space [8], meanshift [14], k-means [4], watershed [12], markov random field [7], and conditional random field [6] methods are popular examples of this type of task. Even FCN [9], which is a popular early scheme for semantic segmentation based on deep learning, has not shown good results in practical scenarios.

The accuracy of semantic segmentation is drastically improved using PSP-Net [16], which was presented in CVPR2017, and originally showed good results in semantic segmentation at ILSVRC2016 [13]. After the emergence of PSP-Net, other schemes showing an improved accuracy have been proposed [3,17], although they still cannot outperform PSPNet under certain conditions.

Whereas some schemes have tried to improve the accuracy of semantic segmentation, other schemes have been developed for an increased speed while maintaining the segmentation. accuracy. ICNet [15] achieves high-speed processing, maintaining nearly the same level of classification accuracy as PSPNet. ICNet was adopted in vision-based road-following [11] and provides sufficiently accurate segmentation results for such tasks.

In this paper, ICNet is adopted for a semantic segmentation scheme because the same scheme should be applied to the road-following and intersection recognition.

2.2   Visual Navigation Using a Topological Map

The overall navigation scheme based on semantic segmentation proposed in [11] mainly consists of road-following and intersection recognition. Road-following is required for a ground vehicle to move from one intersection to another. Intersection recognition enables appropriate route changes at an intersection. By repeating these procedures, a robot can move from the current location to the destination once all information regarding the route is given. The route information can be represented by a list of intersections where a ground vehicle should

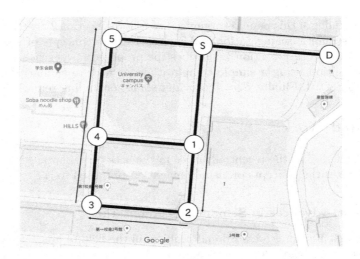

Fig. 1. Example of a topological map composed of a list of intersections.

change its route. Figure 1 shows an example of a topological map having a global path from the start to the destination as a list of intersections.

The navigation scheme can be summarized as follows:

1. Route setting to the destination based on a given topological map.
2. Road-following until an intersection is detected.
3. Classification of the detected intersection.
4. The robot will stop if the intersection is the destination. The robot changes its course and returns to the 2 process if the intersection is the location where the robot should change its route. The moving operation returns to the 2 process if the intersection is not at location where the robot should change its route.

In this study, an attempt was made to construct a computationally effective scheme that enables the accurate and robust classification of intersections required for the navigation scheme described above.

3   Intersection Recognition Using Results of Semantic Segmentation

This section proposes a novel scheme for intersection recognition using the results of semantic segmentation. The computation of the proposed scheme after semantic segmentation can mainly be divided into feature extraction and classification based on a similarity computation. The rest of this section details these operations.

3.1   Feature Extraction

This subsection describes the basic operation of the feature extraction of the proposed scheme, and then introduces an effort to reduce the influence of moving objects at intersections.

Basic Operation. Figure 2 shows an overview of the basic operation of the proposed feature extraction, where mode filtering extracts a feature from an input image. With this process, to obtain an $M \times N$ feature, a segmented image is divided into $M \times N$ rectangular regions before mode filtering. After the grid division, mode filtering is simply applied to each rectangular region. Finally, we can obtain a feature corresponding to an input image the size of which is $M \times N$.

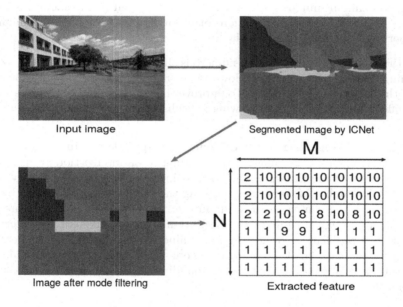

Fig. 2. Overview of feature extraction.

Removal of Moving Objects. In general, there are several types of moving objects at an intersection, including vehicles, bikes, and humans. However, such objects cannot be used as the features of an intersection using our scheme because their locations and their number drastically changes each moment. To reduce the influence of such obstacles for classification that do not belong to an intersection, the objects are removed during the mode filtering: The second most frequent class is selected if it is the most dominant in a sub-window. When only a class to be removed is included in a sub-window, the value corresponding to the sub-window is set according to the surrounding sub-windows.

3.2   Classification

This subsection describes how to classify a target intersection using features extracted using the proposed scheme.

Classification Flow Using Extracted Features. The proposed scheme classifies an image captured at an intersection through the following operations:

1. Feature extraction is applied to reference images corresponding to intersections where a robot should change its route. The features obtained are stored as a reference.
2. Feature extraction is applied to an input image captured at an intersection.
3. Similarities between the features obtained from the input image and the references are computed.
4. If the maximum similarity is greater than a prior defined threshold, the input image is classified as an intersection corresponding to the maximum similarity. Otherwise, the input image is classified as a non-target intersection.

In the similarity computation, numerous different components, such as the Hamming distance, are counted. However, a straightforward computation may reduce the classification accuracy. To improve the accuracy and robustness, some of the efforts described in the following subsubsections are applied.

Appropriate Weighting for Similarity Computation. In general, many of the center regions of an image corresponding to an intersection are runnable or road regions. If the Hamming distance is simply applied to a similarity computation, the number of components having the same value is increased by the road region at the center of an input image. As a result, a middle score can be obtained through a similarity computation even for different intersections. To avoid this problem, a lower weighting value is multiplied to components of features whose class label is the road region. Using weighting for a similarity computation, it is expected that the surrounding objects at an intersection will become dominant.

How to Cope with Misalignment. In actual scenes, a horizontal misalignment occasionally occurs because a robot does not operate along the same course within the road region. A misalignment along the vertical and depth directions also occurs, although this can be resolved when the robot moves forward. Therefore, only a horizontal misalignment should be considered for the similarity computation. To appropriately cope with such a horizontal misalignment during a similarity computation, sub-features are extracted using sliding windows, as shown in Fig. 3.

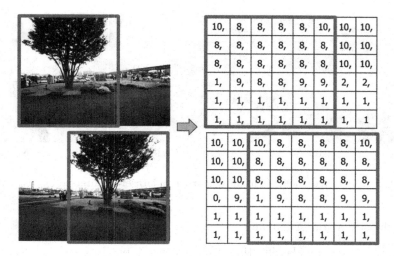

Fig. 3. Example of a horizontal misalignment.

Support Using Location Information of a Robot. A ground vehicle often runs through several intersections with an extremely similar appearance. Of course, the classification accuracy tends to worsen if the appearances of different intersections are similar. To reduce the number of misclassifications caused by a similar appearance, the proposed scheme uses the location information. According to the distance in a given topological map, the weighting coefficients are multiplied as follows: 1.0 for intersections next to the previous intersection, 0.75 for intersections with a distance of 2 on a topological map, and 0.5 otherwise.

4   Evaluation

This section evaluates the accuracy of the proposed scheme using a dataset constructed by the authors and shows the actual running experiment conducted at a university campus.

4.1   A Dataset Used for Evaluation

To evaluate the recognition of an intersection, a dataset was constructed using intersections included in the course at the Tsukuba-Challenge 2018, which is the most popular robot competition on autonomous navigation in an urban environment. The dataset created consists of 1100 images corresponding to 11 intersections: A total of 100 images were labelled for each intersection. Considering the actual conditions during operation, a variety of images under several conditions were captured for this dataset, including in different seasons, under different weather conditions, and taken at different times of the day and from various viewpoints. Figure 4 shows example images of the dataset.

Fig. 4. Example images included in the constructed dataset for intersection recognition.

4.2   Classification Accuracy

Before evaluating the classification accuracy, the input images were converted into segmented images using ICNet, which was trained on the CityScapes dataset [5] and fine-tuned using a dataset for semantic segmentation proposed in [10]. Once the input images were converted into segmented images, the proposed scheme classified them into one of 11 classes.

Tables 1, 2, 3 and 4 show the classification accuracy of the proposed scheme when the sizes of the features were 8×10, 16×15, 32×30, and 64×60, respectively. The results show that a misalignment does not affect the classification accuracy. For all sizes, support based on location information improves the classification accuracy except for case no. 5. Appropriate weighting when considering the road region seems to achieve good results, but does not work well when the size of the features is 64×60. As the results indicate, 32×30 sized features consistently showed the best classification accuracy.

Table 1. Classification accuracy for 8×10 features.

	No. 1	No. 2	No. 3	No. 4	No. 5	No. 6	No. 7	No. 8	No. 9	No. 10	No. 11
Basic	28	100	33	100	100	90	96	100	79	100	100
Align	28	100	33	100	100	90	96	100	79	100	100
Weight	28	100	35	100	100	90	95	100	77	100	100
Neighbor	82	100	100	100	94	100	97	100	79	100	100
All	82	100	98	100	94	100	97	100	79	100	100

4.3   Running Experiment

In addition to a quantitative evaluation of the classification accuracy, the results of a running experiment conducted at a university campus are shown. The course that the robot was required to move along during the running experiment is defined as a topological map, as shown in Fig. 1. During this experiment,

Table 2. Classification accuracy for 16 × 15 features.

	No. 1	No. 2	No. 3	No. 4	No. 5	No. 6	No. 7	No. 8	No. 9	No. 10	No. 11
Basic	50	100	14	100	100	87	96	100	70	100	99
Align	50	100	14	100	100	87	96	100	70	100	99
Weight	51	100	18	100	100	92	96	100	70	100	99
Neighbor	76	100	99	100	94	100	96	100	70	100	100
All	78	100	100	100	94	100	96	100	70	100	100

Table 3. Classification accuracy for 32 × 30 features.

	No. 1	No. 2	No. 3	No. 4	No. 5	No. 6	No. 7	No. 8	No. 9	No. 10	No. 11
Basic	65	100	23	100	100	90	96	100	71	100	100
Align	65	100	23	100	100	90	96	100	71	100	100
Weight	65	100	27	100	100	97	96	100	71	100	100
Neighbor	72	100	100	100	94	100	96	100	71	100	100
All	78	100	100	100	94	100	96	100	71	100	100

Table 4. Classification accuracy for 64 × 60 features.

	No. 1	No. 2	No. 3	No. 4	No. 5	No. 6	No. 7	No. 8	No. 9	No. 10	No. 11
Basic	54	100	28	100	100	88	96	100	71	100	100
Align	54	100	28	100	100	88	96	100	71	100	100
Weight	42	100	31	100	100	93	96	100	70	99	100
Neighbor	71	100	100	100	94	100	96	100	71	100	100
All	60	100	100	100	94	100	96	100	70	99	100

a robot started at location 'S', passed intersections '1', '2', '3', '4', and '5', and finally reached the destination, represented by 'D'. The proposed scheme for intersection recognition with the vision-based road-following approach proposed in [11] worked well, and the robot completed the 500 m long course using only cheap visible cameras.

5 Conclusion

This paper proposed a novel scheme for intersection recognition using the results of semantic segmentation. Using the proposed scheme, mode filtering is applied to segmented images, and feature extraction is conducted for a similarity computation. Similarities between features extracted from an input image and the reference features for classification are computed, such as the Hamming distance, and the number of different components are counted. To improve the accuracy of a similarity computation, the use of the appropriate weighting when considering the road regions, applying offsets according to a misalignment, and providing support through location information are all proposed. Experiment results using a dataset constructed by the authors showed that support from location information improved the classification accuracy and that appropriate weighting also works well. However, offsets according to a misalignment do not affect the classification accuracy despite its apparent importance.

A running experiment held at a university campus showed that the proposed scheme can detect intersections where a robot should change its route. As a result, a robot was able to complete a 500 m long course as represented by a topological map using the proposed scheme with the vision-based road-following proposed in [11].

This scheme showed good accuracy for intersection classification when 11 intersections were used and there were few obstacles. However, it is necessary to classify larger numbers of intersections for practical applications, although it is not clear how doing so will impact the performance of the proposed scheme. In the future, we plan to evaluate the classification accuracy of the proposed scheme using a novel dataset composed of numerous intersections and improve the proposed scheme if its accuracy is deemed insufficient.

Acknowledgements. This research was partially supported by Research Project Grant(B) by Institute of Science and Technology, Meiji University.

References

1. Haneda robotics lab. https://www.tokyo-airport-bldg.co.jp/hanedaroboticslab/
2. Velodyne lidar. https://velodynelidar.com/
3. Chen, L.-C., Zhu, Y., Papandreou, G., Schroff, F., Adam, H.: Encoder-decoder with atrous separable convolution for semantic image segmentation. In: Ferrari, V., Hebert, M., Sminchisescu, C., Weiss, Y. (eds.) ECCV 2018. LNCS, vol. 11211, pp. 833–851. Springer, Cham (2018). https://doi.org/10.1007/978-3-030-01234-2_49
4. Chen, T.W., Chen, Y.L., Chien, S.Y.: Fast image segmentation based on k-means clustering with histograms in HSV color space. In: 2008 IEEE 10th Workshop on Multimedia Signal Processing, pp. 322–325. IEEE (2008)
5. Cordts, M., et al.: The cityscapes dataset for semantic urban scene understanding. In: The IEEE Conference on Computer Vision and Pattern Recognition, June 2016
6. He, X., Zemel, R., Carreira-Perpinan, M.: Multiscale conditional random fields for image labeling, vol. 2, pp. II-695–II-702, 27 June–2 July 2004
7. Kato, Z., Pong, T.C.: A markov random field image segmentation model for color textured images. Image Vis. Comput. **24**(10), 1103–1114 (2006)
8. Li, J., Bioucas-Dias, J.M., Plaza, A.: Spectral-spatial hyperspectral image segmentation using subspace multinomial logistic regression and markov random fields. IEEE Trans. Geosci. Remote Sens. **50**(3), 809–823 (2011)
9. Long, J., Shelhamer, E., Darrell, T.: Fully convolutional networks for semantic segmentation. In: Proceedings of the IEEE Conference on Computer Vision and Pattern Recognition, pp. 3431–3440 (2015)
10. Miyamoto, R., Adachi, M., Nakamura, Y., Nakajima, T., Ishida, H., Kobayashi, S.: Accuracy improvement of semantic segmentation using appropriate datasets for robot navigation. In: Proceedings of the International Conference on Control, Decision and Information Technologies, pp. 1610–1615 (2019)

11. Miyamoto, R., Nakamura, Y., Adachi, M., Nakajima, T., Ishida, H., Kojima, K., Aoki, R., Oki, T., Kobayashi, S.: Vision-based road-following using results of semantic segmentation for autonomous navigation. In: Proceedings of the International Conference of Consumer Electronics in Berlin, pp. 194–199 (2019)

12. Patras, I., Hendriks, E.A., Lagendijk, R.L.: Video segmentation by map labeling of watershed segments. IEEE Trans. Pattern Anal. Mach. Intell. **23**(3), 326–332 (2001)

13. Russakovsky, O., et al.: ImageNet large scale visual recognition challenge. Int. J. Comput. Vis. (IJCV) **115**(3), 211–252 (2015). https://doi.org/10.1007/s11263-015-0816-y

14. Tao, W., Jin, H., Zhang, Y.: Color image segmentation based on mean shift and normalized cuts. IEEE Trans. Syst. Man Cybern. Part B (Cybern.) **37**(5), 1382–1389 (2007)

15. Zhao, H., Qi, X., Shen, X., Shi, J., Jia, J.: ICNet for real-time semantic segmentation on high-resolution images. In: Ferrari, V., Hebert, M., Sminchisescu, C., Weiss, Y. (eds.) ECCV 2018. LNCS, vol. 11207, pp. 418–434. Springer, Cham (2018). https://doi.org/10.1007/978-3-030-01219-9_25

16. Zhao, H., Shi, J., Qi, X., Wang, X., Jia, J.: Pyramid scene parsing network. In: Proceedings of the IEEE Conference on Computer Vision and Pattern Recognition, pp. 2881–2890 (2017)

17. Zhao, H., et al.: PSANet: point-wise spatial attention network for scene parsing. In: Ferrari, V., Hebert, M., Sminchisescu, C., Weiss, Y. (eds.) ECCV 2018. LNCS, vol. 11213, pp. 270–286. Springer, Cham (2018). https://doi.org/10.1007/978-3-030-01240-3_17

Autonomous MAV Navigation in Underground Mines Using Darkness Contours Detection

Sina Sharif Mansouri[1]($\boxtimes$), Miguel Castaño[2], Christoforos Kanellakis[1],
and George Nikolakopoulos[1]

[1] Robotics Team Department of Computer, Electrical and Space Engineering,
Luleå University of Technology, 97187 Luleå, Sweden
sinsha@ltu.se
[2] eMaintenance Group Division of Operation, Maintenance and Acoustics
Department of Civil, Environmental and Natural Resources Engineering,
Luleå University of Technology, 97187 Luleå, Sweden

Abstract. This article considers a low-cost and light weight platform for the task of autonomous flying for inspection in underground mine tunnels. The main contribution of this paper is integrating simple, efficient and well-established methods in the computer vision community in a state of the art vision-based system for Micro Aerial Vehicle (MAV) navigation in dark tunnels. These methods include Otsu's threshold and Moore-Neighborhood object tracing. The vision system can detect the position of low-illuminated tunnels in image frame by exploiting the inherent darkness in the longitudinal direction. In the sequel, it is converted from the pixel coordinates to the heading rate command of the MAV for adjusting the heading towards the center of the tunnel. The efficacy of the proposed framework has been evaluated in multiple experimental field trials in an underground mine in Sweden, thus demonstrating the capability of low-cost and resource-constrained aerial vehicles to fly autonomously through tunnel confined spaces.

Keywords: Micro Aerial Vehicles (MAVs) · Vision-based navigation · Autonomous drift inspection · Otsu's theshold · Moore-Neighborhood tracing

1 Introduction

The deployment of Micro Aerial Vehicle (MAV) is gaining more attention in different applications, such as infrastructure inspection [6], underground mine inspection [4], subterranean exploration [9]. etc. since in general MAVs can reduce service and inspection procedures and increase the overall safety of the personnel by navigating and exploring unattainable, complex, dark and dangerous mining underground locations in production areas.

This work has been partially funded by the European Unions Horizon 2020 Research and Innovation Programme under the Grant Agreement No. 730302 SIMS.

D. Tzovaras et al. (Eds.): ICVS 2019, LNCS 11754, pp. 164–174, 2019.
https://doi.org/10.1007/978-3-030-34995-0_16

In underground mine tunnels, incidents such as rock falls, blasting, drift expansion, fire detection or propagation, etc. require a frequent inspection for collecting information e.g. images, gas levels, dust levels, 3D models, etc. that can be used for ensuring the safety of the human workers and the overall increase of productivity and the reduction of the related down times in production. However, underground mine tunnels are challenging environments for deploying MAVs due to the lack of illumination, the existence of narrow passages, dust, wind gusts, water drops, conductive dust, etc. Additionally, during the autonomous mission of the MAV it should detect obstacles and avoid collisions for providing a successful and safe autonomous navigation. In general, the MAVs are equipped with high-end and expensive sensor suites to provide stable autonomous navigation, nonetheless, in the case of harsh environments inside mines, there is an ongoing trend to utilize low cost hardware and give emphasis on the algorithmic part of the aerial platform in order to enable the consideration of the MAVs as consumable platforms for the completion of specific missions.

Inspired by the mining needs, a low computational complexity image processing method is proposed for correcting the heading of MAVs in dark underground tunnels. In the proposed method, the image stream from the on-board forward looking camera is utilized for extracting the darkest contour of the image. In the sequel, the center of this area is extracted in order to correct the heading of the MAV towards the tunnel axis. In general, due to the lack of natural and external illumination of the environment, as depicted in Fig. 1, the obstacles such as walls are brighter especially when compared to the open areas, a fact that this results to the correction of the MAV's heading towards the open area. Moreover, due to uncertainties in the position estimation, the platform navigates as a floating object with v_x and v_y velocity commands, while the position information on the x and y axes are not used, moreover the potential fields method [4] provides desired velocities $v_{d,x}$, $v_{d,y}$ for avoiding collisions to the walls.

1.1   Background and Motivation

Autonomous navigation in unknown environments is highly coupled with the need for collision avoidance and obstacle detection. Moreover, obstacle detection and navigation, based on vision based techniques for MAVs has received a significant attention the latest years and with a big variety of application scenarios [5]. At the same time, visual stereo or monocular camera systems are able to provide depth measurements for obstacle avoidance, while the obstacle detection methods, based on a monocular camera, in the corresponding literature, are based mainly either on computer vision algorithms or on machine learning methods.

In [10], a mathematical model to estimate the obstacle distance to the MAV was implemented for collision avoidance. However, the method provided poor results at high velocity and low illumination environments. In [13] an obstacle avoidance scheme for MAVs was presented, consisting of three stereo cameras for 360° coverage of the platform's surroundings in the form of pointclouds. However, the proposed method relied on sufficient illumination, landmark extraction and high on-board processing power. In [2], random trees were generated to find

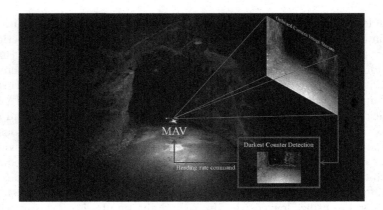

Fig. 1. The proposed approach for extracting the center of the darkest contour from the forward looking camera during autonomous navigation in underground mines. *Supplementary Video:* https://youtu.be/KNWE0BTpALU

the best branch on-line, while building the occupancy map of the perceived environment. This method requires in general a high computation power to process the images.

Moreover, few works using Convolutional Neural Network (CNN) for navigation, such as [1,12], utilized the image frame of on-board camera to feed the CNN for providing heading commands. These works have been evaluated and tuned in out-door environments and with good illumination. Furthermore, preliminary and limited studies of MAV navigation in an underground mine using CNN was presented in [7], however the method was evaluated in off-line collected data-sets from two underground tunnels, without the MAV in the loop.

In general, the performance of the computer vision-based algorithms mainly relies on the surrounding environment with good distinctive features, good illumination and lighting conditions [10] and on a high computation power, factors that could limit the usage of these methods in real-life underground mine applications. Furthermore, when using machine learning techniques a large amount of data and high computation power for the off-line training of a CNN are required. However, the online use of the trained CNN has lower computation power demands and is applicable for the navigation task.

1.2   Contributions

Based on the aforementioned state of the art, the main contributions of this article are provided in this section. The first and major contribution of this work is the development of the low computational complexity method for providing heading rate commands. The proposed method does not require training data-sets, which is a major limitation of most machine learning methods. To the best of our knowledge, this is the first work towards MAV navigation in underground tunnels based on centroid extraction of the darkest contour from on-board looking forward image stream. The method requires low computation power and enables online MAV autonomous navigation.

The second contribution, stems from the evaluation of the proposed method and the low-cost MAV in a dark underground tunnel in Sweden, while accurate pose estimation is not available and the platform operates as a floating object. The experimental results demonstrate the performance of the proposed method in underground tunnels without a natural illumination, while the following link provides a video summary of the system in https://youtu.be/KNWE0BTpALU.

1.3   Outline

The rest of the article is structured as follows. Initially, Sect. 2 presents the system architecture. Then, the algorithm for the darkest contour centroid extraction is presented in Sect. 3. Later, in Sect. 4 the experimental setup and the extended experimental evaluation of the proposed method in an underground tunnel in Sweden are presented. Finally, the article concludes by summarizing the article and future works in Sect. 5.

2   System Architecture

The MAV body-fixed frame is $\mathbb{B}$ and the world frame is denoted by $\mathbb{W}$ in the North-West-Up (NWU) frame. The forward looking camera frame is $\mathbb{C}$ and the image frame is $\mathbb{I}$ with unit vector $\{x^{\mathcal{I}}, y^{\mathcal{I}}\}$. The $p_x \in \mathbf{Z}^+$ and $p_z \in \mathbf{Z}^+$ are the pixel coordinates of the image I, and $\mathbf{Z}$ is the integer set of numbers $\mathbf{Z} = \{-\infty, \ldots, -1, 0, 1, \ldots \infty\}$. Figure 2 depicts the coordinate system.

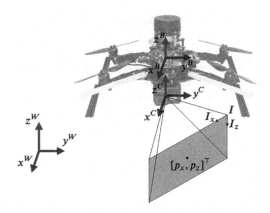

Fig. 2. Coordinate frames, where $\mathbb{W}$, $\mathbb{B}$, $\mathbb{C}$ and $\mathbb{I}$ denote the world, body and camera and image coordinate frames respectively.

The MAV is considered as a floating object, while the state of the system is $X = [z, v_x, v_y, v_z, \phi, \theta]^\top$. The Inertial Measurement Unit (IMU) measurements, which are a_x, a_y, a_z, w_x, w_y, and w_z for the linear and angular accelerations along each axis are passing through an Extended Kalman Filter (EKF) and provides the ϕ and θ. The down-ward optical-flow sensor provides v_x and v_y and the single beam lidar provides altitude z estimation. The image stream from the looking forward camera is denoted by I and the estimated states is indicated with $\hat{\ }$. Additionally, the potential fields method is implemented to generate proper velocity commands $[v_{d,x},\ v_{d,y}]^\top$ to avoid collisions to the walls or obstacles from range measurements $R = \{r_i | r_i \in \mathbf{R}^+, i \in \mathbf{Z} \cap [-\pi, \pi]\}$ of the 2D lidar placed on top of the MAV. The heading rate commands $\dot{\psi}_d$ are provided from the centroid extraction from the darkest contours of the image stream to move towards open spaces. Furthermore, for tracking the desired velocity and altitude commands $[z_{d,x}, v_{d,x}, v_{d,y}]^\top$ the Nonlinear Model Predictive Control (NMPC) [11] is implemented to generate the corresponding thrust and attitude commands $[T_d, \phi_d, \theta_d]^\top$ for the low level controller. The low level controller generates the motor commands $[n_1, \dots, n_4]^\top$ for the MAV. The overall control structure is presented in Fig. 3.

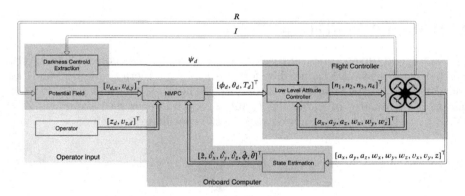

Fig. 3. Control scheme of the proposed navigation module with potential fields for desired velocities generation, while the heading commands are generated from the contour extraction method. The NMPC generates thrust and attitude commands, while the low level controller generates motor commands $[n_1, \dots, n_4]^\top$. The state estimation is based on IMU measurements, optical flow and single beam lidar.

3   Methodology

This section describes the darkness centroid extraction method (see Fig. 3) that receives a camera image as input and delivers the heading correction. The input to the algorithm is an image I taken by the MAV's forward looking camera. The method can be described as follows.

Darkness Centroid Extraction:

– Input: RGB image acquired by the forward looking camera.
Start
• Step 1. Convert the RGB image to binary image using Otsu's threshold.
• Step 2. Extract contour of background objects using Moore-Neighborhood tracing algorithm.
• Step 3. Identify the darkness in the tunnel as the background object with largest area.
• Step 4. Calculate the heading rate correction based on the centroid of the darkness.
End
– Output: Heading rate correction.

It is considered that the images are composed by a set of physical objects (walls, pipes, lights) as well as a background. Step 1 is described in Subsect. 3.1 and deals with the separation of such physical objects from the background, which is done by finding a threshold that separates the data and uses that threshold to create a binary image. Tracing the contour of the dark background objects is performed by using the Moore-Neighborhood tracing method described in Subsect. 3.2. Later, the largest set of background objects is identified and assumed to be the darkness in the tunnel as described in Subsect. 3.3. Finally, the position of the center of the darkness is determined and converted from pixel coordinates to heading rate commands, described in Subsect. 3.4.

3.1 Step 1. Converts the RGB Image to a Binary Image Using Otsu's Threshold

Otsu's method for the threshold selection was introduced in [8] and is widely used to reduce a grayscale image to a binary image. It is therefore necessary that the images have been priory converted to a grayscale. In a grayscale image, a single intensity value is assigned to each pixel and the input pictures are represented by a matrix of intensities $I(p(p_x, p_z))$, where $p(p_x, p_z)$ represents a pixel at image coordinates (p_x, p_z).

Otsu's method assumes that the image is composed by two classes of pixels, namely the foreground pixels and the background pixels. Otsu's method targets the separation of these two classes of pixels that will be named as:

– $\mathcal{T}_1$: class of pixels belonging to the bright foreground.
– $\mathcal{T}_2$: class of pixels belonging to the dark background.

Otsu's threshold is calculated independently for each image, due to the fact that each image has different properties, such as brightness, and they require therefore separate processing for separation between foreground and background. A threshold τ is sought, which determines the belonging of the pixels $p(p_x, p_z)$ to either class. Being a pixel $p(p_x, p_z)$ belonging to the background if the intensity of the pixel $\mathcal{I}(p(p_x, p_z))$ is smaller than the threshold τ.

$$\mathcal{T}_1 = \{p(p_x, p_z) : \mathcal{I}(p(p_x, p_z)) \geq \tau\} \quad ; \quad \mathcal{T}_2 = \{p(p_x, p_z) : \mathcal{I}(p(p_x, p_z)) < \tau\}$$

The goal of Otsu's method is to find the value of the separation threshold τ, which minimizes the intra-class variance $\sigma_w^2(\tau)$:

$$\sigma_w^2(\tau) = \omega_1 \sigma_1^2(\tau) + \omega_2 \sigma_2^2(\tau) \tag{1}$$

where ω_1 and ω_2 are the number of pixels in each class and where $\sigma_1^2(\tau)$ and $\sigma_2^2(\tau)$ are the variance of the elements of each class. Minimizing the intra-class variance is equivalent to maximizing the separation between classes.

The calculation of the variances $\sigma_1^2(\tau)$ and $\sigma_2^2(\tau)$ requires the previous calculation of the means of each class. However, Otsu showed that minimizing the intra-class variance in (1) is equivalent to maximizing the inter-class variance $\sigma_b^2(\tau)$, which only depends on the means of the classes μ_1 and μ_2:

$$\sigma_b^2(\tau) = \omega_1(\tau) \cdot \omega_2(\tau) \cdot (\mu_1(\tau) - \mu_2(\tau))^2 \tag{2}$$

The search of the value of τ, which maximizes the inter-class variance can be performed by considering all the different values of the intensity $\mathcal{I}(p(p_x, p_z))$ and choosing the one which maximizes (2). The threshold τ synthesized using Otsu's method is then used to create a binary image $I_b(p(p_x, p_z))$. In this binary image, the pixels, which correspond to elements in the foreground class $\mathcal{T}_1$ get assigned the value 1 and the pixels that belong to the background class $\mathcal{T}_2$ get assigned the value 0. This binary image therefore classifies the pixels as belonging to the set of foreground pixels or to the set of background pixels.

$$I_b(p(p_x, p_z)) = \begin{cases} 1 \; if \; p(p_x, p_z) \in \mathcal{T}_1 \\ 0 \; if \; p(p_x, p_z) \in \mathcal{T}_2 \end{cases} \tag{3}$$

Step 1 is therefore executed as follows:

Step 1. Convert the RGB image to a binary image using Otsu's threshold:

– Input: RGB image acquired by the forward looking camera.
  Start
• Step 1.1 Convert the RGB image to a gray scale image represented by an intensity matrix $I(p(p_x, p_z))$. This is performed by eliminating the hue and saturation information from the image, while retaining the luminance.
• Step 1.2 Find Otsu's threshold τ by searching through the different intensity values in $I(p(p_x, p_z))$ until the value that maximizes the interclass variance in (2).
• Step 1.3 Use Otsu's threshold τ to convert the image to binary as stated in (3).
  End
– Output: Binary image $I_b(p(p_x, p_z))$.

3.2   Step 2. Finding Boundaries of the Darkness Using the Moore-Neighbor Tracing Algorithm

The goal of using Moore-Neighbor tracing is to find all the sets of the interconnected pixels by using a search based on the Moore-neighborhood concept. The details of Moore-Neighborhood tracing algorithm are omitted here for being widely available in the literature. For details, the reader can refer to [3].

The input to the Moore-Neighbor tracing algorithm is the binary image $I_b(p(p_x, p_z))$. Notice that the usual formulation of the Moore-Neighbor algorithm traces pixels with a binary value of 1 (foreground pixels), but in this example the dark objects (value of 0) has to be traced. This can be done e.g. by simply performing the binary negation of $I_b(p(p_x, p_z))$ before applying Moore-Neighbor tracing.

The output of the algorithm is sequence of sets of ordered points $\mathcal{F}_k$ related to each dark background object, being $\mathcal{F}_k = \{P_{k,1}, P_{k,2}, \ldots, P_{k,N}\}$, where each point $P_{k,i}$ has as coordinates $(x_{k,i}, z_{k,i})$.

3.3   Step 3. Identify the Darkness in the Tunnel as the Background Object with the Largest Area

The area of each set $\mathcal{F}_k$ can then be calculated using the shoelace formula as:

$$area(\mathcal{F}_k) = \frac{1}{2} \sum_{i=1}^{N-1} (x_{k,i} \cdot z_{k,i+1} - x_{k,i+1} \cdot z_{k,i}) \tag{4}$$

where the last vertex given by the coordinates (x_N, z_N) is the same vertex as the original vertex (x_1, z_1).

The set of pixels with the largest area is considered to enclose the darkness in the tunnel. Last, the abscissa $\hat{x}_k$ of the centroid of the largest dark area is calculated as:

$$\hat{x}_k = \frac{1}{6 \cdot area(\mathcal{F}_k)} \sum_{i=1}^{N-1} (x_{k,i} + x_{k,i+1})(x_{k,i} z_{k,i+1} - x_{k,i+1} z_{k,i})] \tag{5}$$

3.4   Step 4. Calculate the Heading Rate Correction Based on the Centroid of the Darkness

Finally we map the value of $\hat{x}$ to heading rate command.

Step 4. Calculate the heading rate correction based on the centroid of the darkness:

– Input: Centroid of the darkness $(\hat{x}, \hat{z})$

Start

- Step 4.1. $\overline{x} = \frac{\hat{x}}{n}$, $\overline{x} \in [0, 1]$ // Linear mapping $\hat{x} \rightarrow [0, 1[$
- Step 4.2. $\dot{\psi}_d = \frac{\overline{x} - 0.5}{2.5}$ // Linear Mapping $[0, 1] \rightarrow [-0.2, 0.2]$ rad/sec
- Output: Heading rate correction $\dot{\psi}_d$.

4 Results

This section describes the experimental setup and the experimental evaluation of the proposed method for sending heading rate commands to the MAV for autonomous navigation in an underground tunnel. The following link provides a video summary of the overall results: https://youtu.be/KNWE0BTpALU.

4.1 Experimental Setup

A low-cost quadcopter has been utilized for underground tunnel navigation, which was developed at Luleå University of Technology based on the ROSflight flight controller. The ROSflight Based quad-copter weights 1.5 kg and provides a $10 - 14$ mins of flight time with a 4-cell 3.7 hA LiPo battery. The Aaeon UP-Board is the main processing unit, incorporating an Intel Atom x5-Z8350 processor and 4 GB RAM. The operating system running on the board is the Ubuntu Desktop 18.04, while Robot Operating System (ROS) Melodic has been also included. The platform is equipped with the PX4Flow optical flow sensor, a single beam Lidar-lite v3 for altitude measurement, two 10 W LED light bars in front arms for providing additional illumination for the looking forward camera and four low-power LEDs looking down for providing illumination for the optical flow sensor. Figure 4 presents the platform, highlighting it's dimensions and the overall sensor configuration.

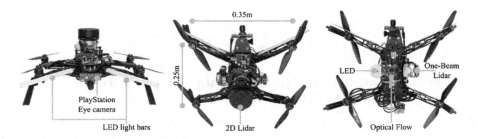

Fig. 4. The developed ROSflight based quad-copter equipped with 2D and one beam lidars, optical flow, PlayStation camera and LED bars.

4.2 Experimental Evaluations

The performance of the proposed method is evaluated in an underground tunnel located at Luleå Sweden with lack of natural and external illumination in the tunnel. The tunnel did not have corrupting magnetic fields, while small particles were in the air. The tunnel morphology resembled an S shape and the dimensions of the area where the MAV navigates autonomously were $3.5(\text{width}) \times 3(\text{height}) \times 30(\text{length})$ m^3. The platform is equipped with a PlayStation Eye camera with a resolution of 640×480 $pixels$ and 10 fps. The front LED bars provide illumination of 460 lx from 1 m distance, while for the optical flow sensor low-power LEDs looking down are provided. The desired altitude and velocities for the MAV were set to 1 m, and $v_{d,x} = 0.5$ m/s, $v_{d,y} = 0.0$ m/s respectively.

In Fig. 5 some examples from the on-board image stream during the autonomous navigation are depicted, while the centroids of the darkest contours are shown. Moreover, it is observed that in case of branches in the tunnel the proposed method cannot recognize them and select the darkest branch as darkest contour or combine both branches.

Fig. 5. Sample images of the on-board forward looking camera, while the boundary of the darkest contour is shown by red color. (Color figure online)

5 Conclusions

This article presented a darkness contours detection for MAV navigation in underground tunnels using a forward looking camera. In the proposed method, the image is first converted to grayscale and later a threshold of the image stream from the on-board camera is obtained by Otsu's method, then the image is converted to binary image and boundaries of the darkness areas are extracted. Later on, the largest dark area is selected and the center of this area is used to generate heading commands for the platform. The proposed method is evaluated in dark underground tunnels in Sweden and provides autonomous navigation, while the heading is corrected towards the center of the tunnel axis.

As expected the method fails to obtain correct heading rate for the platform, in the tunnel with external illumination, as in these cases the tunnel center is not the darkest area.

References

1. Adhikari, S.P., Yang, C., Slot, K., Kim, H.: Accurate natural trail detection using a combination of a deep neural network and dynamic programming. Sensors **18**(1), 178 (2018)
2. Bircher, A., Kamel, M., Alexis, K., Oleynikova, H., Siegwart, R.: Receding horizon next-best-view planner for 3d exploration. In: IEEE International Conference on Robotics and Automation (ICRA), pp. 1462–1468 (2016)
3. Blanchet, G., Charbit, M.: Digital Signal and Image Processing Using MATLAB, vol. 4. Wiley Online Library, Hoboken (2006)
4. Kanellakis, C., Mansouri, S.S., Georgoulas, G., Nikolakopoulos, G.: Towards autonomous surveying of underground mine using MAVs. In: Aspragathos, N.A., Koustoumpardis, P.N., Moulianitis, V.C. (eds.) RAAD 2018. MMS, vol. 67, pp. 173–180. Springer, Cham (2019). https://doi.org/10.1007/978-3-030-00232-9_18
5. Kanellakis, C., Nikolakopoulos, G.: Survey on computer vision for UAVs: Current developments and trends. J. Intell. Robot. Syst. pp. 1–28 (2017). https://doi.org/10.1007/s10846-017-0483-z
6. Mansouri, S.S., Kanellakis, C., Fresk, E., Kominiak, D., Nikolakopoulos, G.: Cooperative coverage path planning for visual inspection. Control Eng. Pract. **74**, 118–131 (2018)
7. Mansouri, S.S., Kanellakis, C., Georgoulas, G., Nikolakopoulos, G.: Towards MAV navigation in underground mine using deep learning. In: IEEE International Conference on Robotics and Biomimetics (ROBIO) (2018)
8. Otsu, N.: A threshold selection method from gray-level histograms. IEEE Trans. Syst. Man Cybern. **9**(1), 62–66 (1979)
9. Rogers, J.G., et al.: Distributed subterranean exploration and mapping with teams of UAVs. In: Ground/Air Multisensor Interoperability, Integration, and Networking for Persistent ISR VIII, vol. 10190, p. 1019017. International Society for Optics and Photonics (2017)
10. Saha, S., Natraj, A., Waharte, S.: A real-time monocular vision-based frontal obstacle detection and avoidance for low cost UAVs in GPS denied environment. In: 2014 IEEE International Conference on Aerospace Electronics and Remote Sensing Technology, pp. 189–195. IEEE (2014)
11. Small, E., Sopasakis, P., Fresk, E., Patrinos, P., Nikolakopoulos, G.: Aerial navigation in obstructed environments with embedded nonlinear model predictive control. In: 2019 European Control Conference (ECC). IEEE (2019)
12. Smolyanskiy, N., Kamenev, A., Smith, J., Birchfield, S.: Toward low-flying autonomous MAV trail navigation using deep neural networks for environmental awareness. arXiv preprint arXiv:1705.02550 (2017)
13. Valenti, F., Giaquinto, D., Musto, L., Zinelli, A., Bertozzi, M., Broggi, A.: Enabling computer vision-based autonomous navigation for unmanned aerial vehicles in cluttered gps-denied environments. In: 2018 21st International Conference on Intelligent Transportation Systems (ITSC), pp. 3886–3891 (2018)

Improving Traversability Estimation Through Autonomous Robot Experimentation

Christos Sevastopoulos, Katerina Maria Oikonomou,
and Stasinos Konstantopoulos$^{(\boxtimes)}$

Institute of Informatics and Telecommunications, NCSR 'Demokritos',
Ag. Partaskevi, Greece
sev_chris@yahoo.com, k.m.oikonomou@gmail.com, konstant@iit.demokritos.gr

Abstract. The ability to have unmanned ground vehicles navigate unmapped off-road terrain has high impact potential in application areas ranging from supply and logistics, to search and rescue, to planetary exploration. To achieve this, robots must be able to estimate the traversability of the terrain they are facing, in order to be able to plan a safe path through rugged terrain. In the work described here, we pursue the idea of fine-tuning a generic visual recognition network to our task and to new environments, but without requiring any manually labelled data. Instead, we present an autonomous data collection method that allows the robot to derive ground truth labels by attempting to traverse a scene and using localization to decide if the traversal was successful. We then present and experimentally evaluate two deep learning architectures that can be used to adapt a pre-trained network to a new environment. We prove that the networks successfully adapt to their new task and environment from a relatively small dataset.

Keywords: Robot vision · Adaptability · Self-assessment

1 Introduction

The ability to have *unmanned ground vehicles (UGV)* navigate unmapped off-road terrain has high impact potential in application areas ranging from supply and logistics, to search and rescue, to planetary exploration. The core concept in outdoors navigation is the computation of a 2D *traversability grid* upon which path planning algorithms can operate. This grid is computed by using knowledge of the physical and mechanical properties and capabilities of a specific UGV in order to estimate if each grid cell can be safely traversed of not.

The two main lines of research in traversability estimate are *appearance-based* and *geometry-based* methods. The former rely on computer vision either directly estimate traversibility from raw images [1, 10] or to extract intermediate information that is then used in a separate traversability estimation step. Such

D. Tzovaras et al. (Eds.): ICVS 2019, LNCS 11754, pp. 175–184, 2019.
https://doi.org/10.1007/978-3-030-34995-0_17

information can be either terrain features such as roughness, slope, discontinuity and hardness [7] or terrain classes such as soil, grass, asphalt, vegetation [2]. *Geometry-based* methods, on the other hand, extract terrain features such as roughness, slope, and discontinuity from *digital elevation maps (DEM)* obtained from stereoscopic or depth sensors [4, 13].

Methods that rely on DEM are more popular in recent literature as they estimate terrain features more directly and accurately than appearance-based methods. However, they lack the ability to distinguish between different classes of obstacles. For instance, a concrete obstacle and a patch of bushy vegetation can be practically impossible to distinguish in the DEM alone, although the distinction is important: a bush that can be easily overrun by the UGV should not be considered an obstacle. To address this, *hybrid methods* combine different types of sensory input to assess different aspects of traversability [5].

Hybrid methods, combined with recent advances in vision brought about by deep learning, can re-introduce appearance-based methods, and computer vision in general, as relevant to the outdoors traversability estimation problem. As demonstrated by recent work [1], a priorly learned network can be efficiently fine-tuned to a new environment using only a fraction of the data needed to initially train it. The fine-tuned network achieved an accuracy of 91% on unseen data, a marked improvement over the 75% achieved by the original network. However, in order to conduct their experiment Adhikari et al. had to manually annotate more than 5000 areas from 10 frames as 'traversible' or 'non-traversible'.

In the work described here, we also develop the idea of fine-tuning a generic network for a specific environment, but without requiring any manually labelled data. Instead, we present an autonomous data collection method that allows the robot to derive ground truth labels for scenes it observes and a network architecture that is fine-tuned from a relatively small, autonomously collected, collection of scenes. We prove the concept that the robot can autonomously assess the accuracy of its visual traversability estimation method, by attempting to traverse a scene and using localization to decide if the traversal was successful (Sect. 2). We then discuss how the outcomes of this autonomous experimentation can be used to adapt the robot's visual traversability estimation models (Sect. 3), present experimental results that validate that where substantial adaptation is achieved from a feasible number of autonomous experiments (Sect. 4). We close the paper with conclusions and future research directions (Sect. 5).

2   Autonomous Data Collection

In order to have the robot autonomously fine-tune its traversability estimation, we want it to be able to measure the traversability of a path after traversing it (or having failed to traverse it). One key concept we are putting forward in this work is that traversability can be measured as the *error in proprioceptive localization*, that is, in localization that relies on the wheel encoder and IMU signals. The rationale is that along an easily traversable path, encoder drift is small and when fused with IMU becomes negligible. The more difficult a path

Fig. 1. Indicative scenes for the localization error experiments.

is, the more the wheels drift and this error increases. In order to calculate this error, we will compare proprioceptive localization against the full 3D localization that fuses the encoders, IMU, and stereoscopic camera signals.

In order to prove this concept, we executed the following experiment using a DrRobot Jaguar rover fitted with a Zed stereoscopic camera: we selected several locations with varying degrees of traversability ranging from paved path, to vegetation that appears as an obstacle in the 3D point cloud created by the stereo camera but can be pushed back and traversed, to a wall (Fig. 1 shows some indicative examples). After using standard ROS navigation to approach an obstacle, the robot synchronizes a secondary proprioceptive localization module with the main localization it uses to navigate, and then circumvents normal obstacle avoidance to push against the obstacle. The velocity is the minimum velocity that the robot can obtain, guaranteeing the safety of the platform even when pushing against a wall.

We have carried out 20 experiments, at varying locations and approaching the obstacle from different angles and have empirically found that for our platform and for our specific sensors a localization error of 21 cm or less signifies that the path is traversable.[1] This simple rule achieves an accuracy of 9/10 in both the traversible and the non-traversible tests (Table 1). The specific threshold is likely to be different for different sensors and will need to be empirically identified for each robotic platform and mix of sensors, but after testing it on different kinds of obstacles we believe that it is a constant with respect to the environment.

We have looked at the outlying experiments and found that:

- The difference of 9 cm on non-traversible path was caused by pushing against the wall from a very small angle, causing the platform to fall back to the paved road and continue moving.
- The difference of 89 cm on a traversible path was caused by uneven slippage between the wheels on the one side and the wheels on the other side of the platform, as one side was on grass and the other on paved road. This does not happen when all wheels are on grass and slip evenly, as it is corrected by the IMU.

The former situation can be easily avoided by ensuring that the robot approaches the obstacle correctly. The latter situation is more difficult to detect and avoid

---

[1] The software used and the data collected is publicly available at https://github. com/roboskel/traversability_estimation.

from the localization and navigation system alone, and a future solution will involve reasoning over a more general situational awareness, so that the robot can determine if slippage is due to trying to push against an obstacle or due to slippery terrain. For the experiments presented below we have included both false datapoints in the training set, to validate the robustness of the overall method against these labelling errors.

As we now have provided the robot with the means to use past experience as ground truth data, we will proceed in the next section to prove the concept that a reasonable amount of such data is sufficient to fine-tune visual traversability estimation for a new environment.

3   Adapting Traversability Estimation

In order to avoid requiring that the robot performs an unreasonable amount of experiments before observing any significant improvement in its traversability estimation capability, we assume as a starting point pre-trained deep learning model and experiment with how to use it as a generic feature extractor for classifiers trained on our environment-specific data.

We are going to use the VGG16 pre-trained model [11] as a feature extractor by exploiting its first five powerful convolutional blocks. In particular, its standard architecture comprises of a total of 13 convolutional layers using 3×3 convolution filters along with max pooling layers for down-sampling. In our case we will load the model with the ImageNet weights and freeze the convolutional blocks. As a result we take advantage of the output from the convolutional layers that represent high-level features of the data. By the process of flattening, this output is then converted to a one-dimensional vector that will be fed to the classifier. A commonly seen approach in literature involves adding one fully-connected (hidden) layer.

We have empirically found two hidden layers to give adequate accuracy while keeping the computational cost low, with 64 units for the first layer and 32 for the second layer. On both layers we use the 'relu' activation function [6]. Moreover, following Baldi and Sadowski [3], a dropout layer with a rate of 0.5 is inserted after each hidden layer. For the final output we add a softmax-activated layer with 2 units for the binary prediction; traversible or non-traversible. The reason we use the softmax activation function for the last layer is because of its ability to output the probability distribution over each possible class label [9,11]

Table 1. The maximum difference (in cm) between full localization and proprioceptive localization, measured along the axis of the robot's movement. Using a threshold of 21 cm, we get two mis-classifications in 22 runs.

Threshold:												21	
Traversible:	0	1	1	2	6	7	7	8	14	16	18		89
Non-traversible:		110	30	28	28	27	26	25	25	22			9

As for the adaptation itself, we have experimented with both *transfer learning* and *fine tuning*. In *transfer learning*, a base network is trained on an initial base dataset for the purposes of an initial task. Then, the learned features are transferred to a target network that is trained on a custom dataset for a different task. The technique we are going to implement is the combination of a *global average pooling layer* [11] one Dropout that tosses out 50% of the neurons and one dense layer that outputs the prediction of the two classes we are investigating. The reason for using the global average technique is that due to its structural regularization nature, it offers an alternative approach in preventing overfitting of the fully connected layers.

In *fine tuning*, the process we followed is to freeze the weights of the low level convolutional layers and then add dense layers on top. We expect the model to be able to map the knowledge represented in the low level convolutional layers to the desired environment-specific output.

4   Experimental Results and Discussion

In order to validate our approach, we will use data collected following the method presented in Sect. 2 in order to adapt the pre-trained VGG16 object recognition network to our task *and* our environment using the two methods presented in Sect. 3. That is, we aim at adapting not only to improve for a new environment, but also at transferring VGG16 knowledge to the new classification task (traversible vs. non-traversible) for which the robot is able to autonomously collect labelled data. We then evaluate the performance of these two adapted networks against a baseline that makes traversability decisions based on the object recognition results from the original VGG16.

4.1   Data Acquisition

Our dataset comprises the RGB frames from the experiments described in Sect. 2. We have randomly split the runs between training, validation, and testing, but forcing the two mis-classified runs to be placed in the training set. This was done in order to (a) estimate the robustness of the approach to the errors made by our autonomous labelling method; and (b) to ensure that testing set us perfectly labelled in order to get accurate evaluation results. The resulting dataset consists of 1300 training images, 150 validation images, and 150 testing images.

It should be noted that the split was done at the level of individual runs. In this manner, although all data is from the NCSR 'Demokritos' campus, the testing data is from *different* locations within this general environment than the training and validation data.

4.2   Baseline

As a baseline, we will assume inferring traversability from the objects recognized by the unmodified, pre-trained VGG16 network. We started by manually

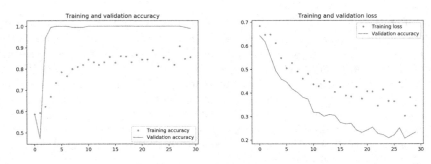

Fig. 2. Accuracy and loss for Model 1

mapping all VGG classes that appear in our dataset to 'traversible' (e.g., vegetation) or 'non-traversible' (e.g., stonewall, vehicle). We then assume the three
most-probably classes, as probabilities after the third prediction become negligible ($\sim 10^{-4}$). We then apply the following rules:

- If all predictions are positive (traversible), the scene is assumed to be positive.
- If the most probable prediction is positive with probability p_{pos}, and the most
  likely negative class has a probability p_{neg}, then the scene is assumed to be
  positive if $p_{pos} - p_{neg} > 0.01$ and negative otherwise.
- If the most probable prediction is negative with probability p_{neg}, and the
  other two predictions are both positive and are both considerably likely, the
  scene is assumed to be positive. Specifically, if the two positive recognitions
  have probabilities p_{pos}^a and p_{pos}^b, then if must be that $p_{neg} - p_{pos}^a < 0.2$ and
  $p_{neg} - p_{pos}^b < 0.5$.
- Otherwise, the scene is assumed to be negative (non-traversible).

We have designed these rules so that they push the balance towards positive
predictions. We have also experimented with more straightforward inferences,
such as simply assuming the traversability label of the single most probable
object, but we have observed that such rules tend to over-generalize in favour
of negative predictions. We have empirically found these rules and thresholds to
give the best possible results for our dataset, setting the baseline par as high as
possible when using the pre-trained VGG16 network.

Qualitatively speaking, observing VGG16 predictions we found some completely erroneous recognitions, as well as some instances where correctly recognized non-traversible objects should not have been the dominant characterization
of the scene but a minor one. But in general, most predictions were correct *and*
meaningful for our task.

4.3 Training

We utilized the Keras deep-learning framework. First, we resized all images to
224×224, which is the original ImageNet format. We also apply *data augmentation* to increase dataset size by rotating, shifting and zooming on the initially
training dataset.

Fig. 3. Accuracy and loss for Model 2

We then built two models with two different front layer architectures. For Model 1 we used the RMSProp optimizer [12] and a learning rate of 10^{-3}; for Model 2 we used the *Adam* optimizer [8] and a learning rate of 10^{-4}. These learning rates were estimated by starting at a value of 10^{-2} and decreasing exponentially. Also, since our classification goal is binary, we are using the *binary_crossentropy* loss function.

Training Model 1 for 30 epochs, we get the accuracy and loss transfer shown in Fig. 2. We can see that the model performs well on both the training and validation sets. More specifically, the Loss plots for both training and validation constantly decrease and stabilize around the same point. As a consequence, we can deduce that the model is learning and generalizing in an adequate manner.

After training Model 2 for 30 epochs, we get the accuracy and loss given in Fig. 3. Overall, the model's performance is satisfactory. However, we can observe certain signs of fluctuation, revealing that potentially there has not been adequate generalization.

4.4   Results

In order to infer a binary traversability decision, comparable to the decision made by our raw-VGG baseline system, we assume as traversible the scenes where the 'traversible' probability is at least 0.075 higher than the 'non-traversible' probability. Table 2 gives the accuracy of the baseline and our two models per traversability category.

It can be noticed that, as already discussed above, VGG errs on the side of safety and its retrieval is low on traversible examples. Our custom models also show higher confidence (Fig. 4), which is to be expected as they have been tuned to our (simpler, binary) task.

4.5   Error Analysis

We observe that both the performances of Model 1 and Model 2 outperform VGG when we are investigating a traversible scene. One common failure, however, is caused by the presence of an non-traversible objects, such as the building on the

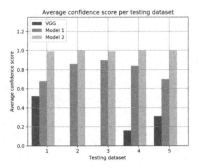

Fig. 4. Average Confidence scores per testing dataset

(a) Traversible vegetation   (b) Non-traversible   (c) Stone wall

Fig. 5. Some characteristic instances

left of Fig. 5a. Although this is a traversible scene, our methods does not focus on the part of the scene directly in front of the robot, but characterizes the scene as a whole.

What is more, there are examples that although the first output of VGG corresponds to a class that has been defined as 'traversible', secondary recognitions lead to mis-classifications. For instance, the scene in Fig. 5b is classified as 'lakeside' and 'park bench', with a small probability difference between the two.

Finally, an example where the baseline is outperformed is given in Fig. 5c. This scene is recognized as 'patio' and 'stone wall', the latter at an exceptionally low probability leading the baseline to classify this as a traversible scene. Model 1

Table 2. Accuracy of the baseline and our two models on traversible and non-traversible training data.

System	Accuracy	
	Traversible examples	Non-traversible examples
VGG	33.3	97.3
Model 1	99.1	100
Model 2	98.2	100

and Model 2 considered the presence of the stone wall and correctly classified the scene as non-traversible.

5   Conclusion and Future Work

We presented a method that tackles the problem of adapting generic vision problems to a new task and environment by only using whatever labelled data the robot is able to collect autonomously and without any human supervision. The core of the learning method is to build CNNs that can take advantage of *transfer learning* and *fine tuning* to keep the convolutional architecture of the VGG intact.

Overall despite the small number of training data, we evaluated our models on the testing dataset and we witnessed a significant improvement over the VGG's weak performance, especially on traversible scenes. Regarding non-traversible scenes where the predictive capabilities of VGG are indisputable, we were able to take advantage of its convolutional features and transfer its obstacle recognition ability to our models. Hence, we showed that the robot can make predictions about its environment by using a reasonable amount of data, and in fact data that is labelled without any manual annotation.

A common problem that our approach cannot completely address is the lack of a sense of specific traversible paths through a generally non-traversible scene, making the overall methodology unnecessarily cautious. In future work we plan to extend the method so that it estimates the traversability of specific paths rather than of the scene as a whole. One approach we will experiment with will be cropping the test images so that traversability is estimated on a narrower field around the selected path; and, accordingly, collecting training data from similarly cropped images around the path tested during autonomous experimentation.

References

1. Adhikari, S.P., Yang, C., Slot, K., Kim, H.: Accurate natural trail detection using a combination of a deep neural network and dynamic programming. Sensors 18(1) (2018). https://doi.org/10.3390/s18010178
2. Angelova, A., Matthies, L., Helmick, D., Perona, P.: Fast terrain classification using variable-length representation for autonomous navigation. In: IEEE Conference on Computer Vision and Pattern Recognition (CVPR 2007), pp. 1–8. IEEE (2007)
3. Baldi, P., Sadowski, P.J.: Understanding dropout. In: Proceedings of the 2013 Conference on Neural Information Processing Systems (NIPS 2013) (2013)
4. Bellone, M.: Watch your step! terrain traversability for robot control. In: Gorrostieta Hurtado, E. (ed.) Robot Control, Chap. 6. InTech (Oct 2016)
5. Bellutta, P., Manduchi, R., Matthies, L., Owens, K., Rankin, A.: Terrain perception for DEMO III. In: Proceedings of the IEEE Intelligent Vehicles Symposium, IV 2000, pp. 326–331. IEEE (2000)
6. Glorot, X., Bordes, A., Bengio, Y.: Deep sparse rectifier neural networks. In: Proceedings of the Fourteenth International Conference on Artificial Intelligence and Statistics, pp. 315–323 (2011)

7. Howard, A., Seraji, H., Tunstel, E.: A rule-based fuzzy traversability index for mobile robot navigation. In: Proceedings 2001 ICRA. IEEE International Conference on Robotics and Automation, vol. 3, pp. 3067–3071. IEEE (2001)
8. Kingma, D.P., Ba, J.: Adam: a method for stochastic optimization. In: Proceedings of ICLR 2015 (2015). arXiv:1412.6980
9. Krizhevsky, A., Sutskever, I., Hinton, G.E.: Imagenet classification with deep convolutional neural networks. In: Advances in Neural Information Processing Systems, pp. 1097–1105 (2012)
10. Rasmussen, C., Lu, Y., Kocamaz, M.: Appearance contrast for fast, robust trail-following. In: Proceedings of the IEEE/RSJ International Conference on Intelligent Robots and Systems (IROS 2009), St. Louis, MO, USA, 10–15 October 2009
11. Simonyan, K., Zisserman, A.: Very deep convolutional networks for large-scale image recognition. In: Proceedings of the 2015 International Conference on Learning Representations (2015). arXiv:1409.1556
12. Tieleman, T., Hinton, G.: RMSProp: divide the gradient by a running average of its recent magnitude. COURSERA: Neural Networks Mach. Learn. 4(2), 26–31 (2012)
13. Wermelinger, M., Fankhauser, P., Diethelm, R., Krüsi, P., Siegwart, R., Hutter, M.: Navigation planning for legged robots in challenging terrain. In: Proceedings of the 2016 IEEE/RSJ International Conference on Intelligent Robots and Systems (IROS 2016), Daejeon, South Korea, October 2016

Towards Automated Order Picking Robots for Warehouses and Retail

Richard Bormann[1]([⊠]), Bruno Ferreira de Brito[2], Jochen Lindermayr[1], Marco Omainska[1], and Mayank Patel[1]

[1] Fraunhofer IPA, Robot and Assistive Systems, 70569 Stuttgart, Germany
{richard.bormann,jochen.lindermayr,marco.omainska,
mayank.patel}@ipa.fraunhofer.de
[2] Delft University of Technology, Department of Cognitive Robotics,
Delft, The Netherlands
bruno.debrito@tudelft.nl
https://www.ipa.fraunhofer.de, https://www.tudelft.nl/en/

Abstract. Order picking is one of the most expensive tasks in warehouses nowadays and at the same time one of the hardest to automate. Technical progress in automation technologies however allowed for first robotic products on fully automated picking in certain applications. This paper presents a mobile order picking robot for retail store or warehouse order fulfillment on typical packaged retail store items. This task is especially challenging due to the variety of items which need to be recognized and manipulated by the robot. Besides providing a comprehensive system overview the paper discusses the chosen techniques for textured object detection and manipulation in greater detail. The paper concludes with a general evaluation of the complete system and elaborates various potential avenues of further improvement.

Keywords: Order picking · Object localization · Robot vision

1 Introduction

Order picking is commonly considered a process that accounts for the largest share of manual labor and the highest costs of up to 55% of total warehouse operating expenses [16]. Operational Research has a long history in optimizing processes in order picking to lower costs, handling time, and reliability [10]. Besides assistive technical tools for the human worker like electronic guidance or virtual glasses, robotics opens up a new dimension of optimizing the order fulfillment process. Many new robotics companies emerged throughout the last years targeting different aspects in warehouse and retail shop automation. Kiva Robotics started with robots capable of relocating shelves inside the warehouse, e.g. for moving the right shelves to pick and pack stations. The Swisslog Auto-Store system designs the warehouse as a large block of columns that can hold a stack of storage boxes each. Small robots operate on top of this rack accessing these boxes by pulling them up and transporting them to a pick and pack station

© Springer Nature Switzerland AG 2019
D. Tzovaras et al. (Eds.): ICVS 2019, LNCS 11754, pp. 185–198, 2019.
https://doi.org/10.1007/978-3-030-34995-0_18

Fig. 1. Left: Automated order picking with a rob@work 3 platform. Right: RViz display of order picking with ordering GUI in upper left corner, detected objects with pose estimates in lower left corner and grasping trajectory to object "toppas_traube" with collision scene on the right side.

likewise. In both cases, however, picking items from the shelves or boxes is still manual labor.

First order picking robots with specific applications have been introduced by Magazino. The Magazino TORU robot can handle cuboid items such as books or shoe boxes whereas the model SOTO can manipulate small load carriers. From Fetch Robotics customers can obtain a research platform for applications such as order picking. In order to accelerate research on automated order picking from nearly unstructured storage bins, Amazon initiated the Amazon Picking Challenge in 2015 [4,6] which has seen lively competition throughout the last years [12,20]. The system introduced in this paper is conceived as a mobile robot for automatic order picking tasks in a wide variety of applications in the retail and warehousing domains. Supermarkets and retail stores begin to offer their products at online shops and ship orders home to their clients. This paper focuses on such an application when our mobile robot is supposed to go shopping in a retail store or warehouse for collecting an online order. Another application directly emerging from a robot with such capabilities is the verification of stock levels and wrongly placed items in the shop.

This paper explains a complete order picking system for retail store or warehouse order fulfillment on textured retail items (Sect. 3). The employed vision system for object detection and localization (Sect. 4) as well as the motion planning procedures for object pick up (Sect. 5) are discussed and evaluated in detail. Eventually, we provide an evaluation on the overall system performance in Sect. 6 and discuss future improvements on system design in Sect. 7.

2   Related Work

Scientific interest in order picking from shelves filled in an unstructured way has been majorly raised by the Amazon Picking Challenge recently [4,6,12,20]. Participants of the Amazon Picking Challenge 2015 indicated that the most challenging components for order picking were vision and manipulation [4]. Consequently, we will discuss these parts of our system in greater detail than the other components.

2.1 Object Perception and Localization

In the past, object detection and localization approaches usually either focused on textured object recognition [8,21] or untextured object recognition [9,13]. The object recognition system utilized in this work is suited for textured objects, which applies to the majority of packaged retail articles, and it extends the local feature point based approach of [8] by the consideration of color data. A good survey on color extensions for feature point descriptors is given in [23]. The RGBSURF method favored in our system was inspired by [11,27]. In the Amazon Robotics Challenge, Convolutional Neural Network based (CNN) methods appear to be prevalent [12,20]. Sometimes the CNN is used to directly estimate the object pose [28] but often it is just employed for bounding box localization of the objects whereas the 3d pose is recovered with e.g. CAD model matching [12] or model-free grasping based on sensed surface geometry [20].

In contrast to those deep learning based methods, our procedure is based on object model matching. We see the major advantage in the simple extensibility of the object database which does not require any further training for adding a model. CNN-based methods struggle to learn new objects within short time periods and may do so even more if large amounts of new items have to be learned [2,20]. The object recording setup described in Sect. 3 is capable of physically recording a new object and adding its recognition model to the detection system within less than 2 min.

2.2 Motion Planning

Motion planners can be categorized into sampling-based, search-based, and optimization-based. The Open Motion Planning Library (OMPL) [25] is one of the most popular planner frameworks used with MoveIt! [3]. The library features several multi-query, single-query, and sampling-based planners. Multi-query planners build a roadmap of the entire environment that can be used for multiple queries. Single-query planners typically construct a graph of states connected by valid motions. On top of the above, sampling-based planners provide some level of optimization as well.

Optimization based planners like Covariant Hamiltonian Optimization for Motion Planning (CHOMP) [29] and Stochastic Optimization for Motion Planning (STOMP) [15] iteratively improve an initial trajectory while minimizing a cost function responsible for smoothness and obstacle avoidance. Both are highly dependent on the initial seed trajectory and hence sometimes fail at finding a valid motion. Dynamical Movement Primitives (DMP) [14] present a time-independent, scalable trajectory representation that allows start and end states to be changed while maintaining the dynamic characteristics of the motion encoded from demonstrations. DMPs either represent a motion in joint-space or Cartesian-space although rotations in the latter case require special attention [17,26]. For the order picking robot we developed an extension of the STOMP framework termed Guided Stochastic Optimization for Motion Planning (GSTOMP) which learns from demonstration to generate a better initial guess by using DMPs as seed generators.

3   System

This section explains the hardware setup and the employed software components. The application has been embedded in a warehouse or shop-like laboratory environment with typical supermarket items sorted into the shelves (see Fig. 1).

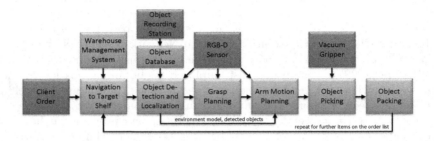

Fig. 2. System overview on the mobile order picking robot.

Robot Hardware. As robotic platform the omnidirectional rob@work 3 manufactured by Fraunhofer IPA is used. It carries a Universal Robots UR 10 arm equipped with a Schmalz Cobot Pump vacuum gripper. The robot offers some payload area where a storage or shipping box can be placed. The camera system for object detection is a combination of an Ensenso N30 projection-assisted stereo camera paired with a 5 Megapixel IDS uEye color camera. The vision system is mounted on a pan-tilt unit next to the manipulator.

Robot Behavior Control. The robot control program is implemented as a Python control script. The overall control diagram is depicted in Fig. 2. It is explained in the order of called components in the following paragraphs.

Order Placement. A simple RViz-based ordering interface emulates the client order, as displayed in Fig. 1 (right). It transfers the types and quantities of ordered articles to the control script, which then queries the warehouse management system for the typical shelf locations where these objects are stored at. The script may choose to optimize the ordering sequence for a short traveling path through the warehouse, but may also consider further constraints such as which objects cannot be placed on top of others in the client box.

Mobile Navigation. The navigation system drives the robot to the next target shelf location while avoiding dynamic obstacles on the way. Our mapping and localization system features multi-feature fusion and multi-robot mapping and localization [5]. Here we utilize the grid map together with line features, i.e. mostly wall segments. The navigation system seamlessly scales up to a fleet of robots with our multi-robot cloud navigation extension [1,19].

Object Detection and Localization. At the shelf the cameras are adjusted to face the putatively correct section in the shelf as told by the warehouse management system. The object detection and localization procedure is outlined in Sect. 4. The script allows for multiple detection trials on failure and also inspects multiple shelf levels to extend the object search. The necessary object recognition models are retrieved from the system's object database, which can be easily extended with new models during operation.

Object Recording Station. In practice, models of thousands of ever-changing retail items and their master data are required. We developed an industrial grade automatic capturing station together with the company Kaptura which can generate colored 3d point cloud and mesh models of arbitrarily textured and untextured objects within 45 s recording time and 45 s 3d modeling time. Required master data such as object weight and bounding box size are captured simultaneously. The station is designed in a modular way and allows for installing defined illumination and multiple cameras such that a reduction in recording time is easily achievable. The obtained 3d data can be used for generating object detection models as well as for showcasing a 3d model at a shop's web page. Figure 3 shows the recording station and some captured 3d models.

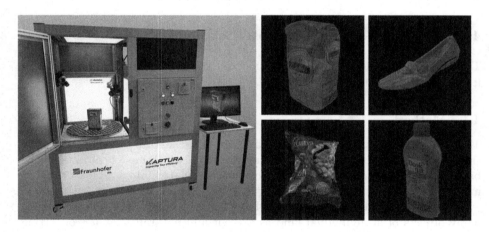

Fig. 3. Kaptura object recording station and four 3d model samples.

Grasp Planning. If the object of interest is available multiple times in the shelf, the script tries with the closest instance first and submits the 3d object pose data to the grasp planner. For cuboid packages and cylinders the planner just computes grasp points with a centered and aligned gripper at the 6 bounding box surface centers. These grasp proposals are usually sufficient in that case. However, on arbitrarily shaped objects we refine or discard the grasp poses based on local surface geometry using a surface-normals approach similar to the method described in [20]. The grasp planner already conducts simple reachability checks based on the captured RGB-D scene data and removes inaccessible grasp pose candidates. Eventually, it outputs a ranked list of possible grasps on the desired object and leaves the motion planner with the choice whether to accept this ranking, revise it or remove grasps due to further scene constraints.

Motion Planning. The motion planning component plans a collision-free path towards the grasp pose based on GSTOMP. This procedure is detailed in Sect. 5. The planner iteratively tries to find a path to any of the provided ranked grasp poses until one attempt is successful.

Object Picking. The found trajectory is then executed by the arm and the vacuum gripper is activated close to the object. Once a contact can be measured by the vacuum system the arm is retracted along a suitable planned trajectory.

Object Packing. Finally, the objects are packed into a storage box using a mixed palletizing method [24] for arranging the objects in a structured and space-saving manner.

If further objects are requested, the procedure continues with step 4 driving to the storage location of the next object on the order list.

4   Object Detection and Pose Estimation

This section explains the object detection and localization system and evaluates its performance on typical retail items.

4.1   Method

The utilized perception system is an extension of our feature-based object detection and localization system from previous work [8] which models objects on the basis of visual feature points which become accumulated into a 3d feature point model.

Due to model matching based on local feature points, this system finds as many objects of an individual kind as present and naturally handles occurring occlusions. Model matching also directly provides a complete 3d object pose estimate with localization in 3d position and angles.

We extended the approach of [8] by a refined model verification procedure and the incorporation of color cues. The original system established object model matches based on a fixed minimum threshold of matching features. This does not generalize well over a diverse set of objects since there is no single optimal number of minimum matching features. The new model verification approach seeks to match a certain percentage of visible model features. We explicitly analyze visibility and occlusions of the model localization hypothesis and only require that the necessary percentage of features is detected on the putative object surface that would be visible from the given perspective. The dynamic feature matching threshold effects that detection is now more accurate in terms of false positives and adapts flexibly to the number of features of the object model, which may range from a few to several hundred features.

Our second enhancement of the object detection system is a color extension of the formerly gray-scale image-based SURF or sORB features. To this end we collected the following set of possible color extensions for feature descriptors.

SURF+HSV Adds a normalized 12-bin color histogram to the SURF descriptor which counts frequencies of surrounding pixels with white, gray, black or one of 9 basic colors. The neighborhood size is adapted with the descriptor's scale. The color assignment follows a fixed comparison scheme of the pixel's hue, saturation, and value in HSV color space.

SURF+OC Adds a 13-bin color histogram to the SURF descriptor, similar to SURF+HSV. The color definitions are derived from the Opponent Color space.

RGBSURF While the SURF feature points are determined from the gray-scale image, the RGBSURF descriptors are computed individually in the color image's R, G, and B-channels [23, 27] yielding a descriptor of length $3 \cdot 64 = 192$.

rgSURF Computes the SURF descriptors on the two channels of the rg color space [23], similarly to RGBSURF.

HSVSURF Computes the SURF descriptors on the 3 channels of the HSV color space [23], similarly to RGBSURF.

OCSURF Computes the SURF descriptors on the three channels of the Opponent Color space (as described in [11, 23]), similarly to RGBSURF.

LabSURF Computes the SURF descriptors on the three channels of the L*a*b* color space, similarly to RGBSURF.

4.2   Evaluation

We evaluated the object detection and localization system on a set of 71 textured supermarket articles[1] containing 16 paper boxes, 11 big paper bags (e.g. flour), 22 paper and plastic bags with limited flexibility (e.g. yeast, pudding), 10 plastic bags with medium flexibility (e.g. sultanas, almonds), 2 plastic bottles, 1 plastic can, 5 plastic wrappings, and 4 plastic blisters. Object sizes are well-distributed ranging from 5 cm to 25 cm side lengths. The data set contains 36 RGB-D perspectives per object captured under 3 conditions. First, the objects were recorded with normal indoor illumination from ca. 65 cm distance, similar to the model recording distance. In the second condition objects were captured from ca. 130 cm distance, which is the maximum camera to object distance in the real application, to evaluate scale invariance. Third, the recording distance was ca. 65 cm and images were captured against varying daylight falling into a window behind the objects (recorded on several consecutive days) and a spotlight illumination was installed at the right side to evaluate illumination invariance. A selection of these objects and recording conditions[2] is displayed in Fig. 4.

---

[1] The dataset is available from the authors upon request (>500 GB).

[2] The data set was captured with a another but similar recording system before the professional Kaptura recording system was available.

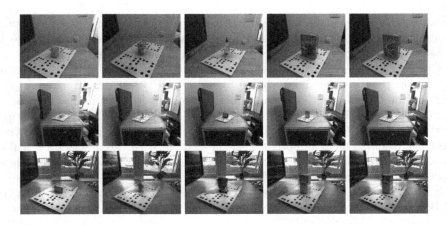

Fig. 4. Exemplary objects from the data set recorded at three conditions.

In the following evaluation we compare SURF and all color descriptor extensions against the SURF descriptor without model verification as described in [8]. For each descriptor, we tuned parameters to achieve the best performance possible. Table 1 provides several performance metrics on all 3 test cases. Reported values are averaged over the 36 views of all the 71 objects. ΔPos is the average translation error, and $\Delta\theta$ is the average angular error. Further we report macro recall, precision and f-score. Recall is the percentage of correctly identified objects throughout the database. For the best methods it is above 90% in test case 1. To interpret this number right, please notice that our database also includes side views of the objects. Almost half of the tested objects are quite flat for which these side views do not give any useful visual information. For those, we have at least 6 difficult side views out of 36 views. Hence, it is nearly impossible to detect the object on about 8.33% of the database images. Consequently, a recall above 90% indicates that nearly every object was correctly found whenever visually possible. Precision is the number of correct predictions divided by all predictions.

While all descriptors perform quite well at test case 1, SURF and RGBSURF are the only methods with high robustness to varying distance and illumination. According to the diagonal offset model [7,18], the SURF descriptor is invariant against uniform and chromatic multiplicative intensity changes and additive

Table 1. Object detection and localization performance for the 3 test cases

Descriptor	test case 1 (65 cm)				test case 2 (130 cm)				test case 3 (65 cm, illumination)			
	ΔPos $\Delta\theta$	recall	prec.	f	ΔPos $\Delta\theta$	recall	prec.	f	ΔPos $\Delta\theta$	recall	prec.	f
SURF+HSV	**5.1 mm 2.84°**	0.928	0.875	**0.901**	26.9 mm 6.58°	0.268	0.497	0.348	**8.2 mm 4.34°**	0.334	0.664	0.444
SURF+OC	5.3 mm 2.97°	**0.932**	0.849	0.888	27.2 mm **6.43°**	0.360	0.559	0.438	**8.2 mm** 4.79°	0.408	0.704	0.517
RGBSURF	6.9 mm 5.13°	0.901	0.868	0.884	27.5 mm 8.00°	**0.546 0.815 0.654**			9.2 mm 6.23°	**0.665 0.798 0.725**		
rgSURF	8.1 mm 5.70°	0.730	0.889	0.801	26.1 mm 7.39°	0.326	0.756	0.455	9.7 mm 6.79°	0.380	0.766	0.508
HSVSURF	9.4 mm 6.85°	0.757	0.841	0.797	**25.1 mm** 6.60°	0.126	0.445	0.196	12.9 mm 7.11°	0.129	0.642	0.209
OCSURF	7.9 mm 5.81°	0.824	0.882	0.852	26.7 mm 7.04°	0.370	0.801	0.506	9.6 mm 6.17°	0.347	0.744	0.473
LabSURF	7.6 mm 5.24°	0.834	**0.899**	0.865	27.0 mm 6.80°	0.328	0.682	0.443	10.1 mm 5.94°	0.315	0.640	0.422
SURF	6.4 mm 4.64°	0.925	0.838	0.880	27.2 mm 7.34°	**0.597 0.828 0.694**			8.3 mm 5.50°	**0.699 0.807 0.749**		
SURF [8]	6.7 mm 4.69°	0.925	0.536	0.678	27.1 mm 7.04°	0.564	0.676	0.615	8.5 mm 5.54°	0.694	0.701	0.698

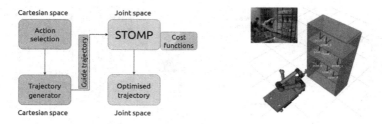

Fig. 5. The architecture of the GSTOMP motion planner (left) and the 8 target poses of the utilized demonstration trajectories in our experiments (right).

diffuse illumination shifts. Only RGBSURF retains the same illumination invariance properties whereas all other extensions are only invariant against uniform intensity variation and/or diffuse shifts and hence perform significantly worse in test case 3. In test case 2, recall is around 55%–60% for the best methods RGBSURF and SURF because half of the database objects are just too small to be detected reliably with the available resolution at this distance (wide angle lens $74° \times 58°$). The real world experiments (Sect. 6) showed that detecting larger objects (>18 cm side length in longest dimension) worked reliably up to this distance.

The value of the model verification step becomes visible by the large improvement on precision between SURF with and without model verification [8]. Although SURF with model verification outperforms the best color extension RGBSURF slightly, our order picking robot was run with RGBSURF because it could still distinguish objects with same texture but different color design. This is especially relevant on partial occlusions which might hide distinguishing texture. Finally, we like to highlight that the reported detection quality is scene-independent in contrast to some CNN-based solutions [20]. We did not observe any influence on detection quality whether 1 or 20 objects were present in the scene.

5   Motion Planning

In this section we explain the motion planning approach GSTOMP which plans arm movements faster by starting from close-by, previously recorded motion primitives that just need to become refined accordingly to the current task.

5.1   Method

The architecture of GSTOMP is presented in Fig. 5 (left). A Dynamical Movement Primitives (DMP) collection is recorded beforehand, encoding for each pick&place task the necessary motion, such as pick, pull, push etc. In our approach we use Cartesian Dynamical Movement Primitives that is the combination of position and rotation component

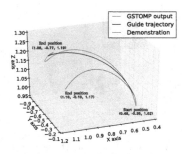 

Fig. 6. Example of demonstrated, *guide,* and optimized trajectories (left) and the execution of an optimized trajectory (right).

$$\tau \ddot{y} = \alpha_z(\beta_z(g_p - y) - \dot{y}) + f_p \tag{1}$$

$$\tau \dot{\eta} = \alpha_z(\beta_z \cdot 2\log(g_o * \bar{q}) - \eta) + f_o \tag{2}$$

where y is the current position, η represents the current orientation, τ is a scaling factor for time, $g = (g_p, g_o)$ is the goal pose, α_z and β_z are scaling terms, $q \in \mathbb{R}^4$ is a unit quaternion representation, $*$ denotes the quaternion multiplication and f_p, f_o are forcing terms for position and orientation defined as

$$f(x) = \frac{\sum_{i=1}^{N} \Psi_i(x)w_i}{\sum_{i=1}^{N} \Psi_i(x)} x(g - y_0) \tag{3}$$

where y_0 stands for the initial pose while g is the goal pose, w_i is a weighting for a given Lagrangian basis function Ψ_i, and x is a unit-free time equivalent, converging monotonically from 1 (initial pose) to 0 (target pose).

The action selection block determines which DMP to use at each planning phase of the task execution. The trajectory generator block takes the selected DMP and generates the initial guess, the *guide trajectory*, based on the previously demonstrated motions. To obtain an optimized GSTOMP trajectory as close as possible to the *guide trajectory* we implemented the Dynamic Time Warping (DTW) [22] method as cost function for the stochastic optimization framework of STOMP, introducing a similarity measure between the *guide* and optimized trajectories.

5.2   Evaluation

We evaluate most of the major state-of-the-art planners implemented in the state-of-the-art software for mobile manipulation, MoveIt! [3]: CHOMP, RRT-Connect provided by OMPL, STOMP, and GSTOMP. We specifically compare our method against all different interpolation methods used in STOMP as they are equivalent in purpose to our *guide trajectories*. For the experiments we recorded 8 demonstration trajectories, randomly sampled within the racks of the utilized shelf type (see Fig. 5, right). The DMPs generate *guide trajectories*

Table 2. Numerical results from 200 experiments with each motion planning algorithm

Algorithm	Planning time (s)		Smoothness		Success rate	Solver Iterations
	Median	Std dev	Median	Std dev		Mean
CHOMP	0.146	2.118	**0.038**	**0.004**	60%	7.6
RRTConnect	**0.071**	**0.049**	0.136	0.057	65%	–
STOMP (cubic)	0.579	2.711	0.185	0.027	70%	2.84
STOMP (linear)	0.575	1.634	0.172	0.023	67%	3.7
STOMP (min c cost)	0.548	3.049	0.188	0.024	58%	2.15
GSTOMP	3.430	1.838	0.049	0.038	**81%**	**1.16**

for the queried arm motion which is in Cartesian space. An example of a *guide trajectory* generated from a demonstrated trajectory as well as the resulting optimized trajectory can be found in Fig. 6.

The planners are compared w.r.t. planning time, number of solver iterations, success rate, and smoothness. The smoothness value is the cumulative function of the linear and angular accelerations. The accelerations are approximated using second order center difference formula. The results of 200 experiments with equally distributed random goal poses in all racks of the shelf are presented in Table 2. It shows that when a better initial guess is provided to the solver the number of iterations to find a solution is reduced, improving the solver performance, and so GSTOMP presents the best result. On the other hand, due to the recent implementation state of our planner the planning time is still suboptimal, resulting in a lower performance in comparison with the fastest planners, RRT-Connect and CHOMP. However, GSTOMP can still achieve reasonable planning times for a pick&place scenario.

In terms of smoothness, the GSTOMP planner is able to find smoother trajectories than all the other STOMP versions, achieving the same levels of performance as CHOMP. Since this smoothness is computed on the Cartesian path this result proves that GSTOMP manages to stay relatively close to an already smooth *guide trajectory*. Finally, the main advantage of the GSTOMP method is the increase of the success rate of planning by 10% compared to the second best, STOMP with cubic polynomial interpolation initial trajectory. This reduces the number of failed planning attempts significantly and hence increases the performance of any manipulation system. GSTOMP is superior to the other approaches since the *guide trajectories* provide a domain-adapted initial guess of the trajectory whereas the others only rely on unspecific heuristic initial guesses, e.g. STOMP with cubic polynomials.

6   System Evaluation

We evaluated the performance of the whole order picking system with 20 random orders, each containing 8–12 target objects out of the 71 database objects, in

Table 3. Performance of the order picking system on 200 picks

Number	%	Event
191	95.5%	successful picks
5	2.5%	object could not be found after 5 attempts
4	2.0%	motion planning did not find a plan after 5 attempts

our supermarket lab environment with 6 different shelf locations. An RViz visualization on object detection and motion planning is provided in Fig. 1 (right). From these 200 picking tasks we counted the statistics as summarized in Table 3.

The objects that could not be found at all were very small objects with few texture. Sometimes, motion planning could not find valid plans when objects were obstructed by others and could not be removed withouth colliding. A complete pick excluding driving the robot to a shelf took 40 s on average (ranging between 35 s and 45 s). Stereo and RGB-D processing took about 5 s, object detection further 3 s, grasp planning accounted for up to 2 s, motion planning for additional 3 s, the grasp execution took 12 s and the packing needed 15 s. Especially the manipulation times can be easily sped up by driving the arm with higher speeds. Likewise, the packing manoeuver is currently suboptimal with turning the arm around its first joint by a full rotation in total.

7   Conclusions and Future Work

In this paper we discussed a mobile order picking robot for order fulfillment applicable to any kind of retail store or warehouse environment. For the initial setup it just needs a navigation map and optionally collision models of the shelves. However, collision scenes for arm motion planning can also be generated online with the onboard 3d sensors. Additionally, a connection to a warehouse management system with knowledge about the usual object storage locations is necessary. The system can handle any textured objects up to 1 kg weight. The objects may be arranged in any ordered or chaotic way. Camera resolution and lens limit the maximum object detection distance. In our case, only objects larger than 18 cm could be detected reliably at 130 cm distance with the 5 MPixel color camera and wide angle lens ($74° \times 58°$). The workspace is only limited by the dexterity of the installed arm, in case of the UR10 we could only reach to objects in the front part for the lowest and topmost shelves. Although performance is already at a good level in a laboratory environment, an industrial application requires even higher dependability. For the system at hand we see the following major steps in achieving that goal.

For reaching to all storage locations in the top and bottom shelves, a different hardware setup with better dexterity is needed. For achieving maximum dexterity we are also working on a GSTOMP extension which can handle full body motion planning, i.e. for mobile platform and arm simultaneously.

We also consider using a gripper-mounted camera for object detection which would allow to get closer to objects of interest and facilitate detection of the smaller items. A light source should be added to the robot for avoiding extreme illumination variance. If needed by the application, the vision system could be complemented with a textured-less object detection method. Last but not least, the implemented error recovery behaviors are not sufficient for grasping obstructed objects. An extension of the order picking system with a reasoning module is necessary for clearing the occluding objects in a task-oriented way.

References

1. Abbenseth, J., Lopez, F.G., Henkel, C., Dörr, S.: Cloud-based cooperative navigation for mobile service robots in dynamic industrial environments. In: Proceedings of the Symposium on Applied Computing, pp. 283–288. ACM (2017)
2. Causo, A., Chong, Z.H., Ramamoorthy, L.: A robust robot design for item picking. In: Proceedings of the IEEE International Conference on Robotics and Automation (ICRA) (2018)
3. Chitta, S., Sucan, I., Cousins, S.: Moveit!. Robot. Autom. Mag. **19**, 18–19 (2012)
4. Correll, N., et al.: Analysis and observations from the first amazon picking challenge. IEEE Trans. Autom. Sci. Eng. **15**(1), 172–188 (2018). https://doi.org/10.1109/TASE.2016.2600527
5. Dörr, S., Barsch, P., Gruhler, M., Lopez, F.G.: Cooperative longterm SLAM for navigating mobile robots in industrial applications. In: 2016 IEEE International Conference on Multisensor Fusion and Integration for Intelligent Systems (MFI), pp. 297–303, September 2016. https://doi.org/10.1109/MFI.2016.7849504
6. Eppner, C., et al.: Lessons from the amazon picking challenge: Four aspects of building robotic systems. In: Proceedings of Robotics: Science and Systems. AnnArbor, Michigan, June 2016. https://doi.org/10.15607/RSS.2016.XII.036
7. Finlayson, G.D., Hordley, S.D., Xu, R.: Convex programming colour constancy with a diagonal-offset model. In: Proceedings of the IEEE International Conference on Image Processing, vol. 3, pp. III-948-51, September 2005. https://doi.org/10.1109/ICIP.2005.1530550
8. Fischer, J., Arbeiter, G., Bormann, R., Verl, A.: A framework for object training and 6 DoF pose estimation. In: Proceedings of the 7th German Conference on Robotics (ROBOTIK), May 2012
9. Fischer, J., Bormann, R., Arbeiter, G., Verl, A.: A feature descriptor for textureless object representation using 2D and 3D cues from RGB-D data. In: Proceedings of the IEEE International Conference on Robotics and Automation (ICRA), pp. 2104–2109 (2013)
10. Gu, J., Goetschalckx, M., McGinnis, L.F.: Research on warehouse design and performance evaluation: a comprehensive review. Eur. J. Oper. Res. **203**(3), 539–549 (2010). https://doi.org/10.1016/j.ejor.2009.07.031
11. Henderson, C., Izquierdo, E.: Robust feature matching in the wild. In: Proceedings of Science and Information Conference (SAI), pp. 628–637, July 2015. https://doi.org/10.1109/SAI.2015.7237208
12. Hernandez, C., et al.: Team delft's robot winner of the amazon picking challenge 2016. In: Behnke, S., Sheh, R., Sarıel, S., Lee, D.D. (eds.) RoboCup 2016. LNCS (LNAI), vol. 9776, pp. 613–624. Springer, Cham (2017). https://doi.org/10.1007/978-3-319-68792-6_51

13. Hinterstoisser, S., et al.: Gradient response maps for real-time detection of textureless objects. IEEE Trans. Pattern Anal. Mach. Intell. **34**, 876–888 (2012)
14. Ijspeert, A.J., Nakanishi, J., Hoffmann, H., Pastor, P., Schaal, S.: Dynamical movement primitives: learning attractor models for motor behaviors. Neural Comput. **25**(2), 328–373 (2013)
15. Kalakrishnan, M., Chitta, S., Theodorou, E., Pastor, P., Schaal, S.: STOMP: stochastic trajectory optimization for motion planning. In: Proceedings of the IEEE International Conference on Robotics and Automation (ICRA), pp. 4569–4574 (2011). https://doi.org/10.1109/ICRA.2011.5980280
16. de Koster, R., Le-Duc, T., Roodbergen, K.J.: Design and control of warehouse order picking: a literature review. Eur. J. Oper. Res. **182**(2), 481–501 (2007). https://doi.org/10.1016/j.ejor.2006.07.009
17. Kramberger, A., Gams, A., Nemec, B., Ude, A.: Generalization of orientational motion in unit quaternion space. In: Proceedings of the IEEE International Conference on Humanoid Robots, pp. 808–813 (2016). https://doi.org/10.1109/HUMANOIDS.2016.7803366
18. Kries, J.v.: Influence of adaptation on the effects produced by luminous stimuli. In: Sources of Color Science, pp. 120–126. The MIT Press, Cambridge, June 1970
19. Lopez, F.G., Abbenseth, J., Henkel, C., Dörr, S.: A predictive online path planning and optimization approach for cooperative mobile service robot navigation in industrial applications. In: Proceedings of the European Conference on Mobile Robots (ECMR), pp. 146–151 (2017)
20. Morrison, D., et al.: Cartman: The low-cost Cartesian Manipulator that won the Amazon Robotics Challenge. In: Proceedings of the IEEE International Conference on Robotics and Automation (ICRA), May 2018
21. Romea, A.C., Torres, M.M., Srinivasa, S.: The moped framework: object recognition and pose estimation for manipulation. Int. J. Robot. Res. **30**(1), 1284–1306 (2011)
22. Sakoe, H., Chiba, S.: Dynamic programming algorithm optimization for spoken word recognition. IEEE Trans. Acoust. Speech Sign. Process. **26**(1), 43–49 (1978). https://doi.org/10.1109/TASSP.1978.1163055
23. Sande, K.V.D., Gevers, T., Snoek, C.: Evaluating color descriptors for object and scene recognition. IEEE Trans. Pattern Anal. Mach. Intell. **32**(9), 1582–1596 (2010). https://doi.org/10.1109/TPAMI.2009.154
24. Schuster, M., Bormann, R., Steidl, D., Reynolds-Haertle, S., Stilman, M.: Stable stacking for the distributor's pallet packing problem. In: Proceedings of the IEEE International Conference on Intelligent Robots and Systems (IROS), pp. 3646–3651, October 2010
25. Sucan, I.A., Moll, M., Kavraki, L.E.: The open motion planning library. Robot. Autom. Mag. **19**, 72–82 (2012)
26. Ude, A.: Filtering in a unit quaternion space for model-based object tracking. Robot. Auton. Syst. **28**(2), 163–172 (1999)
27. Wafy, M., Madbouly, A.M.M.: Increase Efficiency of SURF using RGB Color Space. Int. J. Adv. Comput. Sci. Appl. (IJACSA), 6(8) (2015). https://doi.org/10.14569/IJACSA.2015.060810
28. Zeng, A., Yu, K.T., Song, S., Suo, D., Walker, Jr. E., Rodriguez, A., Xiao, J.: Multiview self-supervised deep learning for 6D pose estimation in the amazon picking challenge. In: Proceedings of the IEEE International Conference on Robotics and Automation (ICRA) (2017)
29. Zucker, M., et al.: Chomp: covariant hamiltonian optimization for motion planning. Int. J. Robot. Res. **32**(9–10), 1164–1193 (2013)

Vision Systems Applications

Tillage Machine Control
Based on a Vision System for Soil
Roughness and Soil Cover Estimation

Peter Riegler-Nurscher[1]([⊠]), Johann Prankl[1], and Markus Vincze[2]

[1] Josephinum Research, Wieselburg, Austria
{p.riegler-nurscher,johann.prankl}@josephinum.at
[2] Automation and Control Institute, Vienna University of Technology,
1040 Vienna, Austria
vincze@acin.tuwien.ac.at

Abstract. Soil roughness and soil cover are important control variables for plant cropping. A certain level of soil roughness can prevent soil erosion, but to rough soil prevents good plant emergence. Local heterogeneities in the field make it difficult to get homogeneous soil roughness. Residues, like straw, influences the soil roughness estimation and play an important role in preventing soil erosion. We propose a system to control the tillage intensity of a power harrow by varying the driving speed and PTO speed of a tractor. The basis for the control algorithm is a roughness estimation system based on an RGB stereo camera. A soil roughness index is calculated from the reconstructed soil surface point cloud. The vision system also integrates an algorithm to detect soil cover, like residues. Two different machine learning methods for pixel-wise semantic segmentation of soil cover were implemented, an entangled random forest and a convolutional neural net. The pixel-wise classification of each image into soil, living organic matter, dead organic matter and stone allow for mapping of soil cover during tillage. The results of the semantic segmentation of soil cover were compared to ground truth labelled data using the grid method. The soil roughness measurements were validated using the manual sieve analysis. The whole control system was validated in field trials on different locations.

Keywords: Stereo camera · Soil roughness · Soil cover ·
Convolutional neural network

1 Introduction

Soil roughness directly influences plant emergence and soil erosion [1]. Soil inhomogeneities in the field make it difficult to get constant soil roughness results during tillage and seeding. To get homogeneous plant emergence while preventing soil erosion methods for estimating soil roughness and controlling a tillage machine are needed. Additionally to soil roughness, soil cover plays an important role in preventing soil erosion and soil cover also influences soil roughness measurements. Measuring soil cover is crucial for informed soil management decisions.

© Springer Nature Switzerland AG 2019
D. Tzovaras et al. (Eds.): ICVS 2019, LNCS 11754, pp. 201–210, 2019.
https://doi.org/10.1007/978-3-030-34995-0_19

Main influences on soil roughness are soil moisture, soil composition and tillage. At tillage with a power harrow parameters influencing the roughness are driving speed, tool speed (PTO- speed), tool geometry and working depth. During work, the driving speed and tool speed can be changed by the tractor. The ISOBUS Class 3 standard gives the tillage implement, in our case the power harrow, the possibility to control tractors functions. To develop a control system, there is the need for a method for real-time soil roughness estimation which can cope with environmental influences like residues, dust and changing light conditions. Advanced residue detection of different soil cover classes can deliver additional important agronomic information for further plant treatments (e.g. site-specific fertilizing).

1.1   Related Work

Different methods for soil roughness measurement have been proposed over the years. Standard methods used mechanical pin-meters to quantify the roughness with a one or two-dimensional array of pins. The measured soil heights from the pins are aggregated to soil roughness indices [3]. Modern methods use laser scanners and cameras to reconstruct the soil surface three dimensional. Methods based on lasers are not as robust, have difficulties to detect residues and are more expensive. Laser scanners are mostly used in academia due to their accuracy and simple use. Methods using structured light have been tested [4], but these do not work well under strong sunlight. First tests with stereo cameras for soil roughness measurement showed promising results [5,6]. Cameras provide additional information which enable the detection of residues and soil cover.

Methods to detect cover from images allow for traditional subjective estimates to be replaced. Proposed methods focus either on dead [7,8] or on living organic matter [9,10]. Over the last years, methods using machine learning are getting more and more popular. For example, Support Vector Machines (SVM) are used in [11], Artificial Neural Networks in [12] or Random Forests in [13]. Artificial neural networks, esp. convolutional neural networks for semantic segmentation are the most promising. State of the art neural networks for semantic segmentation include Enet [14], ESPNet [15] and ERFnet [16]. ESPNet outperforms others in terms of inference speed, but ERFNet still has better classification accuracy, as shown in [15]. The framework bonnet [17] provides implementations of ERFnet in tensorflow.

1.2   Contribution

Our main contributions in this paper are:

- A method for online soil roughness estimation for tillage machine control.
- A method based on convolutional neural networks, to distinguish residues from soil and to detect soil cover during soil roughness estimation, which increases the accuracy for soil cover estimation over state of the art methods.

– Implementation and evaluation of the overall control system, including the embedded vision system and a power harrow, which controls a tractors PTO and driving speed.

2 Online Machine Control

The goal of the overall tillage system is to obtain a homogeneously crumbled soil surface, hence in case the soil is to fine, the driving speed needs to be increased and the PTO speed ($\propto$ tool speed) decreased and the other way around in case the soil is too rough. A proportional controller gets the soil roughness as an input variable and tries to compensate the error to a desired roughness value by sending the commands for varying the speeds to the tractor. An additional hysteresis is used to avoid oscillations and reactions to small roughness changes which avoids integral and derivative terms.

The stereo camera is mounted behind the packer roller of the power harrow. The controller concept is shown in Fig. 1. A single board computer is used for computation of the roughness value. The control algorithm is implemented in the implement ECU on the power harrow.

The controller gets the desired roughness R_{ref} as an input variable from the User Interface. The vision system estimates the actual roughness R_{act} and sends it to the controller. The proportional controller calculates the difference between R_{ref} and R_{act} as error e and sends corresponding speed values v and n_{PTO} to the tractor. Constant environmental changes influence the control loop resulting in a final roughness R on the field. In the following Sects. 3 and 4, the vision system (Module *Camera and Image Processing* in Fig. 1) is described in detail.

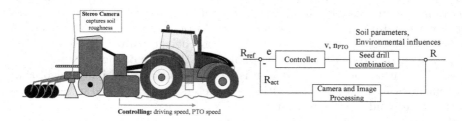

Fig. 1. Left: Depending on the estimated soil roughness, the driving speed and the PTO speed of the tractor are varied. **Right:** The control loop tries to compensate the error e between desired roughness R_{ref} and actual estimated roughness R_{act}

3 Roughness Estimation

In the literature soil roughness is described by one aggregated number, a so called roughness index. Many roughness indices have been proposed over the years. Experiments in previous projects (see [6]) have shown that the RC index and the average angle of surface normals (AoN) are the most suitable choices. The RC index can be computed very efficient, and was therefore selected to be implemented in the machine control system. Equation 1 shows the calculation of the RC index, where the height measurements z at each point (x, y) are its basis. In short, RC is the standard deviation of height measurements $z(x, y)$. Equation 2 shows the formula for the AoN. For the AoN, the average angle to the x-y plane is calculated over all surface normals s_i. Further tests with the controller, only used the RC index.

$$RC = \sqrt{\langle (z(x, y) - \langle z(x, y) \rangle_{XY})^2 \rangle_{XY}} \tag{1}$$

$$AoN = \sum_{i=0}^{N} acos\left(\frac{s_{i,Z}}{\|s_i\|}\right) \tag{2}$$

Stereo cameras have proven suitable for soil roughness estimation, as published in [6] and [5]. As in many proposed methods, the estimation is based on point clouds. The whole estimation process, from camera image, over point cloud to soil roughness, is shown in Fig. 2. The distance of the camera above the soil surface is about 30 cm, which leads to an observed soil surface area of about 60 × 50 cm.

In a first step stereo correspondences between the two RGB images are calculated. The disparity calculation is based on the block matching algorithm implemented in OpenCV [18]. Block matching is suitable in this case due to rich soil textures, which makes the stereo matching very easy, and can be implemented very efficiently. The disparity information and the calibration data from the stereo camera are used to calculate the point cloud. A mean plane of the soil surface is fitted from 5 previous measurements, using the least squares method, to estimate camera angle variations. The point cloud is transformed afterwards on this plane. Residues like straws are filtered from the point cloud (details see in Sect. 3.1). After point cloud processing, the height parameters for the RC index are extracted or the surface normals for estimation of the AoN are calculated. The surface normal calculation is based on the integral image algorithm implement in PCL [19]. This resulting parameters are aggregated in a last step to the corresponding roughness indices. Several measurements are combined to one measurement using a weighted mean for smoothing the measurement.

The roughness estimation is implemented on an Odroid XU4 single board computer. The communication between the computation unit and the implement ECU, where the controller is implemented, is based on CAN 2.0.

3.1   Environmental Influence Handling

Lighting has strong influence on the stereo matching. To reduce illumination variances and to be able to use the system during night, additional headlights were mounted besides the stereo camera. Additionally, the exposure time has to be adapted according to the driving speed of the tractor to avoid motion

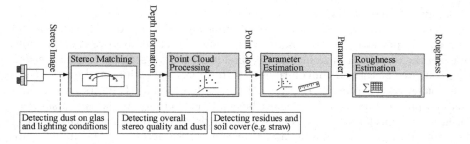

Fig. 2. The roughness estimation pipeline: Input images are from the stereo camera, the pipeline outputs one roughness value

blurring while still getting enough texture for the stereo matching. The necessary velocity value, for calculation of the maximum exposure time, is available over the ISOBUS.

Dust or dirt on the lens can be detected by comparing the left and right image. A simple metric would be the mean grey value. If the difference of the mean grey values between the stereo images exceeds a certain threshold, a message is shown on the implements user interface (ISOBUS) to clean the glass.

Dust in atmosphere reduces the overall matching quality q of the stereo block matching algorithm. If lighting conditions are good, but q is below a certain threshold, it is assumed that strong dust occurs. In that case, a message is sent to the tractor driver via the user interface.

Residues like straw can be filtered by removing speckles and unexpected peaks from the point cloud. However, a more sophisticated and robust solution would be to segment an RGB image from the stereo camera into different soil cover classes and to project them into the point cloud. In the next Sect. 4 we show how to detect residues and other soil cover from the RGB image.

4 Residue and Soil Cover Detection

Residues like straw and plants strongly influence the roughness measurement. To exclude them from the estimation, the covered regions of the soil have to be detected. The RGB images from the stereo camera allow for semantic segmentation of the soil image in different classes. So it is possible to not only detect residues for the roughness estimation but also to map the soil cover of the whole field. The relevant classes for soil cover are soil, living organic matter (like plants), dead organic matter (like straw) and stones. Two different methods were implemented, tested and compared.

The method based on an entangled random forest is described in [13]. This variant of a random forest classifier uses simple pairwise colour difference features and additional maximum a posteriori features, which allow for a certain smoothing while preserving edges. The second approach, based on a convolutional neural net, the Efficient Residual Factorized ConvNet (ERFNet) [16], was implemented using the bonnet framework [17].

Convolutional neural nets for semantic segmentation use an encoder and decoder block consecutive. Down-sampling or encoding in convolutional neural nets lets deeper layers gather more context and therefore improve classification, but has the drawback of reducing the pixel precision for semantic segmentation. ERFNet uses factorized convolutions with residual connections and introduces a non-bottleneck-1D (non-bt-1D) layer. This combination of 1D filters is faster and has less parameters than bottleneck layers. The accuracy remains the same as for non-bottleneck layer. ERFNet uses simple deconvolutions layers with stride 2 to reduce memory and computation requirements for decoding. The architecture for a convolutional neural net for semantic segmentation is shown in Fig. 3.

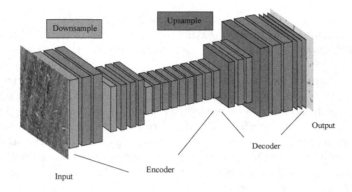

Fig. 3. Encoder-Decoder architecture of a convolutional neural net for semantic segmentation like ERFNet [16].

The model was pre-trained on the Sugar Beets 2016 dataset [20]. For the training on soil cover dataset, we use a trainings set of about 200 different manual labelled images. Additional augmentation like rotation, scaling and gamma modifications were added. The labelled images have a size of about 640×480 pixel and are extracted from high resolution soil images (about 10M pixels).

5   Experimental Results

The whole system was tested in a three step process. First, the roughness estimation was evaluated by sieving soil aggregates. In a second step, the error of soil cover and residue detection was estimated with manually annotated test samples. And last, the effect of the controller on the field was tested in a strip-wise field trial.

5.1   Roughness Estimation

The roughness estimation was evaluated by computing the correlation between soil roughness RC to the mean soil aggregate size. The mean soil aggregate size was determined by sieve analysis [2] and calculation of the mean-weight diameter. The height measurement from the stereo camera where evaluated using a height gauge. The height gauge showed a mean error of 1.55 mm, which is sufficient for our task. The Pearson correlation between MWD and soil roughness index RC was 0.55 (N=52) at the trial field in Gross Enzersdorf (AT) and 0.66 (N = 30) at the trial field in Krummnussbaum (AT). The correlation between soil roughness and mean-weight diameter of soil aggregates was estimated in previous tests, for example in [21]. The work in [21] used a laser scanner for roughness estimation at got a relation of $Roughness = 1.88MWD - 4.02$ and $r^2 = 0.63$. One major factor for good correlation to the is the observed area. Due to mechanical limitations on the power harrow we cannot increase the captured area. We cope with this problem by using the sliding mean window, to increase the observed area.

5.2 Soil Cover Detection

We compared soil cover segmentation results between the entangled random forest method proposed in [13] and the results based on an ERFNet. For comparison the grid method [13] was used. Figure 4 shows the results for both methods, the entangled random forest method in orange and the convolutional neural network based results in blue, compared to the manually annotated ground truth from 114 images. The higher variance for soil and dead organic matter lies in the ambiguity between these classes. Another factor is the low number of trainings images. Increasing the number of training samples could lead to better classification results. Figure 5 shows two examples of stereo images and the corresponding soil cover classification result.

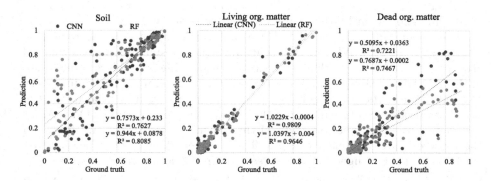

Fig. 4. Comparison between the entangled random forest method (• RF) and the results of the convolutional neural network (• CNN) to manually labelled test samples. The axis go from 0... no cover to 1... full cover. (Color figure online)

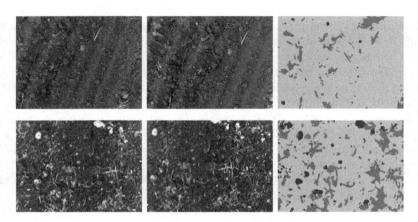

Fig. 5. Examples stereo soil images (left, middle) and the corresponding soil cover classification (right). living plant •, residue •, soil • and stone •. The first example (top) has a roughness value of $RC = 9.50$ and the second example (bottom) of $RC = 7.14$. (Color figure online)

5.3   Machine Control

For testing the machine control, a field at the test site in St. Leonhard am Forst (AT), was tilled with the power harrow at constant speeds, and on alternating tracks, with the controller activated.

The tracks with constant speeds were tilled with a PTO speed of 550 rpm and a constant speed of $1.3\,\mathrm{ms}^{-1}$. At the controlled tracks the target RC value was set to 14. The range of the PTO controller was set between 550 rpm and 1000 rpm. The forward driving speed was set constant at $1.3\,\mathrm{ms}^{-1}$

The mean squared error of the constant tracks to the desired roughness $RC_{ref} = 14$ was 2.16. Whereas, the controlled tracks resulted in a mean squared error of 1.24, which shows that the controller works. Figure 6 shows three tracks of the field trial. The blue curve shows the estimated soil roughness value at the controlled track. Its two neighbouring tracks, with the conventional method of constant speeds, are plotted in grey. The black line presents the PTO speed for the controlled track. The roughness of the controlled track stays mostly within the hysteresis area around $RC = 14$. The corresponding PTO values increase when the soil gets rougher and a decrease at finer field regions at the end. The constant neighbouring tracks, without control algorithm, result in overall high roughness values which go beyond the desired hysteresis area.

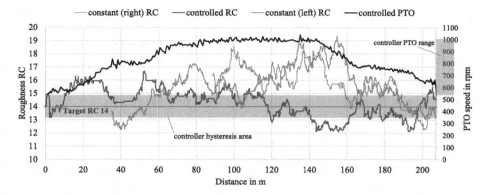

Fig. 6. Three tracks of the field trial, one controlled track with varying PTO speed (blue) and two tracks with constant PTO speeds (grey). The black curve presents the PTO speeds for the controlled track.

6   Conclusions and Future Work

Getting homogeneous soil roughness at changing soil conditions during tillage is an important agronomic goal. The proposed system consists of a stereo camera to estimate soil roughness and soil cover, and a controller to adapt driving speed and PTO speed of a tractor during tillage with a power harrow based on these estimated soil parameters. The effect of the control algorithm could be shown in the field by more homogeneous soil roughness results. The soil cover estimation based on ERFNet showed less error than previous methods, with better distinction between the classes soil and dead organic matter.

Future work could focus on detecting not only soil roughness but find a way to detect soil clods to get a better estimate for the mean clod diameter. At present, no method is known in the literature which can measure the clod size in motion. To further reduce the classification error of the soil cover classifier, the training set has to be increased. Works in [22] have shown that simulated images, from single plant images, could help to pretrain the neural network and therefore improve the detection quality.

Acknowledgement. The research leading to this work has received funding from the Lower Austrian government (WST3-T-140/002-2014). As well as from the Austrian Research Promotion Agency under the program "Bridge 1".

References

1. Adam, K.M., Erbach, D.C.: Secondary tillage tool effect on soil aggregation. Trans. ASAE **35**(6), 1771–1776 (1992)
2. Kirchmeier, H., Geischeder, R., Demmel, M.: Tillage effect and requirements of rotaty harrows with different rotor geometries. Landtechnik **60**(4), 196–197 (2005)

3. Currence, H., Lovely, W.: The analysis of soil surface roughness. Trans. ASAE **13**, 710–714 (1970)
4. Marinello, F., Pezzuolo, A., Gasparini, F., Arvidsson, J., Sartori, L.: Application of the Kinect sensor for dynamic soil surface characterization. Precis. Agric. **16**, 601–612 (2015)
5. Taconet, O., Ciarletti, V.: Estimating soil roughness indices on a ridge-and-furrow surface using stereo photogrammetry. Soil Tillage Res. **93**, 64–76 (2007)
6. Riegler, T., Rechberger, C., Handler, F., Prankl, H.: Image processing system for evaluation of tillage quality. Landtechnik **69**(3), 125–130 (2014)
7. de Obade, V.P.: Review article: remote sensing, surface residue cover and tillage practice. J. Environ. Prot. **3**, 211–217 (2012)
8. Pforte, F., Wilhelm, B., Hensel, O.: Evaluation of an online approach for determination of percentage residue cover. Biosyst. Eng. **112**, 121–129 (2012)
9. Campillo, C., Prieto, M.H., Daza, C., Monino, M.J., Garcia, M.I.: Using digital images to characterize canopy coverage and light interception in a processing tomato crop. HortScience **43**, 1780–1786 (2008)
10. Kırcı, M., Güneş, E. O., Çakır, Y.: Vegetation measurement using image processing methods. In The Third International Conference on Agro-Geoinformatics (2014)
11. Guerrero, J.M., Pajares, G., Montalvo, M., Romeo, J., Guijarro, M.: Support Vector Machines for crop/weeds identification in maize fields. Expert Syst. Appl. **39**(12), 11149–11155 (2012)
12. Mortensen, A. K., Dyrmann, M., Karstoft, H., Nyholm Jørgensen, R., Gislum, R.: Semantic Segmentation of Mixed Crops using Deep Convolutional Neural Network. In: CIGR-AgEng Conference (2016)
13. Riegler-Nurscher, P., Prankl, J., Bauer, T., Strauss, P., Prankl, H.: A machine learning approach for pixel wise classification of residue and vegetation cover under field conditions. Biosyst. Eng. **169**, 188–198 (2018)
14. Paszke, A., Chaurasia, A., Kim, S., Culurciello, E.: Enet: a deep neural network architecture for real-time semantic segmentation. arXiv preprint arXiv:1606.02147 (2016)
15. Mehta, S., Rastegari, M., Caspi, A., Shapiro, L., Hajishirzi, H.: ESPNet: efficient spatial pyramid of dilated convolutions for semantic segmentation. In: Proceedings of the European Conference on Computer Vision (ECCV), pp. 552–568 (2018)
16. Romera, E., Alvarez, J.M., Bergasa, L.M., Arroyo, R.: Erfnet: efficient residual factorized convnet for real-time semantic segmentation. IEEE Trans. Intell. Transp. Syst. (ITS) **19**(1), 263–272 (2018)
17. Milioto, A., C. Stachniss, C.: Bonnet: an open-source training and deployment framework for semantic segmentation in robotics using CNNs. In: Proceedings of the IEEE International Conference on Robotics & Automation (ICRA) (2019)
18. OpenCV library. http://opencv.org/. Accessed 9 May 2019
19. PCL library. http://www.pointclouds.org/. Accessed 9 May 2019
20. Chebrolu, N., Lottes, P., Schaefer, A., Winterhalter, W., Burgard, W., Stachniss, C.: Agricultural robot dataset for plant classification, localization and mapping on sugar beet fields. Int. J. Robot. Res. **36**(10), 1045–1052 (2017)
21. Sandri, R., Anken, T., Hilfiker, T., Sartori, L., Bollhalder, H.: Comparison of methods for determining cloddiness in seedbed preparation. Soil Tillage Res. **45**, 75–90 (1998)
22. Skovsen, S., et al.: Estimation of the botanical composition of clover-grass leys from RGB images using data simulation and fully convolutional neural networks. Sensors **17**(12), 2930 (2017)

Color Calibration on Human Skin Images

Mahdi Amani[1(✉)], Håvard Falk[1,2], Oliver Damsgaard Jensen[1,3],
Gunnar Vartdal[1,4], Anders Aune[1], and Frank Lindseth[1]

[1] Norwegian University of Science and Technology (NTNU), Trondheim, Norway
{mahdi.amani,a.aune,frankl}@ntnu.no
[2] Bekk Consulting AS, Akershusstranda 21, 0150 Oslo, Norway
havard.falk@bekk.no
[3] Itera ASA, Nydalsveien 28, 0484 Oslo, Norway
oliver.jensen@itera.no
[4] Picterus AS, Professor Brochs gate 8A, 7030 Trondheim, Norway
gunnar@picterus.com

Abstract. Many recent medical developments rely on image analysis, however, it is not convenient nor cost-efficient to use professional image acquisition tools in every clinic or laboratory. Hence, a reliable color calibration is necessary; *color calibration* refers to adjusting the pixel colors to a standard color space.

During a real-life project on neonatal jaundice disease detection, we faced a problem to perform skin color calibration on already taken images of neonatal babies. These images were captured with a smartphone (Samsung Galaxy S7, equipped with a 12 Mega Pixel camera to capture 4032×3024 resolution images) in the presence of a specific calibration pattern. This post-processing image analysis deprived us from calibrating the camera itself. There is currently no comprehensive study on color calibration methods applied to human skin images, particularly when using amateur cameras (e.g. smartphones). We made a comprehensive study and we proposed a novel approach for color calibration, Gaussian process regression (GPR), a machine learning model that adapts to environmental variables. The results show that the GPR achieves equal results to state-of-the-art color calibration techniques, while also creating more general models.

Keywords: Color calibration · Color correction · Skin imaging

1 Introduction

Medical imaging is the core of recent medical science and the accuracy of the information highly relies on the environmental parameters such as light sources and camera sensors [1] which cause image inconsistencies. To overcome these issues (i.e., to make images illumination, camera sensors, and actual pixel values independent), we study, implement, and improve the state-of-the-art of color calibration techniques and we propose a novel approach for color calibration;

© Springer Nature Switzerland AG 2019
D. Tzovaras et al. (Eds.): ICVS 2019, LNCS 11754, pp. 211–223, 2019.
https://doi.org/10.1007/978-3-030-34995-0_20

the Gaussian process regression (GPR), a machine learning model that adapts to environmental variables. The results show that the GPR achieves equal results to state-of-the-art color calibration techniques, while also creating more general models. We had faced this problem while working on a dataset of already taken images of neonates for jaundice disease detection.

1.1   Paper Structure

In the rest of this section, an introduction on human skin's light absorption, related color theory, colorcheckers, and color calibration preliminaries is presented. Then, the proposed solutions on color calibration and related machine learning regression techniques are presented in Sect. 2. In Sect. 3, the performance of color calibration techniques are presented and evaluated in terms of $\Delta E*_{a,b}$ using colorcheckers as ground truth and the results are compared. Finally, based on the given results, a conclusion is given in Sect. 4.

1.2   Light Absorption on Human Skin

To understand optical skin diagnosis, we must first understand human skin and how each component's concentration alters the perceived color of skin. The perceived color of an object is due to the light that is reflected off the object. Molecules only reflect a subset of in-coming frequencies [2]. Skin, a layered organ protecting the human organism against environmental stress (e.g. heat, radiation, and infections) contains light-absorbing substances.

Skin consists of three main layers; epidermis, dermis, and the hypodermis. The visual features of skin captured by a camera recording the RGB values and their pattern can reveal a lot of information about the underlying substances, by different lightening and camera sensors. In practice, these RGB values are always dependent to each other (which they should not), hence, having a robust color calibration technique can solve this issue.

1.3   Delta E (ΔE)

While comparing results, a unified metric for color difference is required to express the difference between two colors correctly. The idea is that a color difference of $1\Delta E$ is the smallest color difference a human eye can detect. ΔE was first presented alongside the CIE L*a*b* color space in [3]. Given two colors in the CIE L*a*b* color space, (L_1^*, a_1^*, b_1^*) and (L_2^*, a_2^*, b_2^*), the simplest formula of ΔE is the Euclidean distance between these colors, as follows.

$$\Delta E_{ab}^* = \sqrt{(L_2^* - L_1^*)^2 + (a_2^* - a_1^*)^2 + (b_2^* - b_1^*)^2} \tag{1}$$

It is estimated that a human regards a $\Delta E_{ab}^* \approx 2.3$ difference as 'just noticeable' [3].

1.4   CIE Illuminant

The quality and energy of a *light source* is not always consistent and are often seen as unreliable and cannot technically be *reproduced*. To create light suited for colorimetric calculations, the CIE introduced the concept of standard illuminants. A standard illuminant is a theoretical source of visible light where its spectral power distribution is explicitly defined. Illuminants are divided into series describing source characteristics such as: Incandescent/Tungsten, Direct sunlight at noon, Average/North sky Daylight, Horizon Light, Noon Daylight, etc. Full list can be found on [4] and [5].

1.5   ColorChecker

A *colorchecker*, also known as a *calibration card*, is a physical set of colors defined in the CIE L*a*b* color space. The colors are, thus, defined regardless of light source or image capturing device[1]. If a colorchecker is included in an image, the RGB errors at each color patch can be calculated and provides information about the RGB variations in the image, which again can be used to correct the color errors. Here, we introduce two colorcheckers; the *Macbeth generic colorcheckers* and the *Picterus skin colorchecker*.

Macbeth Generic Colorcheckers. The classic Macbeth colorchecker by [6] (Fig. 1a) is one of the most commonly used reference targets in photographic work. The checker is designed to approximate colors found in nature. Six patches are different neutral gray, from black to white, where the spectral response of each patch is constant at all wavelengths and differ only in intensity. We referred to these six patches as the *grayscale* of the colorchecker. Other adaptations of McCamy's colorchecker have since been produced, here we present Datacolor's **SpyderCHECKR 24**. SpyderCHECKR 24 and its all target values are given in Fig. 2a and c.

Picterus Skin ColorChecker. For jaundice detection on neonatal babies, we designed a custom colorchecker (Fig. 1b) targeted for human skin[2]. We will refer to this colorchecker as the **SkinChecker**, and all target values are given in Fig. 2b and d. The colors on the SkinChecker are based on simulated reflection spectra of neonate's skin with varying skin parameters. These reflection spectra have been printed using spectral printing, a technique which attempts to recreate the whole reflection spectrum instead of just the RGB color values [7].

Figure 3 shows the color diversity of a SpyderCHECKR 24 and a SkinChecker. The figure highlights the difference between general and specialized color correction, drawing triangles to visualize the RGB color sub-spaces defined by the two color checkers. For general color correction, a model must be able to reproduce a

---

[1] However, by specifying a light source from CIE illuminant list, one can approximately define a colorchecker in RGB, but such generic RGB values are not to be fully trusted.

[2] For further information please contact Picterus AS at www.picterus.com.

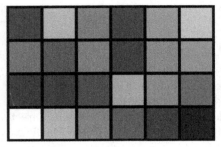
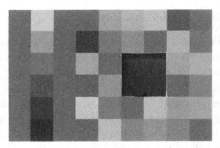

(a) The classic Macbeth ColorChecker   (b) The SkinChecker by Picterus AS

Fig. 1. The Macbeth ColorChecker Classic and the SkinChecker.

wide range of colors, as reflected by the SpyderCHECKR 24. The SkinChecker, on the other hand, contains substantially more data points in its focus area, allowing a model to optimize for more accurate color correction of human skin.

1.6   Color Calibration; Preliminaries and Related Works

Variations in digital cameras' color responses (RGB values), caused by inaccurate sensors, result in device **dependent** values of R, G, and B. The task of correcting errors, in the captured RGB values, is referred to as *color calibration* (or *color correction*[3]). More precisely, *color calibration* refers to adjusting the pixel colors to a default/known/standard color space [9]. Color calibration involves mapping device dependent RGBs to corresponding **independent** color values (e.g. RGB or CIE XYZ) usually by using a colorchecker as reference. The device independent color values are often referred to as tristimulus values, and represent the same color regardless of visual system.

Look-up tables, least-squares linear and polynomial regression and neural networks are some of the methods described in literature regarding the mapping between RGB and tristimulus values.

Look-Up Tables. The trivial *look-up table* is a large collection of camera RGB examples and the corresponding target values, manually created to define a mapping between the two color spaces.

Least-Squares Linear Regression. A linear mapping from camera RGBs to CIE L*a*b* triplets can be achieved through a 3×3 linear transform. If we let ρ define a three element vector representing the three camera responses (R, G, B) and $\mathbf{q}$ define the three corresponding L*, a*, b* values, a simple linear transform can be written as: $\mathbf{q} = \mathbf{M}\rho$,

---

[3] It's worth mentioning that a device (e.g. camera) is to be *calibrated* while images are to be *corrected*, hence, color calibration and color correction are slightly different. However, the concept is the same and here we consider them equivalent.

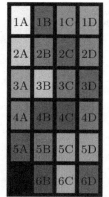

(a) The SpyderCHECKR 24

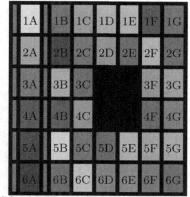

(b) The SkinChecker by Picterus AS

Patch Name		L*	a*	b*	R	G	B
		L*a*b*			sRGB		
1A	Card White	96.04	2.16	2.60	249	242	238
2A	20% Gray	80.44	1.17	2.05	202	198	195
3A	40% Gray	65.52	0.69	1.86	161	157	154
4A	60% Gray	49.62	0.58	1.56	122	118	116
5A	80% Gray	33.55	0.35	1.40	80	80	78
6A	Card Black	16.91	1.43	-0.81	43	41	43
1B	Primary Cyan	47.12	-32.52	-28.75	0	127	159
2B	Primary Magenta	50.49	53.45	-13.55	192	75	145
3B	Primary Yellow	83.61	3.36	87.02	245	205	0
4B	Primary Red	41.05	60.75	31.17	186	26	51
5B	Primary Green	54.14	-40.76	34.75	57	146	64
6B	Primary Blue	24.75	13.78	-49.48	25	55	135
1C	Primary Orange	60.94	38.21	61.31	222	118	32
2C	Blueprint	37.80	7.30	-43.04	58	88	159
3C	Pink	49.81	48.50	15.76	195	79	95
4C	Violet	28.88	19.36	-24.48	83	58	106
5C	Apple Green	72.45	-23.57	60.47	157	188	54
6C	Sunflower	71.65	23.74	72.28	238	158	25
1C	Aqua	70.19	-31.85	1.98	98	187	166
2D	Lavender	54.38	8.84	-25.71	126	125	174
3D	Evergreen	42.03	-15.78	22.93	82	106	60
4D	Steel Blue	48.82	-5.11	-23.08	87	120	155
5D	Classic Light Skin	65.10	18.14	18.68	197	145	125
6D	Classic Dark Skin	36.13	14.15	15.78	112	76	60

(c) Values from the SpyderCHECKR 24 by [8]

Patch	L*	a*	b*	R	G	B
	L*a*b*			sRGB		
1A	88.38	1.51	-3.32	221	221	228
2A	70.14	2.72	-0.74	175	169	172
3A	51.97	1.43	0.66	126	123	122
4A	38.53	-0.15	0.93	91	90	89
5A	30.25	-1.12	2.06	70	71	68
6A	26.02	-1.50	0.71	59	62	60
1B	48.32	1.82	1.49	119	113	112
2B	28.82	12.48	13.19	92	60	48
3B	74.93	8.10	45.14	223	177	100
4B	42.03	39.70	31.89	166	67	48
5B	86.81	2.21	3.52	224	215	210
6B	47.62	0.74	1.01	115	112	111
1C	67.77	15.92	20.13	205	153	129
2C	52.60	24.83	47.00	182	107	42
3C	47.52	2.17	1.81	117	111	109
4C	66.73	8.33	40.28	198	155	89
5C	51.02	32.66	35.69	185	97	61
6C	81.35	4.01	5.79	214	199	191
1D	75.98	5.20	26.58	214	182	138
2D	68.04	8.81	63.77	209	157	39
5D	46.63	1.55	1.91	114	109	107
6D	58.96	19.64	44.55	193	127	62
1E	76.70	16.15	22.70	232	177	148
2E	46.84	2.26	2.76	116	109	106
5E	75.25	20.00	13.09	229	171	162
6E	59.69	29.13	18.90	201	123	112
1F	36.42	24.92	17.63	129	68	58
2F	77.67	10.08	49.47	236	182	99
3F	64.09	31.40	6.45	211	133	145
4F	46.33	2.08	3.07	115	108	104
5F	72.85	15.48	70.20	234	165	35
6F	52.44	38.26	16.14	191	96	99
1G	48.53	2.43	2.41	121	113	111
2G	46.97	26.89	23.61	163	92	73
3G	68.04	26.16	12.62	218	147	144
4G	56.56	32.21	55.49	205	110	33
5G	73.81	14.00	64.45	234	169	55
6G	47.21	1.44	3.02	116	110	106

(d) Values from SkinChecker

Fig. 2. SpyderCHECKR 24 and Picterus' SkinChecker.

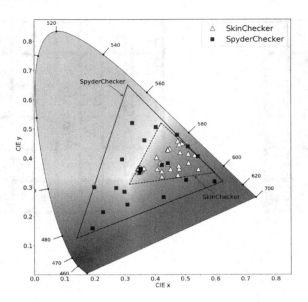

Fig. 3. The CIE xy Chromaticity diagram with the SpyderChecker 24 and SkinChecker's colors individually positioned. The SkinChecker contains skin related colors and resides only in the red-yellow part of the color space while the SpyderChecker 24 is a diverse colorchecker and is spread in a wider range. (Color figure online)

M holds the coefficients (d_{ij}) of the transform that performs the actual color correction. The **linear color correction (LCC)** has proved to perform well in numerous studies, with the advantage of being independent of camera exposure [10], hence, we skip the details. It has also been known to produce significant errors when mapping RGBs to CIE L*a*b* values. Some studies show that LCC can, for some surfaces, generate errors up to $14\Delta E_{ab}^*$ [11]. However, the same study shows that on average the 3×3 LCC transform yields an error of $2.47\Delta E_{ab}^*$ on the 8-bit Professional Color Communicator [12].

Least-Squares Polynomial Regression. A more modern approach for color correction is to assume that the relationship between RGB and target values is polynomial, not linear. This leads to a more complex method for color correction, **polynomial color correction (PCC)**, where the R, G, and B values at a pixel are extended by adding additional polynomial terms of increasing degree.

$$\rho = (R, G, B)^T \xrightarrow{\theta_k} \hat{\rho}_k = (R, G, B, ..., m)^T \tag{2}$$

We denote the kth polynomial extension of an RGB-vector (ρ) by $\hat{\rho}_k$. The extension operator θ_k transforms a three-element column vector to an m-element column vector with a set of added polynomial terms, further derails in [11]. For a simple RGB case i.e. $\rho = (R, G, B)^T$ the polynomial expansions of 2nd, 3rd and 4th degrees are given below:

$$\hat{\rho}_1 = [r, g, b, rg, rb, gb, r^2, g^2, b^2, 1] \tag{3}$$

$$\hat{\rho}_3 = [r, g, b, rg, rb, gb, r^2, g^2, b^2,$$
$$rg^2, rb^2, gr^2, gb^2, br^2, bg^2, r^3, g^3, b^3, rgb, 1] \tag{4}$$

$$\hat{\rho}_4 = [r, g, b, rg, rb, gb, r^2, g^2, b^2,$$
$$rg^2, rb^2, gr^2, gb^2, br^2, bg^2, rgb, r^3, g^3, b^3,$$
$$r^3g, r^3b, g^3r, g^3b, b^3r, b^3g, r^2g^2, r^2b^2, g^2b^2,$$
$$r^2gb, g^2rb, b^2rg, r^4, g^4, b^4, 1] \tag{5}$$

Using the extension operator, the three RGB values recorded in a pixel are extended, represented by 9, 19, and 34 numbers respectively. As apposed to LCC's 3×3 matrix, PCC is carried out by 9×3, 19×3, and 34×3 matrices. Similar to LCC, we find the coefficients that minimize $\mathbf{M}$, the $m \times 3$ PCC matrix.

Root-Polynomial Color Correction. If the correct polynomial fit is chosen, PCC can significantly reduce the colorometric error from LCC. However, the PCC fit depends on sensors and illumination, where exposure alters the vector of polynomial components in a non-linear way. Hence, choosing the right polynomial fit is very important in PCC. To solve the exposure sensitivity of PCC, [10] present a polynomial-type regression related to the idea of fractional polynomials. Their method, named **root-polynomial color correction (RPCC)**, takes each term in a polynomial expansion to its kth root of each k-degree term, and is designed to scale with exposure. The root-polynomial extensions for $k = 2$, $k = 3$ and $k = 4$ are defined as:

$$\bar{\rho}_2 = [r, g, b, \sqrt{rg}, \sqrt{rb}, \sqrt{gb}, 1] \tag{6}$$

$$\bar{\rho}_3 = [r, g, b, \sqrt{rg}, \sqrt{rb}, \sqrt{gb}, \sqrt[3]{rg^2}, \sqrt[3]{rb^2}, \sqrt[3]{gr^2}, \sqrt[3]{gb^2},$$
$$\sqrt[3]{br^2}, \sqrt[3]{bg^2}, \sqrt[3]{rgb}, 1] \tag{7}$$

$$\bar{\rho}_4 = [r, g, b, \sqrt{rg}, \sqrt{rb}, \sqrt{gb},$$
$$\sqrt[3]{rg^2}, \sqrt[3]{rb^2}, \sqrt[3]{gr^2}, \sqrt[3]{gb^2}, \sqrt[3]{br^2}, \sqrt[3]{bg^2}, \sqrt[3]{rgb},$$
$$\sqrt[4]{r^3g}, \sqrt[4]{r^3b}, \sqrt[4]{g^3r}, \sqrt[4]{g^3b}, \sqrt[4]{b^3r}, \sqrt[4]{b^3g},$$
$$\sqrt[4]{r^2gb}, \sqrt[4]{g^2rb}, \sqrt[4]{b^2rg}, 1] \tag{8}$$

We denote a root-polynomial extension of an RGB column vector ρ as $\bar{\rho}$. As with PCC, the RGB-extension naturally increases the size of the transform matrix $\mathbf{M}$. Thus, RPCC is performed by a 7×3, 13×3, and 22×3 matrices.

Machine Learning Approaches. An alternative approach for color correction is the use of machine learning approaches. The most used techniques are Support Vector Machines (SVMs) and fully-connected feed-forward neural networks. Artificial neural network (ANN) have shown to be robust when optimized correctly, and achieve equally good results as a well fitted polynomial approach

218   M. Amani et al.

[13]. However, both SVMs and fully-connected neural networks require extensive hyperparameter-tuning, a tedious process performed through trial and error, or using the computational expensive grid-search with cross-validation.

On the other hand, **Gaussian processes (GPs)** are widely recognized as a powerful, yet practical tool for solving both classification and non-linear regression [14]. A GP has not been applied to the field of color calibration. A GP can be thought of as a generalization of the Gaussian probability distribution over a finite vector space to a function space of infinite dimensions [15]. The processes are probabilistic models of functions and are used for solving both classification and non-linear regression problems. To be more precise, a GP is used to describe a distribution over functions $f(x)$ such that any finite set of function values $f(x_1), f(x_2), ..., f(x_n)$ have a joint Gaussian distribution [16, Chapter 2].

2 Color Calibration Techniques and Results

This section presents the color calibration techniques which are performed on our dataset. The dataset is a collection of 564 images acquired from St. Olav's hospital (https://stolav.no)[4] depicting neonates. In each image, a SkinChecker is placed on the chest, exposing the skin color within the SkinChecker. This dataset is collected from 141 unique neonates by smartphones, where four images were taken with different ranges and lightening (e.g. with or without flash) from each neonate. For each image (Fig. 4a), illumination varies, and inconsistently shifts RGB values. Then the SkinChecker segment is extracted (Fig. 4b). With the SkinChecker as the reference point for the color correction solutions, our goal is to apply pixelwise color correction (Fig. 4c), trying to restore the original colors.

For fieldwork applications, we require the color correction algorithm to be device independent and show high robustness towards illumination. We used SkinChecker[5] to compare these methods using an evaluation method as follows.

2.1 Leave-One-Out Cross-validation Evaluation Method (EM)

For each individual color (on the in-scene colorchecker), we exclude 10% of the sample area on all sides, to eliminate potentially faulty segmentation, and sample the average RGB value. The resulting average RGBs and corresponding CIE L*a*b* values (from the colorchecker) are used to create color correction models.

To evaluate the performance of a model, we perform leave-one-out cross-validation which is an iterative approach where one color patch is withheld every run. For each iteration, the remaining colors are used to build a color correction model, predicting the CIE L*a*b* value of the withheld color patch. By the end of the iteration, all color patches have been withheld and the ΔE_{ab}^* is calculated.

---

[4] NTNU's hospital (https://stolav.no) located in Trondheim, Norway.
[5] We also performed the same evaluation on SpyderCHECKR achieving similar results.

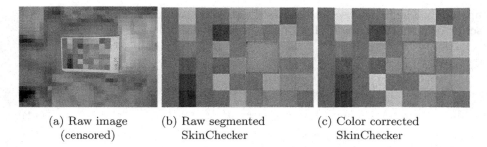

(a) Raw image   (b) Raw segmented   (c) Color corrected
(censored)   SkinChecker   SkinChecker

Fig. 4. Color calibration pipeline. The visible skin patch within the SkinChecker (highlighted by the red square in Fig. 4c), represents the color calibrated image. (Color figure online)

2.2 Color Calibration Frameworks

Literature describes a wide range of polynomial and root-polynomial fits to perform color correction. We are, however, unaware of previous work that proposes a solution as to which extension is to be applied in a given scenario. To overcome the practical issues related to color correction, first, we create two color correction frameworks, polynomial color correction framework (PCCF) and root-polynomial color correction framework (RPCCF), that are implementations of widely used polynomial and root-polynomial extensions. Finally, we test our framework, the novel GPR for color correction. Due to its extreme modeling flexibility, and added noise kernels, in the Sect. 3, we show that the GPR well compete and even outperforms state-of-the-art approaches, creating highly complex models without overfitting. These frameworks are defined by their set of extension operators that transform sensor RGB values to m-element column vectors of an arbitrary combination or power. We will refer to these extension operators as internal methods θ_k (e.g. $(r,g,b)^T \xrightarrow{\theta_k} \hat{\rho}_2(r,g,b,r^2,g^2,b^2)^T$). Each framework builds a color correction model for each of its internal methods (i.e. polynomial expansion) using least-squares regression to solve. Doing so, results in an $m \times 3$ matrix $\mathbf{M}$, mapping

$$\hat{\rho} \xrightarrow{\mathbf{M}} (L^*, a^*, b^*). \tag{9}$$

Polynomial Color Correction Framework Solution (PCCF). To create the set of polynomial extension operators (i.e., the set of internal methods), we apply an iterative scaling approach, increasing the complexity of internal methods for each iteration. We start off by adding a constant term to the LCC:

$$\rho = (r,g,b)^T \xrightarrow{\theta_1} \hat{\rho}_1 = (r,g,b,1)^T \tag{10}$$

While not a significant alteration, the added term gives the least-squares regression some leeway. Continuing, we follow the polynomial orders, and combinations of R, G, and B, and create the following collection of RGB-extension operators. The first subscript denotes the polynomial order, and the second subscript denotes the method id within the polynomial order.

$$\hat{\rho}_{1,1} = [r, g, b, 1] \tag{11}$$

$$\hat{\rho}_{2,1} = [r, g, b, rg, rb, gb, 1] \tag{12}$$

$$\hat{\rho}_{2,2} = [r, g, b, r^2, g^2, b^2, 1] \tag{13}$$

$$\hat{\rho}_{3,1} = [r, g, b, rg, rb, gb, r^2, g^2, b^2, rgb, 1] \tag{14}$$

$$\hat{\rho}_{3,2} = [r, g, b, r^2 g, r^2 b, g^2 r, g^2 b, b^2 r, b^2 g, rgb, 1] \tag{15}$$

$$\hat{\rho}_{3,3} = [r, g, b, rg, rb, gb, r^2, g^2, b^2,$$
$$r^3, g^3, b^3, rgb, 1] \tag{16}$$

$$\hat{\rho}_{3,4} = [r, g, b, rg, rb, gb, r^2, g^2, b^2,$$
$$r^2 g, r^2 b, g^2 r, g^2 b, b^2 r, b^2 g, rgb, 1] \tag{17}$$

$$\hat{\rho}_{3,5} = [r, g, b, rg, rb, gb, r^2, g^2, b^2,$$
$$r^2 g, r^2 b, g^2 r, g^2 b, b^2 r, b^2 g, r^3, g^3, b^3, rgb, 1] \tag{18}$$

Root-Polynomial Color Correction Framework Solution (RPCCF). For a fixed exposure, PCC has shown significantly better results than a 3×3 LCC [11]. However, as pointed out by [10], exposure changes the vector of polynomial components in a nonlinear way, resulting in hue and saturation shifts. Their solution, RPCC, claims to fix these issues by expanding the RGB terms with root-polynomial extensions instead. We adopt their idea and create the RPCCF. The framework includes the root-polynomial extensions suggested in their paper, where the subscript of $\bar{\rho}$ denotes the root-polynomial order.

$$\bar{\rho}_2 = [r, g, b, \sqrt{rg}, \sqrt{rb}, \sqrt{gb}, 1] \tag{19}$$

$$\bar{\rho}_3 = [r, g, b, \sqrt{rg}, \sqrt{rb}, \sqrt{gb}, \sqrt[3]{rg^2}, \sqrt[3]{rb^2}, \sqrt[3]{gr^2}, \sqrt[3]{gb^2},$$
$$\sqrt[3]{br^2}, \sqrt[3]{bg^2}, \sqrt[3]{rgb}, 1] \tag{20}$$

$$\bar{\rho}_4 = [r, g, b, \sqrt{rg}, \sqrt{rb}, \sqrt{gb},$$
$$\sqrt[3]{rg^2}, \sqrt[3]{rb^2}, \sqrt[3]{gr^2}, \sqrt[3]{gb^2}, \sqrt[3]{br^2}, \sqrt[3]{bg^2}, \sqrt[3]{rgb},$$
$$\sqrt[4]{r^3 g}, \sqrt[4]{r^3 b}, \sqrt[4]{g^3 r}, \sqrt[4]{g^3 b}, \sqrt[4]{b^3 r}, \sqrt[4]{b^3 g},$$
$$\sqrt[4]{r^2 gb}, \sqrt[4]{g^2 rb}, \sqrt[4]{b^2 rg}, 1] \tag{21}$$

Gaussian Process Regression Solution. GPR is a very flexible machine learning technique that requires no explicit tuning of hyperparameters (other than the choice of kernel). For color correction, we train three separate GPs, one for each color channel (R, G, B). To predict a new color, the three GPR models take the RGB input color and predicts their respective CIE L*a*b* intensity value (e.g., L* or a* or b*). The three individual results are combined to create the new CIE L*a*b* color coordinate.

The GPR implementation is initiated with a prior's covariance, specified by a kernel object. The hyperparameters of the kernel are optimized using gradient-descent on the marginal likelihood function during fitting, equivalent to maximizing the log of the marginal likelihood (LML). This property makes GPR superior to other supervised learning techniques, as it avoids heavy computational validation approaches, like cross-validation, to tune its hyperparameters. The LML may have multiple local optima, and trail and error testing is employed to verify that the optimal or close to the optimal solution is found by starting the optimizer repeatedly.

Choosing Kernel: GPR is a general method that can be extended to a wide range of problems. To implement GPs for regression purposes, the prior of the GP needs to be specified by passing a kernel. Widely used kernels are empirically tested; *RBF*, *Matérn*, and *rational quadratic (RQ)*. Additionally, combinations of *Constant-* and *White kernel* are added to the kernel function, to find the best performing kernel composition.

3 Experimental Results

In this section, using skinChecker, the proposed color calibration methods are evaluated. The models are evaluated on all images in the jaundice dataset, thus testing each solution in terms of skin color correction. Based on the sampled data and the corresponding CIE L*a*b* triplets, each color correction solution builds a model, evaluated according EM. Figures 5a and b show the results of the experiment and their bar plot visualization, evaluating the LCC, PCCF, RPCCF, and GPR. The dataset is collected through fieldwork, and the light sources are not reproducible in terms of a single CIE illuminant.

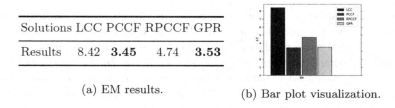

Solutions	LCC	PCCF	RPCCF	GPR
Results	8.42	**3.45**	4.74	**3.53**

(a) EM results.   (b) Bar plot visualization.

Fig. 5. Experimental results of LCC, PCCF, RPCCF, and GPR.

To give insight in the proposed color correction frameworks, we present the results for each internal method. Table 1a shows the internal results for the PCCF, where $\hat{\rho}_{3,1}$ is the best performing internal methods in terms of EM. The RPCCF results are more polarized, where $\bar{\rho}_2$ outperforms the rivaling internal methods by a large margin.

222   M. Amani et al.

Table 1. All color correction evaluation method results for all internal PCCF methods (a) and all internal RPCCF methods (b). The best performing (i.e. lowest ΔE^*_{ab}) internal methods are highlighted in bold.

method	$\hat{\rho}_{11}$	$\hat{\rho}_{21}$	$\hat{\rho}_{22}$	$\hat{\rho}_{31}$	$\hat{\rho}_{32}$	$\hat{\rho}_{33}$	$\hat{\rho}_{34}$	$\hat{\rho}_{35}$
size	4x3	7x3	7x3	11x3	11x3	14x3	17x3	20x3
EM	4.84	**3.63**	**3.69**	**3.53**	**3.77**	4.05	5.04	6.64

method	$\bar{\rho}_2$	$\bar{\rho}_3$	$\bar{\rho}_4$
size	4x3	7x3	7x3
EM	**4.74**	6.04	13.73

(a) EM results for all internal PCCF methods.   (b) Internal RPCCF methods.

Four visualizations of color corrected SkinCheckers using LCC, PCCF, RPCCF, and GPR respectively, are illustrated in Fig. 6. The color corrected image is randomly selected from the dataset.

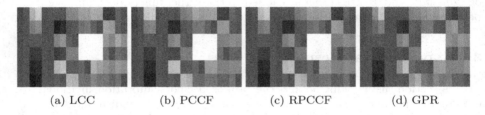

    (a) LCC   (b) PCCF   (c) RPCCF   (d) GPR

Fig. 6. Color correction models are trained on an imaged SkinChecker and evaluated on the same image. Each color patch, in the reconstructed image, is divided into *two parts*: **left** shows the color corrected color, **right** shows the ground truth target color. All images are recreated from CIE L*a*b* coordinates to RGB with CIE D65 illuminant.

4   Conclusion

A reliable color calibration is necessary to avoid using professional image acquisition tools in every clinic or laboratory. To our best knowledge, there is currently no comprehensive study on color calibration methods applied to human skin images, particularly when using amateur cameras. In this work, we made a comprehensive study and we also proposed a novel approach for color calibration, Gaussian process regression (GPR), a machine learning model that adapts to environmental variables. The results indicate that our extended version of Polynomial Color Correction (PCCF) and GPR are viable solutions when color correcting skin, while GPR also creates more general models. We conclude that our solution can be used in a variety of human skin analyses and is an affordable screening alternative to expensive professional image acquisition tools.

References

1. Dougherty, G.: Digital Image Processing for Medical Applications. Cambridge University Press, New Delhi (2009)

2. Kerker, M.: The scattering of light and other electromagnetic radiation: physical chemistry: a series of monographs, vol. 16. Academic press (2013)
3. Sharma, G.: Digital Color Imaging Handbook. CRC Press Inc, Boca Raton (2002). ISBN: 084930900X
4. Pascale, D.: A review of RGB color spaces... from xyy to r'g'b. Babel Color **18**, 136–152 (2003)
5. HunterLab. Equivalent white light sources and cie illuminants, Applications note (2005). https://web.archive.org/web/20050523033826/http://www.hunterlab.com:80/appnotes/an05_05.pdf
6. McCamy, C.S., Marcus, H., Davidson, J., et al.: A color-rendition chart. J. App. Photog. Eng **2**(3), 95–99 (1976)
7. Brito, C.C.: Spectral printing for monitoring jaundice in newborns. Internal article (2016)
8. D Company website. Spydercheckr 24 (2008). https://www.datacolor.com/wpcontent/uploads/2018/01/SpyderCheckr_Color_Data_V2.pdf. Accessed 14 May 2019
9. Lee, H.-C.: Introduction to Color Imaging Science. Cambridge University Press, New York (2005). https://doi.org/10.1017/CBO9780511614392
10. Finlayson, G.D., Mackiewicz, M., Hurlbert, A.: Color correction using root-polynomial regression. IEEE Trans. Image Process. **24**(5), 1460–1470 (2015). https://doi.org/10.1109/TIP.2015.2405336. ISSN: 1057-7149
11. Guowei, H., Ronnier, L.M., Rhodes, P.A.: A study of digital camera colorimetric characterization based on polynomial modeling. Color Res. Appl. **26**(1), 76–84 (2001) https://doi.org/10.1002/1520-6378(200102)26:1h76::AID-COL8i3.0.CO;2-3
12. Park, J., Park, K.: Professional colour communicator-the definitive colour selector. Color. Technol. **111**(3), 56–57 (1995)
13. Cheung, V., Westland, S., Connah, D., Ripamonti, C.: A comparative study of the characterisation of colour cameras by means of neural networks and polynomial transforms. Color. Technol. **120**(1), 19–25 (2004). https://doi.org/10.1111/j.1478-4408.2004.tb00201.x. ISSN: 1478-4408
14. Williams, C.K.I.: Prediction with Gaussian processes: from linear regression to linear prediction and beyond. In: Jordan, M.I. (ed.) Learning and Inference in Graphical Models. NATO ASI Series (Series D: Behavioural and Social Sciences), vol. 89, pp. 599–621. Kluwer, Dordrecht (1997). https://doi.org/10.1007/978-94-011-5014-9_23
15. MacKay, D.J.: Introduction to Gaussian processes. NATO ASI Series FComputer and Systems Sciences 168, 133–166 (1998)
16. Rasmussen, C.E., Williams, C.K.: Gaussian Process for Machine Learning. MIT press, Cambridge (2006). ISBN: 978-0-262-18253-9

Hybrid Geometric Similarity and Local Consistency Measure for GPR Hyperbola Detection

Evangelos Skartados$^{(\boxtimes)}$, Ioannis Kostavelis, Dimitrios Giakoumis, and Dimitrios Tzovaras

Centre for Research and Technology Hellas, ITI,
6th Km Charilaou-Thermi Road, 57001 Thessaloniki, Greece
{eskartad,gkostave,dgiakoum,dimitrios.tzovaras}@iti.gr

Abstract. The recent development of novel powerful sensor topologies, namely Ground Penetrating Radar (GPR) antennas, gave a thrust to the modeling of underground environment. An important step towards underground modelling is the detection of the typical hyperbola patterns on 2D images (B-scans), formulated due to the reflections of underground utilities. This work introduces a soil-agnostic approach for hyperbola detection, starting from one dimensional GPR signals, viz. A-scans, to perform a segmentation of each trace into candidate reflection pulses. Feature vector representations are calculated for segmented pulses through multilevel DWT decomposition. A theoretical geometric model of the corresponding hyperbola pattern is generated on the image plane for all point coordinates of the area under inspection. For each theoretical model, measured pulses that best support it are extracted and are utilized to validate it with a novel hybrid measure. The novel measure simultaneously controls the geometric plausibility of the examined hyperbola model and the consistency of the pulses contributing to this model across all the examined A-scan traces. Implementation details are discussed and experimental evaluation is exhibited on real GPR data.

Keywords: Hyperbola detection · GPR imaging · Soil agnostic detection

1 Introduction

Recently, significant advancements in underground sensing through GPR antennas have been introduced by the respective scientific community that enabled modelling of large subsurface areas [5,10]. The latter, brought a new era in the nondestructive inspection techniques that significantly reduced the construction costs [4]. However, the automatic analysis of GPR data for the detection of subsurface utilities remains an active research topic. This is due to the fact, that even if the synchronous arrays of GPR antennas can produce massive data of the underground environment, most of their annotation is still performed manually or at least requires human verification after the application of a automatic

© Springer Nature Switzerland AG 2019
D. Tzovaras et al. (Eds.): ICVS 2019, LNCS 11754, pp. 224–233, 2019.
https://doi.org/10.1007/978-3-030-34995-0_21

utilities detection method [9]. The most typical processing of GPR data for the detection of underground utilities relies on the image processing techniques applied on B-scans, which are also known as radargrams and are 2D images formulated by gathering A-scan measurements along the GPR antenna motion, for the detection and isolation of reflection signatures [15,16]. In particular, due to specific geometry of scanning sessions, consistent utilities are expected to shape hyperbolic patterns on B-scans. Consequently, utility detection has been mostly tackled as a hyperbola detection problem. In the past, several computer vision techniques, that utilized hand-crafted features, have been utilized to this end and specifically Histograms of Oriented Gradients have been proven robust descriptors of the orientation of recorded reflections [16,18]. Fitting hyperbolic shape has also been treated as an optimization problem, approached by Genetic Algorithms [13] and Neural Networks [1]. Recent works indicate successful application of Sparse Representation to gas pipe localization [17], while state of the art learning techniques such as traditional Convolutional Neural Networks [8] and sophisticated variations optimized towards balancing computational load and accuracy (Faster Region Convolutional Neural Network) [6] have been reported to perform well on detection and classification tasks respectively. Moreover, the work presented in [7] illustrates how a hyperbola detection module can be integrated into a complete underground modelling system. In the current method, a hybrid measure that captures (i) geometric similarity of recorded pulses to a theoretic model and (ii) consistency of those pulses has been developed. At the core of the approach presented herein is the Discrete Wavelet Transform (DWT) decomposition. The DWT is a powerful tool applied in signal processing domains, suitable for the frequency analysis of non-stationary signals, such as GPR signals. Successful application of DWT for filtering and denoising purposes of GPR data has been reported in [3,11], while the authors of [14] applied wavelet analysis in order to perform GPR data classification. In current work DWT decomposition is utilized to extract a feature representation of reflection pulses.

2 Method Description

2.1 Peak Detection and Segmentation

The proposed method commences with the isolation of the most prominent regions on the B-scan that could correspond to a hyperbolic apex. This first processing step is applied in a column-wise manner operating thus on the A-scans, i.e. the separate columns of a B-scan. Each such column corresponds to the time-domain measurements captured by the GPR device at a specific location. A stimulating pulse is transmitted by the GPR antenna and reflections of the pulse originating from underground obstacles travel back to the GPR receiver. Hence, recorded signal is a composition of time-shifted (due to the travel distance) and corrupted (due to noisy nature of underground signal propagation) instances of the initial pulse. This pulse usually closely resembles to a

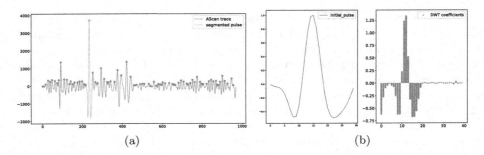

Fig. 1. (a) Highlighted segmented pulse defined by one of the peaks on an A-scan (b) Normalized extracted pulse and the DWT coefficients that constitute its feature representation.

Ricker wavelet, which is also mentioned as Mexican hat wavelet, a name indicative of its waveform, that is centered around a positive peak value. Reflection pulses of an A-scan are expected to have similar waveforms. Therefore, positive peak values along the recorded A-scan are indicators of the existence of waveforms caused by underground reflections of the original pulse. Although the center point is as good as any other point of the Ricker wavelet to characterize the whole waveform, positive peak values are easy to be defined and detected along a signal trace. Peak values of the recorded $y[t]$ trace are computed by transforming the $y[t]$ recorded signal into the frequency domain. Fast Fourier Transform is utilized and $j \cdot \omega \cdot Y[\omega]$ and $-\omega^2 \cdot Y[\omega]$ formulas are applied to calculate first and second derivatives. Finally, first and second derivative checks and a sign check are applied to locate local maxima positive values. It should be noted that most of these values correspond to artifacts of the scanning procedure, underground clutter, etc. and only few of them reflect the existence of actual underground utilities. However, a significant reduction in the search space of candidate hyperbola apexes is achieved by isolating these peak values on each A-scan. In Fig. 1(a) an example of a GPR trace along with the detected positive peaks is illustrated. The initial trace has 970 samples and 83 peaks are detected along it. The set of peaks is more than 10 times smaller then the set of all the time samples. The rest of the pipeline benefits from this reduction, since all the rest steps are applied on the former set. For each peak a pulse of a predefined length, centered around it, is extracted. The length of the extracted pulses is defined in accordance with the working frequency F_o of the GPR scanning device. Frequencies of reflections are expected to fall symmetrically within a band centered around the central frequency and a rule of thumb is to consider the width of the band equal to $1.5 \cdot F_o$. Consequently, the interval of relevant measured frequencies is the following: $[0.25 \cdot F_o, 1.75 \cdot F_o]$. The low frequency of the band defines the largest expected duration of a single reflection pulse in time. This duration in combination with the sampling frequency F_s of the scanning GPR device determines the number of time samples that need to be considered in order to successfully extract a complete waveform of a reflection pulse from the rest of the GPR trace. In our case the original trace is segmented into pulses of length equal to 31 time samples that are centered around each detected peak. An example of an extracted pulse is illustrated in Fig. 1(a).

2.2   Discrete Wavelet Transform Decomposition

For each extracted pulse a feature representation is calculated by utilizing the multilevel Discrete Wavelet Transform (DWT) decomposition of a $1D$ signal. In one level of the DWT the entry signal $y[n]$ (A-scan) is convolved with a high pass filter $h[n]$ and a low pass filter $g[n]$ and is in both cases subsequently down-sampled by two, producing the detail and the approximation coefficients respectively. On the multilevel DWT, approximation coefficients are provided to the next level and the same procedure is repeated. The maximum number of levels the above procedure can be repeated is determined by the length of the original signal and the number of coefficients of the applied filters. From one level to the next one, due to the filtering and the subsampling, the frequency band is reduced to half and at the same time the frequency resolution is doubled. The high-pass filters $h[n]$ comprise the child wavelets of the mother wavelet function and are computed from:

$$\psi_{j,k}[t] = \frac{1}{\sqrt{2^j}}\psi[\frac{t - k2^j}{2^j}] \tag{1}$$

where j is the scale parameter accounting for the level of resolution and k is the shift parameter. Low-pass $g[n]$ filters are produced by a function which is referred as a scaling function. The wavelet family that has been chosen here is Daubechies-6 due to the resemblance of the mother wavelet function to the emitted GPR pulse [2]. Once the decomposition reaches the maximum level, the detail coefficients of all levels along with the approximation coefficients of the maximum level constitute a sparse representation of the signal, which is perfectly invertible and captures information both for frequency and time domain. A feature vector that comprises all detail coefficients and approximation coefficients of the maximum level is formulated for each isolated pulse centered around a peak. In Fig. 1(b) the DWT decomposition of the extracted pulse of Fig. 1(a) is demonstrated.

2.3   Generation of Hyperbolic Templates

Having computed the features for all examined pulses centered around the extracted peak values from all A-scans, the rest of the processing is applied on the 2D space of the B-scan image. The Y axis of a B-scan image represents the time axis of the GPR traces and temporal sampling frequency defines the dy step of this axis (difference in time between two consecutive recorded values). The X axis represents the distance covered by the GPR device along its scanning trajectory and spatial sampling frequency defines the dx step of this axis (distance between two consecutive recorded traces). According to the point reflector

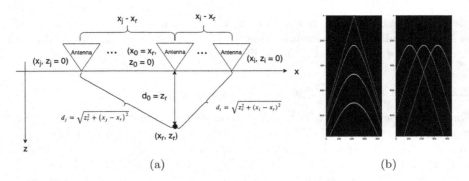

Fig. 2. (a) Travel distances schematic of the GPR signal on consecutive scanning positions. (b) Left: hyperbola templates corresponding to utilities at different depths on the same position along the scanning trajectory, Right: hyperbola templates corresponding to utilities at the same depth on three different positions along the scanning trajectory.

concept [12], the pattern formed on a B-scan image from reflections originating from an underground utility, that lies within the range of the scanning session, will be a hyperbola signature. The point reflector concept introduces a number of assumptions and simplifications which lead to a specific statement: for a single GPR measurement and an isolated point reflector in space, the recorded trace will be the initially transmitted pulse shifted in time according to the time of the two-way travel of the signal from the GPR device to the target and back. With reference to Fig. 2(a), for a reflector located in point $p_r = (x_r, z_r)$ the shortest travel distance d_i will occur when GPR antenna is directly above it at $p_0 = (x_r, 0)$ (for all points first coordinate refers to the scanning direction axis and second coordinate refers to depth). For this scanning position, the time shift of the transmitted pulse is calculated as $t_0 = (2 \cdot d_0)/v = (2 \cdot z_r)/v$ where the 2 factor accounts for the back and forth travel and v is the wave propagation velocity in the examined space. For any other scanning position $p_i = (x_i, 0)$ travel distance d_i is calculated as $d_i = \sqrt{z_r^2 + (x_i - x_r)^2}$. Solving for the corresponding time delay, the following equation is derived:

$$t_i = \frac{2}{v} \cdot \sqrt{z_r^2 + (x_i - x_r)^2} \tag{2}$$

X axis of B-scans images and the scanning trajectory, as depicted on Fig. 2(a), coincide while Z axis of the physical world representation can be transformed to the Y axis of the B-scan image by applying Eq. 2. Unlike a learning approach, in the presented method, hyperbola templates, that will act as models of the class of reflection signatures originating from underground utilities, are not extracted from a set of positive samples but are rather generated based on theoretical Eq. 2. Taking a more careful look on this equation, two statements can be derived: (i) due to the z_r term for a given propagation velocity the shape of the generated template depends on the depth of the candidate utility and (ii) due to the $x_i - x_r$ term, that introduces the notion of relative distance from the x_r coordinate of the candidate utility, the slope of a template for a given depth does not change. Hence, for any given depth z_r coordinate, hyperbola template needs to be computed only once and then it must be simply appropriately translated for each considered x_r coordinate, in order to be centered around it These statements are graphically represented in Fig. 2(b) and lead to an efficient implementation of the proposed template generation. Instead of calculating the corresponding hyperbola template for each candidate position on the 2D image space it suffices to compute a dictionary of all templates for all discrete depths that fall within the device range and simply retrieve and transform the computed template around each considered position.

2.4   Hybrid Measure for Hyperbola Hypotheses

The next step of our method comprises the application of the hyperbola theoretical model on the real GPR measurements. Specifically, each peak detected in Sect. 2.1 constitutes a candidate hyperbola apex and the pre-computed hyperbola template that corresponds to the apex's respective coordinates in B-scan is retrieved from the dictionary. It should be noted that the theoretical model does not take into account any attenuation phenomena or the directionality of the antenna and therefore the hyperbola signature is expanding indefinitely around the apex. Since this is not the case in real data, it is necessary to define a minimum number of GPR traces (B-scan columns) around the candidate apex that the hyperbola signature is expected to be visible. A reasonable number for the following analysis is $N = 60$ traces. For each of those traces the peak with the measured time of occurrence closest to the theoretically computed time of the hyperbola model is selected. Application of the theoretical hyperbola model on real measurements is illustrated in Fig. 3. Then, the two following scores are introduced:

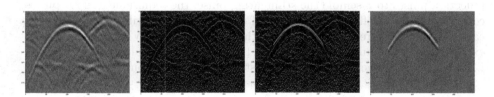

Fig. 3. From left to right: initial B-scan Image, per column detected peaks, hyperbola template computed for a considered apex marked with red color and closest peaks per column marked with green, extracted pulses corresponding to the closest peaks. (Color figure online)

Geometric Similarity Score. The geometric similarity score is computed as the sum of absolute differences between estimated time of occurrence for each trace and the actual detected time of the peak closest to the estimated model among all peaks for the considered trace:

$$score_{similarity} = \sum_{x_i=x_0-N}^{x_0+N} |t_i(x_i, x_0, y_0) - t_{peak_i^*}|$$

$$\text{where } peak_i^* = \underset{peaks[x_i]}{\arg\min} |t_i(x_i, x_0, y_0) - t_{peaks[x_i]}|$$

(3)

The first measure captures only the geometric resemblance of 2D patterns shaped by peaks of actual measurements to automatically generated hyperbola models. However, it fails to discriminate between "positive" peaks corresponding to pulses actually originating from underground targets and "negative" peaks corresponding to pulses originating from underground clutter, artifacts of the scanning procedure, echoes, etc. It can be therefore argued that, if only the geometry measure is considered, patterns on which pulses that are not reflections of a certain underground utility contribute, can be mistakenly classified as reflection signatures.

Pulse Consistency Score. In [17] a strict condition is imposed, demanding the waveforms of reflection pulses contributing to hyperbola atoms to directly resemble the stimulating wavelet initially transmitted by the GPR. This approach does not take into consideration the change of the characteristics (e.g. frequency) of the wavefield as it propagates in any medium. This change differs according to the electromagnetic characteristics of the medium of propagation. Besides, this approach does not account for any of the distortions that the signal might undergo while traveling and impinging on underground targets. Here, a much looser criterion is introduced that can be applied unchanged for signals travelling on various types of soil, reflecting on various types of underground targets and undergoing a reasonable amount of distortion. Instead of asking for reflection pulses to resemble the stimulating pulse, the requirement is for pulses contributing on a reflection signature of an underground utility to resemble to each other. Therefore, we make use of the features (i.e. DWT representations) extracted for each pulse around the selected peaks and we calculate the squared L2 norm between each other. The consistency score based on the calculation of squared L2 norm of the samples is expressed as follows:

$$score_{consistency} = \sum_{x_i=x_0-N}^{x_0+N} \sum_{x_j=x_0-N}^{x_0+N} \left\| dwt(pulse_i^*) - dwt(pulse_j^*) \right\|_2^2$$

$$\text{where } pulse_i^* = \underset{peaks[x_i]}{\arg\min} |t_i(x_i, x_0, y_0) - t_{peaks[x_i]}|$$

(4)

$$\text{and } pulse_j^* = \underset{peaks[x_j]}{\arg\min} |t_j(x_j, x_0, y_0) - t_{peaks[x_j]}|$$

Naive calculation of the consistency score, as presented in Eq. 4, involves two summations among all the pulses participating in the hyperbola signature under consideration. However, squared L2 norm involved in the above calculations can be expressed as a dot product: $\|x - y\|_2^2 = (x - y) \cdot (x - y)^T$. By expanding the right part of this equation, $(x - y) \cdot (x - y)^T = \|x\|_2^2 + \|y\|_2^2 - 2 \cdot x \cdot y^T$ it is derived that the calculation of the squared L2 norm of a vector subtraction can be expressed as two squared L2 norm calculations and one dot product calculation. This deduction, expanded for multiple vector subtractions, can benefit efficient calculation of score consistency. DWT coefficients of all pulses are gathered in a matrix $\mathbf{X} \in \mathbb{R}^{m \times n}$, where m is the number of pulses participating in the considered signature and n is the number of the coefficients in the DWT representation. In this format each row of the matrix is the vector representation of a pulse, $\mathbf{X}[i,:] = dwt(pulse_i)$, and the matrix multiplication $\mathbf{X} \cdot \mathbf{X}^T$ corresponds to calculation of the dot product of all possible combinations between participating pulses. Besides, per row summation of the squared matrix $\mathbf{X}^2$ equals to the squared norm of all the vector representations. Consequently, the two for loops of Eq. 4 can be fully vectorized and implemented as one matrix multiplication and one broadcast sum.

Final Score. The two introduced scores of the hybrid measure need to be combined in a single score that will be eventually utilized to discriminate between positive and negative hyperbola hypotheses. Two empirically defined max thresholds are utilized for the two scores. If a considered hypothesis exceeds any of them, then it is discarded. Remaining scores are normalized with respect to the max thresholds and a final score is computed by the following equation:

$$score_{final} = \sqrt{\left(\frac{score_{similarity}}{max_{similarity}}\right)^2 + \left(\frac{score_{consistency}}{max_{consistency}}\right)^2} \qquad (5)$$

A non max suppression step is finally applied to provide final hyperbola detection. All the overlapping hypotheses are sorted according to their final score and the most prevailing ones are selected.

3   Evaluation

The proposed method has been evaluated on data recorded from a dedicated IDS field site. An example of an input B-scan image along with the acquired results is provided in Fig. 4. As it is illustrated in that figure, the proposed method successfully detects 11 out of 12 reflection signatures with only 2 false positive detections. In the two last images of the same figure, intermediate results of signatures surviving the geometric similarity and the pulse consistency checks are provided. In both images a much bigger rate of false positive detections is examined. This is a result either of peaks of inconsistent pulses that resemble a hyperbola, or of consistent pulses that formulate patterns different from the hyperbola model in shape. As a conclusion, proposed method combines two loose criteria to provide a rather robust detector.

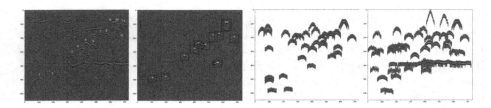

Fig. 4. From left to right: input B-scan image with 12 annotated hyperbola signatures originating from underground utilities, final detections provided by the complete detection pipeline, hypotheses that satisfy only the max similarity score criterion, hypotheses that satisfy only the max consistency score criterion.

4   Conclusions

In this work, a novel hyperbola detection method has been developed based on the fusion of two scores in a single hybrid measure, that on the one hand controls the geometric characteristics of the recorded patterns and, on the other hand demands the consistency of the pulses that participate in the formulation of these patterns. The introduced method does not depend on any other parameters related to the type of soil on which the underground utilities are buried, yet for the creation of the dictionary of the hyperbolic theoretical models, the specifications of the utilized GPR antenna are considered. Proposed implementation makes use of vectorization to diminish computational burden. Overall the proposed method exhibited promising performance when examined on real data, an evaluation that will be extended with various GPR devices and types of soils in our future work.

Acknowledgment. The work funded by the EU-H2020 funded project "BADGER (RoBot for Autonomous unDerGround trenchless opERations, mapping and navigation)" with GA no: 731968.

References

1. Al-Nuaimy, W., Huang, Y., Nakhkash, M., Fang, M., Nguyen, V., Eriksen, A.: Automatic detection of buried utilities and solid objects with GPr using neural networks and pattern recognition. J. Appl. Geophys. **43**(2–4), 157–165 (2000)
2. Baili, J., Lahouar, S., Hergli, M., Al-Qadi, I.L., Besbes, K.: GPR signal de-noising by discrete wavelet transform. Ndt E Int. **42**(8), 696–703 (2009)
3. Baili, J., Lahouar, S., Hergli, M., Amimi, A., Besbes, K.: Application of the discrete wavelet transform to denoise GPR signals. In: International Symposium on Communications, Control and Signal Processing, Marrakech, Morocco, p. 11 (2006)
4. Capineri, L., Grande, P., Temple, J.: Advanced image-processing technique for real-time interpretation of ground-penetrating radar images. Int. J. Imaging Syst. Technol. **9**(1), 51–59 (1998)
5. Daniels, D., Gunton, D., Scott, H.: Introduction to subsurface radar. In: IEEE Proceedings Communications, Radar and Signal Processing, vol. 135, pp. 278–320. IET (1988)

6. Kafedziski, V., Pecov, S., Tanevski, D.: Detection and classification of land mines from ground penetrating radar data using faster R-CNN. In: 26th Telecommunications Forum (TELFOR), pp. 1–4. IEEE (2018)
7. Kouros, G., et al.: 3D underground mapping with a mobile robot and a GPR antenna. In: 2018 IEEE/RSJ International Conference on Intelligent Robots and Systems (IROS), pp. 3218–3224. IEEE (2018)
8. Lameri, S., Lombardi, F., Bestagini, P., Lualdi, M., Tubaro, S.: Landmine detection from GPR data using convolutional neural networks. In: 25th European Signal Processing Conference (EUSIPCO), pp. 508–512. IEEE (2017)
9. Neal, A.: Ground-penetrating radar and its use in sedimentology: principles, problems and progress. Earth Sci. Rev. **66**(3–4), 261–330 (2004)
10. Nguyen, C.: Subsurface sensing technologies and applications II. In: Subsurface Sensing Technologies and Applications II, vol. 4129 (2000)
11. Ni, S.H., Huang, Y.H., Lo, K.F., Lin, D.C.: Buried pipe detection by ground penetrating radar using the discrete wavelet transform. Comput. Geotech. **37**(4), 440–448 (2010)
12. Özdemir, C., Demirci, Ş., Yiğit, E., Yilmaz, B.: A review on migration methods in B-scan ground penetrating radar imaging. Math. Probl. Eng. **2014** (2014)
13. Pasolli, E., Melgani, F., Donelli, M.: Automatic analysis of GPR images: a pattern-recognition approach. IEEE Trans. Geosci. Remote Sens. **47**(7), 2206–2217 (2009)
14. Shao, W., Bouzerdoum, A., Phung, S.L.: Sparse representation of GPR traces with application to signal classification. IEEE Trans. Geosci. Remote Sens. **51**(7), 3922–3930 (2013)
15. Simi, A., Manacorda, G., Miniati, M., Bracciali, S., Buonaccorsi, A.: Underground asset mapping with dual-frequency dual-polarized GPR massive array. In: XIII Internarional Conference on Ground Penetrating Radar, pp. 1–5 (2010)
16. Skartados, E., et al.: Ground penetrating radar image processing towards underground utilities detection for robotic applications. In: 2018 International Conference on Control, Artificial Intelligence, Robotics & Optimization, pp. 27–31. IEEE (2019)
17. Terrasse, G., Nicolas, J.M., Trouvé, E., Drouet, É.: Automatic localization of gas pipes from GPR imagery. In: 24th European Signal Processing Conference, pp. 2395–2399. IEEE (2016)
18. Torrione, P.A., Morton, K.D., Sakaguchi, R., Collins, L.M.: Histograms of oriented gradients for landmine detection in ground-penetrating radar data. IEEE Trans. Geosci. Remote Sens. **52**(3), 1539–1550 (2013)

Towards a Professional Gesture Recognition with RGB-D from Smartphone

Pablo Vicente Moñivar[(✉)], Sotiris Manitsaris, and Alina Glushkova

Center for Robotics, MINES ParisTech, PSL Research University, Paris, France
{pablo.vicente_monivar,sotiris.manitsaris,
alina.glushkova}@mines-paristech.fr

Abstract. The goal of this work is to build the basis for a smartphone application that provides functionalities for recording human motion data, train machine learning algorithms and recognize professional gestures. First, we take advantage of the new mobile phone cameras, either infrared or stereoscopic, to record RGB-D data. Then, a bottom-up pose estimation algorithm based on Deep Learning extracts the 2D human skeleton and exports the 3rd dimension using the depth. Finally, we use a gesture recognition engine, which is based on K-means and Hidden Markov Models (HMMs). The performance of the machine learning algorithm has been tested with professional gestures using a silk-weaving and a TV-assembly datasets.

Keywords: Pose estimation · Depth map · Gesture recognition · Hidden Markov Models · Smartphone

1 Introduction

The role of professional actions, activities and gestures is of high importance in most industries. Motion sensing and machine learning have actively contributed to the capturing of gestures and the recognition of meaningful movement patterns by machines. Therefore, very interesting applications have emerged according to the industry. For example, in the factories of the future, the capabilities of workers will be augmented by machines that can continuously recognize their gestures and collaborate accordingly, whereas in the creative and cultural industries it is still a challenge to recognize and identify the motor skills of a given expert. Therefore, capturing the motion of workers or craftsmen using off-the-shelf devices, such as smartphones, has a great value. New smartphones are equipped with depth sensors and high power processors, which allow us to record data even without very sophisticated devices.

In this work, we aim to create a smartphone application that allows for recording gestures using RGB or RGB-D images by taking advantage from the new smartphone cameras, estimating human poses, training machine learning models by using only a few shots and recognizing meaningful patterns. The motivation of this work is to give the possibility to the users to easily record,

D. Tzovaras et al. (Eds.): ICVS 2019, LNCS 11754, pp. 234–244, 2019.
https://doi.org/10.1007/978-3-030-34995-0_22

annotate, train and recognize human, actions, activities and gestures in professional environments.

2   State of Art

2.1   Qualitative Comparison Between Pose Estimation Frameworks

Different benchmarks of 2D human pose estimation are evaluated through annual challenges that aim to improve various key-parameters, such as the accuracy, how the algorithm performs with partial occlusions or using different keypoints, how to detect the pose of a big number of individuals in the scene, etc. Some examples are the COCO Keypoint Challenge, MPII HumanPose and Posetrack.

The pose estimation in multi-person scenarios can be carried out by using a top-down or a bottom-up approach. In the first approach, a human detector is initiated and both the joints and the skeleton of each person are calculated separately. *AlphaPose* [1] is a top-down method based on regional multi-person pose estimation. Moreover, *DensePose* [2], aims to map each person and extract a 3D surface from it using Mask R-CNN [3].

On the other hand, bottom-up approaches firstly detect and label all the joints candidates in the frame and secondly associate them to each individual person without using any person detector. *DeepCut* [4] is an example of a bottom-up approach, it uses a convolutional neural network (CNN) to extract a set of joint candidates, and then it labels and associates the joints with Non-Maximum Suppression.

Finally, *OpenPose* [5] is a real-time bottom-up approach for detecting 2D pose of multiple people on an image. First, it takes an RGB image and applies a fine-tuned version of the CNN VGG-19 [6] to generate the input features of the algorithm. Second, these features enter a multi-stage CNN to predict the set of confidence maps, where each map represents a joint and the set of part affinities which represent the degree of association between joints. Lastly, bipartite matching is used to associate body part candidates and obtain the full 2D skeletons.

In our work, we have chosen to use OpenPose. The main reason is that top-down approaches suffer from an early commitment when the detector fails, and the computational power increases exponentially with the number of people in the scene. On the other hand, OpenPose includes a hand skeleton estimation, which has been considered an important perspective in our system. Table 1 shows a list of popular open source methods for 2D pose estimation and the classification obtained in their respective challenges.

Table 1. List of popular open-source frameworks for 2D pose estimation

Method	Benchmark	Precision (%)	Rank
OpenPose	COCO Keypoint challenge 2016	60.5	1
AlphaPose	COCO Keypoint challenge 2018	70.2	11
DensePose	Posetrack multi-person pose estimation 2017	61.2	7

2.2   Machine and Deep Learning Frameworks for Mobile Devices

A pose estimation embedded in a smartphone still represents a huge challenge for the scientific community. Neither Apple Store nor Google Play proposes any application that provides pose estimation. The main reason is the computational power needed for running deep learning algorithms, as well as the lack of powerful graphical devices in smartphones. [7] shows different tests when running deep neural networks on Android smartphones. The use of TensorFlow-Lite, Caffe-Android or Torch-Android frameworks is currently possible for implementing CNNs on smartphones.

Nevertheless, at the end of 2017, Apple made a transition in the world of machine learning by launching the CoreML framework for iOS 11 that enable the running of machine learning models on mobile devices. The performance improved in the next version, CoreML2, at the end of 2018.

Today, there are available applications that do eye tracking based on gaze estimation by using CNNs [8]. Nevertheless, pose estimation requires much more computation power than eye tracking or face recognition. For this reason, the use of an external framework has been considered a better solution within the context of this work.

2.3   Gesture Recognition Methods

The implementation of deep learning for gesture recognition has become a common practice and can lead to very good results. The ChaLearn LAP Large-scale Isolated Gesture Recognition Challenge from the ICCV 2017, crowned [9–11] as the best deep learning algorithms for gesture recognition. However, the need for large training databases is not compatible with the constraints of professional gestures where datasets are quite small.

Dynamic Time Warping (DTW) and Hidden Markov Models (HMMs) are machine learning methods that are widely used in pattern recognition. DTW is a template-based approach that is based on a temporal re-scaling of a reference motion signal and its comparison with the input motion signal, such as in [12]. DTW can be good for doing one-shot learning, while HMMs is a robust duration-independent model-based approach.

Thus, in this research, we chose to use the work, described in [13], which makes use of K-means to model the time series of motion data and HMMs for classifying and recognizing the gesture classes by using the Gesture Recognition Toolkit (GRT) [14].

3   Objectives

The general scope of this work is to build the basis of an application for a mobile device, which allows for data recording using its embedded sensors, estimate the human pose, extract the skeleton and recognize (offline) professional gestures. The pose estimation depends on the camera: RGB for 2D, and RGB-D, either infrared or stereoscopic, for 3D skeletons. The whole process can be controlled by the application. More precisely, the annotation, the joint selection, the skeleton visualization and the projection of the recognition results run locally on the phone while the pose estimation and machine learning algorithms are delegated to a GPU server.

4   Overall Pipeline

The smartphone handles the input and output steps of the pipeline. It records the RGB-D frames and shows the results (body skeleton and gesture recognition accuracy). An external motion detection server receives the frames by using WebSockets and, then, it estimates the skeleton and uses the information provided by the body joints to train and test a gesture recognition engine. The overall architecture is shown in Fig. 1.

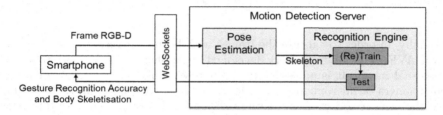

Fig. 1. Architecture of the overall pipeline

4.1   Video Recording Using the Smartphone

A specific module for depth recording has been developed in order to use any iOS device equipped with an RGB and/or RGB-D sensor to record data. More precisely, the new iPhone XS has been used in this work.

On the one hand, iPhone XS uses a dual rear camera, composed by a wide-angle lens and a telephoto lens, to capture the disparity. The normalized disparity is defined as the inverse of the depth. In Fig. 2 on the left is shown how the rear camera captures the disparity. First, the observed object is reflected on the image plane of each camera. Then, the mathematical relation tying the depth (Z), baseline distance (B), disparity (D) and focal length (F), showed in Eq. 1, results in the normalized disparity. Finally, the iPhone needs to filter and post-process the disparity to smooth the edges and fill the holes, which requires a heavy computation.

$$\frac{B}{Z} = \frac{D}{F} \rightarrow \frac{1}{Z} = \frac{D}{BF} \tag{1}$$

On the other hand, the front camera, named True Depth Camera by Apple, is used to measure the depth in meters directly. It has a dot projector that launches over 30,000 dots onto the scene, generally the user face, which is then captured by an infrared camera. To ensure that the system works properly in the dark, there is an ambient light sensor and a flood illuminator which adds more infrared light when needed. The final result is more stable depth images with a higher resolution. Figure 2 on the right shows the architecture of the True Depth Camera.

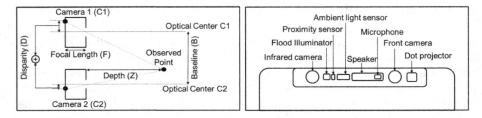

Fig. 2. IPhone XS back camera (Dual Rear Camera) capturing disparity (Left image) and IPhone XS front camera (True Depth Camera) capturing depth (Right image).

Finally, in order to capture the frames with both cameras, it is necessary to use the AVFoundation framework developed by Apple for working with temporal audiovisual media. It is necessary to create a session with at least one capture input (camera or microphone) and one capture output (photo, movie or audio). Then the measured distances (depth or disparity) need to be converted to pixel values in order to visualize depth maps.

4.2   Pose Estimation

Once the data is recorded, it is sent to the GPU server. For compression purposes, the RGB-D data is converted to jpeg format, then WebSockets (RFC 6455) is used to send it. We chose to use the client library Starscream.

The goal of pose estimation is to obtain a series of keypoints that can be used by a gesture recognition engine as input, enabling to train a model that adapts to different situations or environments. OpenPose, in our case, estimates 25 body keypoints, 2×21 hand keypoints and 70 face keypoints. However, some of the estimated keypoints are useless for the recognition, either they are occluded or they do not carry any information about the gesture. Therefore, a joint selection module has been added to the application to give the possibility to the user to select the most appropriate keypoints depending on the use-case.

4.3   Gesture Recognition

The joints obtained with the pose estimation and the data obtained from the depth camera are the input to the gesture classification algorithm. To make the recognition invariant to the position of each person in the frame, the neck joint has been taken as a reference point, and any frame without neck has been discarded.

The gesture recognition engine is based on supervised learning. Before making the gesture recognition, a labeled database can be manually created by the user, by manually selecting starting and ending time stamps of each gesture.

Once the database has been labeled, the gesture recognition engine uses k-Means to obtain discrete-valued observations. Then, Hidden Markov Models is used to train the discrete data and to determine a gesture recognition accuracy. The platform GRT has been used for the entire process.

4.4 Smartphone Application

The graphical user interface of the application consists of four main modules:

- Recording module: allows to record the sequences and visualize depth maps. An example of this module is shown on Fig. 3 right.
- Skeleton Visualization module: allows to visualize the skeleton estimated. Figure 3 left.
- Training module: allows to select the appropriate joints for the gesture recognition.
- Labelling module: interface similar to a photo gallery where you must select the gestures that you want to recognize.

Fig. 3. Example of the smartphone application using the skeleton visualization module and the rear camera (left image) and the recording module of the application enabling the depth visualization with the true depth camera (right image)

5 Datasets

The first dataset used, TV Assembly dataset (TVA), is made up of RGB-D sequences recorded from a top mounted view at a conveyor surface factory. Two different users have been recorded. Each sequence contains around 10.000 RGB

Table 2. Example of frames from the conveyor surface dataset

Gesture 1 (G1)	Gesture 2 (G2)	Gesture 3 (G3)	Gesture 4 (G4)
Take the card from the left side box	Take the wire from the right side box	Connect the wire with the card	Place the card on the TV chassis

frames together with the depth. Table 2 shows the four gestures identified and labeled during the sequences along with the skeleton of each of them.

The second dataset, Silk Weaving dataset (SW), contains sequences recorded from a lateral view and three different positions in a silk weaver museum. Three clear gestures have been identified and showed in Table 3 with their corresponding skeleton.

Table 3. Example of frames from the silk weaver museum dataset

Gesture 5 (G5)	Gesture 6 (G6)	Gesture 7 (G7)
Press the treadle and push the batten	Move the shuttle sideways	Leave the treadle and pull the batten

6 Results

In order to evaluate the performance of our pipeline and the algorithms, we use the 80%–20% evaluation criteria. We randomly divide our dataset in 80% training set and 20% as a testing set, repeating this procedure 10 times and computing the average values to generate the confusion matrix. We also use the Recall (Rc), Precision (Pc) and f-score metrics.

The TVA dataset contains 48 repetitions of each gesture and the SW dataset contains 88 repetitions. Moreover, for the gesture recognition engine, we selected 30 clusters for the K-Means algorithm and 4 states for the HMMs, which follow an ergodic topology. Finally, five different tests have been done in order to compare the gesture recognition accuracy using the following criteria: 2D against 3D, 2 joints against 7 joints, different camera positions and mixing gestures from the two datasets.

6.1 Pose Estimation Comparison Using the TVA and SW Datasets

A comparison of the pose estimation using the two datasets has been made. We compared the number of frames per gestures with 1. the percentage of frames without any estimated skeleton, thus without estimation at all; 2. the percentage of frames without any reference point, thus without the neck and 3. the percentage of frames having the minimum useful estimation, thus at least the neck.

The results are shown on Table 4 and we can affirm that the lateral views provided by the SW dataset have better potential, than the TVA dataset, since a full skeleton has been estimated for all the frames. In the top mounted view of the TVA dataset, the algorithm struggled to estimate any information in 43% of frames for G2 and in 36% of frames for G4, because the user is not captured in many frames. With regard to the TVA dataset, we would expect that these results might have an impact on gesture recognition accuracy. We would also expect that the small duration of the G5 might also affect its recognition accuracy.

Table 4. OpenPose results on the TVA and SW datasets

Dataset	TV Assembly				Silk Weaving		
Gesture	G1	G2	G3	G4	G5	G6	G7
# Frames per sample	80.62	67.89	88.69	164.27	30.85	41.81	66.16
% Frames without any skeleton estimated	0.88	12.27	1.31	14.39	0	0	0
% Frames without reference point (without neck)	2.71	31.05	1.41	22.83	0	0	0
% Frames with at least the minimum useful estimation (at least the neck)	**96.41**	**56.67**	**97.28**	**62.78**	**100**	**100**	**100**

6.2 Gesture Recognition Comparisons Using the TVA Dataset: 2D vs 3D and 2 vs 7 Joints

The joints selected for training the gesture recognition engine with the TVA dataset are the upper-body joints. Table 5 compares the recognition accuracy by using 2 (wrists) or 7 (wrists, elbow, shoulders and head) joints in the 2D or 3D space. The highest results are obtained by using 7 joints in 3D, while the worst with 2 joints in 2D.

Moreover, as we expected, the low number of frames with the minimal useful pose estimation for G2 and G4 impacted the recognition accuracy for these gestures, while G1 and G3, which have data that give good pose estimation, achieved very high accuracy.

Additionally, on one hand, we notice that, while what we gain with the 3D is not so important compared with the 2D, the potential error in the accuracy (standard deviation) decreases for approximately 40% with the 3D. On the other hand, if we use 7 joints instead of 2, we increase the accuracy for more than 10%, meaning that not mostly the hands are involved in effective gestures. In any case, the way the 3rd dimension is extracted is biased by the fact that OpenPose is already pre-trained using only the RGB, meaning that a complete re-training of the OpenPose with the depth might give better results. Finally, the number of joints that give better accuracy really depends on the nature of gestures.

6.3 Gesture Recognition Comparison Using Three Different Camera Positions from the SW Dataset

The accuracy in Table 6 is 100%, meaning that the recognizer works perfectly for the gestures of the SW dataset. We think that the difference between the accuracy of the two datasets is mostly due two main reasons: 1. the top mounted view used in the TVA dataset and 2. the fact that in a number of frames the user is not captured in the TVA dataset, thus there is no any pose estimation. In addition, the three gestures made in the SW dataset have a greater variance in space than in the TVA dataset.

6.4 Comparison Mixing Gestures and Data from the TVA and the SW Datasets

The last experiment we tried is mixing gestures and data from the two datasets. In Table 7, we calculate the precision, recall and f-score for each gesture when we train the gesture recognition engine with samples from both datasets. As a general conclusion, we notice that there is a decrease in accuracy for every

Table 5. Comparison in the gesture recognition using different joints and dimension and TV Assembly dataset

2J-2D	G1	G2	G3	G4	Pr(%)	2J-3D	G1	G2	G3	G4	Pr(%)
HMM1	81	1	3	6	89.0	HMM1	80	1	8	3	87.0
HMM2	0	70	1	29	70.0	HMM2	0	73	1	15	82.0
HMM3	17	12	53	5	60.9	HMM3	1	28	0	85	66.3
HMM4	2	14	0	96	85.7	HMM4	1	28	0	85	74.6
Rc(%)	81.0	72.2	92.3	70.6	76.9 ± 7.9	Rc(%)	84.2	65.0	87.5	77.3	77.2 ± 4.7
7J-2D	G1	G2	G3	G4	Pr(%)	7J-3D	G1	G2	G3	G4	Pr(%)
HMM1	85	1	4	0	94.4	HMM1	98	1	1	0	98.0
HMM2	0	84	0	11	88.4	HMM2	0	82	1	5	93.2
HMM3	4	16	71	8	71.7	HMM3	4	7	68	7	79.1
HMM4	0	6	0	100	94.3	HMM4	6	114	0	94	82.5
Rc(%)	95.5	78.5	94.7	84.0	87.2 ± 6.0	Rc(%)	90.7	78.8	97.1	88.7	88.1 ± 3.4

Table 6. Gesture recognition using all joints and 2 dimensions on Silk Weaving dataset

2J-2D	*G4*	*G5*	*G7*	**Pr(%)**
HMM1	183	0	0	100
HMM2	0	166	0	100
HMM3	0	0	181	100
Rc (%)	100	100	100	100 ± 0

gesture. Nevertheless, this decrease is not important given the fact that we have many users, thus a high variance in the way the gestures are executed, and different camera positions.

Table 7. Gesture recognition mixing gestures from SW and TVA datasets and using 7 joints and 2D

	G1	*G2*	*G3*	*G4*	*G5*	*G6*	*G7*
Pr (%)	81.9	56.7	80.9	62.8	98.3	98.8	99.5
Rc (%)	84.6	77.5	68.5	69.9	96.6	97.6	95.4
f-score (%)	**83.2**	**65.5**	**74.2**	**66.2**	**97.4**	**98.2**	**97.4**

7   Conclusions and Future Work

In this work, we developed the first version of a smartphone application that allows users to record human motion using the RGB or the RGB-D sensors, annotate them, estimate the pose and recognize them. The use of a smartphone in an industrial or professional context is much easier then the use of highly intrusive body tracking systems. The long term goal of this application is to permit industrial actors to record their own professional gestures, to annotate them and to use a user-friendly system for their recognition. We developed a module that extracts the 3rd dimension from a depth. We concluded that the 3rd dimension improves the recognition stability, decreasing by 40% the standard deviation in the accuracy. In addition to this, we observed that while the hand is involved in most cases, using 7 instead of 2 joints can give better recognition results, especially when the camera is top-mounted.

Our future work will be focused not only on improving the application, thus improving the interface, but also on the further development of professional gestures dataset. With a large dataset, we will be able to consider also the use of Deep Learning. Finally, we also plan to extend our pose estimation and gesture recognition system towards the direction of using finger motion as well.

Acknowledgement. The research leading to these results has received funding by the EU Horizon 2020 Research and Innovation Programme under grant agreement No. 820767, CoLLaboratE project, and No. 822336, Mingei project. We acknowledge also the Arçelik factory and the Museum Haus der Seidenkultur for providing as with the use-cases as well as the Foundation for Research and Technology – Hellas for contributing to the motion capturing.

References

1. Fang, H., Xie, S., Lu, C.: RMPE: regional multi-person pose estimation (2016)
2. Güler, R.A., Neverova, N., Kokkinos, I.: DensePose: dense human pose estimation in the wild (2018)
3. Abdulla, W.: Mask R-CNN for object detection and instance segmentation on Keras and TensorFlow. GitHub repository (2017)
4. Pishchulin, L., et al.: DeepCut: joint subset partition and labeling for multi person pose estimation (2015)
5. Cao, Z., Simon, T., Wei, S.E., Sheikh, Y.: Openpose: realtime multi-person 2D pose estimation using part affinity fields (2018)
6. Simonyan, K., Zisserman, A.: Very deep convolutional networks for large-scale image recognition. CoRR, abs/1409.1556 (2014)
7. Ignatov, A., et al.: AI benchmark: running deep neural networks on android smartphones (2018)
8. Krafka, K.: Eye tracking for everyone (2016)
9. Zhang, L., Zhu, G., Shen, P., Song, J.: Learning spatiotemporal features using 3DCNN and convolutional LSTM for gesture recognition. In: ICCV Workshop (2017)
10. Wang, H., Wang, P., Song, Z., Li, W.: Large-scale multimodal gesture recognition using heterogeneous networks. In: ICCV 2017 Workshop, pp. 3129–3137 (2017)
11. Wang, P., Li, W., Liu, S., Gao, Z., Tang, C., Ogunbona, P.: Large-scale isolated gesture recognition using convolutional neural networks (2017)
12. Corradini, A.: Dynamic time warping for off-line recognition of a small gesture vocabulary, pp. 82-89 (2001)
13. Coupeté, E., Moutarde, F., Manitsaris, S.: Multi-users online recognition of technical gestures for natural human-robot collaboration in manufacturing. Auton. Robot. **43**, 1309–1325 (2018)
14. Gillian, N., Paradiso, J.A.: The gesture recognition toolkit. J. Mach. Learn. Res. **15**, 3483–3487 (2014)

Data Anonymization for Data Protection on Publicly Recorded Data

David Münch$^{(\boxtimes)}$ ⓘ, Ann-Kristin Grosselfinger ⓘ, Erik Krempel, Marcus Hebel, and Michael Arens ⓘ

Fraunhofer IOSB, Ettlingen, Germany
`david.muench@iosb.fraunhofer.de`

Abstract. Data protection in Germany has a long tradition (https://www.goethe.de/en/kul/med/20446236.html). For a long time, the German Federal Data Protection Act or Bundesdatenschutzgesetz (BDSG) was considered as one of the strictest. Since May 2017 the EU General Data Protection Regulation (GDPR) regulates data protection all over Europe and it strongly influenced by the German law. When recording data in public areas, the recordings may contain personal data, such as license plates or persons. According to the GDPR this processing of personal data has to fulfill certain requirements to be considered lawful. In this paper, we address recording visual data in public while abiding by the applicable laws. Towards this end, a formal data protection concept is developed for a mobile sensor platform. The core part of this data protection concept is the anonymization of personal data, which is implemented with state-of-the-art deep learning based methods achieving almost human-level performance. The methods are evaluated quantitatively and qualitatively on example data recorded with a real mobile sensor platform in an urban environment.

Keywords: Video data anonymization · Data protection

1 Introduction

Recording visual data in public while abiding by the applicable laws does require special attention concerning data protection. In this paper we present a formal data protection concept and a practical realization of that system with reliable results for handling the recorded data in a secure way. Thereby, a mobile sensor platform is used as underlying practical use case.

MODISSA (Mobile Distributed Situation Awareness) [4] is an experimental platform for evaluation and development of hard- and software in the context of automotive safety, security, and infrastructure applications. The basis of MODISSA is a Volkswagen Transporter T5, which has been equipped with a broad range of sensors on the car roof, see Fig. 1. The necessary hardware for raw data recording, real-time processing, and data visualization is installed inside the vehicle. The battery powered power supply and electronics are installed in the

© Springer Nature Switzerland AG 2019
D. Tzovaras et al. (Eds.): ICVS 2019, LNCS 11754, pp. 245–258, 2019.
https://doi.org/10.1007/978-3-030-34995-0_23

back and allow independent operation for several hours. The sensor equipment can be adapted to the requirements of the current task. To operate the sensors, several state-of-the-art computers are provided in the vehicle. A complete row of seats can be used by up to three people to operate the sensor system or to watch the results on the display devices (screens, virtual reality headsets).

The sensors are mounted on item aluminum profiles, which are mounted on roof racks. The current LiDAR sensor configuration consists of two Velodyne HDL-64E at the front and two Velodyne VLP-16 at the back. Additionally, in each corner of the car roof there are two Baumer VLG-20C.I color cameras which allow to generate a complete 360° panorama around the vehicle. On the middle of the sensor platform a gyro-stabilized pan-tilt unit is mounted and equipped with a thermal infrared camera, a grayscale camera, and a laser rangefinder.

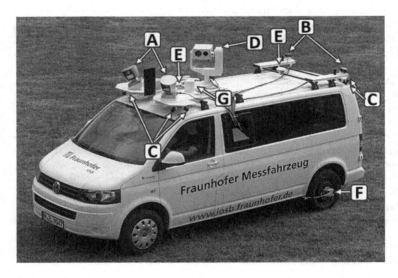

Fig. 1. The external components of the sensor system MODISSA. A: 2x Velodyne HDL-64E LiDAR, B: 2x Velodyne VLP-16 LiDAR, C: Panoramic camera setup 8x Baumer VLG-20C.I. 2 cameras per corner, D: Pan-tilt unit with Jenoptik IR-TCM 640 thermal infrared camera, JAI CM 200-MCL gray scale camera, and Jenoptik DLEM 20 laser rangefinder, E/F: Applanix POS LV V5 520 inertial navigation system with GNSS antennas (E), distance measuring indicator (DMI) (F), inertial measurement unit (IMU) and position computer (both not visible), G: External WiFi antenna. Figure and caption adapted from [4].

2   Data Protection on Publicly Recorded Data

Having the exemplary mobile sensor platform MODISSA described in the paragraph above, we are technically ready to start driving and recording data. Before this can be done, we have to identify applicable law and decide how to implement legal requirements. Processing of personal data in the EU is always regulated by the GDPR which in Germany is further qualified in the BDSG. Depending on the concrete scenario further area-specific regulations may exist.

As we operate with a mobile sensor platform in the public and record data there, it is inevitable that the recorded data contains personal data, such as license plates or persons. A legal opinion [9] and the critical feedback of the federal states data protection commissioner [3] come both to the conclusion that processing is possible when certain regulations and requirements are met.

By default processing of personal data in the EU is prohibited unless at least one of six different reasons legitimate it (Art. 6 GDPR). In short these possible reasons are: (a) data subject has given consent; (b) processing is necessary to fulfill a contract with the data subject; (c) processing is required by obligations of the processor; (d) processing is necessary in order to protect the vital interests of the data subject; (e) processing is necessary for the performance of a task carried out in the public interest; (f) legitimate interest of the processor. In the case of (f) legitimate interest, the data processor has to ensure, that his interest in data processing is not overridden by the interests or fundamental rights and freedoms of the data subject. Therefore, a high level of data protection and data security is mandatory. A data protection concept based on three core principles was developed to achieve this:

- **Avoidance.** Only collect as much data as necessary.
- **Security.** Protection against unauthorized access to the recorded data.
- **Privacy preserving.** Anonymization of personal data.

Fig. 2. Different zones for the recorded data: The red zone handles temporary stored raw data. In the yellow zone we have mostly anonymized data for research purpose with limited access only to people needing the data. In the green zone are few selected and manually checked anonymized data for print and online publication. (Color figure online)

Further we define three different environments where data is processed. In each environment appropriate measures enforce the three core principles. Figure 2 illustrates the three different environments:

- In the **red zone**, we handle temporary stored raw data. It has the highest level of access restrictions. Every access, i.e. using algorithms to anonymize the data and transform it in the yellow zone, is logged and restricted to selected persons. Additionally, physical access to the red zone is restricted.

- In the **yellow zone** we have algorithmically anonymized data for research purpose with limited access by people needing the data for their research.
- In the **green zone** we have selected and manually audited anonymized data for print and online publication.

How to deal with the three core principles in the red zone? **Avoidance:** The need for data recording in public space must exist. The reasons to record are clearly justified and documented in advance in a measurement plan. It is justified in the measurement plan which sensors are needed to achieve the research goals. Data recording is limited to these sensors. As part of the achievement of the research objectives a short time period and a route with a certainly low risk of the occurrence of sensitive data is selected. As a consequence, the responsible person for the data recording is generating a measurement plan and a measurement report.

Security: Access to the mobile sensor platform is limited to a few persons. Every ride on public ground is recorded in the logbook. The group of people involved in data recording is reduced to a necessary minimum. These individuals are aware of their responsibilities for the confidential treatment of sensitive information. Storing of data is minimized and access to this data is restricted at all times by technical measures. As a consequence the mobile sensor platform is never left unattended in public or accessible to third persons. By default, the raw data can be encrypted with AES-256. At the end of a recording, the responsible person has to transfer data stored on the mobile sensor platform to a separate and secured computer and has to delete all recorded data on the mobile sensor platform. All operations are logged in the measurement report.

Privacy preserving: For the planned scientific investigations (described in the measurement plan) personal data is removed from the recorded data. This includes faces and license plates. As a consequence anonymization of faces and license plates takes place automatically on a dedicated and secured computer. After anonymization the recorded and anonymized data is allowed to enter the yellow zone.

Detecting the objects of interest and then blurring them is the most common way in ensuring privacy in video data [1,6,7,15,16,19]. Another possibility instead of blurring sensitive data out, is the total removal of sensitive data, such as removing a whole person and inpainting background instead [20]. Recent works mainly address a street view scenario. In our case, we differ from previous approaches in developing a formal data protection concept abiding the law, gaining almost human performance, and providing a fast running overall system.

In the following we give a detailed description of the process of automatically anonymizing faces and license plates.

3 Data Anonymization

The data anonymization methods address people and license plates. As license plates can be assigned to the vehicle owner, they are subject of the personal right of individuals [3,9]. Thus, they need to be anonymized when recording and

(a)   (b)   (c)   (d)

Fig. 3. After having estimated the facial regions from up to six upper body parts from OpenPose (a) the facial region can be inspected in detail (b) which is used as local mask (c) to anonymize, resulting in an image with blurred face (d). Figure from [8].

processing road scenes [3]. Our implemented processing pipeline is twofold: an automated license plate detection and localization system and all faces of persons in the images need to be anonymized. The source data from MODISSA contains images from cameras mounted on eight different positions around the vehicle, c.f. Fig. 1C. These images are of size 1624×1228 pixels and are stored in a raw Bayer pattern. In addition, images from cameras mounted on a pan-tilt-unit also need to be anonymized, c.f. Fig. 1D.

3.1 Face Anonymization

It has shown, that a fast and robust face detection including the whole person as contextual knowledge yields practical good results to gather the facial regions of the persons to be anonymized.

Towards this end, OpenPose [5, 21], a real-time multi-person system is applied to estimate the keypoints of the persons and their facial region. From up to six positions of upper body parts the face region of the person is estimated. A subset of the facial parts nose, neck, leftEye, rightEye, leftEar, and rightEar, if detected above a given confidence threshold, contribute their weighted centroids to the calculation of a face center. The final result is the average of all passed face position suggestions. The size of the face region is determined from the distances of contributing nodes which are applied with a constant factor [8]. See Fig. 3 for a detailed overview.

3.2 License Plate Anonymization

Automated license plate recognition (ALPR) is well addressed in research, as it is a building block for many real applications, such as automatic toll collection, road traffic monitoring, or intelligent transportation systems. Many proposed systems are hierarchically designed and use a license plate detector in the first stage [2,10,11,13]. License plate detection is a difficult task with high variations in the appearance of license plates, see Fig. 4. The recordings in residential areas does not only contain preceding and following vehicles, but also cars on the roadside or in parking lots. Hence, the variation of orientations, scales, and positions of recorded vehicles and their license plates is large.

(a)   (b)   (c)

Fig. 4. Example images for the license plate anonymization. Challenge: High rotation angle (a), concealed parts (b), small plate sizes (c). Figure from [17].

For training and validation [12] an annotated data set of German license plates would be optimal, but is currently not available. Thus, for creation of an own dataset the following four approaches are taken: First, synthetic images with number plates of the same appearance as German license plates are generated. Secondly, annotated datasets of non-German license plates are used. Thirdly, images of German and European license plates are gathered and annotated. Lastly, the former dataset is expanded by using rectified early test results and data augmentation methods.

As in several related works, the plate detection is performed in two stages: First, a YOLOv3 CNN detects and localizes vehicles and second, another YOLOv3 CNN detects number plates inside the vehicle regions, see Fig. 5. Even

(a)   (b)   (c)   (d)

Fig. 5. License plate detection pipeline. After having localized the cars (a) the license plate region can be detected (b) which is used as local mask (c) to anonymize, resulting in an image with blurred license plates (d).

though the same model is used for both steps, different weights and hyperparameters are set. The main benefit of a two stage detector is the reduction of false positive detections. Obviously, license plates are practically always mounted on vehicles and negligibly rare found isolated in other areas of the image. If the search area is restricted to image parts including vehicles, all false positive detections that are not on a vehicle are avoided. Furthermore, the precision is improved due to a smaller search area.

Properties that make using YOLOv3 a promising approach are the ability YOLOv3 outperforms other object detectors like SSDs in terms of average precision for small objects together with low runtime.

4   Anonymization of Image Regions

A markup language annotation file is used to store the estimated sensitive parts of faces and license plates to be anonymized. An external annotation tool can be used for manual refinement, if necessary. It is possible to add boxes for missing detections, eliminate false positives, or adjust the box boundaries. From the annotation file data, image masks are generated, which are used to determine regions where the image will be changed. Most of the image data to be anonymized is stored as raw Bayer pattern. A modified version of a Gaussian blur is applied to the image as 2D-filter [8].

Anonymization with Gaussian blur can be deanonymized under certain circumstances, see [14]. Since our regions to anonymize cover only few pixels, blur is applied with high size and σ and the blur regions are cropped from wholly blurred images—not phased out—our method can stay for the moment, but alternatives should be kept in mind.

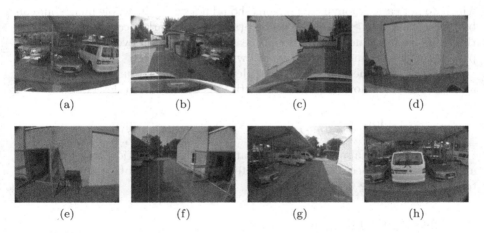

| (a) | (b) | (c) | (d) |

| (e) | (f) | (g) | (h) |

Fig. 6. Examples of MODISSA test dataset. All images are taken at the same time but with different cameras. Figure from [17].

5 Datasets

Three datasets are chosen for testing: A dataset from openALPR[1]. This dataset contains three subsets with Brazilian, European, and US license plates on vehicles. There is one vehicle per image and the perspective variations are little. This dataset is used to compare the performance to other license plate detectors.

For testing the face anonymization a face test dataset with 736 frames was recorded, also using the eight panorama cameras, see Fig. 1C. In this test dataset a person is moving around the MODISSA sensor platform and is visible in up to three cameras at the same time. Ground truth, consisting of 1053 boxes and 91 don't-care boxes is provided. The don't-care areas take into account that the back of a head does not need to be anonymized, but it is no matter if it is anonymized.

For evaluation of the performance for the actual overall anonymization process a third dataset is generated. Representative recorded data from MODISSA is used for evaluation. The dataset consists of 396 fully annotated frames from the eight panorama cameras, see Fig. 1C. A total amount of 1215 faces and 7933 license plates are present in the data. Figure 6 depicts some examples from the MODISSA test dataset.

6 Evaluation

Intersection over Union (IoU) with IoU ≥ 0.5 is a commonly used measure to evaluate detections. In our cases we observed that this measure does neither fit well for face detection nor for license plate detection. The reason is, that there are on the one side regions containing eyes, ears, nose, and mouth which should be anonymized and on the other side there are regions such as hair and forehead which do not care. Anonymized regions should not cover the whole image but can be bigger than IoU ≥ 0.5, if the needed (annotated) part is covered. To address this issue we introduce as additional measure Intersection over Area (IoA), the percentage the ground truth is covered with the detection.

In detail, IoU measures how good detection D and ground truth G match and IoA measures how good ground truth G is covered from detection D.

$$\text{IoU} = \frac{area(D \cap G)}{area(D \cup G)}, \qquad \text{IoA} = \frac{area(D \cap G)}{area(G)}$$

IoU threshold is applied to license plate detection evaluation with IoU ≥ 0.5, and IoU ≥ 0.3 respectively. On face detection evaluation IoU ≥ 0.3 is applied together with IoA ≥ 0.7.

6.1 Face Anonymization

Applying IoU ≥ 0.5 to the face test dataset, we gain weak performance; instead according to the paragraph above using the evaluation metrics IoU ≥ 0.3 and

---
[1] https://www.openalpr.com/.

Fig. 7. Representative test results. High confidences for difficult scenes: Small plate sizes (a), concealed parts (b), high rotation angle (c-e), mirror images (d) and doubled lined plate design (e). False positives for bright surface (e) and wheel rim (f).

IoA $\geq$ 0.7 gives a much more realistic impression of the performance on real data.

With a confidence threshold of 0.575 the face test dataset evaluation achieves the highest precision value 0.998 from 1023 true positives, 2 false positives, 11 used don't-care areas, and 30 false negatives. With confidence threshold 0.32 we get the highest amount of 1039 true positives besides 11 false positives, 43 used don't-care areas and 14 false negatives, from which two are below the IoU threshold and 12 are not detected at all. All false negatives are next to the image border. In a further step the anonymization could be applied to the stiched panorama image instead of the eight single images. Then, the false negatives at the image borders are no longer expected. Thus, we argue, that false negatives at the image border (28 out of 30) can be neglected.

6.2 License Plate Anonymization

The number plate localization system is assessed using the openALPR benchmark dataset. Thereby, the overall license plate detection system is evaluated instead of the two initial stages separately. Using the standard IoU $\geq$ 0.5, an average precision (AP) of 85.36% is yielded. Reducing IoU $\geq$ 0.3, the AP is increased to 98.73%. Considering the fact that the bounding boxes for the test images use another norm than the training images, the smaller threshold is reasonable: While the complete number plate is labeled as ground truth for the benchmark dataset, only the smallest axes-aligned rectangle that includes all characters of the plates states the bounding box for the training images.

Figure 7 shows some example results applying the license plate detection to the MODISSA test dataset.

Even though false positive detections are reduced by the two stage approach, they occur occasionally. Since the distance between the confidence for false positive and false negative detections is large, this is not a problem in general. However, there are few exceptions for false negatives with higher confidences. Mostly, they occur in the following cases:

- periodic, in particular vertical structures, e.g. wheel rims, characters
- bright surfaces, e.g. white cars, overexposed areas
- rectangular objects, e.g. other plates, side mirrors, rear lights.

To improve the license plate detection system, mostly the second stage, i.e. the YOLOv3 for plate detection, should be considered. The training is limited by the number and quality of the training dataset. Thus, further data augmentation and/or suitable annotated datasets might enhance the precision and recall.

6.3 Overall Evaluation

For quantitative results of the MODISSA test dataset, see Fig. 8. Since the dataset was recorded for development and optimized to contain many samples of different challenging situations, the results are correspondingly. The sequences

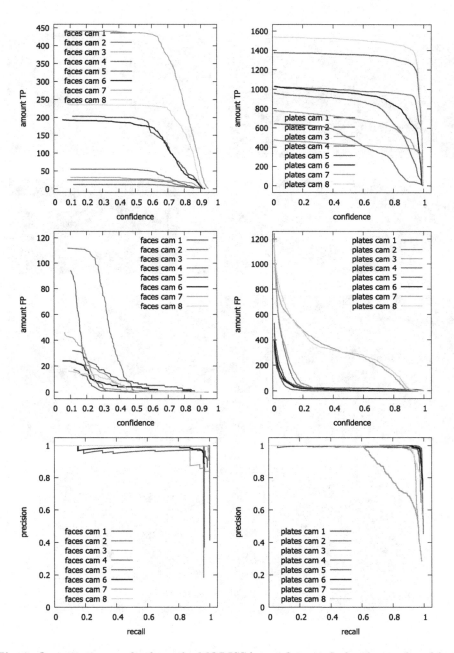

Fig. 8. Quantitative results from the MODISSA test dataset. Left side, results of face anonymization, right side plate anonymization. True positive (TP). False positive (FP).

Fig. 9. Qualitative final automatically anonymized results from the MODISSA test dataset.

of camera 1 to 8 contain a ground truth amount of 206, 26, 32, 57, 12, 199, 446, and 237 faces, as well as 1392, 1032, 485, 647, 978, 1036, 805, and 1558 plates. Besides, there are 53, 16, 9, 2, 4, 301, 148 and 26 don't care faces and 10, 120, 508, 10, 285, 416, 100 and 12 don't care plates annotated. Don't-care plates are too small to show details, that is they do not need to be anonymized, but it does not matter if they are anonymized. During the experiments 17, 0, 0, 0, 2, 32, 21, and 22 don't care faces and 444, 112, 182, 32, 329, 92, 286, and 345 don't care plates are used. Qualitative results are visualized in Fig. 9.

6.4   Runtime

Manually annotating 1000 faces takes about 120 min. Automatic anonymization takes between 0.6 and 1.1 s per camera and frame for license plate detection, depending on car density in the image. Face detection takes about 0.15 s per camera and frame. Applying the Gaussian blur takes about one minute for 1000 frames. The computation time of automatic data anonymization including manual post-treatment is about ten times faster than completely manually anonymizing.

7   Conclusion

In this paper, we addressed recording visual data in public while abiding by the applicable laws. As a consequence, a data protection concept is developed for a mobile sensor platform. The core part of this data protection concept is the anonymization of personal data. Face anonymization is implemented based on OpenPose [5,8,21] and license plate anonymization is realized as a hierarchical approach based on YOLOv3 [17,18]. The quantitative evaluation on example data recorded with a mobile sensor platform in an urban environment shows almost human-level performance and manual post-treatment data anonymization is reduced to a minimum.

References

1. Agrawal, P., Narayanan, P.: Person de-identification in videos. IEEE Trans. Circuits Syst. Video Technol. **21**(3), 299–310 (2011)
2. Arsenovic, M., Sladojevic, S., Anderla, A., Stefanovic, D.: Deep learning driven plates recognition system. In: 17th International Scientific Conference Industrial Systems (IS 2017) (2017)
3. Bayerisches Landesamt für Datenschutzaufsicht: Bundesdatenschutzgesetz (BDSG); Rechtsgutachten zum möglichen Erprobungseinsatz eines mit Laser- und Kameratechnik ausgestatteten Sensorfahrzeuges im Stadtgebiet von Ettlingen, December 2015
4. Borgmann, B., Schatz, V., Kieritz, H., Scherer-Klöckling, C., Hebel, M., Arens, M.: Data processing and recording using a versatile multi-sensor vehicle. ISPRS Ann. Photogramm. Remote Sens. Spat. Inf. Sci. **IV-1**, 21–28 (2018)

5. Cao, Z., Simon, T., Wei, S.E., Sheikh, Y.: Realtime multi-person 2D pose estimation using part affinity fields. In: CVPR (2017)
6. Flores, A., Belongie, S.: Removing pedestrians from google street view images. In: CVPR - Workshops, June 2010
7. Frome, A., et al.: Large-scale privacy protection in Google street view. In: ICCV. IEEE (2009)
8. Grosselfinger, A.K., Münch, D., Arens, M.: An architecture for automatic multimodal video data anonymization to ensure data protection. SPIE Security + Defence (2019)
9. HÄRTING Rechtsanwälte: RECHTSGUTACHTEN zu Rechtsfragen zum möglichen Erprobungseinsatz eines mit Laser- und Kameratechnik ausgestatteten Sensorfahrzeuges im Stadtgebiet Ettlingen durch das Fraunhofer Institut für Optronik, Systemtechnik und Bildauswertung (IOSB), Abteilung Objekterkennung, June 2013
10. Jørgensen, H.: Automatic license plate recognition using deep learning techniques. Master's thesis, NTNU (2017)
11. Laroca, R., et al.: A robust real-time automatic license plate recognition based on the YOLO detector. arXiv preprint arXiv:1802.09567 (2018)
12. Li, H., Shen, C.: Reading car license plates using deep convolutional neural networks and LSTMs. arXiv preprint arXiv:1601.05610 (2016)
13. Masood, S.Z., Shu, G., Dehghan, A., Ortiz, E.G.: license plate detection and recognition using deeply learned convolutional neural networks. arXiv preprint arXiv:1703.07330 (2017)
14. McPherson, R., Shokri, R., Shmatikov, V.: Defeating image obfuscation with deep learning. arXiv preprint arXiv:1609.00408 (2016)
15. Nodari, A., Vanetti, M., Gallo, I.: Digital privacy: Replacing pedestrians from Google street view images. In: ICPR, November 2012
16. Padilla-López, J.R., Chaaraoui, A.A., Flórez-Revuelta, F.: Visual privacy protection methods: a survey. Expert Syst. Appl. $42(9)$, 4177–4195 (2015)
17. Peter, R., Grosselfinger, A.K., Münch, D., Arens, M.: Automated license plate detection for image anonymization. SPIE Security + Defence (2019)
18. Redmon, J., Farhadi, A.: Yolov3 an incremental improvement. arXiv preprint arXiv:1804.02767 (2018)
19. Ribaric, S., Ariyaeeinia, A., Pavesic, N.: De-identification for privacy protection in multimedia content: A survey. Sig. Process. Image Commun. 47, 131–151 (2016)
20. Uittenbogaard, R., Sebastian, C., Vijverberg, J., Boom, B., Gavrila, D.M., et al.: Privacy protection in street-view panoramas using depth and multi-view imagery. In: CVPR (2019)
21. Wei, S.E., Ramakrishna, V., Kanade, T., Sheikh, Y.: Convolutional pose machines. In: CVPR (2016)

Water Streak Detection
with Convolutional Neural Networks
for Scrubber Dryers

Uriel Jost and Richard Bormann[✉]

Fraunhofer IPA, Robot and Assistive Systems, 70569 Stuttgart, Germany
{uriel.jost,richard.bormann}@ipa.fraunhofer.de
https://www.ipa.fraunhofer.de

Abstract. Avoiding gray water remainders behind wet floor cleaning
machines is an essential requirement for safety of passersby and quality
of cleaning results. Nevertheless, operators of scrubber dryers frequently
do not pay sufficient attention to this aspect and automatic robotic clean-
ers cannot even sense water leakage. This paper introduces a compact,
low-cost, low-energy water streak detection system for the use with existing and new cleaning machines. It comprises a Raspberry Pi with an Intel
Movidius Neural Compute Stick, an illumination source, and a camera
to observe the floor after cleaning. The paper evaluates six different Convolutional Neural Network (CNN) architectures on a self-recorded water
streak data set which contains nearly 43000 images of 59 different floor
types. The results show that up to 97% of all water events can be detected
at a low false positive rate of only 2.6%. The fastest CNN Squeezenet can
process images at a sufficient speed of over 30 Hz on the low-cost hardware such that real applicability in practice is provided. When using an
NVidia Jetson Nano as alternative low-cost computing system, five out
of the six networks can be operated faster than 30 Hz.

Keywords: Water streak detection · Cleaning robots · Liquid sensing

1 Introduction

On the task of wet floor cleaning with scrubber dryers a serious problem is
imperfect suction of dirty water after the cleaning. Remaining gray water on the
ground is a threat to humans who may slip, and furthermore it causes visible
dirt streaks on the ground after evaporation. Both must be avoided by the cleaning service provider. Typically, gray water remainders originate from a defective
suction unit or damaged squeegee blades. Especially on ride-on scrubber dryers,
cleaning professionals usually take notice of undesired water remainders too late
or not at all. Likewise, automatic cleaning machines do not check for remaining gray water at all. Hence, an assistive system for the automatic detection

This project has received funding from the German Federal Ministry of Economic
Affairs and Energy (BMWi) through grant 01MT16002A (BakeR).

Fig. 1. Visualization of the water streak detection task behind scrubber dryers (Color figure online) [2].

and warning on gray water remainder events would be a very useful add-on to manually-driven and automatic cleaning machines if it comes at low additional costs. The problem setting is illustrated in Fig. 1 with a good (green frame) and bad (red frame) cleaning result.

This paper presents a compact, low-cost, low-energy visual detection system for water streaks and gray water remainders which can be easily added to existing and newly developed wet cleaning devices. The hardware consists of a Raspberry Pi with attached Intel Movidius Neural Compute Stick for running CNNs efficiently, an illumination stripe for quite stable illumination conditions, and an RGB camera for monitoring the cleaning results behind the machine. The water detection classifier is trained on a self-recorded data set of 59 different floor types with and without water leakage events. We compare Alexnet [16], Densenet [14], Resnet [12], Squeezenet [15], Mobilenet [20], and Shufflenet [17] as CNN architectures for classifying wet against dry floors. The evaluation shows that it is possible to achieve a high recall rate while keeping false alarms very low, which is a core requirement for user acceptance. The water streak detection system has been developed to the request and advising of a major supplier of wet floor cleaning machines. To the best knowledge of the authors this is the first solution to this problem. Summarized, the contributions of this paper are:

1. A description of a low-cost hardware and software system for the reliable detection of water streaks behind floor cleaning machines.
2. A data set of almost 43000 images of real water streak events on 59 different floor types.
3. A thorough evaluation of several classifiers w.r.t. the technical constraints.

The remainder of the paper is structured as follows: Sect. 2 discusses relevant literature, followed by some insights on the requirements on the system and the development process in Sect. 3. Subsequently, the system resulting of the design process is presented in Sect. 4 and evaluated in Sect. 5. We conclude in Sect. 6 and provide some ideas for future work.

2 Related Work

The water streak detection system proposed in this paper is applicable to ordinary manual wet cleaning machines and ride-on scrubber dryers as well as to automatic cleaning machines such as Adlatus, BakeR [1,6], Cleanfix, HEFTER cleantech [9], Intellibot, or TASKI Diversey. The cleaning principle of all these machines is similar: they have one or more rotating brushes which are served with fresh cleaning water and detergents. Their mechanical force and the cleaning chemicals remove dirt from the floor and dissolve it in the water. The gray water is eventually removed by the suction unit at the end of the cleaning tool. If the suction unit has a congestion or defect or the squeegee behind it is blocked by dirt, water leakage may occur behind the cleaning machine. None of the aforementioned devices is currently capable of sensing such water streaks and reacting accordingly, e.g. by lifting the cleaning unit, driving back, and continuing cleaning.

There is currently no system available for detecting water leakage behind a cleaning machine. Similar topics are addressed by various dirt detection systems [3,4,10,11,13] which are supposed to improve the performance of automatic cleaning machines in different ways. The dirt detection system proposed by Bormann et al. [3–5] enables a cleaning robot to sense pollutions at the ground for the task of demand-triggered vacuum cleaning in contrast to systematic cleaning of the whole ground. It is based on a spectral residual novelty detection algorithm which can spot surprising image contents on regular floor surfaces which often coincide with dirt. However, the system is also said to have a high false positive rate and may not be well-applicable on randomly patterned floors. Both properties contradict the requirements of the water detection system at hand (see Sect. 3.1).

The dirt detection system of Grünauer et al. [10,11] focuses on the detection of heavily polluted floor areas to adjust the use of water and detergents by their wet cleaning robot in order to optimize the cleaning result and resource consumption. They propose the unsupervised online learning of a Gaussian Mixture Model representing the floor pattern and detect dirt as novelty regions with good success rates. This system is said to be capable of handling random floor patterns as well and was tested to detect liquids spilled at the floor. It is not reported whether translucent liquids can be handled or whether the system is limited to colored liquids. However, their evaluation shows that recall rates are quite low when a high precision is required.

A third kind of dirt of dirt level estimation system is introduced by Hess et al. [13] which uses a dirt sensor integrated into a vacuum cleaning consumer robot to estimate the most probable distribution of pollutions in the environment. This knowledge is exploited to optimize the robot's driving path for achieving a desired level of cleanliness.

Other related work on the recognition of water is concerned with rain and snow detection and removal from images for weather detection by meteorological centers [7] or for rain removal from images processed in outdoor vision systems [8,24]. The approach of Bossu et al. [7] uses a Gaussian Mixture Model for foreground background separation and a Histogram of Orientation of Streaks to label the individual pixels of rain or snow in order to estimate precipitation amounts. Yang and Chen et al. [8,24] utilize an intricate CNN architecture to detect rain and fog in outdoor images. The work of this domain, however, does not transfer well to liquid sensing at floors because the granularity of objects and the exploited physical effects are very different.

The work of Schenk and Fox [21–23] is focused on perceiving liquids during a pouring process from one container into another. Their approach can even estimate the volume of transferred liquid. The system employs a LSTM-CNN to detect and track liquids during pouring activities. It appears to be optimized to that domain and may not transfer well to the detection of other appearances of liquids such as stains or streaks at the floor.

3   Development Process

This section briefly outlines how the vision system presented in this paper has evolved from the design and development process for solving the task of water streak detection behind scrubber dryers.

3.1   Requirements Analysis

The collection of requirements on the water streak detector has been advised by a major supplier of cleaning machines. The core requirements are:

1. The system should be low-cost at a price of about $100.
2. The processing performance should be compatible with ride-on scrubber dryers at speeds up to 3 m/s.
3. The system should achieve robust performance in the real application.
4. The false alarm rate should be very low, i.e. the system should only alarm when there is really some water streak event. Hence high precision for water event classification is preferred over a high recall rate.
5. The detection sensitivity must be easily adjustable by the end user with a simple turning knob.
6. The water streak detection system must be easily integrated into existing and new cleaning machines.
7. The water streak detection system should be preferably maintenance-free.

Based on these requirements, two reasonable water streak detection approaches remained from the initial brainstorming: an electromechanic device measuring moisture by contact and a vision-based system detecting the specific structure and reflections of water in RGB images. For both options a prototype has been developed and tested with a hand-operated scrubber dryer.

3.2   Electromechanic Measurement System

The electromechanic approach detects water remainders at the floor by measuring the voltage between neighboring copper contact strips which are attached to a second squeegee behind the cleaning unit. Figure 2 illustrates the prototype which is mainly manufactured as a wooden construction. The contact strips are connected to an Arduino microcontroller measuring the different voltage levels which depend on the floor type and the presence of water. The prototype was able to distinguish water leakages from dry floors successfully in most cases, even on metal joints, but had some problems with larger inevitable water remainders in tile joints, which were regularly detected as false positives. For establishing continuous contact this system needs to be pressed towards the ground. However, this is suboptimal because the squeegee may scratch or damage sensitive floor materials with its electric contacts and it is likely not maintenance-free.

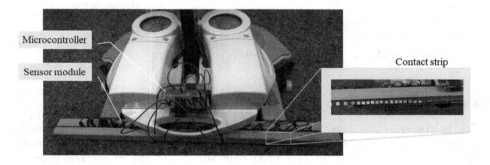

Fig. 2. Prototype of the electromechanic water streak detection device.

3.3   Vision System

The second approach processes RGB camera images with a CNN which classifies the image contents into dry and wet floors (see Fig. 3). The vision system is explained in Sect. 4 and evaluated in Sect. 5. The system achieves robust classification results and is completely maintenance-free. The electromechanic and the vision system are compared in Table 1 according to the requirements. Because of its advantages the vision system was selected for further development on the water detector for cleaning machines.

4   Vision-Based Water Streak Detection System

This section discusses the hardware and software components of the vision-based water streak detection system in greater detail.

Table 1. Comparison of the electromechanic and the vision-based water detection systems with respect to the requirements.

	Electromechanic	Vision
1. Low-cost, price about $100	yes, retail price around $100	almost, retail price around $140
2. Processing speed matches driving speed of 3 m/s	yes	yes
3. Robust performance in real application	problems with water remainders in tile joints, conductive floor materials	very robust, rarely small problems with shiny floor types
4. Low false positive rate	yes, except for wet tile joints	yes
5. Adjustable detection sensitivity	yes (voltage threshold)	yes (classifier confidence threshold)
6. Usability with existing and new cleaning machines	yes	yes
7. Maintenance-free	rather not	yes

4.1 Hardware

The hardware setup is as sketched and displayed in Fig. 3. The vision-based water streak detection system is contained inside a five-sided box open towards the floor. Together with the two LED light sources at the upper box side it establishes mostly stable illumination conditions.

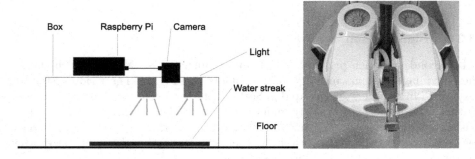

Fig. 3. Schematic drawing (left) and photograph of the core components without enclosing box (right) of the water streak detection device.

The computing hardware consists of a Raspberry Pi 3 B running the control software on a Debian Jessie operating system and an Intel Movidius Neural Compute Stick for the execution of CNN inference computations on the captured images. The images are captured with a Raspberry Pi Camera Rev 1.3 which can capture images with 1920×1080 pixels @ 30 fps, 1280×720 pixels @ 60 fps, or 640×480 pixels @ 60/90 fps. The water detection system can be battery-powered or connected to the power supply of the cleaning device.

4.2 Software

The simple control script of the water detection system was written as a Python script which periodically receives camera images, forwards them to the CNN running on the USB inference stick and retrieves the classification results as softmax confidences for the two classes water and non-water. The confidences can be thresholded with a fixed confidence threshold θ for obtaining a per image label or they can be processed as a time series.

Although our data is a sequential image stream like in the water jet detection system of [21] no LSTM network was tested in this case because they are not supported by the Intel Movidius NCS device with its OpenVINO SDK (ONNX model format) and hence could not be deployed on the target hardware. However, sequence processing has been implemented and analyzed in a simpler way by applying a digital time series filter on the resulting classification confidences according to the following computation scheme:

$$v_n = \begin{cases} 1 \text{ , if } o_{\text{water},n} > \theta_v \\ 0 \text{ , else} \end{cases} \tag{1}$$

$$y_n = \alpha(v_n - y_{n-1}) + y_{n-1} \tag{2}$$

where the CNN confidence on water $o_{\text{water},n}$ is binarized with threshold θ_v in step (1), resulting in binary label v_n. The filter rule (2) updates the filter value y_n with the previous filter value y_{n-1} and the binary label v_n and update rate α. The final filtered classification label is determined by thresholding the current filter value y_n with classification threshold θ. The initial filter value is set to the first CNN confidence output $y_0 = o_{\text{water},0}$.

Beforehand, the six different CNN classifiers for water/non-water classification have been trained on an external PC with Intel Core i7-7700K CPU @ 4.20 GHz and Nvidia GTX 1060 3 GB GPU using PyTorch [18] as deep learning framework. Training was conducted by fine-tuning the networks pre-trained on the 1000 classes of ImageNet Large Scale Visual Recognition Challenge [19] with our data set on water detection (see Sect. 5.1). The classification layer with 1000 neurons was replaced by 2 neurons in our case for representing the water and non-water classification results as softmax confidences. The utilized network architectures are Alexnet [16], Densenet-121 [14], Resnet-18 [12], Squeezenet version 1.0 [15], Mobilenet V2 [20], and Shufflenet V2 [17].

5 Evaluation

The visual water streak detection system has been evaluated on image sequences recorded under real operation conditions with the sensor system mounted on a hand-operated scrubber dryer. This section describes the recorded data set used for evaluating the system, the experimental conditions, and the results obtained in the experiments.

5.1   Water Streak Data Set

For the evaluation of the water streak detection system under controlled conditions we recorded a large data set on 59 different floor types each time with and without water streaks. The image streams were recorded with the hardware of the water streak detector mounted on a hand-operated scrubber dryer in order to generate most realistic and representative data. Each of the 118 image sequences has an approximate duration of 12 s recorded with 30 Hz and hence contains 364 images on average. The hand-operated scrubber dryer allowed very dynamic motions and hence the data set contains all flavors of realistic artifacts such as slowly and fast driven sequences, motion blur effects on the images, image noise, and - to make the data more challenging - motions in all directions, i.e. along and across water streaks and both at the same time.

The original recording resolution of the images is 1920×1080 pixels, however, the images in the data set are downsampled to a resolution of 576×324 pixels. The data set is divided into a training set with 32019 images captured on 45 floor types, a validation set with 3784 images obtained from 5 different floor types, and a test set with 7113 images originating from another 9 different floor types. Furthermore, the detergent type used with the water streaks has been varied on 4 floor types in the test set to increase the difficulty level by changing color and reflection properties of the water streaks. Altogether there are 42916 images in the data set from 59 different floor types. Approximately half of these images are recorded under the condition without water, the other half contains water streaks. Figure 4 illustrates several samples of the data set images.

5.2   Experimental Setup

In all experiments reported we implemented the following experimental procedure. The CNN classifiers tested are Alexnet [16], Densenet-121 [14], Resnet-18 [12], Squeezenet v1.0 [15], Mobilenet V2 [20], and Shufflenet V2 [17] pretrained on the 1000 classes of ImageNet Large Scale Visual Recognition Challenge [19]. We exchanged the last 1000 classes layer against a two-class layer for the water/non-water condition predictions.

All CNN classifiers have been fine-tuned with the data from the training data set in randomized order (i.e. non-sequential) until the loss of the validation set settled and started to increase in order to avoid overfitting. The training has been conducted with mini-batch stochastic gradient descent (SGD) and a learning rate of 0.001. This setting was evaluated best out of learning rates between $[0.001, 0.0005, \ldots, 0.000005]$ used with SGD and Adam. Decreasing the learning rate during training did not improve the performance. Batch sizes were set to 32 for training Alexnet, Squeezenet, Mobilenet, and Shufflenet, and to

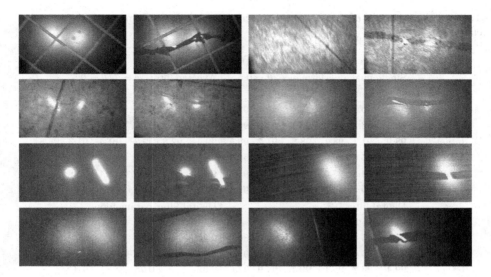

Fig. 4. Exemplary image pairs of the data set showing non-water and water conditions.

8 for Densenet and Resnet. The test set was only used once with each finally trained CNN to evaluate how well the validation set performance generalizes.

5.3   Results

The single image classification performance with respect to mean average precision (mAP) as well as recall and precision obtained with classification threshold $\theta = 0.5$ on the validation and test sets of the water/non-water detection task is provided in the upper left section of Table 2 and in the precision recall plots in Fig. 5. Since the water streak detection system operates on sequential image data two kinds of time sequence filters have been set up, additionally. The first, termed average confidence filter, computes the average over the last w classification confidences and applies classification threshold $\theta = 0.5$ on that average for the binary decision on water/non-water. We tested window sizes between $w = 3$ and $w = 10$ and obtained the best results for $w = 7$. These results are reported in the upper right part of Table 2 and the respective precision recall plots are provided in Fig. 6. The second time series filter is the digital filter as described in Sect. 4.2. The parameters of the digital filter were optimized on the training set via grid search yielding the values reported in the lower right part of Table 2 and the classification performance given in the lower left part. The associated precision recall diagrams for the digital filter are displayed in Fig. 7.

We observe that in each experimental condition and for each of the analyzed CNNs, the validation and test sets have very similar mAP (typically the difference is < 0.4, the maximum difference is 2.5). Especially Resnet, Mobilenet, and Shufflenet retain a nearly constant mAP. This is an indicator that a good generalization performance can be expected from the reported results.

Table 2. Comparison of single image performance and two filtering operations on the classified images.

	mAP	rec.	prec.	mAP	rec.	prec.	mAP	rec.	prec.	mAP	rec.	prec.
	Single image performance, $\theta = 0.5$						Average filter, $w = 7$, $\theta = 0.5$					
	Validation set			Test set			Validation set			Test set		
Alexnet	94.8	89.5	91.5	95.4	85.7	92.9	95.4	90.9	92.3	95.8	85.5	94.9
Densenet	98.1	86.3	97.4	98.4	93.6	98.5	98.5	87.0	97.9	99.2	93.8	98.9
Resnet	98.9	91.1	98.1	98.8	91.2	98.5	99.4	92.8	99.2	**99.4**	91.3	98.8
Squeezenet	98.1	85.6	99.4	96.2	84.3	99.9	98.8	86.3	100	97.0	84.2	100
Mobilenet	**99.2**	94.8	97.0	**98.9**	95.4	95.8	**99.5**	94.7	98.5	**99.4**	95.8	97.7
Shufflenet	98.5	90.7	94.5	**98.9**	95.8	94.1	99.1	91.9	95.8	**99.4**	96.5	94.5
	Digital filter						Digital filter parameters					
	Validation set			Test set			θ_v		α		θ	
Alexnet	93.6	96.8	86.3	96.1	89.9	86.7	0.40		0.25		0.1	
Densenet	98.9	93.6	96.0	99.2	96.9	97.4	0.52		0.25		0.1	
Resnet	99.4	96.7	95.1	**99.5**	94.9	97.6	0.32		0.25		0.3	
Squeezenet	99.3	94.3	98.9	97.4	89.5	99.6	0.52		0.25		0.1	
Mobilenet	**99.6**	96.5	96.7	99.4	97.7	95.3	0.82		0.25		0.1	
Shufflenet	99.2	95.5	93.8	99.4	97.3	92.8	0.51		0.25		0.3	

Furthermore, for all analyzed CNNs both methods of time sequence filtering generally improve mAP performance over the single image performance. Likewise, the digital filter generally achieves slightly higher mAP values than the averaging confidences filter. The precision recall diagrams show that both filtering methods increase the achievable precision at constant recall levels compared to the single image performance. This indicates that false positives for water are usually only sporadically occurring in the sequences.

When comparing the different CNN classifiers, one can see that under the same experimental conditions, Resnet, Mobilenet and Shufflenet always achieve the highest mAP and Densenet comes quite close to that performance (difference often below 0.5%, always below 1.0%). Compared to Densenet, Squeezenet achieves similar or slightly better (up to 0.4%) mAP performance on the validation set but is worse by up to 2.2% on the test set. Alexnet always has the lowest mAP (3.4% to 6.0% worse than the best CNN).

Resnet and Mobilenet constantly provide the best overall performance over all experimental conditions and can be tuned for the highest recall and precision rates of all analyzed methods. In conjunction with the digital filter Resnet obtains a false positive rate below 2.4% and a recall of 94.9%, while Mobilenet achieves a false positive rate below 4.7% and a recall of 97.7% on the test set. Especially Resnet and Mobilenet can be tuned for a very low false positive rate in range $[0\%, 1.0\%]$ at comparably high water streak detection rates well above 90%. These values are very acceptable for the application of the water streak detection system with real cleaning machines. Whereas it could be expected that the larger CNNs achieve better performance it is remarkable that the many times smaller Squeezenet, Mobilenet, and Shufflenet achieve comparable or only slightly worse performance compared to Resnet and Densenet.

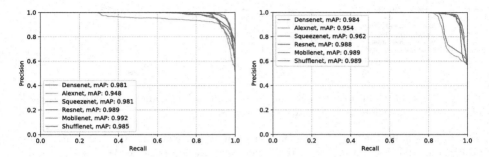

Fig. 5. Precision recall plot for the single image performance of all six classifiers on the validation (left) and test (right) sets.

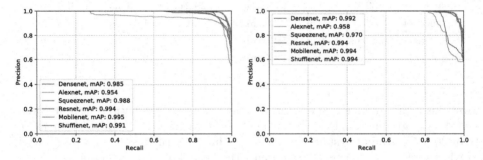

Fig. 6. Precision recall plot for averaging the classification confidences over a sequence of $w = 7$ images and classifying their mean with a threshold of $\theta = 0.5$ on the validation (left) and test (right) sets.

Besides achieving a convincing performance the water streak detection system is also required to run on a low-cost computing device at sufficient speed. Therefore we analyzed the execution times of the considered classifiers to evaluate their applicability to the selected low-cost processing hardware. Table 3 compares the execution times of all CNNs on a Intel Core i7-7700 K CPU @ 4.20 GHz, an Nvidia GTX 1060 3GB GPU, the Intel Movidius Neural Compute Stick, and the newly available NVidia Jetson Nano board for the binary water/no-water classification of this work.

It shows that the performance of all networks is uncritical as long as there is a CPU or even GPU available, however, most models do not run on the Intel

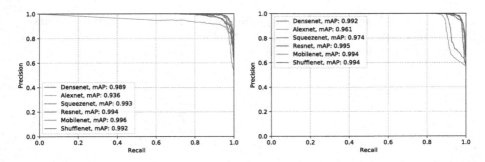

Fig. 7. Precision recall plot for filtering the classification confidences with the digital filter on the validation (left) and test (right) sets.

Table 3. Comparison of measured inference times on executing the tested networks on different devices and driven distance per computation cycle at 3 m/s on the Intel Movidius USB inference system and on the NVidia Jetson Nano processor.

	CPU	GPU	Intel Movidius	distance	Jetson Nano	distance
Alexnet	6.10 ms	1.23 ms	68.80 ms	20.6 cm	7.99 ms	2.4 cm
Densenet	54.21 ms	4.26 ms	184.71 ms	55.4 cm	61.45 ms	18.4 cm
Resnet	15.39 ms	1.54 ms	97.41 ms	29.2 cm	21.93 ms	6.6 cm
Squeezenet	15.64 ms	1.65 ms	27.03 ms	8.1 cm	27.52 ms	8.3 cm
Mobilenet	20.43 ms	1.80 ms	59.30 ms	17.8 cm	22.19 ms	6.7 cm
Shufflenet	7.02 ms	1.27 ms	72.79 ms	21.8 cm	10.12 ms	3.0 cm

Movidius USB inference device at sufficient speed. The NVidia Jetson Nano board instead can run Densenet with 15 Hz and all other models with more than 30 Hz at similar system costs as the solution with Raspberry Pi and Intel Movidius and hence currently represents the best choice of system design[1].

The *distance* columns in Table 3 report the driven distance during one processing cycle at a movement speed of 3 m/s. It can be roughly assumed that the system will not miss any water streak occurrences if images can be recorded every 10 to 20 cm since otherwise the reflections of the active illumination on water areas would not be visible to the camera. Furthermore, there will be some delay in response if the digital filter or the average confidence filter is employed. Consequently, for a timely response and comprehensive monitoring only Squeezenet and possibly Mobilenet would be suitable CNN architectures for usage with the Raspberry Pi and Intel Movidius setup. Using the novel NVidia Jetson Nano processor, all networks could be computed with sufficient speed and the water streak detection system could leverage the full perception power of the favorite CNNs Resnet and Mobilenet.

---

[1] The NVidia Jetson Nano device was not available by the time the project was started. Hence, the Raspberry Pi and Intel Movidius Stick were originally employed as the standard setup for this work.

6   Conclusions and Future Work

This paper addressed the problem of water streak detection behind scrubber dryers. An electromechanical approach was tested but discarded for reasons of potential maintenance effort and damage to the floor materials. The alternative vision-based solution monitors the area behind the scrubber dryer with a camera and evaluates the presence of gray water remainders with a CNN very successfully. The comparison of six CNN architectures has shown that high quality water detection results can even be achieved with low-cost, low-energy hardware at sufficient speed.

The presented vision system can fulfill all requested requirements: the system achieves very good detection rates of water streaks while raising very few false positive alarms. The user can furthermore adjust the sensitivity of the classifier easily with a turning knob by varying the classification threshold θ. The detection speed is compatible with speeds of at least 3 m/s. Due to its small size, the hardware can be easily mounted on existing cleaning machines and integrated into novel developments. The system is basically maintenance-free and its price-performance is very appealing for the use with professional scrubber dryers. The proposed system has a retail price of about $140, hence a manufacturing price below $100 is reasonably achievable. Our collaborating cleaning machine expert acknowledged the great performance of the presented water streak detection system and is considering to offer it as an add-on to the company's scrubber dryers for the benefit of the clients.

As future work, the reduction or removal of the box around the vision system needs to be investigated together with a different design of the active scene illumination for retaining fairly controlled illumination and reflection conditions. Furthermore, the prototype system needs to be developed into a market-ready product. This necessitates a re-design with professional housing and wiring for the processing and sensing units as well as a robust cover for the illumination unit.

References

1. Baker project. https://www.baker-projekt.de
2. Comac Ultra 100 BS AS, product image, KENTER GmbH. https://www.kenter. de/components/com_mijoshop/opencart/image/cache/data/35052-1500x1135. jpg. Accessed 22 July 2019
3. Bormann, R., Weisshardt, F., Arbeiter, G., Fischer, J.: Autonomous dirt detection for cleaning in office environments. In: Proceedings of the IEEE International Conference on Robotics and Automation (ICRA), pp. 1252–1259 (2013)
4. Bormann, R., Fischer, J., Arbeiter, G., Weißhardt, F., Verl, A.: A visual dirt detection system for mobile service robots. In: Proceedings of the 7th German Conference on Robotics (ROBOTIK 2012), Munich, Germany, May 2012
5. Bormann, R., Hampp, J., Hägele, M.: New brooms sweep clean - an autonomous robotic cleaning assistant for professional office cleaning. In: Proceedings of the IEEE International Conference on Robotics and Automation (ICRA), May 2015

6. Bormann, R., Jordan, F., Hampp, J., Hägele, M.: Indoor coverage path planning: survey, implementation, analysis. In: Proceedings of the IEEE International Conference on Robotics and Automation (ICRA). pp. 1718–1725, May 2018
7. Bossu, J., Hautière, N., Tarel, J.P.: Rain or snow detection in image sequences through use of a histogram of orientation of streaks. Int. J. Comput. Vis. **93**(3), 348–367 (2011)
8. Chen, J., Tan, C., Hou, J., Chau, L., Li, H.: Robust video content alignment and compensation for rain removal in a CNN framework. In: Proceedings of the IEEE Conference on Computer Vision and Pattern Recognition (CVPR), pp. 6286–6295, June 2018
9. Endres, H., Feiten, W., Lawitzky, G.: Field test of a navigation system: autonomous cleaning in supermarkets. In: Proceedings of the IEEE International Conference on Robotics and Automation (ICRA), pp. 1779–1781, May 1998
10. Grünauer, A., Halmetschlager-Funek, G., Prankl, J., Vincze, M.: Learning the floor type for automated detection of dirt spots for robotic floor cleaning using gaussian mixture models. In: Liu, M., Chen, H., Vincze, M. (eds.) ICVS 2017. LNCS, vol. 10528, pp. 576–589. Springer, Cham (2017). https://doi.org/10.1007/978-3-319-68345-4_51
11. Grünauer, A., Halmetschlager-Funek, G., Prankl, J., Vincze, M.: The power of GMMs: unsupervised dirt spot detection for industrial floor cleaning robots. In: Gao, Y., Fallah, S., Jin, Y., Lekakou, C. (eds.) TAROS 2017. LNCS (LNAI), vol. 10454, pp. 436–449. Springer, Cham (2017). https://doi.org/10.1007/978-3-319-64107-2_34
12. He, K., Zhang, X., Ren, S., Sun, J.: Deep residual learning for image recognition. In: Proceedings of the IEEE Conference on Computer Vision and Pattern Recognition (CVPR), pp. 770–778 (2016)
13. Hess, J., Beinhofer, M., Burgard, W.: A probabilistic approach to high-confidence cleaning guarantees for low-cost cleaning robots. In: Proceedings of the IEEE International Conference on Robotics and Automation (ICRA) (2014)
14. Huang, G., Liu, Z., van der Maaten, L., Weinberger, K.Q.: Densely connected convolutional networks. In: Proceedings of the IEEE Conference on Computer Vision and Pattern Recognition (CVPR), pp. 2261–2269, July 2017
15. Iandola, F.N., Moskewicz, M.W., Ashraf, K., Han, S., Dally, W.J., Keutzer, K.: Squeezenet: Alexnet-level accuracy with 50x fewer parameters and <1mb model size. CoRR (2016)
16. Krizhevsky, A., Sutskever, I., Hinton, G.E.: Imagenet classification with deep convolutional neural networks. In: Advances in Neural Information Processing Systems (2012)
17. Ma, N., Zhang, X., Zheng, H.-T., Sun, J.: ShuffleNet V2: practical guidelines for efficient CNN architecture design. In: Ferrari, V., Hebert, M., Sminchisescu, C., Weiss, Y. (eds.) Computer Vision – ECCV 2018. LNCS, vol. 11218, pp. 122–138. Springer, Cham (2018). https://doi.org/10.1007/978-3-030-01264-9_8
18. Paszke, A., et al.: Automatic differentiation in PyTorch. In: NIPS Autodiff Workshop (2017)
19. Russakovsky, O., et al.: Imagenet large scale visual recognition challenge. Int. J. Comput. Vis. **115**(3), 211–252 (2015)
20. Sandler, M., Howard, A., Zhu, M., Zhmoginov, A., Chen, L.: MobileNetV2: inverted residuals and linear bottlenecks. In: Proceedings of the IEEE Conference on Computer Vision and Pattern Recognition (CVPR), pp. 4510–4520 June 2018

21. Schenck, C., Fox, D.: Towards learning to perceive and reason about liquids. In: Kulić, D., Nakamura, Y., Khatib, O., Venture, G. (eds.) ISER 2016. SPAR, vol. 1, pp. 488–501. Springer, Cham (2017). https://doi.org/10.1007/978-3-319-50115-4_43

22. Schenck, C., Fox, D.: Visual closed-loop control for pouring liquids. In: Proceedings of the IEEE International Conference on Robotics and Automation (ICRA), pp. 2629–2636 (2017)

23. Schenck, C., Fox, D.: Perceiving and reasoning about liquids using fully convolutional networks. Int. J. Robot. Res. **37**(4–5), 452–471 (2018)

24. Yang, W., Tan, R.T., Feng, J., Liu, J., Guo, Z., Yan, S.: Deep joint rain detection and removal from a single image. In: Proceedings of the IEEE Conference on Computer Vision and Pattern Recognition (CVPR), pp. 1685–1694, July 2017

Segmenting and Detecting Nematode in Coffee Crops Using Aerial Images

Alexandre J. Oliveira[1(✉)], Gleice A. Assis[2(✉)], Vitor Guizilini[3(✉)],
Elaine R. Faria[1(✉)], and Jefferson R. Souza[1(✉)]

[1] Faculty of Computer Science, Federal University of Uberlandia, Uberlândia, Brazil
{alexandre.oliveira,elaine,jrsouza}@ufu.br
[2] Institute of Agricultural Sciences,
Federal University of Uberlandia, Uberlândia, Brazil
gleice@ufu.br
[3] Toyota Research Institute, Los Altos, USA
vitor.guizilini@gmail.com

Abstract. A challenge in precision agriculture is the detection of pests in agricultural environments. This paper describes a methodology to detect the presence of the nematode pest in coffee crops. An Unmanned Aerial Vehicle (UAV) is used to obtain high-resolution RGB images of a commercial coffee plantation. The proposed methodology enables the extraction of visual features from image regions and uses supervised machine learning (ML) techniques to classify areas into two classes: pests and non-pests. Several learning techniques were compared using approaches with and without segmentation. Results demonstrate the methodology potential, with an average for the f-measure of 63% for Convolutional Neural Network (U-net) with manual segmentation.

Keywords: Pest detection · Coffee crops · UAV · Machine learning

1 Introduction

Currently, with a planted area of 10.9 million hectares and world production of 153.6 million bags of 60 kg of coffee benefited in the 2016 harvest, coffee farming represents a significant agricultural activity on the world stage [1].

Although Brazil is the significant coffee producer, accounting for 32.7% of the world's total, the average productivity of 25 bags ha^{-1} is considered low [1], being influenced by abiotic and biotic factors, especially occurrence of pests and diseases. In this sense, the nematodes of the genus *Meloidogyne* carry significant economic losses that vary according to the species present in the area [2], including wide dissemination, high reproductive capacity and aggressiveness [6].

The species *M. paranaensis* has been the most dangerous to crops [13], causing damage to the roots of the plants and leading to symptoms visible from the top such as yellowing and defoliation, with subsequent death of susceptible cultivars. In this sense, we highlight the use of UAVs to detect images of plants

© Springer Nature Switzerland AG 2019
D. Tzovaras et al. (Eds.): ICVS 2019, LNCS 11754, pp. 274–283, 2019.
https://doi.org/10.1007/978-3-030-34995-0_25

with symptoms due to the presence of nematodes in the area, a tool that would reduce production costs and contribute to sustainable coffee production.

For coffee crops, work available on the use of UAVs for the detection of plants with symptoms caused by pests and diseases is scarce, which justifies the importance of this study. Our research was to evaluate the potential of using aerial images with UAVs for the detection of plants with symptoms arising from *M. paranaensis* in coffee crops. The contribution of this paper is a nematodes detection methodology from aerial images, which was selected based on performance-wise evaluation of three ML methods.

The remainder of this paper is organized as follows: Sect. 2 presents related work. Section 3 shows the proposed methodology. The results and analysis are shown in Sect. 4. Section 5 concludes and suggests directions for future work.

2   Related Work

In agriculture, UAVs have been used for several purposes, among them in the diagnosis of diseases. In citrus plantations, the use of multispectral images to detect huanglongbing or greening allowed the classification of diseases and healthy plants with an accuracy of up to 85% with the use of UAVs [7].

In olive trees, the techniques based on the detection of the effects of infection and transmission of *verticillium* wilt were observed with thermal, multispectral and hyperspectral sensors onboard a UAV. It was evidenced that crown temperature, and chlorophyll fluorescence are adequate to indicate physiological damage, and normalized difference vegetation index (NDVI) is designated to evidentiate damage caused by the fungus [5].

For the poppy culture, the use of multispectral and thermal embedded cameras on a UAV indicated that the concentration of chlorophyll, NDVI and cup temperature were parameters adequately related to the physiological stress caused by mildew Peronospora Arborescens, which makes it possible to use them to detect diseases in crops [4].

In the case of nematodes, the use of UAVs has also been promising for localized management of areas from collected images. In soybean areas with *M. incognita*, a hexacopter equipped with two GoPro cameras, one RGB and another modified to the near infrared, with the flight altitude of 80 m for a spatial resolution of the image of 5 cm, has been used with great success. This flight allowed visualization of the damage caused by the nematodes to the area, with the formation of different reefers concerning size and damage [10].

3   Methodology

3.1   Data Collect

To obtain the aerial images, we used an aircraft Phantom 4 Pro (Fig. 2(a)). Aerial images were captured by an onboard camera of 4864 × 3648 pixel resolution and 72 dpi, flying at an altitude of 20 m.

3.2   Pre-processing

Among the 220 images collected by UAV, we selected 10 to the experiments, considering those with high quality, presence of pests, and labeled by the specialist. As the images are collected in the field, several areas corresponding to the soil and shadow, here considered as background, are included in the image. The presence of these background areas makes the problem challenging, which can be seen in Fig. 1(a), so the background of the images were manually removed (manually segment (MS)) - Fig. 1(b), because it is an easy task of distinction for a human being and also allows to reduce the number of non-pest examples as described in Sect. 3.3. The pixels corresponding to the background are set as black and the foreground as the original ones.

(a)   (b)   (c)

Fig. 1. (a) Original image. (b) MS. (c) Ground truth - white areas represent pest.

3.3   Feature Extraction

Using the images obtained by the pre-processing step, a set of features are extracted to generate the instances composing the training and test sets. The set of features are obtained with a block-based approach as proposed in [16], which has the advantage of obtaining an important representation of the region of interest.

Two blocks are admitted, the internal block, also named classification block, represented by a 10×10 square area, and the external block or contextual block, represented by a 20×20 square area. The internal block is used to provide the primary characteristics of the sample. The external block provides the contextual properties of the sample. The blocks have these sizes because after performing some experiments, we noticed that the sizes used in the paper exploit to the maximum the amount of pest areas in the coffee. If we increase these blocks, we could lose pest data. Figure 2(b) illustrates the block-based approach.

Considering the difficulty in reaching an accurate model using a high dimensional data set, we used one method for feature selection, available in the Weka software [8]. The general procedure in applying this technique is, (i) The data set is generated only with the features of the classification block; (ii) Then, we apply a feature evaluator on the data set to obtain a set of most relevant features; (iii) Finally, we generate a new data set with features provided by the

(a)   (b)

Fig. 2. (a) Phantom 4 Pro. (b) Block-based approach proposed in [16].

feature selection algorithms based on the classification block, and then we add the correspondent's features into of the contextual block.

We used the *CorrelationAttributeEval* method [9], correlation of Pearson's between it and the class set of features, while correlation took a few seconds. This method generates a ranking of the best features ordered from highest to lowest correlation value. We selected features using a threshold value of 0.2. All features with a relevance score less than the specified threshold were removed.

The method provides 24 features for the classification block, selecting the following features: (i) mean for each component, RGB, LAB, and XYZ, adding contributing with 9 features; (ii) mean for the following components: S and V from HSV, Y and Cr from YCbCr, Y and Q from YIQ, Y and Q from YUQ, adding 8 features; (iii) min of the components H and V from HSV, adding 2 features; (iv) standard deviation of the components X and Z from XYZ, adding 2 features; (v) mean and max from GRAY, adding 2 features; (vi) The difference between gray-scale image Mean and the block Mean, adding 1 feature. Finally, the same features are selected for the contextual block, resulting in 48 features.

Since pre-processing sets the image background as black, for the feature extraction process only blocks containing all original pixels (i. e. foreground pixels) are considered. We used three classifiers, which are K-Nearest Neighbors (K-NN), Random Forests (RF) and Convolutional Network Architecture (U-net).

3.4   Training and Test Sets

A coffee specialist manually labeled each one of the collected images, identifying pests (Fig. 1(c)). To assign a label to each data instance, when moving the classification block on image to extract features, a label is associated with it as follows. If a percentage P of the pixels of the classification block belong to the same class, then block is labeled as this class. Otherwise, block is discarded. Considering non-pest areas, P is set to 100%, i.e., all 100 pixels of the 10×10 block should be equally classified. However, since the number of pest areas is scarce, P is set to 80%, i.e. if at least 80 pixels of the 10×10 classification block is in an area labeled as pest by the specialist, then the block is marked as a pest.

To evaluate the learning algorithms, we used 10-fold cross-validation to divide the data into training and test sets. Also, aiming to standardize the range of the features, we apply the z-score normalization using information from the training instances. The measures used to evaluate the pixel-wise performance of the classifiers are precision (P), recall (R) and f-measure (F), which are suitable metrics for evaluation in imbalance classes. We used Weka [8] to choose the best classifier, then the complete proposed methodology is developed in python.

3.5   Supervised Learning

Three different supervised learning algorithms are used in our experiments, which are RF, K-NN and U-net. These methods are used in studies of robotics.

Random Forest [3]. It is composed by an ensemble of tree-structured classifiers, which each tree casting a unit vote for the most popular class at input instance. Their popularity is mainly driven by their high computational efficiency during both training and evaluation while achieving state-of-the-art [12]. The classification of a new test example using an RF is based on vote of all individual trees (example, majority or averaging vote). For each tree, a new example traverses it from the root node to a leaf node, testing the feature values in each of the visited split nodes and selecting the corresponding next branch to follow.

K-Nearest Neighbors [14]. It is one of the most straightforward learning algorithms. It is a non-parametric algorithm, i.e. it does not make any assumptions on the underlying data distribution and is also instance-based, which does not build a decision model. K-NN does not have a training phase. A new test instance is classified by a majority vote of its K neighbors. The K nearest neighbors are computed using a distance function, for example, Euclidean distance.

U-Net [11]. It is one Convolutional Neural Network (CNN) architecture that has two phases. First the contraction, it uses several convolutions and max-pooling layers, which decreases the spatial resolution of the input while increasing the number of feature channels. Second the expansion, uses convolutions and transposed convolutions, reverting the effects of the first part, reducing feature channels and increasing the resolution. U-Net uses "skip" connections, connecting channel-wise the results of convolutions performed before each max-pooling to the effects of each transposed convolution. These skips are matched so that the result before the first max-pool layer is concatenated to the result of the last transposed convolution. The advantage of this approach is that they allow for the network to learn an initial result in few training steps without the need of modifying most network parameters.

4   Experimental Results

For image acquisition, the DroneDeploy software can be used to upload a map with a predefined route that must be covered by the UAV. The UAV flies over the route and takes pictures, which are uploaded to a computer after the flight and merged into an image containing the whole area observed, shown in Fig. 3.

Fig. 3. Orthomosaic from DroneDeploy software (data set).

The experimental analysis aims to compare three ML techniques (K-NN, RF, and U-net) using the segmented and non-segmented aerial images. We used 10-fold cross-validation to divide the data into training and test sets. In each round nine of them were used to train a model, while the remaining one served to validate the model (test data). The column "I" of the results Tables represents the image used in the test set, which also corresponds to the ith fold of the cross validation. For all experiments we used K-NN with $K = 1$ (number of neighbors) and *Euclidean distance*, RF with 100 trees and a modified U-Net, where the contraction step is replaced by a pretrain vgg16 network [15] whose parameters have already been found for a very large dataset (e.g. ImageNet, which contains 1.2 million images with 1000 categories). Moreover for U-net training is used 100 epochs.

Table 1 shows the result for K-NN and RF using 48 visual features extracted from blocks (as described in Sect. 3.3) using images without segmentation. Table 2 shows the result using images with segmentation. The U-net was used as a feature extractor and classification algorithm in both tables.

Tables 1 and 2 shows the precision (P), recall (R) and f-measure (F) for each image from test dataset as well as mean and standard deviation for each measure. In each Table, the best results are in bold.

Considering Table 1, U-net achieved the best results for precision in eight of the ten images, while K-NN achieved the worst results. In two images, the U-net obtained low precision (images 1 and 2), which are difficult to classify. The recall is low for most images in all classifiers, meaning that classifiers learn little about the pest concept. It is noteworthy that the number of labeled images available for training are few, and that most of the pixels in these images correspond to areas with healthy plants.

Table 1. Result K-NN, RF, and U-net using images without segmentation.

I	K-NN			RF			U-Net		
	P	R	F	P	R	F	P	R	F
1	0.18	0.73	0.28	0.37	0.77	0.50	0.52	0.70	0.59
2	0.41	0.90	0.57	0.59	0.90	0.71	0.67	0.80	0.73
3	0.01	0.59	0.02	0.03	0.65	0.05	0.88	0.43	0.58
4	0.51	0.43	0.47	0.90	0.23	0.36	0.90	0.35	0.51
5	0.62	0.40	0.49	0.97	0.18	0.30	0.96	0.39	0.55
6	0.56	0.43	0.49	0.88	0.23	0.36	0.88	0.40	0.56
7	0.58	0.58	0.58	0.78	0.52	0.62	0.91	0.43	0.58
8	0.78	0.48	0.60	0.94	0.43	0.59	0.98	0.31	0.48
9	0.64	0.53	0.58	0.93	0.47	0.63	0.97	0.38	0.49
10	0.66	0.45	0.54	0.93	0.37	0.56	0.98	0.32	0.49
M	0.50	0.55	0.46	0.73	0.48	0.47	0.87	0.45	**0.56**
SD	0.24	0.16	0.18	0.31	0.24	0.20	0.15	0.16	**0.07**

Considering Table 2, RF and U-net achieved the best results for precision in ten images. The recall is high for most images in all classifiers, because the soil information was manually removed. The manual segmentation process improved the classifiers' recall and f-measure performance. And again, the standard deviation for U-net is smaller compared to other learning techniques.

Figure 4 shows the results of the prediction of the three machine learning algorithms in two test images (Images 2 and 5). The Figs. 4a, b, and c show the results of K-NN, RF, and U-net for Image 2 without segmentation. While Figs. 4d, e, and f show the results of these algorithms for Image 5 with manual segmentation. The pixels in lilac represent the false positives (FP), the blues are the true positives (TP), and the red ones are the false negatives (FN).

For Figs. 4a, b, and c, it is possible to see a high false positive rate, especially for K-NN. Referring now to Figs. 4d, e and f false positive rates are low since they were manually segmented.

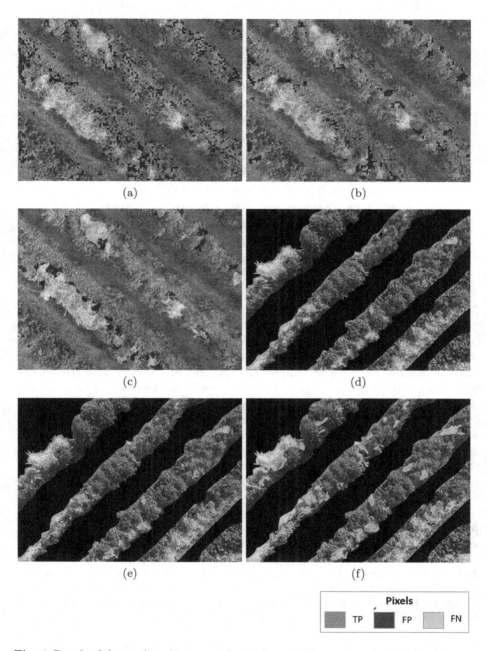

Fig. 4. Result of the predicted image without segmentation and segmented. (a–c) Image 2 without segmentation, predicted respectively by (a) KNN, (b) RF and (c) U-Net; (d–f) Image 5 segmented, predicted respectively by (d) KNN, (e) RF and (f) U-Net

Table 2. Result K-NN, RF, and U-net using segmented images.

	K-NN			RF			U-Net		
I	P	R	F	P	R	F	P	R	F
1	0.23	0.88	0.37	0.30	0.96	0.46	0.50	0.78	0.61
2	0.45	0.92	0.60	0.56	0.92	0.70	0.66	0.77	0.71
3	0.01	0.77	0.02	0.01	0.88	0.02	0.35	0.65	0.46
4	0.83	0.63	0.71	0.97	0.48	0.65	0.73	0.63	0.67
5	0.83	0.71	0.77	0.96	0.69	0.80	0.93	0.47	0.62
6	0.66	0.73	0.69	0.77	0.72	0.75	0.65	0.72	0.68
7	0.76	0.70	0.73	0.89	0.69	0.78	0.88	0.49	0.63
8	0.81	0.50	0.62	0.95	0.45	0.61	0.85	0.53	0.65
9	0.74	0.64	0.69	0.86	0.62	0.72	0.95	0.52	0.67
10	0.78	0.60	0.68	0.87	0.58	0.70	0.96	0.42	0.58
M	0.61	0.71	0.59	0.71	0.70	0.62	0.75	0.60	**0.63**
SD	0.27	0.12	0.22	0.31	0.17	0.22	0.21	0.13	**0.07**

5   Conclusions

A method for segmenting and detecting healthy and diseased plants has been proposed, using RGB images from an UAV. These features were processed by three distinct machine learning techniques (K-NN, RF, and U-net), and the results were compared using f-measure metric. The aim is the detection of nematode, a common coffee crop disease in country Brazil.

The result obtained by the proposed methodology shows that the F-measure average is equal to 63% for U-net with manual segmentation. Besides that, the standard deviation was the lowest compared with K-NN and RF. In this way, the results encourage to search for improvements in the methodology, from segmentation until classification. The major challenge to be faced is to obtain a larger set of labeled data to improve the learning phase of the model.

We believe these results are an essential step towards the advancement of precision agriculture. Further work will focus on applying techniques to work with imbalanced data sets and to improve the accuracy of the prediction model.

Acknowledgements. The Titan Xp used for this research was donated by the NVIDIA Corporation. This work was supported by the Federal University of Uberlandia, CNPq scholarship (process number 163641/2018-8) and CNPq under Grant 400699/2016-8.

References

1. Food and agriculture organization of the united nations (2018). http://faostat.fao.org
2. Boisseau, M., Aribi, J., Sousa, F.R., Carneiro, R.M.D.G., Anthony, F.: Resistance to meloidogyne paranaensis in wild coffea arabica. Trop. Plant Pathol. **34**, 38–41 (2009)
3. Breiman, L.: Random forests. Mach. Learn. 5–32 (2001). https://doi.org/10.1023/A:1010933404324
4. Calderon, R., Montes-Borrego, M., Landa, B., Navas-Cortés, J.A., Zarco-Tejada, P.J.: Detection of downy mildew of opium poppy using high-resolution multspectral and thermal imagery acquired with an unmanned aerial vehicle. Prec. Agric. **15**, 639–661 (2014)
5. Calderon, R., Navas-Cortes, J.A., Lucena, C., Zarco-Tejada, P.J.: High-resolution airborne hyperspectral and thermal imagery for early detection of verticillium wilt of olive using fluorescence, temperature and narrow-band spectral indices. Remote Sens. Environ. **139**, 231–245 (2013)
6. Campos, V.P., Villain, L.: Plant Parasitic Nematodes in Subtropical and Tropical Agriculture. CAB International, Wallingford (2005)
7. Garcia-Ruiz, F., Sankaran, S., Maja, J.M., Lee, W.S., Rasmussen, J., Ehsani, R.: Comparison of two aerial imaging platforms for identification of huanglongbing-infected citrus trees. Comput. Electron. Agric. **91**, 106–115 (2013)
8. Hall, M., Frank, E., Holmes, G., Pfahringer, B., Reutemann, P., Witten, I.H.: The WEKA data mining software: an update. SIGKDD Explor. **11**(1), 10–18 (2009)
9. Hall, M.A.: Correlation-based feature selection of discrete and numeric class machine learning (2000)
10. Otoboni, C.E.M., Gabia, A., Martins, A.S.: Determinação de áreas para o manejo localizado de nematoides em soja com o uso de imagens de vant. In: Congresso Brasileiro de Nematologia (2015)
11. Ronneberger, O., Fischer, P., Brox, T.: U-Net: convolutional networks for biomedical image segmentation. In: Navab, N., Hornegger, J., Wells, W.M., Frangi, A.F. (eds.) MICCAI 2015. LNCS, vol. 9351, pp. 234–241. Springer, Cham (2015). https://doi.org/10.1007/978-3-319-24574-4_28
12. Saffari, A., Leistner, C., Santner, J., Godec, M., Bischof, H.: On-line random forests. In: 12th International Conference on Computer Vision Workshops, pp. 1393–1400 (2009)
13. Salgado, S.M.L., Rezende, J.C., Nunes, J.A.R.: Selection of coffee progenies for resistance to nematode meloidogyne paranaensis in infested area. Crop Breed. Appl. Biotechnol. **14**, 94–101 (2014)
14. Shakhnarovich, G., Darrell, T., Indyk, P.: Nearest-Neighbor Methods in Learning and Vision: Theory and Practice (Neural Information Processing). The MIT Press, Cambridge (2006)
15. Simonyan, K., Zisserman, A.: Very deep convolutional networks for large-scale image recognition. arXiv preprint arXiv:1409.1556 (2014)
16. Souza, J.R., Mendes, C.C., Guizilini, V., Vivaldini, K.C., Colturato, A., Ramos, F., Wolf, D.F.: Automatic detection of ceratocystis wilt in eucalyptus crops from aerial images. In: 2015 IEEE International Conference on Robotics and Automation (ICRA), pp. 3443–3448. IEEE (2015)

Automatic Detection of Obstacles in Railway Tracks Using Monocular Camera

Guilherme Kano[1] 🆔, Tiago Andrade[1] 🆔, and Alexandra Moutinho[2]([✉]) 🆔

[1] Instituto Superior Técnico, Universidade de Lisboa, Av. Rovisco Pais 1,
1049-001 Lisboa, Portugal
{guilherme.kano,tiago.andrade}@tecnico.ulisboa.pt
[2] IDMEC, Instituto Superior Técnico, Universidade de Lisboa, Av. Rovisco Pais 1,
1049-001 Lisboa, Portugal
alexandra.moutinho@tecnico.ulisboa.pt

Abstract. This paper presents an algorithm for automatic detection of obstructions on railway tracks. Based on computer vision techniques, this algorithm extracts the railway tracks from the image feed and automatically detects obstacles that can endanger normal railway system operation, as well as the safety of its users. To segment the railway tracks, two techniques are explored. First, the Hough transform is used to detect straight lines, which proves to be inefficient when dealing with curves. To overcome this problem, an alternative solution is developed based on mathematical morphology techniques and BLOB (Binary Large OBject) analysis, leading to a more robust segmentation. The surrounding terrain is also subject to analysis. The algorithm's performance is evaluated considering different scenarios with and without simulated anomalies, demonstrating the effectiveness of the proposed solution.

Keywords: Railway obstruction detection · Monocular camera ·
Computer vision

1 Introduction

Train accidents caused by obstacles over or close to the railway amount up to 23% of the total number of accidents registered in 2017 [1]. Accidents have a significant social and economical impact as they can delay the normal schedule of passenger and cargo trains. The focus of safety measures is slowly being shifted to automated inspection systems. Commercial solutions [2] exist, which are based on a variety of sensors to scan the railways and feed information to a specialized external station with high-end processing units running machine learning tools, which are very reliable, although they are very costly and complex. For this reason, researchers look for solutions which require less computational resources and are more cost-efficient.

Several proposed systems employ a variety of sensors to acquire information to be used as inputs to machine learning or computer vision algorithms for

© Springer Nature Switzerland AG 2019
D. Tzovaras et al. (Eds.): ICVS 2019, LNCS 11754, pp. 284–294, 2019.
https://doi.org/10.1007/978-3-030-34995-0_26

processing. In [3], a solution is proposed based on image acquisition of the track from thermal camera sensors, which work under any lighting conditions, coupled with computer vision tools to detect obstacles. In [4], a fusion of radar and camera sensors is used to improve the scanning process, which is then fed to a computer vision algorithm. Another approach is presented in [5] where two pairs of unaligned cameras or, alternatively, a pair of laser scanners are used for rail reconstruction and inspection. Alternatives are available for detailed inspection of railways, although these are mostly focused on close rail inspection and are unable to detect obstacles on the line. In [6], a vision based algorithm is applied to a real time camera feed to scan the track for cracks and deformations.

This paper presents a low-cost solution for detection of obstacles on railways using a single monocular camera. The respective algorithm is based on computer vision operations applied to the feed of a single monocular camera mounted on a railway inspection vehicle, detecting obstructions in the railway track and large obstacles in close vicinity of the rails.

After this introduction, Sect. 2 states the working conditions of the system, presenting the assumptions made and the respective requirements. Section 3 presents the proposed algorithm, justifying the choice of each operation used. The algorithm is then tested in Sect. 4 against a benchmark of 60 images to evaluate its performance in quantitative and qualitative terms. Lastly, some concluding remarks are made in Sect. 5.

2   Problem Statement

The purpose of this work is to design a simple, low-cost system capable of detecting railway obstacles employing a single monocular camera. This problem can be reformulated to simplify its solution without changing its initial purpose.

Instead of directly identifying obstacles or anomalies in the track, the proposed solution is designed to detect the presence of an unobstructed set of rails. If possible, the segmentation of a sufficiently long section of a railway is a strong indicator of an unobstructed track.

A set of assumptions is needed as a starting point for the classification algorithm. The solution to the proposed problem assumes the following image acquisition premises:

1. *The monocular camera is mounted on an inspection vehicle travelling at a speed lower than a train's regular speed to ensure a correct analysis due to processing time;*
2. *The camera's perspective is fixed and facing the railway, being centered with the track and capturing its visible portion in its entirety;*
3. *The rails are presumed to span across the majority of the region of interest's area (described in Sect. 3.1) lengthwise, with a specific orientation range;*
4. *No more than one railway track should be identified, corresponding to the track subject to inspection.*

It is not uncommon for railways to be located near mountain ranges or unstable terrain with the risk of rock slides. For this reason, besides detecting obstacles within the railway track, the algorithm must also be capable of detecting any obstacles not directly obstructing the railway but within the train's critical area. Possible danger scenarios are represented in Fig. 1.

Fig. 1. Possible collision (left, middle) and simulated obstacle (right) scenarios (adapted from https://www.railway-technology.com/wp-content/uploads/sites/24/2017/10/cop9.jpg)

As such, a secondary problem arises, requiring the analysis of the terrain directly surrounding the train tracks.

Due to the inexistence of a specialized database and the limited availability of specific images which characterize critical situations, image editing software was used to simulate the image feed in critical scenarios. A set of images was generated through the use of unobstructed rail images and the placement of simulated obstacles within the railway. Both the newly generated as well as the original images were used to build the database.

With the objective guidelines stated in this section, the proposed solution is presented in the following section.

3   Algorithm Breakdown

The algorithm developed to tackle the proposed problem is based on segmentation and morphological operations applied to an image feed of a railway track. The analysis however is limited to processing methods applied to a single monocular camera and thus, no methods such as disparity or proximity analysis are possible. The individual steps of this algorithm are explained in depth in the following sections.

Figure 2 depicts a high-level version of the algorithm proposed.

Fig. 2. Algorithm activity diagram.

3.1 Preprocessing the Image

Prior to the analysis of the image feed, a preprocessing step is required. This step is crucial to the effectiveness of this algorithm and will greatly impact the overall performance of subsequent operations.

Firstly, the image histogram is equalized to enhance the contrast of the image, allowing for a better performance in edge detection. Then, with the intent of attenuating the response of high frequency and undesired features produced by the edge detection algorithm, a smoothing Gaussian filter is applied. Finally, a normalized region of interest (ROI) is extracted, discarding unnecessary sections of the original image (Fig. 3). This approach is developed and tuned based on assumption 2 (Sect. 2).

(a) Captured image   (b) Region of Interest

Fig. 3. Region of interest extraction

Two detection algorithms suited for edge detection are the vertical Sobel Operator [7] and Canny edge detector [7]. Since the Canny method has the ability to detect weak edges if connected to stronger edges, it is expected to output a response with more irrelevant edges.

The results of applying both detection methods are depicted in Fig. 4.

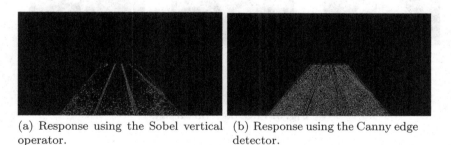

(a) Response using the Sobel vertical   (b) Response using the Canny edge
operator.   detector.

Fig. 4. Edge detection results.

As expected, the Canny edge detector produces a substantially denser edge response than the Sobel operator counterpart. Considering the main objective of the preprocessing phase, the proper detection of the rail's edges, the vertical Sobel operator proves to be better suited for this task.

3.2   Rail Segmentation

With the extraction of the relevant edges in the original image after pre-processing, the resulting image contains only the hard edges associated with distinguishable objects within said image (Fig. 4a). The algorithm must then be able to identify the edges matching the railway tracks.

In order to correctly identify the railway, a set of relevant features is required to distinguish these objects from the rest of the image. Based on rule 3 stated in Sect. 2, the edges are expected to have the same geometry as the rails. With this in mind, the classical Hough transform method can then be applied to identify the straight lines corresponding to the rail edges.

Hough Transform is a computer vision technique specialized in identifying object shapes such as curves and ellipses, as well as their positions based on feature discrimination [8]. In this application, Hough transform is used to identify straight lines.

By applying both an object count threshold to the number of desired Hough peaks and a range threshold to the slopes of the straight lines to be identified, the algorithm is constrained to identifying no more than two lines (and consequently, two Hough peaks) which correspond to the railway tracks (Fig. 5).

Candidate objects, characterized by a set of parameters similar to the Hough peak pairs, are merged to obtain the full length of the detected rail.

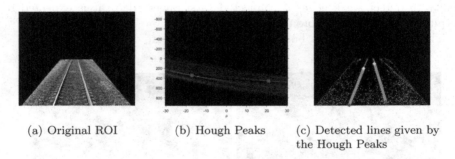

(a) Original ROI   (b) Hough Peaks   (c) Detected lines given by
                                                          the Hough Peaks

Fig. 5. Edge detection results

While slight curvatures in the railway tracks will still be correctly identified to an extent, this methodology is inadequate when dealing with high curvatures (Fig. 6), where the merging algorithm fails to find sufficiently long straight lines.

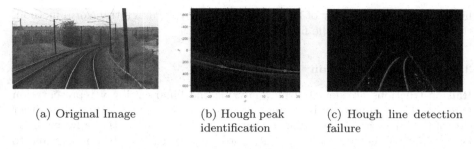

(a) Original Image   (b) Hough peak   (c) Hough line detection
                          identification   failure

Fig. 6. Hough line detection failure in curved track

Kano's Method is a proposed method based on mathematical morphology operations and BLOB analysis developed to tackle the curve limitation of the Hough transform algorithm. While the previously developed algorithm analyzes the rails' edges, this alternate method attempts to extract and reconstruct the rail prior to analysis. It requires an additional preprocessing step due to the high sensibility to small details, in which low pixel count connected components are removed (Fig. 7a).

Then, a closing operation with a small structural element is responsible for reconstructing objects from single edges, resulting in a binary image containing the rails and other objects (Fig. 7b). Objects that do not satisfy the specified orientation (angle measured between the horizontal and the major axis of the ellipses with the same second moment as the objects, Fig. 7(b) of the rails are discarded, leading to Fig. 7c. Lastly, a closing operation with a wider structural element is used to reconstruct the geometry of the rails (Fig. 7d). Due to the

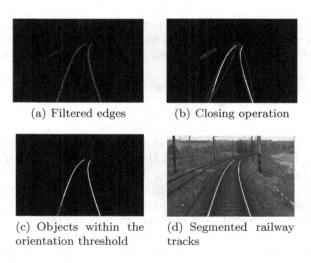

(a) Filtered edges   (b) Closing operation

(c) Objects within the   (d) Segmented railway
orientation threshold   tracks

Fig. 7. Railway segmentation

higher degree of reliability of Kano's method over the Hough transform method, the former is chosen for the final implementation.

3.3   Obstacle Detection

Rail obstructions are detected by analyzing the binary railway representation (Fig. 7d). If the number of blobs (characterized by its centroid) located in either side of the image is different than 1, there is an obstruction in the railway tracks. By computing the centroid's spatial coordinates of each blob it is possible to determine which track is obstructed.

Figure 8 depicts an obstruction caused by a landslide. Since two of the blobs' centroids are located in the left half of the image, the left track is the one to be identified as obstructed.

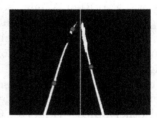

(a) Railway segmentation   (b) Representation of the blobs' centroids (circles).

Fig. 8. Obstruction in the left track caused by a landslide

In situations where the number of identified blobs in either side is equal to 1 (Fig. 9), this is not a sufficient condition to rule out the existence of obstructions. Instead, the vertical distance of each rail is needed, which is computed by

$$VertDist = L_{MajorAxis} \cdot |sin(\theta)| \qquad (1)$$

where $L_{MajorAxis}$ is the major axis' length of the ellipses with the same second moment as the blob and θ is the angle between $L_{MajorAxis}$ and the image's horizontal axis (orientation). If either of the detected rails' length ($VertDist$) is lower than a fraction of the height of the ROI ($Threshold \cdot ROIHeight$), the rail is assumed to be obstructed (Fig. 9).

If the detection of unobstructed rails is successful, the algorithm will then scan the surrounding terrain for any obstacles. It is worth noting that this analysis should not be required if a direct obstruction on the railway is found, which already constitutes a situation in which the passage of trains is no longer safe.

Based on [9], a *k-means* clustering algorithm coupled with a superpixel merging algorithm, is applied to the region of interest of the Gaussian filtered original

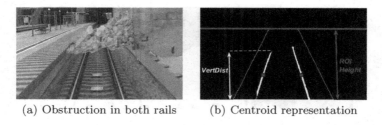

(a) Obstruction in both rails   (b) Centroid representation

Fig. 9. Obstruction identification principle

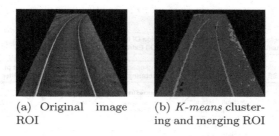

(a) Original image ROI   (b) *K-means* clustering and merging ROI

Fig. 10. Superpixel merging by k-means algorithm

image, grouping the entirety of the ROI into two regions based on color and texture. These distinct regions correspond to the track region and the surrounding terrain. The result of this algorithm composition is represented in Fig. 10.

With the extraction of the rails, the ROI image is segmented based on the positioning of the detected rails. A closing operation with a small structural element applied to each distinct segment is responsible for discarding any smaller discontinuities (gaps or small objects that do not pose a threat to trains). The result is represented in Fig. 11.

Fig. 11. ROI segmentation

An Euler number [7] image analysis is then performed in each distinct region to detect large obstacles and their locations. As such, an opening operation with a larger structural element is needed to allow for the detection of obstacles that do not completely obstruct one of the previous regions. An example of a positive detection is represented in Fig. 12.

The developed algorithm can be represented and summarized by Fig. 13

Fig. 12. Positive obstacle detection with Euler analysis

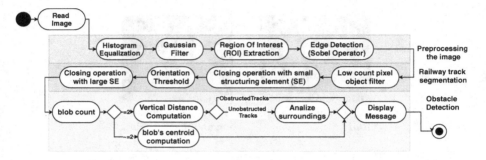

Fig. 13. Algorithm

4 Validation

To test the reliability of this implementation, a set of image examples (Fig. 14) compliant with the specifications of Sect. 2 are used as inputs.

A set of 60 images, 30 images without obstacles (e.g. Fig. 14a) and 30 with obstacles (e.g. Fig. 14b) was used to test the algorithm and some examples are shown in Fig. 14. Images (c) through (f) are correctly classified. However, in the case of figure (g), the classification is not correct due to the drastic variation in color in the vicinity of the track, which leads to an incorrect classification of the existence of an obstacle in the left side of the track.

The confusion matrix M for the validation set is

$$M = \begin{bmatrix} 27 & 12 \\ 3 & 18 \end{bmatrix} \quad \text{where} \quad \begin{bmatrix} TP & FP \\ FN & TN \end{bmatrix} \tag{2}$$

with TF the true positives, FP the false positives, FN the false negatives and TN the true negatives.

An overall accuracy of 75%, a specificity of 60% and a sensitivity of 90% were obtained. When developing prevention systems, a higher cost is assigned to the occurrence of false negatives as opposed to false positives, since false negatives can lead to casualties or economic losses. As such, the algorithm's performance is considered acceptable.

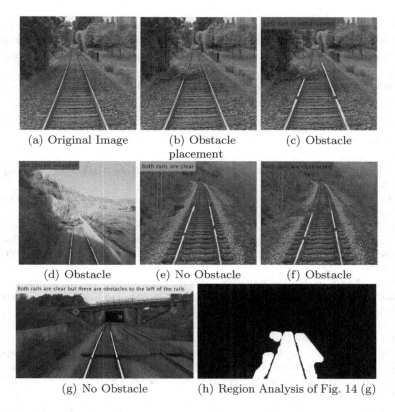

(a) Original Image   (b) Obstacle placement   (c) Obstacle

(d) Obstacle   (e) No Obstacle   (f) Obstacle

(g) No Obstacle   (h) Region Analysis of Fig. 14 (g)

Fig. 14. Algorithm validation. Ground truth in the corresponding sub caption (c, d, e, f)

5   Conclusion

This paper proposes a computer vision-based algorithm for detection of obstacles in railway tracks using a monocular camera. An obstacle is identified when the algorithm is unable to detect two long enough rail lines. Comparing the segmentation techniques, the developed Kano's method led to better results, especially when facing a curve.

Obstacle detection in the vicinity of the track is more prone to fail when these are not as easily distinguishable.

This implementation opens up a whole range of paths to follow for the development of future work. The system can be mounted on a variety of vehicles to perform the scans such as rail carts or drones. The algorithm can be improved upon to allow for the identification of the type of obstacles, classifying them in terms of designation and whether maintenance is needed to remove them in the case of stationary obstacles (such as rocks and land slides) or moving obstacles which may eventually leave the track (cars and other vehicles or people/animals).

Due to the unreliability when detecting obstacles in the vicinity of the railway track, the scanning method can be complemented by the addition of a proximity sensor, fusing the data with the image feed to allow for more accurate readings.

Acknowledgements. This research is funded by FCT, through IDMEC, under LAETA, project UID/EMS/50022/2019.

References

1. UIC Safety Database. https://safetydb.uic.org/IMG/pdf/sdb_report_2018_public. pdf. Accessed 26 May 2019
2. Pleora Technologies. https://www.pleora.com/markets/machine-vision/railway-inspection/. Accessed 26 May 2019
3. Berg, A., Öfjäll, K., Ahlberg, J., Felsberg, M.: Detecting rails and obstacles using a train-mounted thermal camera. In: Paulsen, R.R., Pedersen, K.S. (eds.) SCIA 2015. LNCS, vol. 9127, pp. 492–503. Springer, Cham (2015). https://doi.org/10.1007/978-3-319-19665-7_42
4. Ukai, M., Nassu, B.T., Nagamine, N., Watanabe, M., Inaba, T.: Obstacle detection on railway track by fusing radar and image sensor. In: 9th World Congress on Railway Research, Lille, France (2011)
5. Babenko, P.: Visual inspection of railroad tracks. Ph.D. dissertation, University of Central Florida (2009)
6. Singh, M., Singh, S., Jaiswal, J. and Hempshall, J.: Autonomous rail track inspection using vision based system. In: 2006 IEEE International Conference on Computational Intelligence for Homeland Security and Personal Safety, pp. 56–59 (2006)
7. Pratt, W.: Digital Image Processing. John Wiley and Sons Inc., New York (1991)
8. Illingworth, J., Kittler, J.: A survey of the Hough transform. Comput. Vis. Graphics, Image Process. **44**(1), 87–116 (1988). https://doi.org/10.1016/S0734-189X(88)80033-1
9. Mathworks Documentation. https://www.mathworks.com/help/images/land-classification-with-color-features-and-superpixels.html. Accessed 23 Apr. 2018

A Sequential Approach for Pain Recognition Based on Facial Representations

Antoni Mauricio[1]($\boxtimes$) (iD), Fábio Cappabianco[2], Adriano Veloso[3],
and Guillermo Cámara[4]

[1] Department of Computer Science, Universidad Católica San Pablo, Arequipa, Peru
manasses.mauricio@ucsp.edu.pe
[2] Group for Innovation Based on Images and Signals,
Universidade Federal de São Paulo, São Paulo, Brazil
cappabianco@unifesp.br
[3] Department of Computer Science, Universidade Federal de Minas Gerais,
Belo Horizonte, Minas Gerais, Brazil
adrianov@dcc.ufmg.br
[4] Department of Computer Science, Federal University of Ouro Preto, Ouro Preto,
Minas Gerais, Brazil
guillermo@iceb.ufop.br

Abstract. Pain assessment is a hard subjective problem, but still, it
is critical in many medical situations. Many computational approaches
explore pain detection and estimation using different types of data and
descriptors. Among these, spontaneous facial expressions coded by the
Facial Action Coding System (FACS) have achieved outstanding results
in frame-by-frame stationary analysis, but not in temporal analysis. We
explore spatiotemporal features extracted from video sequences considering pain stimuli as references in the temporal analysis. Our proposal
focuses on guided learning by warping the appearance surround the facial
action units (AUs). The facial features from frames are processed sequentially to extract their temporal correspondences. These sequences are
generated from the original videos and must represent a single-stimulus
effect in a short period, so we develop generation policies. Experimental
results on the publicly available UNBC-McMaster database have demonstrated that our approach yields significant advances over the state-of-
the-art.

Keywords: Spatiotemporal features · Pain recognition · Deep learning

1 Introduction

Over the last decade, the adult population suffering from at least one condition
associated with chronic pain has increased from 20% to 50% concerning the world

The present work was supported by grant 234-2015-FONDECYT (Master Program)
from Cienciactiva of the National Council for Science, Technology and Technological
Innovation (CONCYTEC-PERU).

D. Tzovaras et al. (Eds.): ICVS 2019, LNCS 11754, pp. 295–304, 2019.
https://doi.org/10.1007/978-3-030-34995-0_27

population [15]. These figures place chronic pain as one of the foremost medical issues. Besides, pain management is considered critical in clinical settings, so it is taken as a health condition indicator [5]. Traditionally, pain assessment relies on patients' self-reports through well-known metrics like Visual Analog Scale (VAS) or Observed Pain Intensity (OPI). Though useful, these metrics lack objectivity due to its dependence on the patient's expression capabilities and the doctor's interpretation alongside other external factors [2,3]. Recently, Prkachin and Solomon Pain Intensity (PSPI) metric has arisen as an objective and trusted model based on the face's physiognomy. Moreover, several works [11,12,17,21] have demonstrated that facial expressions show a reflective and spontaneous reaction to pain stimuli, thus making them reliable indicators [18,19].

Many interdisciplinary research works have raised several datasets for pain analysis in different settings. In 2011, Lucey et al. [10] published the UNBC-McMaster Shoulder Pain Expression Archive Database[1], since then, the majority of papers take it as the benchmark. Nevertheless, the temporal correlation of pain is still a hard problem due to variations in the pain stimulus effect over time. A conditioned stimulus generates different tendencies along a sequence so similar to multiple-point stimuli effect [18]. Trends change sporadically and plot a distribution alike to the response cycle of a single-point stimulus. These response cycles are often ignored or considered as noise even though they not only provide information about the patient's condition but also about the stimulus [18].

Like any real-world dataset, the UNBC-McMaster Shoulder Pain Database has several problems that involve image registration, unbalance, and non standardized images. We follow standardized procedures for feature extraction in frames but modifying its parameters [12,17,21]. Then, we used a recurrent architecture similar to those raised in [12,17], but changing GRU instead of LSTM units and adding one more recurrent layer. Furthermore, we tune several hyperparameters, not only the learning ones but those related to the sequences resampling and subsequences size. These changes behalf the analysis of shorter response cycles without affecting the results in longer ones.

In summary, this paper has two significant contributions. (1) A balanced subsequence generation considering the response cycle of a stimulus and its shifts. (2) A novel facial warping model which performs either as a data augmentation method and a guided learning strategy.

2   Related Work

Majority of studies have adopted many handcrafted features to recognize the AUs related to pain including Gabor Wavelets [9,10,20], Histogram of Topographical Features (HoT) [4], Histogram of Oriented Gradients (HOG) [7], Local Binary Patterns (LBP) [6,20], among others. The results and metrics vary depending on the analysis aim and the number of pain levels analyzed. Most research works [7–10,14] use only two categories ("*no pain*" and "*pain*") for pain detection, achieving almost 97% in accuracy and an AUC of 0.86. These results are much

---

[1] http://www.consortium.ri.cmu.edu/painagree/.

better from those obtained by other authors because among more pain categories are considered, the classification is harder [12,16,17].

Florea et al. [4] transfer the topographical facial features from the Cohn-Kanade dataset (CK+)[2] to guide the feature extraction in the UNBC-McMaster Shoulder Pain dataset. Pedersen [14] untangles face identity from rigid deformations by a Discriminative Feature Extractor (DFE). Similarly, Rathee et al. [16] represent the facial deformations using rigid and non-rigid parameters by Thin Plate Spline (TPS) mapping. Then, deformation parameters are mapped into a higher discriminative space to carry out a 16-classes classification. On the other hand, Kaltwang et al. [6] propose a continuous pain estimation using LBP and Discrete Cosine Transform (DCT) as features. Zhao et al. [20] explore the temporal correlations through an Ordinal Support Vector Regression (OSVR) using Gabor coefficients and LBP features as inputs.

Among the feature learning-based methods, CNNs are the most remarkable for many end-to-end applications. In pain analysis, almost every deep learning related efforts focus on the temporal analysis; hence, sequential models run above CNNs. Zhou et al. [21] propose a recurrent CNN to achieve a continuous pain estimation. Face images are flattened into 1D vectors and concatenated along their columns to feed an RCNN considering the last frame's label as output. Nasrollahi et al. [12] propose a CNN-LSTM to achieve a 14-classes classification by using a fine-tuned VGG_faces [13] as the facial feature extractor. They apply resolution changes as a data augmentation policy. Finally, one LSTM layer performs the sequential analysis using the last frame's label as the sequence's label. Following the previous work, Rodriguez et al. [17] add a preprocessing scheme like Zhou et al. [21] for a 5-levels pain classification. Furthermore, they prioritize the frontalization problem and develop a data augmentation module by landmark-based random deformations and vertical flipping.

To conclude this review, the following points should be taken to do a fair comparison: (1) the number of pain levels influences the estimation results; and, (2) the temporal analysis is much more complex to achieve however it has greater medical acceptance than the stationary analysis.

3   Dataset Description

The UNBC-McMaster dataset records 129 patients self-identified with chronic shoulder pain. In each video, the patient moves one arm, if it is the healthy one, then every frame and the whole sequence are labeled as painless. Otherwise, each frame is evaluated by the PSPI metric, while offline parameters (VAS and OPI) appraise the sequences. Additionally, the dataset includes 66 facial landmarks per each frame computed by an Active Appearance Model (AAM). Figure 1 shows a sequence and its distribution regarding pain intensity. Each sequence presents several disturbances in its pain level even though there is just one pain stimulus

---

[2] The Cohn-Kanade AU-Coded Facial Expression Database is labeled for 6 emotions and includes 486 sequences from 97 posers and 66 facial landmarks per frame.

(harmed shoulder motion) – higher the pain intensity, shorter its duration due to the patient's reflexes.

Unbalance: The database presents a huge imbalance in its distribution. At frames level, the first-class "painless" has 82.71% of all data, the second one contains 6.01%, while the last one "extreme pain" reaches only 0.01%. At sequences level, each patient has different amounts of video clips; also, each sequence has different durations, between 84 to 615 frames. Albeit the majority of initial and final frames represent the patients before the stimulus acts or after its effect ends.

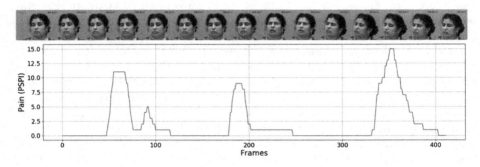

Fig. 1. Above: Original frames. Below: Frames distribution vs. pain intensity.

4   Methodology

Our proposal includes a preprocessing scheme and balancing policies. We present two architectures, one for the stationary analysis and one fo the temporal analysis. The implementation details are given per stage alongside explanations.

4.1   Preprocessing

The preprocessing has three steps: masking, frontalization, and resize. Firstly, we calculate the facial landmarks to apply the convex hull algorithm for masking. The frontalization aims to solve the camera perspective error by estimating the projection matrix [1]. We use frontal-view markers of a 3D-model[3] as a reference to calculate the camera matrix and the projection matrix. The projection matrix maps the original image to get the canonical normalized appearance, which undergoes a smooth symmetry process. We select the landmarks which surround the facial action units of pain to preserve the original features after frontalization; those are the chin, eyebrows and, mouth, and eyes commissures.

---

[3] https://github.com/dougsouza/face-frontalization.

4.2 Data Balancing

The unbalance affects every distribution, including patients' samples and pain levels. We propose a downsampling step and data augmentation policies to face the unbalance depending on the processing level.

Frames Balancing. Let $M_{[p,l]}$ a the data distribution matrix over the patients and pain levels, being $\mathbb{P}$ and $\mathbb{L}$ the set of patients and levels, respectively. $T_{[p,l]}$ is the scaling matrix that balance $M_{[p,l]}$, being τ_{max} its fixed maximum value. When the scaling factor τ is less than 1, then downsampling is done with a probability of τ; else, augmentation policies run with the same odds, except for the facial affine deformations. The policies include rotations ($3°, 6°$ and $9°$), vertical flipping and facial affine deformations.

Facial Affine Deformations: Facial deformations strongly affect the feature vector depending on its magnitude and location. By orienting the distortions, we seek to reinforce the pain features representation. We use Delaunay triangulation and piecewise-affine warping for the facial deformations. Deformations intensity depends on the landmark's relationship with the facial action units of pain proportionally from 0.2% to 10%.

Sequences Balancing. Sequence balancing has two stages; in the first one, the sequences split to generate new sub-sequences. In the second stage, the sub-sequences balancing happens and applies the same frame balancing methods considering that the label of a sub-sequence is its last frame's label. Data augmentation policies work in every frame of a sub-sequences with the same parameters (angles, odds, etc.).

Sub-sequence Generation: The sequences split into a length subsequences, considering a single-stimulus response cycle and the segments that do not express pain. At the beginning of a sub-sequence, there may be b painless frames. The sub-sequences generation steps are: (1) the sequences split into single-stimulus response cycles and painless sub-sequences; (2) in case that a new cycle is a replica then it is attached to the previous period, however, if gradients change abruptly, then it is considered as a new cycle; (3) eliminate the sub-sequences with sizes smaller than a; and, (4) multiple sub-sequences are made using a displacement window over the previous fragments. Figure 2 illustrates the stages in color scale for gradients; being red for ∇^+, blue for ∇^-, yellow for ∇^0 and white for painless.

4.3 Frame-by-Frame Analysis

Pain analysis at frames level is an end-to-end problem that can be managed by a CNN. We fine-tuned the VGG_faces[4] [13] using the UNBC-McMaster and Cohn-Kanade datasets. VGG_faces is designed to accept 224×224 RGB format images

---

[4] http://www.robots.ox.ac.uk/~vgg/software/vgg_face/.

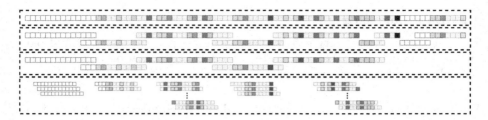

Fig. 2. From top to bottom. **First**: Original sequences. **Second**: Gradient based split. textbf Third: Sub-sequences merge and elimination based on sizes and gradients. **Fourth**: Generation through a 12 length window with an unitarian displacement. (Color figure online)

as input, and it is composed of 16 convolutional layers. We employ an SGD with a momentum of 0.9, a learning rate of 0.001, and a mini-batch size of 100 to fine-tune VGG_faces to the face recognition task. Then, we fix the convolutions to work as a feature extractor. Finally, the fully connected layers train for the pain classification task using the feature map as input.

4.4   Temporal Analysis

The temporal analysis of pain is complicated due to the variability of the stimulus-response cycle. UNBC-Mcmaster dataset holds records of shoulder pain stimulated by arm moves. These movements require dozens of interconnected muscles moves; hence, it is complex to analyze from the punctual-stimulus assumption. The shock registered in a sequence depends on the pain expressed in each of its frames. Nonetheless, every frame express the pain sensation from the previous frames, at least from the most recent. Then, we use the PSPI metric to label the subsequences for balancing and training.

Figure 3 illustrates the spatiotemporal architecture. The prior fine-tuned CNN works as a feature extractor at frames level; then a recurrent network correlates the feature vectors of the sequence. We design a two-layer GRU architecture due to GRU's buffer the gradient vanishing problem in short sequences.

5   Experiments and Results

We employ the Pytorch framework and a PC with the following settings: 3,6 GHz Intel Core i7 processor, 16 GB 3000 MHz DDR4 memory and NVIDIA GTX 1080ti, for every experiment. The upcoming subsections present the results of the proposed steps.

5.1   Balancing Results

We consider 350 samples per level for frame balancing, while for sequence balancing only 250 are contemplated. Both values were set based on the original

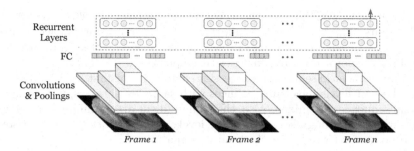

Fig. 3. Overview of the spatiotemporal architecture. A pre-trained CNN performs the spatial extraction of features. Meanwhile, a two-layer GRU network explorers the temporal correspondence among the features vectors.

distribution. As seen in Table 1, some levels have insufficient samples; so those do not meet the data quota, even after the maximum augmentation. We generate 20-frames sequences with five painless frames of tolerance.

Table 1. Data distribution before and after balancing. **Above**: Frame balancing per level. **Below**: Sequence balancing per level.

		Classes															
		0	1	2	3	4	5	6	7	8	9	10	11	12	13	14	15
Samples Distribution	Frames	40029	2909	2351	1412	802	242	270	53	79	32	67	76	48	22	1	5
		48398 original frames															
		350	350	350	350	350	350	350	350	350	350	350	350	350	350	16	80
		4996 balanced frames															
	Sequences	33140	2160	2494	1052	577	122	210	31	61	15	23	46	36	7	1	5
		39980 generated subsequences															
		250	250	250	250	250	250	250	250	250	240	250	250	250	112	16	80
		3448 balanced subsequences															

5.2 Testing

After balancing, we split the data into training (80%) and testing (20%) sets. We chose the one-subject-out strategy for performance measurement. Figure 4 shows an almost diagonal confusion matrix at frames level. The accuracy depends on the dataset size, albeit data augmentation cushions the unbalance the majority of cases. The closest levels tend to have similar feature vectors; then, some blocks appear in the confusion matrix.

At sequences level, we use the `fc6` layer outputs of the `VGG_faces` as the RNN's inputs. The `fc6` layer learns low-level facial descriptors; hence, it works better than `fc7` and `fc8` layers. Table 2 presents our results at frames level, while Table 3 shows the same metrics at sequences level. The results show a slight improvement in sequential processing, mainly in the first levels. One reason is that the first levels have subsequences with same-level frames, while the last-levels subsequences are mixed. Also, the pain gradient draws different trends that complicate the learning process on some levels. Table 4 plots the comparison of the-state-of-the-art results for the temporal analysis of pain.

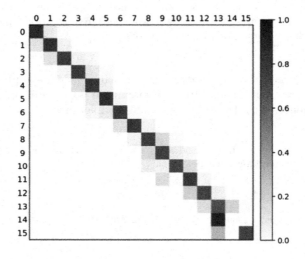

Fig. 4. Confusion matrix for 16-classes classification at frames level.

Table 2. Accuracy per class at frames level.

Frames level - Accuracy																
0	1	2	3	4	5	6	7	8	9	10	11	12	13	14	15	Prom
92.4	88.1	86.9	84.7	83.4	83.3	86.1	82.4	80.8	81.3	79.8	82.5	74.3	74.6	10.7	86.4	**82.72**

Table 3. Accuracy per class at sequences level.

Sequences level - Accuracy																
0	1	2	3	4	5	6	7	8	9	10	11	12	13	14	15	Prom
95.7	90.1	90.6	87.3	86.8	88.9	85.1	84.1	80.9	80.2	83.4	86.3	75.3	73.5	13.2	90.2	**85.37**

Table 4. Comparison of the-state-of-the-art metrics at sequences level.

	MAE	MSE	PCC	ICC	ACC
Zhou et al. [21]	-	1.54	0.65	-	-
Nasrollahi et al. [12]	-	-	-	-	0.619
Rodriguez et al. [17]	0.5	0.74	**0.78**	0.45	-
Our	**0.462**	**0.622**	0.687	**0.615**	**0.854**

6   Conclusions and Future Works

The automatic pain assessment is a significant medic issue that can be analyzed from different approaches using many types of data. Nonetheless, RGB videos are a low-cost solution to capture the facial expressions of pain, considering the spatiotemporal correlation. On the other hand, data distribution has a high-impact over the features representation; hence, data balancing acquires quite an importance. Our proposal faces the unbalance by structured augmentation policies, alongside a fine-tuning step for the spatial analysis; thus, data balancing increases its relevance. We overcome the unbalance by formal augmentation policies, alongside a preprocessing step to ease the spatial analysis.

Based on the results, the sub-sequence generation algorithm shows a good impact on the results in the temporal analysis. Also, the fine-tuning process improves metrics at frame-by-frame level, albeit the classification for closest classes incurs more considerable confusion. Hence, the more degrees of pain, the higher the difficulty in classification. Finally, we surpass most of the-state-of-the-art metrics because we consider the pain cycle responses for the sub-sequence generation. In the future, we are planning to untangle high-level expression features from low-level pain features before the temporal analysis.

References

1. Banerjee, S., et al.: To frontalize or not to frontalize: do we really need elaborate pre-processing to improve face recognition? In: 2018 IEEE Winter Conference on Applications of Computer Vision (WACV), pp. 20–29. IEEE (2018)
2. Bijur, P.E., Silver, W., Gallagher, E.J.: Reliability of the visual analog scale for measurement of acute pain. Acad. Emerg. Med. **8**(12), 1153–1157 (2001)
3. Egede, J.O., Valstar, M.: Cumulative attributes for pain intensity estimation. In: Proceedings of the 19th ACM International Conference on Multimodal Interaction, pp. 146–153. ACM (2017)
4. Florea, C., Florea, L., Vertan, C.: Learning pain from emotion: transferred hot data representation for pain intensity estimation. In: Agapito, L., Bronstein, M.M., Rother, C. (eds.) ECCV 2014. LNCS, vol. 8927, pp. 778–790. Springer, Cham (2015). https://doi.org/10.1007/978-3-319-16199-0_54
5. Glowacki, D.: Effective pain management and improvements in patients' outcomes and satisfaction. Critical Care Nurse **35**(3), 33–41 (2015)

6. Kaltwang, S., Rudovic, O., Pantic, M.: Continuous pain intensity estimation from facial expressions. In: Bebis, G., et al. (eds.) ISVC 2012. LNCS, vol. 7432, pp. 368–377. Springer, Heidelberg (2012). https://doi.org/10.1007/978-3-642-33191-6_36

7. Khan, R.A., Meyer, A., Konik, H., Bouakaz, S.: Pain detection through shape and appearance features. In: 2013 IEEE International Conference on Multimedia and Expo (ICME), pp. 1–6. IEEE (2013)

8. Lo Presti, L., La Cascia, M.: Using hankel matrices for dynamics-based facial emotion recognition and pain detection. In: Proceedings of the IEEE Conference on Computer Vision and Pattern Recognition Workshops, pp. 26–33 (2015)

9. Lucey, P., et al.: Automatically detecting pain in video through facial action units. IEEE Trans. Syst. Man Cybern. Part B (Cybernetics) **41**(3), 664–674 (2011)

10. Lucey, P., Cohn, J.F., Prkachin, K.M., Solomon, P.E., Matthews, I.: Painful data: the unbc-mcmaster shoulder pain expression archive database. In: 2011 IEEE International Conference on Automatic Face & Gesture Recognition and Workshops (FG 2011), pp. 57–64. IEEE (2011)

11. Martinez, D.L., Rudovic, O., Picard, R.: Personalized automatic estimation of self-reported pain intensity from facial expressions. In: 2017 IEEE Conference on Computer Vision and Pattern Recognition Workshops (CVPRW), pp. 2318–2327. IEEE (2017)

12. Bellantonio, M., et al.: Spatio-temporal pain recognition in CNN-based super-resolved facial images. In: Nasrollahi, K., et al. (eds.) FFER/VAAM -2016. LNCS, vol. 10165, pp. 151–162. Springer, Cham (2017). https://doi.org/10.1007/978-3-319-56687-0_13

13. Parkhi, O.M., Vedaldi, A., Zisserman, A., et al.: Deep face recognition. In: BMVC, no. 1, 6 (2015)

14. Pedersen, H.: Learning appearance features for pain detection using the UNBC-McMaster shoulder pain expression archive database. In: Nalpantidis, L., Krüger, V., Eklundh, J.-O., Gasteratos, A. (eds.) ICVS 2015. LNCS, vol. 9163, pp. 128–136. Springer, Cham (2015). https://doi.org/10.1007/978-3-319-20904-3_12

15. Raffaeli, W., Arnaudo, E.: Pain as a disease: an overview. J. Pain Res. **10**, 2003 (2017)

16. Rathee, N., Ganotra, D.: A novel approach for pain intensity detection based on facial feature deformations. J. Vis. Commun. Image Representation **33**, 247–254 (2015)

17. Rodriguez, P., et al.: Deep pain: exploiting long short-term memory networks for facial expression classification. IEEE Trans. Cybern. **99**, 1–11 (2017)

18. Turk, D.C., Melzack, R.: Handbook of Pain Assessment. Guilford Press, New York (2011)

19. Wang, F., et al.: Transferring face verification nets to pain and expression regression. arXiv 1702 (2017)

20. Zhao, R., Gan, Q., Wang, S., Ji, Q.: Facial expression intensity estimation using ordinal information. In: Proceedings of the IEEE Conference on Computer Vision and Pattern Recognition, pp. 3466–3474 (2016)

21. Zhou, J., Hong, X., Su, F., Zhao, G.: Recurrent convolutional neural network regression for continuous pain intensity estimation in video. In: Proceedings of the IEEE Conference on Computer Vision and Pattern Recognition Workshops, pp. 84–92 (2016)

A Computer Vision System Supporting Blind People - The Supermarket Case

Kostas Georgiadis[1,2], Fotis Kalaganis[1,2], Panagiotis Migkotzidis[1],
Elisavet Chatzilari[1(✉)], Spiros Nikolopoulos[1], and Ioannis Kompatsiaris[1]

[1] Information Technologies Institute, Centre for Research and Technology Hellas,
57001 Thermi, Greece
{kostas.georgiadis,fkalaganis,migkotzidis,ehatzi,nikolopo,ikom}@iti.gr
[2] Artificial Intelligence and Information Analysis Lab, Department of Informatics,
Aristotle University of Thessaloniki, Serres, Greece

Abstract. The proposed application builds on the latest advancements
of computer vision with the aim to improve the autonomy of people with
visual impairment at both practical and emotional level. More specifically, it is an assistive system that relies on visual information to recognise the objects and faces surrounding the user. The system is supported
by a set of sensors for capturing the visual information and for transmitting the auditory messages to the users. In this paper, we present
a computer vision application, e-vision, in the context of visiting the
supermarket for buying groceries.

Keywords: Computer vision system · Assistive device · Visually
impaired

1 Introduction

Although computer vision-based assistive devices for vision-impaired people have
made great progress, they are confined to recognising obstacles and generic
objects without taking into account the context of the activities performed by
the person. This context differentiates significantly the functional requirements
of an assistive device. Indeed, in the context of visiting a supermarket, a vision-
impaired person would need different categories to be recognised compared to
going for a walk on the beach, e.g. products and trails vs people, signs and
obstacles respectively. However, existing solutions do not take into account this
context and their design and structure is based on generic categories.

For example, in seeingAI[1], a mobile application that recognises what the
camera sees and narrates it through the speakers, the user must select between
generic categories for recognition, such as reading text, recognising barcodes,
detecting people and describing colours. In a similar vein, applications also utilising the mobile phone sensors (i.e. camera and speakers) Envision[2] and Eye-D[3]

---

[1] https://www.microsoft.com/en-us/ai/seeing-ai.
[2] https://www.letsenvision.com/.
[3] https://eye-d.in/.

© Springer Nature Switzerland AG 2019
D. Tzovaras et al. (Eds.): ICVS 2019, LNCS 11754, pp. 305–315, 2019.
https://doi.org/10.1007/978-3-030-34995-0_28

again provide options between these generic categories, with the latter one also enabling GPS-based functionalities (e.g. where am I, what is around me).

Besides the mobile-based applications assisting the visually impaired, another approach is the systems based on glasses. Prominent examples of this category are OrCam MyEye 2[4], eSight[5], NuEyes[6] and Eyesynth[7]. OrCam MyEye 2 is a small device with a camera that is attached to the users' glasses and is able to recognise up to 100 custom objects inserted by the user (e.g. products, people), read text and recognise barcodes. On the other hand, eSight and NuEyes are glasses that work as digital magnifiers, enhancing the vision of the partially visually-impaired user. Finally, Eyesynth is a pair of glasses accompanied by a portable microcomputer communicates the surroundings of the user through the bones of the head and can be used mainly for avoiding obstacles.

On the other hand, the proposed system (e-vision) is a hybrid approach, combining the movement freedom of an external camera with the processing power and the penetration rate of mobile phones. Besides the system design, its main novelty lies in the context-aware design of the application, i.e. the structure of the application is built based on a specific context. In more detail, we showcase the supermarket context, an activity of daily living that is easy for the seeing people and currently impossible for the visually-impaired. In this context, if we rely on existing solutions, we have two options; first, one could use the text reading option, which would read out loud all the text detected in all of the products that are seen by the camera, providing an annoying experience to the user. Second, one could use the barcode recognition option, which however would mean that the user has to pick each product in the supermarket, find the angle where the barcode is visible to the camera and repeat until they find the product of interest.

In the presented system, towards creating a user friendly experience, we consider the abstraction levels of information needed to be communicated when we visit a supermarket and that is if we are looking at a trail, shelf, product or we are at the entrance/exit of the supermarket. In each abstraction level the user would need different levels of information communicated to them (i.e. in the trail level the system should be able to say you are at the drinks trail, in the shelf level that you are looking at the beers and at the product level that you are holding a specific brand of a beer). As a result, the communication of the system with the user is hassle-free providing a pleasant context-aware experience. Eventually, for the detection between the abstraction levels we rely on computer vision through deep learning, while OCR is used for detecting the text on each image and a supermarket database in combination with the detected abstraction levels is used to refine what will be communicated to the user.

---

[4] https://www.orcam.com/en/.
[5] https://www.esighteyewear.com/int.
[6] https://nueyes.com/.
[7] https://eyesynth.com/?lang=en.

2   Related Work

Lately, towards the new self-service supermarkets (e.g. Amazon Go), there have
been a few efforts on recognising products through object detection. Early efforts
relied on the combination of SIFT-alike descriptors, such as [2], where dense SIFT
is combined with LLC to retrieve the best candidate classes and then a GA-based
algorithm is used to create the final list of products seen in an image. In [1], SIFT-
alike descriptors are combined with deep learning to generate attention maps
from a combination of SIFT, BRISK and SURF features in order to achieve
one-shot deep learning of products, utilising a single image per product. One-
shot recognition of products is also proposed by Karlinsky et al. [5], where a
probabilistic model is employed to generate bounding boxes and then deep fine-
grain refinement based on the VGG-f network is applied to the coarse results. In
a similar vein, in [8], a ROI detector based on Yolo_v2 for detecting the product-
agnostic bounding boxes is combined with global feature matching between the
database of products and the features from a fine-tuned CNN (VGG-16) of the
bounding box. Finally, more closely related to e-vision is the work in [3], where
an application for assistive grocery shopping is presented. More specifically, an
image-based product classification scheme is proposed combining HOG features
with discriminative patches and SVMs that can classify a test image between
26 coarse classes (e.g. coffee, soft drinks). On the contrary, e-vision, combining
OCR and deep learning features, is a fully functional system that can provide
the information that the user is looking for, including the full scale of grocery
products (i.e. more than 100k products), a scalability that is offered due to the
fully unsupervised nature of the proposed system.

3   The Concept of the Proposed Supermarket Application

Our objective is develop an application that can help the visually impaired in
their daily living. In this direction, the proposed system comprises of 3 sen-
sors/devices; (i) a camera that can be attached to the person's head/glasses
or other body part, (ii) a mobile device with an accompanying application that
gets the images from the camera and through the proposed methodology (Sect. 5
extracts meaningful information for what the camera is seeing and (iii) a set of
earphones that connect to the application and provide auditory feedback to the
user with respect to the previously extracted information using text-to-speech
technologies.

    In this work we focus on one activity that is very common and frequent in our
lives, a visit in a supermarket and a session of buying a set of products. In our
case, we consider the Greek case, i.e. a Greek supermarket chain (Masoutis[8]),
with a variety products, a combination of Greek and English labels and a Greek
text-to-speech solution. If we decompose a visit to the supermarket, we can
distinguish between 4 levels of abstraction. More specifically, these levels from
specific to abstract are described below:

---

[8] https://www.masoutis.gr/.

Product: The user has a specific product in their hands and they want the system to tell them what is the product (e.g. in the case of Fig. 1a, the system should say that it is *Edesma charcuterie from chicken fillet*).

Shelf: The user is in front of a shelf with a limited number of relevant products and they want the system to tell them what is the fine-grained category of the products (e.g. in the case of Fig. 1b, the system should say that you are in front of a *Shelf with chocolates*).

Trail: The user is in front of a trail with a large number of relevant products and they want the system to tell them what is the coarse-grained category of the products (e.g. in the case of Fig. 1c, the system should say that you are at the *Trail with frozen products*).

Other: The user is in the entrance/exit of the supermarket (Fig. 1d) and in this case the system informs them that they are not in front of a shelf or trail.

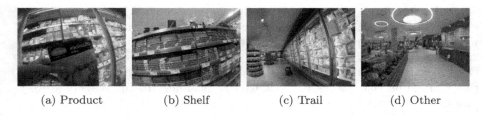

(a) Product   (b) Shelf   (c) Trail   (d) Other

Fig. 1. Example images for each abstraction level

4   Onsite Visits Dataset and Annotation

In order to evaluate our system, we performed onsite visits to two supermarkets with two individuals participating, one with partial and one with complete visual impairment, who were set up with a GoPro camera and were asked to go through a grocery shopping session. Two different stores were selected for the onsite visits, one hypermarket and one small local store so as to have images with varying lightning conditions and also diverse surroundings. The scenarios were identical in both cases and required each participant to complete a full visit in the supermarket, purchasing five distinct pre-defined products. Once the participant acquired all five products, he was instructed to proceed to the cashier where the scenario was completed. Throughout the entirety of the visit the participant had a GoPro camera attached to his chest and was instructed to use the voice command "GoPro Take Photo" in cases where the user wanted the application to provide information to them. This resulted in capturing 39 photos in total, 16 and 23 from the first and second respectively, used for evaluating e-vision.

The acquired images were annotated in two stages: (1) abstraction level (2) category level. In the first case, images were visually inspected to provide a class label (i.e regarding the abstraction level), resulting in 14 product images, 10 shelf images, 8 trail images and 7 images labelled as other. In the second case, we utilised a full product database from the Greek supermarket chain Masoutis[9]. More specifically, the database portrays a detailed categorisation for all the available supermarket products, including product description and product category in three generalisation levels (i.e. category level I-III), starting from the more general category (e.g. frozen product) and ending to the most specific (e.g. frozen cheese pie). Moreover, by inspecting the database, we associate the generalisation levels with the abstraction levels defined earlier in the following way; the generalisation levels I and II correspond to trail and shelf abstraction levels respectively, while the product level images were annotated by the product description level of the database.

5   Methodology

The first step towards the implementation of the supermarket application is to distinguish between the various levels of abstraction in order to identify what type of information the user is expecting as feedback. In this direction, we rely on computer vision and based on the visual features of the images we detect whether the user needs product/shelf/trail/other level of information. Next, in the cases where the images contain products, i.e. in the product/shelf/trail abstraction level, detecting text on the grocery packages and cross-referencing it with the supermarket product catalogue can provide a clear indication of what the user is looking at. Finally, in the other abstraction level, where there are no products in sight, the system informs the user that they are not in front of a trail, shelf or product.

5.1   Distinguishing Between Product, Shelf, Trail and Other

The concept of differentiating among products, shelves, trails and other is of paramount importance in the context of navigation in a supermarket since it guides the level of information abstraction that needs to be communicated to the user. In order to achieve this crucial goal for the proposed system we took advantage of the increased capabilities that Deep Neural Networks (DNNs) combined with Support Vector Machines (SVMs) offer.

Three DNN architectures were compared for the purposes of feature extraction from images, namely VGG16 [6], ResNet [4] and Inception [7] pre-trained on the ImageNet dataset. From each architecture two types of features were extracted and fed to a linear-kernel SVM setting for classifying the images into one of the four categories (product, shelve, trail and other). The first type of features was formed as a concatenation of activations from the intermediate, convolutional, layers. Formally, let $M_l \in \mathbb{R}^{W \times H \times C_l}$ be the feature map of lth layer (after ReLu), where $W \times H$ denotes the spatial dimensions (width and

---

[9] https://www.masoutis.gr/.

height respectively) and C_l the channels of the lth layer. Hence, the first feature vector was formed as the $\mathbf{v} = max\{M(i,j,:), i \in \{1\ldots,W\}, j \in \{1\ldots,H\}\}$ and will be referred to as *intermediate convolutional activations*. In an equivalent aspect, we extracted the features that correspond to the output of the fully connected layers and will be referred to as *FC activations*. Let us denote by $\mathbf{W}$ the weight matrix of the penultimate layer of the employed artificial neural network architecture. Then, the second type of features corresponds to $u = ReLu(\frown)$, where u indicates the output of the layer previous to the penultimate. Then, in order to classify the corresponding images into one of the four categories we employed linear SVMs. The cost parameter C was discovered using cross validation, while we followed the common practice of all-vs-all scheme to achieve multi-class classification.

5.2   Detecting Text in Food Packages

Once the image is categorised based on the abstraction level, an optical character recognition (OCR) mechanism, identifying characters in both English and Greek, is enabled to extract the available text from the product packages (e.g. brand, product description). Depending on abstraction level, two different algorithmic approaches are followed. In the case, the user wants to identify a specific product (i.e. abstraction level: Product) the bounding box with the largest dimensions and the text it encapsulates is only examined, as it will be the one closest to the camera. A search for the identified word(s) in the Masoutis database determines the product description. In the case the user wants to be informed regarding the trail or shelf he is currently standing (i.e. abstraction level: Trail, Shelf) the entirety of the text detected in the whole image by the OCR algorithm is utilised, resulting in an array tabulating each recognised word. A shrinkage of the array is performed by removing the stopwords, any invalid words that cannot be traced in Masoutis database (e.g. frazen) and a number of words that emerge in the database in high frequency but with no discriminative power (e.g. gift, offer), formulating an array that includes only valid words. A search for each valid word is then performed in the product description entries aiming to identify all registrations that encapsulate the selected word and by performing a process equivalent to data binning, associate them with the generalisation level I and II descriptions (depending on the abstraction level). The corresponding unique generalisation level descriptions are encountered as votes and are fed to a majority voting (MV) protocol to determine which shelf or trail is illustrated in the selected image. MV is a decision making protocol that provides a decision (a label in our scenario) when it receives more than half of the votes. More specifically, if the class label corresponds to "Trail", the aforementioned MV protocol will be applied only in the generalisation level I descriptions, while if the image is identified as "Shelf" MV will be performed only in the generalisation level II descriptions.

6   Experiments

In this section, we evalute the proposed system with respect to the accuracy of the detected content utilizing the dataset from the onsite visits (Sect. 4). In this direction, we present two lines of experiments; the first one aims to demonstrate the methodologies in Sect. 5 independently and the second one aims at evaluating the quality of the proposed supermarket application as a whole.

6.1   Distinguishing Between Abstraction Levels

Here, we commence by quantifying the benefits of each feature extraction approach of the 6 presented in Sect. 5.1 regarding their classification capabilities. For classification purposes we employed a suitable machine learning procedure and measure its performance in the quaternary classification task (products, shelves, trails, other). The classifier was a standard linear kernel SVM that operated on the extracted image features. In all cases, classification performance was evaluated through the accuracy metric by means of 10-fold cross-validation. In order to produce comparable results among different feature extraction schemes, the same cross validation partition was used in every case.

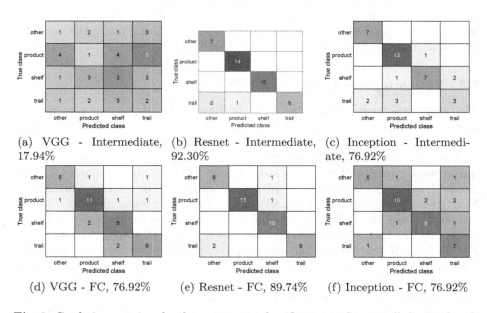

(a) VGG - Intermediate, 17.94%

(b) Resnet - Intermediate, 92.30%

(c) Inception - Intermediate, 76.92%

(d) VGG - FC, 76.92%

(e) Resnet - FC, 89.74%

(f) Inception - FC, 76.92%

Fig. 2. Confusion matrices for the quaternary classification task using all the employed feature extraction schemes (Intermediate activations - top, Fully Connected activations - bottom). The percentages correspond to average accuracy across 10-fold CV.

Figure 2 illustrates the classification performances, by means of confusion matrices, for every feature extraction procedure, while the accuracy of each feature space is denoted in the caption. It is obvious that features obtained using the Resnet architecture manage to achieve significantly better classification performance in all of the four available classes. Then, among the two available Resnet-based feature types, the features obtained from the intermediate layers provide a more robust classification performance compared to the fully connected layer features (92.3% vs 89.74%). It is worth noting that the only class that does not have 100% classification accuracy is the one of Trail, which can be attributed to the variability of the trails visual appearance from the two supermarket store of the onsite visits.

6.2   Classifying Trail, Shelf and Product Images

Based on the abstraction level annotation (i.e. trail, shelf, product), a category level classification is given to each image, with accuracy serving as the evaluation metric. Examining the total accuracy for each abstraction level separately, all levels exceed 80%. The images corresponding to "Shelf" demonstrate the highest accuracy score, as only three images were miss-classified ($Acc_{Shelf} = 89.00\%$). Similar classification scores are obtained in the case of "Trail" images ($Acc_{Isle} = 87.50\%$). The lowest accuracy score is observed, in the "Product" images, with the accuracy being at $Acc_{Product} = 83.30\%$. This is expected due to the various similar products being available (e.g. beer can of 330 ml versus beer can of 500ml), which cannot be distinguished in all product angles. Finally, the system's overall accuracy is 87.0%.

6.3   Evaluating the System as a Whole

The evaluation of the system as a whole, blends the predictions provided in abstraction and category level, resulting in the final version that will be employed for the application's needs. This way a two-stage classification scheme is realised with the first deciding upon the abstraction level and the second based on the initial prediction specifying the category level. The system's accuracy is formulated as the joint probability of correct category and correct abstraction level (i.e. $P(Category, Abstraction)$), while the one presented in the previous section was the conditional probability that the correct category is found given that the image was correctly classified in terms of abstraction level (i.e. $P(Category|Abstraction)$). The system produces high classification scores for both the "Product" and "Shelf" images, with the accuracy being 83.3% and 89.0% respectively. This is attributed to the infallible predictions of the first stage classifier combined with the high rates of the second (refer to $Acc_{Product}$ and Acc_{Shelf}). On the contrary, the accuracy levels for the "Trail" images is significantly lower (i.e. 54.7%) mainly due to the poor performance produced by the first stage classifier that reaches 62.5%. Overall, the system will produce a correct label with a probability of 77.15%, that is considered acceptable considering the task's complexity.

7   The Implementation of the Proposed Supermarket Application

Our final objective is the integration of the aforementioned components in a lightweight and compact mobile application. In this direction, the main tools utilised were the **Unity** game engine (Unity[10]) for the application's graphics, **Tensorflow** for extracting features from deep learning models in order to classify between trail/shelf/product/other (implementing the methodology in Sect. 5.1) and the **Google cloud vision** API for the OCR in order to find the descriptions of trails/shelves/products (implementing the methodology in Sect. 5.2).

Initially, we want to get the image from the camera of the user's device. The first step was to capture the screen of the device so that the user observations could be annotated. To achieve this, the *WebCamTexture* class offered by the Unity game engine was used, capturing the pixels of the device's selected camera and transforming them into a 2D texture. Having captured the image from the camera, the next step is to extract visual features from the intermediate layer of a pretrained ResNet model. For this, we rely on the open source Tensorflow framework, with which pretrained deep learning models can be used. For the integration of the Tensorflow API within the Unity engine we utilised the TensorFlowSharp library. Finally, the images that have been classified as Trail/Shelf/Product are given to the OCR system so as to categorise them based on which Trail/Shelf/Product they are depicting. In this direction, the captured image (texture) is given as input to the Google Cloud Vision API enabling the Optical Character Recognition (OCR). However, because both the interface and back-end architecture where structured and developed through the Unity engine, the communication of those two platforms had to be established. Therefore, a plugin (Unity Google Cloud Vision) was employed that bridged this gap, by using a specified API key that instantly connects to the Cloud Vision API requiring an input and receiving multiple outputs. In our case, the input is the retrieved image from the camera of the device and the outputs are the responses from the Cloud Vision API plugin, which includes labels, products, text, safe search, etc. Afterwards, using the Masoutis database of products, the methodology of Sect. 5.2 is applied to the responses of the OCR system (i.e. text and bounding boxes). Last, after the full classification of an image is performed, the system announces the results as voice information using the **Google text-to-speech** API.

8   Conclusions

In this paper, we presented a novel context-aware application for assisting the visually-impaired in their daily living, and more specifically we showcase the activity of grocery shopping. The presented application has an overall accuracy of 77.15% in voicing exactly what information the user needs. In the future, our

---

[10] https://unity.com/.

plan is to increase this accuracy by increasing the size of the training through either more onsite visits from more stores or by extracting more images from the accompanying video that was shot during the onsite visits. Next, we plan to increase the functionalities of the application in the following ways: first, by enhancing the system narrative in the case "Other" was detected through localising useful objects (i.e. shopping carts, employees and cashiers). As this cannot be facilitated with text detection, we will rely on visual object detection and localisation. Preliminary results with existing object localisation deep networks (e.g. RetinaNet) show promising results, but in order to include the classes of interest (e.g. shopping carts), transfer learning will be employed. As a result, for example, having the location of the shopping carts will enable the application to give directions to the user (e.g. he shopping carts are in your right). Second, detecting non-packaged products such as fruits and vegetables through fine-grained image classification is within our future plans. Finally, we also plan a cashier sub-context, where the application constantly informs the user on what money they are holding or recongises the products as the person puts them on the tray.

Acknowledgements. This work is part of project Evision that has been co-financed by the European Regional Development Fund of the European Union and Greek national funds through the Operational Program Competitiveness, Entrepreneurship and Innovation, under the call RESEARCH – CREATE – INNOVATE (project code: T1EDK-02454).

References

1. Geng, W., et al.: Fine-grained grocery product recognition by one-shot learning. In: Proceedings of the 26th ACM International Conference on Multimedia, pp. 1706–1714. MM 2018, ACM, New York (2018). https://doi.org/10.1145/3240508.3240522, https://doi.acm.org/10.1145/3240508.3240522
2. George, M., Floerkemeier, C.: Recognizing products: a per-exemplar multi-label image classification approach. In: Fleet, D., Pajdla, T., Schiele, B., Tuytelaars, T. (eds.) ECCV 2014. LNCS, vol. 8690, pp. 440–455. Springer, Cham (2014). https://doi.org/10.1007/978-3-319-10605-2_29
3. George, M., Mircic, D., Soros, G., Floerkemeier, C., Mattern, F.: Fine-grained product class recognition for assisted shopping. In: Proceedings of the 2015 IEEE International Conference on Computer Vision Workshop (ICCVW), ICCVW 2015, pp. 546–554. IEEE Computer Society, Washington, DC (2015). https://doi.org/10.1109/ICCVW.2015.77
4. He, K., Zhang, X., Ren, S., Sun, J.: Deep residual learning for image recognition. In: Proceedings of the IEEE Conference on Computer Vision and Pattern Recognition, pp. 770–778 (2016)
5. Karlinsky, L., Shtok, J., Tzur, Y., Tzadok, A.: Fine-grained recognition of thousands of object categories with single-example training. In: CVPR (2017)
6. Simonyan, K., Zisserman, A.: Very deep convolutional networks for large-scale image recognition. arXiv preprint arXiv:1409.1556 (2014)

7. Szegedy, C., Ioffe, S., Vanhoucke, V., Alemi, A.A.: Inception-v4, inception-resnet and the impact of residual connections on learning. In: Thirty-First AAAI Conference on Artificial Intelligence (2017)
8. Tonioni, A., Serro, E., Di Stefano, L.: A deep learning pipeline for product recognition on store shelves. arXiv preprint arXiv:1810.01733 (2018)

High-Level and Learning Vision Systems

Comparing Ellipse Detection and Deep Neural Networks for the Identification of Drinking Glasses in Images

Abdul Jabbar$^{(\boxtimes)}$, Alexandre Mendes, and Stephan Chalup

School of Electrical Engineering and Computing, The University of Newcastle,
Callaghan, NSW 2308, Australia
Abdul.Jabbar@uon.edu.au
{Alexandre.Mendes,Stephan.Chalup}@newcastle.edu.au

Abstract. This study compares a deep learning approach with the traditional computer vision method of ellipse detection on the task of detecting semi-transparent drinking glasses filled with water in images. Deep neural networks can, in principle, be trained until they exhibit excellent performance in terms of detection accuracy. However, their ability to generalise to different types of surroundings relies on large amounts of training data, while ellipse detection can work in any environment without requiring additional data or algorithm tuning. Two deep neural networks trained on different image data sets containing drinking glasses were tested in this study. Both networks achieved high levels of detection accuracy, independently of the test image resolution. In contrast, the ellipse detection method was less consistent, greatly depending on the visibility of the top and bottom of the glasses, and water levels. The method detected the top of the glasses in less than half of the cases, at lower resolutions; and detection results were even worse for the water level and bottom of the glasses, in all resolutions.

Keywords: Deep learning · Computer vision · Ellipse detection · Drinking glasses · Training data · Algorithm tuning · Deep neural networks

1 Introduction

Detection of semi-transparent objects has been a major research problem in the fields of image processing and computer vision [1–3]. Traditional image processing techniques such as edge detection fail on these objects due to their transparency. Numerous attempts have been made in detecting translucent or semi-transparent objects but they are all specific to certain types of objects placed in well-known controlled environments, detected by specialised equipment like Light Field Cameras [1], Time of Flight systems [2], and X-ray tomography [3]. The disadvantages of such systems are that they work in specific settings and require highly specialised and expensive equipment not readily available.

© Springer Nature Switzerland AG 2019
D. Tzovaras et al. (Eds.): ICVS 2019, LNCS 11754, pp. 319–329, 2019.
https://doi.org/10.1007/978-3-030-34995-0_29

The objects under discussion in our research are semi-transparent drinking glasses. Keeping their shape and structure in mind, it was observed that the top and bottom half of most drinking glasses make elliptical shapes if observed from certain angles from the top. Also, if filled with water, the water level can also be viewed as an ellipse. This property was utilised in the first part of this study where a state-of-the-art ellipse detection algorithm was used for glass detection.

There are several parameters required to fully define an ellipse: centre point, orientation, major axis, and minor axis. A generalised Hough transform needs a five dimensional parameter space to detect an ellipse which is both memory and time-consuming. There are certain techniques applied to reduce the time and resource utilisation of the Hough transform [4,5]. Time and resources required are two of the most important factors in any real-time object detection application. One fast ellipse detection approach is presented by Xie et al. [6] which takes advantage of the major axis of an ellipse to find the other ellipse parameters more efficiently. In their approach, they take one pair of pixels and assume them to be two vertices of the major axis. With this assumption and some mathematical work, they find the remaining parameters. If those parameters fulfil the structure of an ellipse, then it is counted as a positive detection and all those pixels within the bounds of the ellipse are removed from the image, and the same steps are repeated for the rest of the image pixels. They tested the accuracy of their proposed algorithm on real-world and synthetic images containing ellipses. No false ellipses were detected for those test images and the resulting detections were almost identical to the actual ellipses present in the image.

The second ellipse detection algorithm tested on our data was proposed by Fornaciari et al. [7]. Their algorithm was developed with real-time applications in mind and yields better time and resource efficiency. Their technique involves a novel strategy for edge points estimation, which reduces the overall number of points to be examined, resulting in faster performance. Only the edge points which are likely to belong to the same ellipse are considered by their technique. They try to find at least three out of four edge segments, which are likely to belong to the same ellipse, and discard all other segments that do not belong to this set. They put some constraints on each edge segment to find out whether it belongs to an ellipse or not, based on its mutual position with respect to other segments and the closeness of the centres of ellipses formed by pairs of two adjacent segments. After detecting all the edge segments that are likely to belong to an ellipse, all other parameters of the ellipse (minor, major, orientation, centre points) are calculated. After identifying all the ellipses, multiple detections of the same ellipse (due to an ellipse having more than 3 edge segments fulfilling the constraints) are discarded using the technique presented by Prasad et al. [8]. They tested their technique on various real life and synthetic images and compared the results to several other state-of-the-art ellipse detection algorithms. Their algorithm performed up to 300 times faster than other algorithms and in most cases gave better accuracy. These results became one of the major motivations for us to use their technique in this research.

The next part of this study uses deep learning for object detection. Deep learning has achieved excellent results in image classification, object detection, and other computer vision and machine learning applications in the recent past [9–12]. For these experiments, we used Tensorflow's [13] object detection API, which contains various object detection models pre-trained on the COCO [14] data set. The COCO [14] data set contains 330 thousand images of objects in 80 categories. Out of those 80 categories, two are related to the type of objects under consideration in this study: cups and wine glasses. Two models from Tensorflow's [13] object detection API were chosen for further testing using our data set. They are discussed in Sect. 2.1. The time performance, type of output, and mean average precision on the COCO [14] data set for both networks are shown in Table 1.

Table 1. Tensorflow's object detection models used in this study and their corresponding information [13].

Model name	Speed (ms)	COCO mAP [14]	Outputs
ssd_mobilenet_v1_coco	30	21	Boxes
faster_rcnn_inception_v2_coco	58	28	Boxes

The networks mentioned in Table 1 were later trained on additional 2,500 and 8,000 images of drinking glasses (generated in our previous work [15]), to determine the effect of this further training on the detection accuracy. After compiling the results of both ellipse detection and deep learning techniques for drinking glass detection, comprehensive discussion and conclusion are presented in Sect. 4.

2   Methodology and Experimental Setup

2.1   Test Data Set

The test data for ellipse detection and deep learning consisted of three different types of glasses, as shown in Fig. 1. The setup in which the glasses were placed was a white table in an office environment. Six pictures were taken of each glass from varying heights and vertical angles using a GoPro Hero 4 camera at 4 K ultra-wide resolution (4000 × 3000 pixels).

2.2   Ellipse Detection

In order to test the accuracy of the ellipse detection method [7] on our test data set, images were re-sized to four different resolutions: 1000 × 750, 1500 × 1125, 2000 × 1500, and 2500 × 1875. These resolutions were selected to check for variations in performance and speed of the algorithm [7].

The detections were divided into three categories: detection of top of the glass, its water level, and the bottom of the glass. In order to produce numerical

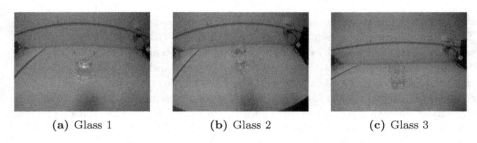

(a) Glass 1   (b) Glass 2   (c) Glass 3

Fig. 1. Images of the three glasses used in our test set, showing the shape and structure of each of them. Glasses 1 and 3 are normal drinking glasses, and Glass 2 is a wine glass.

values for detection accuracy, a new method is proposed, in which the output ellipses were compared with the correct top, water level, and bottom ellipses of the glasses in each image. Ellipses' intersection calculation was performed using the Python package Shapely [16], which provides a function to calculate the overlap area between two ellipses. A valid detection was accounted for when the intersection area was more than 50% of the union area. Based on this method, the valid top, water level, and bottom detections were counted and the complete glass detection was determined based on three possible outcomes:

1 Correct detection of any one of the three (top, water level, and bottom) ellipses.
2 Correct detection of any two ellipses.
3 Correct detection of all three ellipses.

In the first outcome, if any of the three ellipses (top, water level, and bottom) is found correctly then it will be considered as a valid glass detection, whereas in outcomes 2 and 3, any two, and all three ellipses must be found, respectively. The results of all three variations are presented in Sect. 3.1.

2.3 Deep Learning

Most deep neural networks require large amounts of training data in order to produce generalised results for object detection. The networks that we chose to test were pre-trained on the COCO data set [14], which has around 1,000 images containing a cup or wine glass. In our previous work [15], we generated approximately 8,000 images containing drinking glasses. This data set had around 2,500 real-world images compiled using a Canon EOS 60D camera and 5,500 images synthetically generated using Blender [17]. The real-world and synthetic images had six and eight different types of glasses, respectively, which were placed under various lighting conditions, backgrounds and surfaces, as well as camera angles. Some of these images are shown in Fig. 2. Each image was re-sized to the resolution of 370 × 246 before the training of deep neural networks.

The training of each network with our data set was performed for 48 hours using their default configuration, on a Tesla K80 graphics card, in two variations.

(a)   (b)   (c)

(d)   (e)   (f)

Fig. 2. Sample images used in the training of the deep neural networks. Images a, b, c, and d are real-world images; whereas, e and f are synthetic images.

First, they were trained on the 2,500 real-world images, and then on all 8,000 images. In total, four training runs were performed, two for each network.

After training, test images were fed into each network, with the output being the bounding box co-ordinates (if any) of any detected object(s). The actual and detected bounding boxes were compared using the intersection over union method described below.

Intersection over Union (IoU). This technique was originally proposed by Michael et al. [18] and compares two bounding boxes, returning a numeric value representing the overlap between them.

$$IoU = \frac{\text{area of overlap}}{\text{area of union}}$$

This performance measure is well-established and regularly used in object detection competitions, such as the PASCAL VOC challenge [19]. Higher IoU values indicate more similarity between the two bounding boxes, and a value of 1 would mean a perfect overlap. For this study, we considered IoU values greater than 0.5 as a valid detection.

3 Results

3.1 Ellipse Detection

The number of correct detections of glass tops, water levels, and glass bottoms are presented in Table 2. The number of images on each resolution was six for each glass, i.e. 18 images for each resolution. The highest number of detections for the top of the glasses 1 and 2 was achieved with the resolution of 2000×1500 (6 and 5 correct detections, respectively). For glass 3, all 6 tops were detected with the resolutions of 2000×1500 and 2500×1875. For the water level, the highest number of detections for glass 1 was only 2 with all resolutions; for glass 2 there were 4 valid detections with the resolution of 2500×1875; and for glass 3 there was only 1 detection for the resolutions of 2000×1500 and 2500×1875. The highest number of bottom ellipse detections for glass 1 was 3, with the resolution 2500×1875; glasses 2 and 3 had the same number of detections (5 and 1, respectively) with both resolutions of 2000×1500 and 2500×1875. The reason for highly accurate detections of the bottom of glass 2 (Fig. 1) was that it is a wine glass with a well distinguishable bottom, as compared to the other two glasses. Overall, correct detections of the top, water level, and the bottom of the glasses are high with the resolutions of 2000×1500 and 2500×1875. The number of detections of the top of all three glasses is significantly higher than the water level and the bottom because the top ellipse is visually clearer in all images, with the water level and the bottom of the glasses being more difficult to spot.

Table 2. Results for the detection of top, bottom, and water level ellipses for each type of glass (Fig. 1). There are 6 images of each glass, making a total of 18 images at each resolution.

Top of the glass				
	1000×750	1500×1125	2000×1500	2500×1875
Glass 1	4	4	6	5
Glass 2	2	3	5	4
Glass 3	2	5	6	6
Total	8	12	17	15
Water level				
Glass 1	2	2	2	2
Glass 2	2	2	2	4
Glass 3	0	0	1	1
Total	4	4	5	7
Bottom of the glass				
Glass 1	1	2	2	3
Glass 2	1	3	5	5
Glass 3	0	1	1	1
Total	2	6	8	9

For complete glass detection, results on all three variations (one, two, and three ellipses) of detection are presented in Table 3. Having the criteria for detection as one ellipse only, the method detects all glasses with the resolutions of

2000×1500 and 2500×1875. If we change the criterion of detection to two ellipses, then the highest number of detections drops to 8, with the resolution of 2500×1875. Finally, with the most strict criterion of detection of all three ellipses (top, water level, and bottom), the number of detections drops to only 5 with the resolutions of 2000×1500 and 2500×1875.

Table 3. Ellipse detection results with all four resolutions and for 1, 2, and 3 ellipses as the criterion for detection. There are 6 images of each type of glass, thus the maximum number of detections is 18 in any column.

1000×750				1500×1125			
# of ellipses	1	2	3	# of ellipses	1	2	3
Glass 1	4	2	1	Glass 1	4	2	2
Glass 2	4	1	0	Glass 2	4	3	1
Glass 3	2	0	0	Glass 3	5	1	0
Total	10	3	1	Total	13	6	3

2000×1500				2500×1875			
# of ellipses	1	2	3	# of ellipses	1	2	3
Glass 1	6	2	2	Glass 1	6	2	2
Glass 2	6	4	2	Glass 2	6	5	2
Glass 3	6	1	1	Glass 3	6	1	1
Total	18	7	5	Total	18	8	5

Processing Time. The average processing time for the ellipse detection for an image starts at 0.71s, with the resolution of 1000×750, and goes up as the resolution of the images increases. It took 3.76s on 1500×1125, 10s on 2000×1500, and 27.7s on the resolution of 2500×1875. The increase in processing times with the increasing resolution is depicted in Fig. 3.

3.2   Deep Learning

The performance of both networks *faster_rcnn_inception* and *ssd_mobilenet*, pre-trained on COCO data set [14] was tested on our 18 test images in three ways. (1) Without any additional training; (2) with the additional training on the 2,500 real-world images; and (3) with the additional training on all 8,000 images, containing both real-world and synthetically generated ones. The number of detections is reported in Table 4 for the images on their original resolution of 4000×3000 and after re-sizing them down to the resolution of 370×246, in order to closely match the testing speed of the networks given in Table 1.

From the results, it is clear that *faster_rcnn_inception* outperformed *ssd_mobilenet* in all scenarios. The number of detections of both networks without any additional training were 12 and 6, respectively, when the resolution of test images was 370×246. With the original resolution of 4000×3000, the number of detections of *faster_rcnn_inception* increased to 15, whereas, it decreased

Table 4. Results for the number of detections (considering 18 test images) for the two deep neural networks, trained on different configurations of additional images, and tested on two resolutions: 370 × 246 and 4000 × 3000.

| | faster_rcnn_inception_v2_coco | | **ssd_mobilenet_v1_coco** | |
| | Test image resolutions | | | |
Additional training images	370 × 246	4000 × 3000	370 × 246	4000 × 3000
None	12	15	6	5
2,500	18	18	17	17
8,000	17	18	15	16

to 5 for *ssd_mobilenet*, as it classified one of the glasses as a clock in one of the images taken from a top angle.

With the additional training of the networks on our own drinking glasses data set, the number of detections of both networks increased significantly. For the *faster_rcnn_inception*, they increased to 17 and 18 detections on the test images resolutions of 370 × 246 and 4000 × 3000, respectively (when trained on 8,000 real-world and synthetic images). The number of detections reached 18 for both test resolutions when the training of this network was performed on 2,500 real-world images. For the *ssd_mobilenet*, there was also a significant increase in number of detections when trained on 8,000 and 2,500 images (15 and 17, respectively) with the resolution of 370 × 246. The number of detections is mostly higher when the training was performed on only 2,500 images, compared to 8,000. These results are in line with our previous study [15] and are further discussed in Sect. 4. Some of the detection output images are presented in Fig. 4.

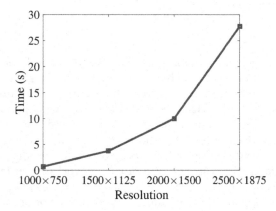

Fig. 3. Average processing time per image (in seconds) for the ellipse detection algorithm with each test image resolution.

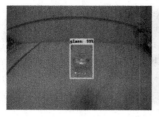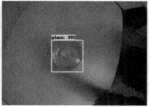

Fig. 4. Sample detection outputs of the deep neural networks.

4   Discussion and Conclusion

In the ellipse detection testing method, having three different variations of glass detection criteria provides more flexibility and a wider range of results depending on the application requirements. For example, one ellipse is enough to identify a glass in an image; two ellipses can be used to draw a bounding box around the glass which would cover most of it; and if an application requires the measurement of the water level, then all three ellipses should be correctly detected. Also, our method can be completely automated once we know the parameters of the correct placement of ellipses and the parameters of output/detected ellipses.

For the ellipse detection method, detection rates go up as the resolution of test images increases, but flattens out at a certain level – i.e. results are very similar on the resolutions of 2000×1500 and 2500×1875. Also, the processing time for an image increases rapidly with the increasing resolution, all the way up to almost 30s at the resolution of 2500×1875, which is often unacceptable in most real-life applications. The number of correct detections becomes lower as we increase the number of ellipses required to fully detect a glass from 1 to 3, e.g. at the resolution of 2500×1875, it went down from 18 to 8 and 5 detections, respectively.

If we compare the ellipse detection results with the deep neural networks' results, the ellipse detection method performed slightly better than deep neural networks if we require only one ellipse to characterise a detection. However, if we require more ellipses, then the deep neural networks outperformed the ellipse detection method by a large margin. The *faster_rcnn_inception* network performed better than the *ssd_mobilenet* in almost all scenarios, but the later network is faster – so there is a speed/accuracy trade-off. After additional training of each network with 2,500 images, both networks performed extremely well, with the average number of detections at 17.5. When further training was performed on 8,000 images, the average number of detections dropped to 16.5. The reason for this is the addition of more training data that does not closely match our test set. Similar results were obtained using deep neural networks in our previous study [15] as well.

Moving over to the ellipse detection algorithm, it performed well for glass identification but if you require more than one ellipse to be detected, then clear visibility of water level and the bottom of the glass is essential. One such example is shown in Fig. 5. This is only possible if the image is taken from certain camera angles (mainly from the top), which is a major limitation for the method. Another limitation of the ellipse detection method is the processing time, as it takes 10–30 s to process an image at higher resolutions (for better results), compared to the deep networks, which once trained, take only a few milliseconds to detect objects in images.

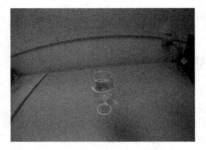

Fig. 5. Example of all 3 ellipses being detected on glass 2. Best viewed in colour.

To summarise the findings of our study, it can be stated that the ellipse detection method is more suitable for glass identification using one ellipse, constraint to less time limitations and visible shape of the drinking glass in the image. No training data or algorithm tuning is required for ellipse detection and it can perform well regardless of any surroundings the drinking glass is placed in – whereas, deep neural networks can perform significantly better than ellipse detection in terms of bounding box around the entire drinking glass, if good amount of training data (at least 1,000 images in our study) is available.

Acknowledgements. Abdul Jabbar was supported by a UNRSC50:50 PhD scholarship from The University of Newcastle.

References

1. Xu, Y., Nagahara, H., Shimada, A., Taniguchi, R.: Transcut: Transparent object segmentation from a light-field image. CoRR abs/1511.06853 (2015)
2. Klank, U., Carton, D., Beetz, M.: Transparent object detection and reconstruction on a mobile platform. In: IEEE International Conference on Robotics and Automation (ICRA), Shanghai, China (2011)
3. Ihrke, I., Kutulakos, K., Lensch, H., Magnor, M., Heidrich, W.: State of the art in transparent and specular object reconstruction. In: EUROGRAPHICS Star Proceedings, pp. 87–108. EG, Crete, Greece (2008)
4. Nair, P., Saunders, A.: Hough transform based ellipse detection algorithm. Pattern Recogn. Lett. **17**(7), 777–784 (1996). http://www.sciencedirect.com/science/article/pii/0167865596000141

5. Yuen, H.K., Illingworth, J., Kittler, J.: Detecting partially occluded ellipses using the hough transform. Image Vis. Comput. **7**(1), 31–37 (1989). https://doi.org/10.1016/0262-8856(89)90017-6
6. Xie, Y., Ji, Q.: A new efficient ellipse detection method. In: 16th International Conference on Pattern Recognition, vol. 2, pp. 957–960. IEEE (2002)
7. Fornaciari, M., Prati, A.: Very fast ellipse detection for embedded vision applications. In: 2012 Sixth International Conference on Distributed Smart Cameras (ICDSC), pp. 1–6 (2012)
8. Prasad, D.K., Leung, M.K.H.: Clustering of ellipses based on their distinctiveness: an aid to ellipse detection algorithms. In: 2010 3rd International Conference on Computer Science and Information Technology, vol. 8, pp. 292–297, July 2010
9. Goodfellow, I., Bengio, Y., Courville, A.: Deep Learning. MIT Press (2016)
10. Krizhevsky, A., Sutskever, I., Hinton, G.E.: Imagenet classification with deep convolutional neural networks. In: Pereira, F., Burges, C.J.C., Bottou, L., Weinberger, K.Q. (eds.) Advances in Neural Information Processing Systems 25 (NIPS), pp. 1097–1105. Curran Associates, Inc. (2012)
11. Simonyan, K., Zisserman, A.: Very deep convolutional networks for large-scale image recognition. CoRR abs/1409.1556 (2014). http://arxiv.org/abs/1409.1556
12. Szegedy, C., et al.: Going deeper with convolutions. In: 2015 IEEE Conference on Computer Vision and Pattern Recognition (CVPR), pp. 1–9 (2015)
13. Abadi, M., et al.: TensorFlow: Large-scale machine learning on heterogeneous systems (2015). https://www.tensorflow.org/. software available from tensorflow.org
14. Lin, T., et al.: Microsoft COCO: common objects in context. CoRR abs/1405.0312 (2014). http://arxiv.org/abs/1405.0312
15. Jabbar, A., Farrawell, L., Fountain, J., Chalup, S.K.: Training deep neural networks for detecting drinking glasses using synthetic images. In: Liu, D., Xie, S., Li, Y., Zhao, D., El-Alfy, E.S. (eds.) ICONIP 2017. Lecture Notes in Computer Science, vol. 10635, pp. 354–363. Springer, Cham (2017). https://doi.org/10.1007/978-3-319-70096-0_37
16. Shapely. https://pypi.org/project/Shapely/
17. Blender Foundation: Blender. https://www.blender.org/
18. Michael, L., David, W.: Distance between sets. Nature **234**, 34–35 (1971)
19. Everingham, M., Van Gool, L., Williams, C.K.I., Winn, J., Zisserman, A.: The pascal visual object classes (voc) challenge. Int. J. Comput. Vis. **88**(2), 303–338 (2010)

Detecting Video Anomaly with a Stacked Convolutional LSTM Framework

Hao Wei$^{(\boxtimes)}$, Kai Li , Haichang Li , Yifan Lyu , and Xiaohui Hu

Institute of Software Chinese Academy of Sciences, Beijing, China
whvino@gmail.com, {likai18,haichang,hxh}@iscas.ac.cn,
lvyifan18@ucas.ac.cn
http://www.is.cas.cn

Abstract. Automatic anomaly detection in real-world video surveillance is still challenging. In this paper, we propose an autoencoder architecture based on a stacked convolutional LSTM framework that highlights both spatial and temporal aspects in detecting anomalies of surveillance videos. The spatial component(i.e. spatial encoder/decoder) uses Convolutional Neural Network (CNN) and carries information about scenes and objects. The temporal component(i.e. temporal encoder/decoder) uses stacked convolutional LSTM and conveys object movement. Specifically, we integrate CNN and the stacked convolutional LSTM to learn normal patterns from the training data, which contains only normal events. With the integrated approach, our method can better model spatio-temporal information than many others. We train our models in an unsupervised manner, and labels are required only in the testing phase. Our method is evaluated on the datasets of Avenue, UCSD and ShanghaiTech Campus. The results show that the accuracy of our method rivals state-of-the-art methods with a faster detection speed.

Keywords: Anomaly detection · Stacked convolutional LSTM · Unsupervised learning

1 Introduction

Automatic anomaly detection is becoming more and more important in real-world video surveillance not only for its potentials in activity recognition and scene understanding but also for its advantages in reducing time and human labor. In the past few years, an increasing number of public video surveillance systems has produced massive data, which brings both chances and challenges.

Anomaly detection in videos refers to the identification of events that do not conform to expected behaviors [2]. However, anomaly remains ill-defined. Abnormal events have an extremely low probability of occurrence in video surveillance and there is no clear boundary between normal and abnormal events for anomalies are highly contextual. That is, the normality of events may change with different scenes. What's more, it's unrealistic to collect all kinds of abnormal

© Springer Nature Switzerland AG 2019
D. Tzovaras et al. (Eds.): ICVS 2019, LNCS 11754, pp. 330–342, 2019.
https://doi.org/10.1007/978-3-030-34995-0_30

events and tackle the problem with a classification method. Therefore, only normal scenarios are given in the training set. The idea is to obtain regular patterns on the training set, and the patterns that deviate from the regular ones would be considered as irregular.

Many efforts have been made in detecting anomalies. Sparse-coding and dictionary-learning based approaches [15,30,31], and autoencoder based approaches [23] have demonstrated their success for anomaly detection and achieved state-of-the-art results. These methods are all based on the assumption that videos contain little or no abnormal events. The main idea for anomaly detection is to learn normal patterns and enforce them to reconstruct normal events with small reconstruction errors in unsupervised manners.

Motivated by the success of CNN for image feature learning and LSTM for modeling the change of sequential data, we propose an autoencoder architecture based on a stacked convolutional LSTM framework. To capture spatial representations of the video, we utilize stacked convolutional layers, which we call spatial encoder/decoder. And to improve the smoothness of prediction over neighboring frames, we employ a stacked convolutional LSTM, which we call temporal encoder/decoder. The convolutional layers and the stacked convolutional LSTM are integrated to pattern normal events. Our method is strong in its ability to characterize spatio-temporal information by taking both CNN and LSTM into consideration. Features corresponding to normal events could be reconstructed by the framework with small errors and vice versa. We can thus identify whether an anomaly occurs. Our method is fully automatic and can be applied to different scenes no matter how complex the foreground is. As the experimental results on real-world datasets show, our method consistently rivals state-of-the-art methods while maintaining a faster speed.

The contributions of this paper can be summarized as follows: (i) We propose an autoencoder architecture based on a stacked convolutional LSTM framework, which models spatio-temporal information for anomaly detection by leveraging only unlabeled training videos. (ii) We obtain spatial and temporal features by learning from data instead of extracting them manually. What's more, video preprocessing, frame resizing and anomaly alerting are all automatically performed.

2   Related Work

Recent years have witnessed significant progress in anomaly detection. State-of-the-art methods can be roughly classified into two categories: anomaly detection with hand-crafted features and anomaly detection with deep learning.

2.1   Anomaly Detection with Hand-Crafted Features

Similar to the traditional machine learning process, anomaly detection methods with hand-crafted features mainly consist of three steps: (i) extracting features; Features are formed and obtained manually with existing knowledge.

(ii) determining the hypothesis space of the model and learn the model; Previous approaches utilized the model to characterize the distribution of normal events or encode regular patterns in videos. (iii) identify the outliers as anomalies. Methods based on trajectory feature extraction [13,20,27,28,32] have long been popular in anomaly detection. However, these methods are based on object tracking and may lose the target when the scenes are crowded and have complex foreground. In the circumstances, failures occur easily. For overcoming the shortcomings of trajectory features, low-level spatio-temporal features, such as the histogram of oriented gradients [5], the histogram of oriented flows [6] and optical flow [21] came into being. Zhang et al. [7] model normal patterns with a Markov Random Field(MRF). Adam et al. [1] utilize an exponential distribution to characterize the regular histograms of optical flow. To model the local optical flow patterns, Kim and Grauman [10] take a mixture of probabilistic PCA into account. Thanks to the low computational cost and the ability to focus on abnormal behavior even in extremely crowded scenes [11] of these approaches, they were popular with researchers. Another popular technique is sparse-coding [4,15,31]. They are all based on the assumption that any regular feature can be linearly represented as a linear combination on the basis of a trained dictionary. Such methods utilize the reconstruction errors to determine whether a pattern is abnormal or not. That's to say, a pattern with significant reconstruction error is considered abnormal and vice versa. However, it's time-consuming to optimize the sparse coefficients in sparse-coding based methods, making real-time anomaly detection impractical. What's more, hand-crafted methods require prior knowledge, which is very difficult to define in case of complex video surveillance scenes.

2.2   Anomaly Detection with Deep Learning

Deep learning has recently achieved remarkable success in computer vision, outperforming former methods for various challenging tasks such as image classification [12], object detection [8,22] as well as anomaly detection [3,14,17,26,29]. Unlike hand-crafted methods, there's no need to define features manually from data. The powerful capacities of artificial neural networks, which contain multiple layers of hidden nodes, make learning hierarchical features automatically possible. In [26], Sultani et al. propose to learn anomaly through a deep MIL framework that contains 3D convolutional neural networks. In another work [9], a 3D convolutional Auto-Encoder is proposed to model regular patterns. Xu et al. [29] design a deep Auto-Encoder for feature learning. In the work [17], Luo et al. propose a sparse-coding based method, which can be mapped to a stacked RNN framework. Convolutional LSTM is applied for learning regular temporal patterns in [19]. Liu et al. [14] take prediction learning based approaches into consideration and apply Generative Adversarial Networks to anomaly detection. This approach predicts future video frames based on previous video frames and compares the prediction with its ground truth. From all the results of these deep learning based approaches we can conclude that CNN is designed for learning spatial patterns and performs well in spatial tasks and meanwhile, RNN including LSTM is well-known for learning temporal patterns and predicting time series data. However, an approach containing only either of CNN and RNN can not characterize both spatial and temporal information well. Inspired by the advantages of CNN and RNN, [3,16] utilize a convolutional LSTM Auto-Encoder to characterize both spatial and temporal information.

3 Our Approach

We propose a novel encoder/decoder architecture with convolutional layers and stacked convolutional LSTM to learn spatial and temporal features in an unsupervised way. The proposed architecture is only trained on normal events and the training objective is to minimize the reconstruction errors between the input video volumes and the output video volumes reconstructed by the model. During the testing phase, normal video volumes are expected to have smaller reconstruction errors while video volumes containing abnormal events are expected to have bigger reconstruction errors. Thus our system is able to detect anomaly by setting a threshold on the reconstruction error. We will discuss more details in the following subsections.

3.1 Network Architecture

Our proposed network architecture is shown in Fig. 1. In order to balance the speed of anomaly detection and the expressive ability of the network, the depth of the CNN is set to 3 and the depth of the stacked convolutional LSTM is set to 4. Three convolutional layers, which are set to extract spatial features of frames at the bottom, take a video volume that contains T consecutive frames as input. The numbers of filters are 256, 128 and 64 from bottom to top and the kernel sizes are all 3*3. The strides are 2, 1 and 1 respectively. In the middle, a stacked convolutional LSTM, whose number of layers is 4, takes the output of convolutional layers as input. The stacked convolutional LSTM maintains the temporal information of the video volume. The numbers of filters are 64, 32, 32, 64, the kernel sizes are all 3*3 and the strides are all set to 1. Three convolutional layers lie at the top, which reconstruct the input video volumes and take the output of the stacked convolutional LSTM as input. The numbers of filters from bottom to top are 128, 256 and 1 and the kernel sizes are all 3*3. The strides are 1, 1, 2 respectively. Inspired by the VGG-NET [25], we set kernel size to 3*3 for all convolutional operations. And as our experimental results show, the model with 3*3 kernels converges faster than those who have larger kernel sizes.

We denote the mapping functions of our proposed network below. In Eq. 1, x_t represents the t-th spatial feature extracted from the t-th frame and the function f_c denotes the convolutional operation (spatial encoding). In Eq. 2, h_t represents the t-th temporal feature extracted from the (t-T)-th to the t-th spatial features, and the function g denotes sequential modeling (temporal encoding/decoding).

In Eq. 3, x_t^R stands for the t-th reconstructed frame and the function f_d^c denotes the convolutional operation (spatial decoding).

$$x_t = f_c(X_t) \tag{1}$$

$$h_t = g(f_c(X_t), f_c(X_{t-1})..., f_c(X_1)) \tag{2}$$

$$X_t^R = f_d^c(h_t) \tag{3}$$

We assume that the normal frames can be well reconstructed while abnormal frames should be reconstructed with bigger reconstruction errors. Thus we can identify whether anomaly occurs or not based on the reconstruction errors. Then we arrive at the loss function where 'B' denotes the size of batch and 'T' denotes the length of an input sequence:

$$L = \frac{1}{2} \sum_1^B \sum_1^T \left|\left| X_t - X_t^R \right|\right|_F^2 \tag{4}$$

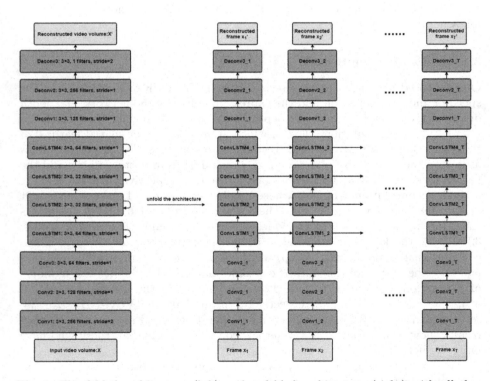

Fig. 1. The folded architecture (left) and unfolded architecture (right) with all the parameters listed.

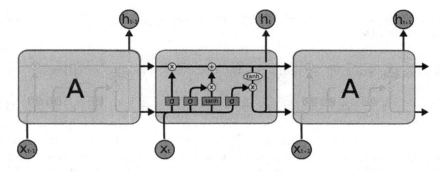

Fig. 2. The inner structure of LSTM.

3.2 Convolutional LSTM

Long Short-Term Memory (LSTM) is a time-cycle neural network specially designed to solve the long-term dependence problem of general RNN. LSTM has made great progress in many fields such as speech recognition, video analysis, and sequence modeling. The inner structure of LSTM is shown in Fig. 2. However, The major drawback of LSTM in handling spatio-temporal data is its usage of full connections in input-to-state and state-to-state transitions, in which no spatial information is encoded. To overcome this problem, the convolutional LSTM [24] is proposed which replaces the matrix multiplication with convolutional operation in LSTM. Compared with LSTM, the convolutional LSTM preserves the spatial information which is of vital importance to anomaly detection. So we introduce the convolutional LSTM into our architecture. The formulations of the convolutional LSTM are shown in Eq. 5, where x_t and h_t represent the input and output of the convolutional LSTM at time t, and i_t, f_t and o_t represent the input gate, the forget gate and the output gate respectively. A memory cell c_t essentially acts as an accumulator of the state information. $\otimes$ denotes the convolution operator, and $\circ$ denotes the Hadamard product and δ is the sigmoid activation function.

$$
\begin{aligned}
i_t &= \delta(w_{xi} \otimes x_t + w_{hi} \otimes h_{t-1} + b_i) \\
f_t &= \delta(w_{xf} \otimes x_t + w_{hf} \otimes h_{t-1} + b_i) \\
o_t &= \delta(w_{xo} \otimes x_t + w_{ho} \otimes h_{t-1} + b_i) \\
c_t &= f_t \circ c_{t-1} + i_t \circ tanh(w_{xc} \otimes x_t + w_{hc} \otimes h_{t-1} + b_i) \\
h_t &= o_t \circ tanh(c_t)
\end{aligned}
\tag{5}
$$

3.3 Anomaly Detection

We employ a stacked convolutional LSTM framework based model in the training phase, and we feed a video volume containing T consecutive frames to the model each time. Then we can utilize Eq. 6 to calculate the reconstruction error of the t-th frame. In Eq. 6, X_t denotes the t-th frame and X_t^R denotes the corresponding reconstructed frame.

$$l(t) = \left|\left|X_t - X_t^R\right|\right|_F^2 \tag{6}$$

Following the work [9], we normalize the reconstruction error as follows:

$$s(t) = 1 - \frac{l(t) - min_t l(t)}{max_t l(t) - min_t l(t)} \tag{7}$$

Here min_t denotes the minimum reconstruction error in T consecutive frames and max_t denotes the maximum reconstruction error in T consecutive frames. And s(t) is used to determine whether an abnormal event occurs at time t. The smaller the value of s(t), the greater the probability of anomaly occurrence.

4   Experiments

In this section, we evaluate our proposed method on 4 real-world anomaly detection datasets and show the results of our experiments.

4.1   Datasets

We train and evaluate our models on four commonly used datasets: Avenue [15], UCSD Ped1 and Ped2 [18], ShanghaiTech Campus Dataset [17].

In Avenue dataset, there are 16 training videos and 21 testing videos with a total of 47 abnormal events, including throwing objects, loitering and running.

In UCSD Ped1 dataset, there are 34 training videos and 36 testing videos with 40 abnormal events. All of these abnormal events are about vehicles such as bicycles and cars.

In UCSD Ped2 dataset, there are 16 training videos and 12 testing videos with 12 abnormal events. The definition of abnormal events is the same as Ped1.

In ShanghaiTech Campus dataset, there are over 270,000 training frames and over 40,000 testing frames with 130 abnormal events. The dataset has 13 scenes with complex light conditions and camera angles.

All of our training datasets contain only normal events and testing datasets contain both normal and abnormal events. For each dataset, we train an individual model because the definition of anomaly varies according to the datasets.

4.2   Data Preprocessing

Firstly, we split videos into frames and resize all frames to 225*225. Secondly, each frame is converted to grayscale. Thirdly, we calculate the mean frame by averaging each pixel value of all frames in the training set and subtract each frame from the mean frame. Finally, we generate the training video volumes and perform data augmentation(White Gaussian Noise and datasets expansion). Earlier we mentioned that each video volume contains T consecutive frames, so the ordinary 1st volume is made up of frame $\{1, 2, 3, ..., T\}$ and so on. We ignore the first three frames and the first five frames in the training set respectively and the transformed 1st volumes are made up of frame $\{4, 5, 6, ..., T + 3\}$ and frame $\{7, 8, 9, ..., T + 6\}$ respectively. We append all the transformed video volumes to the ordinary training set and as a result, the scale of the training set has tripled. What's more, White Gaussian Noise is added to the training data before the data is fed into the network. Batch Normalization is applied to the input of each convolutional layer.

4.3   Implementation Details

We train our models from scratch by minimizing the reconstruction errors of the input video volumes. We use the mini-batch stochastic gradient descent algorithm to learn the parameters of the network, where the batch size is set to 8. The learning rate is set to 0.01 and the activation functions of all convolutional layers are RELU. The network takes T consecutive frames as input and as our experimental results show, when T is set to 8 the network has the best performance. We train the models for a maximum of 500 epochs or until the validation loss hasn't decreased for 5 consecutive epochs. It takes 10 h to train the model on Avenue, 6 h on Ped1, 2 h on Ped2 and 30 h on ShanghaiTech. Once the model is trained, the model can reach a testing speed of up to 145 fps which is faster than most of the considered methods. We conduct all the experiments on a GPU (NVIDIA Titan Xp). For evaluation metrics, we use frame-level Receiver Operating Characteristic (ROC) curve and corresponding Area Under the Curve (AUC) as well as Equal Error Rate (EER). The ROC curve is plotted with True Positive Rate(TPR) against the False Positive Rate(FPR) where TPR is on y-axis and FPR is on the x-axis. The AUC denotes the corresponding area under the ROC curve. And the Equal Error Rate(EER) is obtained when the False Positive Rate(FPR) equals to the False Negative Rate(FNR). A higher AUC and a lower EER correspond to better performance.

4.4   Evaluation with Real-World Anomaly Detection Datasets

The experimental results on 4 real-world datasets show the generality and the robustness of our method. Figure 3 shows the reconstruction error of our proposed model on the UCSD Ped1 training set, and it displays the convergence procedure of our model. Figure 4 shows the regularity scores in Eq. 7 of some video samples in UCSD datasets. In Fig. 4, the red area represents the ground truth of abnormal frames, and here the corresponding regularity score s(t) is low. The part outside the red area represents the ground truth of normal frames, and here the corresponding regularity score s(t) is high. These results demonstrate the effectiveness of the proposed method.

Table 1 shows the frame-level AUC and EER. We can see that on Avenue dataset, the AUC of our method is higher than those of the Conv-AE [9] and ConvLSTM-AE [16]. On UCSD Ped1 and Ped2 datasets, our method achieves the lowest EER. And our method also exhibits high AUC, which is better than or comparable to those of other methods. Compared with the Conv-AE [9], our

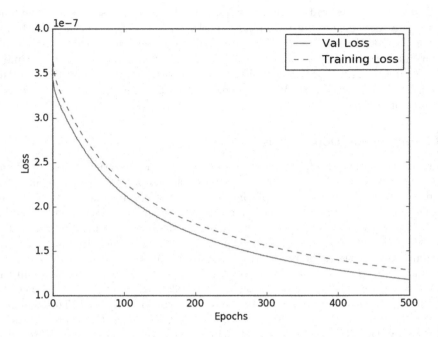

Fig. 3. Reconstruction error on the UCSD Ped1 training set.

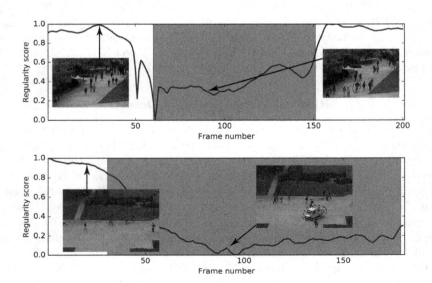

Fig. 4. Regularity scores of video 1 from Ped1 dataset (top) and video 4 from Ped2 dataset (bottom), anomalies are marked with red dotted lines. (Color figure online)

Table 1. Comparison of AUC(%) and EER(%) of our method with other methods. Most papers did not publish their AUC/EER for Avenue.

Method	Avenue	Ped1	Ped2
Adam [1]	N/A	77.1/38.0	-/42.0
MPPCA [18]	N/A	66.8/40.0	69.3/30.0
MPPCA+SF [18]	N/A	74.2/32.0	61.3/36.0
Conv-AE [9]	70.2/25.1	**81.0**/27.9	**90.0**/21.7
ConvLSTM-AE [16]	77.0/-	75.5/-	88.1/-
Ours	**79.7/23.0**	80.6/**18.9**	89.5/**14.2**

Table 2. Comparison of AUC(%) and EER(%) of our method with stacked RNN on ShanghaiTech Campus dataset.

Method	ShanghaiTech
Stacked RNN [17]	68.0/-
Ours	67.2/31.5

method achieves comparable AUC on the three datasets. However, the EER of our method is much lower, which illustrates the accuracy of our method. As ShanghaiTech Campus dataset is a recently released dataset, the compared methods in Table 1 are not tested on this dataset. Therefore, we only compare our method with the stacked RNN [17], in which the dataset was released. Referring to the results shown in Table 2, our method achieves comparable AUC to the stacked RNN [17]. We can draw a conclusion from Tables 1 and 2 that the accuracy of our method rivals state-of-the-art methods owing to the proposed architecture, which is capable of modeling spatial and temporal information. And our method can reach a testing speed of up to 145 fps with only one GPU.

5   Conclusions

In this paper, we proposed an autoencoder architecture based on a stacked convolutional LSTM framework for anomaly detection of surveillance videos. By using convolutional layers, the spatial information of each video frame can be well characterized. And the stacked convolutional LSTM characterizes the temporal information and preserves the spatial information. The experimental results on 4 real-world datasets show that the accuracy of our method rivals the state-of-the-art methods with a faster detection speed. In the future, we will evaluate our proposed method on more datasets and try to introduce parallel algorithm into our method to improve the computational efficiency.

References

1. Adam, A., Rivlin, E., Shimshoni, I., Reinitz, D.: Robust real-time unusual event detection using multiple fixed-location monitors. IEEE Trans. Pattern Anal. Mach. Intell. **30**(3), 555–560 (2008). https://doi.org/10.1109/TPAMI.2007.70825
2. Chandola, V., Banerjee, A., Kumar, V.: Anomaly detection: a survey. ACM Comput. Surv. (CSUR) **41**(3), 15 (2009). https://doi.org/10.1145/1541880.1541882
3. Chong, Y.S., Tay, Y.H.: Abnormal event detection in videos using spatiotemporal autoencoder. In: Cong, F., Leung, A., Wei, Q. (eds.) ISNN 2017. LNCS, vol. 10262, pp. 189–196. Springer, Cham (2017). https://doi.org/10.1007/978-3-319-59081-3_23
4. Cong, Y., Yuan, J., Liu, J.: Sparse reconstruction cost for abnormal event detection. In: CVPR 2011, pp. 3449–3456 (2011). https://doi.org/10.1109/CVPR.2011.5995434
5. Dalal, N., Triggs, B.: Histograms of oriented gradients for human detection. In: Schmid, C., Soatto, S., Tomasi, C. (eds.) International Conference on Computer Vision & Pattern Recognition (CVPR 2005), vol. 1, pp. 886–893. IEEE Computer Society, San Diego, June 2005. https://doi.org/10.1109/CVPR.2005.177
6. Dalal, N., Triggs, B., Schmid, C.: Human detection using oriented histograms of flow and appearance. In: Leonardis, A., Bischof, H., Pinz, A. (eds.) ECCV 2006. LNCS, vol. 3952, pp. 428–441. Springer, Heidelberg (2006). https://doi.org/10.1007/11744047_33
7. Zhang, D., Gatica-Perez, D., Bengio, S., McCowan, I.: Semi-supervised adapted hmms for unusual event detection. In: 2005 IEEE Computer Society Conference on Computer Vision and Pattern Recognition (CVPR 2005), vol. 1, pp. 611–618, June 2005. https://doi.org/10.1109/CVPR.2005.316
8. Girshick, R.: Fast r-CNN. In: 2015 IEEE International Conference on Computer Vision (ICCV). IEEE, December 2015. https://doi.org/10.1109/iccv.2015.169
9. Hasan, M., Choi, J., Neumann, J., Roy-Chowdhury, A.K., Davis, L.S.: Learning temporal regularity in video sequences. In: 2016 IEEE Conference on Computer Vision and Pattern Recognition (CVPR), IEEE Jun 2016. https://doi.org/10.1109/cvpr.2016.86
10. Kim, J., Grauman, K.: Observe locally, infer globally: a space-time MRF for detecting abnormal activities with incremental updates. In: 2009 IEEE Conference on Computer Vision and Pattern Recognition, pp. 2921–2928, June 2009. https://doi.org/10.1109/CVPR.2009.5206569
11. Kratz, L., Nishino, K.: Anomaly detection in extremely crowded scenes using spatio-temporal motion pattern models. In: 2009 IEEE Conference on Computer Vision and Pattern Recognition, pp. 1446–1453, June 2009. https://doi.org/10.1109/CVPR.2009.5206771
12. Krizhevsky, A., Sutskever, I., Hinton, G.E.: ImageNet classification with deep convolutional neural networks. In: Association for Computing Machinery (ACM), vol. 60, pp. 84–90, May 2017. https://doi.org/10.1145/3065386
13. Li, C., Han, Z., Ye, Q., Jiao, J.: Abnormal behavior detection via sparse reconstruction analysis of trajectory. In: 2011 Sixth International Conference on Image and Graphics, pp. 807–810, August 2011. https://doi.org/10.1109/ICIG.2011.104
14. Liu, W., Luo, W., Lian, D., Gao, S.: Future frame prediction for anomaly detection - a new baseline. In: 2018 IEEE/CVF Conference on Computer Vision and Pattern Recognition. IEEE, June 2018.https://doi.org/10.1109/cvpr.2018.00684

15. Lu, C., Shi, J., Jia, J.: Abnormal event detection at 150 fps in matlab. In: 2013 IEEE International Conference on Computer Vision, pp. 2720–2727, December 2013. https://doi.org/10.1109/ICCV.2013.338

16. Luo, W., Liu, W., Gao, S.: Remembering history with convolutional lstm for anomaly detection. In: 2017 IEEE International Conference on Multimedia and Expo (ICME), pp. 439–444, July 2017. https://doi.org/10.1109/ICME.2017.8019325

17. Luo, W., Liu, W., Gao, S.: A revisit of sparse coding based anomaly detection in stacked RNN framework. In: 2017 IEEE International Conference on Computer Vision (ICCV). IEEE, October 2017. https://doi.org/10.1109/iccv.2017.45

18. Mahadevan, V., Li, W., Bhalodia, V., Vasconcelos, N.: Anomaly detection in crowded scenes. In: 2010 IEEE Computer Society Conference on Computer Vision and Pattern Recognition, pp. 1975–1981, June 2010. https://doi.org/10.1109/CVPR.2010.5539872

19. Medel, J.R.: Anomaly detection using predictive convolutional long short-term memory units (2016)

20. Piciarelli, C., Micheloni, C., Foresti, G.L.: Trajectory-based anomalous event detection. IEEE Trans. Circ. Syst. Video Technol. **18**(11), 1544–1554 (2008). https://doi.org/10.1109/TCSVT.2008.2005599

21. Reddy, V., Sanderson, C., Lovell, B.C.: Improved anomaly detection in crowded scenes via cell-based analysis of foreground speed, size and texture. In: CVPR 2011 WORKSHOPS, pp. 55–61, June 2011. https://doi.org/10.1109/CVPRW.2011.5981799

22. Ren, S., He, K., Girshick, R., Sun, J.: Faster r-CNN: Towards real-time object detection with region proposal networks, vol. 39, pp. 1137–1149. Institute of Electrical and Electronics Engineers (IEEE), June 2017. https://doi.org/10.1109/tpami.2016.2577031

23. Sabokrou, M., Fathy, M., Hoseini, M., Klette, R.: Real-time anomaly detection and localization in crowded scenes. In: 2015 IEEE Conference on Computer Vision and Pattern Recognition Workshops (CVPRW), pp. 56–62, June 2015. https://doi.org/10.1109/CVPRW.2015.7301284

24. Shi, X., Chen, Z., Wang, H., Yeung, D.Y., Wong, W.K., Woo, W.C.: Convolutional LSTM network: a machine learning approach for precipitation nowcasting. In: Cortes, C., Lawrence, N.D., Lee, D.D., Sugiyama, M., Garnett, R. (eds.) Advances in Neural Information Processing Systems 28, pp. 802–810. Curran Associates Inc., New York (2015). http://papers.nips.cc/paper/5955-convolutional-lstm-network-a-machine-learning-approach-for-precipitation-nowcasting.pdf

25. Simonyan, K., Zisserman, A.: Very deep convolutional networks for large-scale image recognition. arXiv preprint arXiv:1409.1556 (2014)

26. Sultani, W., Chen, C., Shah, M.: Real-world anomaly detection in surveillance videos. In: 2018 IEEE/CVF Conference on Computer Vision and Pattern Recognition. IEEE, Jun 2018. https://doi.org/10.1109/cvpr.2018.00678

27. Tung, F., Zelek, J.S., Clausi, D.A.: Goal-based trajectory analysis for unusual behaviour detection in intelligent surveillance. Image Vis. Comput. **29**(4), 230–240 (2011). https://doi.org/10.1016/j.imavis.2010.11.003

28. Wu, S., Moore, B.E., Shah, M.: Chaotic invariants of lagrangian particle trajectories for anomaly detection in crowded scenes. In: 2010 IEEE Computer Society Conference on Computer Vision and Pattern Recognition, pp. 2054–2060, June 2010. https://doi.org/10.1109/CVPR.2010.5539882

29. Xu, D., Ricci, E., Yan, Y., Song, J., Sebe, N.: Learning deep representations of appearance and motion for anomalous event detection (2015). https://doi.org/10.5244/c.29.8
30. Yen, S., Wang, C.: Abnormal event detection using HOSF. In: 2013 International Conference on IT Convergence and Security (ICITCS), pp. 1–4, December 2013. https://doi.org/10.1109/ICITCS.2013.6717798
31. Zhao, B., Fei-Fei, L., Xing, E.P.: Online detection of unusual events in videos via dynamic sparse coding. CVPR **2011**, 3313–3320 (2011). https://doi.org/10.1109/CVPR.2011.5995524
32. Zhou, S., Shen, W., Zeng, D., Zhang, Z.: Unusual event detection in crowded scenes by trajectory analysis. In: 2015 IEEE International Conference on Acoustics, Speech and Signal Processing (ICASSP), pp. 1300–1304, April 2015. https://doi.org/10.1109/ICASSP.2015.7178180

Multi-scale Relation Network for Few-Shot Learning Based on Meta-learning

Yueming Ding[1], Xia Tian[1], Lirong Yin[2], Xiaobing Chen[1],
Shan Liu[1,3], Bo Yang[1], and Wenfeng Zheng[1(✉)]

[1] School of Automation, University of Electronic Science and Technology of China,
Chengdu 610054, China
wenfeng.zheng.cn@gmail.com
[2] Geographical and Sustainability Sciences Department, University of Iowa,
Iowa City, IA 52242, USA
[3] Department of Modelling, Simulation, and Visualization Engineering,
Old Dominion University, Norfolk, VA 23529, USA
https://www.overleaf.com/project/5cd418d4628c686a4947e234

Abstract. Deep neural networks can learn a huge function space, because they have millions of parameters to fit large amounts of labeled data. However, this advantage is a major obstacle for few-shot learning, because which has to make predictions based on only few samples of each class. In this work, inspired by multi-scale features methods and relation network which uses neural network to learn metrics, we propose a concise and efficient network, multi-scale relation network. The network consists of a feature extractor and a metric learner. Firstly, the feature extractor extracts multi-scale features by combining features from different convolutional layers. Secondly, we generate the relation feature by calculating the absolute value of the difference between multi-scale features. The results on benchmark sets show that our method avoids the over fitting and elongates the period of learning process, providing higher performance with simple design choices.

Keywords: Few-shot learning · Deep learning · Meta-learning · Multi-scales feature

1 Introduction

Based on a large number of labeled data, deep neural network have succeeded in the field of computer vision. However, in a lot of real time interaction systems, especially in the field of medical engineering and information security, it is impractical to collect a large amount of labeled data because of the absence of data and of the high cost of data collection. Once the amount of data is lacking, the model would have good performances in training processes, but have poor generalization performance. However, human could learn from few samples and

© Springer Nature Switzerland AG 2019
D. Tzovaras et al. (Eds.): ICVS 2019, LNCS 11754, pp. 343–352, 2019.
https://doi.org/10.1007/978-3-030-34995-0_31

draw inferences about other cases from one instance. Inspired by the learning ability of human, more and more researchers focus on the studies of few-shot learning.

Few-shot learning is a kind of method which attempts to abstract information about the object from one or few samples [1]. It is quite challenging, because the model not only need to integrate the prior knowledge and the new information but should avoid the over fitting on the new data sets [2]. However, human could abstract concepts from few samples and extract abundant representations from sparse data sets, whose ability is called meta-learning or learning to learning [3]. Meta-learning could realize fast learning by utilizing prior experience, which could solve the problem whose categories and concepts vary from task to task [4]. In order to make deep learning methods to be used in few-shot learning, researches are in progress to use prior knowledge in deep learning and to make deep learning method to realize self-tuning. Researchers put forward metric learning on wide tasks space to make deep learning to utilize prior knowledge better [1,5–7].

Few-shot learning based on metric learning contains two parts, the feature extractor and the metric learner. The feature extractor is used to extract features from the images. The metric learner is used to measure the distance between different features. The performance of this method is depended on the quality of the extracted features as well as on the selection of the metric learning method. Thus, inspired by the ideas from multi-scale features and relation network which uses neural network to learn metrics, we propose a concise and efficient network, multi-scale relation network(MSRN). Firstly, we employ multi-scale features method in the feature extractor for improving the variation between different categories. Secondly, we design a fusion methods for multi-scale feature and employ a metric learning based on neural network to learn the similarity between the support set and the target set. The multi-scale relation network fuses features from different layers of the convolutional network and utilize the complementary information of different features. The result of the benchmark data sets shows that our approach outperforms a lot of prior methods. This concise method reduces the over fitting of the relation network(RN) [7], and promotes the accuracy.

2   Related Work

With the successful development of deep learning and meta-learning, more and more scholars attempt to utilize those methods for few-shot learning. From 2015 to now, there are four main few-shot learning methods based on meta-learning.

Methods Based on Memory Reinforcement [8–10]. In this method, recurrent neural network with memory is used to accumulate the useful information from the hidden layers and outer memory. However, the defect of this method is the requirement for large memory spaces for the huge historical information.

Learning Optimizer [11–14]. Learning optimizer means that the leaner could learn to update its parameters, which usually learns an update function or update rules for updating the parameters. However, employing Long Short-Term Memory as meta learner would make the model become complex and would limit the performance of the model.

Optimizing the Initial Representation [2,15–19]. In this method, the initial parameters would be optimized to approximate the optimal values in the gradient step. The weights of learner are updated by the gradients, which has no additional parameters and no limits for model structures but needs fine tuning for new tasks.

Methods Based on Metric Learning [1,5,6,20,21]. In this method, a similarity measurement is learned from different tasks, and this measurement is utilized to distinguish samples from unseen categories. In previous researches, there are the matching network [1] which learns an end to end linear classifier, and the prototypical network [6] which learns an end to end nearest neighbor classifier. Both networks benefit from the advantages of parametric structure and of unparametric structure, so that both could learn fast on new samples and have well generalization performances. However, both are based on the constant metric, which assumes that the features should be linearly separable and be based on element-wise compare. As a result, the performance of the method based on constant metric is limited by the quality of the extracted features. Relation network [7] is different from matching network and prototypical network, which does not define a measurement but to learn a measurement. However, the dissimilarities of the extracted features and the generalization could be promoted in the relation network.

3 Model

3.1 Problem Definition

In the few-shot learning, there exist three kinds of data sets, the training set for training the learner on different tasks, the validation set for selecting the classifier model and the test set for testing the classifier. To realize the training purpose of classifying the images from unseen categories through just few samples, there is no intersection among three data sets. As proposed in [1,6,7], each task consists of a support set and a target set. Based on the rules of N-way k-shot learning [1], we select N categories, each one containing k samples, which constitute the support set $S = \{(x_n^i, y_n^i)|i = 1, ..., k; n = 1, ..., N\}$. The support set contains $m_T = k * N$ samples. In the rest samples of each category, b samples are selected to constitute the target set $T = \{(x_n^j, y_n^j)|j = 1, ..., b; n = 1, ..., N\}$. The target set contains samples $n_T = b * N$.

3.2 Model

As is shown in Fig. 1, our multi-scale relation network consists of feature extractor $f_\varphi(\bullet)$, which use to extract image features, and metric leaner $g_\phi(\bullet)$, which use to calculate the similarity between two features. The multi-scale features of the support set and the target set would be extracted by the feature extractor. Then, we combine the multi-scale features of the support set and of the target set to generate relation features. Last, the relation features are used as input for the metric learner to generate the relation scores.

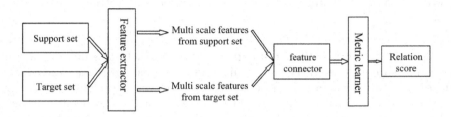

Fig. 1. The structure of the multi-scale relation network

Feature Extractor. As other few-shot learning models [1,6,7], we employ four-layer convolutional neural network as the feature extractor, whose full connected layer(FC) is removed. The multi-scale features would be extracted from different convolutional layers. In muti-scale features, the low-level features contains more details, and the high-level features contains advanced semantic information. Thus, combining the features from different layers could improve the image representation. However combining the features from all layers could generate the redundant information. In addition, combining the features from low layers would be unhelpful or even detrimental because of the over fitting [22,23]. Thus, we combine the features from the third layer and the forth layer to generate the multi-scale features.

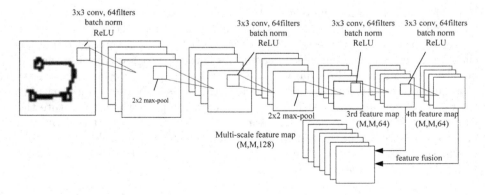

Fig. 2. The structure of feature extractor in the multi-scale relation network

As is shown in Fig. 2, each convolutional block consists of convolutional layer composed of 64 filters with size of 3 × 3, a batch normalized layer and rectified linear unit(ReLU). The first block and the second block are followed by a max pooling layer with size of 2 × 2, as well as the last two blocks are not followed by max pooling layer.

Metric Learner. The input of the metric learner is relation feature, which is generated by combining the mean values of the sample from each category in the support set and the multi-scale features of the sample from the target set. As is shown in Fig. 3, the metric learner of the relation network [7] is employed in our model, which is used to connect features in depth to generate relation feature. However, this combination method would increase the dimension of the relation feature. Thus, after subtracting two multi-scale features, we employ the absolute value as our relation feature to avoid the dimension increasing. In addition, the subtraction method has a better representation of the difference among features, which makes our model performs better.

As proposed in [7], a convolutional neural network composed of two convolutional blocks and two full connected layers is utilized as our metric learner. Each block consists of convolutional layer composed of 64 filters with size of 3 × 3 or 2 × 2, normal patch layers and ReLU layers. Each block is followed by the max pooling layer with size of 2 × 2. The first FC layer employs the ReLU activation function, as well as the second FC layer employs the Sigmoid activation function. The structure of the metric learner is shown in the Fig. 3.

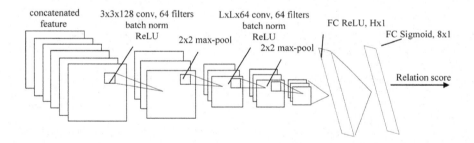

Fig. 3. The structure of metric learner in the multi-scale relation network

Relation score represents the similarities between the samples in the target set and in the support set, which is generated by the metric learner. In one N-way k-shot image classification task, we would get $N^2 \times b$ relation score r_{mn}^j. For every sample in target set, we set the maximum of the relation scores, as the prediction label $\overline{y}_n^j$. Thus, we acquire $N \times b$ classification labels in target set. Specifically, n represents the number of the categories in the target set and in the support set. i represents the number of the samples in the support set. j represents the number of the samples in the target set.

We use function (1) to represent the multi-scale relation network:

$$\hat{x}_n^i = f_\varphi(x_n^i), \hat{x}_n^j = f_\varphi(x_n^j), \overline{x}_m = \frac{1}{k} \sum_{i=1}^{k} \hat{x}_n^i$$
$$r_{mn}^j = g_\phi(|\overline{x}_m - \hat{x}_n^j|), \overline{y}_n^j = max(r_{1n}^j, r_{2n}^j, ..., r_{Nn}^j) \tag{1}$$
$$i = 1, ..., k; j = 1, ..., b; n = 1, ..., N; m = 1, ..., N$$

In function (1), x_n^i and x_n^j represents the original image feature in the support set and in the target set respectively. $\hat{x}_n^i$ and $\hat{x}_n^j$ represents multi-scale feature in the support set and in the target set respectively. $\overline{x}_m$ represents the mean value of the multi-scale features of each category in the support set. $|...|$ represents generating the absolute value after subtracting the 3-dimension vectors.

4   Experiments

4.1   Omniglot

Omniglot [24] contains 50 kinds of international languages and 1623 kinds of categories in total. The training set, the validation set and the test set are divided by 1200:211:212 ratio. The outcomes of experiments in Omniglot are shown in Table 1. FT is the acronym of fine turning.

Table 1. Few-shot classification results on Omniglot.

Model		F	5-way(%)		20-way(%)	
		T	1-shot	5-shot	1-shot	5-shot
Based on memory enhancement	Meta-Learner LSTM [13]	-	-	-	-	-
Learning optimizer	META NETS [9]	N	99.00	-	97.00	-
Optimized initial representation	MAML [2]	Y	98.70 ± 0.40	**99.90 ± 0.10**	95.80 ± 0.30	98.90 ± 0.20
	REPTILE [17]	Y	95.39 ± 0.09	98.90 ± 0.10	88.14 ± 0.15	96.65 ± 0.33
Based on metric learning	Relation Network [7]	N	99.32 ± 0.16	99.64 ± 0.09	96.73 ± 0.15	98.91 ± 0.05
	MSRN(ours)	N	**99.35 ± 0.25**	99.70 ± 0.08	**97.41 ± 0.28**	**99.01 ± 0.13**

The baseline is more than 97% because of the simpleness of the Omniglot. Our model achieves the state-of-the-art performance on 5-way 1-shot setting and 20-way settings. For each task, the best-performing method is highlighted. "-":not reported. In order to compare the same type of methods fairly, we conducted experiments of relation networks under the same experimental conditions.

4.2   MiniImageNet

MiniImageNet [1] is proposed by Vinyals etc, which consists of 60,000 chromatic images with size of $84 \times 84 \times 3$ and includes 100 categories. Because Vinyals did not open the source, Ravi and Larochelle [13] produce a new MniImageNet data set by randomly selecting 100 categories from ImageNet data set and dividing the data set into training set, validation set and test set by 64:16:20 ratio.

Table 2. Few-shot classification results on MiniImageNet database.

Model		F	5-way Acc.(%)	
		T	1-shot	5-shot
Based on memory enhancement	Meta-Learner LSTM [13]	N	43.44 ± 0.77	60.60 ± 0.71
Learning optimizer	META NETS [9]	N	49.21 ± 0.96	-
Optimized initial representation	MAML [2]	Y	48.70 ± 1.84	63.11 ± 0.92
	REPTILE [17]	Y	47.07 ± 0.26	62.74 ± 0.37
Based on metric learning	Relation Network [7]	N	50.16 ± 1.34	64.77 ± 0.34
	MSRN(ours)	N	$\mathbf{50.21 \pm 1.08}$	$\mathbf{65.89 \pm 0.32}$

The results of the experiments on the MiniImageNet are shown on the Table 2. It is shown that our model achieves the state-of -the-art performance on 5-way 1-shot settings and on 5-way 5-shot settings.

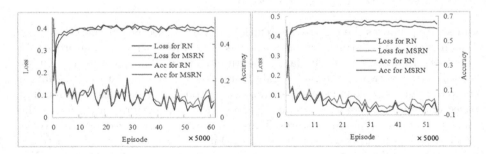

Fig. 4. Accuracy and loss values on MiniImageNet data. The left figure: 5-way 1-shot experiment. The right figure: 5-way 5-shot experiment

Under the definition of meta-learning, the learning process could realize life-long learning [25]. However, in complex tasks, the learning process could not sustain along with the iterations. For example, as is shown in Fig. 4, with the increase of iterations, the accuracy of relation network decreases along with the decline of the loss. But with the increase of the iterations, the accuracy of the multi-scale relation network remains stable after reaching the zenith, which shows the over fitting reduced because of the extension of the learning period.

Table 3. Results of the experiments on the remote sensing image data set and on the fine grit image data set

Model	F	Acc.(%)	
	T	WHU-RS	Standford Dogs
		5-way 1-shot	5-way 1-shot
Relation Network [7]	N	62.07 ± 0.86	30.65 ± 0.44
MSRN(ours)	N	**64.10 ± 0.45**	**32.62 ± 0.49**

We employ the 5-way 1-shot classifier, which is pretrained on the MiniImageNet, on the WHU-RS data set, a remote sensing image data set, and on the Stanford Dogs data set, a fine-grained data set. As is shown in Table 3, our multi-scale relation network outperform relation network both on the WHU-RS data set and the Stanford Dogs data set. In addition, the accuracy in remote sensing data set is 12% higher than the accuracy in MiniImageNet set. The accuracy in fine-grained data set is 18% lower than that in MiniImageNet data set. The result shows that the accuracy of multi-scale relation network is influenced by the distribution of the task, which means that our model is more adaptive for remote sensing image classification.

5   Conclusion and Future Work

Our study proposes the multi-scale relation network based on the meta-learning. This network could utilize the prior knowledge to direct the learning process in new tasks by learning the similarity measurement in tasks space. In the multi-scale relation network, the multi-scale features shows larger category variation compared with the single scale feature. In addition, the metric learning method based on neural network is more flexible, and performs better in abstracting the variation between different features. Thus, our multi-scale relation network works effectively in new tasks without fine tuning. Fusing features from different scales realizes a complimentary advantage, reduces the over fitting on MiniImageNet and improves the classification accuracy. However, the effect of our multi-scale feature connection method needs more considerations when using on fine grit image classification task. Compared with the results on Omniglot data set, the multi-scale relation network performances a little unsatisfactory on miniImageNet. We consider that the reason is the required measurements vary rapidly when there are huge difference among tasks.

In the future, the multi-scale relation network could be promoted in three aspects. First, the multi-scale feature extractor could be optimized to insure the dissimilarity of the features. Second, the influence from the task distribution would be reduced to promote the accuracy. Last, we would attempt to combine our feature extraction method and connection method with deep neural networks to realize some breakthrough results.

References

1. Vinyals, O., Blundell, C., Lillicrap, T., Wierstra, D.: Matching networks for one shot learning. In: Advances in Neural Information Processing Systems, pp. 3630–3638 (2016)
2. Finn, C., Abbeel, P., Levine, S.: Model-agnostic meta-learning for fast adaptation of deep networks. In: Proceedings of the 34th International Conference on Machine Learning, pp. 1126–1135 (2017)
3. Biggs, J.B.: The role of metalearning in study processes. Br. J. Educ. Psychol. **55**(3), 185–212 (2011)
4. Brazdil, P., Carrier, C.G., Soares, C., Vilalta, R.: Metalearning: Applications to data mining. Springer, Heidelberg (2008). https://doi.org/10.1007/978-3-540-73263-1
5. Koch, G., Zemel, R., Salakhutdinov, R.: Siamese neural networks for one-shot image recognition. In: ICML Deep Learning Workshop (2015)
6. Snell, J., Swersky, K., Zemel, R.: Prototypical networks for few-shot learning. In: Advances in Neural Information Processing Systems, pp. 4077–4087 (2017)
7. Sung, F., Yang, Y., Zhang, L., Xiang, T., Torr, P.H., Hospedales, T.M.: Learning to compare: relation network for few-shot learning. In: Proceedings of the IEEE Conference on Computer Vision and Pattern Recognition, pp 1199–1208 (2018)
8. Santoro, A., Bartunov, S., Botvinick, M., Wierstra, D., Lillicrap, T.J.: One-shot learning with memory-augmented neural networks. arXiv preprint arXiv:1605.06065 (2016)
9. Munkhdalai, T., Yu, H.: Meta networks. In: Proceedings of the 34th International Conference on Machine Learning, pp. 2554–2563 (2017)
10. Mishra, N., Rohaninejad, M., Chen, X., Abbeel, P.: A simple neural attentive meta-learner. In: International Conference on Learning Representations (2018)
11. Hochreiter, S., Younger, A.S., Conwell, P.R.: Learning to learn using gradient descent. In: International Conference on Artificial Neural Networks, pp. 87–94 (2001)
12. Maclaurin, D., Duvenaud, D., Adams, R.: Gradient-based hyperparameter optimization through reversible learning. In: International Conference on Machine Learning, pp. 2113–2122 (2015)
13. Ravi, S., Larochelle, H.: Optimization as a model for few-shot learning. In: In International Conference on Learning Representations (2017)
14. Cheng, Y., Yu, M., Guo, X., Zhou, B.: Few-shot learning with meta metric learners. In: NIPS 2017 Workshop on Meta-Learning (2017)
15. Li, Z., Zhou, F., Chen, F., Li, H.: Meta-sgd: learning to learn quickly for few shot learning. arXiv preprint arXiv:1707.09835 (2017)
16. Rusu, A.A., et al.: Meta-learning with latent embedding optimization. arXiv preprint arXiv:1807.05960 (2018)
17. Nichol, A., Achiam, J., Schulman, J.: On first-order meta-learning algorithms. arXiv preprint arXiv:1803.02999 (2018)

18. Gui, L.-Y., Wang, Y.-X., Ramanan, D., Moura, J.M.: Few-shot human motion prediction via meta-learning. In: European Conference on Computer Vision, pp. 441–459 (2018)
19. Zhang, R., Che, T., Ghahramani, Z., Bengio, Y., Song, Y.: MetaGAN: an adversarial approach to few-shot learning. In: Advances in Neural Information Processing Systems, pp. 2371–2380 (2018)
20. Shyam, P., Gupta, S., Dukkipati, A.: Attentive recurrent comparators. In: Proceedings of the 34th International Conference on Machine Learning, pp: 3173–3181 (2017)
21. Oreshkin, B.N., Lacoste, A., Rodriguez, P.: TADAM: task dependent adaptive metric for improved few-shot learning. arXiv preprint arXiv:1805.10123 (2018)
22. Yang, S., Ramanan, D.: Multi-scale recognition with DAG-CNNs. In: Proceedings of the IEEE International Conference on Computer Vision, pp. 1215–1223 (2015)
23. Yu, W., Yang, K., Yao, H., Sun, X., Xu, P.: Exploiting the complementary strengths of multi-layer CNN features for image retrieval. Neurocomputing **237**, 235–241 (2017)
24. Lake, B., Salakhutdinov, R., Gross, J., Tenenbaum, J.: One shot learning of simple visual concepts. In: Proceedings of the Annual Meeting of the Cognitive Science Society (2011)
25. Lemke, C., Budka, M., Gabrys, B.: Metalearning: a survey of trends and technologies. Artif. Intell. Rev. **44**(1), 117–130 (2015)

Planar Pose Estimation Using Object Detection and Reinforcement Learning

Frederik Nørby Rasmussen[1], Sebastian Terp Andersen[1], Bjarne Grossmann[1], Evangelos Boukas[2], and Lazaros Nalpantidis[2]([⊠])

[1] Department of Materials and Production, Aalborg University,
Copenhagen, Denmark
{afnr17,sander17}@student.aau.dk,
bjarne@mp.aau.dk
[2] Department of Electrical Engineering, Technical University of Denmark,
Kongens Lyngby, Denmark
{evbou,lanalpa}@elektro.dtu.dk

Abstract. Pose estimation concerns systems or models dealing with the determination of a static object's pose using, in this case, vision. This paper approaching the problem with an active vision-based solution, that integrates both perception and action in the same model. The problem is solved using a combination of neural networks for object detection and a reinforcement learning architecture for moving a camera and estimating the pose. A robotic implementation of the proposed active vision system is used for testing with promising results. Experiments show that our approach does not only solve the simple task of planar visual pose estimation, but also exhibits robustness to changes in the environment.

Keywords: Pose estimation · Object detection · Reinforcement learning

1 Introduction

A discussion of whether to make an agent intelligent as a subject of its representation of rich information about the environment, or instead as a consequence of a hierarchy of subsumed systems is debated in [15]. End-to-end systems seem to illustrate, that given a sufficient amount of rich data, an agent can display intelligence if it has sufficient representation, i.e. modelling that creates inference from the data.

An example is learning by training a neural network on extensive amounts of data to solve a specific classification task, which is inherently subject to the data on which it is trained. Other solutions try to model the representation, based on knowledge about the desired outcome of a task. An example of such a task is pose estimation. In this paper, we ask, what if the end-to-end system can serve

---

F. N. Rasmussen and S. T. Andersen—The two authors have contributed equally to the work.

© Springer Nature Switzerland AG 2019
D. Tzovaras et al. (Eds.): ICVS 2019, LNCS 11754, pp. 353–365, 2019.
https://doi.org/10.1007/978-3-030-34995-0_32

as a sensory input to a learning system, that is learning actions that solves the task? In this case, the actions required to solve the task is a consequence of exposure to the environment and less attributed to a system of engineered features and decision structures. In such a system, we also question, whether simulation can substitute data and sensory input in phases that traditionally require large amounts of information. In this paper, we revisit theories on subsumption and create an agent, a robot, that uses a sequence of two neural networks which is interfaced by a very limited representation of the environment and test if the system is able to display simple intelligence when solving a simple pose estimation task. The solution is based on simulation and synthesizing of data for learning in a reinforcement learning architecture. The approach is relevant, because it integrates synthetic perception with actions that makes the agent able to solve the task at hand and display some robustness to changes in the learned environment.

2   Related Work

In a 1991 paper [3] Brooks discusses the contemporary views on perception modelling and establishes a *reactive* paradigm that maps actions and perceptions closely. The theory is supported by a series of mobile robots that exhibit insect-like autonomy. The reactive paradigm is based on a principal of subsumption, where no explicit representation is modelled. [15] quotes Brooks' theories as describing intelligence as emergent from the complexity of a hierarchy of task-accomplishing behaviours - without explicit pattern matching or reasoning. The opposite system view can be described as a more *deliberate* paradigm, where information about the world is viewed as rich and useful for modelling a high-level representation of the world, as in e.g. [1,10,12]. In this paper we seek to take a middle stance between the two paradigms, and utilize a rich representation in a subsumption hierarchy for creation of an intelligent agent. This approach is chosen to explore an alternative to rich models dependent on labelled data, representation and feature engineering. An example of object detection from RGB data is [6], while [7] and [8] utilizes cameras with depth perception and visual renderings of images to directly train a 6D pose estimation, albeit using reinforcement learning for some aspects of the system. Another example is [13] that accurately estimates human poses with a cascade of regression based neural networks and extensive use of labelled images - even simple pose estimations, as in [5], also rely on labelling, in this case 45 degree object rotations in images. In this paper, we avoid using images that require the pose to be represented in the labelling of the image. In [10] it is seen how a pose estimation system is used for grasping purposes. The system, however, also relies on stereo cameras and identifies simulation of data as a problem preventing immediate scaling the solution. An example of moving away from this reliance on labelled data is [14] that utilizes synthetic data and domain randomization to bridge the reality gap in state-of-the-art 6D pose estimation and grasping based on simple RGB images. We include the idea of domain randomization, but rather than training the pose into a neural network using the direct perception, we have a neural network learn it using reinforcement learning, and apply it on a simple case of 1D planar pose estimation to identify the direction of an objects rotation.

3   Proposed System Architecture

In order to challenge the more classical end-to-end neural network for pose estimation and rely less on labelled data we split the end-to-end solution in two parts. We utilize the well proven technology from object detection based on convolutional neural networks for the first part. The output of the object detector serves as the input for the second part, a reinforcement learning architecture, which learns the policy that retains the estimate of the objects pose. This means that the object detector can be based on transfer-learning from networks in the Tensorflow environment. This requires a minimal amount of labelled data for fine tuning to output a bounding box for the object. The bounding box coordinates serves as the input to the learning architecture, which means that the learning architecture is not directly dependent on images for learning - an advantage that allows for learning on simulated data. The pose is defined as a rotation around a top view of a tabletop work space, discretized in 8 (0...7) possible poses.

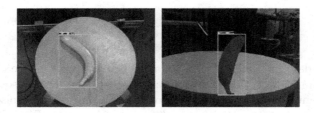

Fig. 1. Left: An image of a plastic banana during in situ object detection. Right: Object detection on the synthetic banana randomly masked on to an image of the work space.

To test the model in a realistic environment it is integrated in a system as illustrated in Fig. 2. The reinforcement learning architecture controls a robotic arm with a simple colour web-camera mounted as an eye-in-hand setup. The action space thus consists of either guessing a discretized pose of the object placed in the robots work space or moving the camera to a new position and receive an observation consisting of a bounding box and a camera position. The feature space for the object detector is an RGB image from the camera. The object detector, based on a Tensorflow model, outputs the coordinates for the bounding box enclosing the detected object. The systems works in a closed loop manner, because the initialization prompts the reinforcement learning policy to step by either moving or guessing. The guess results in the final reward that subsequently terminates the episode.

The model is tested by comparing the pose estimate to the true pose of the object. Furthermore, the system is perturbed by intermittently blocking the cameras view and also rotating the object slightly away from the defined pose. This is done without explicit observation of these situations during learning.

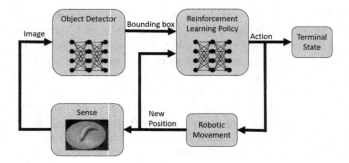

Fig. 2. Proposed model architecture showing how the **reinforcement learning policy**, using **Robotic movement** actions, moves the camera and **sense** images using the **object detector**, until the **reinforcement learning policy** outputs an **action** that results in a pose estimate which triggers the **terminal state**.

3.1   Model

The model is described in Algorithm 1 and illustrates an episode of the model based on a trained object detector and a learned policy. The episode initializes with the accumulated reward set to zero. Once initialized, the episode iterates until the terminal state occurs, i.e. when $Done$ is set to $True$. The episode consists of T discrete steps that are iterated by the policy π reacting to the current state $s_{i,t}$ by outputting an action a. If the action is of the movement type M, the move function $MO()$ is triggered and executes a move by moving the camera to a new position, c_i. In the new position the object detector denoted as the function $OD()$ samples the camera and provides a bounding box $b_{i,t}$ enclosing the object. The state $s_{i,t}$ is updated and the reward for the state-action pair is defined as r_t. If the action in the step is a pose estimate of type P, then no movement and sensing occurs. Rather, the pose estimate p_k is defined and the reward for the state-action pair is defined as r_t. The pose estimate triggers the terminal state condition and breaks the loop. This returns the pose estimate p_k and the accumulated reward for the episode.

Limitations. The model in its current form relies on some conditions. First, the object being estimated is static. Second, the object being tested is elongated which results in bounding boxes that, from some perspectives, appear rectangular.

Model sequence

let $p_k \in P$ be a pose estimate of K possible poses
let $m_n \in M$ be a direction of N possible directions
let $a \in A$ be an action of $A = P \cup M$
let $c_i \in C$ be a camera position of I possible positions
let $b_{i,t} \in B$ be a bounding box from $OD()$ in c_i at step t
let $s_{i,t} \in S$ be a state $(c_i, b_{i,t})$ of I, T possible states
let r_t be the reward in step t for an action a
let $\pi()$ denote a function that takes $s_{i,t}$ and outputs a
let $OD()$ denote a function that takes c_i and outputs $b_{i,t}$
let $MO()$ denote a move that takes m_n and moves to c_i
let $R()$ denote a function that takes $s_{i,t}, a$ and outputs r_t
episode:
$reward = 0$
while *not Done* **do**
    step:
       $a = \pi(s_{i,t})$, generate action
       if $a \in M$ **then**
          $c_i = MO(a)$, move to new position
          $b_{i,t} = OD(c_i)$, generate bounding box
          $s_{i,t} = (c_i, b_{i,t})$, update state
          $r_t = R(s_{i,t}, a)$, obtain reward
       else
          $p_k = a$, map pose estimate to action
          $r_t = R(s_{i,t}, a)$, obtain reward
          $Done = True$
       end
end
$reward = \sum(r_0, r_1, r_2, \ldots, r_t)$, calculate total reward
return$(p_k, reward)$

Algorithm 1: Model overview describing one episode resulting in the pose estimation of the object.

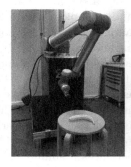
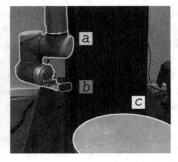

Fig. 3. Experimental setup. Overview (left) and close-up (right) showing the robotic arm (a), the mounted camera (b) and the plane used to place objects onto (c).

4   Experimental Setup and Data Generation

Figure 3 shows an image of the experimental setup.

4.1   Object Detector

The object detectors purpose in this system is to convert the image sensed from the camera into a bounding box. This represents the objects position in the image and reflects the size of the object relative to the camera by ensuring a bounding box that tightly fits the object. The object detector is based on a Python3 implementation of Tensorflow using Cuda on a GPU - an Nvidia GeForce GTX950m with 2GB of VRAM. A trade-off between time and accuracy can be found in Faster R-CNN using Resnet101 or R-FCN using Resnet101 [4]. Tensorflow, however, floods the VRAM during training which results in crashes that denies the use of Resnet101 models due to memory limitations. This can be explained by Resnet101 using more memory than the other models [4]. An alternative is found in using Faster R-CNN with the InceptionV2 feature extractor [4]. Since the original paper on speed and accuracy was published, the figures have changed, and the current advertised performance suggests that Faster R-CNN with the InceptionV2 is a suitable choice - this model is capable of running on the hardware.

Model Training. The Faster R-CNN model is pre-trained on the COCO data set. For this paper, the object for pose estimation is a plastic banana of similar size and appearance to a regular household Cavendish banana. The COCO data set allows for more labels than the banana, but the network is limited to only predict whether this label represented by an object in the image. In order to train the network for the specific task, two data sets are generated and annotated. One data set consists of 336 images of the banana in situ, sampled from 12 different positions of the camera mounted on the robot arm. The other data set consists of 336 images containing only 12 samples of a view in each position. These images does not contain a banana, but rather has a synthetic banana randomly masked into the image. The synthetic banana is sampled from a 3D model of a banana. The image in Fig. 1 shows an example of bounding box detections on two example images. Note that the domain randomization is obtained in the random masking, but does not go beyond this - the background and surface are similar to the real banana samples to allow for comparison. Furthermore, this synthetic generation of the intended object has the benefit of allowing automatic generation of bounding box annotations and minimizes the risk of human errors compared to manual labelling. The synthetic data does not, however, in this case capture the reflections of the plastic banana - this does not significantly affect the accuracy of the object detector. [14] experiences a similar problem, albeit including dynamic lighting conditions for the synthetic data. While training, the previously mentioned hardware limitations means that the batch size has to be of size 1. This essentially means that the training is done as Stochastic gradient decent (SGD) rather than mini-batch SGD as in [4]. The learning rate and beta parameters are set to 2.0^{-4} and 0.89 respectively. Experimentation with changing the learning rate or learning rate decay does not improve the object detector *mean average precision* (mAP) significantly and it settles on a mAP of about 0.94 for data sets with both real or synthetic objects. The described accuracy provides the bounding boxes for the reinforcement learning in this paper.

Model Verification. To verify the model, the Intersection over Union (IoU) measure is used for verification on a test set. This allows for an evaluation of the object detectors performance relative to the ground truth, using a test set consisting of 96 image samples, one from each camera position for each object pose. The test set is used to evaluate both the object detector for real and synthetic data. The original pre-trained object detector based on the COCO data set is used for reference, although it is limited to only output the banana-label predictions. The results of the test are presented in Table 1. Real and synthetic data sets score a very similar score of about 0.92 and a comparable standard deviation. The object detector based on only the COCO data set scores poorly, which might be attributed to the occasional glare from the plastic banana compared to the COCO training data.

Table 1. Accuracy of the classifiers with various amounts of features.

Data	Mean IoU	stddev IoU
Real images	0.922	0.0257
Simulated images	0.920	0.0299
COCO	0.343	0.4360

4.2   Reinforcement Learning

For the reinforcement learning part of the system, the target policy that we want to learn is deterministic, but we want the agent to have a stochastic element to it during learning the target policy. We therefore implement the off-policy model Deep Q-Network (DQN) [9]. Further reasoning for not using an on-policy model is, that in the environment of this experiment, an on-policy agent might avoid going to a state because it has this state associated with a large negative reward, e.g. by having made a wrong prediction from this state, but in this environment, the agent could also make the correct prediction from this state as it can predict all object poses from all camera positions. The reinforcement learning network gets the state $s_{i,t}$ as five inputs containing: the camera position c_i, and the bounding box $b_{i,t}$ (as values: $x_{min}, y_{min}, x_{max}, y_{max}$) detected by the object detector. The reinforcement learning network then has three hidden layers, each with 40 nodes, that use the Rectified Linear Unit (ReLU) activation function and l2 regularization. The output layer uses the identity function to return a single output for each of the 12 actions that the reinforcement learning agent can take. These actions are: moving the robot in one of four directions (N, S, E, W) or making a prediction on the pose of the object ($p_0, p_1, p_2, \ldots, p_7$). Making a prediction ends the episode. Furthermore, to prevent the agent from moving between camera positions indefinitely, the episode will also be terminated if the agent does not make a prediction after 10 steps, which essentially implements reinforcement learning with a finite reward horizon.

Training. As stated above, the bounding box coordinates serves as the input to the reinforcement learning architecture. This makes the simulation of data for this part of the system simpler, as it does not require labelled images for training. However, bounding boxes and the object poses which they represent are still needed, and to train an agent like this, a lot of this type of data is needed. To avoid extracting sets of bounding boxes and object poses from images, the data used to train the model is generated during training. This has been done by first taking five sample pictures of the object from each camera pose for every object pose. These ground-truth bounding boxes have then been extracted from the pictures and used to generate a mean and standard deviation for each bounding box coordinate per object pose and camera position. These are then used to generate a bounding box for the object pose the first time the agent goes to a given camera position during an episode of training. If the agent re-samples an already visited camera position within one episode, the same bounding box will be given again, since the object does not change pose during an episode. However, to accommodate noise encountered during running on the physical setup, each bounding box coordinate has a small amount of Gaussian noise multiplied to it.

To train the DQN agent, a model of the experimental setup has been built with Python3, which allows for running and training of the agent in both the real version with the physical robot and in a strictly simulated implementation. The environment has been structured to conform to a standard environment from the *OpenAI Gym* toolkit, which allows for using the *Keras-rl* library for convenient implementation of the DQN design [2,11]. During training the DQN agent will try to update its Q-values, which map how good it is to perform an action a in a given state $s_{i,t}$. To determine how good an action is, the agent will use the rewards it receives from the environment. These rewards have been set to the following: *Correct prediction: 10, Wrong prediction: −10, Move: −1, Illegal move: −5*. An illegal move is when the agent tries to move in a direction which is not possible, because it is already in the furthest possible position in that direction. The agent uses Boltzmann exploration during training with $\tau = 1$.

Verification. After training for 100,000 episodes the agent is tested for 500 episodes. The result of a 0.94 accuracy appears high enough to solve the task, but low enough to not seem over-fitted (Table 2).

Table 2. Verification results of episodes in simulated environment. Metrics for steps are averages.

Accuracy	Steps per episode	Steps per win	Steps per loss
0.940	2.782	2.583	5.900

5 Experimental Evaluation

The system was tested on the physical setup for 96 episodes with initialization in all camera poses for each object pose. The object detector trained on real images was used along with the DQN agent trained for 100,000 episodes in the simulated environment. In Table 3 the results are shown. These results show, that the agent estimates the correct pose in 92.7% of the episodes tested. Figure 4 shows a confusion matrix of the 96 test episodes. From the matrix it is evident, that the agent only made errors when the object was either in pose 6 or 7. For all the other object poses, the agent estimated the correct object pose 12 out of 12 times. The worst results were when the object was in pose 7. In this scenario, the agent wrongly estimated the object to be in pose 6 four times and had a timeout once. Still, the object was correctly categorized the remaining 7 out of 12 times.

Table 3. Results after running system for 96 episodes with initialization in every camera position for every object pose. Metrics for steps are averages.

Accuracy	Steps per episode	Steps per win	Steps per loss
0.927	2.740	2.494	5.857

To perturb the system, testing was done to see how the agent would handle blockages of the camera. In the first test of camera blockages, the agent would initialize in a random camera position three episodes per object pose. In the first episode, the agent would receive no detection in the first step but afterwards get detections for the remaining steps. In the second episode, the agent would not receive a detection in the second step. Finally, in the third episode per object pose, the agent would not receive a detection in the third step if the episode had not already been completed. In Table 4, an overview of the test is shown. From the table it is evident, that the agent estimated the correct object pose in all trials with no detection in the third step. However, it should be noted, that the episode had already been completed in the first two steps in all but one of the trials. Overall, the agent made three wrong pose estimates, which indicates adequate performance.

To have a result comparable to the 96-episode test of the system, a more comprehensive test with blockages was conducted in a similar fashion to the test with no blockages, but this time the agent had a 25% risk, at each step,

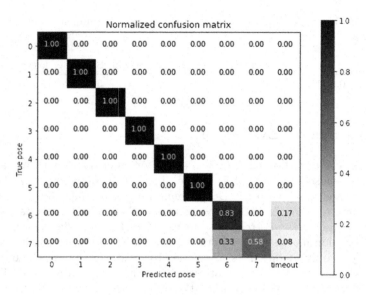

Fig. 4. Confusion matrix for running model on robotic arm with 12 instances per object pose, each initialized in the different camera poses.

of receiving a detection with the camera view blocked. The results are shown in Table 5, and show a drop in accuracy from 94% to 82.3%. Figure 5 shows a confusion matrix of the 96 episodes.

To investigate if the agent would categorize object poses in-between the established poses to the closest known object pose, a test was made where the object was placed to either side of the established poses. The results are presented in Table 6.

Table 4. Number of steps and pose predictions (guesses) with no detections in the first, second and third step respectively. Wrong predictions are written in bold.

Object pose	Block in first		Block in second		Block in third	
	Steps	Guess	Steps	Guess	Steps	Guess
0	2	0	3	0	2	0
1	3	1	3	1	6	1
2	3	2	3	2	2	2
3	2	3	6	3	2	3
4	4	4	1	**0**	2	4
5	3	5	4	5	2	5
6	4	6	5	6	2	6
7	4	**6**	6	**6**	2	7

Table 5. Results after running system for 96 episodes with initialization in every camera position for every object pose and no detections in 25% of taken pictures. Metrics for steps are averages.

Accuracy	Steps per episode	Steps per win	Steps per loss
0.823	4.031	3.354	7.176

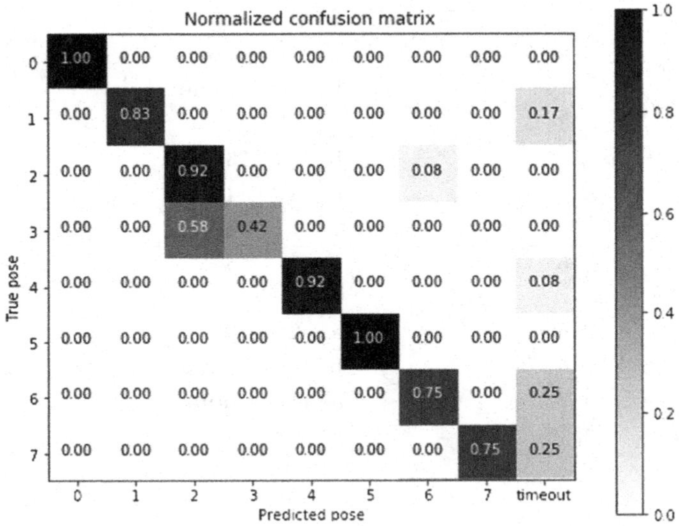

Fig. 5. Confusion matrix for running model on robotic arm with 12 instances per object pose, with a 25% chance of not receiving an object detection.

6   Conclusion

This paper presents our solution to estimating planar object poses using an object detector and a reinforcement learning agent. The system consists of a robotic arm with an eye-in-hand, that uses the object detector to feed a bounding box of the detected object to the reinforcement learning architecture, which uses this information along with the position of the camera to estimate the pose of the object. The advantage of this approach is that the system is able to learn the object pose by training in a simulated environment without the requirement of extensive feature engineering or task specific manual labeling of large amounts of data.

The results shown in Sect. 5 shows that after the agent is trained in the simulated environment for 100,000 episodes, it is able to correctly categorize object poses in 92.7% of episodes run on the physical setup. Perturbations of the system shows that it has the potential to handle occurrences that it has not experienced during training quite well, which can be viewed as beginning stages of intelligence. Even though the environment presented in this paper is simple,

Table 6. Number of steps and pose predictions for when object pose is in-between two of the known discrete poses.

Closest known pose		Steps	Prediction
0	a	11	none
	b	2	0
1	a	11	none
	b	1	1
2	a	11	none
	b	11	none
3	a	2	3
	b	2	3
4	a	1	4
	b	1	4
5	a	2	5
	b	2	5
6	a	2	6
	b	11	none
7	a	11	none
	b	2	7

we believe that this result serves as an indicator that this method has potential to be implemented in more challenging environments, e.g. pose estimation with more degrees of freedom.

References

1. Boukas, E., Gasteratos, A.: Modeling regions of interest on orbital and rover imagery for planetary exploration missions. Cybern. Syst. **47**(3), 180–205 (2016). https://doi.org/10.1080/01969722.2016.1154771
2. Brockman, G., et al.: OpenAI Gym (2016)
3. Brooks, R.A.: Intelligence without representation. Artif. Intell. **47**, 139–159 (1991)
4. Huang, J., et al.: Speed/accuracy trade-offs for modern convolutional object detectors. In: Computer Vision and Pattern Recognition (2017)
5. Jia, Z., Chang, Y.J., Chen, T.: Active view selection for object and pose recognition. In: 2009 IEEE 12th International Conference on Computer Vision Workshops, September 2009, pp. 641–648 (2009)
6. Kostavelis, I., Nalpantidis, L., Gasteratos, A.: Object recognition using saliency maps and HTM learning. In: IEEE International Conference on Imaging Systems and Techniques, pp. 528–532. IEEE, Manchester (2012)
7. Krull, A., Brachmann, E., Michel, F., Yang, M.Y., Gumhold, S.: Learning analysis-by-synthesis for 6D pose estimation in RGB-D images. In: 2015 IEEE International Conference on Computer Vision, December 2015, pp. 954–962 (2015)

8. Krull, A., Brachmann, E., Nowozin, S., Michel, F., Shotton, J., Rother, C.: PoseAgent: budget-constrained 6D object pose estimation via reinforcement learning. In: 2017 IEEE Conference on Computer Vision and Pattern Recognition, July 2017, pp. 2566–2574 (2017)

9. Mnih, V., et al.: Human-level control through deep reinforcement learning. Nature **518**(7540), 529–533 (2015). https://doi.org/10.1038/nature14236

10. Piater, J., Jodogne, S., Detry, R., Kraft, D., Krueger, N., Kroemer, O.: Learning visual representations for perception-action systems. Int. J. Robot. Res. **30**, 294–307 (2015)

11. Plappert, M.: Keras-RL (2016). https://github.com/keras-rl/keras-rl

12. Polydoros, A.S., Boukas, E., Nalpantidis, L.: Online multi-target learning of inverse dynamics models for computed-torque control of compliant manipulators. In: IEEE/RSJ International Conference on Intelligent Robots and Systems (IROS), Vancouver (2017)

13. Toshev, A., Szegedy, C.: DeepPose: human pose estimation via deep neural networks. In: 2014 IEEE Conference on Computer Vision and Pattern Recognition, June 2014, pp. 1653–1660 (2014)

14. Tremblay, J., To, T., Sundaralingam, B., Xiang, Y., Fox, D., Birchfield, S.: Deep object pose estimation for semantic robotic grasping of household objects. Cornell University Library (2018)

15. Wooldridge, M., Jennings, N.R.: Agent theories, architectures, and languages: a survey. In: Wooldridge, M.J., Jennings, N.R. (eds.) ATAL 1994. LNCS, vol. 890, pp. 1–39. Springer, Heidelberg (1995). https://doi.org/10.1007/3-540-58855-8_1

A Two-Stage Approach
for Commonality-Based Temporal
Localization of Periodic Motions

Costas Panagiotakis[1,2]($\boxtimes$) and Antonis Argyros[1,3]

[1] Institute of Computer Science, FORTH, Heraklion, Greece
{cpanag,argyros}@ics.forth.gr
[2] Department of Management Science and Technology, Hellenic Mediterranean
University, Heraklion, Greece
[3] Computer Science Department, University of Crete, Heraklion, Greece

Abstract. We present an unsupervised method for the detection of all
temporal segments of videos or motion capture data, that correspond to
periodic motions. The proposed method is based on the detection of similar segments (commonalities) in different parts of the input sequence and
employs a two-stage approach that operates on the matrix of pairwise
distances of all input frames. The quantitative evaluation of the proposed method on three standard ground-truth-annotated datasets (two
video datasets, one 3D human motion capture dataset) demonstrate its
improved performance in comparison to existing approaches.

Keywords: Periodicity detection · Repetitive motions detection ·
Commonalities discovery · Video segmentation

1 Introduction

Periodic or repetitive motions are very common in natural and man-made environments [3]. Therefore, their detection constitutes an important step towards
the segmentation and the high level interpretation of video and motion capture (mocap) data. This is an interesting problem in computer vision and pattern recognition whose solution has several applications in action and activity
recognition, medical diagnosis [1], detection of machine failures [13], repetition
counting [14], etc.

In this work, we address the problem of temporal segmentation of periodic
segments in videos and mocap data and we propose a solution which neither
requires prior knowledge nor imposes constraints on the speed, the number or the
length of the periods of the periodic segments. Moreover, the method tolerates
variations of the period of the periodic motions. The input to the proposed
method is the $N \times N$ symmetric matrix D that contains the pairwise distances
of all N frames of the input sequence. Figure 1 provides an example of such a
matrix D, where warm (red) and cold (blue) colors correspond to high and low

© Springer Nature Switzerland AG 2019
D. Tzovaras et al. (Eds.): ICVS 2019, LNCS 11754, pp. 366–375, 2019.
https://doi.org/10.1007/978-3-030-34995-0_33

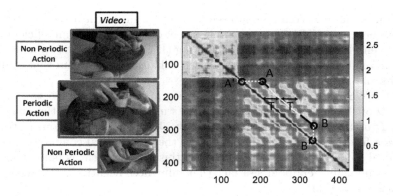

Fig. 1. An example of the distance matrix D of the frames of a video, in which a periodic action appears between two non-periodic actions. The periodic nature of the middle action gives rise to low values in D (i.e., similar frames) along straight line segments that are parallel to the diagonal of D. The goal of the proposed method is to detect such lines and to estimate the period of the corresponding periodic action (i.e., the offset from the diagonal). The periodic segment detected by the first and second stage of the proposed method are plotted with a white and a black straight line, respectively. The projection of the endpoints (A, B) of the black line on the main diagonal, give the start (A') and the end (B') of this periodic segment. (Color figure online)

values in D, respectively. The main diagonal of this matrix contains zeros, since these points hold the distance of each frame of the sequence to itself.

Let us consider that a periodic segment $[A', B']$ with period T starts at frame A' and ends at frame B' of the video. Due to the definition of periodicity, we know that there is a strong similarity between the segment $[A', B']$ and another part of the sequence that is temporally displaced by T (see Fig. 1). This means that D contains a straight line segment $\overline{AB}$ where $A = (A', A'+T)$ and $B = (B'-T, B')$ that is parallel to the main diagonal, along which D has very low values. In practice, due to deviations from perfect periodicity, the path connecting A and B might deviate from straightness and might not be perfectly parallel to the diagonal. Detecting periodic segments and estimating their period, amounts to detecting and localizing straight lines of minimum cost that are located off the main diagonal, as for example the line $\overline{AB}$ in Fig. 1. The sum of the entries of D under such a line/path is inversely proportional to the similarity of the two parts of the sequence. A low (high) cost path corresponds to a high (low) confidence on the existence of a periodic segment. Short paths correspond to periodic segments of a few frames, which are not that interesting. As paths increase in length, their cost also increases. Thus, the trade-off between the length of the path (the duration of the segments) and its cost should be balanced.

A related work [8], developed *P-MUCOS*, a method that reduces the problem of periodicity detection to the problem of finding common sub-sequences in a sequence. This is achieved by employing *MUCOS* [9] a graph-based search algorithm for finding common sub-sequences (commonalities) between two different sequences. *MUCOS* has a complexity of $O(N^2)$ and it is an efficient alternative to employing Dynamic Time Warping (DTW) [10]. As shown in [8,9], the computational complexity of a DTW-based exhaustive algorithm that evaluates all paths and keeps the best one is $O(N^6)$.

P-MUCOS is an one-stage approach that applies *MUCOS* to the distance matrix D of the input sequence. This results in the detection of the main diagonal of D as the major commonality. To avoid this trivial solution, *P-MUCOS* employs a filtering technique that enhances the off-diagonal commonalities that correspond to periodic segments. However, this enhancement is not strong close to the commonality endpoints, a fact that influences negatively the accuracy of periodicity detection.

Thus, in this work, we present *P-MUCOS-S2*, an improvement of *P-MUCOS*, which addresses the drawback of *P-MUCOS* by introducing a two-stage approach for commonality detection. In the first stage, the strongest part of the off-diagonal commonality is detected in an improved version of the enhancement performed by *P-MUCOS*. In the second stage, a hysteresis-thresholding-like operation extends the initial detections by optimizing an appropriately defined objective function. *P-MUCOS-S2* is evaluated quantitatively in comparison to the *P-MUCOS* and another, Fourier-based baseline approach [8] and is shown to improve substantially the accuracy of periodicity detection. The experimental evaluation is performed on the publicly available video datasets employed in [8], but also on a relevant mocap dataset.

In summary, the contributions of this work are: the improvement of the *P-MUCOS* algorithm for temporal localization of periodic segments by (a) improving the commonalities enhancement filtering approach of the first state and (b) by introducing a hysteresis-thresholding-based second stage, as well as the quantitative evaluation of the new algorithm *P-MUCOS-S2* on standard datasets on motion captured and video datasets in comparison to existing approaches.

2   Related Work

The problem of detecting periodic segments in time-series has been well studied. In [4], the problem of periodicity detection in time series is addressed using the time warping algorithm WARP. A given time series is transformed to time-stamped events drawn from a finite set of nominal event types. The main idea of WARP is that if the time series is shifted by a number of elements equal to the period of the time series, then the original time series and the shifted one will be very similar. More recently, Karvounas et al. [5] formulate the detection of a periodic segment as an optimization problem that is solved based on an evolutionary optimization technique. Given a time series representing a periodic signal with a non-periodic prefix and tail, the method estimates the start, the end and the period of the periodic part. The most important limitation of that method is that it assumes a video containing a single periodic segment. Another related challenging problem concerns the problem of periodicity detection from incomplete observations [7]. In [7], the authors propose a probabilistic model for periodic behaviors that was successfully applied on real human movement data.

The periodicity detection in videos is a more challenging problem, due to the high variability of the video content. In [11], Polana and Nelson devise an extension of the Fourier formula to detect periodicity in videos based on normal flow variation between successive image frames. The authors show that periodicity is an inherent low-level motion cue that can be exploited for robust detection of periodic phenomena without prior structural knowledge. Cutler and Davis [3] address the problem of periodicity detection for both stationary and non-stationary periodic signals. For the case of stationary signals, this can be achieved by a Fourier Transform followed by a Hanning filter. For the non-stationary case, Short-Time Fourier Transform is employed to better handle the shifting spectrum. As in [11], the objects are tracked and aligned before the periodicity analysis. The baseline method (see Sect. 4), that is presented in [8], is a natural extension of the power spectrum method [3]. According to this method, a given signal is periodic if the peak of its spectrum is greater than $\mu + 3\sigma$, where μ and σ denote the mean and the standard deviation of the signal spectral power. Such spectral domain methods have the limitation that the action frequency should be almost constant and it would emerge as a discernible peak at a time frequency graph. However, the amount of variation in appearance between repetitions and the variation in action length means that in certain cases, no such clear peak may be identifiable [6].

Wang et al. [17] proposed a method for retrieval of social games that are characterized by repetitions, from unstructured videos. Each frame is mapped to the nearest keyframe, yielding a sequence of keyframe indices that are used to mine recurring patterns. The approach proposed in [15] combines ideas from nonlinear time series analysis and computational topology, by translating the problem of finding recurrent dynamics in video data, into the problem of determining the circularity of an associated geometric space. There exist several supervised techniques that attempt to identify sequences in similarity/distance matrices [2]. In [2], for loop-closure were detected based on a classifier trained on similarity matrices. The proposed methodology can be also applied to such problems.

Levy and Wolf [6] use a deep learning approach to count the number of repetitions of approximately the same action in an input video sequence. In [14], the problem of visual repetition from realistic video has been formulated and solved, improving the results derived by Levy and Wolf [6]. The authors derive three periodic motion types by decomposition of the 3D motion field into its fundamental components and employ the continuous wavelet transform and combine the power spectra of all representations to support viewpoint invariance.

Most of the aforementioned approaches cannot handle the problem of periodicity detection under any video content and without some type of supervision. The method presented in this paper improves the results of [8] that was the first that makes no such assumption and is fully unsupervised. Additionally, the proposed method has been also applied to motion captured data and it can be also used for period tracking and in repetition estimation [14].

3   *P-MUCOS-S2*: **Commonality-Based Periodicity Detection**

Let A be a sequence of N frames and D the $N \times N$ symmetric matrix of the pair-wise distances of these frames. The proposed method can assume a variety of frame representations and corresponding frame distance metrics.

An example of such a distance matrix D of the frames of a sequence, in which a periodic action appears between two non-periodic actions, is shown in Fig. 1. The periodic nature of the middle action gives rise to low values in D (i.e., similar frames) along straight line segments that are parallel to the diagonal of D and at horizontal offset T from the diagonal, where T is the period of the periodic action. This is because the distance between frames f_i and f_{i+T} of a periodic action with period T is expected to be very low. The goal of the proposed method is to detect such lines and to estimate the period T of the corresponding periodic action. Such a straight line is shown in Fig. 1 in black color. The projection of the endpoints (A, B) of the black line on the main diagonal, give the start (A') and the end (B') of the corresponding periodic segment. By detecting such segments, we can segment the periodic parts of a sequence and estimate the period of each of them. This can be achieved by employing a method that detects commonalities between two sequences. The *MUCOS* [9] method suits this purpose, as it discovers all commonalities (similar segments) of two sequences v_1 and v_2, given their distance matrix D. The application of *MUCOS* on the square matrix D of distances of a single sequence will result in the detection of the diagonal as the major commonality. This trivial solution can be excluded from consideration, as performed in [8] where the *P-MUCOS* algorithm was proposed. *P-MUCOS* is improved by introducing, *P-MUCOS-S2*, an approach that operates in two stages.

P-MUCOS-S2, Stage 1: *P-MUCOS* applies the following symmetric filter H_p at point $p = (i, j)$ of the distance matrix D to emphasize the commonalities to be detected:

$$H_p(q) = -a \cdot cos\left(\frac{2\pi d}{\tau}\right) \cdot (\tau - d), \tag{1}$$

where $q = (u, v)$, $d = |v - u|$, $\tau = j - i$, and a is determined by the constraint $\sum_q |H_p(q)| = 1$. The response of filter H_p on point p is given by $D_H(p) = \sum_{q \in R(\tau)} H_p(q) \cdot D(p-q)$ where the square region R is defined as $R(\tau) = \{(u, v)| -\tau/2 \le u \le \tau/2, -\tau/2 \le v \le \tau/2\}$. Intuitively, H operates as follows. At a point $p = (i, j)$ the distance matrix is convolved with a filter whose width is τ and whose coefficients are a sinusoidal pattern, evolving perpendicularly to the main diagonal of D, with a minimum at p. Thus, in case that a commonality path c passes through p, the locally minimum value at p will be further pronounced. Figure 2 illustrates the positioning of filter H at two different points p and p' of a distance matrix. The dotted rectangle denotes the periodic part of the video corresponding to this distance matrix. The rectangles located at p and p' denote the color-coded coefficients of filter H. Finally, the red diagonal lines besides the main diagonal denote commonality paths.

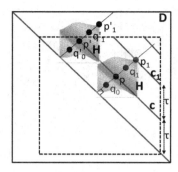

Fig. 2. A schematic illustration of the filtering operations in [8] and in the current work (see text for details).

In this work, we simplify the filter H significantly. Essentially, the realized improvement is that at a certain point p, the response is obtained by combining collinear distance matrix values in a direction perpendicular to the main diagonal. More specifically, for the point p in Fig. 2, the new filter response is given by

$$D_f(p) = 2 \cdot D(p) - D(q_0) - D(q_1) + D(p_1), \qquad (2)$$

where $v = [\frac{\tau}{4}, -\frac{\tau}{4}]^T$, $q_0 = p + v$, $q_1 = p - v$, $p_1 = p - 2 \cdot v$. If p belongs to a commonality path c, then the point p_1 will belong to the next commonality path c_1 that is parallel to c. The point q_0 is located halfway between the main diagonal and commonality c, and q_1 is located halfway between commonalities c and c_1. So, the new filter response can be explained by the fact that D should get low values on p and p_1 and high values on q_0 and q_1. Finally, we subtract from D_f its minimum value, so that D_f becomes a non-negative matrix.

Let S be the part of the upper right triangle of D that is restricted by the minimum (T_m) and the maximum (T_M) duration of a period. In our implementation, we set $T_m = 3$ and $T_M = \lfloor N/3 \rfloor$ frames, so as to ensure that there exist at least three periods of the periodic part of the video. The new filter H is applied to all points in S, to emphasize the commonalities that are close to the diagonal of the distance matrix D (see Fig. 3(b)).

The computational cost of the application of the new filter is constant and equal to 5 operations per point, while the recursive computation of the filter H proposed in [8] has computational cost $O(\tau \cdot N^2) = O(N^3)$ ($4 \cdot \tau$ operations per point). Thus, the new filter results in a significant reduction of the computational cost. Moreover, as demonstrated by experimental results, the application of this filter improves the quantitative metrics of periodicity detection compared to the filter employed in [8].

P-MUCOS-S2, Stage 2: The filtering operation of stage 1 enhances the distance matrix so that a commonality can be detected more easily by *MUCOS*. However, this enhancement fails close to the endpoints of a commonality path. This is because the filtering operation close to the end-points of a commonality

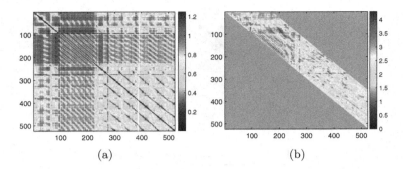

(a)   (b)

Fig. 3. (a) An example of a distance matrix D in which two periodic motions (jumping, waiving) appear after a non-periodic one (stand up). (b) The enhanced distance matrix D_f which is fed into *P-MUCOS-S2*. The two orange triangular parts (top-right, bottom left) are excluded from the commonality search space.

path involves, inevitably, values that are outside the commonality rectangle (see for example points q'_1 and p'_1 when the filter is applied to point p in Fig. 2). To deal with this issue, the second stage of *P-MUCOS-S2* operates as follows. For each detected commonality c of the first stage, its endpoints are extended in the direction that is locally parallel to c, by measuring the following objective function $\omega(c)$

$$\omega(c) = \frac{A(c)}{\sum_{p \in c} D(p) + \epsilon},\qquad(3)$$

where $A(c)$ is equal to the area of the bounding rectangle of the commonality path c, $P(c) = \sum_{p \in c} D(p)$, and ϵ is a small constant preventing division by zero. This objective function captures the trade-off between the terms $A(c)$ and $P(c)$. Specifically, commonalities c with large $A(c)$ are preferable. At the same time, as $A(c)$ increases, $P(c)$ also increases. The selected commonality endpoints are the ones that optimize the objective function $\omega(c)$. Example detections of the second stage are plotted as black lines in Fig. 1.

Finally, the period T of a periodic segment that corresponds to a detected commonality c, can be estimated as the average of the quantities $j - i$, for all points $(i, j) \in c$.

Fig. 4. Snapshots from videos of the PERTUBE dataset.

4   Experimental Results

Datasets: The proposed method was evaluated on one dataset that contains motion capture (mocap) data and on two datasets that contain conventional RGB videos.

- **MHAD202-s dataset**: Contains 202 motion data sequences of the 101 sequence pairs of the MHAD101-s dataset [10]. Each video consists of 3–7 periodic actions (e.g. jumping in place, jumping jacks, bending hands) and non periodic actions (e.g. throwing a ball, sit down). Each frame of a sequence represents the 3D human pose as a 64D vector whose dimensions encode angles of selected body parts. Distance matrices are obtained by estimating the Euclidean distances in pairs of such vectors.
- **MHAD202-v dataset**: Contains the 202 videos that correspond to the 202 motion capture sequences of the MHAD202-s dataset. As suggested in [10] and employed in [8], each frame is represented as a 100D vector that concatenates trajectory shape, HOG, HOF, and MBH descriptors computed on top of Improved Dense Trajectories (IDT) features [16].
- **PERTUBE dataset**: This dataset was introduced in [8] for assessing solutions to the periodicity detection problem. PERTUBE contains 50 videos showing human, animal and machine motions in lab settings or in the wild (see Fig. 4). The representation of video frames is as in MHAD202-v.

Performance Metrics: In order to assess the performance of the proposed methods we employed the standard metrics of precision $\mathcal{P}$, recall $\mathcal{R}$, F-measure score F_1 score and overlap $\mathcal{O}$ (intersection-over-union). The reported metrics were computed in each individual sequence and then averaged across all sequences of a dataset.

Obtained Results: Table 1 summarizes the results of *P-MUCOS-S2*, *P-MUCOS* and *BASELINE* methods [8] on the MHAD202-s, MHAD202-v and PERTUBE datasets. All algorithms run with the same parameters in different datasets. It can be verified that in all datasets, *P-MUCOS-S2* outperforms *P-MUCOS* and the baseline method. The difference in performance is more striking in the more challenging, real-world PERTUBE dataset where *P-MUCOS-S2* achieves a 10% and 7% improvement in overlap and F_1 score, respectively, compared to *P-MUCOS*. Our interpretation is that the MHAD202-s and MHAD202-v datasets contain simpler data, i.e., each sequence contains clearly periodic and non-periodic parts of the motion of a human recorded in laboratory conditions. This is contrasted to the more complex situations encountered in the youtube videos of PERTUBE. As a result, the distance matrices of the MHAD datasets are already of good quality. Therefore, the improved filtering and the two stage approach followed in this paper has more significant impact when applied to the lower-quality distance matrices of the PERTUBE data set.

We have also evaluated the following variant of the proposed method. We kept the original filtering of *P-MUCOS* and on that result, we applied the second stage proposed in this work. This hybrid scheme yields an overlap rate $\mathcal{O} = 70.5\%$

Table 1. Evaluation results on the MHAD202-s, MHAD202-v and PERTUBE datasets.

	Mocap				Video							
Dataset	MHAD202-s				MHAD202-v				PERTUBE			
Metric	$\mathcal{R}(\%)$	$\mathcal{P}(\%)$	$F_1(\%)$	$\mathcal{O}(\%)$	$\mathcal{R}(\%)$	$\mathcal{P}(\%)$	$F_1(\%)$	$\mathcal{O}(\%)$	$\mathcal{R}(\%)$	$\mathcal{P}(\%)$	$F_1(\%)$	$\mathcal{O}(\%)$
P-MUCOS-S2	**90.4**	93.2	**91.2**	**85.4**	87.8	**98.6**	**92.7**	**87.1**	92.5	**80.2**	**83.9**	**75.9**
P-MUCOS	85.8	**95.7**	89.5	82.1	**94.4**	89.5	90.9	84.7	**97.5**	68.0	76.8	65.8
BASELINE	86.3	92.9	88.8	82.4	93.2	86.2	88.9	81.6	79.3	61.1	66.8	57.3

on the PERTUBE dataset. This means that the filtering improvement of this work and the introduction of the second stage contribute approximately equally to the improvement of the P-MUCOS-S2 over the P-MUCOS. Finally, from a computational point of view, P-MUCOS-S2 is about three times faster than P-MUCOS.

5 Conclusions

We proposed a method for discovering periodic segments in motion captured data and videos improving the state-of-the-art method proposed in [8]. The proposed framework is applied to distance matrices of pairwise distances of all frames of a given sequence detecting periodic actions in two stages. The experimental results on challenging datasets showed the effectiveness of the proposed method. As future work, we plan to extend the proposed method in two directions, (a) computation of the number of repetitions of a certain action [14] and (b) monitoring of the variation of the period of periodic motions. Both of these quantitative and qualitative characterizations of periodic motions find important applications in several computer vision applications involving action recognition, anomaly detection (e.g., [12]), performance characterization (e.g. [1]), e.t.c.

Acknowledgments. This work was partially supported by the EU project Co4Robots (H2020-731869).

References

1. Afsar, O., Tirnakli, U., Marwan, N.: Recurrence quantification analysis at work: quasi-periodicity based interpretation of gait force profiles for patients with Parkinson disease. Sci. Rep. **8**(1), 9102 (2018)
2. Bampis, L., Amanatiadis, A., Gasteratos, A.: Fast loop-closure detection using visual-word-vectors from image sequences. Int. J. Rob. Res. **37**(1), 62–82 (2018)
3. Cutler, R., Davis, L.S.: Robust real-time periodic motion detection, analysis, and applications. IEEE Trans. Pattern Anal. Mach. Intell. **22**(8), 781–796 (2000)
4. Elfeky, M.G., Aref, W.G., Elmagarmid, A.K.: Warp: time warping for periodicity detection. In: Fifth IEEE International Conference on Data Mining, 8 pp. IEEE (2005)
5. Karvounas, G., Oikonomidis, I., Argyros, A.A.: Localizing periodicity in time series and videos. In: BMVC (2016)

6. Levy, O., Wolf, L.: Live repetition counting. In: Proceedings of the IEEE International Conference on Computer Vision, pp. 3020–3028 (2015)
7. Li, Z., Wang, J., Han, J.: ePeriodicity: mining event periodicity from incomplete observations. IEEE Trans. Knowl. Data Eng. **27**(5), 1219–1232 (2015)
8. Panagiotakis, C., Karvounas, G., Argyros, A.: Unsupervised detection of periodic segments in videos. In: 2018 25th IEEE International Conference on Image Processing (ICIP), pp. 923–927. IEEE (2018)
9. Panagiotakis, C., Papoutsakis, K., Argyros, A.: A graph-based approach for detecting common actions in motion capture data and videos. Pattern Recogn. **79**, 1–11 (2018)
10. Papoutsakis, K., Panagiotakis, C., Argyros, A.: Temporal action co-segmentation in 3D motion capture data and videos. In: IEEE Conference on COmputer Vision and Pattern Recognition (CVPR) (2017)
11. Polana, R., Nelson, R.C.: Detection and recognition of periodic, nonrigid motion. Int. J. Comput. Vision **23**(3), 261–282 (1997)
12. Ramasso, E., Panagiotakis, C., Pellerin, D., Rombaut, M.: Human action recognition in videos based on the transferable belief model. Pattern Anal. Appl. **11**(1), 1–19 (2008)
13. Ramasso, E., Placet, V., Boubakar, M.L.: Unsupervised consensus clustering of acoustic emission time-series for robust damage sequence estimation in composites. IEEE Trans. Instrum. Meas. **64**(12), 3297–3307 (2015)
14. Runia, T.F., Snoek, C.G., Smeulders, A.W.: Repetition estimation. arXiv preprint arXiv:1806.06984 (2018)
15. Tralie, C.J., Perea, J.A.: (quasi) periodicity quantification in video data, using topology. arXiv preprint arXiv:1704.08382 (2017)
16. Wang, H., Kläser, A., Schmid, C., Liu, C.L.: Dense trajectories and motion boundary descriptors for action recognition. IJCV **103**(1), 60–79 (2013). https://doi.org/10.1007/s11263-012-0594-8
17. Wang, P., Abowd, G.D., Rehg, J.M.: Quasi-periodic event analysis for social game retrieval. In: 2009 IEEE 12th International Conference on Computer Vision, pp. 112–119. IEEE (2009)

Deep Residual Temporal Convolutional Networks for Skeleton-Based Human Action Recognition

R. Khamsehashari[✉], K. Gadzicki, and C. Zetzsche

Cognitive Neuroinformatics, University of Bremen, Bremen, Germany
{rkhamseh,gadzicki}@uni-bremen.de, zetzsche@cs.uni-bremen.de

Abstract. Deep residual networks for action recognition based on skeleton data can avoid the degradation problem, and a 56-layer Res-Net has recently achieved good results. Since a much "shallower" 11-layer model (Res-TCN) with a temporal convolution network and a simplified residual unit achieved almost competitive performance, we investigate deep variants of Res-TCN and compare them to Res-Net architectures. Our results outperform the other approaches in this class of residual networks. Our investigation suggests that the resistance of deep residual networks to degradation is not only determined by the architecture but also by data and task properties.

Keywords: Deep residual networks · Action recognition · Degradation · Hyperparameters

1 Introduction

Human activity recognition has become a prominent research area due to its potential application in video surveillance, human computer interfaces, ambient assisted living, human-robot-interaction, etc. More recently it has also become important for understanding pedestrian behavior in autonomous driving [12]. As in other areas of artificial intelligence, deep learning techniques have also gained exceptional achievements in human activity recognition. Particularly, deep learning methods based on Recurrent Neural Networks (RNN) and Convolutional Neural Networks (CNN) architectures have shown great performance in classification tasks by learning discriminant features from large amounts of data.

Residual networks (e.g. Res-Net [1]) can avoid the degradation problem in deep CNN architectures. Originally developed for image recognition tasks, they have recently been extended to human activity recognition [7]. As to be expected

This work has been supported by the German Aerospace Center (DLR) with financial means of the German Federal Ministry for Economic Affairs and Energy (BMWi), project "OPA³L" (grant No. 50 NA 1909) and by the German Research Foundation DFG, as part of CRC (Sonderforschungsbereich) 1320 "EASE - Everyday Activity Science and Engineering", University of Bremen (http://www.ease-crc.org/).

© Springer Nature Switzerland AG 2019
D. Tzovaras et al. (Eds.): ICVS 2019, LNCS 11754, pp. 376–385, 2019.
https://doi.org/10.1007/978-3-030-34995-0_34

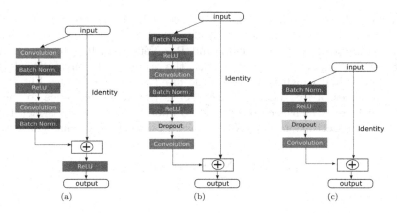

Fig. 1. The basic residual unit in different approaches. (a) Original ResNet [1]; (b) improved ResNet [7]; (c) Res-TCN [4]

for residual architectures, good performance levels could be obtained with quite deep architectures, i.e. 56 to 110 layers [7]. Somewhat surprisingly, however, a recent approach which combined the residual network concept with a temporal convolutional network architecture (Res-TCN [4]) could achieve an almost comparable performance with a comparably "shallow" configuration of 11 layers. And this, although the basic building unit of this approach also is considerably simpler than that of the Res-Net architecture (cf. Fig. 1), and offers in addition a much easier interpretability.

However, while the ResNet approaches have been systematically investigated with respect to the overall depth of the architecture (from 20 up to 110 layers [7]) the Res-TCN approach has only been tested in one single 11-layer variant. This prompted us to investigate whether very deep network architectures for activity recognition can profit from using the structurally simple residual unit of Res-TCN as basic building block, and a representation based on temporal convolutions. We introduce two variants of a deep learning architecture, Deep Res-TCN-3 and Deep Res-TCN-4, with depths ranging from 11 to 152 layers, to learn features from skeleton data and classify them into action classes.

2   Related Work

Skeleton data are of particular interest for human activity recognition as they carry high level information in the form of abstracted 3D joint positions. They can be obtained by optical tracking of body markers, from depth videos (e.g. Microsoft Kinect), or from RGB video data with pose estimation methods [2].

We shortly review the central ideas behind the residual based architectures such as Res-Net [1] and Res-TCN [4]. Res-Net employs injected residual connections between processing streams to allow spatial-temporal interaction between them. Res-TCN redesigned the original TCN [5] by factoring out the deeper layers into additive residual terms that yielded both an interpretable hidden representation and model parameters. Each unit in layer $L + 1$ performs the following computations in Res-TCN and Res-Net, according to Eqs. 1 and 3 respectively. In residual based networks, the traditional convolutional layers calculate a residual which is added to the input of the layers (see Fig. 1). One of the differences of Res-TCN and Res-Net is the direct reference to the first convolution layer, which according to Eq. 2 operates on the raw skeletal input and the created activation map, $X1$, is passed on to the subsequent layers.

$$X_L = X_1 + \sum_{i=2}^{L} W_i * max(0, X_{i-1}) \tag{1}$$

where

$$X_1 = W_1 * X_0 \tag{2}$$

$$X_{L+1} = W_L * max(0, X_L) + X_L. \tag{3}$$

X_L and X_{L+1} are input and output features of the L_{t_h} residual unit, respectively. The architecture of a residual based network is shown in Fig. 2.

The basic residual unit of Res-TCN [4], as compared to that of the original ResNet [1], does not use ReLUs behind the element-wise additions $\oplus$ (see Fig. 1a, c) and can thus provide readily interpretable representations. In addition, such units produce a direct path that enables the signal to be directly propagated in a forward pass through the entire network to any unit and also the gradients can be backwards propagated to any unit (cf. [7]). Finally, the Res-TCN building block is considerably shallower compared to both the original [1] and the improved ResNet [7] (cf. Fig. 1).

3   Methods

In this study we investigate different variants of deep residual architectures. Residual networks are assumed to cope with the degradation phenomenon by allowing for the fusion of all lower-level features of previous layers in the deeper layers, thus enabling more complex mappings to higher level feature maps. Approaches with residual units in both image processing [1] and action recognition [7] indicate that performance can be systematically improved by deeper architectures. For this study, we were particularly interested whether these advantages of deeper architectures can be successfully combined with the temporal representation and the simpler and more shallow nature of the basic Res-TCN building block [4], and how the performance of the resulting architectures compares to the more sophisticated approaches used in [1] and [4] (cf. also Fig. 1).

For our investigation and performance comparisons we wanted to have a broad coverage of the depth dimension, reaching from the relative shallow depth of 11 layers, as used in the original Res-TCN approach [4], up to a quite high depth of 152 layers, the maximum depth used in the original Res-Net study [1].

Table 1. Specification of architectures. The values in the brackets state the filter length and number of features of each building block. The number of stacked building blocks is given after the brackets. Down sampling is performed within `conv1` of each block A to D with a stride of 2. The 11-layer variant corresponds to the original Res-TCN architecture.

Layer name	Output size	11-layer	18-layer	34-layer	50-layer	101-layer	152-layer
(a) Deep Res-TCN-3							
Conv1	300	8,64					
Block A	300	[8,64]*3	[8,64]*5	[8,64]*10	[8,64]*16	[8,64]*33	[8,64]*50
Block B	150	[8,128]*3	[8,128]*5	[8,128]*11	[8,128]*16	[8,128]*33	[8,128]*50
Block C	75	[8,256]*3	[8,256]*6	[8,256]*11	[8,256]*16	[8,256]*33	[8,256]*50
Average pool, fc-60, softmax	1						
(b) Deep Res-TCN-4							
Conv1	300	8,64					
Block A	300	[8,64]*3	[8,64]*4	[8,64]*6	[8,64]*9	[8,64]*33	[8,64]*46
Block B	150	[8,128]*3	[8,128]*4	[8,128]*8	[8,128]*12	[8,128]*33	[8,128]*46
Block C	75	[8,256]*3	[8,256]*4	[8,256]*12	[8,256]*18	[8,256]*22	[8,256]*35
Block D	38	-	[8,512]*4	[8,512]*6	[8,512]*9	[8,512]*11	[8,512]*23
Average pool, fc-60, softmax	1						

Furthermore, we organized our investigations into three, more specific research questions: First, we wanted to find out whether it is possible to use the simplified residual unit of Res-TCN for the design of deeper architectures, in order to obtain an improved classification performance. We stick with the setting of the original Res-TCN architecture [4], Res-TCN-3 in Table 1a, and varied the depth only, in order to have an as direct as possible comparison. We did not use the bottleneck architecture from [1], since we wanted to avoid a change in architecture along the depth dimension. Control tests of a few bottleneck variants of the networks also indicated no advantage in classification performance.

The second research question addresses in how far the simpler structure of the basic residual unit of the Res-TCN architecture [4] will possibly limit performance in comparison to the more sophisticated Res-Net architectures [1,7] (cf. also Fig. 1). Since those architectures make use of a 4-block design, we also designed a 4-block deep Res-TCN-4 architecture (see Table 1b).

The third research question addressed the improvement of the hyperparameters that we found to be optimal for training (see Table 2) which are different from the ones used in the original Res-TCN study [4]. In order to disentangle the influence of the hyperparameters from those of the other properties varied in this study we hence tested a further set of Res-TCN-4 networks which have been trained with the original Res-TCN parameters as described in [4].

We used the residual units of Res-TCN ([4], Fig. 1c), and repeated them within one block multiple times. Between blocks the signal is down-sampled by a factor of 2 by convolution with stride 2 in the first element of each block. The convolutions are 1-dimensional with length 8 throughout the network. The

Table 2. Modified architectures of Deep Res-TCN with hyperparameter tuning.

	Original Res-TCN [4]	Deep Res-TCN-3 Deep Res-TCN-4
Optimizer	SGD	SGD
Nesterov acceleration	True	False
Momentum	0.9	0.01
L-1 regulizer	1e−4	1e−4
Learning rate	0.01	0.01
Batch size	128	128
Dropout	0.5	0.4
Epochs	200	200
Weight initialization	Scratch	Scratch

number of features varies from 64 in the early stages up to 512 in the later ones. Figure 2 shows an example of our architecture.

We evaluate the architectures on the 3D skeleton based human activity recognition dataset NTU RGB+D [8], the currently largest set offering several modalities (RGB video, depth video, IR video and skeleton data) with more than 56k training videos across 60 action classes. The actions were performed by 40 distinct subjects, and recorded from three view points with a Microsoft Kinect v2. The dataset provides two different evaluation criteria: Cross-Subject (CS) and Cross-View (CV). The skeleton data used are the x, y, z coordinates for 25 joints per person. There are maximally two people tracked during the recording.

For validation, we apply two standard testing protocols. One is Cross-Subject, for which half of the subjects are used for training and the rest are used for testing. The other is Cross-View, for which 67% of camera views are considered as training data and the rest as test data. The training on skeleton data provided by NTU RGB+D was performed with the Keras implementation of Res-TCN [4]. The hyperparameters of the proposed architectures are tuned according

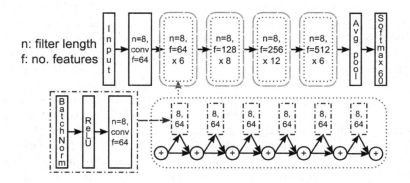

Fig. 2. Deep Res-TCN-4 architecture with 34 layers.

to Table 2 with weight decay of $1e^{-4}$, stochastic gradient descent (SGD), and we adopt the weight initialization and batch normalization [3]. We set the mini-batch size of 128 on one GPU for all structures except for the 152-layer networks for which we used a batch size of 96. The training is started with rate of 0.01, divided by 10 when the testing loss plateaus for more than 10 epochs. Finally, we use sparse softmax cross entropy for loss calculation during training. Evaluation of the loss curves of training and validation set did not show any indication for overfitting problems.

4   Results

In the following we present graphs of the classification performance in dependence on network depth to answer the research questions of Sect. 3. A detailed summary of all results in form of numerical values can be found in Table 3.

(1) *Can we use the simpler residual unit of Res-TCN to design deeper networks with improved performance?* As mentioned we want to keep this comparison straightforward and thus stick here to the 3-block design of the original Res-TCN [4]. Figure 3 shows the performance curve for this 11-layer Res-TCN and for our models with depths between 18 and 151 layers. The main result is that the deeper variants *all* provide a significantly improved classification accuracy in comparison to the original 11-layer Res-TCN architecture. However, the optimum occurs already at comparatively moderate depth levels of 34 and 18 layers for cross-subject and cross-view tests, respectively. But the optimum is a shallow one, and the decrease of accuracy for greater depths is quite moderate.

(2) *How does the simpler architecture of Res-TCN compare to the more sophisticated Res-Net architectures [1,7]? (cf. also* Fig. 1) Comparison is based on a 4-block design (Res-TCN-4), as used in the Res-Net architectures. Classification accuracy curves are shown in Fig. 4. Although the Res-TCN-4

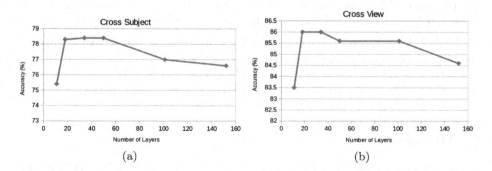

(a)   (b)

Fig. 3. Influence of network depth on accuracy (Deep Res-TCN-3). The leftmost data point corresponds to the original 11-layer Res-TCN architecture.

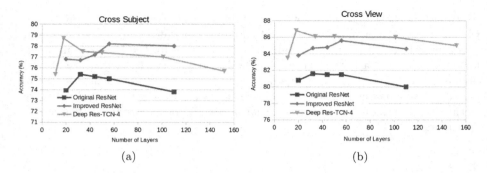

Fig. 4. Accuracy curves of different architectures.

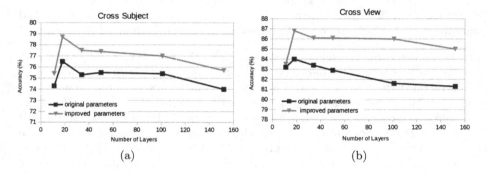

Fig. 5. Influence of hyperparameters.

architecture is simpler than those of the two Res-Net variants its classification accuracy is similar or even superior. In particular it provides the best performance level (78.7% for cross-subject and 86.8% for cross-view tests). Again, this optimum is achieved for a relatively moderate depth of 18 layers (Res-TCN-4-18).

(3) *Influence of hyperparameters.* Figure 5 shows how the classification accuracy is improved by our new hyperparameters, particularly for cross-view.

Table 3 shows the accuracy results for the different deep Res-TCN variants we tested (and for the original Res-TCN-11). A comparison of our models with other ResNet-like architectures is shown in Table 4. The best Res-TCN-4-18 network trained with improved parameters achieved the overall best performance, an accuracy of 78.7% for cross-subject and of 86.8% for cross-iew.

5   Discussion

In the previous section we demonstrated that the proposed deep Res-TCN architecture clearly outperforms the original "shallow" 11-layer Res-TCN [4]. In addition, it outperforms also alternative deep residual approaches of the Res-Net class [1,7], using the same experimental setting and dataset.

Table 3. Recognition accuracy. The best results are highlighted in bold.

Architectures	Original parameters		Improved parameters	
	Res-TCN	Deep Res-TCN-4	Deep Res-TCN-3	Deep Res-TCN-4
(a) Cross-subject accuracy				
Original Res-TCN-11	74.3	–	75.4	–
Res-TCN-18	–	**76.5**	78.3	**78.7**
Res-TCN-34	–	75.3	**78.4**	77.5
Res-TCN-50	–	75.5	**78.4**	77.4
Res-TCN-101	–	75.4	77.0	77.0
Res-TCN-152	–	74.0	76.6	75.7
(b) Cross-view accuracy				
Original Res-TCN-11	83.2	–	83.5	–
Res-TCN-18	–	**84.0**	**86.0**	**86.8**
Res-TCN-34	–	83.4	**86.0**	86.1
Res-TCN-50	–	82.9	85.6	86.1
Res-TCN-101	–	81.6	85.6	86.0
Res-TCN-152	–	81.3	84.6	85.0

Table 4. Relationship between number of layers on ResNet and Res-TCN based architectures and their best performance on the NTU-RGB+D dataset. The numbers are in format cross-subject/cross-view.

Method	11-layer	18-layer	32-layer	34-layer	50-layer	56-layer
Res-TCN [4]	74.3/83.1					
Original ResNet [7]			75.4/81.6			
Improved ResNet [7]						78.2/85.6
Deep Res-TCN-3		–/86.0		78.4/86.0	78.4/–	
Deep Res-TCN-4		78.7/86.8				

A somewhat surprising result is a systematic pattern observed for all architectures (except the improved Res-Net [7]): a strong initial performance gain due to the first step in depth, followed by a rapid leveling-off, or even a shallow decrease, for the deeper network variants. This effect is most expressed for the Res-TCN-4 architecture, where an increase from 11 to 18 layers yields the absolute optimum performance of all architectural variants considered in this study (78.7% and 86.8% accuracy for Cross-Subject and Cross-View, respectively, see also Fig. 4). The same pattern arises also with the Res-TCN-3 architectures (Fig. 3). The effect is not caused by the specific deep Res-TCN architectures since the original Res-Net architecture shows a similar pattern both for Cross-Subject and, less expressed, for Cross-View [7] (Fig. 4). However, the effect is also not solely caused by the dataset, since it does not occur with the improved Res-Net suggested by [7] which produces a systematic slow performance gain with increasing network depth, with an optimum at 56 layers (Fig. 4).

What can we learn from these results? First, the systematic pattern observed for all architectures but the improved Res-Net indicates problems with degradation. Although residual architectures have the basic potential to cope with degradation, they here seem to fail at early network stages. In the case of the Res-TCN variants this may be attributed to the simplified structure of the residual unit with only one convolutional layer, which limits the power for nonlinear approximation. But why do we observe a similar behavior for the original Res-Net, an architecture that has proven to be able to cope with the degradation phenomenon for a variety of datasets [1]? The cause for this could be the format of the representation. The deep Res-TCN architectures are based on purely temporal convolutions, the interrelations between joints being represented by the filters. Although the order of joints has been carefully rearranged in [7], and has proven to be essential for the performance, we assume that the 2-D convolutions are not optimally suited for the representation. This idea is also supported by the fact that our Res-TCN-4 network (with limited power of the basic residual unit) can outperform the improved 56-layer Res-Net architecture by using only as few as 18 layers. Taken together this suggests that the resistance to degradation effects is not solely determined by the specific residual structure of a network but also by a non-trivial interaction of architecture and task/data properties.

It is worth mentioning that the influence of the hyperparameters is quite strong compared to the other effects (Fig. 5). Since the main difference is the avoidance of momentum optimization this might indicate that this common default choice is not optimally suited for the relatively smooth and regular loss landscape of deep residual architectures [6], or will require additional measures to ensure full convergence.

Among the class of models which use only skeleton data and straight-forward deep residual network architectures the deep Res-TCN model suggested here provides the best performance. Other recent approaches exhibit even better performance levels, but these are achieved using quite specialized model structures, e.g. by making use of additional attention modules [9], multi-modal processing streams [11], view point adaptation [10] or other improvements e.g. [13–17]. It has to be expected that further gains can be obtained by a combination of our model with these more sophisticated approaches, for example by using our model as one component in a multimodal architecture.

References

1. He, K., Zhang, X., Ren, S., Sun, J.: Deep residual learning for image recognition. In: IEEE Conference on Computer Vision and Pattern Recognition (CVPR), pp. 770–778 (2016)
2. He, K., Gkioxari, G., Dollár, P., Girshick, R.: Mask R-CNN. In: 2017 IEEE International Conference on Computer Vision (ICCV), Venice, pp. 2980–2988 (2017). https://doi.org/10.1109/ICCV.2017.322
3. Ioffe, S., Szegedy, C.: Batch normalization: accelerating deep network training by reducing internal covariate shift. arXiv preprint. arXiv:1502.03167 (2015)

4. Kim, T.S., Reiter, A.: Interpretable 3D human action analysis with temporal convolutional networks. In: 2017 IEEE Conference on Computer Vision and Pattern Recognition Workshops (CVPRW) (2017)
5. Lea, C., Flynn, M.D., Vidal, R., Reiter, A., Hager, G.D.: Temporal convolutional networks for action segmentation and detection. In: The IEEE Conference on Computer Vision and Pattern Recognition (CVPR) (June 2017)
6. Li, H., Xu, Z., Taylor, G., Goldstein, T.: Visualizing the loss landscape of neural nets. In: CoRR. arXiv:1712.09913 (2017)
7. Pham, H., Khoudour, L., Crouzil, A., Zegers, P., Velastin, S.: Exploiting deep residual networks for human action recognition from skeletal data. Comput. Vis. Image Underst. (CVIU) **170**, 51–66 (2018)
8. Shahroudy, A., Liu, J., Ng, T.-T., Wang, G.: NTU RGB+D: a large scale dataset for 3D human activity analysis. In: Proceedings of the IEEE Conference on Computer Vision and Pattern Recognition (2016)
9. Yang, Z., Li, Y., Yang, J., Luo, J.: Action recognition with visual attention on skeleton images. In: CoRR. arXiv:1804.07453 (2018)
10. Zhang, P., Lan, C., Xing, J., Zeng, W., Xue, J., Zheng, N.: View adaptive neural networks for high performance skeleton-based human action recognition. IEEE Trans. Pattern Anal. Mach. Intell. **41**(8), 1963–1978 (2019)
11. Zhu, J., et al.: Action machine: rethinking action recognition in trimmed videos. In: CoRR. arXiv:1812.05770 (2019)
12. Rasouli, A., Tsotsos, J.K.: Joint attention in driver-pedestrian interaction: from theory to practice. In: CoRR. arXiv:1802.02522 (2018)
13. Liu, M., Hong, L., Chen, C.: Enhanced skeleton visualization for view invariant human action recognition. Pattern Recogn. **68**, 346–362 (2017)
14. Li, C., Wang, P., Wang, S., Hou, Y., Li, W.: Skeleton-based action recognition using LSTM and CNN. In: 2017 IEEE International Conference on Multimedia & Expo Workshops (ICMEW). IEEE (2017)
15. Li, C., Zhong, Q., Xie, D., Pu, S.: Skeleton-based action recognition with convolutional neural networks. 2017 IEEE International Conference on Multimedia & Expo Workshops (ICMEW). IEEE (2017)
16. Ke, Q., Bennamoun, M., An, S., Sohel, F., Boussaid, F.: A new representation of skeleton sequences for 3D action recognition. In: Proceedings of the IEEE Conference on Computer Vision and Pattern Recognition (2017)
17. Yan, S., Xiong, Y., Lin, D.: Spatial temporal graph convolutional networks for skeleton-based action recognition. In: Thirty-Second AAAI Conference on Artificial Intelligence (2018)

Monte Carlo Tree Search on Directed Acyclic Graphs for Object Pose Verification

Dominik Bauer$^{(\boxtimes)}$ ⓘ, Timothy Patten, and Markus Vincze

Automation and Control Institute, TU Wien, 1040 Vienna, Austria
{bauer,patten,vincze}@acin.tuwien.ac.at

Abstract. Reliable object pose estimation is an integral part of robotic vision systems as it enables robots to manipulate their surroundings. Powerful methods exist that estimate object poses from RGB and RGB-D images, yielding a set of hypotheses per object. However, determining the best hypotheses from the set of possible combinations is a challenging task. We apply MCTS to this problem to find an optimal solution in limited time and propose to share information between equivalent object combinations that emerge during the tree search, so-called transpositions. Thereby, the number of combinations that need to be considered is reduced and the search gathers information on these transpositions in a single statistic. We evaluate the resulting verification method on the YCB-VIDEO dataset and show more reliable detection of the best solution as compared to state of the art. In addition, we report a significant speed-up compared to previous MCTS-based methods for object pose verification.

Keywords: Hypotheses verification · Object pose estimation · Monte Carlo Tree Search · Analysis-by-synthesis

1 Introduction

For robots to co-inhabit the human environment, they require an understanding of their surroundings. It is important for robots to detect objects and their location in the environment to fulfill their tasks. To this end, object pose estimation methods are proposed (see SIXD challenge [13]) that enable robotic manipulation, for example, on the YCB object set [22]. While these methods are often able to provide precise estimates among their top-n hypotheses, the difficulty is to identify and select an estimate among those hypotheses. One approach is to cluster similar hypotheses and select the one with the highest consensus [8]. Hypotheses can also be filtered by rejecting those that only partially fit the

Funded by the TU Wien Doctoral College TrustRobots. Partially funded by OMRON Corporation and Aeolus, Inc.

ⓒ Springer Nature Switzerland AG 2019
D. Tzovaras et al. (Eds.): ICVS 2019, LNCS 11754, pp. 386–396, 2019.
https://doi.org/10.1007/978-3-030-34995-0_35

scene or whose silhouette does not align with scene edges [21]. When considering a scene consisting of multiple objects, these heuristics fail to capture the impact of object interrelations. Hypotheses verification frameworks [1,2] combine heuristics with other cues to disambiguate between multi-object hypotheses. However, verification frameworks need to consider different hypotheses combinations and therefore are limited by the number of possible combinations of the top-n hypotheses that grows quickly with an increasing number of objects.

As exhaustive search becomes inhibiting in these large search spaces, verification methods integrating physics simulation with Monte Carlo Tree Search (MCTS) [4,16] exploit the algorithm's any-time nature to determine an optimal solution in a limited number of iterations. Still, the run time of these methods is prohibitive for use in robot vision systems due to the physics simulation being a part of the reward computation.

Drawing from work on general game playing [19] and feature selection [18], we show that, in the domain of object pose verification, the search space can be more efficiently represented as a directed acyclic graph (DAG). Equivalent solutions are only represented once and we circumvent the issue of duplicate subtrees that would otherwise emerge in a tree representation and divide the attention of the search method. In this work, we propose a novel MCTS-based object pose verification method that achieves state-of-the-art performance by:

- modeling the search space of object pose hypotheses as a DAG
- a selection policy that exploits equivalent solutions in this DAG representation using an adapted version of the UCB1-tuned policy, called UCD-tuned
- an expansion policy that combines the idea of a "stopping hypothesis" for subset solutions with a confidence-based heuristic
- a fast, rendering-based evaluation of candidate solutions

We compare our selection policy to the theoretical best-case, worst-case and a baseline set of hypotheses, which are selected solely on an estimated confidence value. We report an improvement over the baseline object pose estimation on the YCB-VIDEO dataset and show that our approach selects hypotheses that are better explained by the observation. Compared to other MCTS-based verification methods, we furthermore report a significantly reduced average run time of 3.5 seconds per frame, enabling the use of the method in real-world vision systems.

2   Related Work

Object Pose Estimation: Methods based on local feature descriptors determine an object's pose from correspondences between the observation and a 3D model. Point Pair Features (PPF) have been shown to be well suited for this task [8,21]. Template-based methods perform well on texture-less objects [12,14]. These methods use pre-computed views of a 3D model of an object that are matched to an observation during inference. Learning-based methods can be separated in two categories: Regressing feature descriptors [15,23] and regressing the pose itself [5, 20]. We use the method by Wang et al. [22] that jointly regresses the pose and an estimation confidence as baseline.

Object Pose Verification: Object pose estimation might yield false positives and picking the best hypothesis among the top-n estimates is not straight-forward. Clustering of similar hypotheses [8], penalizing partial scene fit or penalizing misaligned silhouettes [21] helps to deal with this problem. Aldoma et al. [1] use Simulated Annealing to select a subset of hypotheses that fits the observation. This work proposes a cost function that considers geometrical cues together with cues for clutter and conflicting correspondences to the observation. In principle, fitting the observation can be performed by exhaustively searching through all possible object pose hypotheses in a brute-force manner [17]. This work illustrates the use of the analysis-by-synthesis paradigm to evaluate candidate solutions in terms of depth discrepancy. More efficiently, [4, 16] apply MCTS to search through a pool of pose hypotheses and integrate physics simulation in the evaluation of candidate solutions. While this potentially improves the solution's plausibility and highlights inconsistencies, it dramatically slows down the run time to between 18 s [4] and 30 s per frame [16].

Monte Carlo Tree Search: The basic idea of MCTS is to grow a search tree by repeatedly selecting the best known solution, expanding this solution and sampling its reward to update the solution statistics for the next round. This procedure can be stopped at any time to yield an optimal solution in terms of evaluated possibilities. For an overview of research on MCTS, see [6]. If expanding two different solutions leads to an equivalent solution, this is called a transposition. Childs et al. [7] identify the need to consider transpositions and propose variations to the UCB1 policy to deal with transpositions in the search tree. Transpositions can be handled by using a DAG representation for the search space. However, the policies used in MCTS have to be adapted to maintain convergence. For feature selection, this problem is posed as Best Arm Identification and Best Leaf Identification on an expanding DAG [18]. Parameterized policies are proposed to control the exploitation of the DAG representation [19] in the LeftRight and Hex games. However, no conclusive evidence is provided for selecting the parameterization in other domains, with the best parameters significantly differing between the two considered games. We extend this parameterized policy for our selection policy in this work. In the related domain of feature selection, a so-called stopping feature is proposed to deal with the a-priori unknown number of features to select [10, 18]. We translate this idea to object pose verification to deal with solutions containing only a subset of objects.

3   6D Object Pose Verification

The presented object pose verification method assumes a pool of candidate object pose hypotheses per object instance and a corresponding depth image as input. Furthermore, the object pose estimation method requires a RGB-D image as input and generates the pool of candidate hypotheses. Our proposed verification method's goal is to select the best set of hypotheses from this pool of candidates. This is achieved by identifying the hypotheses for which there is the most evidence in the observation through analysis-by-synthesis, i.e., comparing rendered depth images of the candidate hypotheses with the actual observation. Thereby, we effectively consider the interrelation of the hypotheses of a candidate solution, for example through occlusion. The method is furthermore able to detect false positives and exclude them from candidate solutions. The search for candidate solutions is guided by MCTS. To prevent missing objects from the final solution in cases where search is stopped prematurely, we augment the final solution with the highest confidence hypothesis. In the following, we discuss the proposed pipeline.

3.1   Generation of the Hypotheses Pool

The first part of the pipeline generates a pool of object pose hypotheses. The pool consists of a set of candidate transformations per object and corresponding confidence scores. Many state-of-the-art methods can act as a hypotheses generator to output a set of hypothesis-confidence tuples. We choose the approach by Wang et al. [22], called DenseFusion, that achieves state-of-the-art performance on YCB-VIDEO. But other methods such as PPF-based [21] or template-matching-based methods [14] could be used instead.

DenseFusion requires an RGB-D image and an object instance mask as input. Per-pixel instance masks are efficiently computed using CNN-based methods such as Mask R-CNN [11]. Per object instance, DenseFusion samples a set of pixels from the observation. The method then predicts one object pose hypothesis and a corresponding confidence value per sample. We select the top-n pose hypotheses per object instance, ordered by confidence score, and add them to the hypotheses pool.

3.2   Generation of Candidate Solutions and Verification

Given the hypotheses pool and the observation, the goal of the second stage of the pipeline is to select a subset of hypotheses from the pool that best explains the observation. Previous work [4,16] shows that MCTS can be successfully applied to the domain of object pose verification. The reward function guides the tree search. We define the reward in terms of the discrepancy between the observed depth image and a synthesized depth image of the candidate solution following the analysis-by-synthesis approach.

DAG Representation. Previous work [4,16] define the set of candidate solutions as follows: Starting from the empty set, new candidate solutions are found by iteratively adding a hypothesis of unconsidered object instances until all object instances have been considered exactly once. This is equivalent to a tree with the empty set as root and nodes at level d consisting of d hypotheses for d different object instances.

With this approach, the search tree will contain the same solution several times – so called transpositions. In the worst case this splits the "attention" of

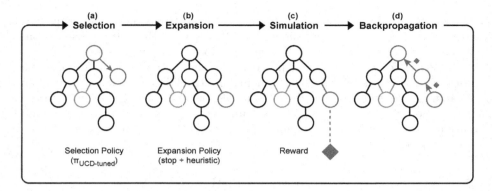

Fig. 1. Overview: Hypotheses Verification using MCTS on DAGs. The subgraphs considered in each step are highlighted in green, transpositions are indicated in gray. See Sect. 3.2(a)–(d) for a detailed description. Adapted from [6].

the search algorithm between the identical subtrees resulting from the transpositions. In general game playing, this problem is tackled by either altering the computation of the node statistics to consider transpositions [7] or changing the search tree to a directed acyclic graph (DAG) [19]. We follow the latter approach; the resulting search graph is illustrated in Fig. 1. The edges represent the added hypothesis, the nodes are the resulting solutions. Transpositions are explicitly encoded by the corresponding solution node featuring several in-going edges – the same solution can be achieved by adding a, potentially different, hypothesis to different ancestor solutions. For algorithmic reasons of the selection policy, we move the reward statistics from the nodes to the edges. Saffidine et al. [19] show that this improves performance.

(a) Selection policy. The DAG representation allows us to jointly consider different paths that lead to the same solution, i.e, transpositions. However, when backpropagating rewards only along the descent path, not all relevant statistics will be updated correctly. For example, if an ancestor of a transposition is not contained in a decent path, the ancestor's statistics will not account for this reward. The assumption of the Upper Confidence bounds for Trees (UCT) algorithm, that all information is locally available, no longer holds in this case as illustrated by Saffidine et al. [19]. To overcome this problem, they propose an adaptation of the UCT algorithm that considers information from a larger neighborhood of descendant edges, called Upper Confidence bound for rooted Directed acyclic graphs (UCD).

Let $R(e)$ be the rewards stored at edge e, $n_0 = n(e)$ their count, $p_0 = p(e)$ the total count of rewards stored at e and its siblings and $\hat{\mu}_0 = \hat{\mu}$ the mean reward at edge e. For larger subscripts, these statistics are computed with information from deeper DAG layers w.r.t. e. Setting one of the subscripts to ∞ is equal to setting it to the maximal DAG depth – the number of object instances in our case. Furthermore, let $c(e)$ be the set of out-going edges from e, i.e., its children. See [19] for a detailed definition. For the sake of clarity, we drop the reference to a specific edge (e) from the notation in cases where all statistics refer to the same edge. Using this notation, the UCD policy at edge e is defined as follows:

$$\pi_{UCD}(c, d_1, d_2, d_3) = \hat{\mu}_{d_1} + c \cdot \sqrt{\frac{\ln(p_{d_2})}{n_{d_3}}} \tag{1}$$

The UCB1-tuned policy presented by Auer et al. [3] considers the empirical variance of rewards. Previous work shows that considering the empirical mean is beneficial in the feature selection domain [10]. We adapt the exploration term in UCB1-tuned to the UCD formulation to arrive at the selection policy:

$$\pi_{UCD\text{-}tuned}(c, d_1, d_2, d_3) = \hat{\mu}_{d_1} + c \cdot \sqrt{\frac{\ln(p_{d_2})}{n_{d_3}} \cdot \min\left(\frac{1}{4}, \hat{\sigma}_{d_1}^2 + \sqrt{\frac{2\ln(p_{d_2})}{n_{d_3}}}\right)} \tag{2}$$

To extend the UCD parameterization to the UCB1-tuned policy, we define the empirical variance $\hat{\sigma}_{d_1}^2$ as the pooled variance at edge e up to depth d_1:

$$\hat{\sigma}_{d_1}^2(e) = \frac{\hat{\sigma}^2(e) \cdot (n(e) - 1) + \sum_{f \in c(e)} (n(f) - 1) \cdot \hat{\sigma}_{d_1-1}^2(f)}{(n(e) - 1) + \sum_{f \in c(e)} (n(f) - 1)} \tag{3}$$

$$\hat{\sigma}_0^2(e) = \hat{\sigma}^2(e) = \frac{1}{n-1} \sum_{r \in R} (r - \hat{\mu})^2 \tag{4}$$

(b) Expansion policy. Our expansion policy is two staged: The first stage implements the "stopping feature" as presented in [10] for the domain of feature selection. The second stage uses the confidence-based heuristic proposed in [4].

The stopping feature – or stopping hypothesis in our case – indicates that no further hypotheses will be added to this solution, i.e., it is a final solution. It is selected in an expansion step at level d with probability:

$$p(stop|d) = 1 - q^d \tag{5}$$

where $q < 1$ is a parameter set to 0.75 in our experiments.

If the stopping hypothesis is not selected during expansion, we apply a confidence-based heuristic as presented in [4]. A currently unused object is selected with a probability proportional to the maximal confidence of its hypotheses. After selecting the object, the same approach is applied for the hypotheses, selecting with probability proportional to their confidence value.

(c) Simulation. The simulation phase of MCTS samples a reward for the expanded solution. As this occurs in each iteration, the reward function should be fast to compute. This rules out the use of complex reward computations as proposed in [1,2]. We want to reward a depth discrepancy below a given threshold ρ. If the discrepancy is above that threshold, we want to penalize the search. This is achieved using the following reward function:

$$r(s, O) = \sum_{\delta \in \Delta(s,O)} \begin{cases} 1 - \frac{\delta}{\rho}, & \text{if } \delta \leq \rho \\ 0, & \text{otherwise} \end{cases} \qquad (6)$$

where δ is one pixel in the depth discrepancy map Δ and ρ is the discrepancy threshold beyond which the reward is 0. The depth discrepancy map $\Delta(s, O)$ for the solution s is computed as follows:

$$\Delta(s, O) = |R_d(s) - O_d| \qquad (7)$$

where $R_d(s)$ is the rendered depth of solution s and O_d is the observed depth.

(d) Backpropagation. Using a search tree representation, the reward is back-propagated to all ancestors. As previous work shows, updating all ancestors in a DAG representation violates the assumption that the mean reward of a node and that of its best child converge towards the same value (see [19] for a counter example). To replicate the behavior of MCTS on a search tree, we instead only update the ancestors along the descent path as suggested in [19].

4   Results

As the proposed method's goal is to select the best set of hypotheses from the pool of given hypotheses, we compare it to the best and worst possible results, i.e., always selecting the best or worst scoring hypotheses, respectively. We compare our MCTS-based approach to the baseline of estimating a confidence value together with the object pose hypothesis and selecting the highest scoring ones.

We choose use YCB-VIDEO for evaluation as it contains scenes of 3 to 9 objects and the baseline method, DenseFusion, achieves state-of-the-art performance on this dataset. The methods are compared on the VSD metric using the default parameters defined in [13]. The average precision, defined in PASCAL VOC [9] as Area Under Curve (AUC), is computed for thresholds $\theta \in [0, 0.3]$. Recall is computed for the threshold $\theta = 0.3$.

For all experiments, we use the following parameters for the presented method referred to as (*Ours*): 5 hypotheses per object instance, a limit of 300 MCTS iterations, threshold of the reward function $\rho = 2cm$, parameter for the stopping probability $q = 0.75$, and $\pi_{UCD\text{-}tuned}$ parameters $(\infty, 0, 0)$. We use the pre-computed instance masks from [24] and the pre-trained weights from [22].

4.1   YCB-VIDEO

To visualize the variation of the hypotheses pool as well as to compare the baseline to the proposed method, we present the results with the baseline's score as 0 in a candlestick chart in Fig. 2. As shown, for most classes the scores

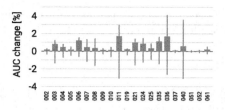 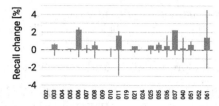

Fig. 2. Results on YCB-VIDEO for the VSD metric. The baseline method is shown as reference value (0.0). The upper and lower shadows give the best and worst case values, respectively. The body indicates the results for our approach.

Table 1. Results on YCB-VIDEO for VSD with $\tau = 0.02$, $\delta = 0.015$ and $\theta = 0.3$.

Class	AUC				Recall			
	Worst	Baseline	Ours	Best	Worst	Baseline	Ours	Best
002_master_chef_can	86.7	87.1	87.2	87.4	100.0	100.0	100.0	100.0
003_cracker_box	82.8	84.2	85.0	85.5	96.8	97.5	98.0	98.2
004_sugar_box	84.0	84.7	85.1	85.5	99.8	99.9	99.9	99.9
005_tomato_soup_can	78.3	78.8	79.0	79.3	95.9	96.1	96.2	96.2
006_mustard_bottle	82.4	83.0	84.2	84.6	95.8	96.6	98.9	99.2
007_tuna_fish_can	60.3	61.5	62.0	62.9	96.3	96.6	96.7	97.1
008_pudding_box	54.7	56.4	56.7	57.8	88.8	89.7	90.2	90.7
009_gelatin_box	74.9	75.2	75.3	75.5	100.0	100.0	100.0	100.0
010_potted_meat_can	69.3	70.0	70.1	70.5	81.3	82.1	82.1	82.3
011_banana	50.6	53.7	55.4	56.7	90.2	93.1	94.7	95.3
019_pitcher_base	91.7	92.2	92.4	92.5	99.8	99.8	99.8	99.8
021_bleach_cleanser	75.4	76.9	77.9	78.4	99.2	99.5	99.9	99.9
024_bowl	40.8	42.1	42.9	43.6	55.9	55.9	55.9	55.9
025_mug	57.6	58.5	58.8	59.4	95.9	96.4	96.9	96.9
035_power_drill	70.9	72.3	73.4	74.0	94.9	95.2	95.7	96.0
036_wood_block	56.7	59.3	61.0	63.4	93.4	94.2	94.6	95.9
037_scissors	2.2	2.4	2.5	2.6	8.3	8.8	11.1	11.1
040_large_marker	40.7	43.8	44.3	47.4	88.7	90.1	90.3	91.5
051_large_clamp	1.8	1.9	1.9	2.0	8.9	9.4	10.0	10.4
052_extra_large_clamp	2.8	3.0	3.0	3.1	10.4	10.4	10.4	10.4
061_foam_brick	3.0	3.3	3.5	3.9	18.1	20.1	21.5	24.7
ALL	62.1	63.1	63.6	64.2	83.1	83.6	83.9	84.2

of the hypotheses in the pool varies less than ±1% and there is, therefore, little variations in the pose hypotheses themselves. Still, the proposed method is able to correctly identify better hypotheses than the baseline method on almost all classes for AUC and is close to the optimum for several classes regarding recall. Overall, this results in a 0.5% increase on the AUC measure and 0.3% on the recall measure, limited by the number of classes where the optimal improvement over the baseline is below 0.5%. The full results are given in Table 1.

4.2   Convergence and Run Time

UCD-tuned starts to converge after 100–200 MCTS iterations as shown in Fig. 3 on a subset of YCB-VIDEO. Note that, still, more iterations can be beneficial as this allows the search to evaluate deeper layers of the DAG. We therefore opted for a maximum number of 300 MCTS iterations.

In our experiments, we observed an average run time of 3.5s per frame with a duration of one MCTS iteration of about 10ms on average. As the main computational bottleneck of the implementation is the rendering of the depth image, we decided to pre-compute a depth image per hypothesis and then compose the solutions' depth images using a simple z-test. This approach is almost a magnitude faster as compared to rendering and, especially, reading back full images from the GPU each iteration.

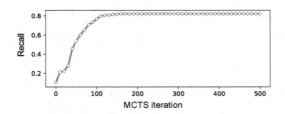

Fig. 3. Convergence behavior of UCD-tuned on a subset of YCB-VIDEO. Recall scores on the VSD metric are computed each 10 MCTS iterations.

4.3   Discussion

Compared to the baseline method [22], the run time increases significantly from 60 ms to 3.5 s per frame. The results, however, support our observation that the top-n hypotheses contain pose estimates that help to improve performance and the proposed method allows for more robust estimates. Still, our method achieves significantly faster run times as compared to the MCTS-based approaches by Bauer et al. [4] with 18 s and by Mitash et al. [16] with 30 s. These methods' run time is heavily increased by physics simulation which we do not apply in order to gain speed. Nevertheless, our method is able to perform better than the baseline method, DenseFusion [22], in terms of VSD on YCB-VIDEO. For future work it would be interesting to find the optimum in terms of run time versus performance by investigating efficient combinations of physics simulation and rendering as found in modern game engines. This will allow an improved method to provide physically plausible scene explanations in addition to the plausibility in terms of depth discrepancy.

5 Conclusion

We present a novel MTCS-based hypotheses verification method for 6D object pose estimation that provides a significant reduction in run time as compared to other MCTS-based approaches. By efficiently representing the search space by a DAG, we are able to identify the best subset among the top-n hypotheses more reliably than the confidence estimation of the baseline method. We report improvement over our baseline object pose estimation on most classes of YCB-VIDEO as well as an increased performance overall using the VSD metric.

References

1. Aldoma, A., Tombari, F., Di Stefano, L., Vincze, M.: A global hypotheses verification method for 3D object recognition. In: Fitzgibbon, A., Lazebnik, S., Perona, P., Sato, Y., Schmid, C. (eds.) ECCV 2012. LNCS, vol. 7574, pp. 511–524. Springer, Heidelberg (2012). https://doi.org/10.1007/978-3-642-33712-3_37
2. Aldoma, A., Tombari, F., Di Stefano, L., Vincze, M.: A global hypothesis verification framework for 3D object recognition in clutter. IEEE TPAMI 38(7), 1383–1396 (2016)
3. Auer, P., Cesa-Bianchi, N., Fischer, P.: Finite-time analysis of the multiarmed bandit problem. Mach. Learn. 47(2–3), 235–256 (2002)
4. Bauer, D., Patten, T., Vincze, M.: 6D object pose verification via confidence-based Monte Carlo tree search and constrained physics simulation. In: OAGM & ARW Joint Workshop, pp. 153–158 (2019)
5. Brachmann, E., Michel, F., Krull, A., Yang, M.Y., Gumhold, S., et al.: Uncertainty-driven 6D pose estimation of objects and scenes from a single RGB image. In: IEEE CVPR, pp. 3364–3372 (2016)
6. Browne, C.B., et al.: A survey of Monte Carlo tree search methods. T-CIAIG 4(1), 1–43 (2012)
7. Childs, B.E., Brodeur, J.H., Kocsis, L., et al.: Transpositions and move groups in Monte Carlo tree search. In: IEEE CIG, pp. 389–395 (2008)
8. Drost, B., Ulrich, M., Navab, N., Ilic, S.: Model globally, match locally: efficient and robust 3D object recognition. In: IEEE CVPR, pp. 998–1005 (2010)
9. Everingham, M., Van Gool, L., Williams, C.K.I., Winn, J., Zisserman, A.: The pascal visual object classes (VOC) challenge. IJCV 88(2), 303–338 (2010)
10. Gaudel, R., Sebag, M.: Feature selection as a one-player game. In: ICML, pp. 359–366 (2010)
11. He, K., Gkioxari, G., Dollár, P., Girshick, R.: Mask R-CNN. In: IEEE ICCV, pp. 2980–2988 (2017)
12. Hinterstoisser, S., et al.: Model based training, detection and pose estimation of texture-less 3D objects in heavily cluttered scenes. In: Lee, K.M., Matsushita, Y., Rehg, J.M., Hu, Z. (eds.) ACCV 2012. LNCS, vol. 7724, pp. 548–562. Springer, Heidelberg (2013). https://doi.org/10.1007/978-3-642-37331-2_42
13. Hodaň, T., et al.: BOP: benchmark for 6D object pose estimation. In: ECCV, pp. 19–34 (2018)
14. Hodaň, T., Zabulis, X., Lourakis, M., Obdržálek, Š., Matas, J.: Detection and fine 3D pose estimation of texture-less objects in RGB-D images. In: IEEE/RSJ IROS, pp. 4421–4428 (2015)

15. Kehl, W., Milletari, F., Tombari, F., Ilic, S., Navab, N.: Deep learning of local RGB-D patches for 3D object detection and 6D pose estimation. In: Leibe, B., Matas, J., Sebe, N., Welling, M. (eds.) ECCV 2016. LNCS, vol. 9907, pp. 205–220. Springer, Cham (2016). https://doi.org/10.1007/978-3-319-46487-9_13

16. Mitash, C., Boularias, A., Bekris, K.E.: Improving 6D pose estimation of objects in clutter via physics-aware Monte Carlo tree search. In: IEEE ICRA, pp. 1–8 (2018)

17. Narayanan, V., Likhachev, M.: Perch: perception via search for multi-object recognition and localization. In: IEEE ICRA, pp. 5052–5059 (2016)

18. Pélissier, A., Nakamura, A., Tabata, K.: Feature selection as Monte-Carlo search in growing single rooted directed acyclic graph by best leaf identification. In: SDM, pp. 450–458 (2019)

19. Saffidine, A., Cazenave, T., Méhat, J.: UCD: upper confidence bound for rooted directed acyclic graphs. Knowl.-Based Syst. **34**, 26–33 (2012)

20. Tejani, A., Tang, D., Kouskouridas, R., Kim, T.-K.: Latent-class hough forests for 3D object detection and pose estimation. In: Fleet, D., Pajdla, T., Schiele, B., Tuytelaars, T. (eds.) ECCV 2014. LNCS, vol. 8694, pp. 462–477. Springer, Cham (2014). https://doi.org/10.1007/978-3-319-10599-4_30

21. Vidal, J., Lin, C.Y., Martí, R.: 6d pose estimation using an improved method based on point pair features. In: ICCAR, pp. 405–409 (2018)

22. Wang, C., et al.: DenseFusion: 6D object pose estimation by iterative dense fusion. In: IEEE CVPR (2019)

23. Wohlhart, P., Lepetit, V.: Learning descriptors for object recognition and 3D pose estimation. In: IEEE CVPR, pp. 3109–3118 (2015)

24. Xiang, Y., Schmidt, T., Narayanan, V., Fox, D.: Posecnn: a convolutional neural network for 6D object pose estimation in cluttered scenes. In: CoRR (2017)

Leveraging Symmetries to Improve Object Detection and Pose Estimation from Range Data

Sergey V. Alexandrov$^{(\boxtimes)}$, Timothy Patten, and Markus Vincze

Vision4Robotics Group, ACIN, TU Wien, Vienna, Austria
{alexandrov,patten,vincze}@acin.tuwien.ac.at

Abstract. Many man-made objects around us exhibit rotational symmetries. This fact can be exploited to improve object detection and 6D pose estimation performance. To this end we propose a set of extensions to the state-of-the-art PPF pipeline. We describe how a fundamental region is selected on symmetrical objects and used to construct a compact model hash table and a Hough voting space without redundancies. We also introduce a symmetry-aware distance metric for the pose clustering step. Our experiments on T-LESS and ToyotaLight datasets demonstrate that these extensions lead to a consistent improvement in the pose estimation recall score compared to the baseline pipeline, while simultaneously reducing computation time by up to 4 times.

1 Introduction

Object detection and pose estimation is an important computer vision problem with applications in service robotics, manufacturing, augmented reality, and other domains [11]. While detection is a challenging task in its own right [17], when it comes to embodied agents, the estimation of 6D pose is necessary to enable interaction with the objects.

Fig. 1. Many household (left) and industrial (right) objects are rotationally symmetric. They may have single or multiple axes of symmetry (blue arrows) of different order (denoted by n). Our extensions to the PPF pipeline support automatic detection of symmetries to enable faster and more accurate pose estimation. (Color figure online)

© Springer Nature Switzerland AG 2019
D. Tzovaras et al. (Eds.): ICVS 2019, LNCS 11754, pp. 397–407, 2019.
https://doi.org/10.1007/978-3-030-34995-0_36

Since the introduction of consumer RGB-D cameras nearly a decade ago, object detection and pose estimation from range data has received a lot of attention. Methods based on both learned and hand-crafted global and local features, classical learning techniques, as well as convolutional neural networks have been proposed. In a recent comprehensive benchmark [9] a method based on Point Pair Feature (PPF) pipeline emerged as a current leader [15].

The original idea of using PPF for pose estimation was introduced by Drost *et al.* [7]. It consists of coupling a global point-based feature with an efficient voting scheme in local coordinate space. Numerous follow-up works mitigated problems related to feature discretization and inefficient point sampling [8], introduced weighted voting and interpolated recovery of pose parameters [1], or added boundary points as an alternative to surface points [3]. Several multimodal extensions were presented, including exploitation of edge information in RGB images [5] and addition of point color to the feature [2]. Most recently Vidal *et al.* [15] reported a major boost in performance by adding carefully designed preprocessing, hierarchical pose clustering, and verification steps to the pipeline.

The basic PPF pipeline and the extensions listed above support detection and pose estimation of free-form objects. However, most man-made objects that surround us exhibit either reflection or rotational symmetry (see Fig. 1). This fact has been exploited in various computer vision areas; algorithms in semantic shape matching [13], geometry completion [14], and segmentation [16] have benefited from information about object symmetries.

In the context of pose estimation, however, object symmetries are troublesome, especially for learning-based methods. Multiple views of symmetric objects may look the same, leading to ambiguities in loss functions and disturbances to the learning process [12]. In the area of PPF-based methods, Drost and Ilic [6] proposed a specialized pipeline for geometric primitives such as spheres and cylinders. Their method that uses an alternative low-dimensional Hough space cannot be applied to free-form objects.

In this paper we introduce a set of extensions to the PPF pipeline that exploit rotational symmetry of man-made objects. Our approach is not limited to primitive shapes and supports arbitrary free-form models with different classes of rotational symmetry. It achieves consistent improvement in pose estimation compared to the baseline pipeline, while simultaneously reducing computation time by up to 4 times. Additionally, we describe a method for automatic extraction of symmetry information from object models. This method is flexible and does not require the model coordinate frame to be aligned with symmetry axes or the center of symmetry.

The rest of the paper is structured as follows. Section 2 provides a brief overview of the original PPF pipeline. Section 3 describes the proposed extensions to the pipeline and Sect. 4 presents our symmetry extraction method. Section 5 reports evaluation results and Sect. 6 concludes the paper.

2 PPF Overview

This section gives a brief overview of the PPF-based object detection and pose estimation pipeline. We first introduce the main ideas behind the approach and then explain how they are put together in a pipeline. An interested reader is referred to an exposition by Drost [4] for a detailed treatment of the topic.

2.1 Building Blocks

Point Pair Feature. A pair of points $\mathbf{m}_1$ and $\mathbf{m}_2$ with associated normals $\mathbf{n}_1$ and $\mathbf{n}_2$ can be described by an asymmetric 4-dimensional feature:

$$\mathbf{F} = \left(\|\mathbf{d}\|_2 , \angle\left(\mathbf{n}_1, \mathbf{d}\right) , \angle\left(\mathbf{n}_2, \mathbf{d}\right) , \angle\left(\mathbf{n}_1, \mathbf{n}_2\right) \right), \tag{1}$$

where $\mathbf{d} = \mathbf{m}_1 - \mathbf{m}_2$ and $\angle : \mathbb{R}^3 \times \mathbb{R}^3 \to [0, \pi]$ is the angle between two vectors. This feature is invariant under rigid transformations and is easy to compute.

Model Description and Search. An object model M can be described by a set of all possible pairs of points together with their corresponding features. This set can be organized into a hash table H, where the features are quantized into integers and are used as keys, and point pairs are stored as values. Given a point pair, one can compute its PPF, quantize, and use H to retrieve a set of all similar model point pairs. This operation can be performed in constant time.

Local Coordinate Space. A submanifold of the $SE(3)$ space can be defined by fixing a reference point in the scene and allowing only transformations that align this point and its normal with *some* model point. Every 3D transformation from this manifold can be parametrized by a model point and a rotation angle, *i.e.* the manifold has two dimensions $M \times [0; 2\pi)$. Due to its low dimensionality, a Hough voting procedure can be carried out efficiently in this space.

2.2 Pipeline

A subset of reference points $\{\mathbf{s}_i\} \subset S$ is sampled from the scene and is processed one-by-one in the following fashion. A reference point $\mathbf{s}_i$ is paired with every other point $\mathbf{s}_j \in \mathcal{N}(\mathbf{s}_i)$ from its spatial neighborhood. For each such pair the hash table H is used to find a set of similar model pairs. Every match defines a transform $\mathbf{T}$ that aligns the scene pair and the model pair. This transform belongs to the submanifold, thus can be expressed with two parameters and used to cast a vote in the Hough space. Once the neighborhood $\mathcal{N}(\mathbf{s}_i)$ is processed, the peak in the Hough voting space is extracted. It represents the most plausible transform that aligns the model with the reference point $\mathbf{s}_i$ and its neighborhood.

Processing the set of reference points yields a set of transforms $\{\mathbf{T}_i\}$, each of which represents a hypothesis about the possible location of the object in the scene. As a last step in the pipeline, pose clustering is performed. The purpose of this step is to group similar poses and improve the estimate through averaging.

3 Our Approach

In this section we describe our proposed modifications to the PPF pipeline. We precede them with an introduction of the symmetry classes that can be exploited.

3.1 Symmetry Classes and Notation

A symmetry is described by a group of transformations G, which we call a *symmetry group*. An object $M \subset \mathbb{R}^3$ is said to be symmetric if it is invariant under the group actions. In other words, any transformation $\mathbf{G} \in G$ can be applied to M unnoticed. Depending on the composition of the group, several symmetry classes can be distinguished [13]. In this work we only consider *rotational symmetries*, *i.e.* those with $G \subset SO(3)$. Such symmetries can be subdivided into *continuous* and *discrete*; we support both.

Continuous Rotational Symmetries. Their symmetry groups are infinite and consist either of rotations around an axis (*cylindrical symmetry*) or rotations around the object center (*spherical symmetry*). The former is a wide-spread symmetry class; 2nd and 4th objects in Fig. 1 are examples. The latter class is less relevant for typical pose estimation applications since the only objects possessing such symmetry are spheres. This class has been considered by Drost and Ilic [6], so we exclude it from the present work.

Discrete Rotational Symmetries. Their symmetry groups are finite and consist of repeated rotations by $360/n$ degrees around a single (*cyclic symmetry*) or multiple (*polyhedral symmetry*) axes, where n is the *order* of the symmetry. Objects 3 and 6 in Fig. 1 are examples of 2nd order cyclic symmetry; object 5 has 4th order. The first object has three symmetry axes and is an example of polyhedral symmetry.

3.2 Modifications of Model Description and Voting Space

Our proposed modifications are motivated by the observation that for a symmetrical model the set of its surface points M is a redundant representation. Indeed, given a model point $\mathbf{m} \in M$, the set of its images $\{\mathbf{G} \circ \mathbf{m} : \mathbf{G} \in G\}$ under the symmetry group action also belongs to the model. This set is called an *orbit* of the action. A subset $F \subset M$ that contains exactly one point from each of the orbits is called a *fundamental region* of the model. Given a fundamental region and a symmetry group one may reconstruct the entire model. In this sense, F is a more concise representation of the model.

Recall that the local coordinate space introduced in Sect. 2.1 has two dimensions, the first one being the dimension of model points. We note that the local coordinates corresponding to the model points from the same orbit encode poses that are equivalent under the symmetry group action. It is therefore sufficient to keep a single representative per orbit. In other words, the local coordinate space

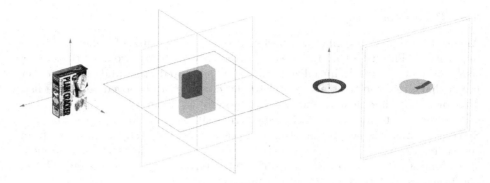

Fig. 2. Selection of the fundamental region for a model with discrete (left) and continuous (right) rotational symmetry. The selected fundamental region F is show in red and the rest of the model $M \setminus F$ is shown in gray. The planes involved in the computation are also presented. (Color figure online)

can be reduced to $F \times [0; 2\pi)$, leading to a more compact Hough space and faster voting. We further note that all hash table entries where the first point of the pair belongs to the set $M \setminus F$ are redundant and can be removed. In practice, we avoid creating them in the first place, *i.e.* we build the hash table only from the pairs in $F \times M$.

Fundamental Region Selection. An important practical consideration is how to select the fundamental region F. There are many ways to choose it, although typically connected subsets are preferred. We propose the following algorithm.

As the first step, we construct a set of planes that pass through symmetry axes. Specifically, for axis $\mathbf{a}_i$ of n_i^{th} order we build a plane that passes through it as well as the next axis $\mathbf{a}_{i+1}$. Then we create $n_i - 1$ images of this plane by applying $360/n_i$ degree rotations around $\mathbf{a}_i$. Figure 2 (left) illustrates the symmetry axes and the constructed set of planes.

As the second step, we choose an arbitrary model point that does not lie on any of the planes. We flip plane normals such that the point is on the positive side of each of the planes. Finally, we select model points that are on the positive side of each plane; this is the fundamental region (show in red in Fig. 2).

The described algorithm is unsuitable for cylindrical symmetries since it would require creation of an infinite set of planes. We propose an alternative approach, where only two planes parallel to the symmetry axis are created. The distance between the planes is set to a certain percentage of the model diameter (in our implementation same as the downsampling resolution). All model points lying in between the planes and having positive x coordinate are selected as the fundamental region. Figure 2 (right) illustrates this. We note that this is not strictly a fundamental region since it may contain multiple points from the same orbit, especially in the areas close to the axis. However, this has no practical implications since the selected set F is still significantly smaller than M.

3.3   Pose Clustering

As in the original pipeline, we perform greedy clustering of generated pose hypotheses. The list of hypotheses is sorted in decreasing order of the voting score. A new cluster is initialized with the top-scored pose and all the other poses that are close to it are added to the cluster and removed from the list. This procedure is repeated until the hypotheses list is empty.

We propose to use a pose distance metric that accounts for object symmetry. Since the translational distance between poses is not affected by rotations from the symmetry group, we propose to use a two-valued metric with separate measures for rotation and translation differences. Given a symmetry group G and two poses P_1 and P_2, the metric is defined as:

$$m_G\left(P_1, P_2\right) = \left(\measuredangle_G\left(\mathbf{R}\right), |\mathbf{t}|\right), \tag{2}$$

where $(\mathbf{R}, \mathbf{t})$ is the difference between the poses $P_1^{-1} \circ P_2$. The translation difference measure is simply the Euclidean norm $|\cdot|$, whereas the rotation difference measure is defined as:

$$\measuredangle_G\left(\mathbf{R}\right) = \begin{cases} \measuredangle\left(\mathbf{R}\right) & \text{no symmetry;} \\ \min\left\{\measuredangle\left(\mathbf{G} \circ \mathbf{R}\right) : \mathbf{G} \in G\right\} & \text{discrete symmetry;} \\ \arccos\left(\mathbf{R}_{3,3}\right) & \text{cylindrical symmetry,} \end{cases} \tag{3}$$

where $\measuredangle\left(\mathbf{R}\right)$ is the angle of rotation that can be computed using the matrix trace. Note that in the case of cylindrical symmetry we only compute the angular distance between rotation axes. Under the assumption that the axis of rotation is aligned with model z-axis, this is simply the bottom-right element of $\mathbf{R}$.

The application of transforms from the symmetry group to either of the input poses does not affect the metric, *i.e.* it is invariant under the group actions. Furthermore, this metric is independent of rigid motions applied to both poses.

4   Automatic Symmetry Extraction

In order to apply the extensions described above, one needs to know the information about symmetry, *i.e.* its type, number of axes and their directions, and order. Most of the pose estimation datasets do not readily provide such information; it needs to be extracted. Manual annotation may be a tedious process, especially if the model coordinate frame is not guaranteed to be aligned with the symmetry axes and/or center. We therefore propose a simple automatic symmetry extraction pipeline that is driven by PPF-based pose estimation.

First we use the object model as both source and target for pose estimation. We configure the PPF pipeline to output multiple pose hypotheses. For each hypothesis the model is transformed and the Hausdorff distance to the original model is computed. The transforms with score below a certain (dataset and application dependent) threshold are retained.

At this point we have a finite set $T \in SE(3)$ of transforms that align the object with itself. However it is not a rotational symmetry group yet. We therefore proceed to convert the rotational part of each transform into an axis-angle representation and compute a point on this axis. Next we perform greedy clustering in this space, keeping record of the rotation angles. The result of this step is a set of rotation axes, each with a discrete set of rotation angles.

The next step is to determine the center of symmetry of the object. In case there is a single axis of rotation, any point on it can serve as the center. We project model points on the axis and take the centroid. In case there are multiple axes of rotation, we find the point where they intersect. Note that due to model imperfections and imprecisions of registration, the extracted axes will not intersect exactly; we thus find a point that is mutually closest to all of them.

The final step is to determine the order of each of the axes. We check the set of rotation angles associated with an axis. If it is larger than a certain threshold (set to 4 in our implementation), we check if this is in fact a continuous symmetry by rotating the model with 12 rotations and computing the Hausdorff distance. If most of the angles are below the threshold, the axis is continuous. Otherwise we find the greatest common divisor of the angles d and assume $n = 360/d$.

5   Experimental Evaluation

In this section we present experiments that evaluate the effect that the proposed modifications have on the feature matching performance, entire detection and estimation pipeline, and the processing time.

We used two publicly available datasets that contain large number of symmetric objects, T-LESS [10] and ToyotaLight [9]. The former includes 30 objects relevant for manufacturing applications, whereas the latter includes 21 common household items. Both datasets provide ground truth pose annotations, however lack information about object symmetries. We used the method described in Sect. 4 to automatically extract symmetry annotations. We created two subsets for each dataset, one with objects with discrete rotational symmetry (17 in T-LESS and 5 in ToyotaLight), and one with cylindrical symmetry (6 in T-LESS and 8 in ToyotaLight).

5.1   Feature Matching Performance

First we demonstrate that our proposed modifications have a positive impact on feature matching performance. To this end we devise the following experiment. Given a scene and an annotated pose of an object, we identify the scene points that belong to the object instance. Next, for each of these points we perform matching and voting steps, arriving at a single pose hypothesis, which we compare to the ground truth annotation. We count the fraction of points that yield a correct pose, which we use as a proxy for feature matching performance.

We compute the described score for each symmetric object in the T-LESS dataset. While the average fraction of matched points varies greatly depending on

Table 1. Recall scores (%) on the discrete symmetry subset of T-LESS.

Method	1	2	5	6	7	8	9	10	16	19	20	23	24	27	28	29	30	Average
Extension	**30**	**39**	**75**	**62**	**78**	64	**87**	65	72	**53**	39	**81**	**57**	56	68	**81**	**85**	**64.23**
Baseline	24	38	70	53	77	64	84	**66**	72	50	39	79	56	56	**70**	77	84	62.29

Table 2. Recall scores (%) on the cylindrical symmetry subset of T-LESS.

Method	3	4	13	14	15	17	Average
Extension	**59**	47	39	**47**	**57**	83	**55.33**
Baseline	48	**49**	39	43	55	83	52.83

the object type, we found that for most objects enabling symmetry optimizations improved the score by around 5%.

5.2 Pose Estimation Performance

We follow the evaluation protocol of Hodan *et al.* [9], *i.e.* computing the visible surface discrepancy (VSD) pose-error function and use the same misalignment and occlusion tolerances and correctness threshold ($\tau = 20\,\text{mm}$, $\delta = 15\,\text{mm}$, and $\theta = 0.3$ respectively). We report recall scores per dataset and per object for our baseline implementation of the PPF pipeline with and without symmetry extensions. This implementation does not include other advanced extensions proposed in the literature, *e.g.* view-dependent hypothesis rescoring or surface boundary verification step [15].

The results over the discrete symmetry subset of T-LESS are presented in Table 1. For most of the objects we observe a slight improvement in the recall score, however there are several objects whose performance boost is 5 or more percent. A similar pattern can be seen in Table 2 for the cylindrical symmetry objects. Overall, on these subsets we observe an average improvement by 2% and 2.5% respectively.

ToyotaLight is a less challenging dataset where scenes typically contain a single isolated object. Tables 3 and 4 show that the baseline pipeline demonstrates a good performance for majority of the objects. Nevertheless, our symmetry extensions improve the performance by 3% and 1.25% for objects with discrete and cylindrical symmetry respectively.

Across the two datasets our proposed extensions demonstrated a consistent performance gains for objects with both discrete and cylindrical symmetries.

Table 3. Recall scores (%) on the discrete symmetry subset of ToyotaLight.

Method	2	14	15	16	19	Average
Extension	**82**	96	100	**46**	**92**	**83.20**
Baseline	76	96	100	39	90	80.20

Table 4. Recall scores (%) on the cylindrical symmetry subset of ToyotaLight.

Method	7	8	10	11	12	13	17	18	Average
Extension	98	100	**82**	**86**	**90**	93	100	100	**93.63**
Baseline	98	100	78	83	87	93	100	100	92.38

5.3 Computation Time

In this experiment we demonstrate time savings resulting from exploitation of symmetries. We use models from the T-LESS dataset with cylindrical and 2nd order discrete symmetries. For each model we uniformly sample a subset of the scenes where it appears and run object detection and pose estimation, both with and without the proposed modifications. We measure the average time spent to generate a hypothesis for a reference scene point.

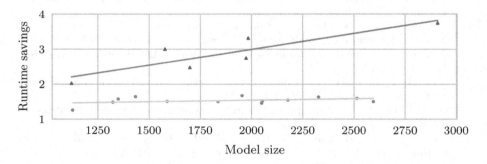

Fig. 3. Runtime comparison, relative improvement when using symmetry information as a function of model size. Objects with discrete rotational symmetry of 2nd order (•) and objects with cylindrical symmetry (▲).

Figure 3 presents the dependency between the size of the object (number of points) and the average time spent on hypotheses generation. We group objects into series according to the symmetry class. A linear model was fit to data points of each of the series. We observe that cylindrical symmetry offers the best improvement; it increases linearly with model size, reaching 4x speedup for the largest object. Discrete symmetry of 2nd order offers what seems to be a fixed improvement of approximately 50%. The datasets have an insufficient number of objects with symmetries of higher order to include in this analysis, however we expect proportionally larger fixed speedups.

6   Conclusion

We have demonstrated that rotational symmetries in objects can be exploited to improve instance detection and pose estimation performance while additionally significantly reducing the runtime of PPF-based methods. These modifications are complimentary to other extensions reported in the literature and thus can be combined to achieve further performance gains. We also presented a method for automatic symmetry extraction from arbitrary free-form objects. In future we would like to exploit mirror symmetries, which is perhaps the largest symmetry class. It would also be interesting to take advantage of partial symmetries, *i.e.* objects that consist of symmetric and non-symmetric parts.

Acknowledgments. The work presented in this paper has been partially supported by Aeolus Robotics, Inc.

References

1. Birdal, T., Ilic, S.: Point pair features based object detection and pose estimation revisited. In: Proceedings of 3DV (2015)
2. Choi, C., Christensen, H.I.: 3D pose estimation of daily objects using an RGB-D camera. In: Proceedings of IROS (2012)
3. Choi, C., Taguchi, Y., Tuzel, O., Liu, M.Y., Ramalingam, S.: Voting-based pose estimation for robotic assembly using a 3D sensor. In: Proceedings of ICRA (2012)
4. Drost, B.: Point cloud computing for rigid and deformable 3D object recognition. Ph.D. thesis (2016)
5. Drost, B., Ilic, S.: 3D Object detection and localization using multimodal point pair features. In: Proc. of 3DV (2012)
6. Drost, B., Ilic, S.: Local hough transform for 3D primitive detection. In: Proceedings of 3DV (2015)
7. Drost, B., Ulrich, M., Navab, N., Ilic, S.: Model globally, match locally: efficient and robust 3D object recognition. In: Proceedings of CVPR (2010)
8. Hinterstoisser, S., Lepetit, V., Rajkumar, N., Konolige, K.: Going further with point pair features. In: Leibe, B., Matas, J., Sebe, N., Welling, M. (eds.) ECCV 2016. LNCS, vol. 9907, pp. 834–848. Springer, Cham (2016). https://doi.org/10.1007/978-3-319-46487-9_51
9. Hodaň, T., et al.: BOP: benchmark for 6D object pose estimation. In: Ferrari, V., Hebert, M., Sminchisescu, C., Weiss, Y. (eds.) ECCV 2018. LNCS, vol. 11214, pp. 19–35. Springer, Cham (2018). https://doi.org/10.1007/978-3-030-01249-6_2
10. Hodaň, T., Haluza, P., Obdrzalek, Š., Matas, J., Lourakis, M., Zabulis, X.: T-LESS: an RGB-D dataset for 6D pose estimation of texture-less objects. In: Proceedings of WACV (2017)
11. Oñoro-Rubio, D., López-Sastre, R.J., Redondo-Cabrera, C., Gil-Jiménez, P.: The challenge of simultaneous object detection and pose estimation: a comparative study. Image Vis. Comput. **79**, 109–122 (2018)
12. Sundermeyer, M., Marton, Z.-C., Durner, M., Brucker, M., Triebel, R.: Implicit 3D orientation learning for 6D object detection from RGB images. In: Ferrari, V., Hebert, M., Sminchisescu, C., Weiss, Y. (eds.) ECCV 2018. LNCS, vol. 11210, pp. 712–729. Springer, Cham (2018). https://doi.org/10.1007/978-3-030-01231-1_43
13. Tevs, A., Huang, Q., Wand, M., Seidel, H.P., Guibas, L.: Relating shapes via geometric symmetries and regularities. ACM Trans. Graph. **33**, 119 (2014)
14. Thrun, S., Wegbreit, B.: Shape from symmetry. In: Proceedings of ICCV (2005)

15. Vidal, J., Lin, C.Y., Lladó, X., Martí, R.: A method for 6D pose estimation of free-form rigid objects using point pair features on range data. Sensors **18**, 2678 (2018)
16. Wang, Y., et al.: Symmetry hierarchy of man-made objects. Comput. Graph. Forum (2011)
17. Zou, Z., Shi, Z., Guo, Y., Ye, J.: Object detection in 20 years: a survey. arXiv preprint (2019)

Towards Meaningful Uncertainty Information for CNN Based 6D Pose Estimates

Jesse Richter-Klug$^{(\boxtimes)}$ ⓘ and Udo Frese ⓘ

University Bremen, 28359 Bremen, Germany
{jesse,ufrese}@uni-bremen.de

Abstract. Image based object recognition and pose estimation is nowadays a heavily focused research field important for robotic object manipulation. Despite the impressive recent success of CNNs to our knowledge none includes a self-estimation of its predicted pose's uncertainty.

In this paper we introduce a novel fusion-based CNN output architecture for 6d object pose estimation obtaining competitive performance on the YCB-Video dataset while also providing a meaningful uncertainty information per 6d pose estimate. It is motivated by the recent success in semantic segmentation, which means that CNNs can learn to know what they see in a pixel. Therefore our CNN produces a per-pixel output of a point in object coordinates with image space uncertainty, which is then fused by (generalized) PnP resulting in a 6d pose with 6×6 covariance matrix. We show that a CNN can compute image space uncertainty while the way from there to pose uncertainty is well solved analytically. In addition, the architecture allows to fuse additional sensor and context information (*e.g.* binocular or depth data) and makes the CNN independent of the camera parameters by which a training sample was taken. (Code available under https://github.com/JesseRK/6D-UI-CNN-Output.)

Keywords: 3D object pose estimation · Uncertainty · gPnP

1 Introduction

1.1 Motivation

Convolutional Neural Networks (CNNs) have immensely improved the capability of state of the art computer vision. This development started with qualitative questions, such as "What is in the image?" [10], over "Where is something in the image?" [17] to metrical questions, such as "Where is an object in space?" [22,23]. The latter is of particular interest for robotics, because a robot normally needs a 6d pose (position and orientation) to grasp an object.

The success of Deep Learning is accompanied by an uneasiness on the lack of transparency, which has many aspects. One aspect is, that it is hard to understand how CNNs come to the solution they output. Another aspect is, that one

ⓒ Springer Nature Switzerland AG 2019
D. Tzovaras et al. (Eds.): ICVS 2019, LNCS 11754, pp. 408–422, 2019.
https://doi.org/10.1007/978-3-030-34995-0_37

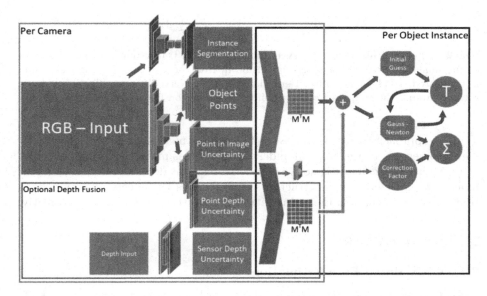

Fig. 1. Proposed architecture overview (cf. Sect. 3): CNNs take the image and option-ally depth data and provide per-pixel instance segmentation, object points and uncer-tainties, as well as a per-object overall uncertainty correction factor (Sect. 4). This data is aggregated into a 13×13 matrix $M^T M$ and fused by a gPnP algorithm (Sect. 5) to a pose estimate T with covariance Σ.

would desire a measure of uncertainty that indicates how precise and/or reliable the output is. Towards this goal, this paper provides such a measure as a 6×6 covariance matrix for 6d object poses estimated by a CNN.

Such uncertainty information offers many opportunities. It could be used to

- safeguard action, *e.g.* only grasp if the object pose is precise enough,
- trigger action *e.g.* look from a different perspective, if necessary,
- fuse several object observations with established probabilistic methods [21],
- fuse with prior information, *e.g.* that objects often rest on a horizontal plane.

1.2 Contribution

Towards this goal, we propose here an architecture (Fig. 1) where an encoder/decoder CNN with bypass connections predicts for every pixel belong-ing to an object "what is seen there" with uncertainty. Concretely it predicts the 3d object point belonging to that pixel in object coordinates as well as the 2D uncertainty of the predicted object point in the image space.

These 3D-2D pairs with uncertainty are then input to a Perspective-n-Point (PnP) solver that computes the pose estimate with uncertainty using probabilis-tic methods and the camera calibration. Actually, the solver handles generalized PnP (gPnP) [9] problems, so available depth information and binocular cameras can be included.

Fig. 2. Object orientation is not even approximately a translation invariant function of the image: In the middle, there is a tube in front of the camera. The right tube is translated, *i.e.* has the same orientation but looks very different. The left tube is rotated around the camera, *i.e.* has a different orientation but looks almost identical.

This approach follows the idea that analytical methods are better at multi-view geometry because it is well understood, whereas machine learning is better at recognition, because it is not well understood what defines an object in a real image. In particular, by construction gPnP handles the effects of camera calibration and occlusion on the estimate and its uncertainty structurally correctly.

2   Related Work

2.1   CNN Based Pose Estimation

From the perspective of this paper, the CNN based pose estimation methods can be categorized according to how the pose is coded in the CNNs output.

Some practical approaches, *e.g.* [25] in the Amazon Picking Challenge use CNNs to segment an object in RGB data and then perform ICP on the segmented depth data, however this ignores most of the information in the RGB data.

Early pose estimation CNNs, such as [18] for outdoor camera pose estimation directly predicted a vector and quaternion. So some object pose CNNs, such as PoseCNN [23], SSD6D [5], AAE [19] and Periyasamy *et al.* [13] evolved from object detection CNNs by adding a pose output stage. Usually they predict the 2D object center (relative to the pixel, often with voting), distance (directly or from bounding box size) and orientation (discretized or quaternion). This wrongly treats orientation as a translation invariant function of the image which is not true as we show in Fig. 2.

Newer works assume, that CNNs can predict image-space quantities best. Examples are YOLO6D [20], BB8 [15], Crivallaro *et al.* [2], Oberweger *et al.* [12] and DOPE [22]. These predict the image position of *e.g.* bounding box corners which are then fed into a PnP-solver (*e.g.* EPnP [11]). This also isolates the CNN from camera calibration handled by the PnP stage.

One can take this idea further and let every pixel predict, which object point it sees. Brachmann *et al.* [1] did this using random-forests, very recently DPOD [24] with CNNs and we also follow this line. The approach handles occlusions well [24, Table 2], arguably because as long as visible object points are correct, PnP tolerates missing points. Crivallaro *et al.* [2] predict points for object parts and Oberweger *et al.* [12] limit the receptive field for the same reason.

AAE [19] considers the problem of symmetries, which come in continuous form (*e.g.* a bottle), discrete form (*e.g.* a cube) and as view dependent ambiguity (*e.g.* a cup with occluded handle). They solve it with a denoising autoencoder that discovers the appropriate rotation representation for the object.

Among the mentioned approaches only [2] considers uncertainty. By propagating an assumed image noise through the Jacobian of the CNN, they predict 2D covariances for the control points. They see these only as a tool to improve the fusion result and neither validate them empirically nor output a pose covariance.

2.2 PnP and Variants

PnP, *i.e.* computing a pose from n 2D-3D point correspondences, is a classical geometric vision problem. Lepetit *et al.* [11] condensed the n points into a fixed size 12×12 matrix, by expressing them as a convex combination of four so-called control points. Their EPnP algorithm globally finds the best fitting control points with four cases depending on the number of small eigenvalues.

Kneip [9] extends EPnP to rays not intersecting in a single point, *e.g.* from multiple cameras (generalized or gPnP). The input is condensed into a 12×12 matrix and a 12 vector and solved using Gröbner bases considering 6 cases. This and other geometric vision algorithms are published in the opengv library [8].

Zheng [26] casts PnP as quaternion minimization, solved by a Gröbner basis.

These algorithms provide global solutions making them fairly complex. Unlike the iterative Gauss-Newton, they don't provide uncertainty information.

The PnP solution in [2] fuses with a pose prior of 9 Gaussians, effectively by Gauss-Newton on 9 initial guesses, while [12,15,20,22,24] use EPnP black-box.

2.3 Gaussian Distributions and Gauss-Newton on $SE(3)$

The space of poses $SE(3) \subset \mathbb{R}^{4 \times 4}$ is a manifold and special care is needed to define Gaussians and perform least-squares there. We use $\boxplus$-manifolds [3] that encapsulate the manifold structure into an operator $\boxplus : SE(3) \times \mathbb{R}^6 \to SE(3)$. It takes a pose and changes it according to a vector in the tangential space, usually using the exponential map. A converse operator $\boxminus : SE(3) \times SE(3) \to \mathbb{R}^6$ axiomatized by $T_1 \boxplus (T_2 \boxminus T_1) = T_2$ gives the vector that goes from one pose T_1 to another T_2. These operators allow also to define Gaussians in $SE(3)$.

$$T \boxplus \delta = T \exp\begin{pmatrix} 0 & -\delta_3 & \delta_2 & \delta_4 \\ \delta_3 & 0 & -\delta_1 & \delta_5 \\ -\delta_2 & \delta_1 & 0 & \delta_6 \\ 0 & 0 & 0 & 0 \end{pmatrix} \overset{\text{linea-}}{\underset{\text{rized}}{\approx}} T \begin{pmatrix} 1 & -\delta_3 & +\delta_2 & \delta_4 \\ +\delta_3 & 1 & -\delta_1 & \delta_5 \\ -\delta_2 & +\delta_1 & 1 & \delta_6 \\ 0 & 0 & 0 & 1 \end{pmatrix}, \tag{1}$$

$$T_2 \boxminus T_1 = \left(L_{32}, L_{13}, L_{21}, L_{14}, L_{24}, L_{34} \right)^T, \quad L = \log(T_1^{-1} T_2) \overset{\text{linea-}}{\underset{\text{rized}}{\approx}} T_1^{-1} T_2 - I \tag{2}$$

$$\mathcal{N}(\mu, \Sigma) = \mu \boxplus \mathcal{N}(0, \Sigma), \quad \mu \in SE(3), \Sigma \in \mathbb{R}^{6 \times 6} \tag{3}$$

$$\mathcal{N}(\mu, \Sigma)(x) \approx |2\pi\Sigma|^{-1/2} \exp\left(-\tfrac{1}{2}(x \boxminus \mu)^T \Sigma^{-1}(x \boxminus \mu)\right) \tag{4}$$

In this framework a squared residual $\|f(T)\|^2$ on $T \in SE(3)$ is minimized with Gauss-Newton [14] by parameterizing the increment with $\boxplus$ in each iteration:

$$\underset{T \in SE(3)}{\arg\min} \|f(T)\|^2 = \lim_k T_k, \quad T_{k+1} = T_k \boxplus \overset{\text{linearized}}{\underset{\delta \in \mathbb{R}^6}{\arg\min}} \|f(T_k \boxplus \delta)\|^2 \tag{5}$$

3   Approach

Figure 1 shows our approach: A CNN recognizes as first stage what is in the images, segments the objects and predicts for every pixel u_i belonging to an object which object point p_i^O this is. It also predicts the uncertainty of this information, expressed as a covariance $C_i \in \mathbb{R}^{2 \times 2}$ in image space. These pieces of information are then fused by the second stage (gPnP) to obtain an object pose with covariance derived from the point covariances.

Actually the uncertainty originates from p_i^O while the pixel coordinates u_i are exact, however the CNN operates on the image, so presumably it can more easily express image uncertainty (e.g. along an image edge). Furthermore, only this way fits into the mathematical structure of gPnP.

To handle multiple cameras, the unknown object pose $T := T_O^R$ is expressed in a reference frame (R). A known transformation $T_R^{C_i}$ locates the reference frame in the frame of the camera to which pixel u_i belongs. We omit the index at C_i in the following. Let p_i^C denote p_i^O in camera-coordinates, u_0 the image center, and f the focal length. With this notation, the camera equation with noise η_i defines the interface between the CNN and gPnP stages:

$$p_i^C = T_R^C T_O^R p_i^O, \quad T := T_O^R, \quad u_i = u_0 + f \frac{p_{i12}^C}{p_{i3}^C} + \eta_i, \quad \eta_i \sim \mathcal{N}(0, C_i) \quad (6)$$

$$C_i = (W_i^T W_i)^{-1} \quad \Longrightarrow \quad W_i \left(u_i - u_0 - f \frac{p_{i12}^C}{p_{i3}^C} \right) \sim \mathcal{N}(0, I_2) \quad (7)$$

The CNN provides C_i indirectly as a 2×2 weight matrix W_i. This should help, as the rows of W_i describe independent image-directions of information with a larger norm meaning more information. In particular, (7) ensures positive semi-definiteness and facilitates expressing anisotropic situations, e.g. at image edges.

Contrary to the usual assumption in a PnP stage (cf. Sect. 5), different p_i^O are not stochastically independent. Therefore the CNN stage predicts a correction factor w_{oj} per object instance j that upscales the computed covariance Σ_{gPnP} assuming every pixel carries the same percentage w_{oj}^{-1} of "new" information.

$$\Sigma = w_{oj} \Sigma_{\text{gPnP}} \quad (8)$$

The proposed method can also incorporate per-pixel depth information z_i if available into the gPnP fusion with a depth-pendant to (6):

$$z_i = p_{i3}^C + \eta_{di}, \quad \eta_{di} \sim \mathcal{N}(0, w_{di}^{-2}), \quad w_{di} = (w_{doi}^{-2} + w_{dsi}^{-2})^{-1/2} \quad (9)$$

$$\Longrightarrow \quad w_{di} \left(z_i - p_{i3}^C \right) \sim \mathcal{N}(0, 1) \quad (10)$$

The noise η_{di} has two sources: The main RGB-CNN gives a weight w_{dpi} describing the noise in p_i^O. A small depth-CNN gives a weight w_{dsi} describing the noise in z_i, identifying e.g. invalid depth data. Both are combined by adding variances. This way the depth data does not contribute to the recognition but is integral part of the gPnP-fusion stage not a separate postprocessing as in PoseCNN.

In the gPnP stage the measurement Eqs. (6) and (9) are converted into 13×13 matrices and added, thereby condensing all information into a globally valid fixed

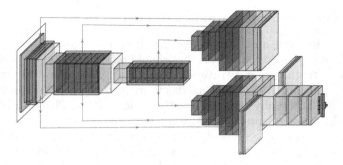

Fig. 3. CNN layers, where outputs are ordered according to Fig. 1: Shades from yellow to orange denote 1×1, 3×3 or 5×5 convolutions; magenta, pairs of 1×1 followed by 3×3 convolutions; green, upsampling; blue, global pool and fully connected layers. All layers with trainable weights are followed by SELU [7] units. Output layers have no activation. For further details see https://github.com/JesseRK/6D-UI-CNN-Output. (Color figure online)

size matrix $M^T M$ without linearization. We obtain an initial guess by removing the translation with Schur-complement and optimizing the rotation-only result on a finite set of 22.5°-spaced rotations. This guess is refined with Gauss-Newton on the $SE(3)$ manifold of poses, giving the pose T and the pose covariance Σ_{gPnP} as the inverse of Gauss-Newton's pseudo-Hessian. $\mathcal{N}(T, \Sigma)$, in the sense of (3), is the then final posterior distribution.

4   CNN Details

The main CNN consists of an encoder-decoder DenseNet-like [4] structure with horizontal connections. Instead of batch-norm-ReLU we exclusively use SELU [7] activations. There are a shared encoder path and two different but identical decoder paths for object points (p_i^O) and uncertainty predictions. The latter forks into heads for W_i, w_{doi} and w_{oj}. All output images have a quarter (x and y) of their input resolution. More details are shown in Fig. 3, however we see the overall architecture, not the layer details, as the paper's contribution.

Instance segmentation can be performed with an arbitrary method since it is independent of the core object estimation algorithm, and was not the main focus here. For prediction, we use the segmentation masks as they are calculated by PoseCNN [23] and omit further description regarding the segmentation procedure. During training we use ground truth segmentation to prevent distracting dependencies. In general we try to prevent those by training the CNN in three stages as declared in Sect. 4.1. In Sect. 4.2 we describe the secondary CNN which is only used in case of depth fusion to estimate the sensor's uncertainty.

4.1   Loss Function and Learning Procedure

Every pose estimation unit (i.e. all except those for segmentation) receives a gradient only if it points at an object that should be learned (i.e. the loss function gets masked with the ground truth segmentation). This allows the network to e.g. predict a p_i^O under the condition that u_i maps to an object point but leaves the decision if this actually is the case to the segmentation. Therefore all outputs are masked by the segmentation before they are used as well.

Currently, we train one network per object as DOPE [22] does. We use Adam [6] with the amsgrad expansion [16] as optimizer and augment our input images with contrast, Gaussian and brightness noise.

Object Points. The p_i^O are exclusively learned first with a mean squared error loss against ground truth $p_i^O{}_{\text{true}}$, which can be obtained by rendering the corresponding model with the ground truth pose. It is supplied with the YCb-data set and for rendered images we obtained it from the depth image and ground truth pose. Each axis is separately scaled to the range $[-1, 1]$ for the CNN and the original scale restored before gPnP.

Pixelwise Uncertainties. The uncertainty weights W_i and w_{doi} are trained with a negative log likelihood (NLL) loss from the Gaussians they represent.

$$\sum_i - \log \mathcal{N}(0, (W_i^T W_i)^{-1})(\eta_i) - \log \mathcal{N}(0, w_{doi}^{-2})(\eta_{doi}), \tag{11}$$

The residuals η_i and η_{doi} are obtained from (6) and (8) resp. by setting $T = T_{\text{true}}$. The NLL becomes minimal when the residual distribution corresponds to the weight-defined covariance.

After freezing the encoder as well as the respective p_i^O decoder we trained W_i and w_{doi} together as a second training stage.

Object Uncertainty. The per object correction factor w_{oj} for the uncertainty obtained by gPnP is also trained with an NLL loss

$$- \log \mathcal{N}(T, \Sigma)(T_{\text{true}}), \tag{12}$$

here of the ground truth pose in the predicted Gaussian posterior. (4) is used, since it is a ⊞-manifold Gaussian.

As third stage, all previously trained layers were frozen and only the w_{oj}-head trained. This training involves the gPnP part. To stabilize the training, Gauss-Newton uses the ground truth pose as initial guess. This approach is sound for training, because the initial guess does not influence the limit, as long as it stays in the basin of convergence for that limit value.

4.2   Depth Sensor Uncertainty CNN

To predict the depth sensor uncertainty we use a separate small network (cf. Fig. 1, Sect. 3), which consists of seven SELU-activated convolutions in a ResNet-fashion (up to 32 channels) with one down and one up sampling layer. The only input for this network is the depth image. Since the resulting weights are later combined with the w_{doi} the input and output are using its resolution, too. The network is trained on the NLL loss

$$\sum_i - \log \mathcal{N}(z_i, w_{dsi}^{-2})(z_{i\,\text{true}}), \tag{13}$$

where the true depth $z_{i\,\text{true}}$ is obtained as $p_{i\,\text{true}}^O$ above.

5   Pragmatic gPnP Algorithm

Overall, the gPnP stage approximates the posterior distribution of T given (6) and optionally (9) as a Gaussian in pose space using (3). It is derived as follows:

5.1   Perspective Measurements

We multiply (7) by $p_{i3}^C W_i$ making it linear in p_i^C with no C_i in the noise.

$$\eta_i' = W_i(p_{i3}^C(u_i - u_0) - f p_{i12}^C) = W_i \begin{pmatrix} -f & 0 & u_{i1}-u_{01} & 0 \\ 0 & -f & u_{i2}-u_{02} & 0 \end{pmatrix} p_i^C \tag{14}$$

$$:= W_i A_i p_i^C = W_i A_i T_R^C T p_i^O \qquad \eta_i' = p_{i3}^C W_i \eta_i \sim \mathcal{N}(0, (p_{i3}^C)^2 I_2) \tag{15}$$

Gaussian measurements can only be fused if their relative covariance is known, however p_{i3}^C, the object point's camera-Z-coordinate, is unknown. Thus we treat the covariances as identical, i.e. $\eta_i' \overset{\text{approx}}{\sim} \mathcal{N}(0, \bar{z}^2 I_2)$, where $\bar{z}$ is the mean p_{i3}^C. This is a acceptable, since the distance camera to object is usually several times its size, so the p_{i3}^C differ only by a small factor (3% avg. on our data). The common factor $\bar{z}$ is still a-priori unknown, but will be recovered later in (25).

The term $T p_i^O$ is linear in T and hence can be expressed as a matrix-vector-product $\bar{p}_i^O \bar{T}$, if T is flattened as $\bar{T} \in \mathbb{R}^{13}$ and p_i^O converted into $\bar{p}_i^O \in \mathbb{R}^{4 \times 13}$:

$$\bar{T} = \begin{pmatrix} T_{11} & T_{12} & T_{13} & T_{21} & T_{22} & T_{23} & T_{31} & T_{32} & T_{33} \mid 1 \mid T_{13} & T_{23} & T_{33} \end{pmatrix}^T, \tag{16}$$

$$\bar{p}_i^O = \begin{pmatrix} p_1 & p_2 & p_3 & 0 & 0 & 0 & 0 & 0 & 0 & 0 & 1 & 0 & 0 \\ 0 & 0 & 0 & p_1 & p_2 & p_3 & 0 & 0 & 0 & 0 & 0 & 1 & 0 \\ 0 & 0 & 0 & 0 & 0 & 0 & p_1 & p_2 & p_3 & 0 & 0 & 0 & 1 \\ 0 & 0 & 0 & 0 & 0 & 0 & 0 & 0 & 0 & 1 & 0 & 0 & 0 \end{pmatrix} \tag{17}$$

The flattened transform $\bar{T}$ consists of 9 rotations entries, followed by a constant 1 and 3 translation entries. Unlike [9,11,26] we directly estimate the transformation matrix instead of control points. The ordering of $\bar{T}$ will allow later to separate the translation as the 3-vector tail.

Now (15) becomes a linear measurement in $\bar{T}$ with i.i.d. Gaussian noise:

$$\eta_i' = W_i A_i T_R^C T p_i^O = (W_i A_i T_R^C \bar{p}_i^O)\bar{T} =: M_i\bar{T}.   \quad (18)$$

Consider the column $M_{\bullet,10}$ which is multiplied by the fixed $\bar{T}_{10} = 1$. From the pattern in A_i and $\bar{p}_i^O$ it is zero, if and only if the translation in T_R^C is zero. This shows, that it is needed in general but not for a single camera ($T_R^C = I$).

Following the usual least-squares approach, the M_i are stacked as $M = (M_{1...n})$ and assuming independence $M\bar{T} \sim \mathcal{N}(0, \bar{z}^2 I_{2n})$. The negative log-likelihood of $\bar{T}$ is now expressed by a $\mathbb{R}^{13 \times 13}$ matrix $M^T M = \sum_i M_i^T M_i$ as

$$-\log p(T|u_{1...n}, p_{1...n}^O) + \text{const} = \bar{z}^2\|M\bar{T}\|^2 = \bar{z}^2 \bar{T}^T (M^T M)\bar{T},   \quad (19)$$

and shall be minimized for $T \in SE(3)$. For several cameras, their M's are stacked, resp. their $M^T M$'s added. So far, the approach resembles [9] except, that we weight by W_i and estimate T directly instead of control points.

5.2   Depth Measurements

The depth measurements z_i are handled similarly multiplying (10) with z_i:

$$\eta_{di}' = z_i w_{di}(z_i - p_{i3}^C) = \begin{pmatrix} 0 & 0 & -w_{di}z_i & w_{di}z_i^2 \end{pmatrix} p_i^C =: B_i p_i^C   \quad (20)$$

$$= (B_i T_R^C \bar{p}_i^O)\bar{T} =: M_{di}\bar{T}, \quad \eta_{di}' = z_i \eta_{di} \sim \mathcal{N}(0, z_i^2)   \quad (21)$$

This makes also $\eta_{di}' \overset{\text{approx}}{\sim} \mathcal{N}(0, \bar{z}^2)$ so the M_{di} can be stacked with M_i or equivalently $\sum_i M_{di}^T M_{di}$ is added to $M^T M$. This fuses perspective and depth measurements into one matrix. Even for $T_R^C = I$ it requires the $M_{\bullet,10}$ column. Indeed, the fixed $\bar{T}_{10}$ entry is necessary whenever scale is defined even without the $T \in SE(3)$ constraint, because without $\bar{T}_{10}$ any linear equation in $\bar{T}$ is scale-ambiguous.

5.3   Gauss-Newton Iteration

Following the approach from [3] specifically (5) to minimize (19) over T by Gauss-Newton we need to minimize $\|M\overline{\check{T} \boxplus \delta}\|^2$ for δ in a linearized way in each iteration, where $\check{T} = T_k$ is the starting point of the current iteration. We therefor write the linearization (1) for $\check{T} \boxplus \delta$ as a matrix-vector product $P\binom{\delta}{1}$.

$$\overline{\check{T} \boxplus \delta} \approx \begin{pmatrix} 0 & -\check{T}_{i3} & +\check{T}_{i2} & 0 & 0 & 0 & \check{T}_{i1} \\ +\check{T}_{i3} & 0 & -\check{T}_{i1} & 0 & 0 & 0 & \check{T}_{i2} \\ -\check{T}_{i2} & +\check{T}_{i1} & 0 & 0 & 0 & 0 & \check{T}_{i3} \\ 0 & 0 & 0 & 0 & 0 & 0 & 1 \\ 0 & 0 & 0 & & \check{T}_{i1} & \check{T}_{i2} & \check{T}_{i3} & \check{T}_{i4} \end{pmatrix} \begin{matrix} \}i=1..3 \\ \\ \\ \\ \}i=1..3 \end{matrix} \begin{pmatrix} \delta \\ 1 \end{pmatrix} =: P\begin{pmatrix} \delta \\ 1 \end{pmatrix}   \quad (22)$$

The first block of 3 rows is repeated for $i = 1 \ldots 3$ (rotation in row-wise ordering) and also the last block (translation). It results a 6-D quadratic minimization

$$\left\|M\overline{\check{T} \boxplus \delta}\right\|^2 \approx \binom{\delta}{1}^T (P^T M^T M P)\binom{\delta}{1} =: \binom{\delta}{1}^T Q\binom{\delta}{1},   \quad (23)$$

$$\delta^* = \arg\min_\delta \binom{\delta}{1}^T Q\binom{\delta}{1} = -Q_{1...6,1...6}^{-1} Q_{1...6,7}, \quad T_{k+1} = \check{T} \boxplus \delta^*   \quad (24)$$

and following (5), $T_{k+1} = \check{T} \boxplus \delta^*$ is the result of one Gauss-Newton iteration.

In Gauss-Newton, the final inverse pseudo-Hessian $Q^{-1}_{1...6,1...6}$ is the covariance of the estimate [14], here in the sense of (3). However, there is still the unknown factor $\bar{z}^2$ from (19). In theory, the obtained minimum of (23) should be measurement dimension minus state dimension on average, i.e. $\text{nRow}(M) - 6$. We use this to estimate a scaling factor for the covariance [14]. It gets rid of $\bar{z}^2$ and any wrong factor in the weights predicted by the CNN. Such a factor arises, because they are trained on *training* data and quantify the prediction error on *training* data which is lower than on test data (Sect. 6, Fig. 6).

$$T_{\text{gPnP}} = \check{T} \boxplus \delta^*, \quad \Sigma_{\text{gPnP}} = \frac{T_{\text{gPnP}} T^T M^T M T_{\text{gPnP}}}{\text{nRow}(M) - 6} Q^{-1}_{1...6,1...6} \tag{25}$$

As a final remark: The Gauss-Newton iterations are very efficient, because they work on the fixed size $M^T M$ instead of the original measurements. During training and for all tests we use ten iterations albeit empiric results indicate that four iterations might already be sufficient for full convergence.

5.4   Calculation of the Initial Guess

Iterative algorithms need an initial guess, however we are far from the minimal case that makes e.g. [9,11,26] so complex. So we decompose into a rotation $\bar{R} \in \mathbb{R}^{10}$ (with the fixed 1) and a translation $t \in \mathbb{R}^3$ solved by Schur-complement [14].

$$\min_{T \in SE(3)} \bar{T}^T M^T M \bar{T} = \min_{R \in SO(3)} \min_{t \in \mathbb{R}^3} \left(\begin{smallmatrix} \bar{R} \\ t \end{smallmatrix} \right)^T \left(\begin{smallmatrix} P & Q \\ Q^T & S \end{smallmatrix} \right) \left(\begin{smallmatrix} \bar{R} \\ t \end{smallmatrix} \right) \tag{26}$$

$$= \min_{R \in SO(3)} \bar{R}^T \left(P - Q^T S^{-1} Q \right) \bar{R}. \approx \min_{R \in SO_d} \bar{R}^T \left(P - Q^T S^{-1} Q \right) \tag{27}$$

$$t^*(\bar{R}) = \arg \min_{t \in \mathbb{R}^3} \left(\begin{smallmatrix} \bar{R} \\ t \end{smallmatrix} \right)^T \left(\begin{smallmatrix} P & Q \\ Q^T & S \end{smallmatrix} \right) \left(\begin{smallmatrix} \bar{R} \\ t \end{smallmatrix} \right) = -Q S^{-1} \bar{R} \tag{28}$$

For (27) we simply try 960 fixed rotations $R \in SO_d$ precomputed by choosing the x-axis on the 60 vertices of a truncated icosahedron (football, $\approx 23°$ spaced), followed by an x-rotation in $22.5°$ steps. With (28) we exclude objects behind the camera $(t^*(\bar{R})_3 \leq 0)$ and get the final translation.

6   Experimental Results

We evaluate our system on the YCB-Video dataset [23]. We exclude symmetric objects, because the network cannot be expected to distinguish symmetric points[1]. Handling this is future work. For training we use the data provided by [23] as well as additionally generated data according to the approach by [22].

---

[1] This is, because the network is trained to distinguish each and every object point based on its appearance (which is the same for multiple points within a symmetrical object).

We follow [23] and utilize the ADD-error as error metric, which is defined for a model with a finite set of object points M and two poses T, T' as

$$ADD(T, T') = \frac{1}{|M|}\Sigma_{p \in M} ||Tp - T'p||. \tag{29}$$

In addition, since we estimate uncertainty of the predicted pose, we also compute the expected ADD-error $(\widehat{ADD})$ from Σ and M by

$$\widehat{ADD} = \frac{1}{|M|}\Sigma_{p \in M}\hat{N}\left(([p]_\times, I_3)\,\Sigma\,([p]_\times, I_3)^T\right), \; a = \sqrt{\lambda_1/\lambda_3}, b = \sqrt{\lambda_2/\lambda_3},$$

$$\hat{N}(C) \approx \sqrt{\lambda_3}\left(0.782 + 0.234(a + b) + 0.246(a^2 + b^2) - 0.126ab\right). \tag{30}$$

$\lambda_1 \leq \lambda_2 \leq \lambda_3$ are the eigenvalues of C. We cannot explain this approximation here for lack of space, but have verified it has 1.3% average (2.9% max) error.

Figure 4 shows object-wise accuracy-threshold curves for ADD-errors in the range of $[0\ m, 0.1\ m]$ both with and without depth-data. In addition to the actual errors we show the expected accuracy-threshold curves based on our predicted uncertainties, which are obtained as the ADD-error from samples from the posterior 6d error distribution $\mathcal{N}(0, \Sigma)$. One can see that the network's prediction about its own accuracy curve looks fairly close to its actual accuracy curve, indicating that the uncertainty information Σ is meaningful. The curves also show that incorporating depth greatly improves the result.

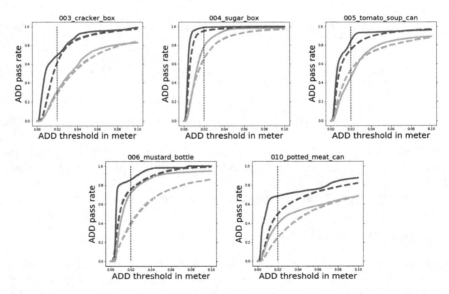

Fig. 4. Accuracy-threshold curves for evaluated objects with RGB input (red solid lines) and with additional depth fusion (green solid lines). Dashed lines describe the expected accuracy based on the predicted uncertainties. The vertical line marks the two centimeter spot. Due to space we depict the object subset which was used in [22]. (Color figure online)

Table 1. Pose evaluation: Area under ADD-accuracy-treshold-curve ([0,10] cm). `Our` method is compared against PoseCNN's [23] and DOPE's [22] results - RGB-only and with depth fusion (^{DF}) or respective ICP. `Seg`$_{gt}$ state comparative results using our's with true segmentation.

YCB-Video dataset Method Object	RGB				RGB-D		
	[23]	[22]	Seg$_{gt}$	Our	[23]^{ICP}	Seg$_{gt}^{DF}$	Our^{DF}
002_master_chef_can	50.9	-	56.3	**54.6**	69.0	78.0	**78.1**
003_cracker_box	51.7	55.9	57.7	**57.6**	80.7	80.9	**82.6**
004_sugar_box	68.6	75.7	85.5	**84.1**	**97.2**	95.8	95.3
005_tomato_soup_can	66.0	**76.1**	69.9	68.3	81.6	87.5	**86.6**
006_mustard_bottle	79.9	**81.9**	79.0	79.0	**97.0**	89.9	90.0
007_tuna_fish_can	**70.4**	-	42.6	43.5	83.1	84.5	**84.4**
008_pudding_box	**62.9**	-	51.0	50.3	**96.6**	91.4	92.0
009_gelatin_box	**75.2**	-	86.2	74.8	**98.2**	96.7	95.5
010_potted_meat_can	**59.6**	39.4	50.9	50.3	**83.8**	72.0	72.1
011_banana	**72.3**	-	8.6	8.2	**91.6**	45.6	45.8
019_pitcher_base	52.5	-	77.7	**77.8**	**96.7**	92.7	92.7
021_bleach_cleanser	50.5	-	60.4	**59.3**	**92.3**	83.1	82.7
025_mug	57.7	-	69.8	**69.1**	81.4	88.4	**91.0**
035_power_drill	55.1	-	71.5	**71.4**	**96.9**	89.3	88.3
AVG	62.4	(65.8)	61.9	60.6	88.3	84.0	84.1

Table 1 compares the performance PoseCNN [23] and DOPE [22]. It lists the area under these ADD-accuracy-threshold-curves for all objects. In addition to our proposed architecture, which utilizes the segmentation of PoseCNN (cf. Sect. 4), we also state comparative results using the true segmentation instead. The intention here is to evaluate the effect of the segmentation algorithm selection.

One can see that there are no big differences between either segmentation use. An exception makes the object 009_gelatin_box where the true segmentation performs much higher, because the segmentation has some outliers.

Our results fluctuate between different objects. In general our detection performance (without the uncertainty estimation) is the higher the better possible it is to distinguish all regions of an object. Otherwise, dissenting p_i^O regions might accrue, disturbing gPnP (which is strongest in case of 011_banana).

Table 1 also shows our results with depth data fusion compared to PoseCNN with iterative closest point (ICP). It can be clearly seen that our results benefit highly from the depth fusion but the more powerful ICP produces overall better results by the cost of much higher computation time. ICP may also dismiss known rotations of objects where the 3D-model (without its texture) is rotational invariant, which results in worse ADD-errors than our depth fusion.

Figure 6 investigates the provided uncertainty information. The χ^2-error measures consistency between actual error and estimated uncertainty. For the training data, both points p_i^O and pose T uncertainties are good. For the test data,

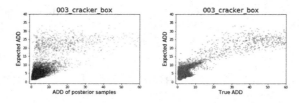

Fig. 5. Example expected ADD vs. posterior samples (left) and vs. true ADD-errors (right).

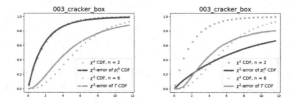

Fig. 6. Cummulative χ^2 curves on training (left) and test images (right). Solid lines show the measured curves for points and poses resp. The dotted lines show the theoretical curves, corresponding to perfectly correct posterior distributions.

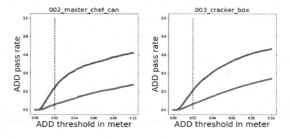

Fig. 7. Example accuracy-threshold curves for monocular (red) and binocular (green) results. (Color figure online)

the CNN underestimates the uncertainty in p_i^O, naturally, since it is trained on training data, where the p_i^O are better. Still the gPnP computes a meaningful uncertainty from that, because it normalizes with the actual residuum in (25).

Figure 5 shows a distribution of actual vs. expected ADD (by (30)) and how that would look if the posterior was perfectly correct. Both match qualitatively, showing that the system mostly knows when it performs badly. Interestingly actual ADD is never small, when expected ADD is very large.

At last we want to show that our system is able to combine multiple cameras for one detection. Therefore we generated a small simulated binocular dataset, whereon we compare our monocular with our binocular results (Fig. 7). One can see that the binocular results are much higher.

7   Conclusion and Future Work

To our knowledge we are the first to present a 6d pose estimation CNN that predicts its own uncertainty (per estimate) with meaningful accuracy. Our network system predicts observed object points per pixel as well as per pixel their uncertainty in the image plane. This information is fused through a gPnP resulting in a pose estimate with covariance. Currently, our pose prediction (not the uncertainty estimation) can be confused by not clearly distinguishable object regions which we want to study in future work including also texture-less objects, rotation invariance and occlusion, as well as, train a single CNN for all objects.

Acknowledgments. The research reported in this paper has been supported by the German Research Foundation DFG, as part of Collaborative Research Center (Sonderforschungsbereich) 1320 EASE - Everyday Activity Science and Engineering, University of Bremen (http://www.ease-crc.org/). The research was conducted in subproject R02 'Multi-cue perception supported by background knowledge'.

References

1. Brachmann, E., Krull, A., Michel, F., Gumhold, S., Shotton, J., Rother, C.: Learning 6D object pose estimation using 3D object coordinates. In: Fleet, D., Pajdla, T., Schiele, B., Tuytelaars, T. (eds.) ECCV 2014. LNCS, vol. 8690, pp. 536–551. Springer, Cham (2014). https://doi.org/10.1007/978-3-319-10605-2_35
2. Crivellaro, A., Rad, M., Verdie, Y., Moo Yi, K., Fua, P., Lepetit, V.: A novel representation of parts for accurate 3D object detection and tracking in monocular images. In: ICCV, pp. 4391–4399 (2015)
3. Hertzberg, C., Wagner, R., Frese, U., Schröder, L.: Integrating generic sensor fusion algorithms with sound state representations through encapsulation of manifolds. Inf. Fusion **14**(1), 57–77 (2013)
4. Huang, G., Liu, Z., Van Der Maaten, L., Weinberger, K.Q.: Densely connected convolutional networks. In: CVPR, pp. 4700–4708 (2017)
5. Kehl, W., Manhardt, F., Tombari, F., Ilic, S., Navab, N.: SSD-6D: making RGB-based 3D detection and 6D pose estimation great again. In: ICCV (2017)
6. Kingma, D.P., Ba, J.: Adam: a method for stochastic optimization. arXiv preprint arXiv:1412.6980 (2014)
7. Klambauer, G., Unterthiner, T., Mayr, A., Hochreiter, S.: Self-normalizing neural networks. In: NIPS, pp. 971–980 (2017)
8. Kneip, L., Furgale, P.: OpenGV: a unified and generalized approach to real-time calibrated geometric vision. In: ICRA, pp. 1–8. IEEE (2014)
9. Kneip, L., Furgale, P., Siegwart, R.: Using multi-camera systems in robotics: efficient solutions to the NPnP problem. In: ICRA, pp. 3770–3776. IEEE (2013)
10. Krizhevsky, A., Sutskever, I., Hinton, G.: Imagenet classification with deep convolutional neural networks. In: NIPS (2012)
11. Lepetit, V., Moreno-Noguer, F., Fua, P.: EPnP: an accurate O(n) solution to the PnP problem. IJCV **81**(2), 155 (2009)
12. Oberweger, M., Rad, M., Lepetit, V.: Making deep heatmaps robust to partial occlusions for 3D object pose estimation. In: Ferrari, V., Hebert, M., Sminchisescu, C., Weiss, Y. (eds.) ECCV 2018. LNCS, vol. 11219, pp. 125–141. Springer, Cham (2018). https://doi.org/10.1007/978-3-030-01267-0_8

13. Periyasamy, A.S., Schwarz, M., Behnke, S.: Robust 6D object pose estimation in cluttered scenes using semantic segmentation and pose regression networks. In: IROS, pp. 6660–6666. IEEE (2018)
14. Press, W.H., Teukolsky, S.A., Vetterling, W.T., Flannery, B.P.: Numerical Recipes in C (2nd ed.): The Art of Scientific Computing. Cambridge University Press, New York (1992)
15. Rad, M., Lepetit, V.: BB8: a scalable, accurate, robust to partial occlusion method for predicting the 3D poses of challenging objects without using depth. In: ICCV, pp. 3828–3836 (2017)
16. Reddi, S.J., Kale, S., Kumar, S.: On the convergence of adam and beyond (2018)
17. Redmon, J., Farhadi, A.: YOLO9000: better, faster, stronger. In: CVPR (2017)
18. Su, H., Qi, C.R., Li, Y., Guibas, L.J.: Render for CNN: viewpoint estimation in images using CNNs trained with rendered 3D model views. In: ICCV (2015)
19. Sundermeyer, M., Marton, Z.-C., Durner, M., Brucker, M., Triebel, R.: Implicit 3D orientation learning for 6D object detection from RGB images. In: Ferrari, V., Hebert, M., Sminchisescu, C., Weiss, Y. (eds.) ECCV 2018. LNCS, vol. 11210, pp. 712–729. Springer, Cham (2018). https://doi.org/10.1007/978-3-030-01231-1_43
20. Tekin, B., Sinha, S.N., Fua, P.: Real-time seamless single shot 6D object pose prediction. In: CVPR, pp. 292–301 (2018)
21. Thrun, S., Burgard, W., Fox, D.: Probabilistic Robotics. MIT Press, Cambridge (2005)
22. Tremblay, J., To, T., Sundaralingam, B., Xiang, Y., Fox, D., Birchfield, S.: Deep object pose estimation for semantic robotic grasping of household objects. arXiv preprint arXiv:1809.10790 (2018)
23. Xiang, Y., Schmidt, T., Narayanan, V., Fox, D.: PoseCNN: a convolutional neural network for 6D object pose estimation in cluttered scenes. arXiv preprint arXiv:1711.00199 (2017)
24. Zakharov, S., Shugurov, I., Ilic, S.: DPOD: dense 6D pose object detector in RGB images. arXiv preprint arXiv:1902.11020 (2019)
25. Zeng, A., et al.: Multi-view self-supervised deep learning for 6D pose estimation in the amazon picking challenge. In: ICRA, pp. 1383–1386. IEEE (2017)
26. Zheng, Y., Kuang, Y., Sugimoto, S., Astrom, K., Okutomi, M.: Revisiting the PnP problem: a fast, general and optimal solution. In: ICCV, pp. 2344–2351 (2013)

QuiltGAN: An Adversarially Trained, Procedural Algorithm for Texture Generation

Renato Barros Arantes$^{(\boxtimes)}$, George Vogiatzis$^{(\boxtimes)}$, and Diego Faria$^{(\boxtimes)}$

Aston University, Aston St, Birmingham B4 7ET, UK
{180178991,g.vogiatzis,d.faria}@aston.ac.uk
https://www2.aston.ac.uk/

Abstract. We investigate a generative method that synthesises high-resolution images based on a single constraint source image. Our approach consists of three types of conditional deep convolutional generative adversarial networks (cDCGAN) that are trained to generate samples of an image patch conditional on the surrounding image regions. The cDCGAN discriminator evaluates the realism of the generated sample concatenated with the surrounding pixels that were conditioned on. This encourages the cDCGAN generator to create image patches that seamlessly blend with their surroundings while maintaining the randomisation of the standard GAN process. After training, the cDCGANs recursively generate a sequence of samples which are then stitched together to synthesise a larger image. Our algorithm is able to produce a nearly infinite collection of variations of a single input image that have enough variability while preserving the essential large-scale constraints. We test our system on several types of images, including urban landscapes, building facades and textures, comparing very favourably against standard image quilting approaches.

Keywords: GAN · Procedural generation · Inpainting

1 Introduction

1.1 Motivation

The quality and quantity of training data is often the most critical part that determines the performance of a model. Once we have the training data, in the needed amount and quality, the rest usually follows. However, to produce a dataset that has the needed features to train a model, in a representative volume, is not an easy task. When data acquisition is costly, the solution typically employed by machine learning practitioners is data augmentation [16] but this is mostly limited to simple randomised manipulation of an existing dataset (e.g. affine warping, rotation or other small perturbations).

This work proposes an investigation on whether GANs [6] can be used to synthesise images, based on a single given image, in a way that the resulting

© Springer Nature Switzerland AG 2019
D. Tzovaras et al. (Eds.): ICVS 2019, LNCS 11754, pp. 423–432, 2019.
https://doi.org/10.1007/978-3-030-34995-0_38

(a) Source image.

(b) Synthesised images.

Fig. 1. Our algorithm in action. Source image used to train our GAN and several different synthesised outputs none of which contains exact copies of the pixels of Fig. 1a

synthesised image is similar to the constraint and shares with it the main features. The proposed method is compared with the results obtained from the *image quilting* (IQ) algorithm [3] in several different variants. We demonstrate that the proposed method can be superior to the simple image quilting algorithm because a GAN can generate new patches that do not exist in the constraint, while the IQ algorithm can only repeat patches from the source image.

1.2 Related Work

The introduction of generative adversarial networks (GAN) [6], provides a new approach for image synthesis in computer vision. Compared with traditional supervised machine learning algorithms, GAN employs the concept of adversarial training, in which two models are simultaneously trained: a generative model, G that generates images from a data distribution, and a discriminative model D, that attempts to determine whether a sample came from the training data, rather than G.

In [14] a type of CNN called deep convolutional generative adversarial networks (DCGANs) was introduced, proposing a specific style of architecture, bridging the gap between the achievement of CNNs for supervised learning and unsupervised learning. According to [5] most GANs today are at least loosely based on the DCGAN architecture, including the proposed method in this paper, where we join the DCGAN architecture and the conditional GAN (cGAN) [11] approach. A conditional GAN (cGAN) was first introduced by [11] where the authors proposed that a GAN can be extended to a conditional model if both the generator and discriminator are conditioned on some extra information **y**. This could be any auxiliary information, such as class labels or data from other modalities. The conditioning can be performed by feeding $\mathbf{y}$ into both the discriminator and generator as an additional input layer. In the generator, the prior input noise z, randomly sampled from a uniform Gaussian, $p_z(z)$, and $\mathbf{y}$ are combined in a joint hidden representation.

In [8], the authors propose a triple network architecture for inpainting: a completion network, a global context discriminator and a local context discriminator. The completion network is fully convolutional and used to complete the image, while both the global and the local context discriminators are auxiliary networks used exclusively for training. These discriminators are used to determine whether or not an image is consistent. The global discriminator takes the full image as input to recognise global consistency of the scene, while the local discriminator looks only at a small region around the completed area in order to judge the quality of more detailed appearance. The authors claim that their method can be used to complete an extensive variety of scenes and also can generate fragments that do not appear elsewhere in the image. Our method can be regarded as an inpainting method because it uses conditional GANs to generate blocks of images that are complementary to a given model. However, differently from the majority of the proposed methods listed here, we aim to create an entirely new image from scratch that is based on a given model. This is the novelty of our proposed approach: instead of generating a patch that fills a hole in an image, the proposed method uses that image as a model to synthesise an entirely new one that is highly related to the model.

An extension of [12] appeared in [8], where the authors proposed a Context Encoder as a convolutional neural network trained to generate the contents of an arbitrary image region conditioned on its surroundings. They found that a context learns a representation that captures not just appearance but also the semantics of visual structures. Their approach is to use an alternate formulation, by conditioning only the generator (not the discriminator) on context. They also found results improved when the generator was not conditioned on a noise vector. Our method also uses that alternate formulation as it only contextualises the generator. Together with [8], our method emphasises the importance of contextualisation to capture the semantics of the visual structures in the constraint image.

PatchGAN was introduced by [9] where a simple framework can be adapted to various image generation problems. Instead of grading the whole image, a PatchGAN slides a window over the input and produces a score that indicates whether the patch is real or fake. We also use this approach of a sliding window to generate the dataset, which is composed exclusively of patches extracted from the constraint image.

Similar to [8], where a triple network architecture is used, in [2] the authors introduce a generative CNN model and a training procedure for the arbitrary hole filling problem. The generator network takes the corrupted image and tries to reconstruct the repaired image. They utilised the ResNet [7] architecture as the generator model with a few alterations. The critical point of their work is that they propose to design a novel discriminator network that combines a global

GAN (G-GAN) structure with PatchGAN approach which they call PGGAN. Regular GANs usually maps a noise input vector to an image. Instead, Spatial GANs (SGAN) [10] extends the input noise distribution space from a single vector to a whole spatial tensor. The authors lately extended SGAN to Periodic Spatial GAN (PSGAN) [1], which extend the structure of the input noise distribution by constructing tensors with different types of dimensions. In PSGAN, the input spatial tensor is composed of three parts: a local independent, a spatially global, and a periodic part. Each piece has the same spatial dimensions but can vary in their channel dimensions. PSGAN can synthesise periodic and high-resolution textures and both SGAN and PSGAN architecture, as the proposed method in this paper, are based on DCGAN [14]. PSGAN authors claim that their model also can be trained using only a single image.

The GAN architecture has brought extra strength to the subject of synthesis and texture transfer, with many papers released that use this approach after its introduction. We have described some of them above. However, more recently [4] introduced a new parametric texture model based only on a convolutional neural network where textures are represented by the correlations between feature maps in several layers of the network. The same problem is tackled in [13], where the authors describe an algorithm for synthesising random images, subject to a statistical model for texture images. These are parameterised by a set of statistics constraints regarding spatial locations, orientations, and scales. Similar to the present work, both [4] and [13] they generate new textures based on a given image using a neural network. The difference in our work is that we use a patch based approach that allows our model to encode both high-frequency textures as well as large scale interactions. We can therefore synthesise larger images with coherent features e.g. the aligned windows in the architectural facades.

1.3   Problem Definition

Using just one image, we would like to train a GAN and obtain as a result a synthesised image that is different from the source image used to train the GAN, but at same time shares its main characteristics. For example, we used the facade [15] in Fig. 1a to train our GANs and we would like, at the end of the process pipeline, to obtain an image like the one in the Fig. 1b. The Fig. 1b was created using our proposed method, as described in Sect. 1.4.

The problem we are considering can be stated as a constrained image generation. Given a $N \times N$ source image S we would like to generate three blocks of size $N \times N$, named $S^{(1)}$, $S^{(2)}$ and $S^{(3)}$, in a way that they are compatible with each other. Figure 2 show how these blocks are conditioned on their generation.

By compatible, we mean an image that can be placed alongside another image to generate a new image that will look like being a single image when evaluated by a person. To achieve this, we need to guarantee semantic compatibility between both images and a continuous transition between them.

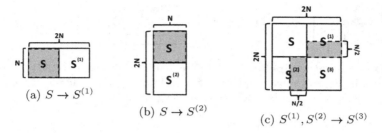

Fig. 2. Conditional patches. The three types of conditional GAN generation are shown here. Shaded grey is the image regions that are used to condition the image. $S^{(1)}$, $S^{(2)}$ and $S^{(3)}$ are the generated patches that should be *compatible* with the condition images.

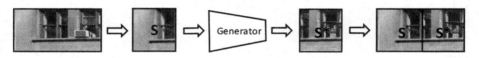

Fig. 3. Generator pipeline. An input $N \times 2N$ image patch is split into two $N \times N$ patches, one of which (S') is used to condition the generator. The $N \times N$ output of the generator S^* is concatenated with S' before it's fed to the discriminator.

1.4   Proposed Method

We propose a new framework to generate constrained images based on deep convolutional generative adversarial networks (DCGAN) [14] and conditional GAN (cGAN) [11].

We propose a cDCGAN that consists of a conditioned generator G and a non-conditioned discriminator D. The generator receives as input a random vector z concatenated with an $N \times N$ input image S. Its output is another $N \times N$ image S^*. Three cDCGANs were trained, as follows:

1. A pairwise cDCGAN, that given the conditioning image S generates $S^{(1)}$, that is compatible to S on the right.
2. A pairwise cDCGAN, that given the conditioning image S generates $S^{(2)}$, that is compatible to S on the bottom.
3. A cDCGAN that receives $S^{(1)}$ and $S^{(2)}$, extracts $S^{(12)}$ as a condition and generates $S^{(3)}$, that is compatible to $S^{(1)}$ above and is compatible to $S^{(2)}$ on the left.

That output image S^* will be concatenated with S, as in Fig. 3, to create a resulting image. The discriminator network D takes both generated images and the real image, see Fig. 4, and strives to distinguish them while the generator network G makes an effort to fool it. Note that our method differs from [11] as we are conditioning only the generator G, letting the discriminator D as proposed by [14], without any constraint.

Fig. 4. Discriminator pipeline. The discriminator is fed a series of $N \times 2N$ patches and tries to distinguishes between those extracted from actual images and those where the right $N \times N$ sub-patch is synthesised by the generator.

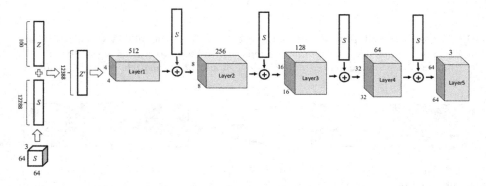

Fig. 5. Conditional GAN architecture. First, the 64×64×3 input image is reshaped into an array and then concatenated with input noise vector z. Subsequently, the output of one layer is concatenated with the condition vector S before it is fed to the next layer.

We condition the generator by concatenating the constraint image S with the input for the layer L_i, where i is the number of the current layer L of the generator G. For the first layer of the generator G, $i = 1$, this means concatenate the constraint image with the input noise vector sampled from a uniform Gaussian, (see Sect. 1.5). For the subsequent layers, $i = \{2, .., n\}$, we concatenate the output of the layer L_{i-1} with the constraint image S and gives this concatenation as the input for the layer L_i. See Fig. 5 for the overall generator network structure.

1.5   Adversarial Loss

Generative adversarial networks are composed of two adversarial models: a generative model G that captures the data distribution, and a discriminator model D that estimates the probability that a sample came from the training data rather than G. To learn a generator distribution p_g, over the data x, the generator builds a mapping function from a prior noise distribution $p_z(z)$ to data space as $G(z; \theta_g)$. Moreover, the discriminator, $D(x; \theta_d)$, outputs a single scalar representing the probability that x came from the training data rather than p_g.

G and D are both trained simultaneously. The parameters θ_g for G are adjusted to minimise $log(1 - D(G(z)))$, and the parameters θ_g for D are adjusted to minimise $log(D(X))$, as if they are following the two-player min-max game with value function $V(G, D)$:

$$\min_G \max_D V(D, G) = \mathbb{E}_{p \sim p_{data}}[\log(D(x))] + \mathbb{E}_{z \sim p_z(z)}[\log(1 - D(G(z)))]. \quad (1)$$

Our proposal is to extend the generative adversarial network model to a conditional model, where only the generator G is conditioned on an constraint image S. Our work differs from [11] where they condition both the generator G and the discriminator D. In that work the proposed objective function is

$$\min_G \max_D V(D, G) = \mathbb{E}_{p \sim p_{data}}[\log(D(x|y))] + \mathbb{E}_{z \sim p_z(z)}[\log(1 - D(G(z|y)))]. \quad (2)$$

In the generator the prior input noise $p_z(z)$, and y, the condition or constraint, are combined. In the discriminator x and y are presented as inputs to the discriminator. The objective of our cDCGAN is simpler, and is expressed as

$$\min_G \max_D V(D, G) = \mathbb{E}_{p \sim p_{data}}[\log(D(x))] + \mathbb{E}_{z \sim p_z(z)}[\log(1 - D(G(z|S)))]. \quad (3)$$

where x is the ground truth, S is the constraint image and z is the noise vector. G tries to minimise this loss function against an adversarial D that seeks to maximise it. The noise vector z is needed because otherwise the generator G would produce deterministic outputs, and therefore fail to meet any distribution of the input data.

1.6   Image Quilting

This section provides a brief review of the *image quilting* algorithm to be found in [3]. Given a set B of $N \times N$ pixels, extracted from a source image, the algorithm below will be executed to generate as output an array of blocks selected from B:

1. Select at random block from B as the top left block of the new synthesised image.
2. Go through the image to be synthesised, from left to right and from top to bottom, in steps of one block.
3. For every new location, search the set B for a block that minimises the overlap constraints, above and left.
4. Compute the minimum cost path1 between the newly chosen block and the old blocks at the overlap region, above and left. Find the minimum cost path along this surface and make that the boundary of the new block. Paste the block.
5. Repeat until the new image is completely synthesised.

2 Experimental Results

In all experiments carried with the IQ algorithm, the block size was 64 × 64 pixels, the size of the DCGAN input and generated image. The width of the overlapping edge was 1/6 of the size of the block. The error was computed using the $L2$ norm on pixel values, and the generated images are 5 × 5 blocks.

The results of the cDCGAN process for a wide range of input images are shown on Fig. 6. Column two on Fig. 6 was created using the simple version of the IQ algorithm as described in [3]. Column three was created with a variant of the IQ algorithm that, to our knowledge has not been evaluated elsewhere. The approach mixes the IQ algorithm and GAN generated patches, as follows: Initially, a set B with five thousands patches of 64 × 64 pixels each was generated using a non-conditional GAN, trained on the source image in the first column. This set B is then used to synthesise the images in column two, as described in Sect. 1.6. Column four was created using our proposed method as described Sect. 1.4.

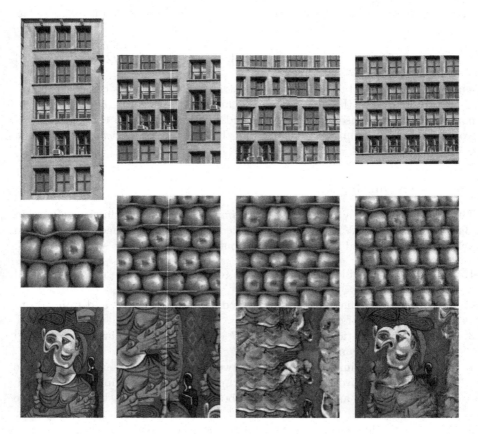

Fig. 6. Experimental results. Column one is the source image. Column two is IQ generated. Column three is IQ + GAN, and column four is cDCGAN

Fig. 7. A typical failure mode. The cDCGAN is applied recursively and linearly through the image, occasionally causing error to accumulate. Fortunately one can detect this automatically by a drop in the discriminator score.

We observe that the IQ generated images are very coherent, but lack variability because they locally repeat patches from the source image. The variant of IQ that uses the GAN for generating a pool of candidate patches does better in terms of variability but lacks coherence. This is because there is nothing to force the IQ algorithm to respect longer-range pixel interactions such as for example the architectural constraint of placing the building windows at regular intervals. Our cDCGAN-based images exhibit the best of both worlds. They do provide randomisation and variability (e.g. none of the building windows is an exact copy of the source image - see Fig. 1b) while also respecting the global image constraints. The algorithm proves capable of also handling non repeating images such as the Picasso painting (Fig. 6, third row). The synthesised painting respects the overall structure of the source image while providing a degree of random variation.

One failure case of our method results in a type of *fading* condition. If one cDCGAN generated patch has some errors, i.e. some non realistic appearance, then the next patch generated which will be conditioned on it will further deteriorate. This will result in a chain of progressively worse image patches leading to an unrealistic image. See Fig. 7 for example. This is an artefact due to the recursive application of the cDCGAN that follows a linear path through the image. One can automatically detect when this happens by a drop in the discriminator score and those image generations can be discarded.

3   Conclusion

In this paper, we introduced a method to synthesise images based on conditioned DCGANs. Our method uses just a single image to train three GANs for generating image patches that can be stitched together to synthesise a larger image. The GANs are conditioned on an adjacent part of the image and learn how to complete it by synthesising a *matching* image region. We compare the results with the *image quilting* algorithm [3] and conclude that the proposed method, although more complex than *image quilting* algorithm, can produce better results and above all can produce patterns that are not in the source image but are constrained by it. In future work we intend to explore the use of our algorithm for data augmentation in semantic segmentation.

References

1. Bergmann, U., Jetchev, N., Vollgraf, R.: Learning texture manifolds with the periodic spatial GAN. In: Precup, D., Teh, Y.W. (eds.) Proceedings of the 34th International Conference on Machine Learning, Proceedings of Machine Learning Research, vol. 70, pp. 469–477. PMLR, International Convention Centre, Sydney, Australia, 06–11 August 2017
2. Demir, U., Ünal, G.B.: Patch-based image inpainting with generative adversarial networks. CoRR abs/1803.07422 (2018). http://arxiv.org/abs/1803.07422
3. Efros, A.A., Freeman, W.T.: Image quilting for texture synthesis and transfer. In: Proceedings of the 28th Annual Conference on Computer Graphics and Interactive Techniques, SIGGRAPH 2001, pp. 341–346. ACM, New York, NY, USA (2001)
4. Gatys, L.A., Ecker, A.S., Bethge, M.: Texture synthesis using convolutional neural networks. In: Proceedings of the 28th International Conference on Neural Information Processing Systems - Volume 1, NIPS 2015, pp. 262–270. MIT Press, Cambridge, MA, USA (2015)
5. Goodfellow, I.: NIPS 2016 Tutorial: Generative Adversarial Networks (2016)
6. Goodfellow, I., et al.: Generative adversarial nets. In: Ghahramani, Z., Welling, M., Cortes, C., Lawrence, N.D., Weinberger, K.Q. (eds.) Advances in Neural Information Processing Systems 27, pp. 2672–2680. Curran Associates, Inc. (2014)
7. He, K., Zhang, X., Ren, S., Sun, J.: Deep residual learning for image recognition. In: 2016 IEEE Conference on Computer Vision and Pattern Recognition, CVPR 2016, Las Vegas, NV, USA, 27–30 June 2016, pp. 770–778 (2016)
8. Iizuka, S., Simo-Serra, E., Ishikawa, H.: Globally and locally consistent image completion. ACM Trans. Graph. **36**(4), 107:1–107:14 (2017)
9. Isola, P., Zhu, J.Y., Zhou, T., Efros, A.A.: Image-to-image translation with conditional adversarial networks. In: 2017 IEEE Conference on Computer Vision and Pattern Recognition (CVPR), pp. 5967–5976 (2017)
10. Jetchev, N., Bergmann, U., Vollgraf, R.: Texture synthesis with spatial generative adversarial networks. CoRR abs/1611.08207 (2016)
11. Mirza, M., Osindero, S.: Conditional generative adversarial nets. CoRR abs/1411.1784 (2014)
12. Pathak, D., Krähenbühl, P., Donahue, J., Darrell, T., Efros, A.: Context encoders: feature learning by inpainting (2016)
13. Portilla, J., Simoncelli, E.P.: A parametric texture model based on joint statistics of complex wavelet coefficients. Int. J. Comput. Vision **40**(1), 49–70 (2000)
14. Radford, A., Metz, L., Chintala, S.: Unsupervised representation learning with deep convolutional generative adversarial networks abs/1511.06434 (2016)
15. Radim Tyleček, R.Š.: Spatial pattern templates for recognition of objects with regular structure. In: Proceedings GCPR, Saarbrucken, Germany (2013)
16. Ratner, A.J., Ehrenberg, H., Hussain, Z., Dunnmon, J., Ré, C.: Learning to compose domain-specific transformations for data augmentation. In: Advances in Neural Information Processing Systems, pp. 3236–3246 (2017)

Automated Mechanical Multi-sensorial Scanning

Vaia Rousopoulou$^{(\boxtimes)}$, Konstantinos Papachristou, Nikolaos Dimitriou,
Anastasios Drosou, and Dimitrios Tzovaras

Information Technologies Institute Centre for Research and Technology Hellas,
Thessaloniki, Greece
{vrousop,kostas.papachristou,nikdim,drosou,Dimitrios.Tzovaras}@iti.gr

Abstract. The 3D reconstruction of Cultural Heritage objects is a significant and advantageous technology for conservators and restorers. It contributes to the proper documentation of CH items, allows researchers, scholars and the general public to better manipulate and understand CH objects and gives the opportunity for remote and enhanced on-site experiences through virtual museums or even personal digital collections. The latest technological advances in computer vision in conjunction with robotics facilitate the development of automated and optimal solutions for digitizing complicated artifacts. In this direction, the current study presents an integrated, portable solution based on a modular architecture, for accurate multi-sensorial 3D scanning via a dedicated motorized mechanical arm and efficient automatic 3D reconstruction of a big variety of cultural heritage assets even in situ. The system is composed of a customized 3D reconstruction module, an automated motion planning module and a physical positioning system built by combining a mechanical arm and a rotary table. The key strength of the proposed system is that it is a cost-effective and time-saving solution applying computer vision and robotic technologies in order to serve Cultural Heritage preservation.

Keywords: 3D reconstruction · Texture mapping · Motion planning

1 Introduction

Cultural heritage is the bridge between the past and the future bearing the sense of history in the present. A plethora of cultural items, ranging from works of art and archaeological findings to buildings and historical monuments, need to be preserved and be accessible to public as they play essential role in society and connect the future generation with its identity. The induction of computer vision technologies in the field of Cultural Heritage is a major contribution towards the preservation of art in future centuries. Computer vision applications can actuate the effective documentation of the items in terms of not only their current state and condition but also the estimation of their deterioration and the prevention of

© Springer Nature Switzerland AG 2019
D. Tzovaras et al. (Eds.): ICVS 2019, LNCS 11754, pp. 433–442, 2019.
https://doi.org/10.1007/978-3-030-34995-0_39

any harmful effects, so that all this information is easily accessible to the public but more importantly to the researchers of this field, towards the stimulation of research and the selection of the optimum methodologies of item restoration and conservation.

The 3D representation is the most complete way to represent the whole structure of an object. However, its is important to introduce new technologies in order to exceed the methods of manual 3D reconstruction and find more time-saving, portable and automated solutions able to manipulate even the most complicated objects. Meanwhile, the use of robotic arms in various fields of application is rising progressively. Undoubtedly, computer vision technologies are inextricably linked with robotics, as they are enabled via cameras mounted on robots.

In the context of utilizing computer vision and robotic technology for the sake of cultural heritage, a mechanical arm application was developed. The proposed system aims to provide an integrated, portable solution based on a modular architecture, for accurate multi-sensorial 3D scanning, via a dedicated motorized mechanical arm. The proposed system combines computer vision techniques with a robotic application supporting 3D model reconstruction and multi-sensorial scanning. It is a cost-effective solution, a useful tool to the preservation, conservation and restoration via automatic digitization of a wide variety of cultural heritage assets even at their demonstration sites.

2   Related Work

There is a multitude of approaches in literature pointing out scientific or technological challenges of the 3D reconstruction process in order to achieve realistic and quality results in the cultural heritage domain. With the advent of range imaging sensors there has been a substantial amount of work that uses computer vision techniques for 3D reconstruction in order to create digital replicas of cultural heritage (CH) asset items. Indicatively, in this category fall the works of [1] and [2] that propose pipelines for the 3D reconstruction and inspect the most relevant methods and technologies classified in four main stages that compose a full 3D reconstruction pipeline with focus in CH: acquisition, registration, surface and texture reconstruction. The prevalent method for solving the registration problem is by implementing EM-type algorithms, with Iterative Closest Point (ICP) algorithm [3] and SoftAssign algorithm [4] being the most well-known solutions. Surface reconstruction can robustly recover fine detail from noisy real-world scans as described in [5]. Regarding texture reconstruction, a large-scale texturing approach is introduced in [6] through a comprehensive texturing framework for large 3D reconstructions with registered images. Furthermore, a model-based color correction method is presented in [7], which considers the distribution of metamers inside the metamer mismatch gamut in a perceptually uniform color space.

The work of Karaszewski, Sitnik and Bunsch [8] combines the technology of 3D digitization and robotics in the field of Cultural Heritage. They present a system for 3D digitization of cultural heritage objects which allows one to perform completely automated shape measurements. The system is composed of a view planning module that calculates the next scanner head position during measurements, an inverse kinematics with collision detection module, and a physical positioning system built by combining an industrial welding robot with a rotary table and a positioning column in order to extend the measurement volume. A rotary table is necessary to avoid problems with positioning the measurement head behind the object. It is a customizable solution as any hardware element can be exchanged easily because no specialized functions of any driver are used and the number of degrees of freedom (DOFs) of the kinematic chain can be easily altered to perform measurements of objects of various dimensions.

3 Motivation and Contribution

This research has been conducted in the context of Scan4Reco project and aims to offer to the scientific community a modular, customizable and cost-efficient solution for the digitization of a plethora of cultural objects, even at their demonstration sites. Various state-of-the-art methodologies where combined in order to achieve an accurate 3D reconstruction pipeline and a mechanically-driven arm control. Existing research activities are enhanced such as evaluation and restoration of cultural heritage objects, 3D scanning and modelling of art objects. This work intends to strongly facilitate efforts to bring heritage to the community through digitization. It is an efficient application to museums and galleries in terms of making the conservation procedures more accessible and effortless.

4 System Implementation

The system architecture is illustrated in Fig. 1 and each individual module is described in the next sections.

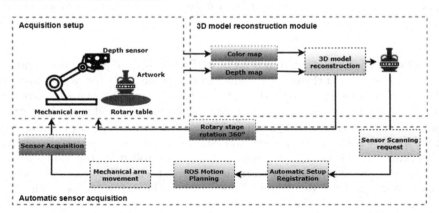

Fig. 1. The system architecture overview including the HW components (green), the reconstruction component (red) and the motion planning module (grey). (Color figure online)

A depth-sensor mounted on a mechanical arm is used to extract color and depth maps from an artwork placed on a rotary table. Next, a 3D model is extracted by a 3D model reconstruction module. Once the 3D model is extracted, a sensor scanning for a specific region of interest on the 3D artwork model is requested. The system setup is registered automatically so as to fit the scanning sensor capabilities (i.e. the artwork object approaches the mechanical arm in a reachable distance) and a motion plan is calculated by the Robotic Operating System. The end-effector of the mechanical arm is moved to the selected point on the actual artwork and a measurement is performed. Once the 3D model is created, various sensor measurements can be performed and repeated as the 3D reconstruction module is independent of the automatic scanning process.

4.1   Acquisition Setup

The proposed system supports automated 3D model digitization and accurate multi-sensorial scanning. The hardware components of the systems are a mechanical arm, a rotary table and various scanning sensors. The setup is presented in Fig. 2. The mechanical positioning system is composed of a 4-DoF mechanical

Fig. 2. The system setup 3D model in the left figure and the real system in the right figure.

arm and a rotary table located in front of it. The artwork object is settled on the rotary table and a depth-sensor is mounted on the end-effector of the mechanical arm. Two PCs are deployed for the software module. The one supports Windows OS and is responsible for the 3D reconstruction module and the other uses the Robotic Operating System in Linux OS for the motion planning software module. Various sensor devices can be mounted on the positioning system's mechanical arm. All hardware components are controlled by the Windows OS PC. The 3D reconstruction procedure is performed with a low-cost commercial depth sensor, the Intel RealSense SR300 device. The depth sensor is mounted on the mechanical arm. The artwork is rotated by a controlled rotary stage with a specific rotation step (e.g., 5°) performing a complete rotation (i.e. 360°).

4.2   3D Model Reconstruction Module

Regarding the 3D reconstruction module, an artwork is initially rotated (360°
rotation) by the controlled rotary stage and is recorded by the depth sensor,
mounted on the mechanical arm, to produce a series of consecutive depth and
color maps. The extracted depth maps are processed so as to discard areas
based on surface normal information. All the recorded views are transformed in
the same coordinate system using the kinematics of the rotary stage.

Then, surface reconstruction [9] is performed on the accumulated point clouds
in order to extract a watertight surface of the object followed by texture mapping
performed on the meshed 3D model so as to reconstruct the texture of the
artwork object. In order to detect areas where the low-cost camera is usually
inaccurate because depth changes abruptly and affects the alignment and surface
reconstruction negatively, [10] we make use of surface normal information. In
particular, for each 3D point corresponding to depth pixel, its normal vector is
estimated as a vector orthogonal to the point cloud of the stage, consisting of
a set of nearest neighbors of the 3D point, using least squares. The 3D point
is removed if the angle between the sensor ray direction and the normal to
the point is bigger than a threshold. When the rotary stage rotates, the input
point cloud is transformed in the same coordinate system. The Poisson Surface
Reconstruction (PSR) algorithm [5] has been incorporated in our 3D scanning
procedure in order to reconstruct the 3D surfaces from the accumulated point
clouds.

After surface reconstruction, texture mapping [11] from a set of color images
on the 3D reconstructed model is performed in order to reconstruct the texture
of the artwork object. The rationale of this step is to find the view where the
image plane of the camera is closer to being parallel to each 3D object's face
and consequently more accurate color information is obtained (i.e. perspective
distortion is minimized in the texture-to-surface mapping and the projections
area is maximized [12]). Texture reconstruction obtained from optimal camera
view selection, as described above, contain many color discontinuities between
patches from different camera views. In order to achieve a smooth color transition
at seams [6], color correction is applied on each vertex belonging at seams. The
optimal correction colours for all vertices at seams are interpolated using their
neighbouring vertices. In order to obtain extra information, apart from colour
images captured by the depth imaging sensor a set of UV-VIS samples obtained
by UV-VIS spectrometer along with corresponding depth sensor RGB responses
are used to compute a mapping utilizing polynomial regression techniques [7].

The results of each step of the 3D reconstruction procedure tested on a
bronze panel are presented in Fig. 3. Specifically, the recorded 72 consecutive
point clouds by the depth sensor have been aligned in the same coordinate system (Fig. 3(2nd row)) using the kinematics of the rotary stage. A watertight
surface (Fig. 3(3rd row)) of the object is extracted then, by performing the Poisson Surface Reconstruction on the accumulated point clouds. Finally, four color
images – corresponding to rotation stage's angles 0°, 90°, 180° and 270° respectively - were used in order to reconstruct the texture of the reference artwork

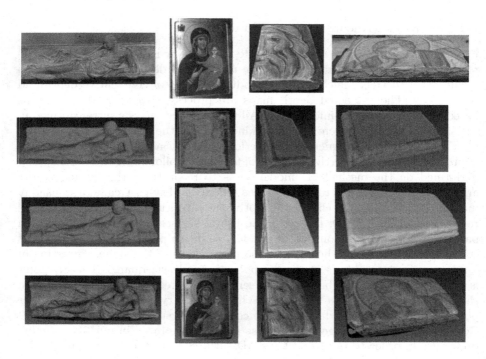

Fig. 3. Artwork samples on the three stages of 3D reconstruction process: one bronze panel (1st column), one paintings on wooded surface (2nd column) and two paintings on plaster surface (3rd and 4th columns).

sample, as illustrated in (Fig. 3(4th row)). As can be seen, our approach achieves a reliable geometric and chromatic representation of the artwork for all the samples.

4.3   Automatic Sensor Acquisition

Sensor Scanning Request. As soon as the 3D model reconstruction procedure is complete, the user can view the model and use the scanning modalities of the mechanical arm in order to scan areas on the artwork. A sensor scanning request is the selection of a specific point on the 3D model of the object to be scanned. This function motivates the process of automatic setup registration and motion planning.

Automatic Setup Registration. In reconstruction phase, the cultural heritage object is placed away from the mechanical arm so as to satisfy the captures of the depth sensor. On the contrary, in the scanning phase the object is brought close to the mechanical arm so as the sensors can reach it. An AprilTag system [13] is used to detect the position of the artwork in reference to the mechanical arm base coordinate system in both states. The system uses a Kinect v2 sensor, fixed across the object to register the actual environment to the simulation setup and locate the exact position of the object. Given the 3D reconstruction position and the desired position of the sensor acquisition, the automatic registration system calculates the translation of the object in these two steps.

Motion Planning. The mechanical arm controllers are handled by the Windows rack PC, the motion planning is performed by the Robotic Operating System (ROS) integrated on the Linux PC and send back to Windows PC and the two systems communicate through TCP Sockets. A motion planning request is sent the moment that the user selects a point to be scanned on the 3D model that is viewed on the Windows platform. The request is a Json object that includes the 3D mesh of the object, the selected point, the current state of the mechanical arm joints and the state of the mechanical arm joints at the 3D reconstruction phase. Concurrently, the object location is detected by the AprilTag system. The 3D mesh model is constructed in reference with the depth sensor coordinate system mounted on the mechanical arm, as well as the target point. The 3D reconstruction state of the mechanical arm joints is used to find the transform of the depth sensor coordinate system to the mechanical arm base system. A translation and rotation matrix is calculated by ROS and the 3D mesh along with the target point are transformed into the arm's coordinate system.

The 3D mesh is inserted in the ROS simulation tool, RVIZ. Next, we calculate the normal of the object surface at the selected point to find the arm' s end effector orientation towards the object. The target point coordinates and the orientation compose a target pose that the mechanical arm is going to reach. The motion planning and movement of the mechanical arm is executed by the ROS tool, MoveIt!. A collision-free path is extracted by the integrated Open Motion Planning Library (OMPL) [14] of MoveIt! Tool. From the library of motion planning algorithms provided by OMPL, various planners have been tested and the Expansive Space Trees (EST) algorithm was chosen as it computes the desired motion plans faster. The algorithm computes a collision-free motion plan for the mechanical arm, which is a set of movements for each of the arm's joints, namely a trajectory sequence. Through TCP sockets, the response is send back to windows PC and the motion planning is executed by the mechanical arm controllers. The targeted point is reached by the arm's end-effector and the measurement is taking place by the selected sensor.

5   Experimental Results

In order to evaluate the accuracy of the proposed method, the coarse 3D modelling was performed on four CH objects: two paintings and two metallic objects. We selected 5 points pairs (see the red lines in Fig. 4) and the Euclidean distances on the physical artworks and the corresponding generated 3D models were calculated. The motion planning module was also tested in these four cultural heritage objects. The Euclidean distance between the end-effector position and

Fig. 4. The accuracy measurements on four artwork objects. The red arrows indicate the regions of measurement for the 3D reconstruction module and the yellow bullets indicate the points measured for the motion planning module. (Color figure online)

Table 1. Average accuracy measurements.

CH item	3D reconstruction (cm)	Position (cm)
Archangel Michael's icon	0.31	0.84
St. Demetrius 's icon	0.18	0.64
Bronze replica of a high-relief	0.14	0.62
Pescatorello statue	042	0.82

the selected point was calculated. The procedure was repeated for five different points on each object (see the yellow bullets in Fig. 4) and the Euclidean distances were measured and are presented below. The results of the 3D reconstruction and motion planning accuracy measurements are presented in Table 1. The 3D reconstruction accuracy varies from 0.14 cm to 0.42 cm, which is quite below the 3D reconstruction accuracy threshold, i.e. 1 cm, confirming that a reliable representation of the artworks' surface is retrieved. Furthermore, the position accuracy of the first object varies from 0.4 to 1.4 cm, with an average at 0.84 cm, which is quite below the threshold, i.e. 2 cm, implying that the measurements are satisfactory. It is worth to mention that the position accuracy is affected by the 3D reconstruction procedure, the more accurate the 3d model is the more accurate is the position of the mechanical arm on the selected point.

6   Conclusions

In this paper an automated mechanical multi-sensorial scanning system is presented, providing an integrated, portable and low-cost solution based on a modular architecture, for accurate multi-sensorial 3D scanning via a dedicated motorized mechanical arm and efficient automatic digitization of a big variety of cultural/heritage assets even in situ. It is composed of a customize 3D reconstruction module, an automated motion planning module and a physical positioning system built by combining a mechanical arm and a rotary table. The rotary table is necessary to perform the 3D reconstruction, as well as to avoid problems with positioning the measurement head behind the object. The 3D reconstruction procedure is performed with a low-cost commercial depth sensor while various sensors are available for surface measurements.

The system asset is the high accuracy established by the capability of repeating the scanning process in a particular region of interest on a Cultural Heritage item and resulting in the same measurement. Overall, this representation of the system architecture denote a modular aspect of the system. In particular, the aforementioned work presents an architecture that inverse kinematics and collision avoidance are separated from the robot drivers and create a custom, software-based, extensible method. The proposed 3D model reconstruction module can reserve satisfactory results providing realistic 3D models and the motion planning module along with the rotary table can achieve automated scanning of CH assets. Furthermore, the strength of the proposed system lies on the fact that various sensor devices can be mounted on the mechanical arm. The system is customizable enough as it can be enhanced with a higher resolution camera and even more sensor probes can be mounted on the mechanical arm.

Acknowledgments. This work has been partially supported by the European Commission through project Scan4Reco funded by the European Union H2020 programme under Grant Agreement no. 665091. The opinions expressed in this paper are those of the authors and do not necessarily reflect the views of the European Commission.

References

1. Vrubel, A., Bellon, O.R.P., Silva, L.: A 3D reconstruction pipeline for digital preservation. In: 2009 IEEE Conference on Computer Vision and Pattern Recognition, pp. 2687–2694 (2009). https://doi.org/10.1109/CVPR.2009.5206586
2. Gomes, L., Bellon, O.R.P., Silva, L.: 3D reconstruction methods for digital preservation of cultural heritage: a survey. Pattern Recogn. Lett. **50**(Depth Image Analysis), 3–14 (2014). https://doi.org/10.1016/j.patrec.2014.03.023. http://www.sciencedirect.com/science/article/pii/S0167865514001032
3. Combès, B., Prima, S.: An efficient EM-ICP algorithm for symmetric consistent non-linear registration of point sets. In: Jiang, T., Navab, N., Pluim, J.P.W., Viergever, M.A. (eds.) MICCAI 2010. LNCS, vol. 6362, pp. 594–601. Springer, Heidelberg (2010). https://doi.org/10.1007/978-3-642-15745-5_73
4. Kent, J.T., Mardia, K.V., Taylor, C.C.: An EM interpretation of the Softassign algorithm for alignment problems. mij **1**, 0 (2010)

5. Kazhdan, M., Bolitho, M., Hoppe, H.: Poisson surface reconstruction. In: Proceedings of the Fourth Eurographics Symposium on Geometry Processing, SGP 2006, pp. 61–70. Eurographics Association, Aire-la-Ville, Switzerland (2006). http://dl.acm.org/citation.cfm?id=1281957.1281965
6. Waechter, M., Moehrle, N., Goesele, M.: Let there be color! Large-scale texturing of 3D reconstructions. In: Fleet, D., Pajdla, T., Schiele, B., Tuytelaars, T. (eds.) ECCV 2014. LNCS, vol. 8693, pp. 836–850. Springer, Cham (2014). https://doi.org/10.1007/978-3-319-10602-1_54
7. Urban, P., Grigat, R.-R.: Metamer density estimated color correction. SIViP **3**(2), 171 (2008). https://doi.org/10.1007/s11760-008-0069-0
8. Karaszewski, M., Sitnik, R., Bunsch, E.: On-line, collision-free positioning of a scanner during fully automated three-dimensional measurement of cultural heritage objects. Robot. Auton. Syst. **60**(9), 1205–1219 (2012). https://doi.org/10.1016/j.robot.2012.05.005. http://www.sciencedirect.com/science/article/pii/S0921889012000619
9. Kazhdan, M., Hoppe, H.: Screened Poisson surface reconstruction. ACM Trans. Graph. **32**(3), 291–2913 (2013). https://doi.org/10.1145/2487228.2487237
10. Khoshelham, K.: Accuracy analysis of kinect depth data. In: ISPRS Workshop Laser Scanning, vol. 38, p. W12 (2011)
11. Callieri, M., Dellepiane, M., Cignoni, P., Scopigno, R.: Processing sampled 3D data: reconstruction and visualization technologies. In: Stanco, F., Battiato, S., Gallo, G. (eds.) Digital Imaging for Cultural Heritage Preservation: Analysis, Restoration and Reconstruction of Ancient Artworks, pp. 105–136. Taylor and Francis, Milton Park (2011)
12. Allene, C., Pons, J.-P., Keriven, R.: Seamless image-based texture atlases using multi-band blending. In: 2008 19th International Conference on Pattern Recognition, ICPR 2008, pp. 1–4. IEEE (2008)
13. Olson, E.: AprilTag: a robust and flexible visual fiducial system. In: 2011 IEEE International Conference on Robotics and Automation (ICRA), pp. 3400–3407. IEEE (2011)
14. Sucan, I.A., Moll, M., Kavraki, L.E.: The open motion planning library. IEEE Robot. Autom. Mag. **19**(4), 72–82 (2012)

Cognitive Vision Systems

Point Pair Feature Matching: Evaluating Methods to Detect Simple Shapes

Markus Ziegler[(✉)], Martin Rudorfer, Xaver Kroischke, Sebastian Krone, and Jörg Krüger

Technische Universität Berlin, PTZ 5, Pascalstr. 8-9, 10587 Berlin, Germany
m.f.z@t-online.de, martin.rudorfer@iat.tu-berlin.de

Abstract. A recent benchmark for 3D object detection and 6D pose estimation from RGB-D images shows the dominance of methods based on Point Pair Feature Matching (PPFM). Since its invention in 2010 several modifications have been proposed to cope with its weaknesses, which are computational complexity, sensitivity to noise, and difficulties in the detection of geometrically simple objects with planar surfaces and rotational symmetries. In this work we focus on the latter. We present a novel approach to automatically detect rotational symmetries by matching the object model to itself. Furthermore, we adapt methods for pose verification and use more discriminative features which incorporate global information into the Point Pair Feature. We also examine the effects of other, already existing extensions by testing them on our specialized dataset for geometrically primitive objects. Results show that particularly our handling of symmetries and the augmented features are able to boost recognition rates.

Keywords: Object detection · Pose estimation · Point Pair Features

1 Introduction

A recent benchmark shows that the original Point Pair Feature Matching (PPFM) algorithm for object detection and 6D pose estimation of Drost et al. [6] still remains competitive even today [10]. Nevertheless, it has seen many modifications over the last years, which attempted to fix some of the algorithm's weaknesses. The most commonly stated drawbacks are: Computational complexity [3,8,9,12,16], problems with noisy scenes [4,9,11,12] and difficulties detecting simple-shaped objects with planar surface patches and rotational symmetries [2,4,5,12,16].

In industrial applications these simple-shaped objects are of particular interest. Workpieces, product parts, tools, etc. tend to have simple geometries, or consist of multiple simple shapes. Those objects are dominated by edges, planes and constant curvature surfaces and often possess discrete or continuous rotational symmetries, also they often lack a distinctive texture.

© Springer Nature Switzerland AG 2019
D. Tzovaras et al. (Eds.): ICVS 2019, LNCS 11754, pp. 445–456, 2019.
https://doi.org/10.1007/978-3-030-34995-0_40

In this paper we aim to extend the original PPFM by modifications which improve the detection of such geometrically simple shapes. We introduce a novel method to detect rotational symmetries. We evaluate and enhance existing methods of Hinterstoisser et al. [9] to handle planar surfaces and sensor noise. Furthermore, we propose a modification of Kim and Medioni's Visibility Context feature [12] to increase the distinctiveness of features for simple-shaped objects. For the evaluation of those methods we also present our 'Geometric Primitives' dataset with geometrically simple model objects.

The paper is structured as follows: First we recap the original PPFM algorithm by [6], which we will refer to as Drost's PPFM. We then review extensions and improvements of the method in Sect. 3. In Sect. 4 we describe our modifications, which are particularly suited for objects with simple geometries, and Sect. 5 presents our evaluation procedure and its results.

2   Point Pair Feature Matching

Drost's PPFM relies on Point Pair Features (PPFs), which describe simple geometric relations between a reference point $\mathbf{p}_r$ and a target point $\mathbf{p}_t$. It is formally described as $\mathbf{PPF}(\mathbf{p}_r, \mathbf{p}_t) = (\|\mathbf{d}\|_2, \angle(\mathbf{n}_{\mathbf{p}_r}, \mathbf{d}), \angle(\mathbf{n}_{\mathbf{p}_t}, \mathbf{d}), \angle(\mathbf{n}_{\mathbf{p}_r}, \mathbf{n}_{\mathbf{p}_t}))$, where $\mathbf{d}$ is the vector from $\mathbf{p}_r$ to $\mathbf{p}_t$ and $\mathbf{n}_p$ the surface normal of point $\mathbf{p}$.

To create the model description the PPF is computed for all permutations of model point pairs. The resulting features are discretized and all PPFs are organized in a hash table. The hash key is the discretized feature itself and the value encodes the feature's pose with respect to the model. This is achieved in an efficient manner by storing the index of $\mathbf{p}_r$ and an angle α that can be derived from a standardized set of transformations. Note that a certain feature can occur multiple times in a model and consequently one hash key can have multiple values.

During matching, every $\mathbf{s}_r{}^{\text{th}}$ scene point is used as reference point with all other points as target points to compute the scene PPFs. The detected scene PPFs are then matched to the corresponding model PPFs and, using the model's hash table, are mapped to a model reference point and an angle, i.e. a pose suggestion. The suggestions are accumulated in one Hough-like voting scheme for each scene reference point. The maxima are extracted along with their number of votes as weight, to form the raw pose hypotheses. These are subsequently clustered and the cluster with the highest accumulated weight is the resulting pose hypothesis.

3   Related Works

Several works tried to improve on Drost's PPFM and developed extensions that cover the following issues and aspects: detection of simple-shaped objects with planar surfaces and rotational symmetries, noisy scenes, computational complexity, and post-processing, i.e. pose refinement and pose verification. Due to the focus of this work we will go into more detail regarding the first issue.

However, since many extensions have side-effects and cannot unambiguously be assigned to one of the four aspects, we also briefly summarize related works on the other issues.

3.1 Planar Surfaces and Rotational Symmetries

For rotationally symmetric objects, and particularly for objects with planar or constant-curvature surfaces, there will be many different point pairs which yield the same feature. In the model description, the hash key corresponding to this feature will thus contain many different values and suggest many different object poses, ultimately rendering the feature indistinctive. This not only weakens recognition but slows it down as well. Even if the object does not exhibit these properties, recognition can be affected by planar surfaces or other simple shapes in the scene. Again, one feature appears multiple times, and we assume it coincidentally matches a model feature. Hence all occurrences of the feature in the scene will cast votes and populate the voting space. This adds noise to the voting space and might skew the voting to yield a high weight for a false hypothesis.

Some authors have addressed that fact by augmenting the PPF with additional features and thus improving its discriminative power. [12] incorporate information about the object's dimensions and the surface's curvature, which is similar to our approach (see Sect. 4.1). [4] and [5] both try to incorporate edge information, as these can be very helpful for simple-shaped objects. It has also been proposed to add color information to the feature [2]. In 2014, [17] trained a SVM to identify the most distinctive features of an object, but this has to be re-trained for each new detection environment.

However, a more distinctive feature does not resolve the issue for rotationally symmetric objects. [7] presented an approach to handle rotational symmetries that yields smaller model descriptions and mainly decreased the average matching time for those objects. However, their approach allowed handling only one axis of symmetry, plus, it had to be known beforehand and could not be detected automatically (in contrast to our approach in Sect. 4.2).

Further ideas to cope with planar surfaces are: Reducing the number of votes a feature can cast [9], adjusting the clustering to limit the weight introduced by ambiguous features [9], or using only a subset of the point cloud which excludes most points that lie on planar surfaces [12,18].

3.2 Noisy Scenes

Another weakness of PPFM is its sensitivity to sensor noise, particularly the estimation of surface normals is affected [11]. In consequence, the same point pairs will generate slightly different features. Promising countermeasures are to also consider adjacent features for both model and/or scene point pairs, and to vote for adjacent poses during matching [9,19], although it comes with additional computational cost.

3.3   Computational Complexity

There are several approaches to lower the computational burden. A consensus is that casting fewer votes will speed up matching, which can be achieved by more discriminative features [2,12], handling rotational symmetries [7], and some other of the already mentioned extensions. However, the computational cost is mainly dominated by the number of scene points and ultimately the number of scene PPFs to be computed and matched. A speed-up can be achieved by segmenting the scene prior to matching [1], or by smart downsampling strategies [12,18,19]. The number of PPFs to be computed is directly reduced e.g. by [9], who introduce voting spheres and disregard all potential target points outside the sphere.

3.4   Post-processing

A final step is typically a pose-refinement using ICP, however, some authors skip it, e.g. [1] use an additional accumulator array during matching to keep track of the continuous rotation angles. They argue that in combination with a subpixel-maximization this already gives satisfactory pose accuracy. Among the resulting list of ranked pose hypotheses, there are typically many false positives. A verification method has been proposed by [15], which, for a limited number of hypotheses, evaluates the numbers of scene points conflicting with the pose vs. scene points that support the pose hypothesis. Many authors deploy verification methods based on this approach [1,9,12,19].

4   Methodology

We describe all extensions along the processing pipeline of the algorithm, based on Drost's original PPFM [6].

4.1   Augmented Feature

In order to make the PPF more discriminative (particularly for geometrically primitive objects) we add further features similar to the visibility context of [12], which adds global information to the PPF. We aim to describe the target point's location within the model. This is done by casting four rays originating from the target point $\mathbf{p}_t$ but pointing into orthogonal directions. The directions are defined by the connecting vector $\mathbf{d}$ and the reference point's normal $\mathbf{n}_{\mathrm{p_r}}$, as depicted in Fig. 1. We now imagine traveling along each ray: Either we are traveling on a surface, and we record the distance it takes to leave the surface, or we are traveling in unoccupied space, and then we measure the distance it takes to intersect with the nearest surface. If the distance is larger than the diameter of the model (or larger than the current voting sphere, see Sect. 4.3), we ignore the feature. See Fig. 1 for a distinction of the possible cases. We add the four acquired distances to the feature vector to yield the augmented feature
$$\mathbf{PPF}_{\mathrm{aug}}(\mathbf{p}_r, \mathbf{p}_t) = \left(\mathbf{PPF}(\mathbf{p}_r, \mathbf{p}_t), \|\mathbf{v}_{\mathrm{out},1}\|_2, \|\mathbf{v}_{\mathrm{out},2}\|_2, \|\mathbf{v}_{\mathrm{out},3}\|_2, \|\mathbf{v}_{\mathrm{out},4}\|_2\right).$$

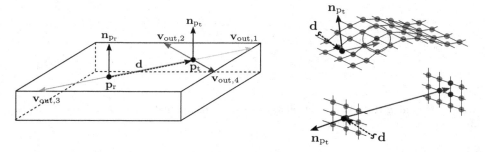

Fig. 1. Left: The four vectors $\mathbf{v}_{\text{out},i}$ add global information to the PPF by describing the position of the target point on the planar surface of the object (image based on [12]). When casting a ray to measure the distance of $\mathbf{v}_{\text{out},i}$, we can distinguish three cases: Left: If reference point and target point lie on the same plane, the rays will follow the planar surface. Right-top: For curved surfaces, the ray will follow the surface until it does not intersect the occupied voxels anymore. It intersects surfaces voxels which are farther away, but this is disregarded as the gap of unoccupied voxels along the ray is larger than the tolerated gap size indicated by the circle. Right-bottom: The normals of reference and target point have different directions, therefore the rays will be cast towards the inside or outside of the object, and the distance to the nearest surface will be measured.

In contrast to [12], who use rendered depth images to compute the visibility context, we apply this method directly to the point cloud to learn features independent of the view. We use an octree with defined voxel dimensions to distinguish occupied and unoccupied space, and ray tracing to determine the intersected voxels. Parameters to tune are the voxel size and the gap size tolerance. The latter defines the tolerated distance between two occupied voxels along a ray. Best results can be obtained with model-specific tuning, however, a voxel-diameter in the range of the point cloud's downsampling distance is a good initial choice.

4.2 Model Creation

During model creation, we automatically detect rotational symmetries of the object and are thus able to deploy the rotational symmetry handling methods by [7] if needed. The process is illustrated in Fig. 2. We detect rotational symmetries by matching the object to itself. The detected pose hypotheses are regarded as equivalent poses, if their weight exceeds a certain threshold. We use the resulting set of equivalent object poses $\mathcal{T}_{\text{eq}}$, which are basically rotations around the axis of symmetry, to merge redundant entries in the model hash table analogous to [7]. In contrast to [7], we do not need any prior knowledge about the axis of symmetry. Furthermore, we can handle discrete symmetries as well as multiple axes of symmetry (e.g. in a cuboid).

Furthermore, as in [9], we also learn adjacent PPFs for more robustness to sensor noise. The original PPF has four elements and therefore $3^4 - 1 = 80$ adja-

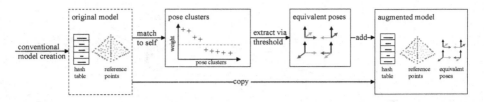

Fig. 2. Matching the object to itself allows us to automatically detect ambiguous poses due to rotational symmetries.

cent PPFs, which we also include in the model hash table. However, we do not apply this enhancement in combination with our augmented PPF, because the PPF then has eight elements and therefore $3^8 - 1 = 6560$ adjacent PPFs, leading to huge amounts of memory consumption for the model description (model with 900 points > 23 GBs).

4.3 Matching

During matching we use four different modifications: Voting spheres, adjacent rotation voting, and flag array, which were all introduced by [9], and the angle refinement of [1].

A voting sphere limits the number of PPF calculations; only those target points which are inside the sphere centered around the reference point are taken into account. Because spherical regions might be bad approximations for the extents of some objects, particularly elongated ones, we use two voting spheres with different diameters. We compute these diameters as described in [9]. To counteract sensor noise, we establish voting for the adjacent rotation angles $\alpha_{1,2} = \alpha \pm \Delta_\alpha$ in the accumulator array, with Δ_α being the quantization step for α. We also use the flag array of [9], which suppresses multiple votes for the same pose by different point pairs which are discretized to the same feature. This is supposed to further reduce the negative effect of planar surfaces.

Lastly, we use a second accumulator array that averages the continuous pose-rotations α as proposed in [1], to get more accurate poses. This second accumulator array also takes into account votes for adjacent rotation angles.

4.4 Pose Clustering

We use two approaches. Our standard approach is based on [4] with poses being processed top-down, forming clusters of similar poses. The similarity check optionally uses the known symmetries $\mathbf{T}_i \in \mathcal{T}_{\mathrm{eq}}$ for each pose as in [7].

Our other clustering approach reduces the negative influence of large planar faces and optionally handles rotational symmetries. The poses are processed bottom-up, i.e. those with the lowest weights are clustered first. Also as in [9], we keep track of the model reference point associated with a pose and only allow this pose to be added to a cluster if the cluster does not already contain a pose with the same model reference point. Otherwise, the pose creates a new cluster. This prevents that several scene points inside a cluster are associated with the same model point, an effect that planar surfaces introduce. For rotationally symmetric models, we then traverse the clusters in descending order and check for each cluster $\mathbf{A}$ if one lower-weighted pose cluster $\mathbf{B}$ is equivalent to $\mathbf{A}$ by applying the rotations $\mathbf{T}_i \in \mathcal{T}_{\mathrm{eq}}$ to $\mathbf{B}$. In case of a match both clusters are merged. By using every $\mathbf{T}_i \in \mathcal{T}_{\mathrm{eq}}$ only once for each $\mathbf{A}$, we uphold the idea to reduce the influence of planar scene surfaces.

4.5   Post-processing

We use a pose verification system which is inspired by [15]. The model points of the estimated pose are analyzed regarding their ability to describe the scene points. Those model points that are close to scene points vote for the pose hypothesis while those that occlude scene points vote against the hypothesis. In addition to [15], we also require the normals of model and scene points to have a similar orientation. We process the n best poses (up to 16) of each voting sphere separately. We do not use ICP pose refinement, as the angle refinement of [1] we use during matching already improves the accuracy of our poses.

5   Experiments

In the following we first introduce the datasets and error metrics that we used in our experiments, and subsequently present and discuss the achieved results.

5.1   Datasets

We use the Linemod Occlusion dataset (LM-O) by [13], with the subset of 200 scenes which were used in the SIXD Challenge 2017 [10], except for the egg-box, glue and holepuncher models. Furthermore, to evaluate our methods with respect to detecting simple shapes, we use our own synthetic dataset 'Geometric Primitives' (GP), whose partially occluded objects exhibit the problematic characteristics: Simple shapes with rotational symmetries and large planar surface patches (see Fig. 3). The 20 scenes of the dataset contain two instances of each of the five objects randomly positioned in a bin. The scenes were created by simulating a Kinect v2 with noise equivalent to 1.1% to 1.8% of the model diameter. The dataset is provided along with our implementation at gitlab[1].

5.2   Metrics

To determine correct matches, we avoid the rotation and translation errors used in [6], because they are erroneous for objects with ambiguous views [10]. Instead, for the GP dataset we use the adaptation of the average distance metric for

---

[1] https://gitlab.com/point_pair_feature_matching.

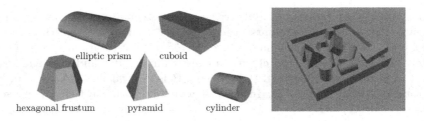

Fig. 3. The simple but challenging models of 'Geometric Primitives' dataset.

symmetric objects e_{ADI} from [8]. For the LM-O dataset we use the visual surface discrepancy e_{VSD} from [10], which correctly handles symmetries and also other ambiguous views. We used the same parameters for computing e_{ADI} and e_{VSD} as in [8] and [10], respectively.

5.3 Results

We perform several different experiments: First, we inspect the effect of our own approach to handle rotational symmetries. Second, we investigate the effects of various combinations of the proposed extensions. While these two experiments focus on detection of the geometric primitives, our third experiment is aimed for comparison with the state of the art and therefore conducted on the LM-O dataset, albeit it does not contain primitive shapes and therefore does not emphasize the strengths of our method. For all our experiments we chose 5% of the model's diameter for downsampling and as distance discretization step and used every 2nd and 5th scene point as reference point for GP and LM-O dataset, respectively.

Rotational Symmetry. We define two variants of handling symmetries: "RS cluster" and "RS cluster&hash". In both variants we automatically detect symmetries and consider equivalent poses during top-down clustering. However, in the second variant ("RS cluster&hash") we also make use of the equivalent poses during model creation to reduce the model hash table. We compare both variants to our baseline implementation, which is Drost's PPFM with angle refinement (see Sect. 4.3) and standard clustering. The results are displayed in Fig. 4. We can observe slightly improved recognition rates, particularly for the cylinder, which is the only object with a continuous rotational symmetry. Both our variants have almost identical recognition rates, but differ drastically in matching times: While considering equivalent poses only during clustering increases the computational effort, the reduction of hash table entries can greatly improve matching time. This is due to a combination of faster hash table look-ups, smaller voting spaces and reduced numbers of poses for clustering.

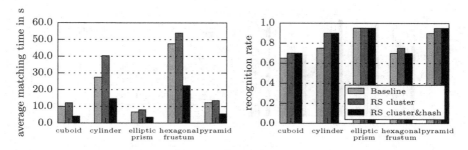

Fig. 4. The differences of handling rotational symmetries only during clustering vs. also during model creation are mainly reflected in the average matching time. Recognition rates are given in %, based on e_{ADI}.

Extension Packs. We made several experiments with various combinations of extension packs (see legend in Fig. 5) to examine their influence on the detection of our GP dataset. As baseline we use the most successful approach from the previous experiment, i.e. Drost's PPFM with angle refinement and full rotational symmetry handling. Figure 5 illustrates the results. Using the sampling and the noise pack alone interestingly can already resolve most of the false detections of the cuboid in the wall of the bin, which is a typical failure case as illustrated in Fig. 6. However, this combination negatively affects the recognition of other objects. Adding the planar surfaces pack only boosts the detection of the cuboid. However, detailed inspection of the results revealed that the higher recognition rate is actually due to a weakness of e_{ADI} error metric: Although two of the estimated poses of the cuboid are rotated by 90° with respect to the ground truth pose, as depicted in Fig. 6 on the right, they are still accepted as correct estimations. With the help of our pose verifier, these wrongly accepted poses could be dismissed, albeit in favor of other wrong estimations. The pose verifier also slightly improves recognition for the other objects. The best improvements are achieved by using the augmented PPF, which suggests that adding global information to the feature really is beneficial for the detection of simple-shaped objects. However, in combination with the other extension packs, particularly the noise pack, the computational complexity prohibitively increases, which is why we did not test further combinations with the augmented feature.

Comparison to State of the Art. Lastly, we check the performance of our implementation on the LM-O dataset. We use our baseline from the previous experiments, but deactivate the rotational symmetry detector as the dataset does not contain symmetric objects. Furthermore, we test a variant with the sampling, noise and planar surfaces packs. This combination is similar to the method of [9]. We compare both variants with recent results from the SIXD Challenge [10], in particular with the original Drost's PPFM [6] and the benchmark winner [19]. The latter is also based on [9], i.e. relatively similar to our variant. Table 1 shows the results. Our approaches perform significantly worse than

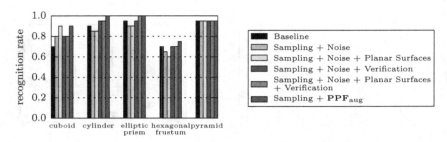

Fig. 5. Comparison of our combinations of extension packs for detecting simple-shaped objects. We report recognition rates in % based on e_{ADI}. Also note the explanation of each pack with the corresponding methods from Sect. 4.

Drost's PPFM. This can not only be explained with the different configurations used in [10] (we use the same as [19], Drost PPFM uses a finer downsampling and distance quantization step of 3% of the model diameter). Our speculation is that the software HALCON 13.0.2 [14], which was used to get the performance of Drost's PPFM, contains an enhanced version and not the original Drost PPFM. However, we can observe that our implementation of the extensions helps to raise the recognition rates particularly for the smaller objects (ape, cat, duck), and thus shows more benefits than for the GP dataset. The PPFM of [19] performs best. We therefore assume that their smart downsampling and more elaborate postprocessing are also very useful extensions.

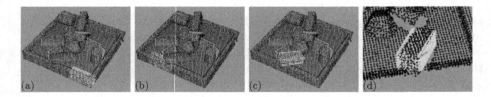

Fig. 6. Selected failure cases. (a)–(c) are the same scene processed with different variants of the algorithm: (a) baseline maximizes surface overlap and therefore detects the cuboid in the wall of the bin; (b) variant with sampling, noise, planar surfaces, and verifier also results in a wall detection, albeit a different one; (c) using the augmented feature avoids the wall detection, but instead leads to confusion of cuboid and elliptic prism. (d) The estimated pose is wrong by 90°, but the e_{ADI} is too permissive and wrongly accepts this pose as correct.

Table 1. Comparison of recognition rates (in %, based on e_{VSD}) on the 200 scenes of the LM-O dataset. We compare the benchmark-results in [10] with our baseline and our variant with the voting spheres and the noise and planar surfaces packs.

Model	Drost [6]	Vidal [19]	Our baseline	Our variant
Ape	62	66	17.7	37.7
Can	75	81	52.8	59.8
Cat	39	46	14.1	21.8
Driller	70	65	46.5	49.5
Duck	57	73	17.9	66.5
Average	60.6	66.2	31.1	48.1

6   Conclusion

We presented a new method for automatic detection and handling of rotational symmetries and showed that it is able to speed up the matching and also increase the recognition rates particularly for models with continuous rotational symmetries. We also integrated further extensions for sampling, noisy scenes or planar surfaces from [9]. Although these did not further improve the recognition of our geometric primitives, we confirmed that they are generally useful on the LM-O dataset. Our pose verification system could eliminate many false detections and slightly improved recognition rates. However, what improved the recognition of geometric primitives most was using the augmented features, which add global information to the PPF. A downside might be that both pose verifier and augmented features have parameters that must be tuned.

Further research could include a more thorough investigation on interactions of different extensions. It is for example unclear if the gained pose accuracy through angle refinement might be impaired by the adjacent rotation voting. Also, naively using the augmented feature in combination with computing adjacent features leads to a prohibitively high memory consumption.

References

1. Birdal, T., Ilic, S.: Point pair features based object detection and pose estimation revisited. In: Brown, M., Kosecka, J., Theobalt, C. (eds.) 2015 International Conference on 3D Vision, 3DV 2015, pp. 527–535. IEEE, Piscataway, NJ (2015)
2. Choi, C., Christensen, H.I.: 3D pose estimation of daily objects using an RGB-D camera. In: 2012 IEEE/RSJ International Conference on Intelligent Robots and Systems, pp. 3342–3349. IEEE (2012)
3. Choi, C., Christensen, H.I.: RGB-D object pose estimation in unstructured environments. Robot. Auton. Syst. **75**, 595–613 (2016)
4. Choi, C., Taguchi, Y., Tuzel, O., Liu, M.Y., Ramalingam, S.: Voting-based pose estimation for robotic assembly using a 3D sensor. In: IEEE International Conference on Robotics and Automation (ICRA), 2012, pp. 1724–1731. IEEE, Piscataway, NJ (2012)

5. Drost, B., Ilic, S.: 3D object detection and localization using multimodal point pair features. In: Second International Conference on 3D Imaging, Modeling, Processing, Visualization and Transmission (3DIMPVT), 2012, pp. 9–16. IEEE, Piscataway, NJ (2012)

6. Drost, B., Ulrich, M., Navab, N., Ilic, S.: Model globally, match locally: efficient and robust 3D object recognition. In: IEEE Conference on Computer Vision and Pattern Recognition (CVPR), 2010, pp. 998–1005. IEEE, Piscataway, NJ (2010)

7. de Figueiredo, R.P., Moreno, P., Bernardino, A.: Efficient pose estimation of rotationally symmetric objects. Neurocomputing **150**(Part A), 126–135 (2015)

8. Hinterstoisser, S., et al.: Model based training, detection and pose estimation of texture-less 3D objects in heavily cluttered scenes. In: Lee, K.M., Matsushita, Y., Rehg, J.M., Hu, Z. (eds.) ACCV 2012. LNCS, vol. 7724, pp. 548–562. Springer, Heidelberg (2013). https://doi.org/10.1007/978-3-642-37331-2_42

9. Hinterstoisser, S., Lepetit, V., Rajkumar, N., Konolige, K.: Going further with point pair features. In: Leibe, B., Matas, J., Sebe, N., Welling, M. (eds.) ECCV 2016. LNCS, vol. 9907, pp. 834–848. Springer, Cham (2016). https://doi.org/10.1007/978-3-319-46487-9_51

10. Hodaň, T., et al.: BOP: benchmark for 6D object pose estimation. In: Ferrari, V., Hebert, M., Sminchisescu, C., Weiss, Y. (eds.) ECCV 2018. LNCS, vol. 11214, pp. 19–35. Springer, Cham (2018). https://doi.org/10.1007/978-3-030-01249-6_2

11. Kiforenko, L., Drost, B., Tombari, F., Krüger, N., Buch, A.G.: A performance evaluation of point pair features. Comput. Vis. Image Underst. **166**, 66–80 (2017)

12. Kim, E., Medioni, G.: 3D object recognition in range images using visibility context. In: Staff, I. (ed.) 2011 IEEE/RSJ International Conference on Intelligent Robots and Systems, pp. 3800–3807. IEEE (2011)

13. Krull, A., Brachmann, E., Michel, F., Yang, M.Y., Gumhold, S., Rother, C.: Learning analysis-by-synthesis for 6D pose estimation in RGB-D images. In: 2015 IEEE International Conference on Computer Vision, pp. 954–962. IEEE, Piscataway, NJ (2015)

14. MVTec Software GmbH: HALCON. https://www.mvtec.com/halcon/. Accessed 9 July 2019

15. Papazov, C., Burschka, D.: An efficient RANSAC for 3D object recognition in noisy and occluded scenes. In: Kimmel, R., Klette, R., Sugimoto, A. (eds.) ACCV 2010. LNCS, vol. 6492, pp. 135–148. Springer, Heidelberg (2011). https://doi.org/10.1007/978-3-642-19315-6_11

16. Rudorfer, M., Kroischke, X.: Evaluation of point pair feature matching for object recognition and pose estimation in 3D scenes. In: 19 Anwendungsbezogener Workshop zur Erfassung, Modellierung, Verarbeitung und Auswertung von 3D-Daten, pp. 27–36. Gesellschaft zur Förderung angewandter Informatik e.V. (2016)

17. Tuzel, O., Liu, M.-Y., Taguchi, Y., Raghunathan, A.: Learning to rank 3D features. In: Fleet, D., Pajdla, T., Schiele, B., Tuytelaars, T. (eds.) ECCV 2014. LNCS, vol. 8689, pp. 520–535. Springer, Cham (2014). https://doi.org/10.1007/978-3-319-10590-1_34

18. Vidal, J., Lin, C.Y., Llado, X., Martí, R.: A method for 6D pose estimation of free-form rigid objects using point pair features on range data. Sensors **18**, 2678 (2018)

19. Vidal, J., Lin, C.Y., Martí, R.: 6D pose estimation using an improved method based on point pair features. CoRR abs/1802.08516 (2018)

Multi-DisNet: Machine Learning-Based Object Distance Estimation from Multiple Cameras

Haseeb Muhammad Abdul[1]([✉]), Ristić-Durrant Danijela[1] [ID], Gräser Axel[1],
Banić Milan[2], and Stamenković Dušan[2]

[1] Institute of Automation, University of Bremen, Otto-Hahn-Allee 1, 28359 Bremen, Germany
{haseeb,ristic,ag}@iat.uni-bremen.de
[2] Faculty of Mechanical Engineering, University of Niš, Aleksandra Medvedeva 14,
18000 Niš, Serbia
{mbanic,dstamenkovic}@masfak.ni.ac.rs

Abstract. In this paper, a novel method for distance estimation from multiple cameras to the object viewed with these cameras is presented. The core element of the method is multilayer neural network named Multi-DisNet, which is used to learn the relationship between the sizes of the object bounding boxes in the cameras images and the distance between the object and the cameras. The Multi-DisNet was trained using a supervised learning technique where the input features were manually calculated parameters of the objects bounding boxes in the cameras images and outputs were ground-truth distances between the objects and the cameras. The presented distance estimation system can be of benefit for all applications where object (obstacle) distance estimation is essential for the safety such as autonomous driving applications in automotive or railway. The presented object distance estimation system was evaluated on the images of real-world railway scenes. As a proof-of-concept, the results on the fusion of two sensors, an RGB and thermal camera mounted on a moving train, in the Multi-DisNet distance estimation system are shown. Shown results demonstrate both the good performance of Multi-DisNet system to estimate the mid (up to 200 m) and long-range (up to 1000 m) object distance and benefit of sensor fusion to overcome the problem of not reliable object detection.

Keywords: Autonomous obstacle detection for railways · Sensor fusion · Machine learning

1 Introduction

Reliable obstacle detection and distance estimation is one of the core problems in safety-critical applications such as autonomous driving in automotive and railway [1, 2]. Use of multiple sensors to overcome the limitations of individual sensors and to combine their benefits has become usual practice for autonomous obstacle detection [3]. Different onboard sensors such as radars, mono/stereo cameras and Light Detection And Ranging - LiDAR, implemented in so-called Advanced Driving Assistance Systems (ADAS), are rapidly increasing the vehicle's automation level [4]. Some of these sensors such as radar

© Springer Nature Switzerland AG 2019
D. Tzovaras et al. (Eds.): ICVS 2019, LNCS 11754, pp. 457–469, 2019.
https://doi.org/10.1007/978-3-030-34995-0_41

and LiDAR can be used for direct distance measurement. However, such sensors do not give enough information on the environment including object's type and size, which is important information for reliable autonomous obstacle detection. In contrast, vision sensors do give enough environment information but they could not be directly used for the distance measurement. Traditionally stereo-vision is used for depth extraction and distance reconstruction, where images from two stereo cameras are used to triangulate and estimate distances to objects, potential obstacles, viewed by cameras [5, 6]. However, as presented in [7], stereo-vision based distance calculation is characterized with the errors as consequences of uncertainties in camera calibration and in images processing of rectified images, in particular of uncertainty in finding the stereo corresponding points. The error is larger for the stereo camera system with a longer baseline, as the calibration error of the system with the longer baseline is bigger than the calibration error for the system with a shorter base line. This is particularly problematic for the applications where long-range obstacle detection is needed, such as railway applications, as longer base line is needed for long-range distance estimation. In order to overcome problems of stereo vision, some authors suggest distance estimation from mono cameras [8]. These methods are beneficial not only because of overcoming stereo-vision problems, but also for their implementation on different camera types, usually used as monocular cameras in obstacle detection, such as thermal cameras. In this paper, a novel machine learning-based method for distance estimation from multiple cameras of different type is presented. The idea behind the multi-sensory vision system is to overcome the limitations of individual sensor and to benefit of system redundancy enabling the use of advantages of individual sensors. Namely, different vision sensors handle light and weather conditions more or less good which causes different object detection results in different cameras' images and so different level of uncertainty in object distance estimation from different cameras. As it can be seen from the evaluation results in [10], object recognition and distance estimation error are different for RGB and thermal vision sensors. Therefore, in case of synchronous and simultaneous, but different object recognition results from different sensors, there is need to combine different sensory data so to get the unique distance estimation such that the resulting information has less uncertainty than would be possible when these sources were used individually. A method for sensor fusion is presented, which is based on novel multilayer neural network named Multi-DisNet that learns the relationship between the distance from the multiple cameras to the detected objects and the corresponding sizes of the objects in both camera images. As a proof-of-concept, the results on integration of two sensors, thermal camera and RGB camera with zooming functionality, and Multi-DisNet distance estimation of objects to both cameras in real-world railways scenarios are shown. The Multi-DisNet network was trained using the first dataset reflecting the real-world railway applications, created within the H2020 Shift2Rail project SMART [9]. This dataset reflects also the need for the long-range obstacle detection (up to 1000 m) due to long train braking distances. Namely, for the usual speed of freight trains, which is in most EU countries limited to 80 km/h (22.22 m/s), the distance of 1000 m has been selected as the relevant distance for detection of obstacles on the rail tracks. The stopping distance of freight trains with mentioned speed is defined by national regulations and it is approximately 700 m for a freight train pulling the 2000 t cargo (the exact stopping distance depends on different factors such as the train speed, the rate of deceleration, gradient of the track). Even though evaluated in the railway scenario, the presented distance estimation system can be applied to different applications where the mid and long-range obstacle detection represents important requirement. In this paper, evaluation results on the obstacle detection are presented assuming humans as possible obstacles. However, presented system can be used for variety of object classes.

The paper is organized as follows. In Sect. 2, an overview of related work on sensor fusion in autonomous obstacle detection with emphasis on the use of machine learning techniques is given. Novel Multi-DisNet system overview including description of long-range dataset generation is presented in Sect. 3. In Sect. 4, the evaluation results are shown. The main conclusions are given in Sect. 5.

2   Related Work

A number of methods from multi-sensor fusion for the purpose of autonomous detection of obstacles in front of vehicles have been developed, from traditional using of Kalman Filter [11] to machine learning, and in particular, deep learning based methods. In this section, examples are presented with the focus on multi-sensor fusion for obstacles (object) detection in autonomous vehicles and in railway applications.

In last decade, there is significant progress in the research and development of machine learning based methods for intelligent transportation mainly in automotive field as well as in robotics field. Significant works in the field starts from proven promising results of vision-based deep learning neural networks [9] and extension of their use for other sensors as well [12]. Most of the current successful object detection approaches are based on a class of deep learning models called Convolutional Neural Networks (CNN). While most existing object detection researches are focused on using CNN with color image data, emerging fields of application such as Autonomous Vehicles (AVs) which integrates a diverse set of sensors, require the processing for multisensory and multimodal information to provide a more comprehensive understanding of the real-world environment. In [13] a multimodal road vehicle detection system integrating data from a 3D-LIDAR and a color camera is proposed. Three modalities (color image, 3D-LIDAR's range and reflectance data). Bounding Box (BB) detections in each one of these modalities are jointly learned and fused by an Artificial Neural Network (ANN) late-fusion strategy to improve the detection performance of each modality. The significant progress of deep-learning based methods for object detection in automotive and robotics field is enabled with existing of number of large datasets such as KITTI dataset [14] and Berkeley Deep Drive dataset [15]. In contrast to automotive field, there is lack of railway specific datasets, which can represent the challenges and real-life scenarios. As one of challenges of deep learning based methods is requirement for massive amounts of annotated data, it is clear that there is delay in deep learning based methods for object detection in railway applications with respect to automotive applications. An example of deep learning based methods for object detection in railway applications is [16]. An automatic object detection system that is based on Feature Fusion Refine neural network (FR-Net) to tackle the real-word railway traffic object detection issue in shunting mode is proposed. Even though proposed on-board object detection device had two sensors, a camera and millimeter-wave radar, the proposed FR-Net was not used for sensor fusion but only for object detection in camera images. Moreover, even though it was said that radar of the proposed system measured the distance between the equipment and the detected object, the measured (estimated) objects distances were not given in the paper. If distances were given they would be rather short-range distances due to considered shunting application. Actually, in all aforementioned work on different sensor fusion methods for object (obstacles) detection, the emphasis was on object detection without regard to the object distances. This is the case in majority of other, if not all, literature on machine learning for sensor fusion for object detection. The particular novelty of this paper is that it presents a machine learning based sensor-fusion method for object distance estimation as the main objective of the presented work is mid (up to 200 m) and long-range (up to 1000 m) obstacle (object) detection.

3 Machine Learning Based Object Distance Estimation from Multi-sensory System

The architecture of the Multi-DisNet-based distance estimation system, with the setup assuming two cameras mounted horizontally parallel, an RGB camera and a thermal camera, is illustrated in Fig. 1.

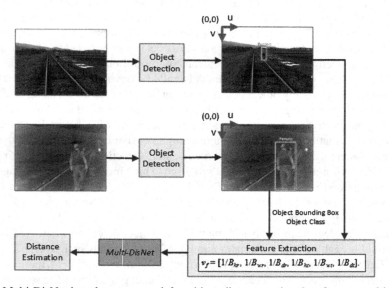

Fig. 1. Multi-DisNet-based system used for object distance estimation from a multi-camera system.

Each of two synchronized images of the same scene, captured by RGB and thermal camera sensors, are inputs to an Object Detector module. Different object detectors can be integrated into the system. Independently of the Object Detector type it is important that its outputs are bounding boxes of detected objects in the image. The resulted corresponding objects bounding boxes are then processed to calculate the features, bounding boxes parameters based on which the trained Multi-DisNet gives as outputs the estimated distance of the object to the multi-camera sensors. In the system architecture illustrated in Fig. 1, an example of the estimation of the distance of a person on the rail tracks is shown.

3.1 Multi-DisNet – Dataset

In order to create dataset which could be used for long-range obstacle detection, the dataset was created by manually extracted 755 bounding boxes of objects in the images of rail scenes recorded by monocular RGB cameras and a thermal camera, mounted on the static test-stand as shown in Fig. 2. The images were recorded in the field tests on the location of the straight rail tracks in length of about 1100 m in different times of the day and night and in different weather conditions in November 2018. The selected sensor RGB camera from "The imaging Source" [17] provides images with resolution up to 2592 × 1944 at 15 fps whereas selected thermal camera from FLIR [18] provides images with resolution up to 640 × 512 at 9 fps. The images were recorded in a synchronized manner to capture same scene from multiple cameras at the same time event. Due to the limitation of network bandwidth and high resolution of images the synchronized images were recorded at 2 fps.

Fig. 2. Field tests performed on the straight rail tracks; Test-stand with the vision sensors viewing the rail tracks and an object (person) on the rail track.

During the performed field tests, for the purpose of dataset generation, two persons imitated potential obstacles on the rail tracks. They were walking along the rail tracks 1000 m away from the cameras and back, 1000 m towards the cameras. At every 5 m, while walking in both directions, they signalized in a particular way, so that frames recorded at moments of the signalization could be used for the dataset generation. These camera frames, corresponding to persons locations at every 5 m from 0 to 1000 m, were extracted from the whole recorded video so that manually drawn bounding boxes of the objects (persons walking along the rail tracks) could be labelled with ground distance. The example RGB and Thermal images recorded in field experiments for the purpose of dataset generation are given in Fig. 3.

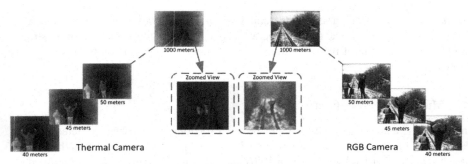

Fig. 3. RGB and thermal images recorded in field experiments performed on the long-range straight rail tracks for the purpose of dataset generation.

Input Feature Vector Extraction. In total 755 bounding boxes of objects were manually extracted from recorded synchronized RGB and thermal camera images using VGG Image Annotator tool [19]. For each object bounding box extracted from RGB images and for each object bounding box extracted from thermal images, a three-dimensional feature vector v was calculated:

$$v = [1/B_{hi}, 1/B_{wi}, 1/B_{di}], \; i = r, t \tag{1}$$

where indexes r and t indicate features extracted from RGB and thermal images respectively, and the coordinates of vector v are following features:

Height, B_{hi} = (height of the object bounding box in pixels/image height in pixels)
Width, B_{wi} = (width of the object bounding box in pixels/image width in pixels)
Diagonal, B_{di} = (diagonal of the object bounding box in pixels/image diagonal in pixels)

The ratios of bounding box sizes and image size were used as the features so to enable use of the generated dataset for any image resolution. The inverse of features, B_{hi}, B_{wi} and B_{di} were finally selected as the best features due to their high correlation with the desired output (distance to the object). Combining the features extracted from both sensor modules, a joint vector, that is a vector of fused data v_f of dimensions 6 × 1 is obtained.

$$v_f = [1/B_{hr}, \; 1/B_{wr}, \; 1/B_{dr}, \; 1/B_{ht}, \; 1/B_{wt}, \; 1/B_{dt}]. \tag{2}$$

3.2 Multi-DisNet – Architecture

In order to find the appropriate number of hidden layers and neurons per layer, the best configuration of hidden layers and neurons obtained from the GridSearch estimator [26] was used for Multi-DisNet architecture. As shown in Fig. 4, the Multi-DisNet network consists of 6 neurons in input layer which represents the features extracted from bounding boxes, followed by the 3 hidden layers with 150 neurons in each layer and one neuron in the output layer which represents the distance estimation. For this analysis and proof-of-concept, a reduced dataset was used. Whereas the dataset was split it into 3 sets for training 80%, validation 10% and testing 10%.

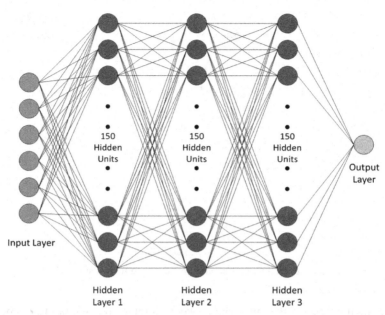

Fig. 4. The structure of Multi-DisNet used for the estimation of object distance from the multi-sensory system.

4 Evaluation

The Multi-DisNet-based system for object distance estimation was evaluated on images recorded in different field tests. As first, testing data of the generated long-range dataset were processed by Multi-DisNet system and the results on testing dataset are shown in Fig. 5. As it can be seen, the estimated object distances vary slightly around the ground truth distance. The error in distance estimation is rather larger with larger distances and the calculated mean absolute error (MAE) was 22.76 m and root means square error (RMSE) was 34.09 m. These results demonstrate the ability of Multi-DisNet to estimate the long-range distance of up to 1000 m with an acceptable error.

Some of the results of the Multi-DisNet long-range object distance estimation in RGB and thermal images are given in Fig. 6. The estimated distances to the objects (persons) detected in the images are given in Table 1. The object bounding boxes in the images were extracted manually so that an ideal object detector, able to extract bounding boxes of far objects of very small size in the images, was assumed.

In order to evaluate performances of Multi-DisNet system with autonomous object bounding boxes extraction, the state-of-the-art computer vision object detector YOLO (You Only Look Once) [21] trained with COCO dataset [20] was implemented in the Object Detector module. The images from above described explained field tests were processed by YOLO and the resulted YOLO objects bounding boxes were processed to calculate the Multi-DisNet input features, bounding boxes parameters. Based on the input features, the trained Multi-DisNet gave as outputs the estimated distances of the objects, persons on the rail tracks, to the camera sensors. It is important to note that images

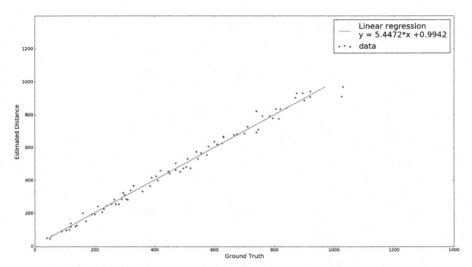

Fig. 5. Ground truth vs estimated object distance from testing dataset.

processed are those recorded while persons were walking along the rail tracks between the points of signalizations done for data set images recording, as explained in Sect. 3.1. Some of the results of the Multi-DisNet object distance estimation in RGB and thermal images are given in Fig. 7. The estimated distances to the objects (persons) detected in the images are given in Table 2. Shown results illustrate the longest range achieved by YOLO, about 395 m, as YOLO detector failed to produce the object bounding boxes for distances larger than 395 m as the YOLO was trained with COCO dataset which contains objects only at short distance to the camera.

Further sensor data for the Multi-DisNet evaluation purposes were recorded during the dynamic field tests [22]. The RGB and thermal cameras were mounted onto the moving locomotive Serbia Cargo type 444 pulling the freight train with 21 wagons on the Serbian part of the Pan European corridor X to Thessaloniki in the length of 120 km with maximal train speed of 80 km/h. The cameras were mounted into specially designed housing as shown in Fig. 8.

The sensors' housing was vibration isolated to prevent transmitting of vibrations from the locomotive onto the cameras as moving vehicle vibration can severely deteriorate quality of acquired images. The vibration isolation system was designed with the rubber-metal springs [9] and it was suitable to suppress low frequency vibrations providing almost 94% isolation efficiency for vehicle primary resonant frequency of 27 Hz. For frequencies above 50 Hz, the isolation efficiency was greater than 98%.

During the train run, onboard cameras recorded the data of the real-world rail tracks scenes in front of the locomotive. Example frames sequence of thermal images is shown in Fig. 9. As it can be seen, the example frames show the scene when a person accidentally crossed the rail track while the train was approaching. These frames were processed by YOLO object detector for person bounding box extraction. The extracted bounding boxes were processed for the extraction of bounding boxes parameters, inputs to Multi-DisNet, so that trained Multi-DisNet estimated distances to the detected object (person crossing

Fig. 6. Manually detected objects in magnified RGB and thermal images from the long-range dataset field tests.

Table 1. Estimated distances vs Ground truth.

Object	Ground truth	Multi-DisNet
Person 1	925 m	937.19 m
Person 2		899.25 m

Fig. 7. Object detected in RGB and Thermal images using YOLO object detector.

Table 2. Estimated distances vs Ground truth.

Object	Ground truth	Multi-DisNet
Person 1	395 m	352.86 m
Person 2		408.25 m

Fig. 8. Vision sensors for obstacle detection integrated into sensors' housing mounted on the frontal profile of a locomotive below the headlights.

the rail tracks). The distance estimation result is given in Table 3. However, as it can be seen from Fig. 9c, YOLO failed to detect object bounding boxes in thermal camera frame. The cause for object detection failure in some of thermal camera images was low images contrast due to lower camera performances influenced with high outside temperature during the dynamic field test of about 38 °C As the Multi-DisNet assumes input feature vectors of six elements where three elements represent features of the object bounding box in thermal image it was necessary to estimate the object bounding box in thermal camera based on detected object bounding box in the RGB camera image. This assumption of bounding box in thermal image was done based on the relationship between the sizes of the objects bounding boxes in RGB and thermal images learned from the generated dataset. The corresponding object bounding boxes between the cameras highly correlate and show a linear relationship, which allows estimation of the missing bounding box in one camera as corresponding to the bounding box detected in the other camera. The results show how Multi-DisNet addresses the problem of object detection module failure to detect object in one of the camera image. The trained Multi-DisNet estimated person distances to the cameras, which were mounted on the moving locomotive using the calculated features of originally detected object bounding boxes in RGB images and the features of estimated object bounding boxes in thermal images, from all subsequent cameras frames. The distance estimation result is given in Table 3. For the sake of better understanding of Multi-DisNet distance estimation accuracy, the ground truth distance in dynamic experiments were calculated offline using the relative GPS position of the train and the approximate GPS position of obstacle on Google Maps.

Table 3. Distance estimation from Multi-DisNet in the dynamic experiment

Frames	Ground truth (GPS)	Multi-DisNet
(a)	155 m	160.92 m
(b)	146 m	141.80 m
(c)	138 m	135.61 m

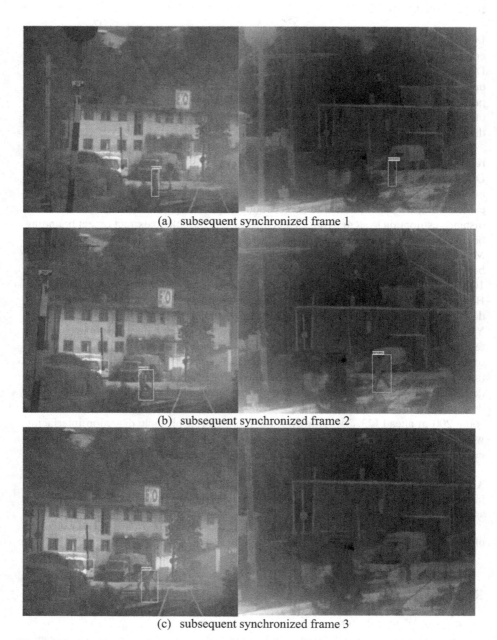

(a) subsequent synchronized frame 1

(b) subsequent synchronized frame 2

(c) subsequent synchronized frame 3

Fig. 9. Three subsequent synchronized frames of RGB and thermal cameras of the scene with a person crossing accidentally the rail tracks, with the YOLO object bounding boxes overlaid. YOLO failed to extract object bounding box in last thermal camera frame.

5   Conclusion

In this paper, a method for long-range obstacle detection from multiple monocular cameras is presented. The method is based on novel multilayer neural network named Multi-DisNet that learns the relationship between the distance from the multiple cameras to the detected objects and the corresponding sizes of the objects in both camera images. The presented work is a part of the research and development of novel integrated on-board obstacle detection system for railway applications. Presented system was evaluated for the setup consisting of a RGB and thermal camera.

The paper presents also the generation of the first long-range dataset reflecting the real-life railway applications, which was necessary for supervised training of the Multi-DisNet network to enable learning of the relationship between the sizes of the bounding boxes of detected object in the camera's images and the distance of the object to the cameras. For the calculation of the object bounding boxes parameters, the Multi-DisNet distance estimation system assumes an Object Detector able to extract object bounding boxes. The evaluation results shown in the paper demonstrate the performance of the Multi-DisNet to reliably estimate the long-range distances in case the object bounding boxes are manually extracted. The system achieved long-range object detection of about 1000 m, which is desired performance in applications such as railway applications. However, the evaluation results in the case of using the YOLO for automatic object detection demonstrated system performance of estimating the distance up to 400 m, as YOLO object detector failed to extract the bounding boxes in long-range images. These results indicate the need for improvement of the object detection system module and need of training of object detector module with datasets consist of objects at longer distance (small in size) in the future work, so to achieve reliable autonomous long-range obstacle detection. Nevertheless, presented system reflects the novelty in long-range obstacle detection as current sensor technology in current land transport research is able to look some 200 m ahead [23, 24]. However, the required rail obstacle detection interfacing with loco control should be able to look ahead up to 1000 m detecting objects on and near track which may potentially interfere with the clearance and ground profile [25].

Acknowledgements. This research has received funding from the Shift2Rail Joint Undertaking under the European Union's Horizon 2020 research and innovation programme under grant agreement No. 730836.

Special thanks to Serbian Railways Infrastructure, and Serbia Cargo for support in realization of the SMART OD Field tests.

References

1. Jiménez, F., Naranjo, J.E., Anaya, J.J., García, F., Ponz, A., Armingol, J.M.: Advanced driver assistance system for road environments to improve safety and efficiency. Transp. Res. Procedia **14**, 2245–2254 (2016)
2. Weichselbaum, J., Zinner, C., Gebauer, O., Pree, W.: Accurate 3D-vision-based obstacle detection for an autonomous train. Comput. Ind. **64**(9), 1209–1220 (2013)
3. Bouain, M., Ali, K.M.A., Berdjag, D., Fakhfakh, N., Atitallah, R.B.: An embedded multi-sensor data fusion design for vehicle perception tasks. J. Commun. **13**(1), 8–14 (2018)

4. Kim, S., Kim, H., Yoo, W., Huh, K.: Sensor fusion algorithm design in detecting vehicles using laser scanner and stereo vision. IEEE Trans. Intell. Transp. Syst. **17**(4), 1072–1084 (2016)
5. Leu, A., Aiteanu, D., Gräser, A.: High speed stereo vision based automotive collision warning system. In: Precup, R.E., Kovács, S., Preitl, S., Petriu, E. (eds.) Applied Computational Intelligence in Engineering and Information Technology, vol. 1, pp. 187–199. Springer, Heidelberg (2012). https://doi.org/10.1007/978-3-642-28305-5_15
6. Bernini, N., Bertozzi, M., Castangia, L., Patander, M., Sabbatelli, M.: Real-time obstacle detection using stereo vision for autonomous ground vehicles: a survey. In: 2014 IEEE 17th International Conference on Intelligent Transportation Systems (ITSC), China, pp. 873–878 (2014)
7. Ristić-Durrant, D., et al.: SMART concept of an integrated multi-sensory on-board system for obstacle recognition. In: 7th Transport Research Arena TRA 2018, Austria, 16–19 April 2018
8. Saxena, A., Sung, H., Ng, A.Y.: 3-D depth reconstruction from a single still image. Int. J. Comput. Vis. **76**(1), 53–69 (2007)
9. Project SMART. http://www.smartrail-automation-project.net
10. Haseeb, M.A., Guan, J., Ristić-Durrant, D., Gräser, A.: DisNet: a novel method for distance estimation from monocular camera. In: 2018 IEEE/RSJ International Conference on Intelligent Robots and Systems - IROS, Spain (2018)
11. Chen, S.Y.: Kalman filter for robot vision: a survey. IEEE Trans. Industr. Electron. **59**(11), 4409–4420 (2012)
12. Caltagironea, L., Bellonea, M., Svenssonb, L., Wahdea, M.: LIDAR-camera fusion for road detection using fully convolutional neural networks. Robot. Auton. Syst. **111**, 125–131 (2019). https://doi.org/10.1016/j.robot.2018.11.002
13. Asvadi, A., Garrote, L., Premebida, C., Peixoto, P., Nunes, U.J.: Multimodal vehicle detection: fusing 3D-LIDAR and color camera data. Pattern Recogn. Lett. **115**, 20–29 (2017). https://doi.org/10.1016/j.patrec.2017.09.038
14. Geiger, A., Lenz, P., Stiller, C., Urtasun, R.: Vision meets robotics: the KITTI dataset. Int. J. Robot. Res. **32**(11), 1231–1237 (2013). https://doi.org/10.1177/0278364913491297
15. Berkeley Deep Drive BDD 100 K dataset. https://bdd-data.berkeley.edu/. Accessed 15 Feb 2019
16. Ye, T., Wang, B., Song, P., Li, J.: Automatic railway traffic object detection system using feature fusion refine neural network under shunting mode. Sensors **18**(6), 1916 (2018). https://doi.org/10.3390/s18061916
17. The imaging source, GigE color zoom camera. https://www.theimagingsource.com/. Accessed 15 Feb 2019
18. FLIR thermal imaging, Tau2. https://www.flir.com/products/tau-2/. Accessed 15 Feb 2019
19. Dutta, A., Gupta, A., Zissermann, A.: VGG image annotator (VIA). http://www.robots.ox.ac.uk/~vgg/software/via. Accessed 15 Feb 2019
20. COCO dataset. https://arxiv.org/pdf/1405.0312.pdf. Accessed 15 Feb 2019
21. Redmon, J., Farhadi, A.: YOLOv3: an incremental improvement. arXiv (2018)
22. Ristić-Durrant, D., et al.: SMART: a novel on-board integrated multi-sensor long-range obstacle detection system for railways. In: RAILCON, Nis, November 2018
23. Duvieubourg, L., Cabestaing, F., Ambellouis, S., Bonnet, P.: Long distance vision sensor for driver assistance. IFAC Proc. Vol. **40**(15), 330–336 (2007)
24. Pinggera, P., Franke, U., Mester, R.: High-performance long range obstacle detection using stereo vision. In: 2015 IEEE/RSJ International Conference on Intelligent Robots and Systems-IROS, pp. 1308–1313 (2015)
25. Shift2Rail Joint Undertaking, Multi-annual Action Plan, Brussels, November 2015
26. Pedregosa, F., et al.: Scikit-learn: machine learning in Python. J. Mach. Learn. Res. **12**, 2825–2830 (2011)

Hierarchical Image Inpainting by a Deep Context Encoder Exploiting Structural Similarity and Saliency Criteria

Nikolaos Stagakis$^{(\boxtimes)}$ ⓘ, Evangelia I. Zacharaki ⓘ,
and Konstantinos Moustakas ⓘ

VVR Group, Department of Electrical and Computer Engineering,
University of Patras, Patras, Greece
nick.stag@ece.upatras.gr
.

Abstract. The purpose of this paper is to present a context learning algorithm for inpainting missing regions using visual features. This encoder learns physical structure and semantic information from the image and this representation differentiates it from simple auto encoders. Such properties are crucial for tasks like image in-painting, classification and detection. Training was performed by patch-wise reconstruction loss using Structural Similarity (SSIM) jointly with an adversarial loss. The reconstruction loss is also augmented using spatially varying saliency maps that increase the error penalty on distinctive regions and thus promote image sharpness. Furthermore, in order to improve image continuity on the boundary of the missing region, distance functions with increasing importance towards the center of the inpainting region are also used either independently or in conjunction with the saliency maps. We also show that our choice of reconstruction loss outperforms conventional criteria such as the L2 norm. This means giving more weight to pixels closer to the border of the missing image parts and also giving more important to salience parts of the image to guide the reconstruction, thus producing more realistic images.

Keywords: Image inpainting · Deep context encoder · Image saliency

1 Introduction

The real world is visually very diverse and difficult to replicate correctly and in a realistic manner. While it truly seems to be impossible to replicate every single detail, due to the continuous efforts of scientists to develop advanced algorithms to synthesize realistic natural images, it has been shown [13] that our world is actually quite structured with structural manifestations being present over different scales in many type of images, like forests or buildings.

The notion of filling missing parts of an image is not new and certainly not limited to convolutional neural networks (CNNs), autoencoders and generative adversarial networks (GANs), although the task has changed names over the

© Springer Nature Switzerland AG 2019
D. Tzovaras et al. (Eds.): ICVS 2019, LNCS 11754, pp. 470–479, 2019.
https://doi.org/10.1007/978-3-030-34995-0_42

years. Algorithms like texture synthesis of Efros *et al.* [6] that create images by inpainting pixels based on a infinitely big texture and utilizing Markov Random Fields, or patch-based approaches [2,9] that evaluate image reconstruction based on two qualities, the completeness and coherency of the image [17], form the backbone of many modern algorithms and are used as benchmarks even to this day.

CNNs represent a very good approach to solving these kind of problems. The advantage of CNNs over these patch-based approach is the latent space, created as part of the training, which contains all the accumulated information from the training about context and basic shapes. This helps to guide the inpainting by not only looking at the contents of the current image, but also at other images which contain similar context and thus fill missing parts with information that is not present in the original image. This is crucial when the missing region covers an area of high context importance, like a photograph where both eyes are missing from the image, and the algorithm has to identify that and create new eyes in the expected places.

GANs and autoencoders in particular are extensively used in image reconstruction with varying success [8]. While in theory having two networks play a game of one trying to deceive the other with fake images and the latter trying to distinguish between true and fake works out very good, in practice GANs are often unstable and difficult to work with. On the other hand, when trained successfully the produced images are very detailed and sharp compared to autoencoders.

Pathak *et al.* [13] proposed a variation of the autoencoder [15], called context encoder, and used it to with great success for image inpainting. They demonstrated that the network is capable of understanding semantics and uses the context of the existent image part to guide the inpainting method. They used L2 norm jointly with an adversarial loss and showed that the two loss functions complement one another. While image inpainting of missing parts when using only L2 norm seemed overall realistic, image content was inherently blurry and distinguishable from the rest of the image. On the other hand, with adversarial loss alone the network would create realistic images for the missing regions but the inpainted image had usually nothing in common with the rest of the image, making it clear to the human eye that the inpainting was a failure. This is why they came up with the idea of the adversarial loss with reconstruction loss (using L2), since it evidently helped to produced sharper and more detailed images, while also containing context relevant to the image. An example of this is illustrated in Fig. 1.

This is why we formulate our approach on this context encoder *et al.* [13] in which an adversarial cost is incorporated in the context encoder's training to combine the advantages of both approaches. In respect to application domains, in contrast to recent CNN architectures that are usually trained and utilized in a very specific context, as for example reconstruction of images of human faces [11] or buildings, we present a deep network architecture for general image inpainting trained on a wide variety of contexts.

Image   (L2)   (Adv)   (L2+Adv)

Fig. 1. Demonstration of the different loss functions and their result when used together

In this paper, we present an improvement on this algorithm augmenting its inpainting capabilities. This is achieved through the incorporation of (i) structural information in the loss function, (ii) saliency criteria guiding the training process through spatially varying weights and (iii) a hierarchical reconstruction mechanism based on the amount of known context in the local neighborhood. The motivation behind those three contributions is described next.

Previous research has shown that image fidelity and reconstruction quality criteria, such as the structural similarity (SSIM) index, that are extracted from local neighborhoods, perform better than pixel-wise approaches (such as L2 norm) commonly used in reconstruction applications [20]. This is because of the tendency of the human vision to extract structural information from an image and overlook individual pixel differences for favor of an overall geometrically coherent and structurally sound image reconstruction. SSIM quantifies the perceived image quality degradation of an image in respect to an other one assumed to be of perfect quality and thus can used as similarity criterion between those two images. There are a lot of cases where images with high values of pixel-based criteria, such as peak signal to noise ratio (PSNR) [18], can still appear unnatural while in contrast, reconstructed images with high SSIM values tend to be more in line with what humans consider realistic. An example of this is shown in Fig. 2, where the first row presents cases where a high score of PSNR does not provide realistic reconstruction and that is reflected on a poor SSIM score, while the second row showcases examples where a high SSIM translates to visually better reconstruction even with a low PSNR score.

Moreover, the idea behind the other two modifications introduced into the baseline architecture is to encode feature distinctiveness and reconstruction confidence into a spatially varying loss function. Aiming at preservation of region boundaries we designed a mechanism that adds a higher penalty of reconstruction errors in the immediate neighborhood of the available content.

Since our aim was not to elaborate deeply in the recognition of image content saliency, we used as criterion a measure of feature distinctiveness such as the standard image gradient quantifying edge intensity, however any other more dedicated approach can be used instead [10]. Additionally to the adjustment of spatial weights based on image distinctiveness, we introduced an influence map to define the image support in each epoch of the loss function minimization process. The influence map aims to give more importance to the parts near the missing area and gradually expand this importance inwards. This serves to avoid discontinuities between the real and the inpainted regions and thus provide a smooth transition. The combination of those two maps determines the overall spatio-temporal contribution of each pixel in the optimization process.

A large variety of methods has been proposed for the extraction of saliency from images or video [3]. Weighting of pixels as a means to incorporate the information content for image reconstruction has been used extensively in computer vision and deep learning applications for image classification, segmentation, registration, clustering and, of course, for inpainting. An example is the work of Erus et al. [7] for domain sampling, while hierarchical approaches based on saliency criteria have also been very successful in reducing ambiguity in image registration [19]. Improved image completion was demonstrated when image gradients were integrated into the distance metric between patches [4]. In contrast to our work in which we use image gradient to define the pixels *contribution* (implemented through a multiplication operation in the loss function, as will be explained in Sect. 3.1), the image gradients in [4] were used as a second term in their objective function (implemented as an additive cost) quantifying patch *similarity*.

Fig. 2. Comparison of PSNR and SSIM for image inpainting quality assessment. First row shows results with high PSNR and low SSIM and the opposite holds for the second row.

2   Context Encoder

In this chapter we will introduce the basic network architecture behind the context encoder created by *Pathak et al.* [13] and explain some of it's inner workings. Later, we will explain how and why the saliency and distance maps were implemented along with results on their effectiveness. The baseline context encoder network takes as input an image with a large central missing part (of size half of the whole image). To be precise, the context encoder expects a 128×128 image with a 64×64 area missing in the middle, and the decoder produces a 64×64 image, same size as the missing area.

2.1   Network Architecture

Encoder. The context encoder created by Pathak *et al.* in [13] is based upon an autoencoder, derived from the AlexNet architecture [12], coupled with an adversarial loss. The encoder receives as input an image with missing parts and produces a feature representation at his last layer. This feature representation is where the semantics information about the image is stored. It is later passed to the decoder which performs the image reconstruction and, essentially, fills in the missing parts of the original image. Due to the huge number of parameters of a fully connected encoder and decoder, a channel wise fully connected layer is used to connect the encoder features to the decoder.

Decoder. The decoder is the other half of the network, playing the role of the generator for the adversarial loss. The decoder architecture is symmetric to the encoder architecture minus one "deconvolutional" layer, which is a convolution that results in an image of higher resolution. The decoder's output is then used to calculate the reconstruction loss while also serving as input, along with the ground truth image, to an adversarial discriminator network. The latter consists of four convolutional layers and the output is evaluated using binary cross entropy.

2.2   Saliency Map

The idea is to use image saliency as criterion for spatial adjustment of the penalty enforced in the reconstruction error of the network, such that more important areas where reconstruction errors are stronger perceived, are weighted more. A saliency map is an image of same size with the original one, that highlights such distinctive regions with rich information context, that can be exploited to better guide the optimization process. A saliency map can be extracted using several sophisticated techniques including detection of interesting visual features or objects [16], but we relied only on distinctive parts, such as edges, employing the standard image gradient. Let $S \in R^2$ be the saliency map. The calculation was restricted on the missing part of the training images where the reconstruction error is evaluated. The saliency map is normalized in the range [0,1] by dividing with the sum of its elements.

And is multiplied element-wise with the output gradient of the loss function in order to act as a spatial weighting function. Through this adaptation the network focuses more in preserving edges during training and thus tends to produce sharper and detailed images.

2.3 Influence Map

We introduced an influence map to perform inpainting gradually starting from regions whose reconstruction is easier to guess. Our hypothesis is that the uncertainty is smaller when there is surrounding information, that is close to the boundary of the missing part and especially in convex regions, while concavities are more difficult to estimate due to smaller amount of information in the neighborhood. For this purpose we calculated a distance map in the form of concentric disks, which expand inwards gradually with the training epochs. This influence map is normalized in the range of [0,1] by dividing with 255 and used control the "temporal" weight of each pixel. This means that at the start of the training the network is more targeted in preserving the continuity of image content near the missing region, providing a realistic inpainting, and slowly work its way inwards. This gradual application of masks to the missing regions helps to control the optimization facilitating convergence.

Overall, the spatiotemporal weight $w_I(x,t)$ for each pixel x at epoch t depends on the value of the influence map at x, the current epoch t and the selected expansion speed that acts as a scaling factor controlling how fast more regions are incorporated into the cost function.

3 Loss Function

This sections defines the loss functions used for training the neural network. In the baseline approach [13] the overall loss function is defined as a weighted sum of reconstruction loss and adversarial loss:

$$L_{total} = \lambda_{rec}L_{rec} + \lambda_{adv}L_{adv}$$

As reconstruction loss, we implemented and incorporated the Structural Similarity criterion replacing the usual mean square error. For the adversarial loss the binary cross entropy criterion is used, as in [13].

3.1 Reconstruction Loss

As stated above, the structural similarity is used for the reconstruction loss due to its demonstrated potential to better express human perception, whereas common criteria, such as the square error, that account for differences in individual pixels without considering the spatial neighborhood, are not as descriptive. This has been mainly illustrated in image restoration tasks, where patch-based criteria showed to outperform the pixel-based metrics [20].

SSIM for pixel p is defined as

$$SSIM(p) = \frac{2\mu_x\mu_y + C_1}{\mu_x^2 + \mu_y^2 + C_1} \cdot \frac{2\sigma_{xy} + C_2}{\sigma_x^2 + \sigma y^2 + C_2} = l(p) \cdot cs(p)$$

where μ_x and μ_y are the mean values of the patches x and y, centered at p, in the ground truth and generated image respectively. The same goes for σ_x and

σ_y which represent the variance while σ_{xy} is the covariance of x and y. The C_1 and C_2 are variables intended to stabilize the division with weak denominator.

The first term, $l(p)$, is defined as the luminance comparison which measures the mean luminance and the second term, cs, is the contrast comparison function multiplied with the structured comparison function. Those two functions measure the closeness and the correlation of the two patches respectively.

Consequently, the reconstruction loss for SSIM in an image I containing patches P each centered at pixel p can be then written as:

$$L_{rec}(I) = \frac{1}{|I|} \sum_{\forall P \in I} \frac{1}{|P|} \sum_{N(p)} 1 - SSIM(p)$$

We compute the output gradient for every pixel q in P centered at p as:

$$\nabla L_{rec}(p) = \frac{\partial L_{rec}(P)}{\partial x(q)} = -\frac{\partial}{\partial x(q)} SSIM(p) - (\frac{\partial l(p)}{\partial x(q)} \cdot cs(p) + l(p) \cdot \frac{\partial cs(p)}{\partial x(q)})$$

while the derivatives of $l(p)$ and $cs(p)$ are:

$$\frac{\partial l(p)}{\partial x(q)} = 2 \cdot G_{\sigma G}(q-p) \cdot (\frac{\mu_y - \mu_x \cdot l(p)}{\mu_x^2 + \mu_y^2 + C_1})$$

Where $G_{\sigma_G}(q-p)$ is the Gaussian coefficient associated with pixel q. And

$$\frac{\partial cs(p)}{\partial x(q)} = \frac{2}{\sigma_x^2 + \sigma_y^2 + C_2} \cdot G_{\sigma G}(q-p) \cdot [(y(q) - \mu_y - cs(p) \cdot (x(q) - \mu_x)]$$

The overall reconstruction loss of the current method use SSIM. If S is the saliency map (adjusting the spatial weights) and I_t the influence map at epoch t (controlling the weights over the optimization epochs), the gradient of the proposed reconstruction loss is spatio-temporally weighted as follows:

$$\nabla L'_{rec}(p) = \nabla L_{rec} \odot [1 + (S \odot I_t)]$$

3.2   Adversarial Loss

The adversarial loss is based around Generative Adversarial Networks (GAN) [8]. The decoder part of the network plays the role of the generator and the adversial role is taken by the adversial network. The training can be described as a game between those two, where the generator tries to create believable images in order to trick the discriminator, and he in turn wants to be able to distinguish between real and forged images. This method while is generally regarded as unstable and hard to control, shows very impressive results and thus it was adopted in [13] as well as in the current implementation. This encourages the generator to produce more realistic images and end up in "sharpening" the image the L2 would normally produce on its own.

4   Results and Future Work

The proposed network was evaluated on the *Learning-Based Image Inpainting Challenge* of ICME [1]. We used the validation dataset of the error concealment track, that simulates the case of transmission error, in the form of missing square blocks at the image center.

Our network was trained with the ImageNet Decathlon Data set [5,14] which contains 48238 different images of different classes to provide the encoder enough variety of features to learn from. Results for both PSNR and SSIM for each approach are shown in Fig. 3 and Table 1.

This work can be used as a first step in implementing an augmented reality system that enables the guidance of patients to manage their own health. For example, patients with respiratory diseases could be educated on the correct use of inhalation devices through a monitoring system that uses diminished reality to remove erroneous holding of the inhalation device (as shown in the bottom right example of Fig. 4) and image inpainting techniques to later augment the scene.

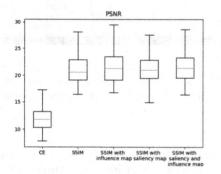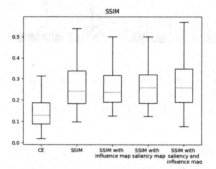

Fig. 3. Network evaluation based on SSIM and PSNR using the baseline context encoder (CE) architecture [13] and the proposed modifications. The red line in the boxlots indicates the median value. (Color figure online)

Table 1. Average values of SSIM and PSNR for the different approaches.

	Context encoder [13]	SSIM	SSIM with influence map	SSIM with saliency map	SSIM with saliency and influence maps
Mean SSIM	0.150	0.275	0.269	0.277	**0.283**
Mean PSNR	12.158	21.388	21.610	21.554	**21.759**

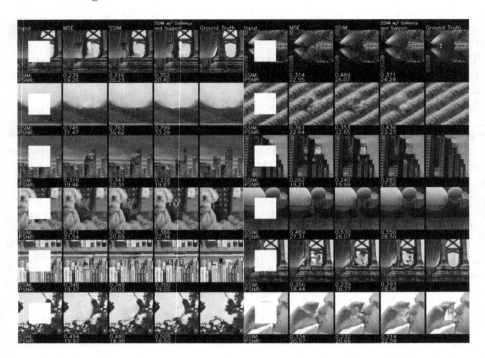

Fig. 4. Twelve examples of image inpainting by the different approaches. For each example from left to right: input image with a central missing part, inpainting by the baseline CE with L2 reconstruction loss [13], with constant SSIM, with spatiotemporaly weighted SSIM (proposed method).

Acknowledgment. This research has been co-financed by the European Regional Development Fund of the European Union and Greek national funds through the Operational Program Competitiveness, Entrepreneurship and Innovation, under the call RESEARCH - CREATE - INNOVATE (project code: T1EDK-03832).

References

1. https://icme19inpainting.github.io/
2. Barnes, C., Shechtman, E., Finkelstein, A., Goldman, D.B.: Patchmatch: a randomized correspondence algorithm for structural image editing. ACM Trans. Graph. **28**, 24:1–24:11 (2009)
3. Chen, L.C., Papandreou, G., Kokkinos, I., Murphy, K., Yuille, A.L.: Deeplab: semantic image segmentation with deep convolutional nets, atrous convolution, and fully connected CRFs. TPAMI **40**(4), 834–848 (2016)
4. Darabi, S., Shechtman, E., Barnes, C., Goldman, D.B., Sen, P.: Image melding: combining inconsistent images using patch-based synthesis. ACM Trans. Graph. (TOG) **31**(4), 82:1–82:10 (2012). Proceedings of SIGGRAPH 2012
5. Deng, J., Dong, W., Socher, R., Li, L.J., Li, K., Fei-Fei, L.: ImageNet: a large-scale hierarchical image database. In: CVPR 2009, pp. 248–255 (2009)

6. Efros, A.A., Leung, T.K.: Texture synthesis by non-parametric sampling. In: Proceedings of the Seventh IEEE International Conference on Computer Vision, vol. 2, pp. 1033–1038, September 1999

7. Erus, G., Zacharaki, E.I., Davatzikos, C.: Individualized statistical learning from medical image databases: application to identification of brain lesions. Med. Image Anal. **18**, 542–554 (2014)

8. Goodfellow, I., Pouget-Abadie, J., Mirza, M., et al.: Generative adversarial nets. In: Advances in Neural Information Processing Systems, vol. 27, pp. 2672–2680 (2014)

9. Herling, J., Broll, W.: High-quality real-time video inpainting with pixmix. IEEE Trans. Visual. Comput. Graph. **20**, 866–879 (2014)

10. Kadir, T., Brady, M.: Saliency, scale and image description. Int. J. Comput. Vis. **45**(2), 83–105 (2001)

11. Karras, T., Laine, S., Aila, T.: A style-based generator architecture for generative adversarial networks. In: CoRR (2018)

12. Krizhevsky, A., Sutskever, I.E., Hinton, G.: Imagenet classification with deep convolutional neural networks. Neural Inf. Process. Syst. **25**, 1097–1105 (2012)

13. Pathak, D., Krahenbuhl, P., Donahue, J., Darrell, T., Efros, A.: Context encoders: feature learning by inpainting. In: CVPR, pp. 2536–2544 (2016)

14. Rebuffi, S.A., Bilen, H., Vedaldi, A.: Learning multiple visual domains with residual adapters. In: NIPS, pp. 506–516 (2017)

15. Schmidhuber, J.: Deep learning in neural networks: an overview. Neural Netw. Off. J. Int. Neural Netw. Soc. **61**, 85–117 (2015)

16. Sharma, G., Jurie, F., Schmid, C.: Discriminative spatial saliency for image classification. In: 2012 IEEE Conference on Computer Vision and Pattern Recognition, pp. 3506–3513, June 2012

17. Simakov, D., Caspi, Y., Shechtman, E., Irani, M.: Summarizing visual data using bidirectional similarity. In: IEEE CVPR, pp. 1–8, June 2008

18. Wang, Z., Bovik, A.C., Sheikh, H.R., Simoncelli, E.P., et al.: Image quality assessment: from error visibility to structural similarity. IEEE Trans. Image Process. **13**(4), 600–612 (2004)

19. Zacharaki, E.I., Shen, D., Lee, S.K., Davatzikos, C.: Orbit: a multiresolution framework for deformable registration of brain tumor images. IEEE Trans. Med. Imaging **27**, 1003–1017 (2008)

20. Zhao, H., Gallo, O., Frosio, I., Kautz, J.: Loss functions for image restoration with neural networks. IEEE Trans. Comput. Imaging **3**, 47–57 (2017)

Online Information Augmented SiamRPN

Edward Budiman Sutanto🆔 and Sukho Lee$^{(\boxtimes)}$🆔

Dongseo University, Busan 47011, Korea
petrasuk@gmail.com
http://kowon.dongseo.ac.kr/~ppddee

Abstract. Recently, many Siamese network based object tracking methods have been proposed and have shown good performances. These method give two images to two identical artificial neural networks as the inputs and find the target area based on the similarity measured by the Siamese network. However, the measure used in the Siamese network is based on the offline training, and therefore, easily fail to adapt to online changes. In this paper, we propose to apply a distance measure which considers the relative position between the objects and the histogram information as additional online information. This additional information prevents the tracking to fail when hard negative cases appear in the scene.

Keywords: Object tracking · Siamese network · Histogram · Deep learning

1   Introduction

The task of video object tracking is to find the most similar patch in the next frame based on the appearance of the target in the previous frame. Video object tracking differ from object detection in the fact that the object to be followed can be any arbitrary target. Object tracking is in fact a class-agnostic task, which make it difficult for well-working classification networks to successfully track the object under consideration. Another problem is the hard negative problem, i.e., there is an object in the video that looks similar to the object we want to chase. To avoid the hard negative problem, a good choice for the similarity measure which measures the similarity between the same object in the previous and the current frame should be made, which is the topic of many Deep-learning based tracking methods.

Deep Learning-based tracking algorithms have come out with good performance since 2016 due to their representation power of the object under consideration. Some important breakthrough algorithms are the MDNET [8] which is one

This work was supported by the Technology development Program (S2644388) funded by the Ministry of SMEs and Startups (MSS, Korea) and the Basic Science Research Program through the National Research Foundation of Korea (NRF-2019R1I1A3A01060150).

D. Tzovaras et al. (Eds.): ICVS 2019, LNCS 11754, pp. 480–489, 2019.
https://doi.org/10.1007/978-3-030-34995-0_43

of the first attempts that learns a shared representation of targets, the GOTURN [4], which uses regression method to track the target, and the ROLO [9], which uses YOLO [10] to get the features of the target region. Siamese network based tracking methods formulate the tracking problem as a cross-correlation problem of convolutional features between the target template and the searching region [2,6]. A representative work is the SiamRPN [6] which tracks the object by measuring the similarity of the features produced by two identical networks.

However, Siamese trackers have still some tracking problems since the highest correlation can also be generated by a false negative object. At this time, introducing online adaption to the output of the SiamRPN can help to avoid the failure in the tracking. In this paper, we do not take the tracking result of the SiamRPN as the final target region, but rather regard the output as a constraint for the target region. That is, we obtain a set of candidate regions from the SiamRPN, and then apply a constraint which regards the positional relationship between the candidates, the past location, and the histogram information to finally predict the current target location. The combined visual understanding of the deep network and the online information based constraint work together to get a more accurate tracking result.

2   SiamRPN

The SiamRPN adapts the Region Proposal Network (RPN), which was proposed by the Faster R-CNN [11], into the Siamese network structure. The Siamese subnetwork extracts features which will used by the RPN. The RPN constitutes of two networks where one is used to differentiate between the background and the foreground, and the other makes proposal regions for the target object. In the training phase, the goal is to find the parameters W for the region proposal subnetwork ζ that minimizes the following loss $\mathcal{L}$:

$$\min_{W} \frac{1}{n} \sum_{i=1}^{n} \mathcal{L} \left(\zeta \left(\varphi \left(x_i; W \right); \varphi \left(z_i; W \right) \right), \ell_i \right) \tag{1}$$

Here, z_i is the template patch, x_i the detection patch, ℓ_i the true label, n the number of training samples, and φ is the function for the Siamese feature extraction subnetwork. After W has been learned, in the test time, the target is passed through the network and the features for the target object is computed. For the next frames coming in, the features are compared with the pre-computed features and the target location is determined to be that with the highest value among the features produced by filtering. The SiamRPN works well when there is a single object due to the good visual understanding. However, if there are many objects of the same class, then the SiamRPN often fails to track the right target and tracks a distractor instead.

3   Proposed Method

In this section, we propose to augment the online color histogram distribution and the history of relative position information into the SiamRPN tracking method so that the combined information helps to avoid tracking failure. Figure 1 shows a video sequence together with their activation maps which are the results of putting them through the SiamRPN. As can be seen from the second row in Fig. 1, the activation maps show large values for all person regions as they are from the same person class, and the SiamRPN has a good visual understanding for these regions. Therefore, it is very likely that the maximum activation switches from the target region to a nearby non-target region which causes the failure in the tracking.

However, the large activation regions which the SiamRPN provides can also be regarded as good candidate regions for the target region. We regard the failure in tracking as a wrong switch between the candidate regions and propose a method that prevents this kind of wrong switch. To obtain the candidate regions, first a non-maximum suppression algorithm is applied on the activation map to eliminate the overlapping regions of the candidates. That is, we erased the boxes which overlap more than 50% with other candidate boxes and ordered them from highest to lowest according to the activation values. The upper row in Fig. 1 shows the candidate regions after applying the non-maximum suppression. Here, the four boxes are the candidate regions numbered by the order of the activation values inside the regions.

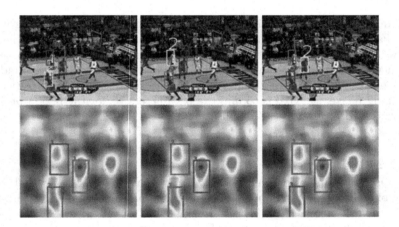

Fig. 1. Showing the candidate regions after applying the non-maximum suppression method on the heatmaps obtained by the SiamRPN

Define by S_p the set of ordered proposal regions. The regions are proposed by the SiamRPN method, which we denote by $p_i, i = 1, 2, ..N$, where the index is ordered in the first frame according to the values of the largest activation in p_i, i.e., p_1 is the region that contains the largest activation value in it, and p_2 is the region containing the second largest activation value. We assume the target region is always in the set S_p. We try to find the set S_p^* in the next frame which solves the following problem:

$$S_p^* = \min_{S_p} \sum_k \sum_i \left\| \mathbf{h}_w \left(p_k^{n-1} \right) - \mathbf{h}_w \left(p_i^n \right) \right\|_2 D \left(\mathbf{c}_{p_k}, \mathbf{c}_{p_i} \right) \tag{2}$$

where p_k^{n-1} denotes the k's region in the previous frame, p_i^n the i's region in the current frame, $\mathbf{h}_w \left(p_k^{n-1} \right)$ the weighted normalized histogram of the region p_k^{n-1}, and $\mathbf{h}_w \left(p_i^n \right)$ the weighted normalized histogram of the region p_i^n. The weights are given by the Gaussian kernel to give a large weight to the center of the region. Furthermore, we give a larger weight for the region containing the largest activation value which is considered in $\mathbf{h}_w \left(p_i^n \right)$. By minimizing the term $\left\| \mathbf{h}_w \left(p_k^{n-1} \right) - \mathbf{h}_w \left(p_i^n \right) \right\|_2$, we try to find the set S_p^* where the candidate regions are ordered so that the magnitude of the total sum of the differences in the weighted histograms between the sets in the previous and the current frames becomes smallest. The term $D \left(\mathbf{c}_{p_k}, \mathbf{c}_{p_i} \right)$ is defined as follows:

$$D(\mathbf{c}_{p_k^{n-1}}, \mathbf{c}_{p_i^n}) = M \left(\tau - \left\| \mathbf{c}_{p_k^{n-1}} - \mathbf{c}_{p_i^n} \right\|^2, 0 \right), \tag{3}$$

where $\mathbf{c}_{p_k^{n-1}}$ and $\mathbf{c}_{p_i^n}$ denote the centers of the regions p_k^{n-1} and p_i^n, respectively, and $M \left(\cdot \right)$ is defined as,

$$M(x) = \begin{cases} T & \text{if } x < 0 \\ \tau - x & \text{if } 0 \le x \le \tau \end{cases} \tag{4}$$

where τ is a threshold value, and T a very large value. The term $D(\mathbf{c}_{p_k^{n-1}}, \mathbf{c}_{p_i^n})$ is large if the distance between $\mathbf{c}_{p_k^{n-1}}$ and $\mathbf{c}_{p_i^n}$ is large, so it tries to prevent to choose the region p_i^n as the matching region for p_k^{n-1} if the distance between these regions is large. The $M(\cdot)$ function applies a very large value T to the distance measure, when the distance between $\mathbf{c}_{p_k^{n-1}}$ and $\mathbf{c}_{p_i^n}$ is larger than the threshold τ, since the object regions cannot move very far in the next frame. Rather than using a distance score for only the target region, we applied a set score which depends on the relative positions of all the candidate regions. In this way, we can constrain the next location of the target region based on the relative positions with other objects together, which we think works as a stronger constraint than a constraint that works on a single object only. For example, let the score value be,

$$Score = \sum_k \sum_i \left\| \mathbf{h}_w \left(p_k^{n-1} \right) - \mathbf{h}_w \left(p_i^n \right) \right\|_2 D \left(\mathbf{c}_{p_k}, \mathbf{c}_{p_i} \right) \tag{5}$$

then, the score values for the first, second, and third image in the upper row of Fig. 1 are 2.3842, 21.5675, and 22.837, respectively. So, the right ordering will be as in the first image, and the next target region becomes the number one region in the first image.

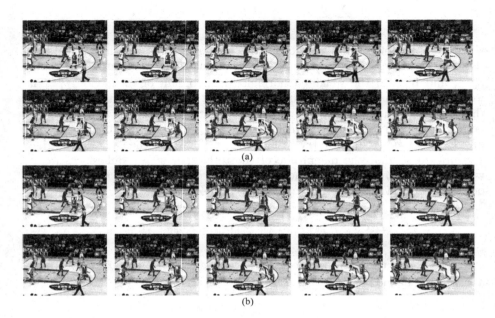

(a)

(b)

Fig. 2. Comparison between the tracking results of the SiamRPN and the proposed method: (a) SiamRPN (b) Proposed

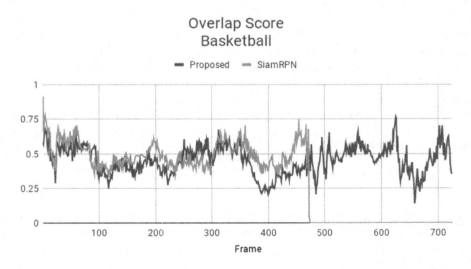

Fig. 3. Graph of overlap scores for the SiamRPN and the proposed methods for the video sequence in Fig. 2

4   Experimental Results

We made a comparison with the SiamRPN on the OTB2015 [12] datasets. To measure the quantitative performance of the tracking algorithms we used the overlap score ϕ_t for each frame t which is defined as follows:

$$\phi_t = \frac{R_t^G \cap R_t^T}{R_t^G \cup R_t^T},\qquad(6)$$

where R_t^G is the ground truth region, R_t^T the estimated region, and $\cap$ and $\cup$ refer to the intersection and the union operators, respectively. Figure 2 shows the tracking results in the case where there are many players in the scene together and overlap with each other. As can be seen from Fig. 2(a), with the SiamRPN the target region changes to a nearby distractor and fails in the tracking. This can be also observed in the graph in Fig. 3, where the overlap score for the SiamRPN drops to zero after the tracking failure. The distractor regards each player in the scene as an object from the same class and gives a high activation value for each region. However, the proposed method successfully avoids the distractor and keeps on tracing the right object. This can be also observed with the football sequence shown in Fig. 4.

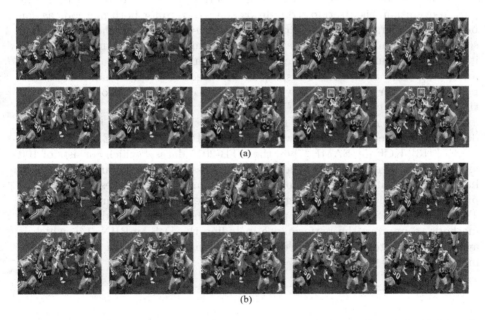

(a)

(b)

Fig. 4. Comparison between the tracking results of the SiamRPN and the Proposed method: (a) SiamRPN (b) Proposed

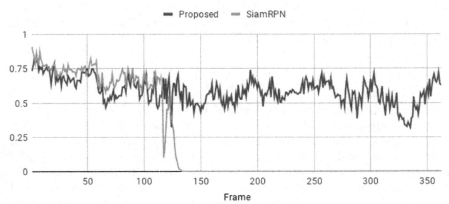

Fig. 5. Graph of overlap scores for the SiamRPN and the Proposed methods for the video sequence in Fig. 4

Figure 6 shows a video sequence where partial occlusion occurs. Due to the partial occlusion, the overlap score drops down to zero for both methods as can be seen in the graph in Fig. 7. However, with the proposed method the overlap score becomes large again after a short term, while with the SiamRPN method it takes a long time. Furthermore, with the SiamRPN, the overlap score is zero for a long time after the occlusion, as the predicted region has moved to a nearby distractor region, while with the proposed method it is only because of the occlusion as can be seen in Fig. 6.

Table 1 compares the success rates of our method with other tracking methods. We compared our method with the SiamRPN [6], STAPLE [1], DSST [3], MEEM [13], KCF [5], SAMF [7] on the OTB-2015 dataset which consists of 100 video sequences, and recorded the average success rates. We regarded the tracking as a success if the overlap score is larger than 0.2 and 0.4, respectively. The proposed method shows in average an improvement over the SiamRPN method. The proposed method can be applied on other neural network detection based methods to improve the performance including the DaSiamRPN [14], which is aware of distractors by using a different training method than the SiamRPN method.

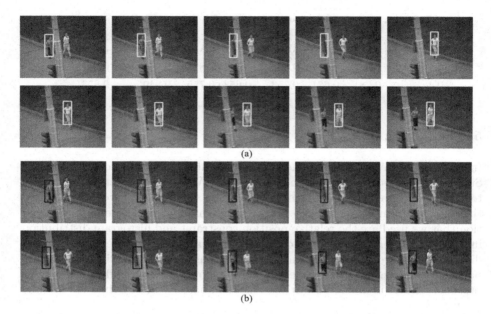

(a)

(b)

Fig. 6. Comparison between the tracking results of the SiamRPN and the proposed method: (a) SiamRPN (b) Proposed

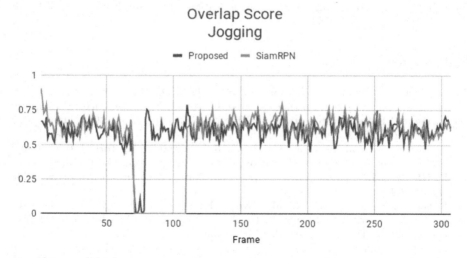

Fig. 7. Graph of overlap scores for the SiamRPN and the Proposed methods for the video sequence in Fig. 6

Table 1. Comparison of success rates with other tracking methods

Threshold	Staple	DSST	MEEM	KCF	SAMF	SiamRPN	proposed
0.2	0.827	0.737	0.851	0.746	0.812	0.874	0.887
0.4	0.781	0.671	0.753	0.653	0.755	0.849	0.853

5 Conclusion

In this paper, we presented a method which enhances the performance of the SiamRPN method by using online information. We used the SiamRPN method to propose several meaningful candidate regions for the target region and used the online information to select between the candidates. The online information that we used are the relative position constraint and histogram distribution information. Interpreting the tracking failure as a false switch between the candidate regions, we tried to prevent this switch based on the relative relationship between these regions. Experimental results show that this additional process prevents failures in difficult cases and results in better tracking than the original SiamRPN.

Acknowledgements. This work was supported by the Technology development Program (S2644388) funded by the Ministry of SMEs and Startups (MSS, Korea) and the Basic Science Research Program through the National Research Foundation of Korea (NRF-2019R1I1A3A01060150).

References

1. Bertinetto, L., Valmadre, J., Golodetz, S., Miksik, O., Torr, P.H.S.: Staple: complementary learners for real-time tracking. In: 2016 IEEE Conference on Computer Vision and Pattern Recognition (CVPR), pp. 1401–1409, June 2016. https://doi.org/10.1109/CVPR.2016.156
2. Bertinetto, L., Valmadre, J., Henriques, J.F., Vedaldi, A., Torr, P.H.S.: Fully-convolutional siamese networks for object tracking. In: Hua, G., Jégou, H. (eds.) ECCV 2016. LNCS, vol. 9914, pp. 850–865. Springer, Cham (2016). https://doi.org/10.1007/978-3-319-48881-3_56
3. Danelljan, M., Häger, G., Shahbaz Khan, F., Felsberg, M.: Accurate scale estimation for robust visual tracking. In: Proceedings of the British Machine Vision Conference. BMVA Press (2014). https://doi.org/10.5244/C.28.65
4. Held, D., Thrun, S., Savarese, S.: Learning to track at 100 FPS with deep regression networks. In: Leibe, B., Matas, J., Sebe, N., Welling, M. (eds.) ECCV 2016. LNCS, vol. 9905, pp. 749–765. Springer, Cham (2016). https://doi.org/10.1007/978-3-319-46448-0_45
5. Henriques, J.F., Caseiro, R., Martins, P., Batista, J.: High-speed tracking with kernelized correlation filters. CoRR abs/1404.7584 (2014). http://arxiv.org/abs/1404.7584
6. Li, B., Yan, J., Wu, W., Zhu, Z., Hu, X.: High performance visual tracking with siamese region proposal network. In: 2018 IEEE/CVF Conference on Computer

Vision and Pattern Recognition, pp. 8971–8980, June 2018. https://doi.org/10. 1109/CVPR.2018.00935

7. Li, Y., Zhu, J.: A scale adaptive kernel correlation filter tracker with feature integration. In: Agapito, L., Bronstein, M.M., Rother, C. (eds.) ECCV 2014. LNCS, vol. 8926, pp. 254–265. Springer, Cham (2015). https://doi.org/10.1007/978-3-319-16181-5_18

8. Nam, H., Han, B.: Learning multi-domain convolutional neural networks for visual tracking. In: 2016 IEEE Conference on Computer Vision and Pattern Recognition (CVPR), pp. 4293–4302, June 2016. https://doi.org/10.1109/CVPR.2016.465

9. Ning, G., et al.: Spatially supervised recurrent convolutional neural networks for visual object tracking. In: 2017 IEEE International Symposium on Circuits and Systems (ISCAS), pp. 1–4, May 2017. https://doi.org/10.1109/ISCAS.2017. 8050867

10. Redmon, J., Divvala, S., Girshick, R., Farhadi, A.: You only look once: unified, real-time object detection. In: 2016 IEEE Conference on Computer Vision and Pattern Recognition (CVPR), pp. 779–788, June 2016. https://doi.org/10.1109/ CVPR.2016.91

11. Ren, S., He, K., Girshick, R., Sun, J.: Faster R-CNN: towards real-time object detection with region proposal networks. In: Cortes, C., Lawrence, N.D., Lee, D.D., Sugiyama, M., Garnett, R. (eds.) Advances in Neural Information Processing Systems, vol. 28, pp. 91–99. Curran Associates, Inc. (2015). http://papers.nips.cc/paper/5638-faster-r-cnn-towards-real-time-object-detection-with-region-proposal-networks.pdf

12. Wu, Y., Lim, J., Yang, M.: Object tracking benchmark. IEEE Trans. Pattern Anal. Mach. Intell. **37**(9), 1834–1848 (2015). https://doi.org/10.1109/TPAMI. 2014.2388226

13. Zhang, J., Ma, S., Sclaroff, S.: MEEM: robust tracking via multiple experts using entropy minimization. In: Fleet, D., Pajdla, T., Schiele, B., Tuytelaars, T. (eds.) ECCV 2014. LNCS, vol. 8694, pp. 188–203. Springer, Cham (2014). https://doi. org/10.1007/978-3-319-10599-4_13

14. Zhu, Z., Wang, Q., Li, B., Wu, W., Yan, J., Hu, W.: Distractor-aware siamese networks for visual object tracking. CoRR abs/1808.06048 (2018). https://arxiv. org/abs/1808.06048

Deep-Learning-Based Computer Vision System for Surface-Defect Detection

Domen Tabernik[1($\boxtimes$)], Samo Šela[2], Jure Skvarč[3], and Danijel Skočaj[1]

[1] Faculty of Computer and Information Science, University of Ljubljana,
Večna pot 113, 1000 Ljubljana, Slovenia
domen.tabernik@fri.uni-lj.si
[2] Kolektor Group d.o.o., Vojkova 10, 5280 Idrija, Slovenia
[3] Kolektor Orodjarna d.o.o., Vojkova 10, 5280 Idrija, Slovenia

Abstract. Automating optical-inspection systems using machine learning has become an interesting and promising area of research. In particular, the deep-learning approaches have shown a very high and direct impact on the application domain of visual inspection. This paper presents a complete inspection system for automated quality control of a specific industrial product. Both hardware and software part of the system are described, with machine vision used for image acquisition and pre-processing followed by a segmentation-based deep-learning model used for surface-defect detection. The deep-learning model is compared with the state-of-the-art commercial software, showing that the proposed approach outperforms the related method on the specific domain of surface-crack detection. Experiments are performed on a real-world quality-control case and demonstrate that the deep-learning model can be successfully used even when only 33 defective training samples are available. This makes the deep-learning method practical for use in industry where the number of available defective samples is limited.

1 Introduction

Reliable visual inspection is one of the key elements of the production processes for ensuring an adequate quality of the manufactured products. Replacing the manual inspection with the automated machine-vision systems has been a trend for many years. By adopting the Industry 4.0 paradigm, the need for advanced machine-vision inspection systems even increases [10]. Increasing demands for customisation of the products, small product series, more complex products and constantly higher quality requirements aiming at zero-defect production call for more general, flexible and complex machine-vision systems.

Machine vision is a well-established engineering discipline that has led to numerous successful machine-vision applications in industrial production lines. A typical machine-vision system is composed of an adequate hardware and software to perform the inspection task that is integrated with the rest of the production line. An appropriate mechanism for positioning the object to be observed is required, and a choice of a suitable illumination and acquisition devices plays a very important role as well. The hardware part of the system should provide as

© Springer Nature Switzerland AG 2019
D. Tzovaras et al. (Eds.): ICVS 2019, LNCS 11754, pp. 490–500, 2019.
https://doi.org/10.1007/978-3-030-34995-0_44

good visual data as possible, so that the software part can reliably extract the required information about the quality of the product. In a classical machine-vision approach to defect detection, an engineer would hand-craft the features adapted to a particular problem at hand based on his previous experience with similar problems. However, this leads to several weaknesses. The hand-crafted features tend to be quite problem-specific so when a new problem arises, an engineer would have to manually adapt the features to specifics of the new problem domain. Additionally, there are several problems that seem to be too difficult or impossible to be solved using the established hand-crafted solutions.

In this paper, we present a different approach to implementation of the software part of the machine-vision system. It is based on deep learning, which has proven to be very successful approach for solving numerous computer vision tasks. Modern computer vision systems heavily rely on deep-learning-based perception, which utilizes more advanced modelling capabilities. Compared to classical machine-vision methods, deep learning can directly learn features from low-level data, and has a higher capacity to represent complex structures. Such an approach addresses previously mentioned weaknesses of the classical machine-vision systems. It enables more general development of machine-vision systems, since they can be adapted to new problems by retraining the systems on the corresponding images, without the need of reprogramming the software. And, due to a better representation capacity, deep learning models are more successful in solving very complex problems as well.

Although the underlying principles are general, we will present a computer vision system for detecting a particular surface defect on a particular industrial product, an electrical commutator shown in Fig. 1. Commutators are an integral part of electrical motor, so the production of such an important component is completely automated. As such, each produced commutator undergoes through complete in-line optical inspection in the acceptable production cycle time. In this paper, we will present the part of the system that inspects the product for the most challenging visual defect.

Fig. 1. Electrical commutator.

The remainder of the paper is structured as follows. The related work is presented in Sect. 2, with details of the optical-inspection system presented in Sect. 3 and evaluation in Sect. 4. The paper concludes with a discussion in Sect. 5.

2 Related Work

A classical machine-vision approaches for defect detection follow more or less the same paradigm. Hand-crafted features are developed for the particular problem domain and classifiers, such as SVM, kNN, decision trees, or similar established computer vision techniques, are utilized to extract the discriminative information from the features. Various filter banks [12], histograms, wavelet transforms [5], morphological operations [6] and others techniques are used to hand-craft the appropriate features, since the classifiers are less powerful than deep-learning methods.

In contrast to the classical machine-vision approach the deep-learning app-roach directly learns the features. Several different works also employed deep-learning methods for optical inspection, which is the main focus of this paper. The work by [7] showed that five-layer convolutional network can outperform classic hand-engineered features on image classification of steel defect. A simi-lar architecture was used by [4] for the detection of rail-surface defects. A more modern network architecture was employed by [2]. They applied the OverFeat [9] network to detect five different types of surface errors and identified a large num-ber of labeled data as an important problem for deep networks. Although they proposed to mitigate this using an existing pre-trained network, however, their method does not learn the network itself on the target domain and is therefore not using the full potential of deep learning.

Full network learning was performed in [11], where authors evaluated several deep-learning architectures with varying depths of layers for surface-anomaly detection. They applied networks ranging from having only 5 layers to a net-work having 11 layers, and, although they showed deep network to outperform any classic method, they demonstrated this only on synthetic dataset. Their method has also shown to be fairly inefficient as it extracted small patches from each image and classified each individual image patch separately. A more efficient network for explicitly performing the segmentation of defects was proposed by [8]. They implemented a fully convolutional network with 10 layers, using both ReLU and batch normalization to perform the segmentation of the defects. Fur-thermore, they proposed an additional decision network on top of the features from the segmentation network to perform a per-image classification of a defect's presence. This allowed them to improve the classification accuracy on the dataset of synthetic surface defects. As opposed to some related works [8,11], the pro-posed network is applied to real-world examples with small number of defective samples instead of using large number of synthetic ones.

3 Deep-Learning-Based Optical Inspection System

Commutators are an integral part of elec-trical motors and as such are often used in various mechanical systems. The production of such an important component is today completely automated, and is subject to a complete in-line optical inspection. In this section, we present the system for fully auto-mated inspection of the compound-body part of the commutator, where a defect is mani-fested as a surface fracture of the material. The whole inspection process is depicted in Fig. 3 and consists of the automatic image acquisition process and the defect detection using the deep learning.

Fig. 2. Optical inspection system

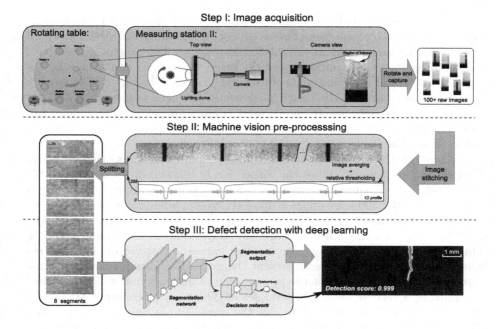

Fig. 3. Surface-defect detection system overview.

3.1 Hardware and Image Acquisition

The optical inspection is performed at the end of the production line, where the commutator enters into the automated machine for a complete optical inspection. The optical inspection machine, as depicted in Fig. 2, consists of 6 measuring stations that inspect 55 distinctive features with 23 different cameras. Manipulation of commutators inside the optical inspection machine is carried out by means of a rotating table with eight separate stations. Two stations are reserved for exit and entry point for the commutators, while six remaining stations perform active measurements of the item in a synchronous manner. Each station has 1.5 s of available time to complete its process. The inspected features range from 2D and 3D measurements of the physical properties of the product to various defects, such as missing material or porosity, mechanical damage on different parts of the commutator or the presence of residue from the production process.

In this work, we focus on the second active measurement station where the circumference of the compound-body of the commutator is being processed for the surface cracks. The whole process of the surface-defect detection consists of acquiring a high-resolution raw image parts of the circumference, combining them into full high-resolution image and then splitting them into eight segments to preserve only the relevant areas for final surface inspection.

In step one, the commutator is rotated in-place for 360° to acquire the whole surface area of the compound-body. The image optics and cameras are synchronized with the rotation of the commutator and the high-power LED strobe-light source. The surface of the commutator is being illuminated with the dome light source that has an opening at the top for the camera. The camera is positioned perpendicular to the circumference of the inspected surface and is viewing the object in a lateral direction as is illustrated in Step I in Fig. 3. The camera observes a larger area of the image as shown in the camera view in Fig. 3, but only smaller region-of-interest area is used in remaining steps. The smaller raw inputs are during the acquisition phase progressively combined to form the entire surface area. The final image size is 11700 × 500 pixels.

In step two, the system removes the dark notches in the image that represent the gaps in the material. Since the inspection is specialized for examining the defects on the surface of the material from which the commutator housing is made, the 11700 × 500 large image is divided into individual segments that do not contain these areas. To detect the horizontal edges of the material, we compute a 1D profile of the image by averaging the gray values in the lateral direction. This results in 11700 × 1 large vector. High values in this vector represent the material, while the low values represent the gaps. The abrupt change in those extreme values then represent the precise edge of the material, which we detect by applying a relative threshold level set at 30% to the computed 1D profile. The edges are used to horizontally cut the image to obtain 8 segments of size 1250 × 500 pixels, which contain only the material needed for the inspection.

3.2   Surface Inspection

Next, each image segment is passed through a deep convolutional network that detects surface defects. We utilize a two-stage network architecture that follows the designs presented in [8], with the first stage implementing a segmentation network to localize the surface defect, while the second stage implementing the binary image classification. The overview is depicted in Fig. 4. The first-stage network is referred to as *the segmentation network*, while the second-stage network, as *the decision network*. We provide a brief description of both stages here and refer a reader to [10] for a more detailed description.

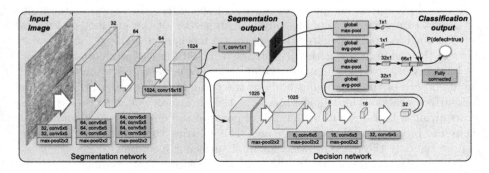

Fig. 4. The proposed architecture with the segmentation and decision networks.

Segmentation Network. The design of the segmentation network focuses on the detection of small surface defects in a large-resolution image. To accomplish this the network contains 11 convolutional layers and three max-pooling layers that each reduce the resolution by a factor of two. Each convolutional layer is followed by a feature normalization and a non-linear ReLU layer, which both help to increase the rate of convergence during the learning. Feature normalization normalizes each channel to a zero-mean distribution with a unit variance. The number of channels and kernel sizes used in each layer are shown in Fig. 4. The final output mask is obtained after applying 1×1 convolution layer that reduces the number of output channels. The resolution of the output map is 8-times smaller than of the input image and is not interpolated back to the original image since this resolution suffices for the problem at hand.

Decision Network. The architecture of the decision network builds on the output from the segmentation network. As the input the decision network takes the output of the last convolutional layer of the segmentation network (1024 channels) concatenated with a single-channel segmentation output map. The input features are processed by a combination of a max-pooling layer and a convolutional layer that are repeated 3 times as shown in Fig. 4. This design effectively results in a 64-times-smaller resolution of the last convolutional layer than that of the original image. Finally, the network performs global maximum and average pooling, resulting in 64 output neurons. Additionally, the result of the global maximum and average pooling on the segmentation output map are concatenated as two output neurons, to provide a shortcut for cases where the segmentation map already ensures perfect detection. This design results in 66 output neurons that are combined with linear weights into the final output neuron.

Learning. Both segmentation and decision networks are trained using the cross-entropy loss, however, the loss is calculated per-pixel for the segmentation network and per-image for the decision network. Both models are initialized randomly using a normal distribution. Networks are trained separately by first training only the segmentation network independently, then freezing the weights for the segmentation network and finally training only the decision network layers. This avoids the issue of overfitting from the large number of weights in the segmentation network. This is more important for the decision layers than for the segmentation layers due to limited GPU memory constraining the batch size to one. The segmentation layers are not effected by this due to pixel-wise loss that effectively increases the number of samples in a batch.

4 Evaluation

In this section, we present the evaluation of the proposed system. The whole system has been evaluated by first utilizing image acquisition and machine-vision pre-processing steps to collect the data. The collected data has then been

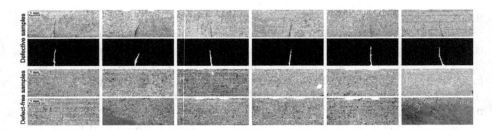

Fig. 5. Several examples of surface images with visible defects and their annotation masks in the top, and defect-free surfaces in the bottom.

used to train and evaluate the deep-learning model. Moreover, the deep-learning model has been compared against the state-of-the-art commercial product, the Cognex ViDi Suite [3].

4.1 Evaluation Setup

Evaluation Data. To collect the data for the evaluation, fifty defective items were passed through the first two stages. This resulted in a total of 399 images; 52 with a visible defect and 347 without any visible surface defects. The defective samples were annotated with a pixel-wise segmentation mask that was additionally dilated with morphological kernel of size 5×5. Several examples of the defective and non-defective surfaces are depicted in Fig. 5. The evaluation is performed with a 3-fold cross validation, while ensuring all the images of the same physical product are in the same fold.

Evaluation Metrics. Three different classification metrics were measured in the evaluation: (a) average precision (AP), (b) number of false negatives (FN) and (c) number of false positives (FP). Note, the positive sample is referred to as an image with a visible defect, and the negative sample, as an image with no defects. We focus mostly on the average precision, since it is more appropriate metric than FP or FN, as it accurately captures the performance of the model under different thresholds in a single value. The number of miss-classifications (FP and FN) are dependent on the specific threshold applied to the classification score. We report FP and FN at a threshold value where the best F-measure is achieved.

Implementation and Learning Details. The network architecture was implemented in the TensorFlow framework [1], using stochastic gradient descend without momentum, a learning rate of 0.1, a batch size of one, and training for 100 epochs. Additionally, positive and negative samples were balanced during the learning by taking images with defects for every even iteration, and images without defects for every odd iteration.

Table 1. Comparison of the defect-detection methods.

	Average precision	False positive	False negative	FP at 100% recall
Seg. and dec. network	**99.9 %**	**0**	**1**	**3**
Cognex ViDi Suite [3]	99.0 %	0	5	7

Commercial Software. We compared the presented model against the Cognex ViDi Suite v2.1 [3]. The training and evaluation model was set to mirror the training and evaluation of the segmentation and decision network. This included using a gray-scale image, learning for 100 epochs and using the same train/test split. This configuration and hyper-parameter setup resulted in the best possible performance that we could achieve with the commercial software on this domain. For more details on the hyper-parameter setup for both the proposed deep-learning model and for the Cognex ViDi Suite the reader is referred to [10].

4.2 Results

The results are presented in Table 1. The segmentation and decision network outperformed the commercial product in all metrics. Observing the number of miss-classification at the ideal F-measure reveals that the segmentation and decision network missed only one defective sample, while the commercial product had 5 miss-classifications. Several miss-classified images are presented in Fig. 6. Both methods performed well on non-defective samples and did not detect and false positives.

Table 1 also shows the number of false positives that would be obtained if a zero-miss rate would be required, i.e., if a recall rate of 100% is required. These false positives then represent the number of items that would be needed to be manually verified by a skilled worker and directly point to the amount of work required to achieve the desired accuracy. Comparing the results in this metric reveals that the presented model introduces only 3 false positives at a zero-miss rate out of all 399 images. This represents 0.75% of all images. On the other hand, the commercial product achieved worse results, requiring the manual verification of 7 images.

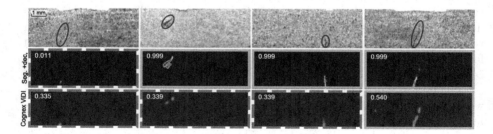

Fig. 6. Examples of true-positive (green solid border) and false-negative (red dashed border) detections with the segmentation output and the corresponding classification (the actual defect is circled in the first row). (Color figure online)

Table 2. Computational cost for the individual processing stages.

Per item processing time	Image capture	Pre-processing and split	Defect detection	Total
GTX 1080 TI 11 GB	61 ms	4 ms	$110 \cdot 8 = 880$ ms	945 ms
GTX 1080 8 GB	61 ms	4 ms	$139 \cdot 8 = 1112$ ms	1177 ms

4.3 Computational Cost

Due to requirements of the production process the processing time for the inspection of the whole item is restricted to 1.5 s. The processing time is composed of the raw image acquisition, cropping of the ROI and combining it into the whole image. This process takes 0.6 ms per raw image part, and 61 ms for all 101 parts. Next, the processing time also accounts for the image splitting, which accounted for 4 ms.

The computationally most demanding part is the actual defect detection with deep-learning. Using the same GPU as used for the learning (NVIDIA GTX 1080 TI 11GB) the processing of a single image takes 110 ms, thus resulting in 880 ms for all eight images and 945 ms for the whole system. Although, deep-learning-based defect detection is computationally most demanding, it is still efficient enough to meet the required criteria. Even with a more cost effective GPU, such as GTX 1080 8GB, the system can still manage to complete the task in the required time, taking 1112 ms for the defect detection on all eight images, and 1177 ms for the whole system (Table 2).

5   Discussion and Conclusion

In this paper, we presented a complete optical inspection system for automated detection of surface defects on electrical commutator. We presented the image acquisition system, which captures the surface of the whole item and converts it into eight non-overlapping images using classical machine-vision processes. We then employed a more powerful deep-learning approach to detect surface-defects using a segmentation and decision networks. We evaluated the deep-learning approach on the problem of surface-defect detection where defects appear as fractures on the compound-body of the electrical commutator. The segmentation and decision network was demonstrated to achieve significantly better results than the related state-of-the-art commercial product, with only one miss-classification for the segmentation and decision network, and five miss-classifications for the commercial product.

The performance of the presented method was achieved by learning the network from only 33 defective samples. This indicates that the presented deep-learning approach is suitable for the studied industrial application with a limited number of defected samples available. The system has also proven to be ready for use in the industrial environment with the required manual inspection rate as low as 0.75% (three images out of all 399 images) when all defective samples need to be found.

Although the main innovation of the presented system is the deep learning approach that has been used for detecting the surface defects, a part of the developed system still uses the classical machine-vision techniques. In the pre-processing step the classical methods are used to stitch and split the images, therefore to produce high-quality images that are latter on used by the learning-based defect-detection method. This part is relatively easy to implement, and robust solutions have already been established, so there is no need to replace it with the data-driven approach. In principle, this could be done, but probably at a cost of a higher required number of training images, which could often be difficult to obtain. A good advice is therefore to use simple machine-vision methods in the pre-processing step to prepare the training data as well as possible, and to use them in the deep-learning approach to solve the more difficult part of the visual inspection problem. Such an approach provides the best results, and it is expected that a reasonable combination of both, classical machine-vision and data-driven learning-based approaches will very often be used in the future machine-vision systems.

Acknowledgements. This work was supported in part by the following research programs: GOSTOP program C3330-16-529000 co-financed by the Republic of Slovenia and the ERDF, ARRS research project J2-9433 (DIVID), and ARRS research programme P2-0214.

References

1. Abadi, M., et al.: TensorFlow: large-scale machine learning on heterogeneous systems (2015). https://www.tensorflow.org/
2. Chen, P.H., Ho, S.S.: Is overfeat useful for image-based surface defect classification tasks? In: IEEE International Conference on Image Processing, pp. 749–753 (2016)
3. Cognex: VISIONPRO VIDI: deep learning-based software for industrial image analysis (2018). https://www.cognex.com/products/machine-vision/vision-software/visionpro-vidi
4. Faghih-Roohi, S., Hajizadeh, S., Núñez, A., Babuska, R., Schutter, B.D.: Deep convolutional neural networks for detection of rail surface defects deep convolutional neural networks for detection of rail surface defects. In: International Joint Conference on Neural Networks, pp. 2584–2589, October 2016
5. Ghazvini, M., Monadjemi, S.A., Movahhedinia, N., Jamshidi, K.: Defect detection of tiles using 2D-wavelet transform and statistical features. Int. Schol. Sci. Res. Innov. **3**(1), 773–776 (2009)
6. Mak, K.L., Peng, P., Yiu, K.F.: Fabric defect detection using morphological filters. Image Vis. Comput. **27**(10), 1585–1592 (2009). https://doi.org/10.1016/j.imavis.2009.03.007
7. Masci, J., Meier, U., Ciresan, D., Schmidhuber, J., Fricout, G.: Steel defect classification with max-pooling convolutional neural networks. In: Proceedings of the International Joint Conference on Neural Networks (2012). https://doi.org/10.1109/IJCNN.2012.6252468

8. Rački, D., Tomaževič, D., Skočaj, D.: A compact convolutional neural network for textured surface anomaly detection. In: IEEE Winter Conference on Applications of Computer Vision, pp. 1331–1339 (2018). https://doi.org/10.1109/WACV.2018.00150

9. Sermanet, P., Eigen, D.: OverFea: integrated recognition, localization and detection using convolutional networks. In: International Conference on Learning Representations (2014)

10. Tabernik, D., Šela, S., Skvarč, J., Skočaj, D.: Segmentation-based deep-learning approach for surface-defect detection. J. Intell. Manuf. 1–18 (2019). https://doi.org/10.1007/s10845-019-01476-x

11. Weimer, D., Scholz-Reiter, B., Shpitalni, M.: Design of deep convolutional neural network architectures for automated feature extraction in industrial inspection. CIRP Ann. - Manuf. Technol. **65**(1), 417–420 (2016). https://doi.org/10.1016/j.cirp.2016.04.072

12. Zheng, H., Kong, L.X., Nahavandi, S.: Automatic inspection of metallic surface defects using genetic algorithms. J. Mater. Process. Technol. **125–126**, 427–433 (2002). https://doi.org/10.1016/S0924-0136(02)00294-7

Color-Guided Adaptive Support Weights for Active Stereo Systems

Ioannis Kleitsiotis[(✉)], Nikolaos Dimitriou, Konstantinos Votis,
and Dimitrios Tzovaras

Information Technologies Institute, Centre of Research and Technology Hellas,
6th km Harilaou - Thermi, 57001 Thessaloniki, Greece
{ioklei,nikdim,kvotis,Dimitrios.Tzovaras}@iti.gr

Abstract. In this paper we present a color-guided adaptive support weight scheme for the cost aggregation of active stereo matching systems. These systems work by stereo matching two images using the texture provided by infrared pseudo-random dot pattern projectors. This method might fail in regions where the pattern is absent, due to the geometry of the scene and/or the reflectivity properties of the test material. However, leveraging the texture information provided by a separate color sensor might uncover details otherwise unseen by the infrared sensors. We propose a cost aggregation method that utilizes both an infrared and a color image of the scene, making smart aggregation choices depending on the underlying texture information provided by the two separate images. We use our cost aggregation method with the fully self-supervised real-time architecture presented in [14], having in mind the usage of low-cost commercial active stereo matching sensors, like the Intel Realsense D435 sensor, in industrial applications demanding high-quality depth maps. We evaluate our results on our own dataset comprised by vehicle surface data, and give qualitative evidence of the disparity estimation improvements.

Keywords: Active stereo · Depth estimation · Adaptive support weights · Color-guided · Cost aggregation · Self-supervised

1 Introduction

The problem of acquiring accurate depth maps using commercial sensors is important for the realization of low-cost data acquisition systems that can be used in high precision industrial applications. Deep learning solutions are known to provide impressive results in estimating the depth of a scenery when trained with high-quality ground truth depth data. But the abscence of large ground truth depth datasets, particularly for specialized industrial settings, means that the depth acquisition system has to rely for its accuracy solely on the depth sensor quality. Therefore, the usage of cheaper commercial sensors in industrial applications is, in most cases, not a viable option, due to the low quality of the depth maps they provide.

© Springer Nature Switzerland AG 2019
D. Tzovaras et al. (Eds.): ICVS 2019, LNCS 11754, pp. 501–510, 2019.
https://doi.org/10.1007/978-3-030-34995-0_45

On the one hand, low-cost active depth sensors, like Microsoft Kinect v2, underperform in highly illuminated scenes, and cannot adequately capture specular surfaces. On the other hand, low-cost passive sensors suffer from wrong stereo matches between their left and right cameras, stemming mainly from textureless and occluded regions. Active stereo sensors sought to solve these problems by combining infrared stereo images, and an infrared projector that creates texture by projecting a pseudo-random light pattern on the scene. A relatively new commercial sensor that uses this technology is Intel Realsense D435. The authors of [14] improved the output depth maps of Intel Realsense D435 by training their neural architecture to learn the disparity between the left and right images in a fully self-supervised way, and with real-time usage of the network in their mind. The self-supervised nature of their network makes it ideal for industrial applications where the test data are more predictable than arbitrary indoor-outdoor scenery. Moreover, the machine learning approach has the advantage of the continuous improvement of the algorithm, as soon as new data are available.

However, apart from producing left and right infrared images, the Intel Realsense D435 sensor also captures a color image of the scenery, which might provide rich details that can enhance the resulting depth map. Furthermore, in an industrial setting, high-quality color images can be captured from distinct color sensors. Based on these observations, we propose an improvement on the active stereo matching algorithm of [7], incorporating color information on the training procedure, and thus helping in cases where color texture is more representative of a region than the corresponding infrared one. Our method uses color information only in the training phase, and thus there is no additional overhead in the real-time usage of the algorithm.

2   Related Work

There is a great amount of literature dedicated to the problem of depth enhancement. The authors of [3] presented a variational shape-from-shading method to solve the underdetermined problem of depth map super resolution. In [13], a method for depth inpainting is proposed, leveraging information from the color image to predict surface normals and occlusion boundaries, and using the Poisson-reconstructed input data as ground truth. The authors of [12] create denoised ground truth through dynamic fusion and train a neural network to learn the denoising function. They afterwards refine the denoised depth data in an unsupervised manner using a shading-based loss guided by color information.

A related problem is the estimation of a depth map from a corresponding color image. In [5], the authors use a novel upsampling block to recover vertically flat or horizontal depths from the color images, and they train a reliability network whose output is used in a CRF refinement method. In [15], depth estimation from real-world color images is tackled by a GAN that utilizes synthetic depth-color ground truth pairs. The authors of [10] extract depth information in different frequency scales by cropping the color image with different cropping ratios, and then combine them in the frequency domain to output the final depth map. Different from the above, the authors of [16] take as input a video sequence and simultaneously train a relative pose estimation network and a depth estimation network. The nearby images are inverse warped to a target image, and the network is trained in an unsupervised manner through a photometric reconstruction loss.

Leveraging the information given by a pair of rectified, synchronized stereo images can provide impressive results in estimating the disparity map and, in a straightforward manner, the depth map, of a scene. The authors of [6] use the geometric knowledge for the stereo problem by developing differentiable layers for the major parts of traditional stereo matching algorithms. In [2], the disparity map is predicted from a pair of stereo images in an unsupervised manner through a left-right consistency loss. A real-time solution is presented in [8], where an estimate of the disparity is produced through a low resolution cost map, and then refined by a color-guided refinement network. Based on [8], the authors of [14] presented an architecture that predicts the disparity of stereo images viewing an actively illuminated scene. They trained their network in a fully self-supervised manner, and thus they gave special care in constructing a loss function that faithfully captures the active stereo matching problem.

3   Proposed Method

3.1   Network Architecture

The model architecture we use is the disparity prediction network presented in [14]. As is common in deep stereo matching architectures [6,8], the authors of [14] use a siamese tower to extract low resolution features from the rectified, synchronized left and right high resolution images, that will afterwards be used in the formation of a low resolution cost volume. With the utilization of a soft argmin operator on the cost volume a low resolution disparity map is produced. This disparity map is then upsampled by a bilinear filter to the original resolution, and refined through a residual refinement network.

3.2   Color Information Preprocessing

In the following sections, we will use color information made available to us by a color sensor, in order to improve the disparity map predicted by the network introduced in [14]. The field of view of the color sensor overlaps with the field of view of the left infrared sensor, and we asume that their relative positions, as well as their internal calibration parameters, are known. Our preprocessing algorithm takes as input the left image of the active stereo system, the color image, and the noisy depth map produced by a computationally cheap off-the-shelf stereo matching algorithm. In our experiments, we used the depth maps provided by the algorithm [7] that the Intel Realsense D435 sensor uses to output depth. Using the calibration parameters and the noisy depth map, we register

Initial Color Image +   Registered Color Image   Inpainted Registered Color Image
Left IR Image

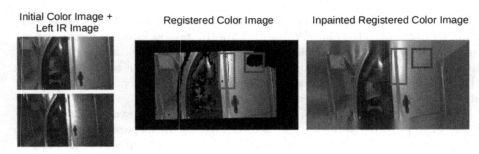

Fig. 1. The outputs of different stages of the color information preprocessing pipeline - from left to right, we present the initial color image of a vehicle, the image registered to the corresponding left infared image, and the inpainted registered color image using the method presented in [11]. Notice how the inpainting algorithm faithfully reconstructs the texture of the metal surface in small missing vegions (marked in red boxes) of the registered color image, but might also introduce smooth color variations in larger missing sections (marked in blue boxes). (Color figure online)

[4] the color image to the left image. This process results in the color image having missing values due to the inaccuracies of the noisy depth. To alleviate this problem, we use the inpainting method described in [11], which produced adequate results in our experiments. In Fig. 1 we observe that sufficiently small regions are adequately inpainted and preserve all the information present on the original color image. On the other hand, in larger regions inpainting produces some smooth color artifacts, which might have an adverse effect on disparity estimation. This necessitated the introduction of a tunable hypermarameter in our loss function. The hyperparameter will control the degree by which we include color information in our training process.

3.3   Color-Guided Adaptive Support Weights

A stereo matching algorithm must alleviate the ambiguity in disparity estimation arising from many left-right pixel matches having the same costs. The traditional method to achieve this is by aggregating the per pixel costs C_{ij} in a rectangular window around every pixel. In a supervised setting, the network can learn by itself a complex aggregation method by passing the cost volume through a series of 3D convolutions and non linear activations [8]. In a fully self-supervised setting like the one we use, it is beneficial to incorporate a cost aggregation method in the learning process through the loss function.

The cost aggregation method used by the authors of [14] is the adaptive support weights (ASW) scheme [9]. Assuming a square $2k \times 2k$ window, the per pixel aggregated cost they use is:

$$\hat{C}_{ij} = \frac{\sum_{x=i-k}^{i+k-1} \sum_{y=j-k}^{j+k-1} w_{xy} C_{xy}}{\sum_{x=i-k}^{i+k-1} \sum_{y=j-k}^{j+k-1} w_{xy}} \tag{1}$$

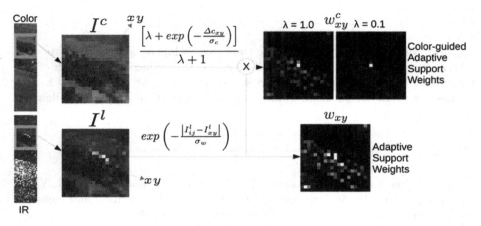

Fig. 2. Pictorial representation of the data and calculations involved in the two weighting schemas, as well as their qualitative difference. On the left hand side, we present the subsampled by a factor of 8 color and infrared images. We calculate the cost aggregation weights inside the green boxes, for the pixel lying in the center of the box. We have zoomed these windows, and highlighted in yellow the central pixel. The pipeline at the bottom of the figure depicts the traditional adaptive support weights calculation, and outputs an intensity map that represents the value of each weight. One can observe that it only utilizes infrared information. On the other hand, the upper pipeline additionally incorpotates color information (through the multiplication of the color based term) to produce the color-guided adaptive support weights that reduce the strength of related pixels in a principled way. The output of this pipeline is depicted for two λ values, namely 1.0 and 0.1. It is worthwhile noticing that the central pixel retains its strength, while the surrounding relevant pixels diminish in intensity, especially for lower values of λ. (Color figure online)

where $w_{xy} = exp\left(-\frac{|I_{ij}^l - I_{xy}^l|}{\sigma_w}\right)$, $\sigma_w = 2$, and C_{ij} is the weighted local contrast normalization costs described in [14]. The ASW scheme uses information from the left infrared image I^l to decide which pixels are similar with the central pixel of the window, and adds their costs. In the loss minimization process, this will lead to lesser ambiguity in cases where a pixel from the left image matches with several pixels from the right image, because it enforces the algorithm to match structures rather than single pixels.

We hypothesized that the incorporation of color information to the adaptive weights would enhance the performance of the stereo matching algorithm. This hypothesis turned out to be true for the loss function of the coarse subsampled disparity map (a loss function is calculated both for the low resolution and the high resolution disparity maps, as will be described in Sect. 3.4). The low resolution part of the loss function seeks to match large, coarse structures of the scenery. If we rely only on subsampled infrared information, the ASW scheme might make a rough agglomeration of these areas, creating ambiguity about the underlying ground truth disparity. Our insight was that a color-guided approach could leverage color information to reduce an overly-aggresive cost aggregation scheme in a principled way.

More concretely, we multiplied the traditional adaptive support weights with a term that discourages color dissimilarity, and can reduce the strength of the traditional weight as low as $\frac{\lambda}{\lambda+1}$. In our experiments we used $\lambda = 0.1$. The complete formula for our color-guided adaptive support weights is the following:

$$w_{xy}^c = \frac{exp\left(-\frac{|I_{ij}^l - I_{xy}^l|}{\sigma_w}\right)\left[\lambda + exp\left(-\frac{\Delta c_{xy}}{\sigma_c}\right)\right]}{\lambda+1} \tag{2}$$

where Δc_{xy} is a distance measure defined in the colorspace (e.g RGB, YUV, CIELab, etc.) of I^c, and σ_c controls the range of Δc_{xy} values that define a region with the same texture. We utilized the sum of L_1 distances between each color in the CIELab colorspace and $\sigma_c = 2.0$. Thus, the per pixel color-guided cost becomes:

$$\hat{C}_{ij}^c = \frac{\sum_{x=i-k}^{i+k}\sum_{y=j-k}^{j+k} w_{xy}^c C_{xy}}{\sum_{x=i-k}^{i+k}\sum_{y=j-k}^{j+k} w_{xy}^c} \tag{3}$$

where we used odd-sized $2k+1 \times 2k+1$ windows instead of even-sized. A pictorial representation of the calculations involved in the cost aggregation process and the differences between the two weighting schemas can be found in Fig. 2. We ephasize that our method only adds overhead in the training process, and thus does not affect the real-time application of the algorithm. Moreover, one can control the usage of color data through the tunable hyperparameter λ, as was stated in Sect. 3.2.

3.4   Loss Function

After incorporating a hard left-right consistency check mask in conjunction with a cross-entropy regularization term $\mathcal{L}_{reg}$ [14], the loss function $\mathcal{L}$ that we propose becomes:

$$\mathcal{L} = \frac{1}{N}\sum_{ij}\hat{C}_{ij}^c + \mathcal{L}_{reg} \tag{4}$$

where the summation is done for the valid pixels ij, and N is the number of valid pixels, as determined by the invalidation mask. Our contribution lies in the utilization of the color-guided aggregated cost $\hat{C}_{ij}^c$ instead of $\hat{C}_{ij}$. The loss function is calculated both for the high resolution disparity output of the network, as well as for the low resolution disparity map produced by the cost volume. For the weighted local contrast normalization computation, we used 9×9 windows in the original resolution, and 3×3 windows in the subsampled resolution, because a 9×9 window in the low resolution disparity would have a global effect on the normalization process.

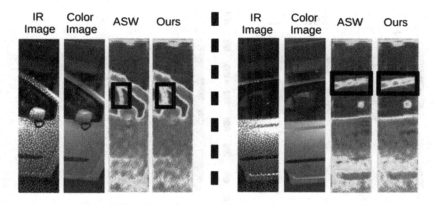

Fig. 3. Two tedrads depicting, from left to right, the cropped left IR image of a car, the corresponding registered and inpainted cropped color image, the disparity output of the network using traditional ASW cost aggregation, and the disparity output of the color-guided cost aggregation method (ours). We have drawn black boxes in regions where the color guidance helps the algorithm correctly estimate relatively thinner structures. (Color figure online)

4 Experiments

For our experiments, we captured both sides of 7 cars with the Intel Realsense D435 sensor, saving the left and right infrared images, as well as the color images and the depth maps provided by the sensor. Due to bandwidth constraints, which will be stricter in the case where multiple sensors are connected to the same computer, we reduced the frame resolution from the maximum available to 848×480, which is the optimal for the operation of the sensor's depth prediction algorithm. Analogous to [14], where the network is fed with 1024×256 images, we used 640×160 images, retaining the $4:1$ aspect. We normalised our images, confining the pixel values to the interval $[-1, 1]$. In [14], the authors mention that 100 images are enough to produce reasonable results. As our images have approximately 2.5 less pixels than the ones used in [14], we train the network with a dataset of 300 infrared pairs to ensure convergence. Due to the small dataset, the estimated disparity for the background of the cars did not always converge to a reasonable value. Moreover, we found the invalidation mask to be a major source of instability for the network, and similar to [14], we applied it after an approximate convergence had been reached. Taking into consideration the height and width of our original (not cropped) images in relation to the size used in [14], we used 21×21 aggregation windows. We implemented our architecture in tensorflow [1]. We precomputed the cost aggregation weights both for the case of ASW and the color-guided ASW. Finaly, we trained our network using the RMSProp optimizer, with a constant learning rate of 10^{-4}, and with a batch size equal to 4.

Fig. 4. Each row contains two tedrads presenting from left to right, the cropped left IR image of a car, the corresponding registered and inpainted cropped color image, the disparity output of the network using traditional ASW cost aggregation, and the disparity output of the color-guided cost aggregation method (ours). We have marked with boxes the regions where the color-guided approach achieves better results than the traditional ASW. Black boxes signify regions where color information more faithfully reproduces relatively thinner structures. We have drawn purple boxes in the tires of a car, where the color-guided approach approximates better the underlying depth of the scenery. (Color figure online)

We qualitatively present the results of our experiments in Fig. 3, comparing the disparity output of the traditional ASW approach used in [14] with the color-guided ASW, and highlight cases where the our method performs better than [14]. We notice that there are regions, like the upper metal parts of the doors, or plastic bar-like constructs, where the traditional ASW scheme cannot reproduce the underlying depth of the scene. On the contrary, our color-guided approach leverages the information made available by the color image, and produces a better disparity map of the vehicle. Additional results where our color-guided approach qualitatively outperforms the traditional ASW scheme are presented in Fig. 4.

5 Conclusions

We presented a color-guided cost aggregation method based on adaptive support weights, that is well suited for the problem of active stereo matching. A color image viewing the same scenery as the left and right infrared images encodes important details that help the self-supervised network of [14] uncover structures that would otherwise be left out of the resulting depth map. The commercial Intel Realsense D435 sensor has a separate color camera mounted on it, making it ideal for our improved active stereo matching method. The main limitation of our method is due to the computational requirements of the preprocessing phase: the precomputation and loading of the color-guided adaptive weights, in order for the training to initiate, requires significant execution time and computer memory, even in the case of a relatively small dataset of 300 data points and 21×21 aggregation windows. However, in our future work we plan to apply our method on larger datasets that have greater geometric variability, and to compare our results with other stereo matching methods as well. Finally, we would like to thank the anonymous reviewers for their constructive comments.

Acknowledgements. This work has received funding from the German Federal Ministry of Education and Research (BMBF) and the Greek General Secretariat for Research and Technology (GSRT) through project INVIVO in the context of the Greek-German Call for Proposals on Bilateral Research and Innovation Cooperation, 2016.

References

1. Abadi, M., et al.: TensorFlow: a system for large-scale machine learning. In: 12th USENIX Symposium on Operating Systems Design and Implementation (OSDI 16), pp. 265–283 (2016). https://www.usenix.org/system/files/conference/osdi16/osdi16-abadi.pdf
2. Godard, C., Mac Aodha, O., Brostow, G.J.: Unsupervised monocular depth estimation with left-right consistency. In: The IEEE Conference on Computer Vision and Pattern Recognition (CVPR), July 2017
3. Haefner, B., Quéau, Y., Möllenhoff, T., Cremers, D.: Fight ill-posedness with ill-posedness: single-shot variational depth super-resolution from shading. In: 2018 IEEE/CVF Conference on Computer Vision and Pattern Recognition, pp. 164–174, June 2018. https://doi.org/10.1109/CVPR.2018.00025

4. Hartley, R., Zisserman, A.: Multiple View Geometry in Computer Vision, 2nd edn. Cambridge University Press, New York (2003)

5. Heo, M., Lee, J., Kim, K.R., Kim, H.U., Kim, C.S.: Monocular depth estimation using whole strip masking and reliability-based refinement. In: The European Conference on Computer Vision (ECCV), September 2018

6. Kendall, A., Martirosyan, H., Dasgupta, S., Henry, P.: End-to-end learning of geometry and context for deep stereo regression. In: IEEE International Conference on Computer Vision, ICCV 2017, Venice, Italy, 22–29 October 2017, pp. 66–75 (2017). https://doi.org/10.1109/ICCV.2017.17

7. Keselman, L., Iselin Woodfill, J., Grunnet-Jepsen, A., Bhowmik, A.: Intel realsense stereoscopic depth cameras. In: The IEEE Conference on Computer Vision and Pattern Recognition (CVPR) Workshops, July 2017

8. Khamis, S., Fanello, S., Rhemann, C., Kowdle, A., Valentin, J., Izadi, S.: StereoNet: guided hierarchical refinement for real-time edge-aware depth prediction. In: The European Conference on Computer Vision (ECCV), September 2018

9. Yoon, K.-J., Kweon, I.S.: Adaptive support-weight approach for correspondence search. IEEE Trans. Pattern Anal. Mach. Intell. **28**(4), 650–656 (2006). https://doi.org/10.1109/TPAMI.2006.70

10. Lee, J.H., Heo, M., Kim, K.R., Kim, C.S.: Single-image depth estimation based on fourier domain analysis. In: The IEEE Conference on Computer Vision and Pattern Recognition (CVPR), June 2018

11. Telea, A.: An image inpainting technique based on the fast marching method. J. Graph. Tools **9**, 23–34 (2004). https://doi.org/10.1080/10867651.2004.10487596

12. Yan, S., et al.: DDRNet: depth map denoising and refinement for consumer depth cameras using cascaded CNNs. In: The European Conference on Computer Vision (ECCV), September 2018

13. Zhang, Y., Funkhouser, T.: Deep depth completion of a single RGB-D image. In: The IEEE Conference on Computer Vision and Pattern Recognition (CVPR), June 2018

14. Zhang, Y., et al.: ActiveStereoNet: end-to-end self-supervised learning for active stereo systems. In: The European Conference on Computer Vision (ECCV), September 2018

15. Zheng, C., Cham, T.J., Cai, J.: T2Net: synthetic-to-realistic translation for solving single-image depth estimation tasks. In: The European Conference on Computer Vision (ECCV), September 2018

16. Zhou, T., Brown, M., Snavely, N., Lowe, D.G.: Unsupervised learning of depth and ego-motion from video. In: The IEEE Conference on Computer Vision and Pattern Recognition (CVPR), July 2017

Image Enhancing in Poorly Illuminated Subterranean Environments for MAV Applications: A Comparison Study

Christoforos Kanellakis(✉), Petros Karvelis, and George Nikolakopoulos

Robotics Team, Department of Computer, Electrical and Space Engineering,
Luleå University of Technology, 97187 Luleå, Sweden
chrkan@ltu.se

Abstract. This work focuses on a comprehensive study and evaluation of existing low-level vision techniques for low light image enhancement, targeting applications in subterranean environments. More specifically, an emerging effort is currently pursuing the deployment of Micro Aerial Vehicles in subterranean environments for search and rescue missions, infrastructure inspection and other tasks. A major part of the autonomy of these vehicles, as well as the feedback to the operator, has been based on the processing of the information provided from onboard visual sensors. Nevertheless, subterranean environments are characterized by a low natural illumination that directly affects the performance of the utilized visual algorithms. In this article, an novel extensive comparison study is presented among five State-of the-Art low light image enhancement algorithms for evaluating their performance and identifying further developments needed. The evaluation has been performed from datasets collected in real underground tunnel environments with challenging conditions from the onboard sensor of a MAV.

Keywords: Low light imaging · Image enhancement · Subterranean MAV applications

1 Introduction

The technology of Micro Aerial Vehicles (MAVs) has shown increased levels of robustness in well defined and constrained lab environments. Recently, there is a large effort to operate these platforms outside of a lab and in real world, on site field scenarios. Until now, the majority of the consumer grade MAVs has been directed towards the photography-cinematography industry, taking advantage of their payload capacity and the stable flight characteristics. Nevertheless, an emerging trend is to integrate MAVs in the operation cycles of underground mines, as well as for search and rescue missions in general subterranean environments, trying to reduce the operating costs, increase the overall efficiency, while

This work has been partially funded by the European Unions Horizon 2020 Research and Innovation Program under the Grant Agreement No. 730302 SIMS.

© Springer Nature Switzerland AG 2019
D. Tzovaras et al. (Eds.): ICVS 2019, LNCS 11754, pp. 511–520, 2019.
https://doi.org/10.1007/978-3-030-34995-0_46

removing the human operator away from the harsh operating conditions, thus providing advanced tools for more effective decision making and overall mining operations [11].

In these application scenarios, multiple challenges arise before employing autonomous MAVs, mainly regarding the perception, planning and control of these air vehicles, such as the narrow passages, the reduced visibility due to rock falls, the dust existence, the uncertainty in localization, wind gusts and the overall lack of proper illumination. Within the related literature of MAVs in subterranean mine operations, several research efforts have been reported trying to address these challenging tasks. In [18] a visual inertial navigation framework has been proposed, to implement position tracking control of the platform. In this case, the MAV was controlled to follow obstacle free paths, while the system was experimentally evaluated in a real scale tunnel environment, simulating a coal mine, where the illumination challenge was assumed solved. In [3] a more realistic approach, compared to [18] regarding underground localization, has been performed. More specifically, a hexacopter equipped with a Visual Inertial (VI) sensor and a laser scanner was manually guided across a vertical mine shaft to collect data for post-processing. The extracted information from the measurements have been utilized to create a 3D mesh of the environment and localize the vehicle. Finally, in [12] the estimation, navigation, mapping and control capabilities for autonomous inspection of penstocks and tunnels using aerial vehicles has been studied by utilizing IMUs, cameras and lidar sensors.

Figure 1(a) depicts a typical example of an underground tunnel with limited natural illumination. Inspired by the challenge of poor visibility in such environments, this work tries to identify the performance of the current state of the art methods in realistic and degraded environments with restricted access. Studying methods to enhance the quality of image data recorded on board of the MAV during the mission has a value for both the mission accomplishment as well as the data post processing for generating 3D models, identifying degradation in the area and general acquiring a direct visual feedback for the asset owners to contextualize the location of the damages found during the inspection task.

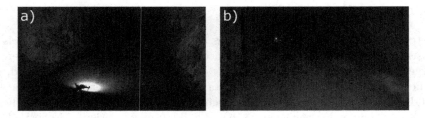

Fig. 1. (a) MAV with onboard illumination in a subterranean area with complete lack of natural illumination sources. (b) Image of low light conditions in a mine captured by the camera on board a MAV.

Generally, the quality of an image or a video sequence largely depends on the light conditions of the environment. For example strong light produces images with a wash out effect and weak light produce images that are not visible due to the darkness. For both of the cases, the contrast of the two images is extremely low and needs further modification in order to reveal the details of the images. These typical low light conditions occur in mines, usually due to the lack of proper illumination and in other underground or indoor dark environments, e.g. factories. In all these cases, even if it was possible and realistic to utilize a light on board of the MAV, in order to illuminate the surrounding area of interest, this would have drawbacks for the mission design due to: (a) the limited weight that the MAV can lift, and (b) the limited power supply that the MAV can support with the corresponding impact on the flight duration. A characteristic example of such a low light image capturing conditions acquired from a camera on a MAV is displayed in Fig. 1(b). Image enhancement techniques [7] have received great attention mostly because of their simplicity as also their effectiveness. Even nowadays in the era of Deep Learning (DL), efforts are being made to enhance images especially in the case of low light conditions [2].

This work is aligned with the vision of deploying aerial robotic platforms in subterranean environments, and it focuses on the study and understanding on how the low-level image enhancement mechanisms can be incorporated in the application scenarios of MAVs flying in low light environments. The contribution of the proposed work is two-folded. Initially, this article presents a thorough performance analysis of State-of-the-Art methods on datasets collected from real underground environments with limited visibility. The second contribution stems from the collected datasets, which include images with varying illumination levels that correspond to real environment, compared to artificially generate low light images. They have been generated by using a low cost aerial platform equipped with a single camera that can be considered as a consumable and easily replaceable in areas that have limited access to the public and there are not many similar datasets available in the literature, where the presented analysis provides insights to bring these algorithms a step closer to real life applications.

The rest of this article is organized as it follows. In Sect. 2 the image processing methods for low light image enhancement are summarized, while in Sects. 3 and 4 the data collection procedure and the study evaluations of the State-of-the-Art methods in field trials are described respectively. Finally, the concluding remarks and future work are presented in Sect. 5.

2 Methodology

In this article five methods have been utilized for the image low light enhancement and in the sequel each one of them is going to be briefly described.

LIME [4] is a method that belongs to the Retinex-based category [1,6], which intends to enhance a low-light image by estimating its illumination map. However one of the fundamental differences from other Retinex based methods that are based on decomposing the recorded image into the reflectance and illumination component, is the fact that it estimates one of the two components, namely the illumination. This has the advantage that it reduces the solution space and thus reduces the overall computation cost. The first step of the method is to estimate the illumination map computing the maximum intensity for each one of the pixels of the three channels (Red, Green, Blue) of the color image. Then, a refinement step is employed to better estimate the illumination map by introducing an Augmented Lagrangian Multiplier.

The goal of the Joint Low-Light enhancement and denoising method [15] is to enhance a low light image, while in parallel tries to minimize the inherent noise. The method is based on the Retinex category and decomposes the image into a reflectance and illumination map and a smoothed illumination map, with a noise free reflectance map, is computed. This is achieved by initially estimating the illumination map and then independently estimating the reflectance map. After the extraction of the smoothed illumination map, a large fraction of the noise is still present in the reflectance map, which is suppressed in the sequel by using weighted matrices.

The LECARM [16] method uses the nonlinear Camera Response Function (CRF) in order to reduce distortions to the enhancement and thus it leads to a significant reduction in the visual quality. In this approach, initially it is employed a fixed response curve in order to obtain the parameters of the CRF function. Then the second step in this methodology is the estimation of the illumination map by using the LIME method [4] and in the sequel the estimated exposure ratio map can be extracted from that, since it is inversely proportional to the illumination map.

The MSR [13,14] method is another method based on the Retinex model [8]. It combines the retinex dynamic range compression and color constancy with a color restoration filter that provides excellent color rendition.

Finally, in the last method [20] a two level exposure fusion approach in order to compute, as accurate as possible, the contrast and light enhancement is introduced. The method first computes a weight matrix for the image fusion step by using the illumination map that will be estimated. The camera response model is then computed and the best exposure ratio is computed for regions where the original images are under exposured. The final image is fused with the enhanced image by using the weight matrix.

3   Dataset Overview

The proposed study of image processing techniques in dark environments has been evaluated by using datasets collected from actual flights of two custom designed aerial platforms shown in Fig. 2 and from two different locations in Sweden. The rest of the section provides a thorough description of the visited underground areas.

Fig. 2. Aerial platforms used for the dataset collection. On the left column is the platform used for the 1st dataset and on the right column is the platform used on the 2nd dataset.

3.1 Description of Datasets

The first dataset was collected inside a tunnel under Mjölkuddsberget mountain located at Luleå, Sweden. The selected environment, which resembles an underground mine tunnel, was pitch dark without any external illumination, while the tunnel surfaces consisted by uneven rock formations. The dimensions of the testing tunnel area were $100 \times 2.5 \times 3\,\mathrm{m}^3$, capturing the camera sequences, while the MAV was following the path along the tunnel. Furthermore, the tunnel lack the presence of strong magnetic fields, while small particles were floating in the air during the flights. The aerial platform has been equipped with one 10 W LED light bar pointing towards the field of view of the camera to illuminate its surroundings. In more details, this light bar was set to different illumination levels (luminous flux per unit area or lux) varying from 2000 lux to 400 lux. The camera used in the dataset sequences was the FOXEER Box[1] that was recording with a resolution of 1920×1080 at 60 fps. Figure 3 depicts representative snapshots of the dataset, where the dominating darkness in the surrounding environment is evident.

The second dataset was collected in an area 790 m deep in an underground mine in Sweden without any natural illumination sources. Furthermore, the underground tunnels did not have strong corrupting magnetic fields, while their morphology resembled an S shape environment with small inclination. Overall, the field trial area had width, height and length dimensions of 6 m, 4 m, and 150 m respectively. The MAV is equipped with two 10 W LED light bars in both front arms for providing additional illumination for the looking forward camera. In more details, this light bar was set to different illumination levels (luminous flux per unit area or lux) varying from 400 lux to 40 lux. During the data collection, the GoPro7 visual sensor has been used with resolution of 3840 × 2160 pixels with 60 fps. Figure 3 depicts representative snapshots of the dataset, where the dominating darkness in the surrounding environment is evident.

---

[1] http://foxeer.com/Foxeer-4K-Box-Action-Camera-SuperVision-g-22.

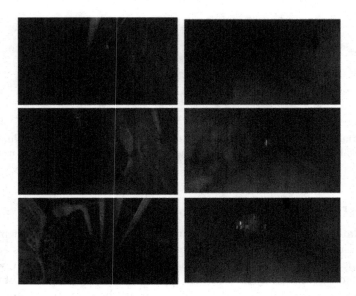

Fig. 3. In the left column are depicted representative snapshots from the first Dataset, while in the right column are depicted representative snapshots from the second Dataset.

3.2   Evaluation Strategies

The merit of the dataset is the fact that it considers non ideal conditions captured in real low light environments with sensors that are commonly used in MAV applications. The datasets have been captured from an aerial vehicle flying along subterranean tunnels. In the performed field trials, it was not possible to capture images with ground truth illumination to match the actual images recorded in the dataset that have low illumination levels, since the MAV motion was slightly different in every run. The popular full-reference PSNR/SSIM [5] metrics found in the literature, for evaluating the enhancement methods, have limited evaluation applicability in practice. Therefore, to complement the evaluation process this work refers to two no-Reference IQA models: (1) BLind Image Integrity Notator using DCT-Statistics (BLIINDS2) [17] and (2) Naturalness Image Quality Evaluator (NIQE) [10], (3) Blind/Referenceless Image Spatial QUality Evaluator (BRISQUE) [9], (4) Perception based Image Quality Evaluator (PIQE) [19]. The no-Reference method provides as an outcome a score $\in [0, 100]$, where the lowest value indicates a better performance.

The described datasets have been used for the method evaluation. More specifically, the first set of evaluation images is described as $Set1$ and is derived from part of $dataset2$. The $Set1$ includes 200 frames captured with 40lux illumination measured from 1 m distance. The MAV was flying manually following a straight path. The second set of evaluation images is described as $Set2$ and is derived from part of $dataset1$. The $Set2$ includes 200 frames captured with

100lux illumination measured from 1 m distance. The MAV was flying manually following the tunnel. Finally, another part from the *dataset*2 has been extracted into the *Set*3 where it includes 200 frames captured with 200lux illumination measured from 1 m distance. The MAV was flying autonomously following the S-shaped tunnel path. All images in the datasets have been exported from the videos recorded onboard and have been saved in a JPEG compression format. Additionally, all frames have been down-sampled to 960*pixels* × 540*pixels* for reducing the overall computation time.

For a fair comparison, all methods have been evaluated in Windows 10 MAT-LAB environment in a machine with 2.69 GHz CPU and 12G RAM.

4 Evaluation

This Section describes the evaluation results of the five State-of-the-Art methods available online, namely: (1) BIMEF[2], (2) LECARM[3], (3) JED[4], (4) MultiscaleRetinex[5], and (5) LIME (See footnote 5).

The evaluation process is divided in two parts: (a) quantitative evaluation, and (b) subjective-qualitative evaluation. Table 1 presents the BLIINDS2 and NIQE metrics for all methods for a number of 10 images from Set1, Set2 and Set3, while Table 2 presents the BRISQUE and PIQE metrics for all methods for 200 images from Set1, Set2 and Set3. By inspecting the tables it is observed that for different metrics different methods perform better. The performance is also varying among different datasets for the same method. When it comes to no-reference metrics the results become less consistent. Such an inconsistency between no-reference evaluations is a point raised in this work and should be addressed in a task oriented manner for different application scenarios.

Table 1. Average no-Reference evaluation results of low light enhancement algorithms

	Set1		Set2		Set3	
	Bliinds2	Niqe	Bliinds2	Niqe	Bliinds2	Niqe
BIMEF	**29.8421**	19.2128	87.3000	21.0303	**33.9500**	19.1589
LECARM	30.9000	19.0942	**16.1500**	20.8437	34.7000	18.8645
JED	34.6500	19.3555	21.9500	25.2385	36.3500	19.5826
MSR	34.6250	**16.5409**	37.6000	19.8599	40.9000	**16.9791**
LIME	33.4500	18.4797	19.9000	**19.3274**	36.9000	18.3015

---

[2] https://github.com/baidut/BIMEF.
[3] https://github.com/baidut/LECARM.
[4] https://github.com/baidut/OpenCE/tree/master/others.
[5] https://github.com/baidut/BIMEF/tree/master/lowlight.

Table 2. Average no-Reference evaluation results of low light enhancement algorithms

	Set1		Set2		Set3	
	Brisque	Piqe	Brisque	Piqe	Brisque	Piqe
BIMEF	**32.2073**	16.9229	37.0642	29.8044	45.9678	12.8824
LECARM	35.6660	15.3142	36.1718	**29.1353**	**46.0225**	10.2393
JED	51.1892	39.2722	39.7473	48.1605	50.2417	42.7230
MSR	40.8146	16.2568	**19.4552**	32.5144	46.6340	10.5247
LIME	40.0704	**13.6910**	39.3943	40.2375	46.8235	**9.4447**

Although the methods have been evaluated using the above no-Reference metrics it is not evident how the algorithms perform in reality and their applicability in various MAV missions. To this end, Fig. 4 visualize the resulting outcome when employing the methods using the collected datasets. Three different frames have been selected from each set of images (Set1, Set2 and Set3), which are representative from such tunnel environments. From this Figure it is shown that all methods reveal further details from the environment by showing the tunnel width. Although MultiscaleRetinex adds to many artifacts and noise, its applicability fits for visual feedback to the operator, thus extending the surroundings perceived from the aerial vehicle. Moreover, they fit for navigation tasks that process the darkness of the tunnels, by revealing the local geometry of the tunnels showing walls and floors. However, the images are noisy for 3D reconstruction and place recognition methods.

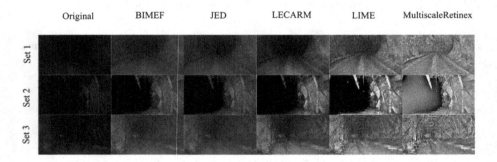

Fig. 4. Qualitative results for the human visual perception.

4.1   Future Directions

From the presented results it is evident that there is room for further developments in the fields of aerial robotics and the low-level image processing. More specifically, since in real applications it is hard to capture ground truth data, there is a need to propose new metrics that capture the performance of the methods within the desired application scenarios. The use of low light image enhancement algorithms should be task oriented, where different algorithms could fit better in different applications. Few application scenarios that are currently pursued include enhanced visual feedback to the operator, 3D reconstruction, object detection in dark environments, visual place recognition for navigation tasks, as well as general vision based navigation and planning tasks. A major parameter that affects the performance of the methods is the geometry of the tunnels. In the case of narrow tunnels the illumination is spread in the local vicinity (walls, floor and ceiling) of the platform, assisting the usage of the methods. On the contrary, wide tunnels are more challenging since usually, the onboard illumination is absorbed from the void in the front of the aerial vehicle and the local vicinity of the platform is weakly illuminated. In this work the methods have been applied in the most challenging case of wide tunnels.

5   Conclusions

This work focuses on a comprehensive study and evaluation of existing low-level vision techniques for low light image enhancement, targeting MAV applications in underground environments. Subterranean environments are characterized by low natural illumination, which directly affects the performance of visual algorithms. In this work, an extensive comparison study has been presented among five State-of the-Art low light image enhancement algorithms to evaluate their performance and further developments. The evaluation has been performed from datasets collected in real underground tunnel environments with challenging conditions from the onboard sensor of a MAV. From the presented results it is shown that there is no single-best low-light enhancement method for all criteria. The inconsistency in the results is complicated depicting that the method performance depends of the prior it uses and the design choices. Future works should focus and optimize on evaluation criteria that are task oriented aligned with higher level tasks.

References

1. Chandra, M.A., Acharya, B., Khan, M.I.: Retinex image processing: improving the visual realism of color images (2011)
2. Chen, C., Chen, Q., Xu, J., Koltun, V.: Learning to see in the dark. arXiv preprint arXiv:1805.01934 (2018)
3. Gohl, P., et al.: Towards autonomous mine inspection. In: 2014 3rd International Conference on Applied Robotics for the Power Industry (CARPI), pp. 1–6. IEEE (2014)
4. Guo, X., Li, Y., Ling, H.: LIME: low-light image enhancement via illumination map estimation. IEEE Trans. Image Process. **26**(2), 982–993 (2017). https://doi.org/10.1109/TIP.2016.2639450
5. Hore, A., Ziou, D.: Image quality metrics: PSNR vs. SSIM. In: 2010 20th International Conference on Pattern Recognition, pp. 2366–2369. IEEE (2010)

6. Jobson, D.J., Rahman, Z., Woodell, G.A.: A multiscale retinex for bridging the gap between color images and the human observation of scenes. IEEE Trans. Image Process. **6**(7), 965–976 (1997). https://doi.org/10.1109/83.597272

7. Lal, S., Narasimhadhan, A., Kumar, R.: Automatic method for contrast enhancement of natural color images. J. Electr. Eng. Technol. **10**(3), 1233–1243 (2015)

8. Land, E.H.: An alternative technique for the computation of the designator in the retinex theory of color vision. Proc. Natl. Acad. Sci. USA **83**(10), 3078–3080 (1986). http://www.jstor.org/stable/27444

9. Mittal, A., Moorthy, A.K., Bovik, A.C.: No-reference image quality assessment in the spatial domain. IEEE Trans. Image Process. **21**(12), 4695–4708 (2012)

10. Mittal, A., Soundararajan, R., Bovik, A.C.: Making a "completely blind" image quality analyzer. IEEE Signal Process. Lett. **20**(3), 209–212 (2012)

11. Nikolakopoulos, G., Gustafsson, T., Martinsson, P., Andersson, U.: A vision of zero entry production areas in mines**this work has been partially funded by the sustainable mining and innovation for the future research program. In: 4th IFAC Workshop on Mining, Mineral and Metal Processing MMM (2015)

12. Özaslan, T., et al.: Autonomous navigation and mapping for inspection of penstocks and tunnels with MAVs. IEEE Robot. Autom. Lett. **2**(3), 1740–1747 (2017)

13. Petro, A.B., Sbert, C., Morel, J.M.: Multiscale retinex. Image Process. Line **4**, 71–88 (2014). https://doi.org/10.5201/ipol.2014.107

14. Rahman, Z.U., Jobson, D.J., Woodell, G.A.: Multi-scale retinex for color image enhancement. In: Proceedings of 3rd IEEE International Conference on Image Processing, vol. 3, pp. 1003–1006. IEEE (1996)

15. Ren, X., Li, M., Cheng, W., Liu, J.: Joint enhancement and denoising method via sequential decomposition. CoRR abs/1804.08468 (2018). http://arxiv.org/abs/1804.08468

16. Ren, Y., Ying, Z., Li, T.H., Li, G.: LECARM: low-light image enhancement using the camera response model. IEEE Trans. Circ. Syst. Video Technol. **29**(4), 968–981 (2019). https://doi.org/10.1109/TCSVT.2018.2828141

17. Saad, M.A., Bovik, A.C.: Blind quality assessment of videos using a model of natural scene statistics and motion coherency. In: 2012 Conference Record of the Forty Sixth Asilomar Conference on Signals, Systems and Computers (ASILOMAR), pp. 332–336. IEEE (2012)

18. Schmid, K., Lutz, P., Tomić, T., Mair, E., Hirschmüller, H.: Autonomous vision-based micro air vehicle for indoor and outdoor navigation. J. Field Robot. **31**(4), 537–570 (2014)

19. Venkatanath, N., Praneeth, D., Bh, M.C., Channappayya, S.S., Medasani, S.S.: Blind image quality evaluation using perception based features. In: 2015 Twenty First National Conference on Communications (NCC), pp. 1–6. IEEE (2015)

20. Ying, Z., Li, G., Gao, W.: A bio-inspired multi-exposure fusion framework for low-light image enhancement. CoRR abs/1711.00591 (2017). http://arxiv.org/abs/1711.00591

Robust Optical Flow Estimation Using the Monocular Epipolar Geometry

Mahmoud A. Mohamed$^{(\boxtimes)}$ⓘ and Bärbel Mertschingⓘ

GET Lab, University of Paderborn, Pohlweg 47–49, 33098 Paderborn, Germany
{mohamed,mertsching}@get.upb.de

Abstract. The estimation of optical flow in cases of illumination change, sparsely-textured regions or fast moving objects is a challenging problem. In this paper, we analyze the use of a texture constancy constraint based on local descriptors (i.e., HOG) integrated with the monocular epipolar geometry to estimate robustly optical flow. The framework is implemented in differential data fidelities using a total variation model in a multi-resolution scheme. Besides, we propose an effective method to refine the fundamental matrix along with the estimation of the optical flow. Experimental results based on the challenging KITTI dataset show that the integration of texture constancy constraint with the monocular epipolar line constraint and the enhancement of the fundamental matrix significantly increases the accuracy of the estimated optical flow. Furthermore, a comparison with existing state-of-the-art approaches shows better performance for the proposed approach.

Keywords: Optical flow · Epipolar geometry · HOG descriptor · Fundamental matrix · Texture constraint

1 Introduction

An optical flow field describes the apparent motion of a scene and involves the camera ego-motion and the motion of the objects themselves. To this day, many differential optical flow approaches are mainly dependent on the brightness constancy assumption by assuming that the intensity of a pixel will not be change if objects or the camera move. Unfortunately, not every object motion yields a change in the gray values and not every change in gray values is generated by body motion. Challenging outdoor scenes (e.g. the sequence shown in Fig. 1) which have poorly textured regions, illumination changes, shadows, reflections, glare, and the inherent noise of the camera image may yield intensity changes while the depicted object remains stationary. Especially in poor illumination conditions the number of photons, which are collected by a photo cell, may vary over time. Therefore, other assumptions such as gradient and texture constraint are used to estimate robust optical flow.

To estimate robust optical flow in the above-mentioned challenging cases, many methods have presented different optical flow models that become suitable for deployment in real-world vision systems. For instance, the normalized

© Springer Nature Switzerland AG 2019
D. Tzovaras et al. (Eds.): ICVS 2019, LNCS 11754, pp. 521–530, 2019.
https://doi.org/10.1007/978-3-030-34995-0_47

Fig. 1. KITTI sequence number 178 [19], a challenging example which contains repeated patterns, regions with little texture, and poor illumination conditions.

cross-correlation is used for a data conservation [1], which leads to increasing the robustness of the estimated optical flow. In turn, the problem of poorly textured regions, occlusions, and small-scale image structure is tackled in [2] by incorporating a low-level image segmentation process that has been used in a non-local total variation regularization term in a unified variational framework. Besides, an optical flow estimation method based on the zero-mean normalized cross-correlation transform is introduced by [3]. The use of the normalized cross-correlation increases the robustness but results in a complicated optimization problem. Moreover, the accuracy of the optical flow estimation is adversely affected in the case of regions with only little texture.

Recently local descriptors have been used to estimate dense optical flow fields to gain robustness for the estimated optical flow in real-world vision systems. For instance, census transform [4], histogram of oriented gradient (HOG) [5], distributed average gradient [6], and modified local directional pattern (MLDP) [7,8] have been incorporated directly into the classical energy minimization framework. Moreover, a local optical flow framework taking into account an illumination model to deal with varying illumination and a prediction step based on a perspective global motion model to deal with long-range motions is proposed in [9]. Furthermore, in [10] a new matching-based algorithm is proposed to estimate dense optical flow fields for the parts whose motion is induced by the moving observer only. In addition, a framework for estimating optical flow in the case of complex motions, large displacements, and difficult imaging conditions is proposed in [18]. Unfortunately, these algorithms fail in the case of poorly textured regions. Moreover, such algorithms require heavy computationally processing.

In case that camera motion only induces an optical flow, i.e., the objects are stationary, several methods use the monocular epipolar geometry to estimate the optical flow estimation and increase the robustness against outdoor effects [11–13]. The epipolar geometry is used to obtain one more extra constraint for the flow of a pixel. In this regard, the fundamental matrix related to the motion of the camera between two frames is firstly estimated. Consequently, given each point in the first frame, the place of the correspondent point in the next consecutive will be constraint over a line known as the epipolar line. Fundamental matrices can be estimated using 8-point or 7-point [14] methods. The estimation of optical flow using the epipolar line can be divided into two categories sparse and dense optical flow.

For sparse optical flow, [11] introduces a method to find the correspondent points by searching over epipolar lines and used a semi-global block matching technique. Unfortunately, the method provide inaccurate match points due to the limited accuracy of the fundamental matrix, the usage of calibration information, and the use of approximation of rotation matrix which is valid only for small rotations. In turn, in [15] a new epipolar flow approach with low computational complexity is proposed to estimate semi-dense optical flow. Furthermore, in [16], the epipolar line constrained is integrated with the data term, and no smoothness term is used which limits the approach to only some features points.

For the estimation of dense optical flow, a joint estimation of the fundamental matrix and the optical flow is discussed in [12]. However, the formulation of the method is not presented for the multi-resolution pyramid analysis which limits the algorithm to small motions. It can be concluded that the simultaneous estimation of the fundamental matrix and the flow looks theoretically interesting. However, the method can diverge due to the sensitivity of fundamental matrix estimation due to the accumulation of small calculation errors. Such a situation occurs when the flow is not finely calculated. In [17], an approach to dense motion estimation from a single monocular camera is proposed. The algorithm segments the optical flow field into a set of motion models, each with its epipolar geometry. In turn, the work in [13] introduces the integration of smoothness constraint with a data term using a texture and the monocular epipolar line constraints. The algorithm integrates the epiploar line constraint directly to the data term using the total variation model which yields a better accuracy in case of texture-less regions.

In this paper, we investigate the extensibility of the optical flow estimation algorithm [13]. Therefore, the concept of integrating the monocular epipolar line is extended by refining the fundamental matrix using the estimated optical flow in a coarse to fine scheme. Consequently, the optical flow accuracy has been significantly improved. Finally, several experiments in various environments are provided and the results are discussed.

2   Optical Flow Model

To estimate dense optical flow, [13] developed a regularized cost function including a data term based on a texture constraint together with the monocular epipolar line constraint and a smoothness term as follows:

$$\min_{u,v} E(u,v) = \sum_{\Omega} \left(\lambda E_{data} + \gamma E_{epip} + \eta E_{smooth} + E_{dual} \right), \tag{1}$$

where

$$E_{data} = \lambda \rho(x, y, u, v)^2 \qquad (2)$$

$$E_{epip} = \| a_l u + b_l v + d \| \qquad (3)$$

$$E_{smooth} = (\| \nabla u \| + \| \nabla v \|) \qquad (4)$$

$$E_{dual} = \frac{1}{\theta}[(u - \hat{u})^2 + (v - \hat{v})^2] \qquad (5)$$

where λ, η and γ are the weight of the data term, the smoothness term, and the monocular epipolar line constraint. θ is a threshold. The epipolar line equation is written as described in [13]:

$$au + bv = -ax - by - c \qquad (6)$$

where a, b, and c are the coefficients of the epipolar constraint.

The cost function is optimized based on the combination of local and global costs [13] as follows:

$$\min_{u,v} \ E_d(u, v) = \lambda \rho(x, y, u, v)^2 + \gamma(a_l u + b_l v + d)^2 + \frac{1}{\theta}(u - \hat{u})^2 + \frac{1}{\theta}(v - \hat{v})^2, \ (7)$$

$$\min_{\hat{u},\hat{v}} \ E_s(\hat{u}, \hat{v}) = \eta \left(\| \nabla \hat{u} \| + \| \nabla \hat{v} \| \right) + \frac{1}{\theta}(u - \hat{u})^2 + \frac{1}{\theta}(v - \hat{v})^2, \qquad (8)$$

The similarity function ρ in Eq. (7) used in between the two images can be rewritten as:

$$\rho(x, y, u, v) = \sum_{i=1}^{N} [S_2(x + u, y + v)_i - S_1(x, y)_i], \qquad (9)$$

where $S_1(x, y)$ and $S_2(x+u, y+v)$ are the two brightness or descriptors extracted from the two images $I_1(x, y)$ and $I_2(x + u, y + v)$, respectively. N is the number of channels of the descriptor. In the case of brightness constraint $S_1 = I_1$ and $S_2 = I_2$ and $N = 1$. Assume $\boldsymbol{w} = [u, v]^T$. Thus, $S_2(x + u, y + v)$ or $S_2(x, y, \boldsymbol{w})$ can be linearized around the starting value of $\boldsymbol{w}$ using the Taylor expansion by considering only the first order as proposed in [4]:

$$S_2(x, y, \boldsymbol{w}) \approx S_2(x, y) + \nabla^T S_2(x, y, \hat{\boldsymbol{w}})(\boldsymbol{w} - \hat{\boldsymbol{w}}). \qquad (10)$$

The derivative $\nabla^T S_2(x, y, \hat{\boldsymbol{w}}) = \left[\frac{\partial S}{\partial x} = S_x, \frac{\partial S}{\partial y} = S_y \right]^T$ can be computed by applying a derivative mask to the resulted image (i.e. Sobel) in the x and y directions. Thus, the similarity function can be written as:

$$\rho(x, y, \boldsymbol{w}) \approx \tilde{\rho}(x, y, \boldsymbol{w})$$
$$= S_2(x, y) - S_1(x, y) + \nabla^T S_2(x, y, \hat{\boldsymbol{w}})(\boldsymbol{w} - \hat{\boldsymbol{w}}). \qquad (11)$$

For every pixel, the similarity is calculated using the differences between two HOG descriptors. Assume $S_1(x, y)$ and $S_2(x + u, y + v)$ are the N channel HOG descriptors extracted from the two images $I_1(x, y)$ and $I_2(x + u, y + v)$, respectively. In practice, the residual function between the two N-channel descriptors can be represented as:

$$\rho(x, y, u, v)^2 = \sum_{i=1}^{N} [S_{2,i}(x + u, y + v) - S_{1,i}(x, y)]^2$$

$$= \sum_{i=1}^{N} \rho_i(x, y, u, v)^2, \qquad (12)$$

In practice, the summation over all ρ_i^2 measures the distance between the two HOG descriptors. Using the dual TV-L1 algorithm, Eq. (7) can be optimized by solving for (u, v) as follows:

$$\frac{\partial}{\partial u} \left(\lambda \tilde{\rho}(x, y, \boldsymbol{w})^2 + \gamma(a_l u + b_l v + d)^2 + \frac{1}{\theta}(\boldsymbol{w} - \hat{\boldsymbol{w}})^2 \right) = 0,$$

$$\frac{\partial}{\partial v} \left(\lambda \tilde{\rho}(x, y, \boldsymbol{w})^2 + \gamma(a_l u + b_l v + d)^2 + \frac{1}{\theta}(\boldsymbol{w} - \hat{\boldsymbol{w}})^2 \right) = 0. \qquad (13)$$

Equation (12) is a linear equation in (u, v) and can be solved as a linear system $A\boldsymbol{w} = B$. Hence, A and B matrices of the linear system can be written as:

$$A = \begin{bmatrix} \frac{1}{\lambda\theta} + \sum S_x^2 + 2\gamma(a_l)^2 & \sum S_x S_y + 2\gamma a_l b_l \\ \sum S_x S_y + 2\gamma a_l b_l & \frac{1}{\lambda\theta} + \sum S_y^2 + 2\gamma(b_l)^2 \end{bmatrix}.$$

and

$$B = \begin{bmatrix} \left(\frac{1}{\lambda\theta} + \sum S_x^2 \right) \hat{u} + \sum S_x \sum S_y \hat{v} - \sum S_x \sum S_t + 2\gamma a_l(a_l x + b_l y + d) \\ \left(\frac{1}{\lambda\theta} + \sum S_y^2 \right) \hat{v} + \sum S_x \sum S_y \hat{u} - \sum S_y \sum S_t + 2\gamma b_l(a_l x + b_l y + d) \end{bmatrix}.$$

The solution of (u, v) is obtained using a lest square minimization scheme together with the optimization of the smoothness term using the fixed point iteration approach to find solutions for $(\hat{u}, \hat{v})$ (see [7] and [13] for more details about the dual variational optical flow approach).

3   Enhancement of the Fundamental Matrix

Theoretically, it sounds plausible that fundamental matrix can also be enhanced based on the flows obtained at each level. To this end, we discuss briefly, how the fundamental matrix can be iteratively at each iteration enhanced and in experimental result section, we evaluate how it affects the results practically. One method uses the epipolar distances between each point and its corresponding epipolar line. Given N matched point as (x_i, y_i) and (x_i', y_i'), $i = 1, ..., N$, the following error function should be minimized:

$$\epsilon = \sum_{i=1}^{N} \frac{(a_i x_i' + b_i y_i' + c_i)^2}{a_i^2 + b_i^2} + \frac{(a_i' x_i + b_i' y_i + c_i')^2}{a_i'^2 + b_i'^2} \qquad (14)$$

where $[a_i \ b_i \ c_i]^T = F[x_i, y_i, s^l]^T$ and $[a'_i \ b'_i \ c'_i]^T = F^T[x'_i, y'_i, s^l]^T$. To force the rank deficiency of the fundamental matrix, it can be parametrized as follows:

$$F = \begin{bmatrix} f_1 & f_2 & f_3 \\ f_4 & f_5 & f_6 \\ \alpha f_1 + \beta f_4 & \alpha f_2 + \beta f_5 & \alpha f_3 + \beta f_6 \end{bmatrix} \tag{15}$$

As a result the parameters vector which should be enhanced is:

$$\mathbf{f} = [f_1, f_2, f_3, f_4, f_5, f_6, \alpha, \beta]^T \tag{16}$$

The error function in Eq. (14) can be rewritten as multi-objective error functions to form a nonlinear equation system containing N equations. The equation system is solved iteratively in a regularized form, e.g., using Marquardt-Levenberg method.

4   Experiments

For quantitative evaluation, we tested the proposed approach on the public KITTI 2012 dataset [19], which provides a challenging testbed for the evaluation of optical flow algorithms (image resolution 1240×376). Pixel displacements in the dataset are generally large, exceeding 250 pixels in some sequences. Furthermore, the images exhibit homogeneous regions, strongly varying lighting conditions, and many non-Lambertian surfaces, especially translucent windows and specular glass and metal surfaces. Moreover, the high speed of the forward motion creates large regions on the image boundaries that move out of the field of view between frames, such that no correspondence can be established.

4.1   Comparison to the State-of-the-art Methods

At May 2019 we evaluated the algorithm on KITTI benchmark two times. The first time we evaluate the algorithm using the HOG descriptor, the epipolar line constraint, and the enhancement of the fundamental matrix. In the second

Table 1. The effect of the use of the enhancement of the fundamental matrix on the percentage of the outliers and the average end-point error using KITTI 2012 online evaluation benchmark on May 2019.

| | Using Epipolar line | | | | Without the Epipolar line | |
| | With enhancement of the fundamental matrix | | Without enhancement of the fundamental matrix | | | |
	% Outliers	AEE	% Outliers	AEE	% Outliers	AEE
2 px	9.01%	1.56 px	9.51%	1.8 px	13.41%	2.9 px
3 px	6.75%	1.56 px	6.95%	1.8 px	10.87%	2.9 px
4 px	5.63%	1.56 px	5.79%	1.8 px	9.58%	2.9 px
5 px	04.60 %	1.56 px	5.06%	1.8 px	08.71%	2.9 px

Table 2. Comparison among the proposed approach and some of the state-of-the art methods on the KITTI 2012 benchmark.

Method	% Outliers	AEE
RLOF (IM-GM) [9]	37.49%	8.2 px
FSDEF [15]	36.85%	8.8 px
GC-BM-Bino [10]	18.83%	5.0 px
TF+OFM [18]	10.22%	2.0 px
MLDP-OF [7]	08.67%	2.4 px
MEC-Flow [13]	06.95%	1.8 px
Proposed method	**06.75%**	**1.56 px**

evaluation, we used only the HOG descriptor without the epipolar line constraint. We got the results shown in Table 1. It can be seen, that the average end-point error AEE is significantly decreased using the proposed method while the percentage of outliers is slightly decreased when we used the monocular

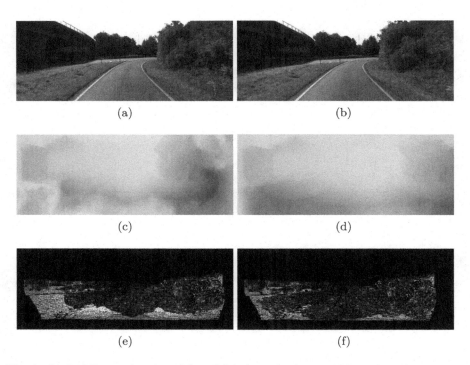

(a)   (b)

(c)   (d)

(e)   (f)

Fig. 2. Optical flow estimation. (a) and (b) show the frames 000025_10 and 000025_11 of sequence 25 of the KITTI training dataset. (c) and (e) depict the estimated optical flow and AEE error map without the monocular epipolar line constraint, while (d) and (f) are with it.

epipolar line constraint integrated with the enhancement of the fundamental matrix. A comparison with state-of-the art methods on the KITTI dataset is shown in Table 2. The proposed method outperformes other methods that used the epipolar geometry to estimate the optical flow. The monocular epipolar line constraint succeeds to estimate a better optical flow. The AEE and the percentage of outliers are significantly reduced. Some images of these sequences are shown in Fig. 2.

4.2 Evaluation of the Fundamental Matrix Re-estimation

We also conducted an enhancement of the fundamental matrix starting at different levels of the coarse-to-fine scheme. In this regard, we used the method which produces the minimum errors which is the 7-point method, LK feature trackers, and HOG descriptor. In Fig. 3, the errors concerning the starting level to

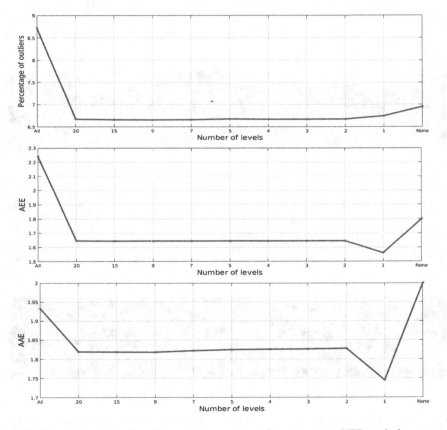

Fig. 3. The percentage of outliers, the average end-point error AEE and the average angular error AAE based on the updating fundamental matrix starting at different levels.

apply enhancement is presented. We can see that the update of the fundamental matrix resulted in functional improvement if it was applied in the fine levels (i.e. level one). We also see that if the matrix was updated from the coarsest level, the errors noticeably increased. One crucial issue is that if no good initial guess of the fundamental was fed to the method, it diverged in the first steps. The reason is that the estimation of the fundamental matrix is generally sensitive to measurement noise. The noise more than two pixels for the 7-point method can result in significant errors in the estimation of the fundamental matrix.

5   Conclusion

We have explained the integration of the monocular epipolar line constraint for an uncalibrated camera in a variational method for the calculation of the optical flow. We have also described an efficient way to enhance the fundamental matrix using the estimated optical flow. The proposed algorithm was evaluated with different sequences of the KITTI datasets providing the correct flow fields and increased the robustness. The performance of the proposed method was evaluated based on the KITTI dataset. The performance could be enhanced by about 15% reduction in the average end-point error less than the baseline method [13] and about 28% less than the method [5].

References

1. Molnár, J., Chetverikov, D., Fazekas, S.: Illumination-robust variational optical flow using cross-correlation. Comput. Vis. Image Underst. **114**(10), 1104–1114 (2010)
2. Werlberger, M., Pock, T., Bischof, H.: Motion estimation with non-local total variation regularization. In: 2010 IEEE Computer Society Conference on Computer Vision and Pattern Recognition, pp. 2464–2471 (2010)
3. Drulea, M., Nedevschi, S.: Motion estimation using the correlation transform. IEEE Trans. Image Process. **22**(8), 3260–3270 (2013)
4. Mueller, T., Rabe, C., Rannacher, J., Franke, U., Mester, R.: Illumination-robust dense optical flow using census signatures. In: Joint Pattern Recognition Symposium, pp. 236–245 (2011)
5. Rashwan, H.A., Mohamed, M.A., García, M.A., Mertsching, B., Puig, D.: Illumination robust optical flow model based on histogram of oriented gradients. In: Weickert, J., Hein, M., Schiele, B. (eds.) GCPR 2013. LNCS, vol. 8142, pp. 354–363. Springer, Heidelberg (2013). https://doi.org/10.1007/978-3-642-40602-7_38
6. Mirabdollah, M.H., Mohamed, M.A., Mertsching, B.: Distributed averages of gradients (DAG): a fast alternative for histogram of oriented gradients. In: Behnke, S., Sheh, R., Sarıel, S., Lee, D.D. (eds.) RoboCup 2016. LNCS (LNAI), vol. 9776, pp. 97–108. Springer, Cham (2017). https://doi.org/10.1007/978-3-319-68792-6_8
7. Mohamed, M.A., Rashwan, H.A., Mertsching, B., García, M.A., Puig, D.: Illumination-robust optical flow using a local directional pattern. IEEE Trans. Circ. Syst. Video Technol. **24**(9), 1499–1508 (2014)

8. Mohamed, M.A., Rashwan, H.A., Mertsching, B., Garcia, M.A., Puig, D.: On improving the robustness of variational optical flow against illumination changes. In: Proceedings of the 4th ACM/IEEE International Workshop on Analysis and Retrieval of Tracked Events and Motion in Imagery Stream, pp. 1–8 (2013)

9. Senst, T., Geistert, J., Sikora, T.: Robust local optical flow: long-range motions and varying illuminations. In: 2016 IEEE International Conference on Image Processing (ICIP), pp. 4478–4482 (2016)

10. Kitt, B., Lategahn, H.: Trinocular optical flow estimation for intelligent vehicle applications. In: 2012 15th International IEEE Conference on Intelligent Transportation Systems, pp. 300–306 (2012)

11. Yamaguchi, K., McAllester, D., Urtasun, R.: Robust monocular epipolar flow estimation. In: Proceedings of the IEEE Conference on Computer Vision and Pattern Recognition, pp. 1862–1869 (2013)

12. Valgaerts, L., Bruhn, A., Weickert, J.: A variational model for the joint recovery of the fundamental matrix and the optical flow. In: Rigoll, G. (ed.) DAGM 2008. LNCS, vol. 5096, pp. 314–324. Springer, Heidelberg (2008). https://doi.org/10. 1007/978-3-540-69321-5_32

13. Mohamed, M.A., Mirabdollah, M.H., Mertsching, B.: Monocular epipolar constraint for optical flow estimation. In: Liu, M., Chen, H., Vincze, M. (eds.) ICVS 2017. LNCS, vol. 10528, pp. 62–71. Springer, Cham (2017). https://doi.org/10. 1007/978-3-319-68345-4_6

14. Hartley, R., Zisserman, A.: Multiple View Geometry in Computer Vision. Cambridge University Press, Cambridge (2003)

15. Garrigues, M., Manzanera, A.: Fast semi dense epipolar flow estimation. In: 2017 IEEE Winter Conference on Applications of Computer Vision (WACV), pp. 427–435 (2017)

16. Mohamed, M.A., Mirabdollah, M.H., Mertsching, B.: Differential optical flow estimation under monocular epipolar line constraint. In: Nalpantidis, L., Krüger, V., Eklundh, J.-O., Gasteratos, A. (eds.) ICVS 2015. LNCS, vol. 9163, pp. 354–363. Springer, Cham (2015). https://doi.org/10.1007/978-3-319-20904-3_32

17. Ranftl, R., Vineet, V., Chen, Q., Koltun, V.: Dense monocular depth estimation in complex dynamic scenes. In: Proceedings of the IEEE Conference on Computer Vision and Pattern Recognition, pp. 4058–4066 (2016)

18. Kennedy, R., Taylor, C.J.: Optical flow with geometric occlusion estimation and fusion of multiple frames. In: Tai, X.-C., Bae, E., Chan, T.F., Lysaker, M. (eds.) EMMCVPR 2015. LNCS, vol. 8932, pp. 364–377. Springer, Cham (2015). https:// doi.org/10.1007/978-3-319-14612-6_27

19. Geiger, A., Lenz, P., Stiller, C., Urtasun, R.: Vision meets robotics: the KITTI dataset. Int. J. Robot. Res. 32(11), 1231–1237 (2013)

3D Hand Tracking by Employing Probabilistic Principal Component Analysis to Model Action Priors

Emmanouil Oulof Porfyrakis[1,2], Alexandros Makris[1(✉)],
and Antonis Argyros[1,2]

[1] Institute of Computer Science, FORTH, Heraklion, Greece
{porfyrak,amakris,argyros}@ics.forth.gr
[2] University of Crete, Heraklion, Greece

Abstract. This paper addresses the problem of 3D hand pose estimation by modeling specific hand actions using probabilistic Principal Component Analysis. For each of the considered actions, a parametric subspace is learned based on a dataset of sample action executions. The developed method tracks the 3D hand pose either in the case of unconstrained hand motion or in the case that the hand is engaged in some of the modelled actions. The tracker uses gradient descent optimization to fit a 3D hand model to the available observations. An online criterion is used to automatically switch between tracking the hand in the unconstrained case and tracking it in the case of learned action subspaces. To train and evaluate the proposed method, we captured a new dataset that contains sample executions of 5 different grasp-like hand actions and hand/object interactions. We tested the proposed method both quantitatively and qualitatively. For the quantitative evaluation we relied on our dataset to create synthetic sequences from which we artificially removed observations to simulate occlusions. The obtained results show that the proposed method improves 3D hand pose estimation over existing approaches, especially in the presence of occlusions, where the employed action models assist the accurate recovery of the 3D hand pose despite the missing observations.

1 Introduction

The problem of effectively inferring the 3D pose of human parts and understanding human actions is a challenging topic in computer vision. In real life, human hands support important functions by executing complex tasks such as object manipulation and sign-based human-to-human communication. By developing technical systems that are able to observe and understand the configuration of human hands we can support applications such as sign language recognition, interactive games or virtual reality environments, robotic arm tele-operation and many others. Such applications typically require high accuracy and robustness. To meet these requirements, many challenges must be addressed such as

© Springer Nature Switzerland AG 2019
D. Tzovaras et al. (Eds.): ICVS 2019, LNCS 11754, pp. 531–541, 2019.
https://doi.org/10.1007/978-3-030-34995-0_48

occlusions, uncontrolled environments and fast hand motions. We focus on the problem of 3D hand pose estimation and gesture recognition based on modeling and exploiting action priors. More specifically, the goal is to exploit prior knowledge on the hand actions to estimate the hand pose and the performed gesture. Taking into account the high dimensionality of hand models, we use Probabilistic Principal Component Analysis [23], a linear dimensionality reduction technique, combined with gradient based optimization. The input to our approach is RGB image sequences. Prior knowledge in the form of kinematic constraints (average size of an articulated structure, degrees of freedom for each articulation), or motion dynamics (physical laws ruling the object movements, assumptions on grasp movements) may provide rich information and facilitates the solution of the aforementioned problem. In our case, prior knowledge is based on the modeling of a set of predefined actions. The main assumption is that the finger motions are correlated given a particular hand action such as an object grasp. In other words, we assume that a grasp that concerns a particular object type will be performed similarly regardless of the subject that performs it.

1.1   Related Work

A large number of methods have been proposed for solving the 3D hand pose estimation and gesture recognition problems using markerless RGB-D or RGB observations. Several works employ prior information on the hand motion to facilitate and speed-up pose estimation, and/or to deal with missing observations.

Model based approaches use 3D hand models and local optimization to estimate the hand pose. Several optimization algorithms have been proposed, such as Particle Swarm Optimization [13,18], hierarchical particle filters [11], or the quasi-Newton method [4]. Methods that estimate the shape of the hand in addition to the pose by using deformable hand models have also appeared [10,21].

Discriminative approaches attempt to regress the pose directly from observations. Hybrid methods use a discriminative component to extract high level features which are then fed into a generative component. Over the last years, Convolutional Neural Network (CNN) based approaches dominate this category. One direction is to estimate 2D keypoints which are then lifted to 3D [2,16]. The downside of passing through a 2D representation is the presence of projection ambiguities which can be overcome by employing suitable priors. Approaches that rely on RGB-D provide good accuracy and avoid the projection related ambiguities [17,20]. Another approach is to directly estimate the 3D pose from RGB images [12].

For gesture recognition, recent methods rely mostly on CNNs. Liang [9] proposed a multi-view framework for recognizing hand gestures using point clouds captured by a depth sensor. They used CNNs as feature extractors followed by an SVM classifier to classify hand gestures. In [15] they utilized a CNN and stacked a denoising auto-encoder for recognizing 24 hand gestures of the American Sign Language. In [8], they used two CNN architectures, one lightweight

CNN architecture for detecting hand gestures and a deep CNN for classifying them.

Several approaches use prior motion information and dimensionality reduction to facilitate hand pose estimation. In [5,7], they employ Principal Component Analysis (PCA) to learn a lower dimensional space that describes compactly and effectively the human hand articulation, thus reducing the computational effort needed for hand poses estimation. In [14], they use PCA to learn subspace models from cyclic motions. Nonlinear dimensionality reduction techniques have also been used, as, for example, in [6] where ST-Isomap is used. However, Isomap and LLE do not provide mapping between the latent space and the data space. Gaussian Process Dynamical Model (GPDM), a nonlinear reduction method, had been applied [19] for 3D human body tracking. Urtasun et al. [24] use a form of probabilistic dimensionality reduction with a GPDM to formulate the tracking as a nonlinear least-squares optimization problem. Tian et al. [22] use Gaussian Process Latent Variable Models (GPLVM) for 2D pose estimation.

Our Contribution: This work aims at exploiting prior knowledge about particular hand motions to reduce the dimensionality of the hand pose estimation problem, which (a) speeds-up the tracking and (b) provides robustness to noise and missing observations. The first contribution is the coupling of the state of the art hybrid approach of [16] with probabilistic PCA dimensionality reduction. An additional contribution is the compilation of a dataset comprised of several actions (mostly grasping) executed by multiple actors. The dataset has been used for the training of our method and will become publicly available.

2 Method Description

2.1 Hand Model

The hand model we use (see Fig. 1) is comprised of a kinematic skeleton and a 3D mesh that represents the geometry of its surface. It has 26 degrees of freedom (DOFs), 6 for global position and rotation and 20 for finger articulation. Specifically, the kinematics of each finger is modeled using four parameters, two for the base joint of the finger and one for each of the two remaining joints.

2.2 Action PPCA Training

The Probabilistic Principal Components Analysis (PPCA) requires a dataset of example executions (RGB sequences) of a set of actions. The hand pose in each frame of the dataset is annotated. As described in Sect. 2.1, the hand pose is comprised of a global translation and rotation and the hand joints articulation. The action modeling concerns only the articulation part, so in the following we stripped the global transform DOFs from the hand state. Given the articulation pose sequences, a small number of key poses specific for each motion is identified. Subsequently, the motions are time wrapped so that the key poses are temporally aligned. Furthermore, the number of poses of each sequence is reduced to a

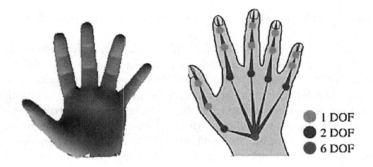

Fig. 1. Hand model: left: hand shape/geometry, right: hand kinematics.

predetermined value N. The state of a pose is denoted by t_n where the index n is the action phase with $n \in [1..N]$.

We used the Expectation-Maximization (EM) algorithm for training a PPCA model for each action. As input, the algorithm takes the state trajectories of a set of sample action executions. Each trajectory state t results from the concatenation of the N hand pose states t_n. The output of EM is the estimated weight matrix W and the variance of noise σ^2. Using these we can convert full dimensional states t_n to reduced dimensional states x_n and vice versa using:

$$x_n = Y^{-1}W^T(t_i - \Theta), \tag{1}$$

$$t_n = x_n * W + \Theta, \tag{2}$$

where Θ is the training states mean and $Y = \sigma^2 I + W^T W$.

2.3 PPCA Hand Tracking

The input to the proposed tracking algorithm is an RGB image and the M action PPCA models. From the image we extract the 2D hand joint locations. These locations are used in an optimization algorithm (see following paragraph) to estimate the hand pose. The optimization is performed $M + 1$ times i.e. on the full dimensional space, and on each of the M modeled sub-spaces. Each optimization provides a candidate solution and the best solution is selected using a method described later in this section. All the steps of the proposed method are summarized in Algorithm 1.

Optimization Algorithm: 3D hand pose estimation is treated as an optimization problem, as in [16]. The input to the optimizer is a set of 2D hand keypoints which are localized in the input image using OpenPose [3]. Typically, the optimization is performed on the full hand state. We also follow this approach to track free hand motion. However, for the pre-modeled actions, we exploit dimensionality reduction to perform optimization on a lower dimensional space.

Algorithm 1. Hand pose estimation.

1: Initialization: t_{f_0}
2: **for <each frame RGB_f> do**
3:    $[t_t^0, S_t^0] = solver(RGB_f, t_{f-1}^0)$ # Sect. 2.3
4:    **for <every model m> do**
5:       $[x_t^m, S_f^m] = solver(RGB_f, x_{f-1}^m)$ # Sect. 2.3
6:       $t_t^m = x_{f-1}^m * W + \Theta_n^m$
7:    **end for**
8:    $m_{sel} = select\_model([x_f^m, S_f^m]_{m=0}^M)$ # Sect. 2.3
9:    Solution: $t_f^{m_{sel}}$
10: **end for**

Given a hand pose and its forward kinematics function, we compute the positions of the joint keypoints $m_i = (u_i, v_i)$, $i \in [1, I]$, $I = 18$, on the image plane. Let $o_i = (u_i, v_i)$, $i \in [1, I]$, represent the detected 2D joints (using Open-Pose) and f_i be a binary flag taking the value of 1 if the i-th keypoint is actually detected and 0 otherwise. For a given pose, the total discrepancy $S(\mathbf{x}, o)$ between the observed and the model joints is given by:

$$S(\mathbf{x}, o) = \sum_{i=1}^{I} f_i \|m_i - o_i\|. \tag{3}$$

The 3D pose x^* that is most compatible with the available observations can be estimated by minimizing the objective function of Eq. 3:

$$\mathbf{x}^* = \arg\min_x \{S(\mathbf{x}, o)\}. \tag{4}$$

This is achieved using the Levenberg-Marquardt optimizer that minimizes this objective function after the automatic differentiation of the residuals. In our implementation, optimization has been performed by employing the Ceres Solver [1].

Model Selection: The selection of the appropriate low dimensional model to be used is performed automatically and on-line. The selection relies on the optimization score but for stability we propose a model locking mechanism based on the action phase and the model likelihood.

For each frame we perform the optimization procedure using all the available models (including the full dimensional model aiming at recovering free hand motion). The optimization score S_f^m for each model m approximates the degree of fit of each model to the observations. We select the model with the minimum score value: $m_{sel} = \arg\min_m \{S_f^m\}_{m=0}^M$. The pose estimation of this model t_f^m is thus the output of the algorithm for the frame f.

The selection procedure based solely on the optimization score is unstable. This is mainly due to the fact that the optimization algorithm relies only on the visible keypoints whose number fluctuates during tracking. To achieve model

(a)   (b)   (c)   (d)   (e)

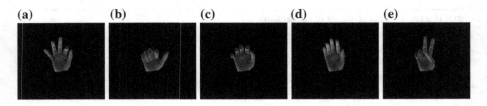

Fig. 2. Grasp actions that have been used in the developed dataset. (a) Pincer grasp, (b) palm grasp, (c) spherical grasp, (d) parallel extension grasp, (e) ring pinch grasp.

selection that is robust to the score fluctuations, we use a model locking approach. By this approach, we lock to a specific model if two criteria are met: (i) the model likelihood L^m is above a threshold value and (ii) the action phase n is above a threshold value. Essentially, these two criteria ensure that if a particular action is detected with a high likelihood and the action execution has advanced considerably, the algorithm will lock the selection to that action model until action completion. The model likelihood is given by:

$$L^m(x_n) = \exp - \frac{((x_m - \Theta_n^m)C^{m-1}(x_m - \Theta_n^m)^T)}{2}, \tag{5}$$

where Θ_n^m is mean value for model m and C^m is the covariance matrix.

3   Experiments

Dataset: For the purposes of this work, we created a new dataset for training and testing the proposed method. The dataset contains 5 grasping actions performed by 6 subjects, 2 females and 4 males. Every subject repeated each action 6 times. The instructions that had been given to all subjects was the verbal description of the actions they had to perform. This gave the opportunity to have action executions with considerable variability. Characteristic snapshots of the specified set of actions are shown in Fig. 2. For every action in the dataset the hand starts from a neutral (open) configuration.

To enable the quantitative evaluation of the proposed method, we used the real world dataset to create a synthetic one. To do so, we tracked the hands in the real dataset to obtain 3D hand poses that we considered as ground truth. We then used the known camera parameters to project the 3D joint locations extracted from the aforementioned ground truth poses back to the image. We provide these 2D image locations as input to the method. To simulate occlusions, we selectively removed some of the 2D keypoints from the input that is provides to the evaluated methods.

Evaluated Methods: We implemented and evaluated the following methods:

- **LEV:** Levenberg-Marquardt optimization without dimensionality reduction.

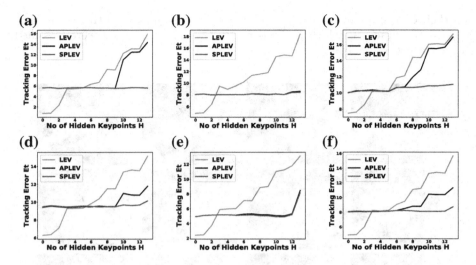

Fig. 3. The tracking error E_t as a function of the occlusion ratio. Each plot concerns sequences of a particular action: (a) pincer grasp, (b) palm grasp, (c) spherical grasp, (d) parallel extension grasp, (e) ring pinch grasp, (f) average over all grasps.

Table 1. Average hand action classification accuracy A_c as a function of the number H of hidden keypoints.

H	0	3	6	9	12
Pincer grasp	0.99	0.98	0.95	0.64	0.46
Palmer grasp	0.99	0.97	0.93	0.89	0.80
Spherical grasp	0.78	0.71	0.50	0.38	0.37
Parallel extension grasp	0.79	0.71	0.63	0.57	0.41
Ring pinch grasp	0.86	0.79	0.77	0.69	0.54
Average	**0.86**	**0.84**	**0.75**	**0.64**	**0.52**

- **APLEV:** Proposed optimization, exploiting the dimensionality reduction with automatic selection of the modeled actions.
- **SPLEV:** Proposed optimization exploiting the dimensionality reduction assuming knowledge (from the ground truth) of the performed actions.

For the low dimensional sub-spaces of **APLEV** and **SPLEV** methods we used 10 dimensions, 7 for global position and rotation, 2 for the articulation, and 1 for the action phase. Moreover, the likelihood threshold for the experiments are 0.55 and the phase threshold 0.31.

Quantitative Evaluation: We evaluated the methods quantitatively using the synthetic dataset described in Sect. 3. We measured the tracking error E_t which is defined as the average 3D distance between the estimated 3D joint locations

538   E. O. Porfyrakis et al.

and their corresponding ground truth values. We also measured the action classification accuracy A_c which is the percentage of of frames that were classified to the correct action class.

Fig. 4. Qualitative Results for grasping objects using **APLEV**. Every row represents a motion model in different phases.

In order to assess the ability of the methods to deal with occlusions, the tracking error was measured for different occlusion ratios. As mentioned in Sect. 3, to simulate occlusions we ignore a number H of 2D keypoints. In Fig. 3, we compare the performance of the methods for an occlusion percentage range from 0% to 60% which corresponds to $H = 0$ up to $H = 12$ of hidden 2D keypoints. The results show that the error of the proposed methods is smaller compared to the baseline method **LEV** if 4 or more keypoints are hidden. For the majority of the modeled actions, automatic model selection performs well and therefore the error of **APLEV** is on par with that of **SPLEV**. In two of the action classes, model selection does not perform so well, so **APLEV** has inferior performance to **SPLEV**. Nevertheless, it still compares favourably to the performance of the **LEV** baseline method.

The primary goal of the proposed **APLEV** method is to leverage prior knowledge about the performed actions in order to perform better tracking. To do so, it classifies each frame either into one of the modeled actions or as free hand motion. In Table 1 we present the action classification accuracy A_c results for different occlusion ratios. We observe that the classification accuracy remains high even in the presence of considerable occlusions. At the same time, classification as a function of occlusion vary considerably among different actions.

Qualitative Evaluation: For the qualitative evaluation we used a real world dataset where a hand performed object manipulation of various objects such as books, paper, bottle, small balls and pens. The obtained videos have a length between 250 and 600 frames and each of them contained at least 2 actions. As it can be verified in Fig. 4, the proposed approach captures the configuration of the hand correctly, despite the considerable occlusions between the hand and the manipulated object. More qualitative results are available as a youtube video[1].

4   Summary

We presented a method for markerless, model-based tracking of human 3D hand pose using dimensionality reduction based on action priors. We developed a dataset that contained instances of 5 action models and performed Probabilistic Principal Component Analysis to model them. The obtained quantitative and qualitative results demonstrate that the proposed approach manages to track the 3D pose of a hand robustly, even in the presence of considerable occlusions due to hand-object interactions. We intend to increase the grasp type action models so as to have a more complete relevant dataset. We also plan to incorporate object detection methods and enrich our method by exploiting fingertip/object contact points as location priors. Another future research direction is the exploitation of the proposed approach for 3D human body tracking.

Acknowledgements. This work was partially supported by the EU H2020 project Co4Robots (Grant No 731869).

References

1. Agarwal, S., Mierle, K., et al.: Ceres solver (2012)
2. Ballan, L., Taneja, A., Gall, J., Van Gool, L., Pollefeys, M.: Motion capture of hands in action using discriminative salient points. In: Fitzgibbon, A., Lazebnik, S., Perona, P., Sato, Y., Schmid, C. (eds.) ECCV 2012. LNCS, vol. 7577, pp. 640–653. Springer, Heidelberg (2012). https://doi.org/10.1007/978-3-642-33783-3_46

---

[1] https://youtu.be/L09qeohuJ9k.

3. Cao, Z., Simon, T., Wei, S.E., Sheikh, Y.: Realtime multi-person 2D pose estimation using part affinity fields. In: Proceedings of the IEEE Conference on Computer Vision and Pattern Recognition, pp. 7291–7299 (2017)
4. de La Gorce, M., Fleet, D.J., Paragios, N.: Model-based 3D hand pose estimation from monocular video. IEEE Trans. Pattern Anal. Mach. Intell. **33**(9), 1793–1805 (2011)
5. Douvantzis, P., Oikonomidis, I., Kyriazis, N., Argyros, A.: Dimensionality reduction for efficient single frame hand pose estimation. In: Chen, M., Leibe, B., Neumann, B. (eds.) ICVS 2013. LNCS, vol. 7963, pp. 143–152. Springer, Heidelberg (2013). https://doi.org/10.1007/978-3-642-39402-7_15
6. Jenkins, O.C., Matarić, M.J.: A spatio-temporal extension to isomap nonlinear dimension reduction. In: Proceedings of the Twenty-first International Conference on Machine Learning, p. 56. ACM (2004)
7. Kato, M., Chen, Y.W., Xu, G.: Articulated hand tracking by PCA-ICA approach. In: 7th International Conference on Automatic Face and Gesture Recognition (FGR06), pp. 329–334. IEEE (2006)
8. Köpüklü, O., Gunduz, A., Kose, N., Rigoll, G.: Real-time hand gesture detection and classification using convolutional neural networks. arXiv preprint arXiv:1901.10323 (2019)
9. Liang, C., Song, Y., Zhang, Y.: Hand gesture recognition using view projection from point cloud. In: 2016 IEEE International Conference on Image Processing (ICIP), pp. 4413–4417. IEEE (2016)
10. Makris, A., Argyros, A.: Model-based 3D hand tracking with on-line shape adaptation, pp. 77.1-77.12. British Machine Vision Association (2015)
11. Makris, A., Kyriazis, N., Argyros, A.A.: Hierarchical particle filtering for 3D hand tracking. In: 2015 IEEE Conference on Computer Vision and Pattern Recognition Workshops (CVPRW), pp. 8–17. IEEE, June 2015
12. Mueller, F., et al.: Ganerated hands for real-time 3D hand tracking from monocular RGB. In: The IEEE Conference on Computer Vision and Pattern Recognition (CVPR), June 2018
13. Oikonomidis, I., Kyriazis, N., Argyros, A.A.: Efficient model-based 3D tracking of hand articulations using kinect. In: BmVC, vol. 1, p. 3 (2011)
14. Ormoneit, D., Sidenbladh, H., Black, M.J., Hastie, T.: Learning and tracking cyclic human motion. In: Advances in Neural Information Processing Systems, pp. 894–900 (2001)
15. Oyedotun, O.K., Khashman, A.: Deep learning in vision-based static hand gesture recognition. Neural Comput. Appl. **28**(12), 3941–3951 (2017)
16. Panteleris, P., Oikonomidis, I., Argyros, A.A.: Using a single RGB frame for real time 3D hand pose estimation in the wild. In: IEEE Winter Conference on Applications of Computer Vision (WACV 2018), also available at Arxiv, pp. 436–445. IEEE, lake Tahoe, March 2018
17. Poier, G., Schinagl, D., Bischof, H.: Learning pose specific representations by predicting different views, April 2018
18. Qian, C., Sun, X., Wei, Y., Tang, X., Sun, J.: Realtime and robust hand tracking from depth, pp. 1106–1113, June 2014
19. Raskin, L.: Dimensionality reduction for 3D articulated body tracking and human action analysis. Technion-Israel Institute of Technology, Faculty of Computer Science (2010)
20. Sun, X., Wei, Y., Liang, S., Tang, X., Sun, J.: Cascaded hand pose regression. In: The IEEE Conference on Computer Vision and Pattern Recognition (CVPR), June 2015

21. Tan, D.J., et al.: Fits like a glove: rapid and reliable hand shape personalization. Microsoft Research, June 2016
22. Tian, T.P., Li, R., Sclaroff, S.: Tracking human body pose on a learned smooth space. Technical report, Boston University Computer Science Department (2005)
23. Tipping, M.E., Bishop, C.M.: Mixtures of probabilistic principal component analyzers. Neural Comput. **11**(2), 443–482 (1999)
24. Urtasun, R., Fua, P.: 3D human body tracking using deterministic temporal motion models. In: Pajdla, T., Matas, J. (eds.) ECCV 2004. LNCS, vol. 3023, pp. 92–106. Springer, Heidelberg (2004). https://doi.org/10.1007/978-3-540-24672-5_8

Cross-Domain Interpolation for Unpaired Image-to-Image Translation

Jorge López[1]([✉])(iD), Antoni Mauricio[1](iD), Jose Díaz[2], and Guillermo Cámara[3]

[1] Department of Computer Science, Universidad Católica San Pablo, Arequipa, Peru
{jorge.lopez.caceres,manasses.mauricio}@ucsp.edu.pe
[2] Universidad Nacional de Ingeniería, Lima, Peru
jcdiazrosado@uni.edu.pe
[3] Department of Computer Science, Federal University of Ouro Preto, Ouro Preto, Minas Gerais, Brazil
guillermo@iceb.ufop.br

Abstract. Unpaired Image-to-image translation is a brand new challenging problem that consists of latent vectors extracting and matching from a source domain A and a target domain B. Both latent spaces are matched and interpolated by a directed correspondence function F for $A \rightarrow B$ and G for $B \rightarrow A$. The current efforts point to Generative Adversarial Networks (GANs) based models due they synthesize new quite realistic samples across different domains by learning critical features from their latent spaces. Nonetheless, domain exploration is not explicit supervision; thereby most GANs based models do not achieve to learn the key features. In consequence, the correspondence function overfits and fails in reverse or loses translation quality. In this paper, we propose a guided learning model through manifold bi-directional translation loops between the source and the target domains considering the Wasserstein distance between their probability distributions. The bi-directional translation is CycleGAN-based but considering the latent space Z as an intermediate domain which guides the learning process and reduces the inducted error from loops. We show experimental results in several public datasets including Cityscapes, Horse2zebra, and Monet2photo at the EECS-Berkeley webpage (http://people. eecs.berkeley.edu/~taesung_park/CycleGAN/datasets/). Our results are competitive to the state-of-the-art regarding visual quality, stability, and other baseline metrics.

Keywords: Image-to-image translation · Generative Adversarial Network · Cross-domain interpolation

The present work was supported by grant 234-2015-FONDECYT (Master Program) from Cienciactiva of the National Council for Science, Technology and Technological Innovation (CONCYTEC-PERU) and the Vicerrectorate for Research of Universidad Nacional de Ingeniería (VRI - UNI).

D. Tzovaras et al. (Eds.): ICVS 2019, LNCS 11754, pp. 542–551, 2019.
https://doi.org/10.1007/978-3-030-34995-0_49

1   Introduction

Style transfer or image-to-image translation is a well-known problem in image processing, which consists of transforming an image A into another image B with a different style but preserving its semantic meaning through a transfer function [7]. Such a problem benefits a wide range of computer vision applications like realistic image synthesis [14], image stylization [3], image editing [17] and image super-resolution [10]. Some authors disentangle the latent features of domains to model a simpler transfer function [2] while others propose to update the transfer function iteratively until converging to the desired style [11, 16–18]. Cross-domain GAN-based methods are hot-trending due to GANs unravel the latent space and learn how to generate new samples from any latent vector.

In this paper, we develop a novel CycleGAN-based architecture considering the latent space as a transferable domain. As most GAN-based architectures, our proposal suffers some phenomena related to the vanishing gradient and learning divergence. These are overcome by including new loss functions in every control-loop and applying a spectral normalization during generation. Our results are visually compared with those obtained by the CycleGAN in several databases, in addition to the metrics proposed in [12]. The paper structure is as follows: Sect. 2 provides a consist resume about related works, their innovations, and results. Section 3 explores hardy the essential methods which support our proposals. Section 4 develops the paper's cores: the architecture, loss functions, and training algorithm. Section 5 details the most relevant experiments and results. Finally, Sect. 6 discusses the results, current challenges, and limitations.

2   Related Works

Hertzmann et al. [7] pioneered image-to-image translation focusing on image analogies. Isola et al. [9] exploit pixel-level reconstruction constraints to build connections between domains, although efficient, this framework requires a broad set of paired data for training. Unpaired image-to-image translation proposals [3, 11, 16–18], avoid pixel-level supervision requirements. Gatys et al. [3] propose an iterative process of encoding a domain into its feature vectors to decode them over white noise. While training, the reconstructed domain gets closer to the original one. Zhu et al. [18] propose a cyclic model GAN-based for unpaired data, known as CycleGAN, to support a bi-directional prediction between the source and target domain using adversarial loss in each new iteration and cycle.

Several works have proposed new improvements and modifications over the CycleGAN. Yi et al. [16] develop a dual-GAN mechanism inspired by dual learning from natural language translation. Each generator maps one real and one generated sample to the other domain in a cross order; *ergo* each generator feeds the other. Li et al. [11] argue CycleGANs based methods achieve low-quality results when the image resolution is high, or domains are significantly different. They propose a Stacked Cycle-Consistent Adversarial Networks (SCANs) to decompose a single translation into multi-stage transformations, boosting translation quality and image resolution. Hiasa et al. [8] introduce the gradient-consistency loss to the CycleGAN to enhance the resolution at the boundaries.

3   Background

In this section, we explore basic stuff related to our proposal enfolding the matching and interpolation of the latent vectors extracted from unpaired domains.

3.1   Generative Adversarial Networks - GANs

Proposed by Goodfellow et al. [5], GAN is considered as the pioneering method in adversarial learning and photo-realistic image synthesis. GANs re-sample the probability distribution by pitting a sample generator G against a binary discriminator D. The discriminator has to maximize the classification performance about if a sample x is synthetic ($x \sim \rho_g$) or real ($x \sim \rho_r$). Meanwhile, the generator maps a vector z from a random distribution ρ_z to synthesize a sample $G(z)$. G minimizes D performance to detect fakes by generating better samples. The interaction between $G(z, \theta_g)$ and $D(x, \theta_d)$ establishes a min-max adversarial game, where θ_g and θ_d are G and D hyperparameters, respectively. $D(x, \theta_d)$ is trained to maximize $log(D(x))$ whereas $G(z, \theta_g)$ is trained to minimize $log(1 - D(G(z)))$. The adversarial loss function L_{GAN} is expressed in Eq. 1.

$$\min_{G} \max_{D} \mathcal{L}_{GAN}(D, G) = \mathbb{E}_{x \sim \rho_r(x)}[log(D(x))] + \mathbb{E}_{z \sim \rho_z(z)}[log(1 - D(G(z)))] \quad (1)$$

According to several authors [1,6,13,14], the learning process is slow and unstable due to $\mathcal{L}_{GAN}$ represents a non-cooperative game with continuous agents updating hence it is hard to achieve Nash equilibrium. Furthermore, Mao et al. [13] mention that the sigmoid cross entropy loss function leads to the vanishing gradients problem during the learning process. To overcome such a problem, they propose to adopt the least squares loss function for D. As a result, two different loss functions are assigned for D and G as shown in Eqs. 2 and 3.

$$\max_{D} \mathcal{L}_{LSGAN}(D) = \frac{1}{2}\mathbb{E}_{x \sim \rho_r(x)}[(D(x) - 1)^2] + \frac{1}{2}\mathbb{E}_{z \sim \rho_z(z)}[(D(G(z)))^2] \quad (2)$$

$$\min_{G} \mathcal{L}_{LSGAN}(G) = \mathbb{E}_{z \sim \rho_z(z)}[(D(G(z)) - 1)^2] \quad (3)$$

Arjovsky et al. [1] stabilize L_{GAN} by using Wasserstein distance as the distribution similarity metric. Given G distribution over X (ρ_g) and real data distribution (ρ_r), $\mathcal{W}(\rho_r, \rho_g)$ measures the Wasserstein distance between ρ_g and ρ_r (see Eq. 4), thus Eq. 5 presents Wasserstein-GAN (WGAN) loss function.

$$\mathcal{W}(\rho_r, \rho_g) = \inf_{\gamma \sim \prod(\rho_r, \rho_g)} \mathbb{E}_{(x,y) \sim \gamma}[\|x - y\|] \quad (4)$$

$$\mathcal{L}_{WGAN}(\rho_r, \rho_g) = \mathcal{W}(\rho_r, \rho_g) = \max_{D} \mathbb{E}_{x \sim \rho_r}[D(x)] - \mathbb{E}_{z \sim \rho_z(z)}[D(G(z))] \quad (5)$$

CycleGANs. The image-to-image translation problem is complex a depends on data configuration. Paired data allow establishing an implicit correspondence function between domains, while unpaired data require a function that links up both domains. As Fig. 1 shows, given a pair of domains A and B, being G_A and G_B their generators, respectively. Then, each generator transfers the style of one domain to a sample from the other domain while each discriminator tries to recognize the real samples (X) from transferred ones/fakes (X^*). Thereby, G_A and G_B are cross-domain autoencoders defined by $G_B(A) = d_B(e_A(A)) = B^*$ and $G_A(B) = d_A(e_B(B)) = A^*$, while reconstructed samples (X^r) can be obtained by the following expressions: $d_A(e_B(B^*)) = A^r \approx A$ and $d_B(e_A(A^*)) = B^r \approx B$. Equation 7 shows CycleGAN loss function which includes both $\mathcal{L}_{GAN}$ and the cycle reconstruction error $\mathcal{L}_{cycle}$ (Eq. 6) pondered by factor λ.

$$\mathcal{L}_{cycle}(G_A, G_B) = \mathbb{E}_{A \sim \rho_r(A)}[\|G_A(G_B(A)) - A\|_1]$$
$$+ \mathbb{E}_{B \sim \rho_r(B)}[\|G_B(G_A(B)) - B\|_1] \tag{6}$$

$$\mathcal{L}_{CycleGAN} = \mathcal{L}_{GAN}(D_A, G_A) + \mathcal{L}_{GAN}(D_B, G_B) + \lambda\mathcal{L}_{cycle}(G_A, G_B) \tag{7}$$

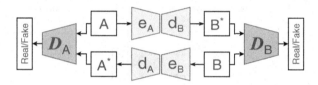

Fig. 1. CycleGAN architecture, $A \rightarrow B$: e_A encodes a sample $a \in A$ into a latent vector, later d_B decodes it to a transferred style sample $b^* \in B^*$. The second loop does the same but in reverse. Finally, discriminators try to recognize the real samples from transferred ones in both domains.

4   Cross-Domain Architecture

Our cross-domain architecture is a CycleGAN based composed of several ResNet blocks, inside generators, used to overcome the vanishing gradient problem. Like vanilla CycleGAN [18], we use feedback loops to control the transfer quality but considering the latent space as a new domain. As illustrated in Fig. 2, an encoder e links both pixel-domains (A and B) to the latent-domain (Z) while their respective decoders (d_A and d_B) reverse the encoding. Whereby, $Z_A = e(A)$, $Z_B = e(B)$, $d_B(Z_A) = B^*$ and $d_A(Z_B) = A^*$, being A^* and B^* transferred style samples. The loops close to compute the reconstructed samples by following in reverse the previous step; then, $Z_A^r = e(B^*)$, $Z_B^r = e(A^*)$, $d_A(Z_A^r) = A^r$ and $d_B(Z_B^r) = B^r$, being A^r, Z_A^r, B^r and Z_B^r reconstructed data. Discriminators D_A and D_B work over their respective domains such as vanilla CycleGAN meanwhile D_Z tries to distinguish Z_A from Z_B (see Fig. 3) aiming to homogenize their latent vectors and unfold the hidden information.

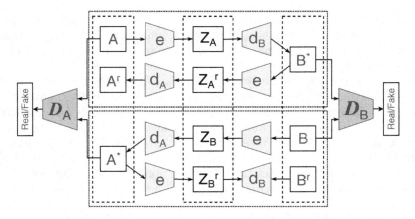

Fig. 2. Proposed architecture. e encodes a sample $a \in A$ into a latent vector $z_A \in Z_A$, then d_B decodes z_A to a transferred style sample $b^* \in B^*$. The loop closes to get the reconstructed sample $a^r \in A^r$ and reconstructed latent vector $z_a^r \in Z_A^*$ using the same encoder e but a different decoder d_A. A second loop is applied although in the opposite direction ($B \to A$) following the same statements. Finally, discriminators try to recognize the real samples from transferred ones in both domains.

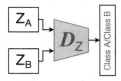

Fig. 3. Discriminator D_Z tries to differentiate the latent vectors from both domains.

4.1   Loss Function

We integrate the WGAN loss function (Eq. 5) into the CycleGAN loss function (Eq. 7) aiming to surpass stability and lack-generalization problems. Lastly, we apply a spectral normalization [15] to each GAN loss functions to generate smooth and uniform images. Equations 8 and 9 present discriminator and generator loss functions, respectively.

$$\mathcal{L}_{GAN}(D) = \mathbb{E}_{A \sim \rho_r(A)}[min(0, -1 + D(A))] + \mathbb{E}_{z \sim \rho_z(z)}[min(0, -1 - D(G(z)))] \quad (8)$$

$$\mathcal{L}_{GAN}(G) = -\mathbb{E}_{z \sim \rho_z(Z)}[D(G(z))] \quad (9)$$

If we consider $G_A(x) = e(d_A(x))$ and $G_B(x) = e(d_B(x))$ then Eqs. 8 and 9 can be rewritten such as in Eqs. 10, 11 and 12. Due to the hierarchy, the latent variables present the vanishing gradient problem, so we adopt the LSGAN loss for the training of D_Z and e (Eqs. 13 and 14). $\mathcal{L}_{ress}$ is the autoencoder reconstruction error [4] for both decoders (d_A and d_B) and the encoder e. Finally, we join all loss functions into one (Eq. 15). Parameter β allows to alternate the training process between the encoder e and decoders (d_A and d_B). As can be appreciated in Algorithm 1, training is primarily focused on latent vector synthesis and then in realistic image generation.

Algorithm 1: Training Algorithm

    Input: A and B: Datasets. K: Epochs. λ factor. α:learning rate. m:
            Batch size.
    Output: D_A, D_B, D_Z, e, d_A and d_B trained.

1 **for** $c = 1, 2, ..., K$ **do**
2      $j \leftarrow 0$
3   **for** $i = 1, 2, ..., |A|$ **do**
4   Sample a tuple $< a, b >; a \in A \wedge b \in B$
5   **if** $j < n_{critic}$ **then**
6   Equations 13, 11 and 10: Update D_Z, D_A and D_B
7   Equation 15: Update e, d_A and d_B with $\beta = 1$
8             $j \leftarrow j + 1$
9   **else**
10   Equation 15: Update e, d_A and d_B with $\beta = 0$
11         $j \leftarrow 0$

$$\mathcal{L}_{GAN}(D_A) = \mathbb{E}_{A \sim \rho_r(A)}[min(0, -1 + D_A(A))]$$
$$+ \mathbb{E}_{B \sim \rho_r(B)}[min(0, -1 - D_A(d_A(e(B))))] \tag{10}$$

$$\mathcal{L}_{GAN}(D_B) = \mathbb{E}_{B \sim \rho_r(B)}[min(0, -1 + D_B(B))]$$
$$+ \mathbb{E}_{A \sim \rho_r(A)}[min(0, -1 - D_B(d_B(e(A))))] \tag{11}$$

$$\mathcal{L}_{GAN}(d_A, d_B, e) = -\mathbb{E}_{A \sim \rho_r(A)}[D_B(d_B(e(A)))] - \mathbb{E}_{B \sim \rho_r(B)}[D_A(d_A(e(B)))] \tag{12}$$

$$\mathcal{L}_{LSGAN_z}(D_Z) = \frac{1}{2}\mathbb{E}_{A \sim \rho_r(A)}[(D_Z(e(A)) - 1)^2] + \frac{1}{2}\mathbb{E}_{B \sim \rho_r(B)}[(D_Z(e(B)))^2] \tag{13}$$

$$\mathcal{L}_{LSGAN_z}(e) = \mathbb{E}_{A \sim \rho_r(A)}[(D_Z(e(A)))^2] + \frac{1}{2}\mathbb{E}_{B \sim \rho_r(B)}[(D_Z(e(B)) - 1)^2] \tag{14}$$

$$\mathcal{L}_{Total}(D) = \lambda\mathcal{L}_{ress}(e, d_A, d_B) + \lambda\mathcal{L}_{cycle}(e, d_A, d_B)$$
$$+ \beta\mathcal{L}_{LSGAN_z}(e) + (1 - \beta)\mathcal{L}_{GAN}(d_A, d_B, e) \tag{15}$$

5 Experiments and Results

In this section, we show the experimental tests to contrast our results against CycleGAN results for different datasets and hyperparameters. Performed tasks include both paired (Fig. 4) and unpaired (Fig. 5) datasets. The Cityscapes dataset is paired and consists of tuples <real image, segmented image>, then we untangle them by considering each one as different domains. Figure 5 shows results over two unpaired datasets: Monet2photo and Horse2zebra, both contain two domains. The results show improvements over the CycleGAN model in terms of resolution and object boundary, although it varies according to image size and the number of objects.

| Input | Ground-truth | CycleGAN | Ours |

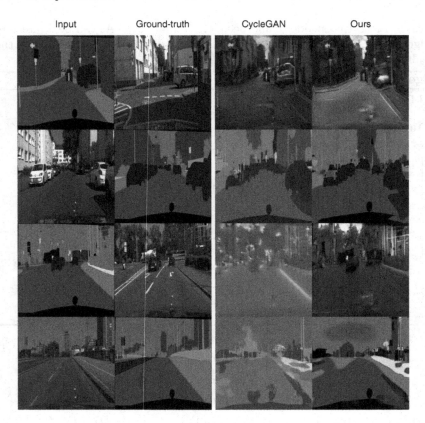

Fig. 4. Comparison of the results using Cityscapes dataset. The input and ground-truth are considered as different domains.

5.1 Implementation Details

For the implementation, the Pytorch library and an 8GB GTX1080ti GPU were used. For training, we use the Adam optimizer with a learning rate of $\alpha = 0.0002$, a decay factor of 50% over 200 epochs and a batch size $m = 18$, this configuration allows for training using less memory else GPU memory overflows. Lastly, we consider a $\lambda = 10$, and a $n_{critic} = 5$ in Eq. 15 because of its visual results.

5.2 Metrics

In addition to visual results, we use three metrics from conventional semantic segmentation and scene parsing evaluations [12]. Suppose there are n_{ij} pixels of class i labeled as class j from k different classes, and let $t_i = \sum_j n_{ij}$ be the total number of pixels of class i. Then, we compute: (1) pixel accuracy: $\sum_i n_{ii} / \sum_i t_i$; (2) mean accuracy: $\frac{1}{k} \sum_i n_{ii}/t_i$; and, (3) mean of region intersection over unions (UI): $\frac{1}{k} \sum_i n_{ii}/(t_i + \sum_j n_{ji} - n_{ii})$ (Table 1).

Table 1. Metric comparison.

	Pixel accuracy	Mean accuracy	Mean UI
CycleGAN	0.388406	0.054650	0.022456
Ours	0.387113	0.055757	0.023924

Fig. 5. Comparison of the results using Monet2photo and Horse2zebra datasets.

6   Conclusions and Future Works

The image-to-image translation takes advantage of the latent space manipulability, however, results interpretation is unclear, and the current metrics are not good enough yet. The proposed method explores latent space as an intermediate domain instead of account it as just an end-to-end problem. Our proposal uses a one-single encoder for all operations; thus, latent vectors from both domains get closer, and the transfer improves in precision. The loss function normalizes the training parameters to overcome generation phenomena. Nonetheless, hyperparameters must be tuned according to the image complexity; otherwise, overfitting or underfitting problems appear. About limitations, our model requires a high computational capacity per epoch proportional to the batch size.

Based on the results, we obtain sharper and more stable images than the CycleGAN results; however, like other proposals, our model worsens for multiple-labels problems due to its complexity. In the future, we will attempt to disentangle the latent vectors to obtain a prior mapping of domains to improve the quality of the results and surpass the image size limitations. Besides, we will prove our method in new datasets for different tasks related to medical image generation due to its social impact and computational complexity. Finally, we have to mention the necessity to raise better evaluation metrics because the current ones are inadequate and uninterpretable in a semantic evaluation context.

References

1. Arjovsky, M., Chintala, S., Bottou, L.: Wasserstein GAN. arXiv preprint arXiv:1701.07875 (2017)
2. Benaim, S., Wolf, L.: One-shot unsupervised cross domain translation. In: Advances in Neural Information Processing Systems, pp. 2104–2114 (2018)
3. Gatys, L., Ecker, A.S., Bethge, M.: Texture synthesis using convolutional neural networks. In: Advances in Neural Information Processing Systems, pp. 262–270 (2015)
4. Goodfellow, I., Bengio, Y., Courville, A.: Deep Learning. MIT Press, Cambridge (2016)
5. Goodfellow, I., et al.: Generative adversarial nets. In: Advances in Neural Information Processing Systems, pp. 2672–2680 (2014)
6. Gulrajani, I., Ahmed, F., Arjovsky, M., Dumoulin, V., Courville, A.C.: Improved training of Wasserstein GANs. In: Advances in Neural Information Processing Systems, pp. 5767–5777 (2017)
7. Hertzmann, A., Jacobs, C.E., Oliver, N., Curless, B., Salesin, D.H.: Image analogies. In: Proceedings of the 28th Annual Conference on Computer Graphics and Interactive Techniques, pp. 327–340. ACM (2001)
8. Hiasa, Y., et al.: Cross-modality image synthesis from unpaired data using CycleGAN. In: Gooya, A., Goksel, O., Oguz, I., Burgos, N. (eds.) SASHIMI 2018. LNCS, vol. 11037, pp. 31–41. Springer, Cham (2018). https://doi.org/10.1007/978-3-030-00536-8_4
9. Isola, P., Zhu, J.Y., Zhou, T., Efros, A.A.: Image-to-image translation with conditional adversarial networks. In: Proceedings of the IEEE Conference on Computer Vision and Pattern Recognition, pp. 1125–1134 (2017)
10. Ledig, C., et al.: Photo-realistic single image super-resolution using a generative adversarial network. In: Proceedings of the IEEE Conference on Computer Vision and Pattern Recognition, pp. 4681–4690 (2017)
11. Li, M., Huang, H., Ma, L., Liu, W., Zhang, T., Jiang, Y.: Unsupervised image-to-image translation with stacked cycle-consistent adversarial networks. In: Proceedings of the European Conference on Computer Vision (ECCV), pp. 184–199 (2018)

12. Long, J., Shelhamer, E., Darrell, T.: Fully convolutional networks for semantic segmentation. In: Proceedings of the IEEE Conference on Computer Vision and Pattern Recognition, pp. 3431–3440 (2015)
13. Mao, X., Li, Q., Xie, H., Lau, R.Y., Wang, Z., Paul Smolley, S.: Least squares generative adversarial networks. In: Proceedings of the IEEE International Conference on Computer Vision, pp. 2794–2802 (2017)
14. Mauricio, A., López, J., Huauya, R., Diaz, J.: High-resolution generative adversarial neural networks applied to histological images generation. In: Kůrková, V., Manolopoulos, Y., Hammer, B., Iliadis, L., Maglogiannis, I. (eds.) ICANN 2018. LNCS, vol. 11140, pp. 195–202. Springer, Cham (2018). https://doi.org/10.1007/978-3-030-01421-6_20
15. Miyato, T., Kataoka, T., Koyama, M., Yoshida, Y.: Spectral normalization for generative adversarial networks. arXiv preprint arXiv:1802.05957 (2018)
16. Yi, Z., Zhang, H., Tan, P., Gong, M.: DualGAN: unsupervised dual learning for image-to-image translation. In: Proceedings of the IEEE International Conference on Computer Vision, pp. 2849–2857 (2017)
17. Zhu, J.-Y., Krähenbühl, P., Shechtman, E., Efros, A.A.: Generative visual manipulation on the natural image manifold. In: Leibe, B., Matas, J., Sebe, N., Welling, M. (eds.) ECCV 2016. LNCS, vol. 9909, pp. 597–613. Springer, Cham (2016). https://doi.org/10.1007/978-3-319-46454-1_36
18. Zhu, J.Y., Park, T., Isola, P., Efros, A.A.: Unpaired image-to-image translation using cycle-consistent adversarial networks. In: Proceedings of the IEEE International Conference on Computer Vision. pp. 2223–2232 (2017)

A Short-Term Biometric Based System for Accurate Personalized Tracking

Georgios Stavropoulos[1]($\boxtimes$), Nikolaos Dimitriou[2], Anastasios Drosou[2], and Dimitrios Tzovaras[2]

[1] Electrical and Computer Engineering Department, University of Patras, Patras, Greece
stavrop@iti.gr
[2] Centre for Research and Technology, Information Technologies Institute, Thessaloniki, Greece

Abstract. Surveillance systems have long been in the focus of the research community. Although the accurate detection of the human presence in the scene is now possible even under extreme environmental conditions via the advanced modern camera sensors, efficient personalized tracking is still an open issue and a significant challenge for researchers addressing. Moreover, personalized tracking will not only enhance the tracking robustness but it can also find useful application in several commercial surveillance use-cases, ranging from security to occupancy statistics (i.e. per building, per space and per human). In this respect, this paper introduces a novel the biometric approach for enhanced privacy preserving human tracking based on a novel soft-biometric feature of humans. The moving blobs in the recorded scene can be easily detected in the colour images, while the human silhouettes are detected from the corresponding depth ones. The state-of-the-art $3D$ Weighted Walk-throughs ($3DWW$) transformation is applied on the extracted human $3D$ point cloud, forming thus, a short-term soft biometric signature. The re-authentication of the humans is performed via the comparison of their last valid signature with current one. A thorough analysis on the adjustment of the system's optimal operational settings has been carried out and the experimental results illustrate the promising robustness, accuracy and efficiency on human tracking performance.

Keywords: Motion detection · Human tracking · Surveillance · Geometric identification

1 Introduction

Among a wide range of applications related to video surveillance such as human motion understanding, content-based video retrieval and object-based video compression, the tracking of moving objects, including human silhouettes, has been one of the essential and fundamental tasks in computer vision.

© Springer Nature Switzerland AG 2019
D. Tzovaras et al. (Eds.): ICVS 2019, LNCS 11754, pp. 552–561, 2019.
https://doi.org/10.1007/978-3-030-34995-0_50

During the past decades, numerous and various approaches have been endeavoured to improve its performance, and there is a fruitful literature in tracking algorithms development that reports promising results under various scenarios. However, human tracking in surveillance applications [1] still remains a challenging problem in tracking the non-stationary appearance of humans undergoing significant pose, illumination variations and occlusions as well as shape deformation for non-rigid objects.

Such tracking system mainly concern about two essential issues: the detection algorithm that should be applied to locate the target (e.g. particle filtering approach [16]) and what type of features should be used to represent the object (e.g. mean-shift algorithm [5]).

1.1 Current Approaches

Modelling objects' appearance in videos is a problem of extracting features and is of utmost significance in the overall tracking procedure. Developing a robust system able to adaptively model [6,11] the object appearance changes has been concern the scientific community the last decades. In general, there is a large number of tracking algorithms in the literature, that can be divided into three main categories:

Blob-based methods: Mimik et al. [13] exploit Kalman filtering that is applied on $3D$ points. Similarly, Black et al. [4] utilize a Kalman filter to simultaneously track in $2D$ and $3D$. Focken et al. [7] compared a best-hypothesis and a multiple-hypotheses approaches to find people tracks from 3D locations obtained from foreground binary blobs extracted from multiple calibrated views.

Colour-Based Methods: Mittal et al. [14] propose a system that segments, detects and tracks multiple people in a scene using a wide-baseline setup several synchronized cameras. Kang et al. [12] introduce a method which tracks humans both in image planes and top view using $2D$ and $3D$ motion models derived from a Kalman filter.

Occupancy map methods: Recent techniques explicitly use a discredited occupancy map into which the objects detected in the camera images are back-projected. Beymer [3] relies on a standard detection of stereo disparities which increase counters associated to square areas on the ground via the utilization of Gaussians mixtures. In the same concept, Yang et al. [17] computed the occupancy map with a standard visual hull procedure. One originality of the approach is to keep for each resulting convex component an upper and lower bound on the number of objects it can contain.

The human tracking approach introduced in this paper, utilizes a state-of-the-art non-invertible transformation, based on the so-called $3D$ Weighed Walk-throughs ($3DWW$) that was originally proposed by Beretti et al. in [2] and achieved impressive accuracy in $3D$ face recognition. In this current work the $3DWW$ algorithm has been accordingly modified and applied on corporal surfaces, as captured by commercial low-cost depth sensors of moderate resolution capabilities and average frame rate (i.e. 30 fps), like Microsoft Kinect 2. The main assumption towards this approach is the fact that within very short time intervals (i.e. the differences between 2–3 frames), both the human posture and the camera's recording view angle will not significantly change. In other words, it is expected that the intra-similarities of one's corporal surface (see Fig. 3) within time-difference of few sequentially captured frames would be significantly higher than their inter-similarities among different people.

The rest of the paper is organized as follows. Initially, the motivation behind our efforts along with the gap indicated in the literature review are elaborated in Sect. 1.2. Section 2 presents a high level, building block diagram of the workflow to be followed approach, while Sect. 2.1 describes the preparation steps, including the human silhouette extraction process, required for the application of the core $3DWW$ algorithm in Sect. 3. After a detailed discussion regarding the application of the $3DWW$ algorithm on corporeal surfaces, the proprietary dataset and the carried out experiment are exhibited while the derived accuracy results are reported. The current paper concludes in Sect. 5 with a short summary and possible future extensions.

1.2   Motivation

Existing methods for real-time surveillance of individuals focus on monitoring the existence or the number of individuals in a specific area, rather than their identities. Identification is accomplished using extra knowledge, such as which person usually occupies a specific sub-area. Such methods fail to accurately track the identity of individuals in cases of occlusion or close proximity to other persons. On the other hand, methods to accurately identify persons based on biometric features already exist in the literature (i.e. biometrics). However, they usually require either long enrolment sessions, training, increased processing power, certain enrolment and recognition scenarios, etc. [9]. On the other hand, soft biometrics have proved their applicability and their recognition capacity [8], that can be further improved when exploited in under the correct scenarios.

Motivated by the weaknesses of the existing methods [10], the proposed approach combines soft biometric characteristics, in order to provide real-time identification of persons in a monitored environment. In order to avoid identifying between several different poses, a so-called "short-term" biometric identification procedure is followed, where each person is re-authenticated continuously in short time-intervals. This way, the pose of a person can be considered fixed, thus allowing successful biometric identification, even under the short occlusion cases or proximity false positives where the tracked identity can be lost with traditional methods. Moreover, it should be also noted that the proposed method exploits only the depth information, while color information is discarded. This way, contrary to most existing approaches, significant ethical and privacy issues are addressed and preserved.

This way, the main novelties of the proposed human tracking algorithm be summarized in (i) the introduction of a novel corporal soft biometric feature of geometric nature, (ii) the utilization of only privacy preserving depth images, (iii) the efficient application of the non-invertible $3DWW$ transformation and (v) the lifting of the occlusions via the introduction of a virtual camera in the preprocessing step (see Sect. 2.1).

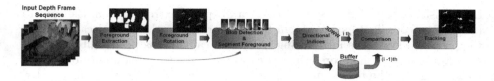

Fig. 1. System overview: the recorded frames are initially processed so as to extract the foreground information. The latter is then rotated to a top view in order to omit occlusions. Using the blobs from the rotated image the foreground image is segmented and the Directional Indices are extracted. The Directional Indices across sequential frames are matched to produce tracking results.

2   System Architecture

The architecture of the proposed system is displayed in Fig. 1. The input of the system is a sequence of depth frames. The first step is the extraction of the foreground, which is achieved using the methodology described in [18]. Then the 3D point cloud of the foreground is extracted, and transformed to the space of a virtual camera placed over the monitored area. The latter is achieved using the calibration information of the camera. The virtual camera is used to overcome the occlusions induced by the real camera. From the image of the virtual camera, the positions of the occupants in the monitored area are estimated, and transformed back to the real camera. These positions are then used to segment the 3D point cloud of the foreground and extract the 3DWW features (see Fig. 2). The extracted features of the current frame are compared with the features of the previous frame, assigning tracking labels to the occupants.

Fig. 2. Left: Input depth frame. Middle: Rotated foreground, where the separation of the occluded people is viewed. Right: Segmented foreground. Each person is marked with different color. Using the information from the rotated image, the individuals that are occluded can be separated. (Color figure online)

2.1   Preprocessing

Initially, and before the tracking can be started, the utilized camera is calibrated with respect to the architectural map of the monitored area. This way, the transformation matrices the convert a 3D point extracted from the camera (in its own coordinate system) can be transformed to the virtual camera that the proposed system is introducing. After the camera has been calibrated, the background of the monitored area is captured, with the area empty of people. Any changes in the background during the monitoring period are compensated using the methodology from [18].

During the runtime, and in order for the features to be calculated, the 3D point cloud of each individual in the monitored are has to be extracted from the depth frame. Towards this end, at first the foreground is extracted using the pre-calculated background and the background is updated if necessary. From the foreground the 3D point cloud is extracted, and transformed to the space of a virtual camera placed on top of the monitored area, thus omitting any occlusions, as can be seen in Fig. 2, where in the depth image (left) people occlude each other, while in the rotated one (middle) all people are clearly separated. Using the view of the virtual camera, the number and positions of the occupants are calculated (see middle part of Fig. 2). Finally the calculated positions are transformed back to the real camera, and used as centres in order to perform K-Means segmentation on the 3D point cloud (see right part of Fig. 2, where all individuals are labelled with different colors). The segments of the point cloud are used for the calculation of the features.

3   3D Weighted Walkthroughs as Short-Term Biometric Feature

The intra-person Directional Indices of a $3D$ surface (i.e. human body) are extracted by estimating the $3D$ Walking Walkthroughs ($3DWW$) on it, as described below. Initially, the shortest geodesic distances of each point on the bodial surface f with respect to the detected highest head location $N(x_0, y_0, z_0)$ is estimated via the Dijkstra algorithm. This way, isogeodesic stripes (i.e. 10) of equal width (i.e. 50 mm) are formed, concentric and centered on head peak (see point N in Fig. 3). Thereafter, the so-called $3DWW$ are computed between all pairs of isogeodesic stripes (interstripe $3DWW$) and between each stripe and itself (intrastripe $3DWW$), as explained in Fig. 5 and as proposed by Beretti et al. [2]. In particular, the $3DWWs$ are computed as aggregate measures (i.e. Directional Indices) that provide a representation for the mutual displacement between the set of points of two spatial entities (i.e. isogeodesic stripes). Finally, these Directional Indices are cast to a graph representation x_{3DWW}, where stripes are used to label the graph nodes and $3DWWs$ to label the graph edges.

In order to form the isogeodesic stripes and compute the $3DWW$ descriptor, only the top 20% (Fig. 3) of the body is considered. This includes the head and the upper part of the shoulders [19, 20] (Fig. 4). The rest of the body is not used since it makes a highly non-rigid motion, which is activity related, something that would introduce big fluctuations in the $3DWW$ descriptor computation.

The 3DWW encodes the relevant positions of the voxels between two stripes, or within a single stripe. By comparing the relevant position between all pairs

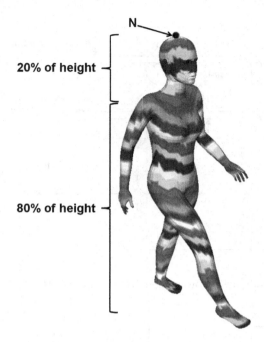

Fig. 3. Isogeodesic stripes over the human body. In the proposed system the top 20% of the body in considered, since the rest performs highly non-rigid motions.

of voxels, a $3 \times 3 \times 3$ matrix is filled with the accumulated number of voxels in each direction. The procedure for calculating the 3DWW is described in Fig. 5 and in more detail in [2].

4   Experimental Results

4.1   Dataset

The proposed method was tested using a proprietary dataset captured with a Microsoft Kinect 2 sensor, which provides depth images at 30 fps, at 512×424 resolution. The dataset consists of $\sim$3200 frames recorded in an indoor environment. During the recording, 6 individuals were walking in the monitored area with no specific pattern. The challenging nature of this dataset lies in the fact that several occlusions and collisions between humans when walking in very close proximity occurred.

4.2   Results

The captured dataset was manually annotated and the results of the proposed method were compared with the annotated ones. Also, the method described in [18] was used for comparison.

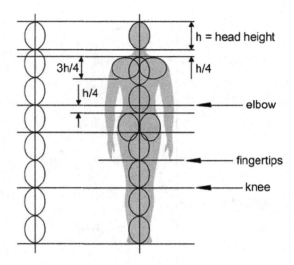

Fig. 4. Human body height proportions with respect to the size of the head [19].

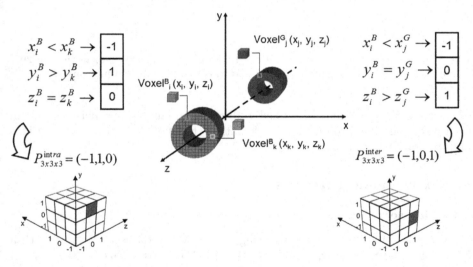

Fig. 5. The basics of the $3DWW$ algorithm are illustrated herein. *Center*: Two $3D$ objects (i.e. stripes) are depicted, each consisting of a certain amount of voxels. The procedure for producing the intra-signature of the each stripe is depicted on the *Left* side of the image, while the inter-signature of the stripes is depicted on the *Right* side. In particular, the relevant position between two voxels indicates a $3D$ position (i.e. directional index) in a $3 \times 3 \times 3$ cube. Once all voxels have been taken into account the accumulation of directional indices in the cube forms the biometric signature for a human in a certain time instance (i.e. frame). More details about the $3DWW$ can be found in [2].

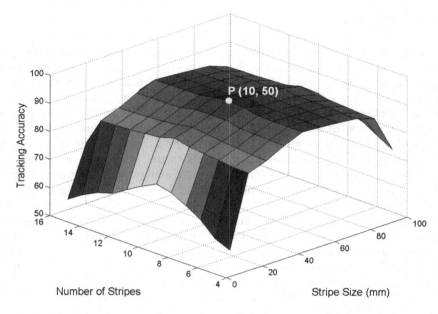

Fig. 6. Effect of the number and size of the Isogeodesic Stripes in the tracking accuracy. The optimal results are achieved using 10 stripes of 50 mm size.

In order to evaluate the effect of the number of stripes and the size of the stripe in performance, a number of experiments was run with a range of values for both variables. More in detail, the number of stripes was tested using 5 to 16 stripes with a step of 1, while the stripe size from 10 to 100 mm with a step of 10 mm. The results of these experiments can be seen in Fig. 6. The best results were obtained using 10 stripes with a geodesic width of 50 mm (point P in Fig. 6).

Using the aforementioned settings, the obtained results can be seen in Table 1. The proposed method clearly outperforms the one from [18]. The accuracy of the tracking was calculated by counting the number of correctly labelled persons between frames, with respect to the manually annotated labels of the same frame.

Table 1. Tracking accuracy

Method	Accuracy
Isogeodesic stripes	91.98%
Proximity	68.73%

5   Conclusion

In this paper, a novel soft-biometric approach for personalized human tracking from a low cost depth camera has been presented based on robust geometric features of the human body. Before the geometric feature extraction a pre-processing step has been applied that concerned the background subtraction, the view-angle transformation (i.e. scene rotation to a virtual top-view camera), the detection of the human blobs and the accurate segmentation of the human silhouettes via k-means algorithms in the $3D$ space. The extraction of the proposed $3DWW$ geometric feature leads to the generation of a biometric signature, that remains valid as long as the relevant position between the user and the camera does not alter significantly. The approach has been evaluated on a proprietary dataset of few people moving around in a closed space and the obtained results exhibited high accuracy and robustness even under difficult conditions. Moreover, it should be noted that the proposed approach supports the preservation of the human privacy by both processing only the depth images and by producing a secure biometric template via the application of an non-irreversible transformation. The proposed approach can be easily adjusted and integrated in privacy preserving cross-calibrated camera networks for usage in multi-space environments.

Acknowledgment. This work is co-funded by the European Union (EU) within the SMILE project under grant agreement number 740931. The SMILE project is part of the EU Framework Program for Research and Innovation Horizon 2020.

References

1. Yun, Y., Song, C., Katsaggelos, A.K., Yanwei, L., Yi, Q.: Wireless video surveillance: a survey. IEEE Access **1**, 646–660 (2013)
2. Berretti, S., Bimbo, A.D., International Continence Society, Pala, P.: 3D Face recognition using isogeodesic stripes. IEEE Trans. Pattern Anal. Mach. Intell. **32**(12), 2162–2177 (2010)
3. Beymer, D.: Person counting using stereo. In: Workshop on Human Motion, pp. 127–133 (2000)
4. Black, J., Ellis, T., Rosin, P.: Multi-view image surveillance and tracking. In: IEEE Workshop on Motion and Video Computing (2002)
5. Comaniciu, D., Ramesh, V., Meer, P.: Kernel-based object tracking. IEEE Trans. Pattern Anal. Mach. Intell. **5**, 564–575 (2003)
6. Fleuret, F., Berclaz, J., Lengagne, R., Fua, P.: Multicamera people tracking with a probabilistic occupancy map. IEEE Trans. Pattern Anal. Mach. Intell. **30**(2), 267–282 (2008)
7. Focken, D., Stiefelhagen, R.: Towards vision-based 3D people tracking in a smart room. In: IEEE International Conference on Multimodal Interfaces (2002)
8. Drosou, A., Tzovaras, D., Moustakas, K., Petrou, M.: Systematic error analysis for the enhancement of biometric systems using soft biometrics. IEEE Sig. Process. Lett. **19**(12), 833–836 (2012)

9. Drosou, A., Ioannidis, D., Tzovaras, D., Moustakas, K., Petrou, M.: Activity related authentication using prehension biometrics. Pattern Recogn. **48**(5), 1743–1759 (2015)
10. Xu, X., Tang, J., Liu, X., Zhang, X.: Human behavior understanding for video surveillance: recent advance. In: 2010 IEEE International Conference in Systems Man and Cybernetics (SMC), pp. 3867–3873 (2010)
11. Jia, X., Lu, H., Yang, M.: Visual tracking via adaptive structural local sparse appearance model. In: Proceedings of IEEE Conference on Computer Vision and Pattern Recognition (CVPR), pp. 1822–1829 (2012)
12. Kang, J., Cohen, I., Medioni, G.: Tracking people in crowded scenes across multiple cameras. In: Proceedings of Asian Conference on Computer Vision (2004)
13. Mikic, I., Santini, S., Jain, R.: Video processing and integration from multiple cameras. In: Image Understanding Workshop (1998)
14. Mittal, A., Davis, L.: M2Tracker: a multi-view approach to segmenting and tracking people in a cluttered scene. Int. J. Comput. Vis. **51**(3), 189–203 (2003)
15. Otsuka, K., Mukawa, N.: Multi-view occlusion analysis for tracking densely populated objects based on 2D visual angles. In: Proceedings of the Conference on Computer Vision and Pattern Recognition (CVPR) (2004)
16. Salih, Y., Malik, A.: 3D tracking using particle filters. In: Proceedings of IEEE Instrumentation and Measurement Technology Conference (I2MTC), pp. 1–4 (2011)
17. Yang, D., Gonzales-Banos, H., Guibas, L.: Counting people in crowds with a real-time network of simple image sensors. In: International Conference on Computer Vision, pp. 122–129 (2003)
18. Krinidis, S., Stavropoulos, G., Ioannidis, D., Tzovaras, D.: A robust and real-time multi-space occupancy extraction system exploiting privacy-preserving sensors. In: International Symposium on Communications, Control and Signal Processing (2014)
19. De Silva, L.: Audiovisual sensing of human movements for home-care and security in a smart environment. Int. J. Smart Sens. Intell. Syst. **1**, 220–245 (2008)
20. Scataglini, S., Andreoni, G., Gallant, J.: Smart clothing design issues in military applications. In: Ahram, T.Z. (ed.) AHFE 2018. AISC, vol. 795, pp. 158–168. Springer, Cham (2019). https://doi.org/10.1007/978-3-319-94619-1_15

Workshop on: Movement Analytics and Gesture Recognition for Human-Machine Collaboration in Industry 4.0

Real-Time Gestural Control of Robot Manipulator Through Deep Learning Human-Pose Inference

Jesus Bujalance Martin$^{(\boxtimes)}$ and Fabien Moutarde

MINES ParisTech, PSL Research University, Center for Robotics,
60 Bd St Michel, 75006 Paris, France
jesus.bujalance_martin@mines-paristech.fr

Abstract. With the raise of collaborative robots, human-robot inter-action needs to be as natural as possible. In this work, we present a framework for real-time continuous motion control of a real collaborative robot (cobot) from gestures captured by an RGB camera. Through deep learning existing techniques, we obtain human skeletal pose information both in 2D and 3D. We use it to design a controller that makes the robot mirror in real-time the movements of a human arm or hand.

Keywords: Collaborative robots · Robot manipulator · Motion control · Real time · Pose estimation · Deep learning · ROS

1 Introduction

The first generation of manufacturing robots were always operating in human-free areas, for safety reasons. But during recent years, new types of robots have been designed for deployment in direct contact, and even cooperation, with human workers. These *collaborative* robots create the opportunity and interest for *gesture-based* control of robots by humans, for seamless and natural Human Robot Interaction.

Gestural control can mean either launching particular robot actions by just executing some predefined gestures interpreted as commands, or direct and continuous motion control of the robot by human movements. In this work, we focus on the latter. Until recently, robust servoing of robot motion on human movement was possible only using wearable sensors (e.g. ElectroMyoGram sensor, EMG, cf. [9] or [4]), or thanks to a depth camera (such as Kinect or Real-sense) allowing real-time human skeletal pose estimation and tracking (e.g. [1]).

Meanwhile, recent deep learning methods have achieved great results in both 2D and 3D human skeletal pose estimation from RGB cameras in real time. Inspired by this progress, we have designed a robot motion continuous controller based on pose estimation. To clarify, there is no gesture recognition module in this work. The objective is to have the robot mimic the movements of the human arm in real time, without any underlying understanding of these movements.

© Springer Nature Switzerland AG 2019
D. Tzovaras et al. (Eds.): ICVS 2019, LNCS 11754, pp. 565–572, 2019.
https://doi.org/10.1007/978-3-030-34995-0_51

In our framework, an RGB camera captures the movements of the user. We extract the 2D and 3D poses in real-time and process them to control the robot motion. We present two implementations, one based on forward kinematics and one on inverse kinematics.

2   Pose Estimation

Pose estimation is the problem of determining the position and orientation of a person from RGB images or videos. Namely, we want to obtain coordinates of keypoints such as joints, eyes, or fingers. In this section we will discuss the methods we used for both 2D and 3D pose estimation.

2.1   2D Pose Estimation

Multi-person pose estimation models can be categorized as either top-down or bottom-up. For instance, AlphaPose [3] is a top-down approach since it detects every person in the scene then extracts their pose individually. Openpose [2] is a bottom-up approach since it detects all the keypoints in the scene and then puts them together to form skeletons. Other bottom-up approaches, like [8], do not separate the detection and grouping stages, obtaining a single-stage network.

We chose OpenPose over other methods because of its proven real-time performance. It also has the most active support and has been regularly updated with new features since its release. Particularly, in this work we use the hand detection module.

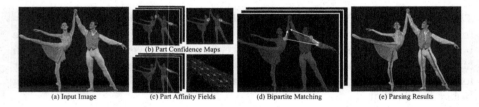

Fig. 1. OpenPose [2] pipeline.

OpenPose. Figure 1 shows the general pipeline of OpenPose. First, an input image enters a multi-stage CNN which jointly predicts the set of confidence maps, one for each part, and the set of part affinities which represent the degree of association between parts. Then, bipartite matching is used to associate body part candidates and obtain the full 2D skeletons. A part refers to a keypoint such as the left elbow or the right eye.

2.2 3D Pose Estimation

2D pose estimation allows us to control the robot in a 2D plane. In order to add a third dimension we need the 3D pose. There are numerous open-source methods that perform this task. We chose Human Mesh Recovery (HMR) [5], but we concede that it might be excessive to compute a full mesh when we only care about the pose. There are other simpler yet effective approaches such as [7] that could have been used as well. Nevertheless HMR provides good results and performs well in real-time (Fig. 2).

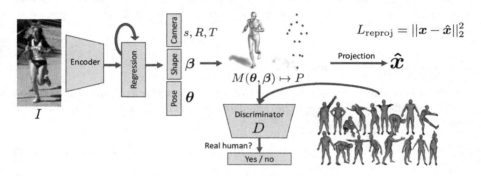

Fig. 2. HMR [5] pipeline.

Human Mesh Recovery. HMR is a recent approach that directly predicts the 3D pose and shape of a human SMPL model [6], along with the camera configuration, from a single image.

The proposed framework is as follows: An image I is sent through a convolutional encoder into a 2048D latent space. The latent features are then decoded by an iterative regression to produce the pose, the shape, and the camera configuration. These 3D parameters are sent to the discriminator D, whose goal is to tell whether they come from a real human.

HMR uses a weakly-supervised adversarial framework that allows the model to be trained on images with only 2D pose annotations, without any ground truth 3D labels. Indeed, the reprojection loss used for training only requires 2D joint locations. Note also that there is no need to make a priori assumptions about the joint limits, bone ratios or other physiognomic constraints. Instead, the 3D mesh pool acts as a data-driven prior.

3 Real-Time Robot Control

The robot we used is a Universal Robot, model UR3. Shown in Fig. 4, it is a 6 DOF robot manipulator designed for collaborative tasks. The easiest way to control Universal Robots directly is through the URScript programming language.

However, we chose to implement our controller in python within the Robot Operating System (ROS) framework. ROS is a middleware which provides hardware abstraction, device drivers, visualizers, message-passing, and other useful tools to manage multi-modular projects like ours.

As a side note, although the UR3 robot is controlled in position (at the hardware level), we should see more velocity-controlled robots in the future as the need for real-time applications increases [10]. The same author of this paper made an open-source ROS package, *jog_arm*, which allows to control the speed of the joints or the end-effector in real time. In this section, we will discuss the advantages (and disavantages) of this package as well as the two different implementations that we developed to control the robot from a sequence of human poses.

3.1   Inverse Kinematics (IK)

The most straight-forward way to control a robot is to control only the position of the end-effector. We recall the equation tying the coordinates in cartesian space x and joint space q:

$$\dot{x} = J(q)\dot{q} \tag{1}$$

For our robot, the Jacobian matrix J is a 6 by 6 matrix (generally 6 by #DOF). The package simply inverts the Jacobian matrix to recover the desired joint angles from the desired end-effector xyz position, corresponding to the 3D hand poisition provided by HMR. The other 3 coordinates correspond to the orientation of the end-effector, which we maintain constant (similar to fixing the 3 wrist joints). The main advantages of this Jacobian method are its speed and that the resulting trajectories can be altered in real time. It is also deterministic, avoiding unexpected behaviours. This makes it very suitable for real-time environments.

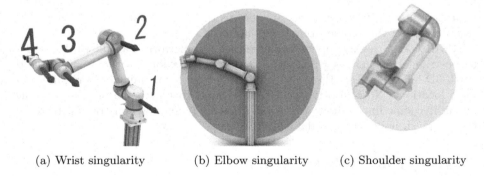

(a) Wrist singularity   (b) Elbow singularity   (c) Shoulder singularity

Fig. 3. Singularities of UR robots (source: www.universal-robots.com)

Singularities. When dealing with an IK approach, one must be aware of singularities and how to handle them. As shown in Fig. 3 the UR3 robot has three different types of singularities. At a singularity, the mobility of the manipulator is reduced. They occur when $det(J) = 0$, and can produce undesired behaviours such as infinite solutions (wrist singularity) or a solution with infinite joint speeds (shoulder singularity).

The *jog_arm* package is able to reverse out of singularities. Indeed, the robot will decrease its speed as it is approaching a singularity and halt before reaching it. More complicated planners should be able to avoid them altogether by taking more sophisticated IK solutions, but this simple Jacobian inversion method cannot plan around them (or obstacles, joint limits, etc). Because of these limitations, this technique is generally only useful over short distances.

3.2 Forward Kinematics (FK)

If we opt for a FK approach, the pose information allows us to control the entire robot, not just the end-effector. However, we come across the problem of mapping the joints between a human arm and the robot manipulator, which don't share the same amount of DOF. Our hand-crafted joint mapping is as follows. The shoulder, elbow and wrist 1 mappings come out naturally. The wrist rotation around the forearm axis (wrist 3 joint) is computed from the angle between the forearm and the thumb of the skeleton. We only require the 2D skeleton from OpenPose here, as shown in Fig. 4.

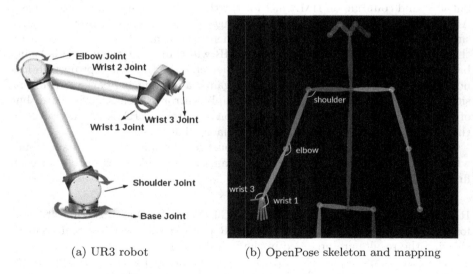

(a) UR3 robot   (b) OpenPose skeleton and mapping

Fig. 4. Joint mapping for forward kinematics.

The base joint is the only one operating outside of the 2D plane (the wrist 2 joint remains fixed). A calibration step measures the length in pixels of the upper arm whenever a human is detected for the first time. When we detect a shorter forearm, we compute a base joint angle accordingly. To distinguish between the arm coming towards or moving away from the camera, we could compare the depth of the elbow and shoulder provided by HMR, but it comes with extra computation time.

The main limitation of this method is that it only works properly if the person is facing the camera and the movements are mostly planar, which limits the range of motion of the base joint to approximately $\pm45°$. Because of this limited range, we choose not to compensate for the differences between the projected angles given by OpenPose and the actual angles of the user. A future version should correct these distortions.

Gripper. For both IK and FK we include a gripper controller which distinguishes between two states: fully open and fully closed. Based on the OpenPose hand detection, we detect a closed hand if the vectors wrist-knuckle and knuckle-fingertip have opposite directions for all fingers (excluding the thumb).

4   Experiments

When testing our controller, we ran into a few issues, most coming from HMR. As discussed in Sect. 2.2, the network is optimized to provide a credible human mesh, not just the pose. Therefore, its precision regarding joint positions is not as good as that of OpenPose. We avoid this issue by simply prioritizing OpenPose outputs, and counting on HMR just for the third dimension even in the IK case. Also, a momentum-like exponential average of the commands proved to give smoother results. Another issue comes when both arms of the user are within the 2D plane of the body. Quite often, HMR will think that the person is facing backwards. Indeed, if the face is not well detected, neither the reprojection loss nor the adversarial loss can discriminate against a mesh facing backwards in this situation. To solve this issue we could simply mirror the backwards mesh, but often it won't be satisfying. Another option is to retrain HMR adding a loss to the discriminator which penalizes meshes facing backwards.

OpenPose is much more precise and robust to lightning. The only issue comes when the hand is perpendicular to the camera, but this is expected since all fingers but one are partially occluded.

Results. In our setup with a GEFORCE GTX 1080 Ti, OpenPose runs at 30 fps for 640×480 images. OpenPose and HMR run together at 6 fps. Some videos showing the results are provided in the linked Google Drive.

Figure 5 compares, in the FK case, command angles and resulting robot angles. It shows that the delay is very short. The commands present some noise but the dynamics of the robot act as a filter and the result is a smooth trajectory. In the IK case, since only the position of the end-effector is controlled, the target/result comparison cannot be made on angles. Figure 6 therefore compares target and obtained position of the end-effector. It shows that the robot

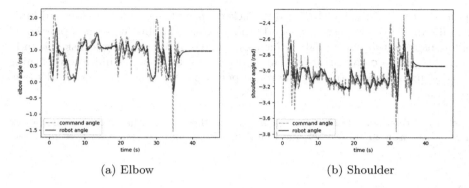

(a) Elbow   (b) Shoulder

Fig. 5. Command angle and actual robot angle (FK)

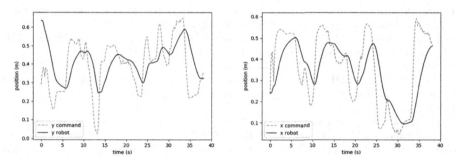

Fig. 6. Position of the end-effector in the 2D plane (IK)

doesn't follow the command as tightly and the delay is bigger, equal to 1 s. This behaviour is expected since the frequency of the commands is much lower and the post-processing is heavier. Note that in our setup with good lightning and only one person, OpenPose performs extremely well and its output is very close to the ground truth. Therefore the command curves could be interpreted as the ground truth corrupted with some additive centered noise.

Overall, the FK solution proved to be more precise and satisfying for the user. Still, the IK solution is also valuable since it allows to control the robot in the entire 3D space with no calibration.

5   Conclusions and Future Work

The raising trend in bringing workers and industrial robots together needs an efficient human-machine interaction. This work shows that a gestural motion control is possible in a simple scenario and it represents a first step towards a richer human-robot collaboration.

The inverse kinematics approach wasn't responsive enough for a real-time application, but 3D pose estimation can still be useful for other tasks such as pointing to a particular object to be picked up by the robot. The forward kinematics approach was responsive and robust. A more sophisticated mapping could allow us to obtain joint angle sequences capable of reproducing a task shown in a human video demonstration. Future work will be done on imitation learning from demonstrations to have the robot perform a variety of complex tasks (e.g. pick and place).

Acknowledgements. The research leading to these results has received funding by the EU Horizon 2020 Research and Innovation Programme under grant agreement No. 820767, CoLLaboratE project.

References

1. Paravati, G., Manuri, F., Sanna, A., Lamberti, F.: A Kinect-based natural interface for quadrotor control. Entertain. Comput. **4**(3), 179–186 (2013)
2. Cao, Z., Hidalgo, G., Simon, T., Wei, S.E., Sheikh, Y.: OpenPose: realtime multi-person 2D pose estimation using part affinity fields (2018)
3. Fang, H., Xie, S., Lu, C.: RMPE: regional multi-person pose estimation (2016)
4. Shin, H.S., Lee, K.H., Ganiev, A.: Study on virtual control of a robotic arm via a myo armband for the selfmanipulation of a hand amputee. Int. J. Appl. Eng. Res **11**(2), 775–782 (2016)
5. Kanazawa, A., Black, M.J., Jacobs, D.W., Malik, J.: End-to-end recovery of human shape and pose. CoRR, abs/1712.06584 (2017)
6. Loper, M., Mahmood, N., Romero, J., Pons-Moll, G., Black, M.J.: SMPL: a skinned multi-person linear model. ACM Trans. Graph. **34**, 248 (2015)
7. Martinez, J., Hossain, R., Romero, J., Little, J.J.: A simple yet effective baseline for 3D human pose estimation. In: Proceedings of the IEEE International Conference on Computer Vision, pp. 2640–2649 (2017)
8. Newell, A., Huang, Z., Deng, J.: Associative embedding: end-to-end learning for joint detection and grouping. In: Advances in Neural Information Processing Systems, pp. 2277–2287 (2017)
9. Xu, Y., Yang, C., Liang, P., Zhao, L., Li, Z.: Development of a hybrid motion capture method using MYO armband with application to teleoperation. In: 2016 IEEE International Conference on Mechatronics and Automation, pp. 1179–1184 (2016)
10. Zelenak, A., Peterson, C., Thompson, J., Pryor, M.: The advantages of velocity control for reactive robot motion. In: ASME 2015 Dynamic Systems and Control Conference, pages V003T43A003-V003T43A003. American Society of Mechanical Engineers (2015)

A Comparison of Computational Intelligence Techniques for Real-Time Discrete Multivariate Time Series Classification of Conducting Gestures

Justin van Heek, Gideon Woo, Jack Park, and Herbert H. Tsang[✉]

Applied Research Lab, Trinity Western University, Langley, BC, Canada
Justin.vanHeek@mytwu.ca, herbert.tsang@twu.ca

Abstract. Gesture classification is a computational process that can identify and classify human gestures. More specifically, gesture classification is often a discrete multivariate time series classification problem and various computational intelligence solutions have been developed for these problems. It is difficult to determine which existing techniques and approaches to algorithms will produce the most effective solutions for discrete multivariate time series classification problems. In this study, we compare twelve different classification algorithms to report which techniques and approaches are most effective for recognizing conducting beat pattern gestures. After performing 10-fold cross-validation tests on twelve commonly used algorithms, the results show that of the algorithms tested, the most accurate were RNN, LSTM, and DTW; all of which had an accuracy of 100%. We found that in general, algorithms which can take in a dynamic sequence input and classification algorithms that are discriminative performed consistently well, while their counterparts varied in performance. From these results we determine that when selecting a computational intelligence technique to solve these classification problems, it would be advantageous to consider the top performing algorithms along with furthering research into new dynamic input and discriminative algorithms.

Keywords: Gesture recognition · Classification · Conducting · Computational intelligence · Neural networks · Machine learning

1 Introduction

Gesture recognition has been a long standing area of pursuit in the computing world. As the research and development in gesture recognition have continued, the gestures being detected have opened up to include more and more complex movements. Gestures can range from simple movements like the raising of the

Supported by The Natural Sciences and Engineering Research Council of Canada (NSERC).

D. Tzovaras et al. (Eds.): ICVS 2019, LNCS 11754, pp. 573–585, 2019.
https://doi.org/10.1007/978-3-030-34995-0_52

hand to more complicated and nuanced ones like the movements of a martial artist. In response to this, more advanced computational intelligence methods have been used to tackle these expanded gesture recognition workloads. These methods include simple mathematical algorithms like Dynamic Time Warping to highly complex neural networks like Gated Recurrent Units. This paper looks to explore the differences between these computational intelligence methods and comparing their effectiveness by applying them to a musical conducting gesture recognition use-case.

2   Literature Review

There are many approaches that have explored the problem of gesture recognition with the help of computational intelligence (CI) in literature. This section will briefly explore how the data has been collected and what CI algorithms have been used to classify gestures.

2.1   Data Collection

In order to use CI algorithms, one must first decide the form of data collection the CI will make use of. Currently, the two main forms of data collection are vision-based and sensor-based.

Vision-Based. Vision-based systems use cameras to extract motion data in one of two ways. The first method has the moving subject placed in front of the camera to capture the movement using computer vision techniques [8]. Tao et al. considered using this approach, but decided to look elsewhere because of the drawbacks [13]. The second way is to extract positional data by moving the camera itself and extrapolating the movement of the phone by comparing details from each video frame [3]. In previous research, van Heek et al. implemented this method to help with tracking orchestral conducting gestures and produced good results [3].

Sensor-Based. Triaxial data from sensors like accelerometers and gyroscopes is the other commonly used form of gesture data collection. Lu et al. built a custom wrist-mounted sensor array solution comprised of accelerometers, gyroscopes and surface electromyography sensors [7]. Other studies have used existing sensory devices like Nintendo's Wii remote which has an accelerometer and gyroscope built-in [5]. A similar third option that researchers have used to gather motion data is to use the built-in sensors inside modern smartphones due to their easy accessibility [6].

2.2   Computational Intelligence (CI) Implementations

After the collection of motion data, it must be run through some form of computational intelligence for detection and classification. There are many possible algorithms out there that are suited for this kind of classification problem.

Dynamic Time Warping (DTW). DTW is an algorithm used for measuring similarity between two different sequences of data and is commonly used for speech recognition [12]. Fahn Chin-Shyurng et al. did research on musical conducting gesture recognition using DTW which produced a total accuracy of 89% for classifying two, three, and four beat conducting gesture patterns [1].

Hidden Markov Model (HMM). HMM is a method that predicts the probability of a given sequence of observations occurring. Classification of a given sequence is determined by the model that outputs the highest probability of the sequence. HMM has been implemented in speech recognition, gesture recognition, and bioinformatics [4]. Kolesnik et al. used HMM for conducting gesture classification with 2D positional data as the input and were able to classify the gestures in real-time with 94.6% accuracy [4].

Naive Bayes Classifier (NBC). NBC is a classification algorithm that makes use of Bayes' theory as its main component. NBC assumes that all the variables in the data are independent of each other. Ziaie et al. implemented NBC to recognize images of gestures and were able to get 93% accuracy at identifying the correct one from three different gestures [9].

Neural Networks. There are two main neural networks that have been used in gesture recognition. The first and older of the two is called Long-Short Term Memory (LSTM), which learns from sequential data in a more thorough manner by saving important information in memory [2]. The idea of giving neural networks memory is further developed with the Gated Recurrent Unit (GRU), which was created to be more efficient compared to LSTM [2]. Both LSTM and GRU were compared by Chung et al. who discovered that there were no significant differences in performance between the two neural networks and the choice of which to use should be evaluated per use-case [2]. Both of these neural networks were further developed with BiDirectional-LSTM and BiDirectional-GRU (BiLSTM and BiGRU) which gives the neural network the ability to look at the past and the future simultaneously, giving it more contextual information. Using these two neural networks, Li et al. evaluated the performance of BiLSTM and BiGRU in detecting alphabet and number writing gestures using data from a smartphone [6]. The results were promising with accuracy above 98% for both neural networks [6].

3   Methods

In order to compare various computational intelligence techniques for classifying conducting gestures, a dataset is required to be used for input to train and/or test each algorithm. To generate this dataset, a method of data collection must be implemented. Once the dataset has been assembled, it is then pre-processed before going through the algorithms. Each algorithm is evaluated by 10-fold cross validation on the pre-processed data and the results are recorded. This section will go into the individual aspects of our workflow in more thorough detail.

3.1 Data Collection

Any gesture recognition application must start with the collection of gesture data. Wanting to make this gesture classification research as reproducible as possible, an Android smartphone was the data collection hardware of choice. The abundance and popularity of smartphones makes them readily available tools for real-time data collection. The built-in Android tools were used to access the phone's sensor data which it would then transmit to the computer over Bluetooth. The inertial sensors in the smartphone are comprised of an accelerometer and gyroscope, each outputting three values (x, y and z axes) at regular time intervals creating a total of 6 features. This defines our classification problem as a discrete multivariate time series classification problem.

A total of 372,256 of raw data was collected from the conducting gestures of three individuals. The data was then grouped into chunks of 128 time steps for dynamic input algorithms and 200 for static input algorithms. The window is larger for the static input ones due to the segmentation pre-processor which isolates a full segment within that chunk; therefore, to isolate the full segment, a larger window must be given and then the excess is trimmed. This results in a dataset of 910 instances for the dynamic input algorithms, and 570 instances for the static input algorithms.

Gestures. In this study, three musical conducting gestures were used to generate and test the algorithms. These gestures are two, three, and four beat patterns as shown in Fig. 1. To keep the data consistent and eliminating an extraneous variable, the tempo of all conducting gestures was set to 80 beats per minute. Since the gesture of conducting can be looped indefinitely with no clear start and end points, it is classified as a dynamic gesture.

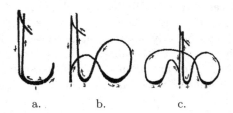

a.   b.   c.

Fig. 1. The conductor's beat patterns for (a) two beats, (b) three beats, and (c) four beats.

3.2 Pre-processing

Some computational intelligence algorithms may require or perform more efficiently with input data in different formats. In order to accommodate the various data formats required, the raw data is run through a series of pre-processors before being passed to the algorithms. Table 1 shows the pre-processing used for each algorithm.

Table 1. Pre-processing for each algorithm

Algorithm	Scaling	Segmentation	Flattening
LSTM	x		
GRU	x		
RNN	x		
HMM	x		
DTW	x	x	
Linear SVM	x	x	x
RBF SVM	x	x	x
Polynomial SVM	x	x	x
KNN	x	x	x
Gaussian NBC	x	x	x
Multinomial NBC	x	x	x
Complement NBC	x	x	x

Scaling. All data is scaled prior to being used as input into the algorithms. The purpose of this scaling is to adjust the range of each of the six features form the inertial sensors so that they retain a proportional level of importance as the numerical distance between values can affect performance of the algorithms.

Segmentation. The segmentation pre-processing method, as shown in Fig. 2, allows computational intelligence methods designed for static gesture recognition to work in a dynamic gesture workflow. It separates the data into chunks that would roughly encompass a single gesture, essentially converting the incoming dynamic data into static data. Due to the difficulties caused by fluctuations in the data from sensor noise,a Butterworth low pass filter with an order of 5 and a cutoff frequency of 0.1 is run over a copy of the data. This filtered data is then used to detect when the acceleration along the z-axis crosses an upwards threshold of 0, constituting a beat. The z-axis of the acceleration was chosen in particular due to its distinct pattern that correlated with the beats of the conducting gesture. The minimum point between the crossed thresholds is considered to be the starting index of that beat. After finding four consecutive beats as that size would contain all three types of gestures, the raw data segment is selected and is then linearly stretched along the time axis to a defined output size and passed to the next pre-processor or the classification algorithm. It is important to note, that the Butterworth filter is only applied to aid in the detection of the segment, afterwards the filtered data is discarded. This approach gives higher consistency in segmentation while still retaining all of the original data from the phone's inertial sensors.

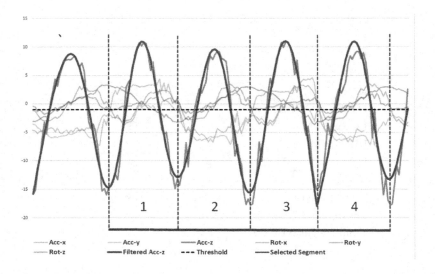

Fig. 2. Beat segmentation

Flattening. The raw inertial sensor data comes in a two-dimensional form due to the data being a time series. For some algorithms, the input data must be one-dimensional. In these cases, a flattening pre-processor is used to convert the two-dimensional data into one dimension.

3.3 Computational Intelligence Techniques

A variety of computational intelligence techniques were tested in this study. The algorithms were categorized by type of classifier (generative or discriminative) and expected input type (dynamic or static). Table 2 displays these categorizations for each algorithm. Each algorithm takes in the pre-processed data as input and outputs an integer corresponding to the beat pattern it has classified the input as.

Dynamic Time Warping. DTW compares its input data to a set of labelled data. The labelled data with the lowest distance compared to the input data is selected as the classification. The distance cost between two time sequences, X and Y, is calculated by finding the sum of the minimum euclidean distances between each index of X to every index of Y and vice versa [10].

Table 2. Algorithm type and expected input type

Algorithm	Generative/Discriminative	Input type
LSTM	Discriminative	Dynamic
GRU	Discriminative	Dynamic
RNN	Discriminative	Dynamic
Linear SVM	Discriminative	Static
RBF SVM	Discriminative	Static
Polynomial SVM	Discriminative	Static
DTW	Discriminative	Static
KNN	Discriminative	Static
HMM	Generative	Dynamic
Gaussian NBC	Generative	Static
Multinomial NBC	Generative	Static
Complement NBC	Generative	Static

K-Nearest Neighbor (KNN). The K-Nearest Neighbor algorithm runs on the assumption that similar things are close together. When training this algorithm, data points of the incoming data are mapped onto a plane where data from the same class should naturally group together. Prediction is done by calculating the distance between the new data point from the current input to the various groups already on the plane and choosing the closest group. This study implemented KNN with a k value of 5.

HMM. HMM is comprised of three main elements as follows:

1. A, the transition probability matrix representing the probability of undergoing the specified transition.
2. B, the output probability matrix representing the probability of outputting the symbol from a finite set of symbols.
3. π, the initial state probability matrix.

For the training phase, a model for each class of gesture is initialized with random values for A B, and π. With the given data, both A and B are updated with the Viterbi algorithm repeatedly until optimized. Incoming data can then be tested with the forward algorithm measuring the forward probability of each class and the model with the highest probability output is predicted as the correct gesture classification.

NBC. There are three different types of NBC implemented in this research: Gaussian, Multinomial, and Complement. For all of these variations, the main component used is the Bayes' theory, which calculates the likelihood of an event,

y, happening given the observation of another event, x. The theory can be represented as follows [11]:

$$P(y|x_1, ...x_n) = \frac{P(y)P(x_1, ...x_n|y)}{P(x_1, ...x_n)}$$

The main difference between these variations is the assumed distribution of $P(x_i|y)$. The Gaussian NBC distribution is assumed to be Gaussian, the Multinomial NBC (MNBC) is based on relative frequency, and the Complement NBC (CNBC) expands on MNBC by weighing the likelihood between classes.

Support Vector Machine (SVM). This computer classification algorithm works by taking the data points given and mapping them to a constructed n-dimensional space based on the number of features the data has (in this case, $n = 6$). When training an SVM, optimal vectors that separate the groups to be classified are calculated so that the distance between the groups and the vectors is maximized. Prediction is then based on which side of the vectors that the new incoming data point lands. Specifically, this research explores three different kernel functions of SVM: Linear, RBF, and Polynomial.

With this particular classification problem, the data has a large number of dimensions which would require multiple hyperplanes that separate the data points in order to perform accurate classification. Theoretically non-linear kernels like RBF and Polynomial should outperform a linear kernel in this case.

After some preliminary testing, the penalty parameter C and gamma value chosen for the implementation of the SVMs in this study was 1000 and the inverse of the number of features respectively. The polynomial SVM was given a degree of 3.

Recurrent Neural Network (RNN). The basic RNN is a fairly basic form of neural network. It improves upon the classic feed forward neural network by allowing internal nodes of the hidden layer to loop back on itself. This unique feature of an RNN is designed to learn the interdependencies of sequential data. This study's implementation of RNN and the following advanced RNNs have their parameters set to 128 internal nodes with a learning rate of 0.0001.

LSTM. Developed in the 90's, LSTM is an extension of a basic RNN. As the name suggests, an LSTM neural network has access to memory where it can store information it deems useful, giving it context for future data analysis. Every LSTM unit can be viewed as a gated cell that has 3 gates: input, forget, and output. The input gate determines whether to let new input be stored into memory. The forget gate is responsible for deleting the data stored if it is deemed unimportant. If the data stored is important, the output gate is responsible for letting that data affect the output.

GRU. Another extension of an RNN is GRU, which gives it the ability to store important information in memory similar to LSTM, but uses one less electronic gate. A single GRU cell is composed of two gates: reset and update. The update gate is responsible for determining what information coming into the cell it should throw away or add. It accomplishes approximately what the forget and input gates in an LSTM cell does, but only with one gate. The reset gate determines how much of the stored past irrelevant information to forget. Since a GRU cell has one less gate, it takes less time to train and process compared to LSTM.

4 Results and Discussions

Using the modular system that we developed, we trained and tested the various algorithms. The individual results of these tests are contained within Table 3. Number of instances is the size of the dataset that was used to train/test the algorithms. This number varies between dynamic and static input algorithms due to the segmentation pre-processing required. Processing time is the mean time in seconds that the entire algorithm takes to process a single input. While the exact values of this would range between machines, the relative order of efficiency between the algorithms would not. Overall accuracy, also known as classification accuracy, is the sum of the correct classifications over all ten folds divided by the total number of instances. Cohen's kappa is a measure of the classifier's actual classification performance by controlling for the accuracy of randomly guessing the classification. Standard deviation of accuracy is an indicator of how consistent the algorithm's accuracy is to the overall accuracy.

4.1 Individual Algorithms

A brief analysis of the results show that the LSTM, RNN and DTW algorithms performed with perfect accuracy; however, our implementation of DTW had a processing time that is too slow for most real-time applications. Multi-threading and additional optimization could benefit the DTW algorithm. Most algorithms tested resulted in an accuracy higher than 90% which is very promising. KNN is another algorithm of interest due to its extremely high accuracy and one of the fastest processing times making it a suitable candidate for more complex real-time gestures. In addition to analyzing the individual top performing algorithms, the different categorizations of algorithms are also compared to discover if certain types of algorithms tend to outperform others when presented with this type of classification problem.

4.2 Generative Vs Discriminative

Classification algorithms are often categorized as either generative or discriminative; therefore, we analyzed the mean averages of the overall accuracy and

Table 3. Results of algorithm performances

Algorithm	Number of instances	Processing time (sec)	Overall accuracy	Cohen's kappa	Standard deviation of accuracy
RNN	910	0.0579	100%	1.000	0.00000
LSTM	910	0.0874	100%	1.000	0.00000
DTW	570	127.7438	100%	1.000	0.00000
GRU	910	0.0782	99.5%	0.992	0.01187
KNN	570	0.0037	99.1%	0.987	0.00014
Polynomial SVM	570	0.0028	92.8%	0.892	0.00077
HMM	910	0.0056	92.5%	0.888	0.00038
RBF SVM	570	0.0026	90.4%	0.855	0.00082
Linear SVM	570	0.0024	85.6%	0.784	0.00147
Gaussian NBC	570	0.0026	75.1%	0.626	0.00263
Complement NBC	570	0.0026	70.0%	0.550	0.00169
Multinomial NBC	570	0.0025	69.6%	0.545	0.00450

the Cohen's kappa along with the standard deviation of algorithms in both categories. Table 4 compares the results and as shown by the table, discriminative algorithms perform better overall with an advantage of 19.1%. They also have a lower standard deviation which shows the algorithms in this category are all performing well. The higher standard deviation and lower overall accuracy from the generative algorithms can be attributed to the fact that of the four algorithms in that category, three are variations of NBC that did not perform well whereas the other was HMM which achieved 92.5% accuracy.

We observe that the discriminative algorithms hold a distinct advantage over their generative counterparts as far as the results show in the study. Perhaps this is due to the large amount of similarities between each gesture. Discriminative algorithms would be capable of identifying the unique aspects of the data that discriminates the gestures between each other. Generative algorithms on the other hand do not take the other classes into context and thus due to the similarities, all three classes would have high probabilities.

4.3 Dynamic Vs Static

While the goal of this experiment is to compare time series classification algorithms that can perform classification in real-time, some algorithms require the input data to be static. This means that the motion data must start and end at

Table 4. Results of comparison between generative and discriminative algorithms

Classification	Mean overall accuracy	Mean cohen's kappa	Standard deviation of algorithm accuracy
Generative	76.8%	0.652	0.10765
Discriminative	95.9%	0.939	0.05601

Table 5. Results of comparison between dynamic and static input algorithms

Input type	Mean overall accuracy	Mean cohen's kappa	Standard deviation of algorithm accuracy
Dynamic	98.0%	0.970	0.03654
Static	86.3%	0.795	0.11965

similar fixed points of time in the gesture rather than feeding the most recent data directly as is into the algorithm (dynamic). To facilitate static algorithms, a pre-processor is used to segment the incoming data. Similar to the previous analysis, we compared the same features between dynamic and static input algorithms. The results displayed in Table 5 show that the dynamic input algorithms have a very high accuracy and low standard deviation. The static input algorithms on the other hand, also have a relatively high accuracy, but have a much larger standard deviation.

From these findings we determine that dynamic input algorithms consistently perform well with discrete multivariate time series data whereas static input algorithms show no significant advantage or disadvantage to performance in this study. This is most likely due to the data being a time series and the variation in movement that can occur in a gesture. A dynamic input algorithm is designed to handle each step in the sequence independently and therefore may be more immune to the variations in the time series data.

5   Conclusions

In attempt to classify real time discrete multivariate time series data of different conducting beat pattern gestures, we have accumulated existing CI algorithms and compared their performances in solving this classification problem. The algorithms can be categorized as either generative or discriminative, as well as categorized by whether the algorithm uses dynamic or static input data.

The top three performing algorithms which all scored 100% accuracy are RNN, LSTM, and DTW. For more complex problems, KNN should also be a consideration due to its faster processing time while maintaining high accuracy. In terms of the findings between the different algorithm categories, it is found that discriminative algorithms, including all of top performing algorithms mentioned above, perform better than generative algorithms with generally lower standard deviation.

Furthermore, we have observed that algorithms using dynamic input data consistently perform well, most of which are neural network algorithms. On the other hand, algorithms using static input data do not show enough consistency in accuracy from each other to conclude their general effect on performance.

From our results, we conclude that it would be worthwhile for further research into new gesture classification techniques to have a focus on discriminative algorithms with dynamic input.

References

1. Chin-Shyurng, F., Lee, S.E., Wu, M.L.: Real-time musical conducting gesture recognition based on a dynamic time warping classifier using a single-depth camera. Appl. Sci. **9**(3), 528 (2019). https://doi.org/10.3390/app9030528
2. Chung, J., Gülçehre, Ç., Cho, K., Bengio, Y.: Empirical evaluation of gated recurrent neural networks on sequence modeling. CoRR abs/1412.3555 (2014)
3. van Heek, J., Park, J., Yu, X., Tsang, H.H.: An evaluation study of recognizing conducting gesture using computational intelligence techniques. In: 2018 IEEE Symposium Series on Computational Intelligence (SSCI), pp. 1005–1012, November 2018. https://doi.org/10.1109/SSCI.2018.8628906
4. Kolesnik, P., Wanderley, M,M.: Recognition, analysis and performance with expressive conducting gestures. In: In Proceedings of the International Computer Music Conference (2004)
5. Lee-Cosio, B.M., Delgado-Mata, C., Ibanez, J.: ANN for gesture recognition using accelerometer data. In: the 2012 IberoAmerican Conference on Electronics Engineering and Computer Science, Procedia Technology, vol. 3, pp. 109–120 (2012)
6. Li, C., Xie, C., Zhang, B., Chen, C., Han, J.: Deep fisher discriminant learning for mobile hand gesture recognition. Pattern Recogn. **77**, 276–288 (2018). https://doi.org/10.1016/j.patcog.2017.12.023
7. Lu, Z., Chen, X., Li, Q., Zhang, X., Zhou, P.: A hand gesture recognition framework and wearable gesture-based interaction prototype for mobile devices. IEEE Trans. Hum.-Mach. Syst. **44**(2), 293–299 (2014). https://doi.org/10.1109/THMS.2014.2302794
8. Ma'asum, F.F.M., Sulaiman, S., Saparon, A.: An overview of hand gestures recognition system techniques. In: 2012 IOP Conference Series: Materials Science and Engineering, vol. 99, no. 01, November 2015. https://doi.org/10.1088/1757-899x/99/1/012012
9. Paalanen, P.: Bayesian classification using gaussian mixture model and EM estimation: implementations and comparisons. Information Technology Project (2004)
10. Ratanamahatana, C., Keogh, E.: Everything you know about dynamic time warping is wrong. In: 3rd Workshop on Mining Temporal and Sequential Data SIGKDD, January 2004
11. Rish, I.: An empirical study of the naive bayes classifier. In: IJCAI 2001 Work Empircal Methods Artifical Intelligence, vol. 3, January 2001

12. Yadav, M., Alam, A.: Dynamic time warping (DTW) algorithm in speech: a review. Int. J. Res. Electron. Comput. Eng. **6** (2018)
13. Yuan, T., Wang, B.: Accelerometer-based Chinese traffic police gesture recognition system. Chin. J. Electron. **19**, 270–274 (2010)

A Deep Network for Automatic Video-Based Food Bite Detection

Dimitrios Konstantinidis[1]([✉]), Kosmas Dimitropoulos[1]([✉]), Ioannis Ioakimidis[2],
Billy Langlet[2], and Petros Daras[1]

[1] CERTH-ITI, 6th km Harilaou-Thermi, 57001 Thessaloniki, Greece
{dikonsta,dimitrop,daras}@iti.gr
[2] Karolinska Institutet, Blickagången 16, 14183 Huddinge, Sweden
{Ioannis.Ioakimidis,billy.langlet}@ki.se

Abstract. Past research has now provided compelling evidence pointing towards correlations among individual eating styles and the development of (un)healthy eating patterns, obesity and other medical conditions. In this setting, an automatic, non-invasive food bite detection system can be a really useful tool in the hands of nutritionists, dietary experts and medical doctors in order to explore real-life eating behaviors and dietary habits. Unfortunately, the automatic detection of food bites can be challenging due to occlusions between hands and mouth, use of different kitchen utensils and personalized eating habits. On the other hand, although accurate, manual bite detection is time-consuming for the annotator, making it infeasible for large scale experimental deployments or real-life settings. To this regard, we propose a novel deep learning methodology that relies solely on human body and face motion data extracted from videos depicting people eating meals. The purpose is to develop a system that can accurately, robustly and automatically identify food bite instances, with the long-term goal to complement or even replace manual bite-annotation protocols currently in use. The experimental results on a large dataset reveal the superb classification performance of the proposed methodology on the task of bite detection and paves the way for additional research on automatic bite detection systems.

Keywords: Deep learning · Bite detection · Video analysis · Motion features

1 Introduction

Food intake is the aggregate of a complex array of eating behaviors, such as bites, chews and inter-bite pauses [1]. In this work, we are primarily concerned with the identification of bite instances that occur when a person opens his mouth for food intake. Studies have shown that increased food intake rate is directly linked to obesity both in children and adults [2, 3]. Thus, bite detection mechanisms that allow food bite quantification and meal analysis can be invaluable for nutritionists, dietary experts and food scientists in order to evaluate individuals and help them avoid health problems related to obesity [4].

Currently, the detection of bite instances is usually performed by human experts, who have to watch hours of videos in order to successfully annotate eating behaviors

© Springer Nature Switzerland AG 2019
D. Tzovaras et al. (Eds.): ICVS 2019, LNCS 11754, pp. 586–595, 2019.
https://doi.org/10.1007/978-3-030-34995-0_53

[1, 5]. Although the annotation of human experts can be really accurate, the annotation procedure is time-consuming and prone to introduce errors due to the repetitive nature of the task. Thus, the need for the automation of the procedure has often been emphasized by experts in the field [1, 6] in order to overcome these problems and speed up the bite detection procedure. In the past several methodologies have been proposed to achieve that, with their majority being based on weight, inertial, motion and visual sensors that facilitate the recognition of bite instances [7].

However, existing methodologies face various challenges that limit their use and potential for large scale evaluation deployments or real life settings. More specifically, the mediocre accuracy of existing weight scales can significantly affect bite detection results. Furthermore, a person can eat with both hands either simultaneously or interchangeably and therefore, wearable sensors should be placed on both hands or else they fail to detect all bite instances. Additionally, sensors without visual feedback are prone to errors as instances of wiping mouth with handkerchief or raising hand to scratch head can be erroneously recognized as bites. Moreover, sensors that monitor jaw movements can become obtrusive, while visual sensors, such as cameras can pose challenges to a bite detection method due to the variety in appearance of people and kitchen utensils used for food intake and the occlusions of body parts. Finally, although the combination of multiple sensors can lead to more sophisticated solutions, it can also reduce the usability of the proposed systems in everyday life scenarios [8].

To overcome the aforementioned limitations of current automatic bite detection methodologies, we propose a novel non-obtrusive deep learning based approach that is capable of achieving highly accurate bite detection results. To this end, we initially employ a deep network [9, 10] to extract human motion features from video sequences. Subsequently, we propose a novel two-steam deep network that processes body and face motion data and combines the extracted information, thus taking advantage of both types of features simultaneously. We evaluate the proposed method on a large bite detection dataset and validate its bite detection performance. The main contributions of this work are summarized below:

- This is the first video-based deep learning approach that utilizes body and face features for the task of automatic bite detection.
- We propose a sophisticated deep network that extracts and combines spatiotemporal information from its inputs using convolutional neural networks (CNNs) and Long Short-Term Memory (LSTMs) units.
- We perform optimization of the hyper-parameters of the proposed deep network and explore data augmentation techniques to further improve its accuracy.

The remainder of the paper is organized as follows. Section 2 reviews work related to ours with respect to automatic bite detection. Section 3 presents our proposed methodology, while Sect. 4 presents the experimental results from the evaluation of our method. Finally, Sect. 5 concludes the work.

2   Related Work

So far, there is limited work in the literature about automatic bite detection as obsolete sensor technology put great challenges to automatic bite detection systems. However, recent research works [2, 3] revealed the strong correlation between food intake and obesity, thus intensifying the efforts towards the development of methods that study eating behaviors in order to prevent health related problems. Moreover, the development of modern sensors and the advancements in the classification techniques has sparked interest towards automatic bite detection methodologies that overcome the tediousness of having experts manually annotate bite instances.

In [11], the authors used a piezoelectric strain gauge sensor to detect the movement of the lower jaw that can characterize the eating behavior. On the other hand, in [5] and [12], the authors used high-precision food weight scales that can model the reduction of food from a subject's plate, while in [13], the authors employ smart-glasses to detect bite instances. More recent studies employ inertial sensors, such as accelerometer and gyroscope, located in wearable devices, in order to automatically extract bite instances [8, 14]. Finally, numerous studies develop methodologies that rely on the combination of audio and motion sensors [15, 16] or the combination of multiple motion and gesture sensors [17, 18] in order to robustly monitor eating behavior.

As far as classification techniques are concerned, early methods employ spectral segmentation and Random Forest classification [19] and Hidden Markov Models that are able to capture the temporal dependencies between hand gestures and bite instances [20]. Most recent bite detection systems take advantage of the success of deep learning on several classification tasks in order to propose more accurate and robust solutions. More specifically, the authors in [8] and [14], employ CNNs and recurrent neural networks in order to capture temporal dependencies between the outputs of inertial sensors and bite instances.

Our proposed methodology attempts to overcome the challenges of current state-of-the-art methods by proposing a video-based deep learning approach that extracts, processes and combines body and face motion data from video sequences. To our knowledge, our work comprises the first attempt to process videos and, more importantly, combine or fuse information extracted from body and face motion data to achieve accurate and robust automatic bite detection results.

3   Proposed Methodology

In this section, we analyze the proposed automatic bite detection methodology. Initially, we employ OpenPose [9, 10] that is able to process videos and extract body and face features from each video frame, along with the confidence of the algorithm on its predictions. Afterwards, we propose a deep network that takes as input only the most relevant features for the task of bite detection out of those computed in the first step.

The proposed two-stream deep network gets as input 2 types of features: (a) upper body and b() face features, as shown in Fig. 1. More specifically, we employ the nose and hand features' x- and y-coordinates and the distances between the selected numbered features. For the face features, we employ only the x- and y-coordinates of the mouth features that we believe are the most relevant for the modelling of bite instances. We perform averaging between similar neighboring mouth features and end up with 3 features that describe the middle points of the upper and lower lips and the point where the lips converge (i.e., corner of mouth). Due to the fact that the videos of our dataset depict people eating from a side view, we select only the mouth corner with the highest confidence (A or C in Fig. 1). Furthermore, we also compute the distances between the 3 mouth features that are shown either as the triangle ABD or BCD in Fig. 1, based on the selected mouth corner. These mouth features are adequate in describing the basic movements of mouth during eating.

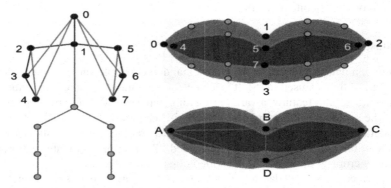

Fig. 1. Upper body (left) and mouth (right and down) human motion features that are used as input to the proposed deep network.

The proposed deep network, shown in Fig. 2 processes the input features and extracts new discriminative ones that can better model the underlying spatiotemporal information. More specifically, the first two blocks of the proposed deep network employ stacked convolutional layers to compute spatial features (i.e., interactions between neighboring features both in time and space). The max-pooling layer between blocks 1 and 2 is responsible for down-sampling the feature space and improve the robustness of the network. Furthermore, the upper parts of blocks 1 and 2 are shortcut connections that tackle the vanishing gradient problem and improve the convergence of the network. The third block of the network employs a series of LSTM units that extract temporal features from the entire sequence of input features. A fusion of the two-streams of spatiotemporal features is performed in the end using a fully connected layer that combines the information from the hands, head and mouth and computes the probability of an input video sequence to describe a bite instance.

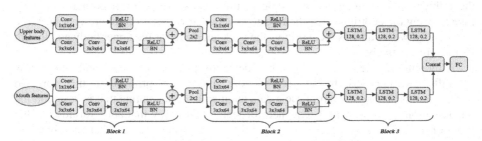

Fig. 2. Proposed deep network for automatic bite detection.

4   Experimental Evaluation

In this section, we initially describe the dataset used for the evaluation of the proposed automatic bite detection method and then present the preprocessing of features before their introduction to the proposed deep network. Moreover, we present the experiments for the hyper-parameter optimization of the proposed network, prior to its evaluation in the task of continuous bite detection.

4.1   Dataset

The evaluation dataset consists of 85 videos that depict people eating from a side view and it is part of the one used in [5]. It is a challenging dataset as there are occlusions of hands and mouth and features a variety of eating patterns and different subjects, types of meals (i.e., soup, breakfast and dinner) and kitchen utensils.

The dataset is manually annotated by human experts (i.e., nutritionists) and ground truth annotations of bite instances are provided. To train and validate our proposed network, we cropped a total of 12121 video clips with a duration of 2 s or 50 frames each and formed an isolated dataset. In this dataset, there are 4149 positive samples (i.e., annotated bite instances) and 7972 negative samples (i.e., randomly cropped non-bite instances). Additionally, this dataset is randomly split in a training set that contains 90% of samples and is used for training the proposed deep network and a validation set that contains the remaining 10% of the samples and is used for the optimization of the hyper-parameters of the proposed network.

4.2   Data Preprocessing and Augmentation

There are 25 body and 70 face features extracted from each video frame. We select only the 8 most relevant body (i.e., nose, neck, left and right hands) features and the 8 most relevant face (i.e., mouth) features as shown in Fig. 2. Furthermore, due to body parts' occlusions, there are features that are not detected or have abnormal values (i.e., outliers) and even frames with no detected features at all. To overcome this problem we apply two preprocessing stages to the extracted features. The first stage fills the empty values with values from the previous temporal instance (i.e., frame), while the second stage performs spline interpolation to the temporal sequences of features in order to remove outliers and smooth feature values across time. Moreover, in order to diminish the influence of the location of people in the videos, we perform normalization by transforming the coordinates of the selected features to a local coordinate system. More specifically, we assume as local origins the neck and nose for the upper body and mouth features respectively.

Although the size of the isolated bite detection dataset may seem quite large, it is not sufficient to properly train a deep network, like the one proposed in this work. To overcome this problem, we propose a data augmentation technique that is based on the manipulation of the temporal sequences of features in a way that adds variation to the input of the network and assists it in identifying the real difference between bite and non-bite instances and achieving higher accuracy in the task of bite detection.

More specifically, we propose two augmentation operations to the input temporal sequences of features. The first operation concerns the addition of a small displacement in the input, thus affecting the global location of features in the videos, while the second operation concerns the circular wrapping of the input by a certain small value in order to add variation to the temporal order of the feature values.

4.3 Hyper-parameter Optimization

The optimization of the hyper-parameters of the proposed deep network is performed based on the performance of the network on the validation set. In our case, the parameters that are optimized are the number of CNN and LSTM layers since all the other parameters (i.e., number and size of kernels and filters and dropout percentages) are kept fixed to optimal values chosen based on our knowledge on the task and after initial experimentation.

The optimization of the number of layers is performed using the following approach. We consider a maximum number of 3 stacked layers and start by introducing one-by-one the recurrent layers (LSTMs) of block 3 and then add convolutional layers for blocks 1 and 2 as long as the performance of the network is increased. Finally, we add the shortcut connections of blocks 1 and 2. The experiments are performed on the initial isolated dataset (i.e., no data augmentation) as the experiments with data augmentation are performed on the optimized network. The performance of the proposed deep network on the validation set for different number of convolutional (Conv) and LSTM layers is presented in Table 1.

Table 1. Experiments with respect to the proposed network architecture.

Proposed network architecture			Performance on validation set
Block 1	Block 2	Block 3	
–	–	1 LSTM	0.81
–	–	2 LSTMs	0.807
–	–	3 LSTMs	0.847
1 Conv	–	3 LSTMs	0.857
2 Conv	–	3 LSTMs	0.864
3 Conv	–	3 LSTMs	0.875
3 Conv	1 Conv	3 LSTMs	0.882
3 Conv	2 Conv	3 LSTMs	0.894
3 Conv	3 Conv	3 LSTMs	0.922
3 Conv + shortcut	**3 Conv + shortcut**	**3 LSTMs**	**0.927**

From Table 1, we conclude that deeper networks with more parameters and thus processing capabilities and with the ability to extract both spatial and temporal information from their inputs achieve higher performance in the task of bite detection. Furthermore,

592   D. Konstantinidis et al.

the shortcut connections are beneficial to the proposed network by slightly improving both its robustness and its classification accuracy.

As far as data augmentation is concerned, we test both data augmentation operations (i.e., displacement and circular wrapping). For the displacement operation, we compute a random displacement in the range [0, 0.1] for both x and y coordinates and add it to all features across time, thus not affecting their relative position. On the other hand, for the circular wrapping operation, we consider either 1 value (i.e., original sequence of features) or 3 values, namely {−5, 0, 5}, where 0 corresponds to the original sequence, while −5 and 5 corresponds to a circular wrapping of the temporal sequence by 5 positions (i.e., frames) to the left and right respectively. Table 2 presents the results of the experimentation with the data augmentation operations, where we can see that augmenting data with the displacement operation improves the performance of the proposed network. However, there is a threshold over which further increase of the displacement augmentation factor leads to a drop in the performance of the proposed network. This can be attributed to the fact that additional samples do not improve the requested variation of the input and the network has already learned the differentiation between bites and non-bites. On the other hand, the combination of the displacement and circular wrapping augmentation operations leads to a significant increase in the performance of the proposed network in the task of bite detection.

Table 2. Experiments with respect to the data augmentation operations.

Data augmentation operations			Performance on the validation set
Displacement factor	Circular wrapping factor	Combined factor	
1	1	1	0.927
3	1	3	0.939
5	1	5	0.937
3	**3**	**9**	**0.947**

4.4 Results on Continuous Bite Detection

To evaluate the ability of the proposed methodology to identify bite instances in a continuous fashion, we test our deep network on the 85 continuous videos of the provided dataset [5]. To achieve this, we employ an overlapping sliding window of 60 frames, which is slightly larger than the clips of 50 frames used for training and as we want smoother output probabilities from our network. Additionally, we employ a step of 1 frame, meaning that a bite detection probability is computed for each frame, taking also into account its neighboring frames.

Since the output of the proposed network is a continuous signal of bite detection probabilities, post-processing should be applied to detect exact timestamps of bite instances and remove false alarms. Initially a nth-order median filter is applied to smooth signal and remove small and abrupt changes (i.e., sawtooth effect and outliers). Afterwards, the mean m and standard deviation s of the signal are computed and all predictions below the threshold of $m + s$ are zeroed. Then, all local maxima (i.e., peaks) of the signal are detected and all peaks with width below a threshold T_W are removed. Finally, the distance between peaks is computed and for peaks having distance smaller than a threshold T_D, we preserve only the peak with the highest probability to belong to the bite instance. We set the order of the media filter n to 40 so as to achieve heavy smoothing and the distance threshold T_D to 50 frames as we believe that under normal circumstances a person cannot receive two bites in less than 2 s time difference. Finally, after experimentation, we set the width threshold T_W of the peaks to 22 frames, which describes the duration of a clear bite instance and corresponds to almost 1 s.

Table 3 presents the overall performance of the proposed method in the task of continuous bite detection. We can observe that our method achieves a bite detection rate (i.e., recall) of 91.71% with a false alarm rate of 8.25%, which means that our method is a very accurate and robust bite detector. Furthermore, Fig. 3 shows a histogram of the distribution of F1-scores (i.e., harmonic averages of recall and precision) for the tested videos. This figure shows that our proposed method achieves superb results on most videos, while there are only a few videos, in which our method achieves mediocre results. Finally, two examples of the predicted bite detection probabilities that our method outputs overlaid on the ground truth are presented in Fig. 4. The proposed bite detector run on a desktop PC having i7-5820 K CPU with 6 cores @ 3.3 GHz, RAM of 64 GB and a Tesla K40 GPU card.

Table 3. Experimentation with continuous automatic bite detection.

Recall	Precision	F1-score
0.9171	0.9175	0.9173

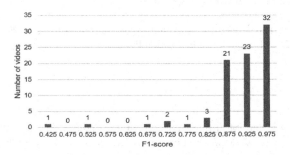

Fig. 3. Distribution of videos based on their F1-score. Total number of videos: 85

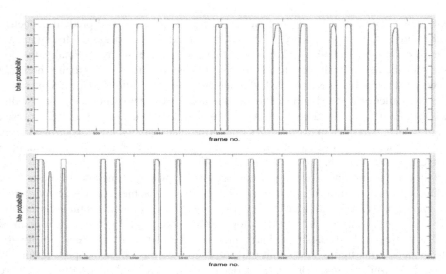

Fig. 4. Bite prediction results (blue signal) overlaid on ground truth (red signal) for two videos of the dataset. (Color figure online)

5 Conclusions

In this work, we present a novel methodology for accurate bite detection results based on the extraction of human body and face motion features from videos. Experiments on both isolated and continuous datasets show the superb performance of the proposed bite detector. Current limitations of the algorithm include noisy features extracted from OpenPose and human moves that can be easily mistaken with bite instances, such as wiping of mouth or scratching of head. These limitations can be overcome with the addition of more advanced feature preprocessing techniques, elaborate network architectures and descriptive features, such as image features.

Acknowledgement. This work was supported by the European Project: PROTEIN Grant no. 817732 with the H2020 Research and Innovation Programme.

References

1. Ioakimidis, I., Zandian, M., Eriksson-Marklund, L., Bergh, C., Grigoriadis, A., Södersten, P.: Description of chewing and food intake over the course of a meal. Physiol. Behav. **104**, 761–769 (2011)
2. Fogel, A., et al.: A description of an "obesogenic" eating style that promotes higher energy intake and is associated with greater adiposity in 4.5 year-old children: results from the GUSTO cohort. Physiol. Behav. **176**, 107–116 (2017)
3. Ohkuma, T., Hirakawa, Y., Nakamura, U., Kiyohara, Y., Kitazono, T., Ninomiya, T.: Association between eating rate and obesity: a systematic review and meta-analysis. Int. J. Obes. **39**(11), 1589 (2015)

4. Fagerberg, P., Langlet, B., Glossner, A., Ioakimidis, I.: Food intake during school lunch is better explained by objectively measured eating behaviors than by subjectively rated food taste and fullness: a cross-sectional study. Nutrients **11**(3), 597 (2019)

5. Langlet, B., Tang Bach, M., Odegi, D., Fagerberg, P., Ioakimidis, I.: The effect of food unit sizes and meal serving occasions on eating behaviour characteristics: within person randomised crossover studies on healthy women. Nutrients **10**(7), 880 (2018)

6. Hermsen, S., Frost, J.H., Robinson, E., Higgs, S., Mars, M., Hermans, R.C.J.: Evaluation of a smart fork to decelerate eating rate. J. Acad. Nutr. Diet. **116**(7), 1066–1067 (2016)

7. Theodoridis, T., Solachidis, V., Dimitropoulos, K., Gymnopoulos, L. Daras, P.: A survey on AI nutrition recommender systems. In: 12th International Conference on Pervasive Technologies Related to Assistive Environments Conference, Rhodes, Greece (2019)

8. Kyritsis, K., Diou, C., Delopoulos, A.: Food intake detection from inertial sensors using LSTM networks. In: Battiato, S., Farinella, G.M., Leo, M., Gallo, G. (eds.) ICIAP 2017. LNCS, vol. 10590, pp. 411–418. Springer, Cham (2017). https://doi.org/10.1007/978-3-319-70742-6_39

9. Simon, T., Joo, H., Matthews, I. Sheikh, Y.: Hand keypoint detection in single images using multiview bootstrapping. In: IEEE Conference on Computer Vision and Pattern Recognition (CVPR), Honolulu, HI, pp. 4645–4653 (2017)

10. Cao, Z., Simon, T., Wei, S. Sheikh, Y.: Realtime multi-person 2D pose estimation using part affinity fields. In: IEEE Conference on Computer Vision and Pattern Recognition (CVPR), Honolulu, HI, pp. 1302–1310 (2017)

11. Sazonov, E., Fontana, J.: A sensor system for automatic detection of food intake through non-invasive monitoring of chewing. IEEE Sens. J. **12**, 1340–1348 (2012)

12. Papapanagiotou, V., Diou, C., Langlet, B., Ioakimidis, I. Delopoulos, A.: A parametric probabilistic context-free grammar for food intake analysis based on continuous meal weight measurements. In 37th Annual International Conference of the IEEE on Engineering in Medicine and Biology Society (EMBC), pp. 7853–7856 (2015)

13. Zhang, R., Amft, O.: Monitoring chewing and eating in free-living using smart eyeglasses. IEEE J. Biomed. Health Inf. **22**(1), 23–32 (2018)

14. Kyritsis, K., Diou, C., Delopoulos, A.: End-to-end learning for measuring in-meal eating behavior from a smartwatch. In: 40th Annual International Conference of the IEEE Engineering in Medicine and Biology Society (EMBC), Honolulu, HI, pp. 5511–5514 (2018)

15. Papapanagiotou, V., Diou, C., Zhou, L., Van Den Boer, J., Mars, M., Delopoulos, A.: A novel chewing detection system based on ppg, audio, and accelerometry. IEEE J. Biomed. Health Inf. **21**(3), 607–618 (2017)

16. Mirtchouk, M. Merck, C. Kleinberg, S.: Automated estimation of food type and amount consumed from body-worn audio and motion sensors. In: Proceedings of the 2016 ACM International Joint Conference on Pervasive and Ubiquitous Computing, pp. 451–462 (2016)

17. Doulah, A., et al.: Meal microstructure characterization from sensor-based food intake detection. Front. Nutr. **4**, 31 (2017)

18. Fontana, J.M., Farooq, M., Sazonov, E.: Automatic ingestion monitor: a novel wearable device for monitoring of ingestive behavior. IEEE Trans. Biomed. Eng. **61**, 1772–1779 (2014)

19. Zhang, S., Stogin, W., Alshurafa, N.: I sense overeating: motif-based machine learning framework to detect overeating using wrist-worn sensing. Inf. Fusion **41**, 37–47 (2018)

20. Ramos-Garcia, R.I., Muth, E.R., Gowdy, J.N., Hoover, A.W.: Improving the recognition of eating gestures using intergesture sequential dependencies. IEEE J. biomedical and health informatics **19**(3), 825–831 (2015)

Extracting the Inertia Properties of the Human Upper Body Using Computer Vision

Dimitrios Menychtas[(✉)], Alina Glushkova, and Sotirios Manitsaris

Centre for Robotics, MINES ParisTech, PSL Université Paris, Paris, France
`dimitrios.menychtas@mines-paristech.fr`

Abstract. Currently, biomechanics analyses of the upper human body are mostly kinematic i.e., they are concerned with the positions, velocities, and accelerations of the joints on the human body with little consideration on the forces required to produces them. Tough kinetic analysis can give insight to the torques required by the muscles to generate motion and therefore provide more information regarding human movements, it is generally used in a relatively small scope (e.g. one joint or the contact forces the hand applies). The problem is that in order to calculate the joint torques on an articulated body, such as the human arm, the correct shape and weight must be measured. For robot manipulators, this is done by the manufacturer during the designing phase, however, on the human arm, direct measurement of the volume and the weight is very difficult and extremely impractical. Methods for indirect estimation of those parameters have been proposed, such as the use of medical imaging or standardized scaling factors (SF). However, there is always a trade off between accuracy and practicality. This paper uses computer vision (CV) to extract the shape of each body segment and find the inertia parameters. The joint torques are calculated using those parameters and they are compared to joint torques that were calculated using SF to establish the inertia properties. The purpose here is to examine a practical method for real-time joint torques calculation that can be personalized and accurate.

Keywords: Computer vision · Biomechanics · Upper human body · Inertia properties

1 Introduction

The science of motion and movements of the human body is called biomechanics and it has wide application to sports, ergonomics, and rehabilitation. As a field it shares a lot of the underlying principles with mechanical devices and specifically articulated robots. However, its most distinguishing characteristic is that it needs to consider the anthropometrics of each individual. Anthropometrics are the measurements of the human body (height, limb lengths, weight etc.) and they are different for every person. It is generally challenging and

© Springer Nature Switzerland AG 2019
D. Tzovaras et al. (Eds.): ICVS 2019, LNCS 11754, pp. 596–603, 2019.
https://doi.org/10.1007/978-3-030-34995-0_54

time-consuming to get these measurements, as a result it is more efficient to average them based on certain characteristics (gender, race etc). However, quite often, the interest is in the individual performance, and therefore using standardized parameters obscures this individuality. There quite a few studies that provide such generic data. Both kinematic and kinetic average measurements are publicly available [1–3]. Those measurements have provided scaling factors (SF) to allow for the regression of the inertial properties based on a person's height and weight. The earlier attempts to calculate the inertia parameters used cadavers. Once the technology became available, medical imaging was used to reconstruct the shape of the human body and calculate the inertia properties ([1–3], among others). Obviously, neither method is particularly practical to be incorporated in an experiment with larger scope or in a working environment where the need for real-time, or at least reasonably quick, results is crucial.

Though the SF can be reasonably accurate, no ground truth exists and the validation of the SF can be challenging [4]. On a more technical level, the more dependent a calculation is on the inertia parameters, the less personalized the results are going to be. This can explain, to a certain degree why joint torques are not used as often as one might expect.

This paper examines the possibility of using computer vision (CV) to extract the volumes of the segments of the human body to find the inertia properties. The joint torques during two simple motions of the right arm will be calculated using inertia properties from CV and SF.

2   Methods

The inverse dynamics problem is when the joint motions of the human body are known and the forces that cause them are unknown. In general, a kinematic model of the human body and the tensors of inertia are required to solve the problem. The most efficient algorithm is the recursive Newton-Euler [5]. The basic components will be briefly discussed here, but note that each aspect has its own intricacies and it is impossible to cover all the material here. Figure 1 shows the basic steps of a biomechanical study.

2.1   Data Collection

For this work, the joint angles of an expert glassblower during crafting were recorded using an inertial measurement units (IMUs) suit (Nansense, Baranger Studios, Los Angeles, CA, USA). Standard RGB video was also recorded. At this point, the concept is shown. The recordings from the IMUs suit will provide the kinematic data, while the RGB video frames will be used to calculate the volumes of the body segments.

2.2   The Kinematic Model of the Human Body

The human body is a highly sophisticated structure that synchronizes different organ systems to generate motion. Those systems interact on a cellular level

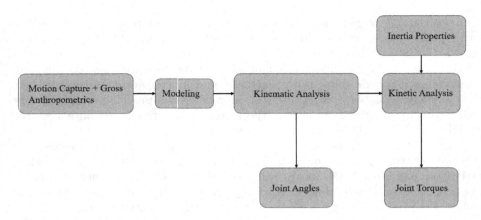

Fig. 1. The flow chart of the standard steps of a biomechanical analysis. "Gross Anthropometrics" can mean just the height and the weight of a person, or more specific measurements depending on the project

and produce an effect on a macroscopic scale. Each joint can move in up to three Cartesian axes, with each axis contributing a degree of freedom (DoF) to the system. As such, it is important to establish the scope of the project and the DoFs the model needs to have. Certain segments of the body (e.g. shoulder complex) can be modeled in different ways, while still keeping the important aspects for each application.

In its most essential definition, the human body can be described mathematically as a kinematic chain. The most elegant way to define complex kinematic chains is with the use of screw theory and the product of exponentials (POE). Proposed by Brockett [7], and described, among others, by Murray et. al [6], Selig [8], and Park and Lynch [9], this methodology defines each DoF as an axis, with respect to a global coordinate frame, about which a rigid body can rotate. The complexity of the mathematics involved does not increase with the number of DoFs. Even though it initially appears to be more complicated than other approaches (e.g. Denavit-Hartenberg parameters), more advanced robotic concepts, such as the Jacobian or the mass matrix, become easier to calculate and incorporate. For the scope of this work, such a model will allow for the calculation of the joint torques using the same methodologies that are applied on manipulators.

In this paper, the model that is used is similar to the one Menychtas used [10]. The upper body is separated into seven segments. The torso, two upper arm segments, two forearm segments, and two hand segments, one in each side. The weight and the centers of mass for each segment are added for the kinetics calculations.

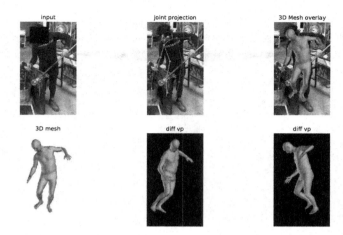

Fig. 2. A glassblower during the creation of an item. HMR identifies the volume of his body

2.3   Calculating the Tensors of Inertia

In this work, the tensors of inertia for each segment of the upper human body will be calculated in two ways. The first method will be to use the person's height and weight along with the SF reported by Dumas et al. [2]. This method will interpolate the inertial properties of the participant with respect to the general population. The mass of each segment of the body is considered to be a percentage of the total weight, and the tensors of inertia are scaled according to the height and the segments' weight.

The second method uses CV to extract the volumes of the body's segments. More specifically, the Human Mesh Recovery (HMR) [11] is employed to produce a 3D mesh of the body. Static images are used to fit the mesh on the person's body shape and posture. Figure 2 shows an example.

HMR uses the skinned multi-person linear model (SMPL) [12] as its foundation and performs its calculations on it. Since the kinematic data related to posture have been recorded from the IMUs suit, there is no need to try and extract the joint angles from the mesh the HMR creates, only the volume is used here. Figure 3 shows the mesh that was extracted in a T-pose.

Now that the 3D mesh is not distorted by the posture, each segment of the upper human body can be separated and the volume can be calculated. Parallelepiped bounding boxes are used to encapsulate the mesh's volume (Fig. 4). To increase accuracy, each segment uses more than one box. The density is assumed to be constant with a value of $985\,\mathrm{kg/m^3}$, therefore the integral of the volume along the density gives the mass of each segment. The tensors of inertia are calculated based on the center of mass and the weight that is measured. Using this method doesn't require any additional input because all relevant data have been extracted from the video recordings.

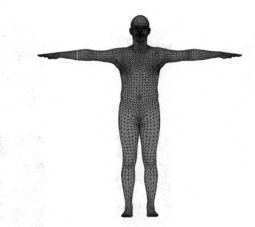

Fig. 3. The 3D mesh extracted from HMR projected on a T-Pose

Fig. 4. The bounding boxes on the upper body

2.4   Experiment

A subject of 65 kg and 1.73 m participated in a series of motions to create an item made of glass. As it is standard practice, the person performed range of motion (ROM) tasks before the actual recordings, to create reference points for his motions. Analyzing the kinematic and kinetic results of the complete movements falls outside the scope of this paper. To limit the discussion on the appropriate way to calculate the joint torques, only two ROM tasks will be examined.

For the first task, the person stood straight up and flexed his elbows all the way up and then down until he returned to the neutral posture. The process was repeated three times. The second task focused on shoulder rotation. The participant stood with his upper arm abducted parallel to the ground and the elbows flexed at 90° angle and the palms open. He rotated his upper arm until his fingers were pointing upwards and then reversed the rotation until his fingers were pointing to the ground and then returned to the initial position. This task was repeated three times as well.

3 Results and Discussion

The torques are calculated using the inertia properties extracted by the two methods previously described but the motion files (joint centers and angles) are the same as well as the kinematic model. As a result, the differences on the resulting data are due to different inertia properties only.

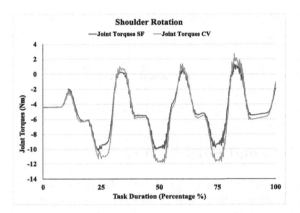

Fig. 5. Shoulder rotation torques

Figure 5 shows the torques of the shoulder rotation calculated using SF and CV. As the arm rotates upwards, the torque is increased because the lever arm moves against gravity. As it goes down, the torque shows that the arm is moving in the negative sense. Both methods yield similar results even though the CV method appears to yield higher peaks and deeper valleys. Since the kinematics are the same, the only explanation is that the inertia properties are higher because the segments' volumes and masses are larger.

Figure 6 highlights this even further. While the profile of the torques is the same in both cases, the magnitude is different. Unfortunately, neither methodology can be considered ground truth. The SF uses average scaling factors with the assumption that the inertia properties will be close enough. The CV method, on the other hand is constrained by how accurately it can identify the human body and fit the 3D mesh on each person. Basically, both methods are almost accurate enough. As a result, it is very difficult to favor one method over the other without further experimentation.

However, using CV, despite its limitations, results were extracted that were very close to the SF method. Considering the experimentation that is required to calculate the SF, using 3D meshes for volumetric measurements appears to have great potential. What's important to note is that using CV, the accuracy of the joint torques is only limited by how well the system can measure the volume from the image. When the SF method is employed, there's not much that context on the error they introduce.

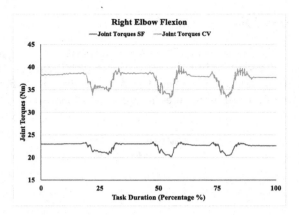

Fig. 6. Elbow flexion torques

4   Conclusions and Future Work

In this paper, the concept of finding the inertia properties of the human body using a 3D mesh that was fitted on an actual person was examined briefly. The volumes of the body segments were found from the mesh and, using the body's density, the mass was calculated. Initial results showed that this approach can give comparable results with the more standard practice of using average SF.

The results from the CV method had a higher magnitude than the respective ones from the SF. The implication here is double, on the one hand, the scaling factors might be underestimating the inertia properties of the subject. This kind of error is expected when using average values and not personalized ones. However, there is not much that can be done about it. Even if an updated version is used, the method is still constrained by the fact that it relies on averaged values.

On the other hand, the bounding boxes the CV uses to identify the volumes might have been too large. This resulted in heavier segments to be identified. It is also possible that the 3D mesh wasn't fitted accurately enough. Of course, the assumption of consistent density for the whole upper body can also be questioned. However, by using CV, the comparable results to the SF method were acquired with a fraction of effort. More importantly, there's a clear direction on what needs to be done in the future to create more accurate tools. Most computer vision projects try to identify the joint angles of multiple persons in a scene, however, there is no dire need for another tool to calculate kinematic data. At the same time, there is no practical way to measure the inertia properties of the human body and computer vision seems to be the only technology that has the potential to address the issue.

To sum up, calculating the inertia parameters is challenging with the use of scaling factors because they may not generalize well across different subjects.

Initial results of using computer vision look promising, but a more rigorous experiment needs to be used to establish the ground truth and the relevancy of the method.

Acknowledgements. The research leading to these results has received funding by the EU Horizon 2020 Research and Innovation Program under grant agreement No. 822336 Mingei project.

The ROM tasks were performed by an expert glass blower at CERFAV Centre Européen de Rercherches et de Formation aux Arts Verriers, Vannes-le-Châtel, France.

References

1. Zatsiorsky, V.: Kinetics of Human Motion (2002)
2. Dumas, R., Chèze, L., Verriest, J.P.: Adjustments to McConville et al. and Young et al. body segment inertial parameters. J. Biomech. **40**, 543–553 (2007)
3. Ma, Y., Kwon, J., Mao, Z., Lee, K., Li, L., Chung, H.: Segment inertial parameters of Korean adults estimated from three-dimensional body laser scan data. Int. J. Ind. Ergon. **41**, 19–29 (2011)
4. Hansen, C., Venture, G., Rezzoug, N., Gorce, P., Isableu, B.: An individual and dynamic body segment inertial parameter validation method using ground reaction forces. J. Biomech. **47**, 1577–1581 (2014)
5. Featherstone, R.: Rigid Body Dynamics Algorithms. Springer, Berlin (2008). https://doi.org/10.1007/978-1-4899-7560-7
6. Murray, R., Li, Z., Sastry, S.: A Mathematical Introduction to Robotic Manipulation (1994)
7. Brockett, R.W.: Robotic manipulators and the product of exponentials formula. In: Mathematical Theory of Networks and Systems, vol. 58, pp. 120–129 (1984). https://doi.org/10.1007/BFb0031048
8. Selig, J.: Geometric Fundamentals of Robotics. Springer, Berlin (2005). https://doi.org/10.1007/b138859
9. Lynch, K., Park, F.: Modern Robotics Mechanics, Planning, and Control. Cambridge University Press, Cambridge (2017)
10. Menychtas, D.: Human body motions optimization for able-bodied individuals and prosthesis users during activities of daily living using a personalized robot-human model, Ph. D. dissertation, University of South Florida (2018)
11. Kanażawa, A., Black, M, Jacobs, D., Malik, J.: End-to-end recovery of human shape and pose. In: 2018 IEEE/CVF Conference on Computer Vision and Pattern Recognition (2017)
12. Loper, M., Mahmood, N., Romero, J., Pons-Moll, G., Black, M.J.: SMPL: a skinned multi-person linear model. ACM Trans. Graph. (Proc. SIGGRAPH Asia) **34**, 248:1–248:16 (2015)

Single Fingertip Detection Using Simple Geometric Properties of the Hand Image: A Case Study for Augmented Reality in an Educational App

Nikolaos Nomikos[✉] and Dimitris Kalles

Hellenic Open University, 26335 Patra, Greece
nomikos.development@gmail.com, kalles@eap.gr

Abstract. We propose a fingertip detection method suitable for portable devices' applications where the user can interact with objects or interface elements located in the augmented space. The method has been experimentally implemented in the context of an application that uses a board containing ArUco markers [1]. The user can press virtual buttons laid on the board or drag items along predetermined paths on the board level by extending the index finger (or thumb) and placing its edge on the object of interest, while the other hand holds the device so that both the object and the hand are visible. We present brief but indicative results of our technique.

Keywords: Fingertip detection · Computer vision · Augmented reality

1 Introduction

In recent years the expanding and growing use of augmented reality has begun to find its way to the education. In the context of the development of a physics educational application, the need for interaction with objects of the augmented space emerged. After we explored possible ways to achieve this interaction, we focused on making use of the fingertip, so that the user can select, move, and drag objects of the augmented space. In order to locate the tip of the finger we devised the algorithm that we will present here.

The subject of digital finger identification has been extensively studied in the literature and three main categories of approaches have been formed based on the material or equipment that is used by the application.

The first category of approaches regards the use of equipment attached to the hand, such as ring-shaped sensors on each finger [2], colored markers attached to the fingers [3] and fiducial markers attached at several hand points in order to determine the basic hand gesture from their detection [4].

The second category of approaches involves the use of additional equipment, such as Kinect or RGB-D cameras, in order to extract the depth image and the fingertip can be determined for example by considering it to be the point of the hand that is farthest away from the arm [5].

© Springer Nature Switzerland AG 2019
D. Tzovaras et al. (Eds.): ICVS 2019, LNCS 11754, pp. 604–611, 2019.
https://doi.org/10.1007/978-3-030-34995-0_55

The third category of approaches concerns the use of recognition methods that are based on the processing of an image of the hand, or the processing of a sequence of images in which the hand is detected. In the latter case, for example, the curvature of the binary image's contour of the hand can be calculated, local maxima are found as candidate fingertips, and after separating them from the valleys, candidate edges are selected as those that have the highest rate of appearance in five consecutive frames [6]. In the same case, for example, we can perform shape filtering on binary images and gradually exclude (fast rejection filter) from the segmented image of the hand regions having geometric features of a non-fingertip [7] so that only the fingertips are left. Alternatively we can find strong peaks along the hand blob perimeters of the vectors' angles defined by each point of the contour of the hand with two other points (n points before and n points after) [8] or by using shadow based methods with heuristics that return a single finger-tip per hand [9]. Our approach belongs to the third category and is based on the detection of simple geometric features. The detection of these features in a binary image of the hand allows us to locate the fingertip.

The aim of the method that we have developed is the location of the fingertip in applications that fall within the range of augmented - mixed reality applications [10]. This localization is useful in situations where it is desirable for the user to be able to interact via his finger with objects of the augmented space either by touching them, selecting them or dragging them with his fingertip.

For this process to succeed it is necessary to specify a correspondence between points of the real environment and their 2D-projection as it is captured by the camera. This allows us to estimate the camera pose, and to specify the portion of the image stream where our algorithm is applicable. In doing so, we shall use binary square fiducial markers, as provided by the ArUco library, namely ArUco markers [11]. The detection of a single marker can give us enough information, by specifying its corners, for estimating the camera pose. Instead of a single marker, we shall use a combination of ArUco markers, in the form of a custom ArUco board. This custom ArUco board acts as a single marker regarding the pose estimation but can provide us with additional information since we can access information for each individual detected marker. Therefore, we can obtain the position of each of theirs corners and obtain the location of areas such as the interior of the ArUco board.

As an example of the concepts underlying our proposed algorithm, Fig. 1 depicts the initial image as captured by the camera and the geometric characteristics which arise during the successive stages of the processing of this image, the determination of which led us from a well-defined area of the hand's image (the interior of the ArUco board) to the geometric abstraction which represents the fingertip.

The rest of this paper is structured in three sections. The next section focuses on the motivation and the need for fingertip detection algorithms. Section 3 sets out the proposed algorithm. The last section presents our conclusions.

2   Interacting with Real, Virtual and Augmented Space

In several applications implemented with augmented reality capabilities, the user's interaction with augmented space's objects is realized using the touch screen of the mobile device. In other cases interaction with the augmented space is accomplished through special devices such as leap motion [12], through markers [13], through camera and depth camera combination in order to better track the hand and achieve this kind of interaction [14, 15]. The algorithm we propose is a way of finding the fingertip position so the interaction with the augmented space becomes possible without the use of any extra equipment. Indicatively, we can interact with virtual buttons, virtual sliders, and generally move virtual objects the motion of which is constrained on predetermined lines or curves. The entrance and the departure of the fingertip in and from these lines or curves indicates the beginning or the termination of the interaction with these objects.

The need for interface elements on the augmented space arises from the fact that on a mobile screen the existing space is limited so the number of them that can be displayed is inversely related with their overall usability. On the other hand, augmented space offers the ability to place a larger number of interactive elements than the number of interactive elements that can be placed on the screen of a mobile device. Additionally, placement of interactive elements in the augmented space allows us to access different elements of the interface by moving the handheld device or by changing the viewing angle, and even allow us to enlarge or reduce the area of interface's elements that are visible, by moving the mobile device closer to or away from the objects of the augmented space.

3   Description of the Algorithm

The proposed algorithm can be applied when the available image contains only the hand or when we are able to locate the area of the hand in the available image. In the first case, the algorithm is applied on the entire binary image resulting from a hand segmentation process while in the second case the method is applied to the binary sub-image of the area where the hand has been detected. Several methods have been proposed for extracting the binary image of the hand and in the following we assume that one of these methods have already been applied and that we have at our disposal the resulting binary image.

The proposed algorithm is based on the observation that if we have an image of a hand with one of the fingers pointing and then we draw the convex hull of the hand, then we can observe that right and left of the pointing finger there exist points of the convex hull that do not belong to the contour of the hand's binary image. For each such deviation of the contour of the hand from its convex hull we can find a triad of points A, B, C such that A is the point where the deviation begins, as we cross the contour clockwise, B is the point where the deviation ends, as we cross the contour clockwise, and C is the point of the contour between A and B that has the maximum distance from the line segment AB. So, we have the triplet of points A_1, B_1, C_1 left of the finger, the triplet A_2, B_2, C_2 right of the finger and also the corresponding triangles $ABC_1 \equiv A_1B_1C_1$ and $ABC_2 \equiv A_2B_2C_2$. Our algorithm is based on the assumption that right and left of a pointing

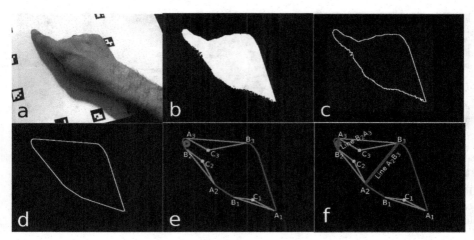

Fig. 1. (a) Original image of the hand placed on an ArUco board, (b) Segmented hand image (using K-Means clustering on channels Cb, Cr in YcbCr color space) restricted in the ArUco's board interior area, (c) Contour of hand image, (d) Convex hull of hand image, (e) Convexity defects and the respective triangles, (f) Candidate line segments (blue segments). (Color figure online)

finger these regions are maximum, in the sense that the corresponding ABCL and ABCR triangles that can be formed with the previously explained procedure have areas larger than any other's triangle ABC_i area that is formed by any other triad A_i, B_i, C_i of points. Once we have identified these two triangles $A_1B_1C_1$ and $A_2B_2C_2$ the fingertip is in the middle of A_1B_2 if $A_1B_1C_1$ is right of the finger, and $A_1B_2 < A_2B_2$ holds, or it is in the middle of A_2B_1 if $A_1B_1C_1$ is left of the finger and $A_1B_2 > A_2B_1$ holds.

The first step of the algorithm is to find the points of the contour of the hand (Fig. 1c). Based on these points, we calculate the convex hull of the hand (Fig. 1d). Then for each deviation of the contour from the convex hull (Fig. 1d) we find the corresponding points A_i, B_i, C_i (Fig. 1e). For each of the defined triangles ABC_i, we calculate their area and we select the two triplets of points that define triangles having the largest areas ($A_2B_2C_2$ and $A_3B_3C_3$ in Figs. 1e and f). Based on these two triplets A_2, B_2, C_2 and A_3, B_3, C_3, we calculate the lengths of the line segments A_3B_2 and A_2B_3. If $A_3B_2 > A_2B_3$ then the middle of A_3B_2 is the fingertip (Fig. 1f) and if $A_3B_2 < A_2B_3$ then the mean of A_2B_3 is the fingertip's position.

The robustness of the proposed algorithm is depicted in Fig. 2 where various realistic hand configurations have been examined, realistic in the sense that their respective images include a finger pointing, or interacting with something, being extended or nearly extended.

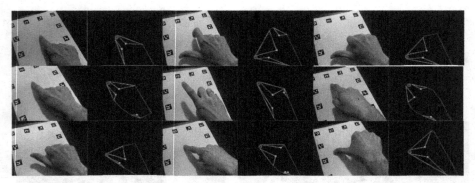

Fig. 2. Robustness of the algorithm: the left image of each pair depicts the image captured by the camera while the right one depicts convex hull, triangles and estimated fingertip position for the hand of the left image.

4   Technology Necessities and the Proposed Algorithm

In the implementation of our algorithm we used the computer vision library OpenCV [16] and the library ArUco [17], which is a library for augmented reality applications based on OpenCV (Fig. 3). Using ArUco library we constructed an ArUco board [18], within the boundaries of which is the arena of fingertip's motion and detection. This ArUco board is printed on a sheet of paper and on the boundary of this board ArUco markers have been placed. With the aid of these markers an area is defined in which there are elements of the augmented space that the user can interact with them using his fingertip. The Board contains 14 markers from one of the ArUco's available dictionaries of markers (Fig. 4).

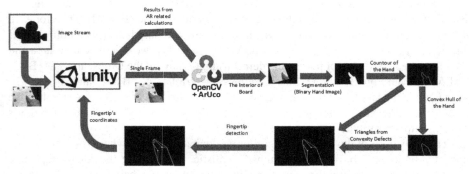

Fig. 3. The main stages of our application that are related with the realization of augmented reality and the detection of the fingertip.

Using an ArUco Board instead of just one marker is advantageous because, on the one hand, tracking is more accurate due to the combined information obtained from the tracking of more than one markers and on the other hand it is still possible to track the board even when some markers are not visible. In the first step of the demonstration of the proposed algorithm's functionality, the part of the image that corresponds to ArUco Board's interior is located. This is done by determining the homography transformation that maps the ArUco board's image to the board image detected by the camera input. Based on this transformation, we specify the points that correspond to the interior of the ArUco Board on the camera's input stream and thus we specify the portion of the input image stream on which we will apply the algorithm. Next, we perform color segmentation in the YCbCr color space using k-means clustering with three centers in order that one of them will be the interesting one in terms of color (away as possible from black or white values) and its cluster will contain pixels that belong to the hand's image. Once the segmentation of the image and the extraction of the binary image of the hand are completed, we calculate the contour and the convex hull of the binary image. The contour and the convex hull do

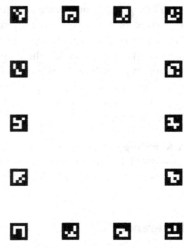

Fig. 4. The custom ArUco Board: Fingertip detection takes place inside the rectangle defined by the four markers on the corners

not much everywhere and the regions where they do not match are convexity defects. For each such defect, three points of the contour are obtained, one for the beginning of the defect, another for its end and the third one is the point of the contour having the maximum distance from the convex hull.

5 Conclusion

We propose an algorithm for fingertip detection using areas of triangles extracted from the comparison of the contour and the convex hull of the binary hand image. Robust and precise performance of our proposed algorithm has been presented through extensive experiments (Fig. 2) even when we change the pointing finger of the hand. Restrictions have been taken into account regarding the natural positioning of the hand with the pointing finger and the handheld device held by the other hand.

The algorithm we propose was created/conceived during the implementation of a physics educational application in which students could interact with elements of the augmented space. These elements could be representations of real objects or 3-D visualizations of concepts used in physics such as force and speed.

During the development of this application and while we were trying to exploit the algorithm, we encountered performance related issues. These problems are related to the fact that the image processing part was taking place in the device's CPU, meaning that we are experiencing the bottleneck of transferring image data from the GPU to the CPU. Deploying the image processing part and the algorithm in the GPU would provide a major performance boost. A lot of effort has been made towards this direction, such as the opencl API, which unfortunately is not available on all mobile devices.

The algorithm in order to be functional and give results requires that the hand is within an ArUco board. Within this region the geometric features that allow the detection of the fingerprint are calculated and processed. It also requires that we make use of the board's markers to realize the augmented reality functionality of the application. This restriction regarding the use of an ArUco board could be removed by first detecting the area where the hand lies (using, for example, haar classifier), and then by following the steps of the algorithm, but this time inside this detected area instead of the interior of an ArUco Board.

Acknowledgment. This research has been co-financed by the Operational Program "Human Resources Development, Education and Lifelong Learning" and is co-financed by the European Union (European Social Fund) and Greek national funds.

References

1. Garrido-Jurado, S., Muñoz-Salinas, R., Madrid-Cuevas, F., Marín-Jiménez, M.: Automatic generation and detection of highly reliable fiducial markers under occlusion. Pattern Recogn. **47**(6), 2280–2292 (2014)
2. Fukumoto, M., Tonomura, Y.: Body coupled FingerRing: wireless wearable keyboard. In: ACM SIGCHI Conference on Human factors in computing systems (CHI '97), New York, USA (1997)
3. Kerdvibulvech, C., Saito, H.: Vision-based guitarist fingering tracking using a bayesian classifier and particle filters. In: Mery, D., Rueda, L. (eds.) PSIVT 2007. LNCS, vol. 4872, pp. 625–638. Springer, Heidelberg (2007). https://doi.org/10.1007/978-3-540-77129-6_54
4. Buchmann, V., Violich, S., Billinghurst, M., Cockburn, A.: FingARtips: gesture based direct manipulation in augmented reality. In: 2nd International Conference on Computer Graphics and Interactive Techniques in Australasia and South East Asia (GRAPHITE '04). Suntec City, Singapore (2004)
5. Feng, Z., Xu, S., Zhang, X., Jin, L., Ye, Z., Yang, W.: Real-time fingertip tracking and detection using Kinect depth sensor for a new writing-in-the-air system. In: 4th International Conference on Internet Multimedia Computing and Service (ICIMCS '12), Wuhan, China (2012)
6. Shah, S.A.H., Ahmed, A., Mahmood, I., Khurshid, K.: Hand gesture based user interface for computer using a camera and projector. In: 2011 IEEE International Conference on Signal and Image Processing Applications (ICSIPA), Kuala Lumpur, Malaysia (2011)
7. Letessier, J., Bérard, F.: Visual tracking of bare fingers for interactive surfaces. In: 17th Annual ACM Symposium on User Interface Software and Technology (UIST '04), Santa Fe, New Mexico (2004)
8. Malik, S., Laszlo, J.: Visual touchpad: a two-handed gestural input device. In: 6th International Conference on Multimodal Interfaces (ICMI '04), State College, PA, USA (2004)
9. Wilson, D.A.: PlayAnywhere: a compact interactive tabletop projection-vision system. In: 18th Annual ACM Symposium on User Interface Software and Technology (UIST '05), Seattle, WA, USA (2005)
10. Milgram, P., Takemura, H., Utsumi, A., Kishino, F.: Augmented reality: a class of displays on the reality-virtuality continuum. In: Telemanipulator and Telepresence Technologies, Boston, MA, United States (1995)
11. OpenCV: Detection of ArUco Markers. https://docs.opencv.org/3.1.0/d5/dae/tutorial_aruco_detection.html

12. El Kabtane, H., El Adnani, M., Sadgal, M., Mourdi, Y.: Toward an occluded augmented reality framework in E-learning platforms for practical activities. J. Eng. Sci. Technol. **13**(2), 394–408 (2018)

13. Sinclair, P.A.S., Martinez, K., Millard, D.E., Weal, M.J.: Augmented reality as an interface to adaptive. New Rev. Hypermedia Multimedia **9**, 117–136 (2003)

14. Cai, S., Chiang, F.-K., Sun, Y., Lin, C., Lee, J.J.: Applications of augmented reality-based natural interactive learning in magnetic field instruction. Interact. Learn. Environ. **25**, 778–791 (2017)

15. Martin-Gonzalez, A., Chi-Poot, A., Uc-Cetina, V.: Usability evaluation of an augmented reality system for teaching Euclidean vectors. Innov. Educ. Teach. Int. **53**, 627–636 (2016)

16. OpenCV library. https://opencv.org/

17. ArUco: a minimal library for Augmented Reality applications based on OpenCV. University of Córdoba. https://www.uco.es/investiga/grupos/ava/node/26

18. OpenCV: Detection of ArUco Boards. https://docs.opencv.org/3.4.0/db/da9/tutorial_aruco_board_detection.html

Leveraging Pre-trained CNN Models for Skeleton-Based Action Recognition

Sohaib Laraba(✉)📧, Joëlle Tilmanne, and Thierry Dutoit

TCTS Lab, Numediart Institute, University of Mons (UMONS), Mons, Belgium
{sohaib.laraba,joelle.tilmanne,thierry.dutoit}@umons.ac.be

Abstract. Skeleton-based human action recognition has recently drawn increasing attention thanks to the availability of low-cost motion capture devices, and accessibility of large-scale 3D skeleton datasets. One of the key challenges in action recognition lies in the high dimensionality of the captured data. In recent works, researchers draw inspiration from the success of deep learning in computer vision in order to improve the performances of action recognition systems. Unfortunately, most of these studies do not leverage different available deep architectures but develop new architectures. Most of the available architecture achieve very high accuracy in different image classification problems. In this paper, we use these architectures that are already pre-trained on other image classification tasks. Skeleton sequences are first transformed into image-like data representation. The resulting images are used to train different state-of-the-art CNN architectures following different training procedures. The experimental results obtained on the popular NTU RGB+D dataset, are very promising and outperform most of the state-of-the-art results.

Keywords: Motion capture · Action recognition · Convolutional Neural Networks · Finetuning

1 Introduction

Human action recognition is an important and challenging research area in computer vision that has received much attention in the research community. It has numerous applications such as video surveillance, human-computer interaction, gaming, robotics, health, etc. Skeleton-based human action recognition has attracted increasing attention in the last decade due to wide accessibility of motion capture devices and large-scale datasets. Skeleton-based human representation considers a human body as articulated system of rigid segments connected by joints in 2D or 3D space depending on the motion capture device. Several studies have been proposed to classify 3D skeleton-based sequences [9,33,37]. Early approaches tend to build classifiers based on hand-crafted features designed manually to extract relevant information from these 3D sequences, such as relative distance between joints [10], joints orientations [29], geometric features [28], etc. These classifiers suffer from a lack of automation because of

© Springer Nature Switzerland AG 2019
D. Tzovaras et al. (Eds.): ICVS 2019, LNCS 11754, pp. 612–626, 2019.
https://doi.org/10.1007/978-3-030-34995-0_56

the dependency on hand-crafted features. These features are time consuming to design and extract and depend on the type of actions and data. Deep Learning (DL) has been adopted recently by the computer vision community thanks to its results that outperformed the state-of-the-art in several domains and to the availability of high-performance processing units able to train the DL models. The main advantage of deep learning, particularly in computer vision, is its ability to directly exploit raw data without any hand-crafted features. Deep learning classifiers are end-to-end systems that extract features in a fully automated way. Two deep learning methods are used in the field of action recognition: Recurrent Neural Networks (RNN) and Convolutional Neural Networks (CNN). RNNs are adopted to capture temporal information from extracted spatial skeleton features. RNNs have shown their strength in language modeling [41], image captioning [38], video analysis [34] and RGB-based activity recognition [25]. CNNs in the other hand were successfully introduced for image and video classification. They learn to recognize patterns across space. A CNN will learn to recognize components of an image (lines, curves, etc.) and then learn to combine these components to recognize larger structures (objects, faces, etc.). Inspired by this success, particularly for RGB image-based action recognition, there is a growing trend of using CNNs for skeleton-based action recognition [24,39,40]. This new trend produces more accurate classifiers compared to traditional machine learning approaches. Deep learning techniques learn by creating a more abstract representation of data as the network grows deeper. As a result, the model automatically extracts features and yields higher accuracy results. Despite these good results, deep learning research in action recognition remains immature and requires more attention to produce practical systems. For example, most of the researchers explore new architectures developed specifically for this task while many new successful deep architectures are not tested in the context of action recognition. In this work, we use a spatio-temporal representation of skeleton sequences presented under the form of a form of 2D images to train CNN classifiers. We use the process of fine-tuning to leverage models already trained for classical image classification for our specific skeleton-based action recognition task. We evaluate these architectures on the public benchmark dataset NTU RGB+D [31].

2   Related Works

Several research works, addressing human action recognition from skeletal data, have been published in the last decades. In recent years, deep learning methods have been widely addressed in many fields including human action recognition. Deep learning has enabled the replacement of hand-crafted features by learned features, and the learning of whole tasks in end-to-end way. Several approaches of deep learning have been proposed for skeleton-based human action recognition and can be categorized into two main categories: Recurrent Neural Networks (RNNs) methods and Convolutional Neural Networks (CNNs) methods.

RNNs are adopted to capture temporal information from spatial sequences. Basic RNN architectures are notoriously difficult to train [4,30], and more elaborate architectures are commonly used instead, such as the LSTM (Long Short-Term Memory) [15] and the GRU (Gated Recurrent Unit) [6]. Applications of these networks have shown promising results in skeleton-based action recognition. Du et al. [8,9] designed an end-to-end hierarchical RNN architectures for skeleton-based action recognition. They divided the human skeleton into five main parts in terms of body physical structure and fed them into five independent bidirectional RNNs for local feature extraction in the first layer. In the following layers, the outputs of the RNNs were concatenated to represent the upper body and lower body, then each was further fed into another set of RNNs. The global body representation was obtained and fed to the next RNN layer. These features are fed into a fully connected layer followed by a softmax layer for classification. Veeriah et al. [36] present a differential RNN that extends LSTM structure by modeling the dynamics of states evolving over time. They proposed to add a new gating mechanism for LSTM to model the derivatives of the memory states and explore the salient action patterns. In this method, all of the input features were concatenated at each frame and were fed to the differential LSTM at each step. Shahroudy et al. [31] propose a part-aware extension of LSTM to utilize the physical structure of the human body. They split the LSTM memory cell to sub-cells to push the network towards learning the context representations for each body part separately. These methods only model the motion dynamics in the temporal domain and neglect the spatial configurations of articulated skeletons.

RNN architectures are generally used for modeling sequential data. However, one of their major drawbacks is the exploding and vanishing gradient problem and the difficulty to parallelize their training. Bai et al. [2] show that convolutional networks can perform as well as or even better that recurrent networks in many tasks such as speech recognition [42], some tasks of NLP like neural machine translation [12], classification of long sentences [1], etc.

The challenge for CNN-based methods is to effectively capture the spatio-temporal information of a skeleton sequence using image-based representation. In fact, it is inevitable to lose temporal information during the conversion of 3D information into 2D information. For example, Wang et al. [40] encode joint trajectories into texture images and utilized HSV (Hue, Saturation, Value) space to represent the temporal information. However, this method cannot distinguish some actions such as "knock" and "catch" due to the trajectory overlapping and the loss of past temporal information. Li et al. [24] propose to encode the pairwise distances of skeleton joints into texture images and encoded the pairwise distances between joints over skeleton motion sequences into color variations to capture temporal information. But this method cannot distinguish some actions of similar distance variations such as "draw-circle-clockwise" and "draw-circle-counterclockwise" due to the loss of local temporal information.

Most of these approaches transform motion capture sequences into the 2D space and develop CNN architectures for classification. However, CNNs are very successful in image classification and a lot of pre-trained models exist and can be adapted for other tasks using fine-tuning and transfer learning techniques. These methods are very successful in many tasks such as, and in addition to image classification [13], object detection [19], person re-identification [11], biomedical image analysis [5], etc. Our proposed approach consists of transforming motion capture sequences into a simple spatio-temporal image-like representation. Then, fine-tuning of different successful state-of-the-art image classification models is applied for our task of skeleton-based action recognition.

3   Background

In this section, we cover the necessary materials for the rest of the paper. We first review Convolutional Neural Networks (CNN) and different techniques used for fine-tuning these models. Then, we propose a method to transform motion capture sequences into images. These images are fed into classical CNN models, and by the process of fine-tuning we adapt these models to our data.

3.1   Review of Convolutional Neural Networks (CNNs)

CNNs are biologically inspired models for computer vision. They have been applied for several image classification tasks such as face identification, object recognition, tumors detection and classification, etc. The architecture of a Convolutional Neural Network consists of various layers between the input and the output. Convolutional layers are the core building block of a convolutional network. They comprise of a list of n^*n filters. They are generally followed by Pooling layers that progressively reduce the spatial size of the representation and reduce the number of parameters. At the end of a CNN, there is generally a "Fully-connected" layer, similar to a multi-layer perceptron (MLP), followed by a classification layer. The classification layer can be implemented as a general-purpose classifier (for example SVM), but generally a SoftMax function is used which transforms the final vector values into probability scores between zero and one.

CNN models are trained using backpropagation algorithm. This algorithm aims to minimize a cost function that measures the total error of the model on the training dataset. Backpropagation is used to compute the gradients of the error with respect to all the weights in the network and gradient descent algorithm updates all parameters to minimize the output error. [3] gives technical details about backpropagation and gradient descent algorithms. In many situations, the amount of training data is not sufficient to adjust all parameters which causes an overfitting. In this case, transfer learning and fine-tuning are used. In these two processes, an initial step of model pre-training is used. It consists of training the CNN model on a large dataset, like ImageNet, before training on the desired dataset, and initial weights are obtained. These weights are used instead of random once and a transfer learning or fine-tuning is then applied.

- Transfer-learning (or shallow retraining) consists in freezing the pre-trained weights and only the parameters of the last classification layers need to be inferred from scratch using the new training set. This is very useful if the pre-training dataset and the final dataset are similar.
- Fine-tuning (or deep retraining) consists in freezing only few first layers or no layer, depending on the similarity of the original and final datasets, and updating all the other parameters.

In this work, six architectures (AlexNet [21], InceptionV3 [35], VGGNet [32], ResNet [14], DenseNet [16] and SqueezeNet [18]) are used with their other varieties. In total, 12 pre-trained models are used and modified to fit our classification problem (i.e. changing the last fully connected layer and the classification layer to fit the number of action classes. We compare both strategies (shallow and deep retaining) in addition to training the models from scratch (i.e. with random values).

3.2   Data Representation

In this work, we transform motion sequences, consisting of 3D joint coordinates, into RGB images to be able to use CNN models pre-trained for image classification, following the approach proposed by [22]. This approach is based on a simple transformation in which joints coordinates are normalized between 0 and 255. The normalization is done using the formula 1.

$$
\begin{cases}
r_i(f) = 255 * \dfrac{(x_i(f) - min(X)}{(max(X) - min(X)} \\[2ex]
g_i(f) = 255 * \dfrac{(y_i(f) - min(Y)}{(max(Y) - min(Y)} \\[2ex]
b_i(f) = 255 * \dfrac{(z_i(f) - min(Z)}{(max(Z) - min(Z)}
\end{cases}
\tag{1}
$$

where:
$(x_i(f), y_i(f), z_i(f))$ are the 3D coordinates of joint i at frame f.
$min(X), min(Y), min(Z)$ are the minimum values for all joints on the 3 axes.
$max(X), max(Y), max(Z)$ are the maximum values for all joints on the 3 axes.
$(r_i(f), g_i(f), b_i(f))$ are hence the normalized values of every joint i at the frame f, where each value corresponds to the format of one channel of the RGB space.

By stacking the normalized values of all joints on all frames we obtain a long image-like representation of the motion sequence where rows correspond to joints and columns correspond to frames. The image is resized to a square representation to be easily used by different CNN architectures. Figure 1 shows some results from the NTU RGB+D dataset. The colored triangle on the bottom right of the figure can be considered as a dictionary to interpret these images. The red color represents the X coordinate, the green color represents the Y coordinate and the blue color represents the Z coordinate. For example, the greener the part of an image from left to right means that the Y value is getting higher. We can see for example at the images corresponding to the action "Throw" the lines on the top and in the middle (corresponding to right and left arms joints) are getting in particular more of the green color at a certain part of the image then it comes back to the original color. This is the translation of the arms moving up to throw the object before going down again.

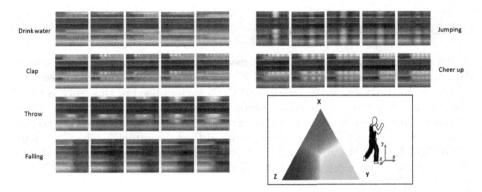

Fig. 1. Examples of motion capture sequences represented into RGB images. (Color figure online)

In the images corresponding to "Falling", we can see, starting from the middle, that the images are getting redder to purple. This can be explained by the fact that, when the person falls, all the joints Y values are getting smaller (hence less green). In order to be invariant to global position, all 3D joint coordinates are normalized with respect to the "SpineBase" joint (the most stable joint from

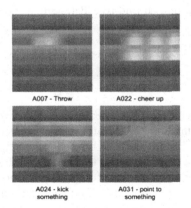

Fig. 2. Examples of the motion capture sequences transformed into images after pre-processing.

the Kinect 2). 3D joint coordinates are hence relative coordinates. The order of the joints is also modified in order to have more present visual patterns, and 4 very noisy joints have been discarded (thumbs and hand tips joints). The final order of joints is the following: 3, 2, 20, 1, 0, 8, 9, 10, 11, 4, 5, 6, 7, 16, 17, 18, 19, 12, 13, 14, 15 (the joints' numbers are shown in Fig. 4.). Figure 2 shows some examples of the final representation. These images will be fed to multiple CNN models to select the best one in terms of accuracy.

4   Experimental Results

In this section, the proposed method is evaluated on the NTU RGB+D dataset. We evaluate different state-of-the-art CNN architectures using different training strategies An overview of the end-to-end CNN system used in this work is illustrated in Fig. 3. After the skeletons are pre-processed, motion capture sequences are transformed into image data representations. These images are fed into different CNN architectures to be trained using different strategies. The CNN part of our models generate automatically features (feature maps) that are fed into a fully connected network for classification.

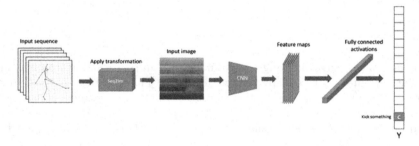

Fig. 3. Illustration of the proposed End-to-end system for skeleton-based action recognition.

4.1   Data Structure

In order to compare our results with existing works, we evaluate our method on the NTU RGB+D dataset. This dataset was collected using three Kinect V2 sensors at the same time covering three views (45°, 0°, 45°) and contains more than 56,000 action sequences. A total of 60 different action classes are performed by 40 subjects aged from 10 to 35 years. Among these action classes, 49 are performed by single persons and 11 are interactions between two people. Only the 49 single person actions were used in our tests. In addition to depth maps, RGB frames, and infrared (IR) sequences, information of 25 3D joints are available. This dataset is challenging because of the large intraclass and viewpoint variations; however, due to its large scale, it is highly suitable for deep learning. The captured Kinect V2 skeletons have 25 joints in total. The configuration and the given order of joints is shown in Fig. 4.

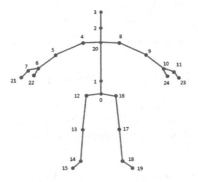

Fig. 4. Configuration of the skeleton given by the Kinect V2.

4.2   Training the Deep Network

In this experiment, 12 state-of-the-art architectures (AlexNet, Inception V3, VGG11, VGG16, VGG19, ResNet34, ResNet50, ResNet152, DenseNet121, DenseNet169, DenseNet201 and SqueezeNet) are trained on the dataset described in the previous section. The numbers after the architectures' names represent the number of layers. This will allow us also to analyze the effect of the depth of the models as well. We trained these architectures using three different strategies. The first strategy consists in training the CNN from scratch starting from random weights. The other two strategies are based on transfer learning using pre-trained networks. The first transfer learning approach, called shallow retraining approach, consists in fine-tuning only the last added fully connected layer, while the rest of the network is used as feature extractor. The second approach, called deep retraining approach, fine-tunes all the network layers. All the 36 training configurations use the same hyperparameters (momentum 0.9, weight decay 0.0005, learning rate 0.001, batch size 30 and a number of epochs of 15). The dataset is divided following two evaluation protocols:

- Cross-subject evaluation: the 40 subjects are split into training and testing groups. Each group consists of 20 subjects.
- Cross-view evaluation: the samples of one camera (corresponding to one viewpoint) are used for testing and samples of the two other cameras are used for training.

All experiments are performed on a machine having the specifications summarized in Table 1.

Table 1. Training and testing machine specifications.

	Characteristics
Memory (RAM)	32 GB
Processor (CPU)	Intel® Core™ i7-7800X CPU @ 3.50 GHz 12
Graphics (GPU)	2 * GeForce GTX 1080 Ti/PCIe/SSE2 (11 GB)
Operating system	Linux Ubuntu 16.04 64 bits

4.3 Models Evaluation: Results and Discussion

In this section, we discuss the results obtained for different experiments. The accuracy and the training time for all the 12 models are shown in Tables 2 and 3 for cross-subject and cross-view evaluation protocols respectively.

Cross-Subject Evaluation Protocol. From Table 2, we get the highest scores of accuracy with the models ResNet50 and DenseNet201, around 82%. This accuracy is obtained using the deep retraining approach. Analyzing this table, we notice that the lowest scores are obtained using the shallow retraining approach. Such low scores are linked to the fact that we retrain only the last added layer and we keep the rest of the model untouched. The convolutional layers in this situation are used as feature extractors. This can work and give good results in cases where we have images that are similar to the original dataset that was used to pre-train the model (ImageNet). However, our images are more abstract and completely different from the original dataset. The features extracted are hence meaningless in regard to our data.

Table 2. Comparison of different proposed models for cross-subject evaluation.

Model	From scratch		Retrain shallow		Retrain deep	
	Time (h)	Acc	Time (h)	Acc	Time (h)	Acc
AlexNet	1.17	0.6544	1.09	0.2371	1.76	0.7319
InceptionV3	1.78	0.6607	0.61	0.2524	1.68	0.7953
VGG11	2.33	0.6841	1.95	0.2627	3.29	0.7691
VGG16	2.82	0.6674	2.15	0.2696	3.87	0.7760
VGG19	3.07	0.6490	2.27	0.2200	3.70	0.7514
ResNet34	1.05	0.7202	0.85	0.3768	1.05	0.8020
ResNet50	1.31	0.6809	0.97	0.3835	1.30	0.8207
ResNet152	2.45	0.6756	2.09	0.3910	2.43	0.8118
DensNet121	1.29	0.7468	1.53	0.4228	1.28	0.8096
DensNet169	1.56	0.7610	1.49	0.4380	1.53	0.8172
DeseNet201	1.86	0.7622	1.84	0.4422	1.82	0.8200
SqueezeNet	0.61	0.6573	0.55	0.3145	0.62	0.7209

We can notice relatively higher scores, compared to the shallow retraining, in the case of the retraining from scratch (from random weights). Retraining the whole model allowed it to develop features that are adapted to our data. Nevertheless, the scores are not as high as the deep retraining approached because retraining from scratch may need more data and more time to converge. Transfer learning using deep retraining allowed a quicker convergence. We can also conclude that more layers do not automatically mean higher accuracy. It can be the opposite in many cases like VGG and ResNet. VGG architecture with 19 layers for example gives lower accuracy than the one with 11 and 16 layers in a deep retraining approach. This can be explained by the fact that the more layers we have, the more risk we have of overfitting. We can finally notice that SqueezeNet gives relatively high scores, particularly in a deep retraining approach. SqueezeNet is a very small network with few parameters. It has a total size of less than 0.5 MB. Compared to AlexNet for example that has a total size of 240 Mb, it makes it easy and practical to fit into embedded systems and smartphones. Evaluation of the models' performances comparing training time and accuracy is displayed in Fig. 5.

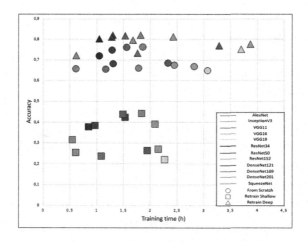

Fig. 5. Cross-subject evaluation: accuracy vs. training time (h).

Cross-View Evaluation Protocol. From Table 3, we get the highest scores in the "Cross-View" evaluation protocol with the models ResNet34, ResNet152 and DenseNet201 of about 86%. The exact same conclusions as in "Cross-Subject" evaluation can be drawn.

Table 3. Comparison of different proposed models for cross-view evaluation.

Model	From scratch		Retrain shallow		Retrain deep	
	Time (h)	Acc	Time (h)	Acc	Time (h)	Acc
AlexNet	0.92	0.6400	1.18	0.2267	1.15	0.7486
InceptionV3	1.73	0.6482	0.61	0.2573	1.54	0.8046
VGG11	1.75	0.7573	2.13	0.2693	2.27	0.8144
VGG16	2.23	0.7635	2.18	0.2530	2.75	0.8269
VGG19	2.50	0.7643	2.25	0.2160	2.99	0.8001
ResNet34	1.01	0.6997	0.83	0.3672	1.03	0.8600
ResNet50	1.27	0.6457	0.96	0.3774	1.29	0.8592
ResNet152	2.67	0.7375	1.61	0.3906	2.38	0.8611
DensNet121	1.30	0.7877	0.94	0.4104	1.27	0.8550
DensNet169	1.64	0.7767	1.10	0.4302	1.51	0.8489
DeseNet201	1.95	0.7924	1.27	0.4422	1.80	0.8654
SqueezeNet	0.4	0.7021	0.57	0.3065	0.62	0.7760

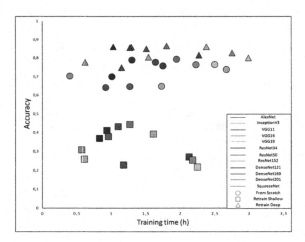

Fig. 6. Cross-view evaluation: accuracy vs. training time (h).

Evaluation of the models' performances by comparing training time and accuracy is also displayed in Fig. 6, which is similar to the previous evaluation scenario (cross-subject). The obtained results can be compared to several works in the state of the art that used the same dataset (Table 4). The highest score we obtained is 82,07% using ResNet50 in the cross-subject evaluation protocol and 86,54% using DenseNet201 in the cross-view evaluation protocol. Our results outperform most of state-of-the-art methods in both cross-subject and cross-view protocols. Our results are obtained by transforming motion sequences

Table 4. Comparison of different proposed models for cross-view evaluation.

Method	Cross-subject	Cross-view
Using CNNs		
two streams 3DCNN [26]	66,85%	72,58%
conversion into image + CNN [20]	79,57%	84,83%
trajectories maps + CNN [40]	76,32%	81,08%
conversion into image + CNN [23]	75,20%	82,10%
conversion into image + CNN [22]	74,27%	75,74%
conversion into image + CNN [7]	**83.2%**	**89.3%**
Using other DL methods		
HBRNN [9]	59,07%	63,97%
Deep RNN [31]	56.29%	64.09%
Deep LSTM [31]	60.69%	67.29%
PA-LSTM [31]	62.93%	70.27%
LieNet [17]	61.37%	66.95%
ST-LSTM [27]	69.20%	77.70%
Our method	**82.07%**	**86.54%**

into images and using pre-trained models without developing new architectures. Specifically, we improve accuracy by 3 to 8% compared to other techniques using CNNs with image-like transformation of motion sequences. This proves that adapting the weights from pre-trained models for a new task improves the performance of deep learning networks and gives high scores even if the original task is very different.

The work done by Li et al. [7] have significantly higher results. Authors have developed a new architecture dedicated for this task, using two-stream CNNs and achieved 83.2% of accuracy for cross-subject evaluation and 89.3% for cross-view evaluation.

5 Conclusion

In this work, we studied the use of pre-trained convolutional neural networks for skeleton-based action recognition. In order to handle the high dimensionality of skeleton sequences, we transformed them into RGB images so they can be fed directly to different CNN models that are designed for image classification. We exploited different state-of-the-art pre-trained architectures and used a process called "fine-tuning" to adapt the weights to our task. Even though the obtained images are abstract and completely different from original dataset (ImageNet) used to pre-train different models, we achieved high classification scores and outperformed most of action classification state-of-the-art results on the NTU RGB+D dataset. This proves that, firstly, the image-like representation of motion capture skeleton sequences can be well interpreted by different CNN architectures. This also proves that using already trained classification models helps in other classification tasks and accelerates the model's convergence. Nonetheless, that are still quite a few open issues. For example, studying the use of rotation in addition to 3D joint locations and exploiting other pre-trained architectures that use both CNNs and RNNs for our task. Besides, building a CNN architecture for our specific task, pre-training it on different motion capture datasets, then apply fine-tuning for the desired dataset, could be more effective. Last but not least, studying the behavior of different CNN models using visualization techniques to set the most optimized parameters would also be an interesting direction.

References

1. Adel, H., Schütze, H.: Exploring different dimensions of attention for uncertainty detection. arXiv preprint arXiv:1612.06549 (2016)
2. Bai, S., Kolter, J.Z., Koltun, V.: An empirical evaluation of generic convolutional and recurrent networks for sequence modeling. arXiv preprint arXiv:1803.01271 (2018)
3. Bengio, Y., Goodfellow, I., Courville, A.: Deep learning, vol. 1. Citeseer (2017)
4. Bengio, Y., Simard, P., Frasconi, P., et al.: Learning long-term dependencies with gradient descent is difficult. IEEE Trans. Neural Netw. **5**(2), 157–166 (1994)
5. Broadwater, D.R., Smith, N.E.: A fine-tuned inception v3 constitutional neural network (CNN) architecture accurately distinguishes between benign and malignant breast histology. Technical report, 59 MDW San Antonio United States (2018)
6. Cho, K., Van Merriënboer, B., Bahdanau, D., Bengio, Y.: On the properties of neural machine translation: encoder-decoder approaches. arXiv preprint arXiv:1409.1259 (2014)
7. Du, Y., Fu, Y., Wang, L.: Skeleton based action recognition with convolutional neural network. In: 2015 3rd IAPR Asian Conference on Pattern Recognition (ACPR), pp. 579–583. IEEE (2015)
8. Du, Y., Fu, Y., Wang, L.: Representation learning of temporal dynamics for skeleton-based action recognition. IEEE Trans. Image Process. **25**(7), 3010–3022 (2016)
9. Du, Y., Wang, W., Wang, L.: Hierarchical recurrent neural network for skeleton based action recognition. In: Proceedings of the IEEE Conference on Computer Vision and Pattern Recognition, pp. 1110–1118 (2015)
10. Efros, A.A., Berg, A.C., Mori, G., Malik, J.: Recognizing action at a distance. In: Null, p. 726. IEEE (2003)
11. Fan, H., Zheng, L., Yan, C., Yang, Y.: Unsupervised person re-identification: clustering and fine-tuning. ACM Trans. Multimedia Comput. Commun. Appl. (TOMM) **14**(4), 83 (2018)
12. Gehring, J., Auli, M., Grangier, D., Dauphin, Y.N.: A convolutional encoder model for neural machine translation. arXiv preprint arXiv:1611.02344 (2016)
13. Han, D., Liu, Q., Fan, W.: A new image classification method using CNN transfer learning and web data augmentation. Expert Syst. Appl. **95**, 43–56 (2018)
14. He, K., Zhang, X., Ren, S., Sun, J.: Deep residual learning for image recognition. In: Proceedings of the IEEE Conference on Computer Vision and Pattern Recognition, pp. 770–778 (2016)

15. Hochreiter, S., Schmidhuber, J.: Long short-term memory. Neural Comput. **9**(8), 1735–1780 (1997)
16. Huang, G., Liu, Z., Van Der Maaten, L., Weinberger, K.Q.: Densely connected convolutional networks. In: Proceedings of the IEEE Conference on Computer Vision and Pattern Recognition, pp. 4700–4708 (2017)
17. Huang, Z., Wan, C., Probst, T., Van Gool, L.: Deep learning on lie groups for skeleton-based action recognition. In: Proceedings of the IEEE Conference on Computer Vision and Pattern Recognition, pp. 6099–6108 (2017)
18. Iandola, F.N., Han, S., Moskewicz, M.W., Ashraf, K., Dally, W.J., Keutzer, K.: SqueezeNet: AlexNet-level accuracy with 50x fewer parameters and <0.5 MB model size. arXiv preprint arXiv:1602.07360 (2016)
19. Kang, K., et al.: T-CNN: tubelets with convolutional neural networks for object detection from videos. IEEE Trans. Circ. Syst. Video Technol. **28**(10), 2896–2907 (2017)
20. Ke, Q., Bennamoun, M., An, S., Sohel, F., Boussaid, F.: A new representation of skeleton sequences for 3D action recognition. In: Proceedings of the IEEE Conference on Computer Vision and Pattern Recognition, pp. 3288–3297 (2017)
21. Krizhevsky, A., Sutskever, I., Hinton, G.E.: ImageNet classification with deep convolutional neural networks. In: Advances in Neural Information Processing Systems, pp. 1097–1105 (2012)
22. Laraba, S., Brahimi, M., Tilmanne, J., Dutoit, T.: 3D skeleton-based action recognition by representing motion capture sequences as 2D-RGB images. Comput. Anim. Virtual Worlds **28**(3–4), e1782 (2017)
23. Li, C., Sun, S., Min, X., Lin, W., Nie, B., Zhang, X.: End-to-end learning of deep convolutional neural network for 3D human action recognition. In: 2017 IEEE International Conference on Multimedia & Expo Workshops (ICMEW), pp. 609–612. IEEE (2017)
24. Li, C., Hou, Y., Wang, P., Li, W.: Joint distance maps based action recognition with convolutional neural networks. IEEE Signal Process. Lett. **24**(5), 624–628 (2017)
25. Li, Q., Qiu, Z., Yao, T., Mei, T., Rui, Y., Luo, J.: Action recognition by learning deep multi-granular spatio-temporal video representation. In: Proceedings of the 2016 ACM on International Conference on Multimedia Retrieval, pp. 159–166. ACM (2016)
26. Liu, H., Tu, J., Liu, M.: Two-stream 3D convolutional neural network for skeleton-based action recognition. arXiv preprint arXiv:1705.08106 (2017)
27. Liu, J., Shahroudy, A., Xu, D., Wang, G.: Spatio-temporal LSTM with trust gates for 3D human action recognition. In: Leibe, B., Matas, J., Sebe, N., Welling, M. (eds.) ECCV 2016. LNCS, vol. 9907, pp. 816–833. Springer, Cham (2016). https://doi.org/10.1007/978-3-319-46487-9_50
28. Müller, M.: Information Retrieval for Music and Motion, vol. 2. Springer, Heidelberg (2007). https://doi.org/10.1007/978-3-540-74048-3
29. Ohn-Bar, E., Trivedi, M.: Joint angles similarities and HOG2 for action recognition. In: Proceedings of the IEEE Conference on Computer Vision and Pattern Recognition Workshops, pp. 465–470 (2013)
30. Pascanu, R., Mikolov, T., Bengio, Y.: On the difficulty of training recurrent neural networks. In: International Conference on Machine Learning, pp. 1310–1318 (2013)
31. Shahroudy, A., Liu, J., Ng, T.T., Wang, G.: NTU RGB+ D: a large scale dataset for 3D human activity analysis. In: Proceedings of the IEEE Conference on Computer Vision and Pattern Recognition, pp. 1010–1019 (2016)

32. Simonyan, K., Zisserman, A.: Very deep convolutional networks for large-scale image recognition. arXiv preprint arXiv:1409.1556 (2014)
33. Song, S., Lan, C., Xing, J., Zeng, W., Liu, J.: An end-to-end spatio-temporal attention model for human action recognition from skeleton data. In: Thirty-first AAAI Conference on Artificial Intelligence (2017)
34. Srivastava, N., Mansimov, E., Salakhudinov, R.: Unsupervised learning of video representations using LSTMs. In: International Conference on Machine Learning, pp. 843–852 (2015)
35. Szegedy, C., Vanhoucke, V., Ioffe, S., Shlens, J., Wojna, Z.: Rethinking the inception architecture for computer vision. In: Proceedings of the IEEE Conference on Computer Vision and Pattern Recognition, pp. 2818–2826 (2016)
36. Veeriah, V., Zhuang, N., Qi, G.J.: Differential recurrent neural networks for action recognition. In: Proceedings of the IEEE International Conference on Computer Vision, pp. 4041–4049 (2015)
37. Vemulapalli, R., Arrate, F., Chellappa, R.: Human action recognition by representing 3D skeletons as points in a lie group. In: Proceedings of the IEEE Conference on Computer Vision and Pattern Recognition, pp. 588–595 (2014)
38. Vinyals, O., Toshev, A., Bengio, S., Erhan, D.: Show and tell: a neural image caption generator. In: Proceedings of the IEEE Conference on Computer Vision and Pattern Recognition, pp. 3156–3164 (2015)
39. Wang, P., Li, W., Li, C., Hou, Y.: Action recognition based on joint trajectory maps with convolutional neural networks. Knowl.-Based Syst. **158**, 43–53 (2018)
40. Wang, P., Li, Z., Hou, Y., Li, W.: Action recognition based on joint trajectory maps using convolutional neural networks. In: Proceedings of the 24th ACM International Conference on Multimedia, pp. 102–106. ACM (2016)
41. Xu, K., et al.: Show, attend and tell: neural image caption generation with visual attention. In: International Conference on Machine Learning, pp. 2048–2057 (2015)
42. Zhang, Y., Pezeshki, M., Brakel, P., Zhang, S., Bengio, C.L.Y., Courville, A.: Towards end-to-end speech recognition with deep convolutional neural networks. arXiv preprint arXiv:1701.02720 (2017)

Workshop on: Cognitive and Computer Vision Assisted Systems for Energy Awareness and Behavior Analysis

An Augmented Reality Game for Energy Awareness

Piero Fraternali and Sergio Luis Herrera Gonzalez[✉]

Dipartimento di Elettronica, Informazione e Bioingegneria, Politecnico di Milano,
Piazza Leonardo da Vinci 32, Milan, Italy
{piero.fraternali,sergioluis.herrera}@polimi.it

Abstract. Energy efficiency requires a behavioral shift towards sustainable consumption. Such a change can be supported by persuasive IT applications, which employ a variety of stimuli to increase the energy literacy and awareness of consumers. We describe FunergyAR, an Augmented Reality digital game targeting children and their families. FunergyAR incorporates Computer Vision and Augmented Reality components within traditional game mechanics and can be used either in a standalone manner or together with Funergy, a card game designed for improving energy savvy behaviors in children.

Keywords: Augmented Reality · Serious games · Energy awareness

1 Introduction

Consuming natural resources responsibly is a key objective for achieving the 2030 UN Sustainable Development Goals [1], which aim at ensuring access to affordable, reliable, sustainable and modern energy and at fostering sustainable consumption and production patterns.

It is now commonly understood that achieving the objective of more sustainable energy (and in general resource) consumption is a multi-faceted effort, which requires the joint use of complementary stimuli, including economic incentives, education towards efficiency, social pressure, and elicitation of personal values. One promising approach to dispatch such stimuli is the use of persuasive digital applications, especially mobile applications, which have the potential of impacting the daily habits of users and thus of nudging their behavior towards sustainability goals. Within the broad spectrum of persuasive IT applications, Games with a Purpose (GWAPs) have emerged as a potentially effective approach. A GWAP exploits the well-proved engagement power of games to attain a non-gaming ("serious") objective. However, designing a GWAP that can effectively attract, engage and retain people, especially young players, who are the most promising target of a behavioral change effort, remains a challenge. GWAPs must face the competition of a menagerie of commercial, professionally developed games, which target their audience with enormous marketing investments.

© Springer Nature Switzerland AG 2019
D. Tzovaras et al. (Eds.): ICVS 2019, LNCS 11754, pp. 629–638, 2019.
https://doi.org/10.1007/978-3-030-34995-0_57

Therefore, it is imperative that GWAPs, which are mostly developed for non-commercial purposes, find an alternative way to reach their audience, based on an original mix of technology, game mechanics, and content.

In this paper, we describe the design of FunergyAR, a digital game for promoting energy efficient behavior in kids, which aims at finding a new way to attract, engage and retain players, by exploiting an original mix of technology (Augmented Reality, boosted by Deep Learning Computer Vision methods for real time object recognition on low-power devices), game mechanics (a digital game with a well-proved mechanics coupled to a "traditional" card game), and content (collaboratively developed energy quizzes and saving tips).

The paper is organized as follows: Sect. 2 surveys the related work and provides the background of the project; Sect. 3 illustrates the design of FunergyAR; Sect. 4 describes the evaluation plan of the game and discusses the ongoing and future work.

2   Related Work and Background

2.1   Games for Energy Awareness and Sustainability

Games have been applied in different ways for environmental awareness, their motivational and engaging characteristics are key elements to retain user attention while delivering information, e.g., about energy saving. An example of such games is Power House, an online game in which the player must assist the other characters in their day-to-day activities. The player oversees turning on and off appliances (lights, TV set, computer, etc.) and keeps track of the activities of every member of the family to reduce waste. The point system is based on the ability to minimize the amount of electricity consumed by the family [2]. A similar approach is used in ecoPet [3], where the player must take care of a virtual pet's needs in a energy efficient way. Social Power [4] is another mobile game application blending social interactions and game mechanics to steer people towards long term behaviour changes in energy consumption. When a player registers to the game she is automatically assigned to a team, and given a set of individual and collective challenges, such as energy saving goals for which progress can be tracked. Some tasks require coordination with the team members. The application delivers information about how to make efficient use of the energy in the households and in shared spaces, such as schools and libraries. The users receive points and badges for their achievements and contributions to the teamwork. EnerGAware [5] is a mobile simulation game in which the objective is to reduce the energy consumption of a virtual house with respect to previous week. The player can execute actions such as changing the location of the lamps in a room or turning off appliances, or actions that are specific for a given time, such as during the World Environment Day. At the end of every week, the players receive points based on the energy saved. Virtual money can be used to buy more efficient devices and continue saving energy. The players can visit their neighbours' houses to help those that are less efficient or to learn from those that are better at saving energy.

FunergyAR differs in the use of Augmented Reality to engage players in the exploration of the surrounding world, in search of energy-related tips and knowledge collectively produced by the community of participating schools, and in the possibility of playing the game in an integrated manner with the Funergy non-digital, card game.

2.2 Augmented Reality Applications for Consumers

Mobile augmented reality (Mobile AR) has increased its popularity thanks to the emergence of consumer grade AR-enabled devices and to the possibility of using AR apps also on plain mobile phones. Also the offer of frameworks for building AR browsers and applications is growing, including such platforms as Google ARCore [6], Wikitude [7], and blippAR [8]. The survey in [9] reviews AR applications for maintenance and assembly in several industries: aviation, mechanical maintenance, consumer technology and production plants. It illustrates the most popular technologies, among which mobile apps total the 30% of the reviewed solutions, and pinpoints the most relevant technical limitations. The retail industry has also started exploring AR. Examples include Ikea Place [10] and Amazon Home [11], which enable the users to visualize furniture catalogues and virtually place pieces on their room space. They also provide identification features for searching products that are similar to the furniture identified in pictures taken with the app. Several AR applications for tourism have been developed. Specifically for trekking tourism apps such as Peaklens [12] provide a location-based mobile outdoor solution that provides users with information (name, altitude, etc.) about the mountains they are looking at through the camera. PeakLens exploit a computer vision module, the location and orientation sensors of the phone, and the Digital Elevation Model (DEM) of the Earth to estimate the visible panorama from the user viewpoint and retrieve the information about the mountains in view. The fashion industry exploits AR and computer vision to let customers virtually try makeup [13], clothes [14] and shoes [15].

FunergyAR applies the AR paradigm to the design of a GWAP for energy awareness and literacy, exploiting two popular mechanics (treasure hunt like search and trivia quizzes) to engage young players towards sustainability.

3   The FunergyAR Augmented Reality Game

FunergyAR is an Augmented Reality game for low-power devices, which exploits a well-proved game mechanics, trivia quizzes, to deliver content aimed at promoting energy literacy and awareness and suggest practical ways to save energy.

The design of FunergyAR is based on the following principles:

- The game should leverage an emerging technology trend, Augmented Reality boosted by object recognition tools on mobile devices, to attract the attention of millennials and stand up within the myriad of digital games available.

- The game should be playable in a standalone manner, as any other digital game, or in a hybrid configuration, coupled to a non-digital game, to leverage the increasing interest and market growth of tabletop and card games [16].
- Support should be given not only to the single-player mode, but also to a collaboration-competition interaction by stakeholders, e.g., by schools.

3.1   Game Mechanics

FunergyAR blends the Augmented Reality interaction paradigm with the trivia quiz game mechanics. The idea is to let players, typically school pupils in the 6–12 age range, to explore the world surrounding them and "query the objects" for hints on how to save energy and for knowledge tips about energy in general.

Figure 1 shows the home page of the game, which comprises a menu listing two play options: the former (*Take a picture*) starts the AR exploration game play, whereas the latter (*Single Player*) permits the user to use the game in a standalone fashion, by responding to a sequence of quizzes with increasing complexity.

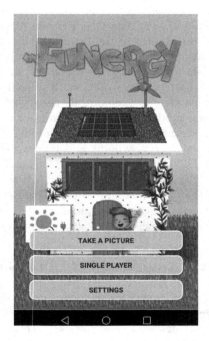

Fig. 1. The home page shows the two game modes of the game: take a picture and single player

Selecting the *Take a picture* option opens the Augmented Reality interface, shown in Fig. 2(a). The player frames the environment around him in the camera view and the Computer Vision module of the game detects objects and displays their class at the bottom of the screen; when some energy quiz is associated to a detected object, a *Continue* button appears, whereby the user can obtain an energy quiz related to the framed object type. Quizzes are binary (Yes/No, true/false) questions, delivered with an increasing level of complexity along the game play session (Fig. 2(b)). After responding to the quiz, the game displays the outcome of the response (see Fig. 3(a)) and the user can ask for an explanation, which is a short text with information on the topic of the question (Fig. 3(b)).

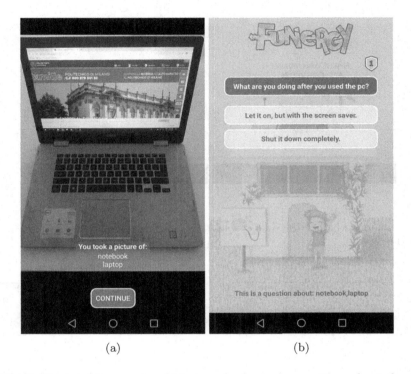

(a)   (b)

Fig. 2. FunergyAR recognizes objects and asks energy questions about them

3.2 Game Implementation

FunergyAR is implemented as a client side app for Android devices, interacting with a server-side backed that exposed REST services for content retrieval. The game is coupled to a gamified web application, whereby schools can cooperate and compete in the creation of content for the game. Figure 4 shows the overall architecture of FunergyAR and of the companion Content Management System.

The Android client application uses the Camera 2 framework to handle the camera sensor. Images are captured continuously at a 640 × 480 resolution to reduce the processing time. When a frame is delivered from the camera

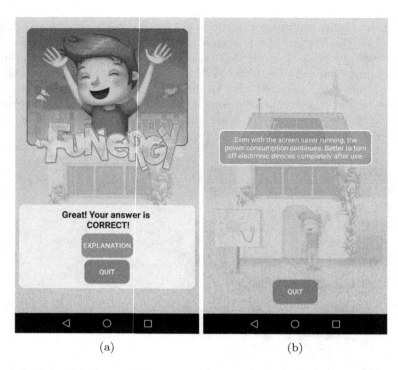

(a)   (b)

Fig. 3. Funergy whether the answer was right or wrong, and a concise explanation about the topic.

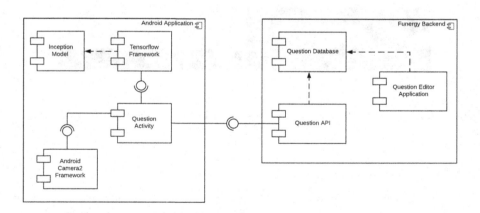

Fig. 4. The architecture of FunergyAR

to the application, it is rotated depending on the device and camera orientation, because the object detection model performs better on vertical images. Each frame is processed with the TensorFlow[1] Image Classifier Framework. The model used for object detection is the Inception model [17][2] version 3, trained with 1000+ classes. The classifier returns the classes with the highest confidence and the application discards the classes with confidence under a threshold and retains the top 3 classes. Then, the application queries the back-end for a question related to the identified classes. The question retrieval API is a server-side component that exposes a set of REST services to retrieve questions based on parameters, such as the level of difficulty, the language, the related classes, etc. The last component is the Funergy content management system, a gamified web application that enables the users to create, edit and translate questions to be used in the game. The users climb positions on a leader board depending on the question that they have created and translated. The registration to the system is done on a group basis, to encourage collaboration among the member of a school class and convey the message that energy saving is a collective effort.

3.3   Coupling FunergyAR with a Traditional Card Game

FunergyAR can be used in conjunction with the Funergy card game, to attain a hybrid game play that mixes the paradigms of card games and of digital games. Funergy (shown in Fig. 5) is a card game aimed at explaining the value of the European Energy efficiency scale. The game is divided into seven rounds, one for every Energy Scale Level. The game begins with the G level (the lowest one) and finishes when players reach the A level (the highest one). At the beginning of the game each player receives seven cards and the rest of the cards form the drawing deck. The objective is to form a combination of cards numbered from 1 to 7 by drawing from the deck and by exchanging cards with other players. The first player completing the right sequence is the winner of the round. For every round, there is a small pack of 5 cards showing a piggy-bank with an increasing value. The winner of the round takes the highest card and distributes the remaining ones to the other players. The values of the piggy-bank increase level by level, so winning the last one can be crucial for determining the final winner. The game contains bad card, showing old appliances, which must be discarded for closing the hand; it also exploits wild cards, which can replace any number and help completing the sequence. At the end of the rounds, all players sum up the points of their Energy Scale Cards and add 3 points for every wild card. The winner is the player with the highest score.

FunergyAR is used to "unlock" the value of a wild card. For the player to retain her wild card, she must locate an object of the type chosen by the other players, scan it, and respond correctly to the question attached to it. If a wrong response is given, the wild card and its associated points must be "donated" to a competitor player.

---

[1] https://www.tensorflow.org/.
[2] https://github.com/tensorflow/models/tree/master/research/inception/inception.

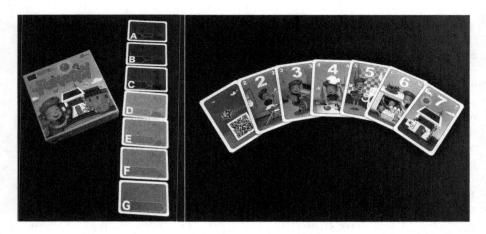

Fig. 5. The Funergy card game: box, Energy Scale score cards (the piggy-bank is on the reverse side of the card) and game play cards

The interplay between the non-digital and the digital part of the game is designed carefully. The use of the FunergyAR app is non-intrusive with respect to the flow of the card game, but at the same time increases the fun of the play by exploiting a treasure hunt like search and a trivia quiz game mechanics.

4   Experience, Discussion and Conclusions

FunergyAR has been developed in collaboration with several schools in Italy and Switzerland, where requirement elicitation has been conducted to understand from teachers the best approach to integrate the game in the education programs and to engage pupils and their families.

A key design choice to better integrate the game within the educational programs of schools about sustainability is to open the content of the game to the classes, who can submit their own questions, associate them with the object classes recognized by FunergyAR, and thus compete to become "top contributors" of the game. To support this feature, a Web application has been developed, whereby classes can register and input their questions and explanations[3]. The application records the number of quizzes that have been accepted by the editorial board of the game and assign scores to the contributing classes, who can check their status in a leader board.

The evaluation of the impact of Funergy and FunergyAR will be assessed by a structured activity with schools. The idea is to verify if and how playing the energy game changes kids' (and family's) energy saving knowledge and attitude. The intervention in schools is structured in a sequence of activities: (1) an initial questionnaire is distributed for determining knowledge and attitude before the

---

[3] The application is published at: http://funergy.ifmledit.org/funergy.

experiment; (2) an activity involving the game is performed; (3) a final questionnaire is gathered, for determining if and how knowledge and attitude have changed after the experiment. To better understand the role of the game play and of the collateral education activities performed in the school, the treatment consists of alternative activities, performed in distinct classes:

- None: only the initial and final questionnaires are filled in; this "no activity" helps us understanding the difference made by the game.
- Playing: the activity consists of a visit to the school by researchers, an explanation of energy sustainability issues and a round of the game play (2 h). Pupils are then assigned a homework, consisting in playing the game with their relatives and friends.
- Creating content: the activity consists of a visit to the school and homework assignment (as in the "Playing" case), followed by a period in which the class creates energy quizzes for the game and inserts them into the web application, before and after the treatment.

The experiment will permit us to assess the impact of gaming, and specifically of hybrid game schemes, on the change of attitude and behavior (at least at the intentional level) towards sustainable energy use. In the future, we plan to partner with a utility company, to integrate the experiment with the analysis of smart meter energy consumption data, so to understand the impact of the proposed game not only on the intention to save, but also on the actual consumption by households.

Acknowledgements. This work is partially supported by the "enCOMPASS - Collaborative Recommendations and Adaptive Control for Personalised Energy Saving" project funded by the EU H2020 Programme, grant agreement no. 723059 and by the "PENNY - Psychological, social and financial barriers to energy efficiency" project funded by the EU H2020 Programme, grant agreement no. 723791.

References

1. Un sustainable development goals. https://www.un.org/sustainabledevelopment/sustainable-development-goals/. Accessed 13 June 2019
2. Reeves, B., Cummings, J.J., Scarborough, J.K., Flora, J., Anderson, D.: Leveraging the engagement of games to change energy behavior. In: 2012 International Conference on Collaboration Technologies and Systems (CTS), pp. 354–358. IEEE (2012)
3. Yang, J.C., Chien, K.H., Liu, T.C.: A digital game-based learning system for energy education: an energy conservation pet. TOJET: Turk. Online J. Educ. Technol. **11**(2), 27–37 (2012)
4. De Luca, V., Castri, R.: The social power game: a smart application for sharing energy-saving behaviours in the city. FSEA **2014**, 27 (2014)
5. Casals, M., Gangolells, M., Macarulla, M., Fuertes, A., Vimont, V., Pinho, L.M.: A serious game enhancing social tenants' behavioral change towards energy efficiency. In: 2017 Global Internet of Things Summit (GIoTS), pp. 1–6. IEEE (2017)
6. Google arcore. https://developers.google.com/ar/. Accessed 13 June 2019

7. Wikitude. https://www.wikitude.com/. Accessed 13 June 2019
8. Blippar. https://www.blippar.com/. Accessed 13 June 2019
9. Palmarini, R., Erkoyuncu, J.A., Roy, R., Torabmostaedi, H.: A systematic review of augmented reality applications in maintenance. Robot. Comput.-Integr. Manuf. **49**, 215–228 (2018)
10. Ikea place. https://highlights.ikea.com/2017/ikea-place/. Accessed 13 June 2019
11. Amazon home AR. https://www.amazon.com/adlp/arview. Accessed 13 June 2019
12. Frajberg, D., Fraternali, P., Torres, R.N.: Heterogeneous information integration for mountain augmented reality mobile apps. In: 2017 IEEE International Conference on Data Science and Advanced Analytics (DSAA), pp. 313–322. IEEE (2017)
13. Sephora virtual artist. https://sephoravirtualartist.com. Accessed 13 June 2019
14. Gap virtual dressing room. https://corporate.gapinc.com/en-us/articles/2017/01/gap-tests-new-virtual-dressing-room. Accessed 13 June 2019
15. Nike fit. https://news.nike.com/news/nike-fit-digital-foot-measurement-tool. Accessed 13 June 2019
16. Arizton: Board games market - global outlook and forecast 2018–2023, August 2018. https://www.reportlinker.com/p05482343
17. Szegedy, C., et al.: Going deeper with convolutions. CoRR, abs/1409.4842 (2014)

Energy Consumption Patterns of Residential Users: A Study in Greece

Aristeidis Karananos[1]($\boxtimes$), Asimina Dimara[2]($\boxtimes$), Konstantinos Arvanitis[1]($\boxtimes$),
Christos Timplalexis[2]($\boxtimes$), Stelios Krinidis[2]($\boxtimes$), and Dimitrios Tzovaras[2]($\boxtimes$)

[1] WATT AND VOLT Anonomi Etairia Ekmetalleysis Enallaktikon Morfon
Energeias, Leoforos Kifissias 217A, 15124 Marousi, Greece
{a.karananos,k.arvanitis}@watt-volt.gr

[2] Centre for Research and Technology Hellas/Information Technologies Institute,
6th Km Charilaou-Thermis, 57001 Thermi-Thessaloniki, Greece
{adimara,ctimplalexis,krinidis,Dimitrios.Tzovaras}@iti.gr

Abstract. Electricity is an integral part of our lives and is directly linked to all areas of indoor human activity. In order to achieve good management of household electricity consumption, it is first necessary to make a correct and detailed measurement of it. Based on that aspect, this paper utilizes smart meters to monitor the electricity consumption of 120 different houses for a year in Greece. The measurements are saved and analyzed in order to gain a perspective of energy consumption patterns in comparison to temperature and personal energy profiling. The results and information of this paper could be used by current and future users as a guide to shift electricity behavior towards energy saving and also create new standardized profiles regarding demand response management to achieve energy efficiency.

Keywords: Energy consumption · Electricity · Behavioral consumption patterns · Outdoor environmental conditions · Smart meters

1 Introduction

The need to limit climate change by keeping global warming at a stable level and reducing greenhouse gases is, considering especially nowadays, an imperative necessity. This, combined with the idea that pervasive technologies – such as smart metering, home automation, sensing, and mobile devices - can enable the collective and individual change, is fundamental to activate energy efficiency policies.

The use of studies on the residential electricity profile is becoming more and more crucial. Temperature plays a very prominent role in this subject. Giannakopoulos and Psiloglou [3] on their research for Athens (capital City of Greece), highlights the sensitivity of energy demand to weather conditions and seasonal variation. A relative study was conducted by Pardo *et al.* [7] for Spain.

© Springer Nature Switzerland AG 2019
D. Tzovaras et al. (Eds.): ICVS 2019, LNCS 11754, pp. 639–650, 2019.
https://doi.org/10.1007/978-3-030-34995-0_58

Hekkenberg *et al.* [5] present the respective results for a colder climate, such as the one in the Netherlands. Gouveia *et al.* [4] demonstrate profiles of electricity consumption for 230 households in Evora, Portugal. Comparative residential profiles between Kenya, Germany and Spain were developed by Stoppok *et al.* [9]. This paper presents results of a study aimed to explore residential energy patterns in Greece, which emanated after an effort to educate household occupants on energy saving behavior as well as how to exploit smart metering equipment for this cause.

In our approach, a newer and more effective approach to energy saving was developed, leveraging data generated from smart sensors and provided energy recommendations by applying advanced consumer behavior models. The aim was to improve the energy consumption habits of people, by always keeping the individual comfort level high and achieve a reduction of resource consumption and a more sustainable society.

To begin with, smart meter technology is described along with the way it is utilized for energy consumption metering (Sect. 2). Afterward the relationship between energy consumption and outdoor temperature and different behavioral patterns of randomly chosen users are presented and evaluated in Sect. 3.

2   Smart Meters

The first type of electricity consumption meters was manufactured in the 1980s [11]. These meters were hardware driven (electromechanical) and were a very good solution for recording and monitoring consumer habits in energy distribution networks. Although the manual monitoring and restoration, the limited control and the often mechanical failures led over the next decade to the development of a new type of meters, known as electronic meters. Those had more features, in addition to simple energy consumption measurements and could calculate more complex parameters such as the use of energy, power demand in Kilowatt (kW) or Kilovolt-Ampere (kVA), electrical voltage, current, reactive power and other parameters related to energy.

Nowadays, a new kind of meters have emerged, the so-called "smart" meters [10]. Smart Metering or Advanced Metering Infrastructure (AMI) is an integrated system of smart meters, communication networks, and data management systems that enables bidirectional communication between utilities and customers. The smart meter is an advanced energy meter that obtains information from the end users' load devices and measures the real-time energy consumption of the consumers and then provides added information to the system operator (utility access points). These utility access points are commonly located on power poles or street lights. The entire system is referred to as a mesh network allowing for continuous connections among the network [13]. The goal of an AMI is to provide utility companies with real-time data about power consumption and allow customers to make informed choices about energy usage, based on the price at the time of use [6].

Smart meters measure, collect and report data back to utilities in short intervals throughout the day. They can also perform advanced functions such as sending signals to home appliances. A smart meter allows not only remote monitoring but also gives the user the ability to detect possible issues (extreme usage of electricity consumption, theft of electricity, etc.). The basic function of a smart meter is to gather and transmit data. Many methods are used for the transmission of data, such as:

(a) Wired via cables such as fiber-optic and copper phone lines;
(b) Wired via power lines such as power line communication or broadband over power lines;
(c) Wireless via antennas (e.g. GPRS, GSM, ZigBee, WiMax) [8].

2.1 Hardware and Software Infrastructure

Measurement equipment is installed to record the energy consumption of each household. This equipment is connected to each central electrical panel of each user. The data is transferred from the measurement point (meter) to a logger/poster using a radio frequency (RF) signal. The poster connects to a modem via an Ethernet port that allows the contents of its memory to be transferred through a wire to an online database using a secure service. At specific set time stamps, the recorded data is transferred from the database to a computer using a suitable computer program. The measurements and the collection of all the data are processed and analyzed in order to produce graphs and indicators for the electrical energy consumption of each dwelling, as well as the whole of the sample. An automation service is retrieving the data from the intermediate server, formats the data correctly and delivers the data via secure FTP protocol to the endpoint. The conceptual architecture of the method described above is shown in the following Fig. 1.

Fig. 1. Conceptual architecture for data integration

The metering devices used for monitoring energy consumption (Fig. 2) are clamps. These devices are utilized for the measurement of a live-conductor circuit without damaging its continuity. The "clamp" device as the name suggests uses a clamp or jaw like object, for the measurement of current flow. It is used for measurement of alternative current (AC)/direct current (DC) voltage and AC/DC current, resistance, frequency, capacitance, etc. In general AC clamps operate on the principle of current transformer (CT) used to pick up magnetic flux generated as a result of current flowing through a conductor. Assuming a current flowing through a conductor to be the primary current, one can obtain a current proportional to the primary current by electromagnetic induction from the secondary side (winding) of the transformer which is connected to a measuring circuit of the instrument. This permits to take an AC current reading on the digital display (in the case of digital clamp meters).

Fig. 2. Smart meter hardware

A meter using clamps to measure electricity is similar to a digital panel meter. A digital panel meter has advanced features, such as the detachable display part for easy user handling. Low pass filters provide accurate and stable measurements and use of advanced signal processing software. Some of these measurements include the calculation of power factor, max demand power and total harmonic distortion. Unlike other meters available on the market, this type has Pulse and Modbus communications built in, eradicating the requirement of purchasing extra modules.

For the current study, the meters with the clamp hardware were selected as the most valid option, since the cost can be kept low and also the error in measurements was negligible.

3   Case Study: Greek Residential Houses

A random number of Greek residential houses were selected for participation in this study; therefore, they can be considered as a representative sample of Greek energy consumers. The sample size consists of 120 unique houses all situated in Thessaloniki, Greece. All of the participants gave initial information about their profiles, where basic information such as number of occupants, type of household, heating type, as well as the number of rooms was provided. Additionally, they signed a consent form, giving us the approval to use their data.

3.1   User Profiles

The demographics of the 120 participating users are available in Table 1. The average persons living in a household is around 2 and most of them have 1 to 2 children.

Table 1. Main user information

Number of houses	120
Average persons in the household	2,08
Average number of children under 16	1,55
Average number of pets per household	1,26

The users are mostly people living in apartment buildings such as multistory buildings (Table 2). This is a very common living arrangement for families in Greece, as it provides plenty of space but also keeps the monthly rent cost down, compared to independent houses.

Table 2. Type of households

Type of households	
Apartment	95
Independent house	8
Semi-detached	2
Terraced house	15

Moreover, according to the users' profiles, most of the houses in this study consist of 3 available rooms but some variation is also apparent (Table 3). Based on this information, the number of rooms can be correlated with the surface area of each household in question, consequently to the amount of energy use.

Table 3. Number of rooms per house

Number of rooms	Number of houses
Two (2)	17
Three (3)	60
Four (4)	28
More than five (5)	15

Additionally, in Greece and particularly for Thessaloniki, there are three main heating type methods. The one of those that has become more available because of its efficiency, is heating using natural gas furnace and normal radiators that dissipate heat. Another method is heating oil furnace, likewise, combined with radiators. This is an older version and not as efficient as natural gas. Other methods consist of using a fireplace or more modern ones, like air conditioning systems. This study reveals that most of the users, use either natural gas or heating oil (Table 4).

Table 4. Type of heating method

Heating type	Number of houses
Electricity	22
Wood (fireplace)	2
Natural gas	58
Home heating oil	38

Each of the above group of users has its own characteristics as well as a unique energy behavior. The focus of this study is to understand how these different groups of users behave in their everyday lives, considering electricity consumption.

3.2 Climate in Thessaloniki

Greece is located in Southern and Southeast Europe. The mainland consists of mountains, hills, and sea. It has the 11th longest coastline in the world. The climate of Greece is predominantly Mediterranean, featuring mild, wet winters and hot, dry summers. However, there is a range of micro-climates because of the country's geography. Generally, there are four main types of climate in Greece: Mediterranean, Alpine Mediterranean, transitional continental – Mediterranean and Semi-arid climate. Thessaloniki is the second largest city of Greece with a population over 1.000.000 and the capital of Macedonia. Thessaloniki is located on the Thermaikos Gulf, at the northwest corner of the Aegean Sea. Because of the sea, Thessaloniki's climate is affected by it. Its climate is humid subtropical that borders to the Mediterranean climate and also semi-arid climate. During winter the climate is dry with morning frost. The coldest month is January. During the summer the climate is hot with humidity during the night. The hottest month is July.

3.3 Use of Energy Related to the Outside Temperature

The current sample is compared with the outside temperature. An average value of the temperature per month has been acquired for Thessaloniki from an external weather provider application programming interface (API) for the sampling period of April 2018 to March 2019. Data is also available for the previous year of 2017 which shows a similar condition of the weather (Fig. 3).

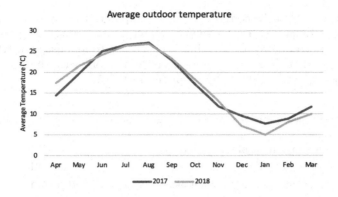

Fig. 3. Average outdoor temperature for Thessaloniki, GR

According to other studies and published papers [1, 12], there is an outdoor temperature "comfort zone" between 15 and 18° (°C) where the average energy consumption is at its lowest.

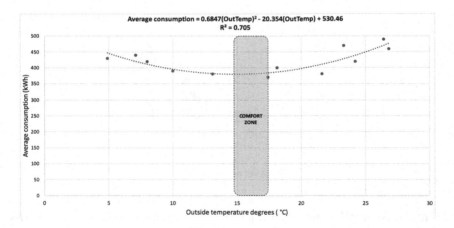

Fig. 4. Relationship/correlation of average consumption in regard to outdoor temperature

The correlation between outdoor temperature and monthly average consumption is depicted in Fig. 4. A nonlinear regression [2] was used to test the relationship. The R-squared (R^2) is a statistical measure that represents the proportion of the variance for a dependent variable that is explained by an independent variable or variables in a regression model. The R^2 analysis reveals that the model fits well the data. Within the comfort zone, which is 15 to 17.5 °C it is apparent that energy consumption is at its lowest. The heater it gets, the higher the heating ventilation and air conditioning systems (HVAC) system usage. On the other hand, within the comfort zone, it is apparent that energy consumption is at its lowest. In such cases, households maintain an average temperature that makes the users feel more comfortable, which leads to less usage of HVAC.

The statistical analysis of average consumption along with the standard deviation is as described in Table 5 and depicted in Fig. 5.

Table 5. Statistical analysis of the average consumption of residential houses

Average monthly consumption	419,8 kWh
Average daily consumption	13,5 kWh
Average monthly consumption per person	201,9 kWh
Average daily consumption per person	6,5 kWh
Standard monthly deviation per household	38,5 kWh
Standard daily deviation per household	3,6 kWh

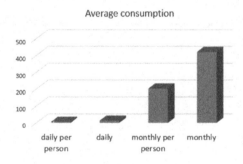

Fig. 5. Average consumption

The change in the relationship between consumption and temperature does not follow a seasonal pattern. The assumption that consumption changes per season is not eligible according to Fig. 6. For example, during the month of April, which is Spring, the average consumption is the same as in the case of the month of November, where the temperature is similar, but the season is different (Autumn). The demand of energy consumption is not season dependent. In case the consumption was affected from season, then the same season the consumption would be the same no matter what the outdoor temperature was.

3.4 Energy Consumption Patterns

Individual energy consumption patterns may differentiate on regards to daily average consumption, demand during peak hours and motifs. Some characteristic

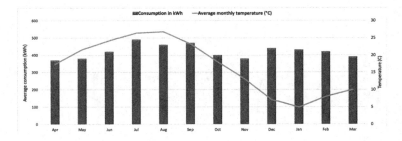

Fig. 6. Outside temperature compared to consumption for one year

behavioral patterns of families with one to three children are as depicted in Fig. 7. The number of occupants within a house affects the amount of total daily energy used. In the figure below, the user with the higher peak value (green line) has three children.

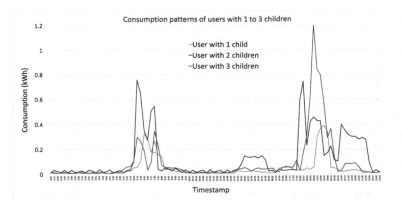

Fig. 7. Daily consumption patterns (Color figure online)

Another pattern that is studied is the relationship between consumption (in kWh) and the number of rooms in the household. In Fig. 8 it is apparent that the differentiation in the number of available rooms does not have a big impact on the total amount of energy consumption during a single day. Nonetheless, a small increase does occur where the number of rooms is more than three.

A significant divergence in energy consumption is visible when comparing two users with non-identical types of heating in the household. In Fig. 9 one of the users has the option to heat the house using a natural gas heater and an air conditioning unit (A/C), whereas the other one can only use natural gas. The user with the extra A/C unit has an average daily consumption of 3,65 kWh more if compared to the user without A/C. Moreover, this user's daily peak

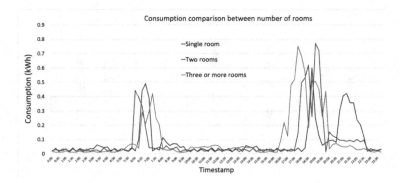

Fig. 8. Patterns of consumption between different number of rooms

value is calculated at 0,30 kWh over the analogous peak value of the second user in question.

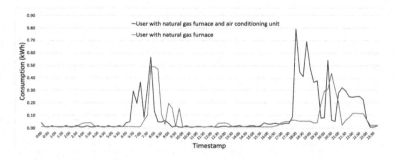

Fig. 9. Patterns of consumption between different heating type

In general (Fig. 10), it may be seen that consumption is never zero, because there are always devices, either active or on standby, for example, a working fridge or LED lights. There are also several time stamps during the day, where a higher consumption is apparent, that concurs with the users' morning routines as well as evening occupancy of the household. Moreover, lower consumption may be due to user absence.

Contrarily, none of the daily patterns overlap with any other. This case may be observed throughout the whole sample. There is a time shift on morning routines, household occupancy and absence. Additionally, electricity consumption is shifted quantitatively, depending on each users' energy consumption profile.

As it may be observed, in Fig. 11 each day has a unique consumption instance, both in the total kWh used and the time of usage. A deeper look reveals how the user is consistent during the weekdays, such as the sleeping routine, the absence during work hours and the evening occupancy. Furthermore, from Friday

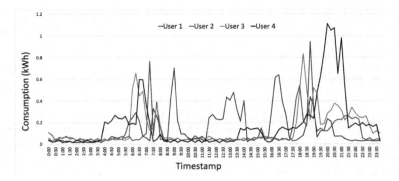

Fig. 10. Daily consumption patterns of four random users

Fig. 11. Weekly consumption patterns

afternoon to Saturday evening the consumption is at its highest, because the user might be using home appliances, like washing machine and dryer.

4   Conclusions

This study aims to establish a better understanding of the patterns of residential energy consumption. The relationship between energy demand and the temperature is examined, which in turn reveals that there is an apparent correlation between them, at least for average monthly measurements. An analysis of the average values concludes that even though the monthly average standard deviation (stdev) is low, the daily stdev value is really significant. This, in turn, leads to the assumption that consumption should be analyzed through individual profiling of users. To this extent, both outdoor temperature and indoor temperature are to be taken into account while analyzing data.

Results showed that consumption is affected basically from daily routines. Energy behavioral patterns are not only affected by outdoor temperature but also depend on the day of the week. To sum up, consumption monitoring can help users to observe their energy usage and improve their behavior towards more efficient load management.

Acknowledgements. This work is partially supported by the "enCOMPASS - Collaborative Recommendations and Adaptive Control for Personalised Energy Saving" project funded by the EU H2020 Programme, grant agreement no. 723059.

References

1. Blázquez, L., Boogen, N., Filippini, M.: Residential electricity demand in Spain: new empirical evidence using aggregate data. Energy Econ. **36**, 648–657 (2013)
2. Erdman, D., Little, M.: Non-linear regression analysis and non-linear simulation models. Survey of SAS System Features. SAS Institute Inc., Cary, NC (1998)
3. Giannakopoulos, C., Psiloglou, B.E.: Trends in energy load demand for Athens, Greece: weather and non-weather related factors. Clim. Res. **31**(1), 97–108 (2006)
4. Gouveia, J.P., Seixas, J., Luo, S., Bilo, N., Valentim, A.: Understanding electricity consumption patterns in households through data fusion of smart meters and door-to-door surveys
5. Hekkenberg, M., Benders, R., Moll, H., Uiterkamp, A.S.: Indications for a changing electricity demand pattern: the temperature dependence of electricity demand in the netherlands. Energy Policy **37**(4), 1542–1551 (2009)
6. Le, T.N., Chin, W.L., Truong, D.K., Nguyen, T.H.: Advanced metering infrastructure based on smart meters in smart grid. In: Smart Metering Technology and Services-Inspirations for Energy Utilities. IntechOpen (2016)
7. Pardo, A., Meneu, V., Valor, E.: Temperature and seasonality influences on spanish electricity load. Energy Econ. **24**(1), 55–70 (2002)
8. Sierck, P.: Smart meter-what we know: measurement challenges and complexities (2011). http://hbelc.org/pdf/resources.SmartMeter_Sierck.pdf. Accessed 1 Nov 2013
9. Stoppok, M., Jess, A., Freitag, R., Alber, E.: Of culture, consumption and cost: a comparative analysis of household energy consumption in Kenya, Germany and Spain. Energy Res. Soc. Sci. **40**, 127–139 (2018)
10. Temneanu, M., Ardeleanu, A.S.: Hardware and software architecture of a smart meter based on electrical signature analysis. In: 2013 8th International Symposium on Advanced Topics in Electrical Engineering (ATEE), pp. 1–6. IEEE (2013)
11. Weranga, K., Kumarawadu, S., Chandima, D.: Evolution of electricity meters. In: Weranga, K., Kumarawadu, S., Chandima, D. (eds.) Smart Metering Design and Applications, pp. 17–38. Springer, Singapore (2014). https://doi.org/10.1007/978-981-4451-82-6_2
12. Yi-Ling, H., Hai-Zhen, M., Guang-Tao, D., Jun, S.: Influences of urban temperature on the electricity consumption of shanghai. Adv. Clim. Change Res. **5**(2), 74–80 (2014)
13. Zheng, J., Gao, D.W., Lin, L.: Smart meters in smart grid: an overview. In: 2013 IEEE Green Technologies Conference (GreenTech), pp. 57–64. IEEE (2013)

Overview of Legacy AC Automation for Energy-Efficient Thermal Comfort Preservation

Michail Terzopoulos[1,2], Christos Korkas[1], Iakovos T. Michailidis[1(✉)], and Elias Kosmatopoulos[1,2]

[1] Information Technologies Institute, Center for Research and Technology Hellas (ITI-CERTH), 57001 Thessaloniki, Greece
{miketerzo,michaild}@iti.gr
[2] Department of Electrical and Computer Engineering, Democritus University of Thrace, 67100 Xanthi, Greece

Abstract. The rapid maturity of everyday sensor technologies has had a significant impact on our ability to collect information from the physical world. There are tremendous opportunities in using sensor technologies (both wired and wireless) for building operation, monitoring and control. The key promise of sensor technology in building operation is to reduce the cost of installing data acquisition and control systems (typically 40% of the cost of controls technology in a heating, ventilation, and air conditioning (HVAC) system). Reducing or eliminating this cost component has a dramatic effect on the overall installed system cost. With low-cost sensor and control systems, not only will the cost of system installation be significantly reduced, but it will become economical to use more sensors, thereby establishing highly energy efficient building operations and demand responsiveness that will enhance our electric grid reliability.

Keywords: Sensors · Building automation · Energy and thermal comfort management

1 Introduction

Due to the mounting global energy crisis, there is a continued focus on improving building design and engineering to reduce energy consumption and enable "smart" energy use. Such "smart" buildings may incorporate a number of dynamic systems that can react to changing environmental conditions in order to minimize overall energy use while maintaining user comfort [1–4].

In order for a building energy management system to intelligently respond to changing building conditions, a network of sensors is required to provide the necessary feedback-data to the control system in order to be able to react appropriately, as well as, to "feed" the user with the necessary internal conditions of the building. Thus, a variety of sensors is required to collect the measurements for indoor factors such as: temperature, relative humidity, occupancy, luminance and CO_2 sensors. Despite the fact that

© Springer Nature Switzerland AG 2019
D. Tzovaras et al. (Eds.): ICVS 2019, LNCS 11754, pp. 651–657, 2019.
https://doi.org/10.1007/978-3-030-34995-0_59

such sensors, are readily available, the tuning and installation process of these elements, can create a large and complicated process. Any way to reduce the time investment and the cost of the installation of such elements (sensors and actuators) could significantly reduce the overall complexity of the system [5, 6].

In this work a system of air-conditioning units control is presented, for household use with the aim of optimizing its efficiency, translated as a reduction of the consumed energy and increasing the satisfaction of the occupants. Key components of this system except the air conditioning unit are microcontrollers, various sensors and actuators and the graphical user interface developed. Surveys and measurements for the effectiveness of the proposed system took place during the summer months. One of the key features missing from the legacy air conditioning units is their fully automated operation. The proposed system offers the required automation without, however, depriving the user of the possibility of intervention when deemed necessary. Our goal in this work was to create a system that can produce optimized results in a relatively short period of time by using conventional computing resources. Also important was the ability to use the system on already installed air-conditioning units that are located in dwellings as well as the easy scalability of the system with new units and new functions.

2 Sensor Equipment and Control Platform

As stated in Introduction, the main goal of this work was to install, operate and evaluate various types of microcontrollers, sensors and actuators for the automated control of an A/C operated room. For the purposes of this work various sets of sensors and microcontrollers were used in order to assess different sets of market products. In Table 1, all the equipment used in this work is presented.

2.1 Decoding Process

We have succeeded in sending various signals from the remote control of the air conditioner to Arduino through the receiver to determine the heading, the duration of the pulses for each logical state, logic 0 and 1, and the magnitude of the signals. A typical A/C signal consists of a series of bits (Fig. 2). Each bit is encoded differently depending on the device manufacturer. The air conditioner used encodes logic 0 as pulse for 500 ms and pause duration of 560 ms. Correspondingly, logic 1 is encoded as a 500 ms pulse and a 1560 ms pause (Table 1). So the signal 00101, looks something like this: 500, 560, 500, 560, 500, 1560, 500, 560, 500, and 1560.

We have therefore come to the conclusion that this control unit works by sending 170 bit signals for start and stop commands and 114 bit for other signals when the device is in operation. By taking a larger number of signals and processing them properly, we managed to decode the role of each bit and then produce our own signal according to our requirements (Fig. 1).

Table 1. Equipment list

Arduino Uno microcontroller: 	Arduino is an open-source hardware and software company. The boards are equipped with sets of digital and analog input/output (I/O) pins that may be interfaced to various expansion boards or breadboards (shields) and other circuits. The boards feature serial communications interfaces, including Universal Serial Bus (USB) on some models, which are also used for loading programs from personal computers.
Raspberry Pi 3 b+ microcontroller: 	The Raspberry Pi 3 Model B is the earliest model of the third-generation Raspberry Pi. Pi 3 has Quad Core 1.2GHz Broadcom BCM2837 64bit CPU ,1GB ram, 2.4GHz and 5GHz IEEE 802.11.b/g/n/ac wireless LAN, Bluetooth 4.2, BLE Gigabit Ethernet over USB 2.0 (maximum throughput 300 Mbps),Extended 40-pin GPIO header, 4 USB 2.0 ports
DHT21 Temperature-humidity sensor: 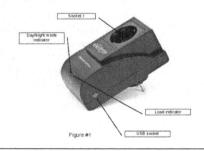	The DHT21 is a basic, low-cost digital temperature and humidity sensor. It uses a capacitive humidity sensor and a thermistor to measure the surrounding air, and spits out a digital signal on the data pin (no analog input pins needed).
USB Energy meter EGM- PWML	The AC energy consumption was measured with USB Energy meter EGM- PWML which can measure energy, voltage, current, frequency, resistance, power factor and is able to transfer the measurements to pc with usb cable. Also EGM-PWML has software which monitoring and save the measurements to pc .The energy meter can measure max 2500w power (10 Ampere RMS.

(*continued*)

Table 1. (*continued*)

Aeotec-Z-Stick Gen5 :	Z-Stick Gen5 lets you build your own gateway. A gateway that is locally hosted. A gateway that is cloud free. One with the features and security your home need. And one that can communicate with over 230 Z-Wave devices as you gradually expand your smart home. It's been designed to bring Z-Wave to every computer platform. You can use it with a PC, HTPC or laptop, with Mac, Linux or Windows, with single-board computers such as Raspberry Pi, and even select NAS systems such as the Asus' Asustor.
AEOTEC-MultiSensor 6:	MultiSensor 6 has 6 sensors Motion sensor, Temperature sensor, Light sensor, Humidity sensor, Vibration sensor, UV sensor
IR receiver and transmitter:	We used IR Receiver 38kHz (TSOP38238) to scan and decode the a/c ir remote control signals in order to send this signal with a IR diode and control remotely the A/C.
Irdroid USB IR Transceiver	Use as a Remote control for your computer, scan infrared remotes and store in a lirc.conf file. The USB infrared Transceiver unit has both transmit and receive functionalities. The unit is based on PIC18F2550, it communicates via USB (ACM) with the host system and it will allow you to send and receive infrared signals via Winlirc or Lirc (LinOS).

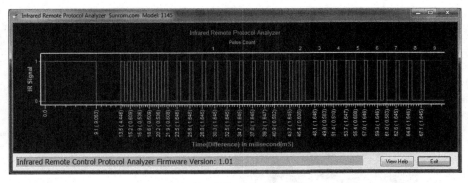

Fig. 1. A/C control signal

3 User Graphical Interface

The Graphical User Interface (GUI) developed in this work was necessary to meet its core needs. In addition to the fact that a GUI is considered necessary for any modern application, it offers easy control over the system even to unskilled users. It allows monitoring of the experiment and the operation of the system and it is offered to display useful information on the conditions inside and outside the room and the measurements made. It also allows the user to intervene manually and at will at the system's operation at any time.

In Fig. 2, the GUI is presented. More specifically, the first, upper left axle system is used to draw graphs of humidity inside and out of the room over time. The second axle system, left and in the middle, displays the graphs of Dew Point temperatures and dew point in and out of the room in respect with time. The three viewing frames of the system parameters and states consist of Readings, A/C Status, and Power Consumption. The Readings box displays the weather conditions prevailing in the area but also the internal conditions of the room like: temperature, humidity and dew point. Next to the Readings box is the A/C Status box. Here are all the indications for the operation of the air conditioner. It consists of a value input field and a sliding bar for operating temperature and an indication of how it works (Off, Heat, Cool, Dry, Auto, Fan). The user is able to change these values if he finds it necessary, either by typing the desired temperature in the corresponding field or by moving the sliding bar and selecting the mode from the popup menu. This is done by pressing the "Set Manually" button.

Finally, a Check Box informs about whether or not a tenant is in the room. Although at first sight this information can be considered useless, it is not. We note that it can be used in the case of remote control as well as for better supervision during experiments.

Last but not least, the last information panel is the Power Consumption panel that gathers the indications for the power consumption of the air conditioner. From the top to the bottom is the "Total Power Consumption" showing the cumulative consumption throughout the experiments, the "Power Consumption for the Day" showing consumption up to the time step and the "Measured Power Consumption", i.e. the measurement for that step. It also allows the user to enter the measurement of the current step manually by typing the value in the corresponding field and pressing the Submit button.

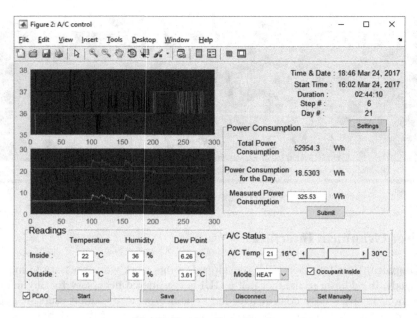

Fig. 2. Graphical user interface

4 Conclusions

The ultimate goal of the current work aims at employing AC control strategies either for cooling or heating which preserve indoor thermal-comfort while minimizing the respective total energy consumption. To help bridge the gap, different controllers aim to enable smartness and automation in the use of old school legacy ACs by emulating their IR-RF remote commands, such as: setting a temperature, fan speed, mode and timer. These controllers have the potential to evolve from IR Remotes (infrared) blasters, to Smartness-enabling AC remote controllers and ultimately AI-enabling AC remote controllers.

In specific, the current study presented the potential of low-cost home automation devices enabling automatized control of legacy air conditioners. The devices considered come from commercially available solutions emerged during the recent past years in the European market. It is proven that using freeware APIs which comply with the architecture and scope of open-platforms for building automation, a reliable control channel can be established in order to automatically manage the operational state of almost any legacy AC appliance having an IR-RF-based remote controller.

Acknowledgements. This work was partially supported by the "Plug-N-Harvest - Plug-n-play passive and active multi-modal energy Harvesting systems, circular economy by design, with high replicability for Self-sufficient Districts & Near-Zero Buildings" project funded by the EU H2020 Programme, grant agreement no. 768735.

References

1. Korkas, C.D., Baldi, S., Michailidis, I., Kosmatopoulos, E.B.: Occupancy-based demand response and thermal comfort optimization in microgrids with renewable energy sources and energy storage. Appl. Energy **163**, 93–104 (2016)
2. Wang, L., Wang, Z., Yang, R.: Intelligent multiagent control system for energy and comfort management in smart and sustainable buildings. IEEE Trans. Smart Grid **3**(2), 605–617 (2012)
3. Michailidis, I.T., Korkas, C., Kosmatopoulos, E.B., Nassie, E.: Automated control calibration exploiting exogenous environment energy: an Israeli commercial building case study. Energy Build. **128**, 473–483 (2016). https://doi.org/10.1016/j.enbuild.2016.06.035. ISSN 0378-7788
4. Michailidis, I.T., Baldi, S., Pichler, M.F., Kosmatopoulos, E.B., Santiago, J.R.: Proactive control for solar energy exploitation: a german high-inertia building case study. Appl. Energy **155**, 409–420 (2015). https://doi.org/10.1016/j.apenergy.2015.06.033. ISSN 0306-2619
5. Conklin, J.A., Hammond, S.R.: U.S. Patent No. 9,772,260. Washington, DC, U.S. Patent and Trademark Office (2017)
6. Agarwal, Y., Balaji, B., Gupta, R., Lyles, J., Wei, M., Weng, T.: Occupancy-driven energy management for smart building automation. In: Proceedings of the 2nd ACM Workshop on Embedded Sensing Systems for Energy-Efficiency in Building, pp. 1–6. ACM, November 2010

Can I Shift My Load? Optimizing the Selection of the Best Electrical Tariff for Tertiary Buildings

Oihane Kamara-Esteban$^{(\boxtimes)}$![ORCID], Cruz E. Borges ![ORCID], and Diego Casado-Mansilla ![ORCID]

DeustoTech - Universidad de Deusto, Avda. Universidades, 24, 48007 Bilbao, Spain
{oihane.esteban,cruz.borges,dcasado}@deusto.es

Abstract. Sustainability is strongly related to the appropriate use of available resources, being an important cornerstone in any company's administration due to the direct influence on its efficiency and ability to compete in the global market. Therefore, the intelligent and proper management of these resources is a pressing matter in terms of cost savings. Among the possible alternatives for optimisation, the one regarding electricity consumption stands out due to its strong influence on the expenses account. In general, this type of optimisation can be carried out from two different perspectives: one that concerns the efficient use of energy itself and the other related to the proper adjustment of the electricity contract so that it meets the infrastructure needs while avoiding extra costs derived from poorly sized bills. This paper describes the application of an artificial intelligence based methodology for the optimisation of the parameters contracted in the electricity tariff in the Spanish market. This technique is able to adjust the power term needed so that the global economic cost derived from energy consumption is significantly reduced. The papers discusses the impact that this proposal may have on a demand response scenario associated to load shifting practices within university buildings. Furthermore, the role of human beings, specifically university employees, and their actions towards reducing the overuse of power consumption at the same time is also addressed.

Keywords: Demand response · Energy costs · Forecasting · Flexibility · Genetic algorithms

1 Introduction

The climate crisis has motivated the need to audit all expenses in order to identify the key areas where saving techniques can be applied to reduce global warming [1]. Nowadays, most of these techniques focus on minimising costs within the electric scope, not only in terms of reducing the overall consumption [2,3], but also in configuring a specific electric contract that meets the energy needs of the infrastructure while avoiding overruns [4]. Smart contracts based on blockchain technologies [5,6] are the summon of this concept.

© Springer Nature Switzerland AG 2019
D. Tzovaras et al. (Eds.): ICVS 2019, LNCS 11754, pp. 658–669, 2019.
https://doi.org/10.1007/978-3-030-34995-0_60

In this sense, electricity contracts can be challenging to understand for all the variables that are involved. In the era of the economic liberalism, new electric companies are entering into the markets with a plethora of terms and conditions associated to their offers and billings which usually residential consumers and even business' managers struggle to understand [7]. The most important endeavour for them should be to understand what type of service they really need and how to get the best rates for such necessity. The result of this misinformation is that most companies and end-users choose contracts based on what a sales representative from the electric company advised them rather than taking into account their own energy needs to decide a type contract. Therefore, the optimisation of the energy bill requires a technical and thorough analysis of both the parameters that influence it and the facility's consumption patterns, as well as the possibilities of negotiation within the electric market.

Therefore, the knowledge on how the energy is consumed within an infrastructure provides a great advantage when negotiating contracts with the electricity supplier. For this, most companies still turn to *energy managers* whose labour is to regulate and monitor energy consumption with the aim of improving its efficiency by implementing new policies and changes where necessary. Traditionally, these managers are responsible for knowing in detail all the variables that come into play within the electric market such as the methodology for the calculation of electricity prices and the advantages and disadvantages of the current electric tariffs. However, new proposals related to artificial intelligence in smart grids are arising to assist or even to replace the work done by these experts [8]. One example of that is the automatic forecast of next-day energy usage [9] and the adjustment of flexible loads from one slot to another [10].

To successfully complete a demand response actions the following steps should be followed:

1. Perform and energy audit and correct any obvious energy losses [11].
2. Optimize your power rate.
3. Introduce behavioural change actions to foster the reduction of energy consumption [12].
4. Identify potential loads that could be shifted in order to reduce the peak power or to move them to periods with lower energy tariff.
5. Introduce behavioural change actions to foster these changes [13–15].

While automatic energy tariff optimization is a hectic research field [16], the optimization of the power contracted is seldom addressed in the literature even when it is a pre-requirement for any Demand Response action, since until recently, the objective of these actions was closely related to the reduction of the power requested [17]. In order to cover this gap in the state of the art, we present the first step towards an automatic power rate optimizer in the smart grid.

The rest of the paper is structured as follows: Sect. 2 describes the tariff model currently operating in Spain, Sect. 3 presents an analytical analysis of the rate, Sect. 4 contains the experimental setup, Sect. 5 shows the results of applying this methodology at an office building in Spain where a behavioural change pilot is being carried out, and finally, Sect. 6 draws out a summary of the main findings and next steps.

2 Tariff Model

In Spain, the concepts defined in any electricity bill are classified under two main components: regulated and non-regulated

Regulated. These refer to those concepts which price or methodology of calculus is defined in the Boletín Oficial del Estado (BOE), the official gazette of the Government of Spain, and they are applicable to every supply contract regardless of the electricity supplier.

Power term: When negotiating an electric contract, companies choose the type of tariff depending on the voltage of their electric infrastructure and the amount of power they expect to consume. Some of these tariffs split the days in *periods* which are assigned an annual power term price by the BOE. In order to calculate the monthly bill, the amount of contracted power in every period is multiplied by the price for the days invoiced.

Power excess: Penalty applied each time the company demands more power than the amount originally contracted.

Others concepts: Rental of equipment and taxes.

Non-Regulated. These include those concepts which price can be negotiated with the electricity supplier and, therefore, can vary among contracts.

Energy term: It is important to select when energy consumption is necessary and when it is superfluous. Energy prices are different depending on the time of the day and month, so it is not irrelevant to consume at night than during daylight. The energy term of the monthly bill is calculated by multiplying the price of energy negotiated with the electricity supplier by the amount of energy measured by the meter.

Complement for consumption of reactive energy: The billing of high electricity consumption in industrial sectors includes a charge for the generation of reactive energy. Its purpose is to promote energy efficiency by issuing an economic penalty for inefficient consumption and is billed according to the power factor.

Based on the aforementioned parameters and the type of contracted electricity tariff, energy optimisation can be achieved in three different ways:

Optimisation of the power term: Adjust the contracted power to the real amount of power expected to be demanded by the infrastructure so as to minimise the penalty for power excesses.

Optimisation of the reactive term: Compensate the generation of this type of energy, derived from the use of some electrical and mechanical devices, by installing capacitor batteries.

Reduction of power and energy consumed: Implement policies and best practices such as replacing lighting fixtures and electric equipment with more efficient alternatives.

The configuration of an appropriate electricity contract is as important as knowing the specific needs and energy capabilities of the infrastructure. The BOE establishes the voltage steps, power ranges and billing characteristics that define each of the available tariffs. These tariffs are classified into two groups: low voltage rates ($v \leq 1\,\mathrm{kV}$), oriented to small and medium energy consumers, and high voltage rates ($v > 1\,\mathrm{kV}$) for large industrial consumers with high energy needs. The selection of the proper tariff is determined by the amount of voltage on which the infrastructure is connected to the electric grid and the minimum and maximum power to be demanded.

The study presented in this article focuses on the optimisation of the regulated cost of the energy bill in terms of power for 6.1 high voltage electric tariffs, though the same methodology can be applied to other tariffs. When contracting energy supply under this type of tariff, the energy manager has to estimate the global power expected to be demanded p_i on each period P_i, configuring them in ascending order:

$$p_1 \leq p_2 \leq p_3 \leq p_4 \leq p_5 \leq p_6. \tag{1}$$

The definition of the hours belonging to each period depends on the global energy demand within the country. For example, the hours belonging to P_1 are the ones in which the global demand reaches its maximum peak and so the cost of demanding power or consuming energy on this period has a higher cost than doing so on P_6.

Nowadays, the power optimisation technique most commonly used by energy managers consists on analysing the power consumption of the installation and identifying the maximum amount of power demanded on each one of the six periods. These values are then modified so as they the fulfil the Inequality (1) and conform the power term of the electric contract. This approach reduces the cost derived from the power term significantly, however there is a great margin of improvement since the existence of a power peak can overstate the power requirements of the facility, incurring in penalties for power excesses. This paper proposes the use of evolutionary algorithms as a technique to optimise the regulated part of the electricity bill by finding the best combination of power values and therefore, minimise the annual electricity cost.

3   Analytical Optimisation of the 6.1 Electric Tariff

This section analyses the mathematical properties defined by the regulated components of the 6.1 tariff. As stated in the previous section, this function only takes into account the power term and power excesses.

Let $p := (p_1, \ldots, p_6)$ denote the power limit assigned to every period P_i. Please note that p is a vector of integers in which at least one of the values must be greater than $450\,\mathrm{kW}$ and should also fulfil Inequality 1. On the other hand, let $\pi_j^{i,m}$ denote the maximum power demanded in j quarter (which corresponds to an hour in the period P_i on month m). Now, for every month m, the cost function ϕ_m of the power term and excesses would be:

$$\phi_m(p) := \sum_{i=1}^{6} \left(c_i p_i + 1.4064 k_i \sqrt{\sum_{j \in J}(\pi_j^{i,m} - p_i)^2} \right),$$

where c_i denotes the power cost of the period P_i, $k := (1, 0.5, 0.37, 0.37, 0.37, 0.17)$ is a vector of coefficients used to assign importance to power excesses on the most critical periods (please note that these two latter quantities are defined in the BOE) and J denotes the set of quarters where the power measured $\pi_j^{i,m}$ is higher than the power limit p_i contracted for this period. Namely, J is the set of quarters j such that:

$$p_i < \pi_j^{i,m}. \tag{2}$$

This function can have as much as 35 040 jump discontinuities depending on the cardinality of J. This results in a highly non-continuous non-linear integer optimization problem in which the use of classical integer programming algorithms is nearly impossible. However, a more graphical approach can give us additional information (see Fig. 1 for help in following the next statements). As can be seen, $\phi_m(p_i)$ is a monotonic increasing linear function until one p_i fulfils Inequality (2) (α label in Fig. 1). In this situation we have the first jump discontinuity. Now, $\phi_m(p_i)$ could be monotonic decreasing or increasing depending on whether c_i is bigger or not than $1.4064 k_i$. The latter case means that the minima has been reached and the process can stop. In the former case, however, the process should continue until a quarter (or a discontinuity jump) $k \in J$ is found so that (β label in Fig. 1):

$$c_i \leq 1.4064 k_i \frac{\sum_{k \geq j}(\pi_k^{i,m} - p_i)}{\sqrt{\sum_{k \geq j}(\pi_k^{i,m} - p_i)^2}}.$$

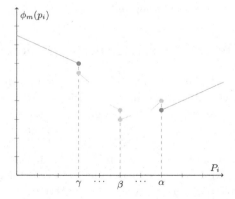

Fig. 1. Outline of the cost function $\phi_m(p_i)$.

Please note that this process only finds the best set of p_i for a single month. Since these coefficients are shared among the twelve months, the problem becomes combinatorial. A brute-force approach would help in testing all the possibilities, however, this would be better accomplished by using more efficiency techniques such as evolutionary algorithms.

4   Experimental Section

4.1   Datasets

The public infrastructure which electricity consumption was evaluated is comprised of six buildings: five located in Bilbao and one in Donosti; both cities situated on the north of Spain. Currently, each building feeds electricity from its own supply point and therefore is managed under a particular electric contract. Five of the buildings (four in Bilbao and the one in Donosti) have a 6.1 tariff (six periods) while the remaining one has a 3.1 tariff (three periods).

The behaviour of the demand curves of four of the buildings with the 6.1 tariff is quite similar:

- There are three day-types, weekdays, Saturdays and holidays.
- The highest energy consumption takes place on weekdays.
- The peak hours being between 10:00 and 18:00.
- On Saturday, the demand curve follows a similar pattern, with lower consumption and the peak hours finishing at 14:00.
- On holidays, the curve is almost flat.

The only difference of the fifth building with the previous ones is the existence of a fourth day type, bank holidays, where the shape of the load profile is similar to a weekday but the peak load is similar to a Saturday. Figure 2 shows the four daytimes of this building.

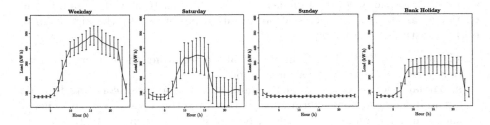

Fig. 2. Mean load over a year for Building E. Errors bar denote $\pm\sigma$.

Based on this configuration, two case studies are proposed:

Individual case: Analyse the electric consumption per building and apply a power term optimisation algorithm to each dataset separately.

Joined case: Simulate that the buildings with the 6.1 tariff located in Bilbao are all connected to a single supply point and are all managed under the same electric contract. This means that the consumption values of the four buildings will be added together to generate a global dataset on which to apply the genetic algorithm.

In both cases, the purpose of applying the genetic algorithm to the consumption datasets is to find the best combination of power values that meets the demand of the infrastructure while minimising the economic cost derived from a contract configured with higher power values than what the facility needs. The analysis of the best solution will depend not only on the degree of optimisation achieved in each case, but also on the possibility of modifying the current electric configuration if the best optimisation is given by a different layout of the supply points.

The energy consumed by the whole infrastructure is monitored by an automated system, which periodically collects the data recorded by the buildings' meters and stores it in a centralised database. The monitoring module manages all aspects of telemetering, from the identification of missing or corrupted data up to the management of events in case of meter malfunction or communication failure. The centralised database also stores additional data regarding the electricity contracts associated with each building, the definition of the tariff's periods, the cost of the power term established by the BOE, and the distribution of bank holidays within the year. All this information feeds the system's optimisation module, which replicates the annual bills in order to simulate scenarios with different sets of contracted power values to find the combination that meets the infrastructure needs while minimising the global cost.

5   Results and Discussion

The implementation of the genetic algorithm and the bill replication module altogether called for a deep study of the parameters and characteristics influencing the electricity contract of each building in order to identify common and particular aspects. The process of optimisation performed by the algorithm in both case studies is as follows:

- The system collects data from the database regarding the annual consumption, contracts, and tariff parameters per building.
- A partial calculus of the annual billing is done for those concepts that do not influence the power term such as cost of active energy, reactive energy and equipment rental. This partial invoice is added as input to the core of the optimisation algorithm. It is important to note that in the joined study use case, in which all buildings are evaluated as a single facility, the values of energy consumption are added together to form a single dataset.
- The genetic algorithm then generates a set of candidate solutions, i.e, combinations of six power values that satisfy the constraints set by the type of tariff (upper and lower limit values of power that can be demanded; in the case of the 6.1 tariff, one of the values must be higher than 450 kW).

– The candidate solutions undergo a series of transformation through the use of genetic operators.

Using the previous approximation, three different experiments were carried out in order to asses the quality of the proposed method.

5.1 Tariff Optimisation

The first experiment assesses the percentage of reduction achieved in the electricity bill using the traditional method (TM), i.e. the optimisation technique explained in Sect. 1 in comparison with the proposed method (GA), i.e. the method of the evolutionary algorithm. Namely, the objective is to look for the power limits $p_1, \ldots, p_6$ of every period that minimises the electric bill of the previous year. The results can be seen in Table 1. Columns $\Delta T\ TM$ and $\Delta T\ GA$ show the *percentage of reduction* on the period of time comprised by the dates in columns *From* and *To* for the traditional algorithm and the proposed method respectively. Please note that negative values denote an increment in the electric bill while positive values denote reductions.

As can be seen, the proposed method almost always finds a better combination of parameters that minimises the annual cost than the traditional procedure. As expected, the improvement is scarce, except in building D. However, please note that these results imply that a simple phone call represent more than 1000€ of savings. Finally, the savings can be particularly important if all the loads are combined. In this case the savings could be up to 30%. Please note that this situation can not be always possible to contract due to physical or legal restrictions but should always be considered in the design phase of a complex of buildings such as a university campus.

5.2 Savings Forecasting

We tried to asses the ability to forecast future savings for the two methods. Namely, in this experiment we suppose that we have changed the power limits using the data comprised between *From* and *To* dates for the period comprised between the *To* date until 2014-03-31. Columns $\Delta V\ TM$ and $\Delta V\ GA$ of Table 1 contain the percentage of reduction over this period for the traditional algorithm and the proposed method respectively. Please note that, just like before, negative values denotes an increment in the electric bill while positive values denote reductions.

As can be seen, the proposed method achieves a reduction in the annual electricity cost and, moreover, this improvement is bigger in all cases. Please note that the results suggest that the savings of the proposed method are kept during long periods.

5.3 Convergence

Since the proposed method is stochastic by nature, its convergence properties have to be tested empirically. Please note that, in this context, convergence

Table 1. Comparison or the results achieved by PM and GA optimisation methods.

B.	From	To	Δ Test			Δ Validation		
			TM (%)	GA (%)	Days	TM (%)	GA (%)	Conv. (%)
A	2012-09-01	2013-08-31	−1.68	0.03	211	−1.69	0.59	0.03
A	2012-10-01	2013-09-30	−1.66	0.09	181	−1.95	0.88	0.03
A	2012-11-01	2013-10-31	−1.51	0.18	150	−1.78	1.01	0.04
A	2012-12-01	2013-11-30	−1.24	0.22	120	−1.50	1.61	0.03
A	2013-01-01	2013-12-31	−0.63	0.91	89	−0.47	2.44	0.04
A	2013-02-01	2014-01-31	0.06	1.47	58	−1.23	1.23	0.04
A	2013-03-01	2014-02-28	0.10	2.09	30	6.49	6.95	0.03
B	2012-09-05	2013-08-31	1.56	3.02	211	5.92	9.54	0.04
B	2012-10-01	2013-09-30	1.97	3.45	181	6.52	10.18	0.04
B	2012-11-01	2013-10-31	2.25	3.88	150	6.39	10.49	0.05
B	2012-12-01	2013-11-30	2.54	4.43	120	6.34	10.31	0.04
B	2013-01-01	2013-12-31	3.82	5.69	89	7.39	11.86	0.04
B	2013-02-01	2014-01-31	4.37	6.61	58	8.01	13.62	0.03
B	2013-03-01	2014-02-28	5.76	8.18	30	15.79	26.04	0.04
C	2012-09-01	2013-08-31	−1.18	0.31	211	−0.31	2.48	0.05
C	2012-10-01	2013-09-30	−1.06	0.45	181	0.29	3.66	0.05
C	2012-11-01	2013-10-31	−0.69	0.76	150	0.62	3.90	0.03
C	2012-12-01	2013-11-30	−0.87	0.95	120	−0.35	4.07	0.03
C	2013-01-01	2013-12-31	−0.41	1.62	89	0.19	4.59	0.03
C	2013-02-01	2014-01-31	0.03	2.18	58	−0.57	3.78	0.03
C	2013-03-01	2014-02-28	0.33	2.88	30	3.34	11.22	0.03
D	2012-09-10	2013-08-31	10.40	12.68	211	9.36	10.39	0.09
D	2012-10-01	2013-09-30	10.80	13.03	181	7.79	9.50	0.07
D	2012-11-01	2013-10-31	11.25	13.35	150	5.16	8.22	0.07
D	2012-12-01	2013-11-30	11.46	13.54	120	1.45	6.69	0.09
D	2013-01-01	2013-12-31	9.77	12.83	89	2.72	10.52	0.08
D	2013-02-01	2014-01-31	9.59	13.03	58	5.77	12.70	0.09
D	2013-03-01	2014-02-28	9.09	12.94	30	34.00	29.11	0.09
E	2012-09-10	2013-08-31	−2.26	0.48	211	−6.06	3.38	0.03
E	2012-10-01	2013-09-30	−1.91	0.74	181	−4.64	4.73	0.03
E	2012-11-01	2013-10-31	0.33	1.22	150	0.90	5.18	0.05
E	2012-12-01	2013-11-30	0.37	1.50	120	0.92	4.99	0.03
E	2013-01-01	2013-12-31	0.41	1.75	89	0.95	6.44	0.03
E	2013-02-01	2014-01-31	0.52	2.19	58	1.27	8.10	0.05
E	2013-03-01	2014-02-28	0.57	2.53	30	1.95	12.50	0.03
JOIN	2012-09-10	2013-08-31	38.91	39.53	211	35.30	37.87	0.03
JOIN	2012-10-01	2013-09-30	38.38	39.05	181	37.05	39.42	0.05
JOIN	2012-11-01	2013-10-31	36.80	28.61	150	40.04	33.65	0.03
JOIN	2012-12-01	2013-11-30	36.22	28.55	120	42.47	35.69	0.03
JOIN	2013-01-01	2013-12-31	36.08	28.72	89	43.45	36.87	0.05
JOIN	2013-02-01	2014-01-31	35.76	28.88	58	44.26	37.45	0.03
JOIN	2013-03-01	2014-02-28	35.48	29.09	30	47.37	42.50	0.03

refers to the ability of the proposed method to find a local minima. In order to measure this ability, the previous experiment was repeated 100 times for each period of time defined by the *From* and *To* columns. The variability of the results was examined by using the Coefficient of Variation [18] which normalises the standard deviation with respect to the mean value of the sample. In this sense, it represents the percentage of central variation of the sample. Namely, a low value would mean that the method always finds the same local minimum which suggests it gets the absolute minimum. On the other hand, high variability may mean that the function has several different local minima or that the method fails to converge. Column *Conv* of the results in Table 1 suggest that in all cases the method converges to the same local minimum so it is likely that this is also the absolute one.

6 Conclusions

The optimisation of the electricity bill is a secure alternative in terms of cost savings given the high energy consumption in which business and particulars alike incur today. Of the possible alternatives for optimisation, the one that equilibrates contracted power values is the most affordable, since its application does neither involve a change in the electric system nor a change of supplier, and is of immediate application when renewing electric contracts. This paper proposes an artificial intelligence based methodology, namely evolutionary algorithms, to find the best combination of power values that minimizes the annual electricity cost while meeting the infrastructure needs.

In this sense, the first two steps towards applying Demand Response actions have been covered by analyzing the degree of tariff optimisation that can be achieved by a set of buildings on two cases: (a) individually and (b) connected to the same single supply point. Results show that the optimisation achieved with the evolutionary algorithm is greater than that obtained with the traditional method.

The future work in this area will regard the remaining phases in the Demand Response approach. On the one hand, the improvement of the model to include variables directly related to the equipment available in the building so as to deepen in the last step of the methodology, namely, to provide knowledge to the building managers about the type and kind of loads that can be shifted along time and to perform behavioural change actions towards its optimization. Loads such as those related to the charging of batteries for electric equipment, the use of printing machines, or the execution of heavy experiments are prone to be deferrable and provide an interesting context to test demand-side management in the workplace. On the other hand, it is of great importance to analyze how this load shifting affect users. If the response to this change is positive and the impact is low, further analysis can be carried out to devise adequate persuasive strategies in order to engage and motivate users in the workplace to take action and engage in load shifting.

Acknowledgments. We acknowledge the support of the Spanish government for SentientThings project under Grant No.: TIN2017-90042-R.

References

1. Ratnatunga, J.: Carbon cost accounting: the impact of global warming on the cost accounting profession. J. Appl. Manag. Acc. Res. **5**, 01 (2007)
2. Kamal, W.: Improving energy efficiency–the cost-effective way to mitigate global warming. Energy Convers. Manag. **38**(1), 39–59 (1997)
3. Allouhi, A., El Fouih, Y., Kousksou, T., Jamil, A., Zeraouli, Y., Mourad, Y.: Energy consumption and efficiency in buildings: current status and future trends. J. Clean. Prod. **109**, 118–130 (2015)
4. Vecchiato, D., Tempesta, T.: Public preferences for electricity contracts including renewable energy: a marketing analysis with choice experiments. Energy **88**, 168–179 (2015)
5. Zhang, J., et al.: Blockchain based intelligent distributed electrical energy systems: needs, concepts, approaches and vision. Zidonghua Xuebao/Acta Automatica Sinica **43**(9), 1544–1554 (2017)
6. Zhao, S., Wang, B., Li, Y., Li, Y.: Integrated energy transaction mechanisms based on blockchain technology. Energies **11**(9), 2412 (2018)
7. Wain, N.: Households still baffled by energy bills and vote them the most difficult paperwork to understand despite rules to make them clearer. Thisismoney.co.uk, October 2014
8. Kuster, C., Rezgui, Y., Mourshed, M.: Electrical load forecasting models: a critical systematic review. Sustain. Cities Soc. **35**, 257–270 (2017)
9. Borges, C.E., Penya, Y.K., Fernández, I.: Optimal combined short-term building load forecasting. In: 2011 IEEE PES Innovative Smart Grid Technologies, pp. 1–7, November 2011
10. Raza, M., Khosravi, A.: A review on artificial intelligence based load demand forecasting techniques for smart grid and buildings. Renew. Sustain. Energy Rev. **50**, 1352–1372 (2015)
11. Abdelaziz, E., Saidur, R., Mekhilef, S.: A review on energy saving strategies in industrial sector. Renew. Sustain. Energy Rev. **15**(1), 150–168 (2011)
12. Casado-Mansilla, D., et al.: A human-centric context-aware IoT framework for enhancing energy efficiency in buildings of public use. IEEE Access **6**, 31444–31456 (2018)
13. Kessels, K., Kraan, C., Karg, L., Maggiore, S., Valkering, P., Laes, E.: Fostering residential demand response through dynamic pricing schemes: a behavioural review of smart grid pilots in Europe. Sustainability (Switzerland) **8**(9), 929 (2016)
14. Quintal, F., Jorge, C., Nisi, V., Nunes, N.: Watt-I-See: a tangible visualization of energy. In: Proceedings of the International Working Conference on Advanced Visual Interfaces, ser. AVI '16, pp. 120–127. ACM, New York, NY, USA (2016)
15. Sugarman, V., Lank, E.: Designing persuasive technology to manage peak electricity demand in ontario homes. In: Proceedings of the 33rd Annual ACM Conference on Human Factors in Computing Systems, pp. 1975–1984. ACM (2015)
16. Mohsenian-Rad, A.-H., Wong, V., Jatskevich, J., Schober, R., Leon-Garcia, A.: Autonomous demand-side management based on game-theoretic energy consumption scheduling for the future smart grid. IEEE Trans. Smart Grid **1**(3), 320–331 (2010)

17. Kreith, F., Goswami, D.Y.: Energy Management and Conservation Handbook. CRC Press, Boca Raton (2007)
18. Reed, G.F., Lynn, F., Meade, B.D.: Use of coefficient of variation in assessing variability of quantitative assays. Clin. Diagn. Lab. Immunol. **9**(6), 1235–1239 (2002)

Occupancy Inference Through Energy Consumption Data: A Smart Home Experiment

Adamantia Chouliara$^{(\boxtimes)}$, Konstantinos Peppas, Apostolos C. Tsolakis$^{(\boxtimes)}$, Thanasis Vafeiadis, Stelios Krinidis, and Dimitrios Tzovaras

Information Technologies Institute, Centre for Research and Technology - Hellas, 6th km Harilaou - Thermi, 57001 Thessaloniki, Greece
{achouliara,tsolakis}@iti.gr,
http://www.iti.gr

Abstract. This work is addressing the problem of occupancy detection in domestic environments, which is considered crucial in the aspect of increasing energy efficiency in buildings. In particular, in contrast with most previous researches, which obtained occupancy data through dedicated sensors, this study is investigating the possibility of using total consumption solely obtained from central smart meters installed in the examined buildings. In order to evaluate the feasibility of this simplified approach, the supervised machine learning classifier Random Forest was trained and tested on the experimental dataset. Repeated simulation tests show encouraging results achieving a high average performance with accuracy of 85%.

Keywords: Occupancy inference · Energy consumption · Smart meters · Machine learning · Random forest

1 Introduction

Recent years have seen an increased effort to produce methodologies for achieving higher energy efficiency in buildings as well as continuous research into assisting users to adopt a less energy consuming behaviour. Various energy saving methodologies are therefore being explored and the subject of detecting whether a house is occupied or empty is raised, as most of them have occupancy monitoring as a basic requirement. For instance, resolving this subject would be of great value to building automation systems, typically used to describe technologies which monitor Heating, Ventilation, and Air Conditioning (HVAC) systems and manage building facilities in general [1,14] which have recently been gaining ground. At the same time, it has been proved that energy-unaware behaviour can add significantly to a building's overall energy consumption and various approaches have been tested to aid occupants into optimizing their energy consumption habits in the long run [10].

© Springer Nature Switzerland AG 2019
D. Tzovaras et al. (Eds.): ICVS 2019, LNCS 11754, pp. 670–679, 2019.
https://doi.org/10.1007/978-3-030-34995-0_61

This is the overall goal of the SIT4Energy project, which provides a smart energy management solution considering efficiency potentials in the local energy production and consumption while implementing an intelligent mobile recommendation service to provide personalized insights and suggestions to the end users. The general idea of optimizing buildings and by extension users' energy behaviour requires occupancy monitoring as an initial step [2]. In the scope of the SIT4Energy project, the inferred occupancy patterns can be integrated into a more efficient recommendation system which can guide inhabitants to minimize the unneeded energy expenditure during unoccupied building states, while they can also be incorporated in optimized scheduling of building automation systems [5].

Extending previous work performed by Vafeiadis et al. [23], this manuscript is investigating the use of total energy consumption indication obtained from electricity meters as an occupancy sensing infrastructure in residential settings. To this end, an extensive experiment has been run to collect electricity consumption data of a Smart Home, serving as a pilot site for the SIT4Energy project, for over 10 months. Supervised machine learning algorithms were used to evaluate the occupancy detection accuracy achievable on this dataset. The results show that occupancy classification from electricity consumption data is feasible, as an average detection accuracy of 85% can be obtained. Due to the novel nature of the subject and the sparsity of performance data from smart meter installations [8], this is one of the first studies that analyze the possibility of occupancy inference using a single value of electricity consumption and including a variety of features as added value information for improving results.

Related work in the literature is discussed in Sect. 2, followed by the methodology proposed in Sect. 3. The overall pilot setup, the project's data collection and the results obtained are demonstrated in Sect. 4. Finally, the manuscript is concluded in Sect. 5.

2 Related Work

Several authors have dealt with evaluation of approaches to detect occupancy in both commercial and residential buildings. A comprehensive overview of the subject focusing on user activity is provided by Nguyen and Aiello [10] who rate the most effectual activities and behaviours on energy saving potential and highlight the significance of occupancy detection in building automation systems.

Recent works [2] distinguish occupancy monitoring systems into direct and indirect methods. While direct methods demand installation of extra physical devices in the buildings [11,12], indirect approaches use other forms of existent information to extract conclusions [1,5,8]. Most used devices in direct methods include wireless motion sensors, CO_2 sensors, door sensors [3,13] and Passive Infrared (PIR) sensors with reed switches [14]. Several strategies involve interrogating sensors carried by the residents, such as Radio-frequency identification (RFID) tags, dedicated wireless transmitters or GPS modules embedded in mobile phones [14], some others also combine PIR sensors with call monitoring

[16] or microphones [17], while another solution incorporates dedicated camera networks [15]. All these high-fidelity sensors have the disadvantage of intrusiveness and are often accused of creating confusion between occupancy detection and surveillance.

On the contrary, indirect methods offer a much more practical solution, limited in requiring off-the-shelf digital electricity meters and are eventually considered as far less invasive to the user. In this respect, an upcoming, quite promising potential has been recently explored, considered as Non-Intrusive Occupancy Monitoring (NIOM) which uses indirect information to infer occupancy [3]. This approach seems more beneficial and financially affordable compared to the conventional sensor-using common techniques as it only needs electricity consumption measurement through smart meters which are already widely deployed. Most of the related researches explore the possibility of having device-level readings leading to estimations about appliance usage mining, activity detection and subsequently occupancy prediction [18, 19]. Some of them [6, 7] obtain occupancy detection using combinations of individual circuits and selected plugs from multiple households. Kleiminger et al. (2015) [4] used classification techniques on the publicly available Electricity Consumption and Occupancy (ECO) dataset and utilizing features that relate to the operation of occupancy-relevant appliances achieved accuracy between 83% and 94% on occupancy detection at households. In the research of Vafeiadis et al. (2017) [22], a combined dataset of electricity and water consumption measurements is evaluated under various machine learning techniques for occupancy detection. Kleiminger et al. (2013) [8] gathered data from digital electricity meters, smart plugs and PIR sensors as well as ground-truth occupancy data in domestic environments, resulting in accuracy of over 80%. The strategy of inferring the disaggregated device activity information from aggregate electricity data consists a field in non intrusive load monitoring (NILM) and has been initially introduced by Hart [20]. However, later studies have shown that limited appliances are able to be detected by current algorithms, while also perplexing the situation requiring additional instrumentation, more training and higher data collection resolution [1].

On that account, recent researches are trying to simplify the overall procedure exploring the possibility of occupancy detection using only the installation of a single digital electricity meter [1]. Chen et al. [3] proved that even a simple threshold based NIOM algorithm performs well in inferring occupancy through average power, although noting that probably better conclusions would be drawn with the use of machine learning techniques over a larger dataset collection. Another study [2] showed that it is achievable to predict occupancy in multiple households by using solely electricity consumption data of 6 months and explored several possible combinations of features to extract with an accuracy of up to 78%. Evidently, the assumption that electricity consumption measurements might indicate the presence of residents in a household is intuitive and has already been examined [8, 9]. Although the viability of this indirect approach is encouraged by the above efforts, any comprehensive, quantitative analyses of the accuracy in occupancy detection through electricity meters, especially with the limitation of a single indicative measurement hardly exist, probably due to the difficulty of finding large datasets including both electricity consumption data and ground-truth occupancy information [8]. As a consequence, the research so far is limited and still remains an area for further experimentation. This study is building and extending upon previous work [23] and attempts to explore an improved solution regarding the predictive model to apply and the features to select for achieving better results, analyzing extensive data input provided by a Smart Home testbed.

3 Methodology

Recent literature [2,22,23], through repeated simulation results, indicates that decision tree and more specifically Random Forest machine learning classifiers achieve the highest predictive performance compared to other tested classification techniques on the occupancy classification scenario. Due to its superiority over other techniques on similar cases, random forest was selected for the present experiment. Random forest is defined as a classifier consisting of a collection of tree structured classifiers $\{h(x, \Theta_k), k = 1, ... \}$ where the $\{\Theta_k\}$ are independent identically distributed random vectors and each tree casts a unit vote for the most popular class at input x [21]. Being an ensemble of decision trees, random forest is one of the best among classification algorithms suitable for large amounts of data. Each decision tree is constructed by using a random subset of the training data, while the output class is the average of the classes decided by each decision tree [21].

In order to evaluate the selected machine learning algorithm's performance over a model tested on the given dataset, an experiment was performed using CERTH/ITI's installed equipment. The workflow followed, as depicted in Fig. 1, is composed of several steps. After the initial data collection, the procedures of pre-processing of raw data for converting it to time-series, re-indexing to stable time intervals and cleaning missing values were required. The basic conditions under which the pre-processing phase was set were to retain a 15-minute time interval between recordings and to fill missing values taking into account neighbouring existing measurements. Weekends, during which occupants' presence was uncertain were excluded from the final dataset. The data cleaning procedure was followed by new feature extraction. Concerning the search of the optimal set of features to select, there were suggestions provided by similar works as well. According to them, occupancy tends to be correlated with the absolute value of the energy consumption, variability of the energy consumption and time [1,2,8]. Based on these studies, the following features were created as depicted in Table 1. The principal feature of the set is the mean energy estimated on a short sized rolling window as it usually conforms to short intervals of data. Similarly, regarding the standard deviation and the maximum values of consumption, the respective features are also calculated over the same 3-hour sized rolling window.

The rest of the features were extracted as proposed in the literature [2] by the following equations.

$$c\_mean\_sd\_neighbour_i =$$

$$= \sqrt{\frac{1}{3}\left[(c\_mean_{i-1} - \mu)^2 + (c\_mean_i - \mu)^2 + (c\_mean_{i+1} - \mu)^2\right]} \qquad (1)$$

Where μ represents the mean of three neighbouring elements, given by the equation:

$$\mu = \frac{c\_mean_{i-1} + c\_mean_i + c\_mean_{i+1}}{3} \qquad (2)$$

$$c\_mean\_sad_i = |c\_mean_i - c\_mean_{i-1}| + |c\_mean_i - c\_mean_{i+1}| \qquad (3)$$

Table 1. Features extracted over 15-minute time intervals

Feature	Description
c_mean	Rolling mean consumption over a 3-hour rolling window
c_sd	Rolling standard deviation of the consumption
c_mean_sd_neighbour	Standard deviation of the mean consumption between neighbours
c_mean_sad	Sum of absolute differences of the mean consumption
c_max	Rolling maximum consumption

Next, the normalization method was applied to address the heterogeneity of fields in terms of scale and range. At this stage, the entire range of values from min to max were mapped to the range 0 to 1 to keep features more consistent with each other and allow for better evaluation. The normalization method is given by the equation:

$$X' = (X - min(X))/(max(X) - min(X)) \qquad (4)$$

Finally, the feature set was completed by the inclusion of occupancy ground truth information as implied by known daily schedule.

For the validation of the results, a training set and a testing set is generated randomly, in a percentage of 70% and 30%, respectively of the given dataset. Moreover, 100 Monte Carlo iterations were executed for different parameter scenarios. Performing 100-fold cross validation is aiming to eliminate the bias and derive more credible results. The random forest classifier is provided with an ensemble of 100 estimators standing for the corresponding decision trees.

4   Experimental Results

4.1   Pilot Setup and Data Collection

As a pilot at Smart Home testbed at CERTH/ITI's premises, an experiment was employed and the information required was obtained with the use of energy smart meters. The measurements are recorded in units of kwh at 15-minute time intervals during a 10 month period from June 2018 to March 2019. Another available estimation is occupancy, which indicates which of the two phases, absence (0) or presence (1) is valid at each timestamp and is extracted in accordance to the daily schedule of regular working hours, approximated as 09:00–17:00. This measurement is used as ground truth for training the machine learning model. To provide a clearer data visualization, a representative figure is displayed below (Fig. 2) including a chart of an indicative day's consumption and another one illustrating the typical changes in consumption over several days.

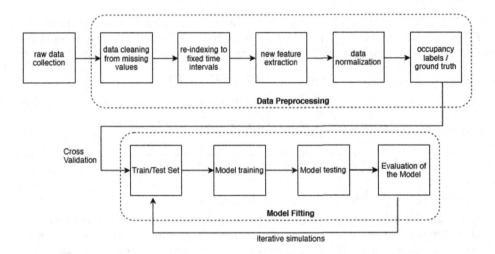

Fig. 1. Experimental workflow

The main objective of the study is to find an inference model that can estimate efficiently occupancy given the energy consumption data. On that account, the simulation schema tested is based on the implementation of the previously explained random forest classifier.

4.2   Evaluation Metrics

Many evaluation metrics can be used to evaluate a classifier. As the current experiment concerns a two-class classification scenario, in order to assess the model, the measures of precision, recall, accuracy and F-measure are used, which

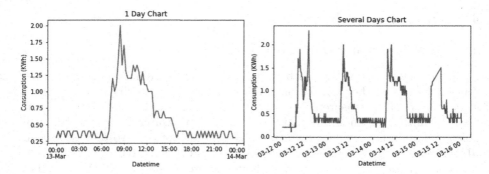

Fig. 2. The daily energy consumption indicates an increase between the hours of occupancy (09:00–17:00).

are computed from the contents of the confusion matrix of the classification predictions.

The above metrics are taking into account the predicted true positive cases (TP), the false positives (FP), the true negatives (TN) and false negatives (FN) and are given respectively by the following equations:

$$Precision = TP/(TP + FP), \tag{5}$$

$$Recall = TP/(TP + FN). \tag{6}$$

$$Accuracy = (TP + TN)/(TP + FP + TN + FN). \tag{7}$$

$$F\_measure = \frac{2 * Precision * Recall}{Precision + Recall} \tag{8}$$

4.3   Evaluation Results

From simulation results presented in Table 2 it can be seen that random forest classifier achieves high predictive performance regarding occupancy inference. More specifically, after 100 iterations for different parameter scenarios, the average precision was 87% while the accuracy value reaches 85%. The recall and F-measure metrics are estimated as an average of 93% and 90% respectively.

Table 2. Experimental Results. The values of metrics are on average from 100-fold cross validation.

Precision	Recall	Accuracy	F-measure
87%	93%	85%	90%

Equally important is the value of the Area under the Receiver Operating Characteristic (ROC) Curve which provides an aggregate measure of performance across all possible classification thresholds and was estimated as 86%.

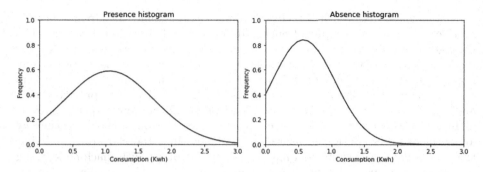

Fig. 3. As opposed to the case of "Presence", the majority of consumption values during "Absence" are distributed on the left side of the histogram.

Similar conclusions, supporting the hypothesis that states of occupancy are related to an increase in energy consumption, are also derived by the difference between the average consumption observed at each state. As demonstrated in the Table 3 below, the mean consumption of the presence state is much greater than the average in the state of absence. A similar difference is also implied by the two central values indicated by the corresponding medians. Using a Two-Tailed Test, there seems to be a statistically significant difference between the two samples' averages in energy consumption ($t = 51.58 > 1.96$(t-critical), $p < 0.0001$, 95% confidence level). Regarding the distributions of the two samples, in the case of absence, the consumption values display a peak on the left side of the histogram, whereas in the case of presence the consumption distribution is shifted in higher values (Fig. 3).

Table 3. Mean, Median and Standard Deviation.

	Absence	Presence
Mean	0.5745	1.0599
Std	0.4732	0.6778
Median	0.4	1.0

5   Conclusions

The present study is focusing on the emerging problem of identifying whether a residential environment is occupied or not, commonly known as occupancy detection. Its purpose is to explore a simple methodology, based solely on the overall energy consumption of the smart home and the use of well-known supervised machine learning classification techniques. In such a way, it evaluates the feasibility of utilizing entirely indirect monitoring methods in order to decrease intrusiveness as well as avoid impractical implications of additional device installments as well as financial unaffordability.

The proposed approach is trying to predict smart home's occupancy per 15-minutes in a day. The input of energy data is provided by smart meters. The simulation is based on previous works' conclusions and experiments on the prediction efficiency of random forest classifier and improves its efficiency by exploiting a large dataset of ten months' measurements. Results show that the random forest machine learning classifier, provided with extensive data pre-processing and including a variety of features as added value information, displays an overall high predictive performance achieving an average of 85% and 90% on accuracy and F-measure evaluation measures, respectively. It is therefore observed that the overall energy consumption measurements combined with machine learning techniques have great potentials regarding the problem of occupancy detection. It is in the project's future plans to develop a more context-aware approach by taking into account parameters additional to the pilot's experimental data, such as integration of billing data, monthly energy consumption and environmental conditions.

Aknowledgement. The SIT4Energy project has received funding from the German Federal Ministry of Education and Research (BMBF) and the Greek General Secretariat for Research and Technology (GSRT) in the context of the GreekGerman Call for Proposals on Bilateral Research and Innovation Cooperation, 2016.

References

1. Kleiminger, W., Beckel, C., Santini S.: Household occupancy monitoring using electricity meters, Japan, vol. 25, no. 3, pp. 273–291. ACM (2015). https://doi.org/10.1145/2750858.2807538
2. Pereira, D.F.: Occupancy Prediction from Electricity Consumption Data in Smart Homes. Instituto Superior Técnico, Universidade de Lisboa, Portugal (2017)
3. Chen, D., Barker, S., Subbaswamy, A., Irwin, D., Shenoy, P.: Non-Intrusive Occupancy Monitoring using Smart Meters. University of Massachusetts Amherst Vanderbilt University (2013)
4. Kleiminger, W.: Occupancy Sensing and Prediction for Automated Energy Savings. ETH Zurich (2015). https://doi.org/10.3929/ethz-a-010450096
5. Ardakanian, O., Bhattacharya, A., Cullere, D.: Non-intrusive occupancy monitoring for energy conservation in commercial buildings. Energy Build. **179**, 311–323 (2018). https://doi.org/10.1016/j.enbuild.2018.09.033
6. Kolter, J.Z., Johnson, M.J.: REDD: a public data set for energy disaggregation research. ACM, USA (2011)
7. Beckel, C., Kleiminger, W., Cicchetti, R.: The ECO data set and the performance of non-intrusive load monitoring algorithms. ACM, USA (2014)
8. Kleiminger, W., Beckel, C., Staake, T., Santini, S.: Occupancy detection from electricity consumption data. ACM, Italy (2013)
9. Molina-Markham, A., Shenoy, P., Fu, K., Cecchet, E., Irwin, D.: Private memoirs of a smart meter, pp. 61–66. ACM, Switzerland (2010)
10. Nguyen, T.A., Aiello, M.: Energy intelligent buildings based on user activity: a survey. Energy Build. **56**, 244–257 (2012). https://doi.org/10.1016/j.enbuild.2012.09.005

11. Agarwal, Y., Balaji, B., Gupta, R., Lyles, J., Wei, M., Weng, T.: Occupancy-driven energy management for smart building automation. In: Proceedings of the 2nd ACM Workshop on Embedded Sensing Systems for Energy-Efficiency in Building, pp. 1–6. ACM, Switzerland (2010)
12. Akbar, A., Nati, M., Carrez, F.: Contextual occupancy detection for smart office by pattern recognition of electricity consumption data. In: IEEE International Conference on Communications, UK (2015)
13. Lu, J., et al.: The smart thermostat: using occupancy sensors to save energy in homes, pp. 211–224. ACM, Switzerland (2010)
14. Kleiminger, W., Mattern, F., Santini., S. : Predicting household occupancy for smart heating control: a comparative performance analysis of state-of-the-art approaches. Energy Build. **85**, 493–505 (2014)
15. Erickson, V.L., Achleitner, S., Cerpa, A.E.: POEM: power-efficient occupancy-based energy management system. In: Proceedings of the 12th International Conference on Information Processing in Sensor Networks (IPSN 2013), pp. 203–216. ACM/IEEE, Germany (2013)
16. Dodier, R.H., Henze, G.P., Tiller, D.K., Guo, X.: Building occupancy detection through sensor belief networks. Energy Build. **38**(9), 1033–1043 (2006)
17. Padmanabh, K., et al.: iSense: a wireless sensor network based conference room management system. In: Proceedings of the 1st ACM Workshop on Embedded Sensing Systems for Energy-Efficiency in Buildings (BuildSys 2009). ACM, USA (2009)
18. Monachi, A.: GREEND: an energy consumption dataset of households in Italy and Austria. In: Proceedings of IEEE SmartGridComm, Italy (2014)
19. Froehlich, J., Larson, E., Gupta, S., Cohn, G., Reynolds, M., Patel, S.: Disaggregated end-use energy sensing for the smart grid. IEEE Pervasive Comput. **10**(1), 28–39 (2011)
20. Hart, G.W.: Nonintrusive appliance load monitoring. Proc. IEEE **80**(12), 1870–1891 (1992)
21. Breiman, L.: Random forests –random features. Statistics Department, University of California, Berkeley (1999). ftp://ftp.stat.berkeley.edu/pub/users/breiman
22. Vafeiadis, T., et al.: Machine learning based occupancy detection via the use of smart meters. In: International Conference on Energy Science and Electrical Engineering (2017). https://doi.org/10.1109/ISCSIC.2017.15
23. Vafeiadis, A., et al.: Energy-based decision engine for household human activity recognition. In: IEEE International Conference on Pervasive Computing Pervasive Computing and Communications Workshops (PerCom Workshops), Greece (2018)

A Dynamic Convergence Algorithm for Thermal Comfort Modelling

Asimina Dimara, Christos Timplalexis, Stelios Krinidis$^{(\boxtimes)}$,
and Dimitrios Tzovaras

Centre for Research and Technology Hellas/Information Technologies Institute,
6th Km Charilaou-Thermis, 57001 Thermi-Thessaloniki, Greece
{adimara,ctimplalexis,krinidis,Dimitrios.Tzovaras}@iti.gr

Abstract. This paper attempts to utilize experimental results in order
to correlate clothing insulation and metabolic rate with indoor temperature. Inferring clothing insulation and metabolic rate values from
ASHRAE standards is an alternative that totally ignores environmental
conditions that actually affect human clothing and activity. In this work,
comfort feedback regarding occupants' thermal sensation is utilized by
an algorithm that predicts clothing insulation and metabolic rate values.
The analysis of those values reveals certain patterns that lead to the
formulation of two non-linear equations between clothing – indoor temperature and metabolic rate – indoor temperature. The formulation of
the equations is based on the experimental results derived from the thermal comfort feedback provided by actual building occupants. On trial
tests are presented and conclusions regarding the method's effectiveness
and limitations are drawn.

Keywords: Thermal comfort · Metabolic rate · Clothing insulation ·
Indoor temperature · User feedback

1 Introduction

The main European objectives require an alternation in energy consumption
behaviour by energy saving. The most challenging task, is that the energy saving
must be achieved with comfortable approaches for the residents. In order to
determine the subjective comfort levels of the occupants there must be a method
for the detection of both indoor climate conditions and occupants estimation.
This indoor climate data is based on sensor generated inputs and it includes
the metering of indoor humidity, indoor temperature and indoor luminance.
The occupants' estimation is based on users' feedback about their comfort level
which can be provided by a mobile or web application.

Thermal comfort indicates the human satisfactory perception of the indoor
environment. Environmental and personal conditions must be estimated for people to feel comfortable. Thermal comfort is provided by the predicted mean vote
(PMV) [2]. The Standard ISO 7730 [2] defines that the PMV is affected by four

© Springer Nature Switzerland AG 2019
D. Tzovaras et al. (Eds.): ICVS 2019, LNCS 11754, pp. 680–689, 2019.
https://doi.org/10.1007/978-3-030-34995-0_62

physical variables, which are air temperature, mean radiant temperature, air humidity and relative air velocity and two personal variables, that are metabolic rate and clothing insulation.

The physical variables needed for thermal comfort estimation can be effortlessly given by technological means like indoor air sensors and indoor humidity and temperature sensors. Diversely, the personal factors are laborious to be estimated. The most precise method to calculate clothing insulation is by thermal manikins [1]. Another approach for clothing insulation estimation is by using scientific questionnaires [11].

Metabolic rate is also a difficult factor of the thermal comfort function to be evaluated. Many studies estimating thermal comfort simply compute PMV by utilizing activities having low metabolic rates (like seating, relaxing and standing) [4,7,13] when calculating PMV, those activities are ranked based on ASHRAE tables [6]. This kind of approach is inaccurate as it excludes main basic indoor activities. Another way to measure metabolic rate is by wearable or portable metabolic devices. Those devices are expensive and are proven not accurate enough [9,12]. As a consequence, they are rarely used.

To overcome all the above issues, initially the thermal comfort is calculated using an assumption for clothing insulation and metabolic rate based on the tables provide by ASHRAE [6]. Afterwards, these two factors are predicted based on user's feedback for thermal comfort. Consequently, PMV is calculated using the updated values of clothing insulation and metabolic rate. The assumption is that indoor temperature affects the way we dress and act in indoor spaces. We propose a dynamic convergence algorithm, which in case of lack of user feedback, updates clothing insulation and metabolic rate values progressively according to indoor temperature.

2   Clothing Insulation and Metabolic Rate Estimation

There are 7 points at the thermal sensation scale, according to ASHRAE thermal comfort scale (in Table 1) [6].

Table 1. ASHRAE thermal comfort scale

Value	Sensation
+3	Hot
+2	Warm
+1	Slightly warm
0	Neutral
−1	Slightly cool
−2	Cool
−3	Cold

The most commonly used method for computing thermal comfort has been suggested by Fanger [3]. The final PMV is calculated by a set of equations. All the equations are described below.

Identification of a skin temperature and sweating rate required for comfort conditions [10]:

$$T_{sk,req} = 96.3 - 0.156 q_{met,heat}. \tag{1}$$

$$q_{sweat,req} = 0.42(q_{met,heat} - 18.43). \tag{2}$$

$$q_{met,heat} = M - \dot{w}, \tag{3}$$

where T_{sk} is the average skin temperature ($^\circ F$), M is the rate metabolic generation per unit DuBois surface area ($Btu/h\ ft^2$), and w is the human work per unit DuBois surface area ($Btu/h\ ft^2$).

Upon those conditions Fanger, corelated PMV as a function to the thermal load L on the body.

$$\begin{aligned}
L = {} & q_{met,heat} \\
& - f_{cl} h_c (T_{cl} - T_a) \\
& - f_{cl} h_r (T_{cl} - T_r) \\
& - 156(W_{sk,req} - W_a) \\
& - 0.42(q_{met,heat} - 18.43) \\
& - 0.00077 M(93.2 - T_a) \\
& - 2.78 M(0.0365 - W_a),
\end{aligned} \tag{4}$$

where clothing temperature is calculated from the required skin temperature:

$$\frac{T_{sk,req} - T_{cl}}{R_{cl}} = f_{cl} h_c (T_{cl} - T_a) + f_{cl} h_r (T_{cl} - T_r), \tag{5}$$

$$T_{cl} = \frac{T_{sk,req} + R_{cl} f_{cl} (h_c T_a + h_r T_r)}{1 + R_{cl} f_{cl} (h_c + h_r)}, \tag{6}$$

where

$$f_{cl} = \begin{cases} 1.0 + 0.2 I_{cl} & I_{cl} < 0.5 clo \\ 1.05 + 0.1 I_{cl} & I_{cl} > 0.5 clo \end{cases}, \tag{7}$$

$$h_c = \max \begin{cases} 0.361(T_{cl} - T_a)^{0.25} \\ 0.151\sqrt{V} \end{cases}, \tag{8}$$

$$h_r = 0.7 Btu/h f t^{2^\circ} F. \tag{9}$$

The final equation is given by the correlation between PMV and the thermal Load [3] and is given by:

$$PMV = 3.155 \left(0.303 e^{-0.114 M} + 0.028\right) L. \tag{10}$$

2.1   Clothing Insulation and Metabolic Rate Estimation

Fanger's PMV equation [3] needs four main factors to be computed: temperature (T_a), humidity (RH), clothing insulation (I_{cl}) and metabolic rate (M). Thus, thermal comfort is dependent upon those factors:

$$PMV = (T_a, RH, I_{cl}, M). \tag{11}$$

Temperature and humidity are received by indoor metering sensors. Clothing insulation and metabolic rate are initially assumed based on ASHRAE [6] tables. Consequently, PMV is calculated with two different approaches based on the existence or lack of the users' feedback.

In case a user gives feedback, clothing insulation and metabolic rate values are predicted based on the PMV feedback value. The values of I_{cl}, M are calculated by solving Fanger's equation [3], where indoor temperature and humidity are estimated from the time the user feedback is provided. The new I_{cl}, M are calculated from a pre-trained model that utilizes as inputs the thermal comfort feedback, the temperature and the humidity. This model calculates I_{cl}, M for selected values of T_a and RH and was trained using Fanger's "Comfort Equation" [3]. The formulated problem requires the prediction of multiple continuous variables $y_i = (M, I_{cl})$ from a vector of k input variables $x_i = (PMV_{feedback}, T_a, RH)$. This is a multi-target regression (MTR) problem so extremely randomized trees were selected [5] for I_{cl}, M prediction.

The total observations of clothing insulation and metabolic rate values that are predicted from feedback are correlated to indoor temperature using a non-linear regression model [5]. The outcome of this regression model is two equations for clothing insulation and metabolic rate which are both indoor temperature dependent. The new clothing insulation value is estimated by:

$$I_{cl} = f(T_a) = 89.279(T_a)^{-1.592}. \tag{12}$$

And the new metabolic rate value is given by:

$$M = f(T_a) = 3081.9(T_a)^{-1.173}. \tag{13}$$

Whenever there is a feedback thermal comfort is estimated utilizing the predicted clothing and activity values. On the other hand, thermal comfort is estimated utilizing I_{cl} and M that results from the correlation of indoor temperature to clothing insulation and metabolic rate. The overall flow-chart of the algorithm is presented in Fig. 1:

2.2   Indoor Building Study

The study takes place in 157 households. The number of the active users is in average 157. Users are asked "how they feel" regarding their thermal comfort scale and their feedback is saved along with the exact time it is given. The indoor conditions from the users' room are monitored by humidity and temperature

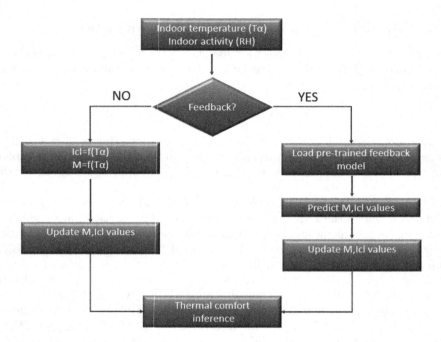

Fig. 1. Thermal comfort flow chart

Table 2. Number of users and feedbacks.

Month	Number of users	Number of feedbacks
September 2018	12	58
October 2018	22	101
November 2018	29	201
December 2018	24	218
January 2019	30	191
February 2019	19	207
March 2019	13	263

sensors and saved every 15 min. As soon as the feedback is given, along with the indoor environmental values, the values of clothing insulation and metabolic rate that are predicted are updated for each user (Table 2).

Average indoor temperature is depicted in Fig. 2 and outdoor average temperature is shown in Fig. 3. Indoor temperature is not affected by the outdoor temperature as it has less fluctuations and smaller range of values.

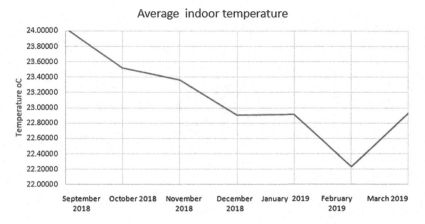

Fig. 2. Average indoor temperature in °C.

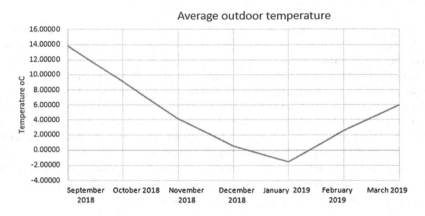

Fig. 3. Average outdoor temperature

3   Results

The relationship between temperature and clothing insulation has been thoroughly examined by Morgan [8], but the indoor environment examined was not domestic. Moreover, the indoor environment tested (shopping mall, offices) is an indoor environment that has a dress-code (casual, formal) and clothes worn are chosen based on the fact that people have to go outside before arriving to the destination the experiment is done. Our sample refers to indoor clothing insulation and metabolic rate in households. The results from the calculation of clothing insulation compared to indoor temperature monitored, are depicted in Fig. 4.

The relationship between clothing insulation and Indoor temperature is as expected, inversely proportional. The colder it gets the more clothes someone

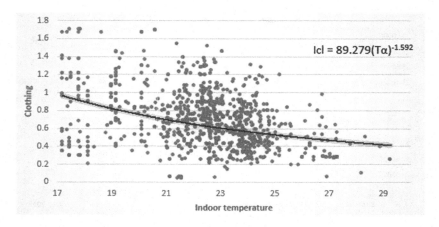

$$Icl = 89.279(T\alpha)^{-1.592}$$

Fig. 4. Clothing related to indoor monitored temperature.

is wearing. Statistical analysis of the correlation clothing – indoor Temperature is given in Table 3. Adjusted R-Square (Table 3), is a statistical measure of how close the data are to the fitted regression line and the p-value tests the null hypothesis, that the coefficient is equal to zero. When p-value is below the confidence interval it suggests strong evidence against the null hypothesis.

As seen in Table 3, adjusted R-square's value is relatively low (0.14) but this is justified by the fact that clothing is subjective and may differ from person to person. Morgan's study [8] adjusted R-square value is 0.24 which is also low but reveals the same pattern observed in this study, as viewed in Fig. 5. The inclusion of more variables in the model could probably improve R-square but this is beyond the scope of the current study which attempts a more precise clothing inference utilizing only temperature measurements.

Table 3. Non-linear regression statistics for clothing insulation.

Nonlinear regression statistics Clothing-indoor temperature	Value
Adjusted R Square	0.137879952
P-value	≤ 0.01
Observations 2018	1288

The equation that came as a result for the relationship between clothing worn inside buildings [8] by Morgan compared to the equation in Fig. 4, is depicted in Fig. 5. It is observed that there is a small deviation between the two lines. This is a remarkable outcome considering the fact that Morgan's study was situated in a different continent, type of indoor environment, number of observations,

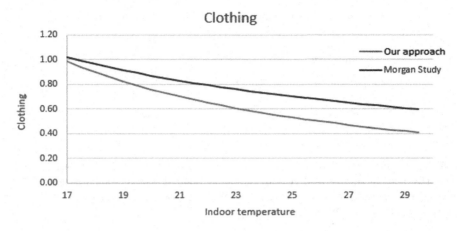

Fig. 5. Comparison of our approach and Morgan's study

season and year. The main assumption is that there is a significant relationship between indoor temperature and clothing insulation.

Likewise, the relationship between indoor temperature and metabolic rate is examined in Fig. 6. The statistical results are better than the clothing insulation as the R-square is almost 30% (Table 4).

The proposed equations are tested for their performance for calculating thermal comfort using real time indoor measurements. Two months were selected. One with "high" Indoor temperatures and one with "low" indoor temperature. Afterwards, a random day was selected for both of them. Thermal comfort was calculated at 15 min frequency, and the results are depicted in Fig. 7. As both

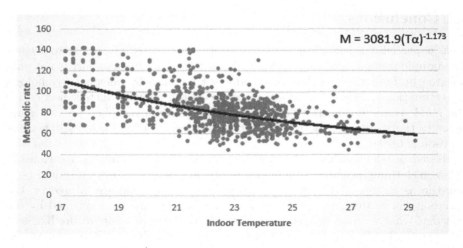

Fig. 6. Metabolic rate related to indoor monitored temperature

Table 4. Non-linear regression statistics for metabolic rate.

Nonlinear regression statistics Metabolic rate-indoor temperature	Value
Adjusted R Square	0.288332
P-value	≤0.01
Observations 2018	1288

clothing insulation and metabolic rate factors are correlated to indoor temperature the results of thermal comfort are the ideal. For this exact reason PMV values for high indoor temperatures are close to positive 0 and PMV values for low temperatures are close to negative 0.

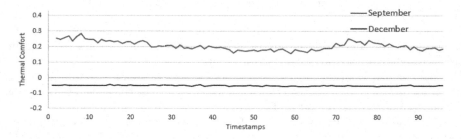

Fig. 7. Thermal comfort

4   Conclusions

This paper emphasizes on a dynamic algorithm that estimates thermal comfort in indoor environments. Based on the fact that thermal comfort is not only affected by indoor micro climatic parameters but also by personal psychological estimation the model concentrates more on the personal factors needed for the PMV computation. Both clothing insulation and metabolic rate values are proven a thorny task for thermal comfort evaluation, so a flexible but still feasible solution is tested.

Based on the feedback observations the non-linear regression relationship between clothing insulation and indoor temperatures advocates that indoor temperature is an essential factor of clothing worn inside buildings. Moreover, the variability in clothing insulation values in the sample may be explained by the fact that what we wear is affected by many factors except temperature like gender, age, and cold and heat tolerance. Furthermore, the non-linear regression relationship between metabolic rate and indoor temperatures also reveals that temperature is a dominant component of metabolic rate.

It is strongly believed that if the non-linear regression model is fitted individually and distinctly for every user the relationship between indoor temperature and personal factors will grow stronger and R-square will be better. After creating a user profile for even a small range of temperature values, clothing insulation and metabolic rate would be fitted to the ideal thermal comfort for each user.

Acknowledgements. This work is partially supported by the "enCOMPASS - Collaborative Recommendations and Adaptive Control for Personalised Energy Saving" project funded by the EU H2020 Programme, grant agreement no. 723059.

References

1. Fan, J., Chen, Y.: Measurement of clothing thermal insulation and moisture vapour resistance using a novel perspiring fabric thermal manikin. Meas. Sci. Technol. **13**(7), 1115 (2002)
2. Fanger, P.: Moderate thermal environments determination of the PMV and PPD indices and specification of the conditions for thermal comfort. ISO 7730 (1984)
3. Fanger, P.O., et al.: Thermal Comfort. Analysis and Applications in Environmental Engineering. Danish Technical Press, Copenhagen (1970)
4. Gao, S., Zhai, Y., Yang, L., Zhang, H., Gao, Y.: Preferred temperature with standing and treadmill workstations. Build. Environ. **138**, 63–73 (2018)
5. Geurts, P., Ernst, D., Wehenkel, L.: Extremely randomized trees. Mach. Learn. **63**(1), 3–42 (2006)
6. American Society of Heating, Refrigerating and Air-Conditioning Engineers: ASHRAE Standard: Thermal Environmental Conditions for Human Occupancy, vol. 55. ASHRAE (2010)
7. Luo, M., Wang, Z., Ke, K., Cao, B., Zhai, Y., Zhou, X.: Human metabolic rate and thermal comfort in buildings: the problem and challenge. Build. Environ. **131**, 44–52 (2018)
8. Morgan, C., de Dear, R.: Weather, clothing and thermal adaptation to indoor climate. Clim. Res. **24**(3), 267–284 (2003)
9. Mukhopadhyay, S.C.: Wearable sensors for human activity monitoring: a review. IEEE Sens. J. **15**(3), 1321–1330 (2015)
10. Rohles, F.: The nature of thermal comfort for sedentary man. ASHRAE Trans. **77**, 239–247 (1971)
11. Watanabe, K., Rijal, H.B., Nakaya, T.: Investigation of clothing insulation and thermal comfort in Japanese houses. In: PLEA (2013)
12. Yang, C.C., Hsu, Y.L.: A review of accelerometry-based wearable motion detectors for physical activity monitoring. Sensors **10**(8), 7772–7788 (2010)
13. Zhai, Y., Elsworth, C., Arens, E., Zhang, H., Zhang, Y., Zhao, L.: Using air movement for comfort during moderate exercise. Build. Environ. **94**, 344–352 (2015)

Thermal Comfort Metabolic Rate and Clothing Inference

Christos Timplalexis, Asimina Dimara, Stelios Krinidis$^{(\boxtimes)}$,
and Dimitrios Tzovaras

Centre for Research and Technology Hellas/Information Technologies Institute,
6th Km Charilaou-Thermis, 57001 Thermi-Thessaloniki, Greece
{ctimplalexis,adimara,krinidis,Dimitrios.Tzovaras}@iti.gr

Abstract. This paper examines the implementation of an algorithm for the prediction of metabolic rate (M) and clothing insulation (I_{cl}) values in indoor spaces. Thermal comfort is calculated according to Fanger's steady state model. In Fanger's approach, M and I_{cl} are two parameters that have a strong impact on the calculation of thermal comfort. The estimation of those parameters is usually done, utilizing tables that match certain activities with metabolic rate values and garments with insulation values that aggregate to a person's total clothing. In this work, M and I_{cl} are predicted utilizing indoor temperature (T), indoor humidity (H) and thermal comfort feedback provided by the building occupants. The training of the predictive model, required generating a set of training data using values in pre-defined boundaries for each variable. The accuracy of the algorithm is showcased by experimental results. The promising capabilities that derive from the successful implementation of the proposed method are discussed in the conclusions.

Keywords: Thermal comfort · Metabolic rate · Clothing insulation

1 Introduction

In modern societies people spend almost 90% of their time indoors [6]. Studies have shown that indoor thermal conditions may impact on the occupants' attendance and cognitive performance [10]. Consequently, indoor thermal conditions should be regulated so that they do not have any negative effect to the occupants' feeling or execution of activities. The American Society of Heating, Refrigerating and Air-Conditioning Engineers (ASHRAE) defines thermal comfort as "the condition of the mind in which satisfaction is expressed with the thermal environment" [1]. The definition emphasizes to the fact that it refers to a state of mind and not a standard condition. As such, it is different for every person and it is influenced by many factors such as age, gender, mood or culture.

Generally, comfort occurs when human temperature remains between a certain range, skin moisture stays low and human body makes a minimal effort for regulation. Lack of comfort is noticed when changes in human behavior are

© Springer Nature Switzerland AG 2019
D. Tzovaras et al. (Eds.): ICVS 2019, LNCS 11754, pp. 690–699, 2019.
https://doi.org/10.1007/978-3-030-34995-0_63

observed [2]. Changing clothes, changing posture, altering activity or just complaining are that type of behaviors. The parameters that affect thermal sensation were defined by MacPherson in 1962 [9]: air temperature, air speed, humidity, mean radiant temperature, metabolic rate and clothing insulation. Those six factors were later incorporated into a steady state heat transfer model developed by Fanger, utilizing experimental results from 1296 human subjects in a controlled climate chamber [4]. In these studies, participants were dressed in standardized clothing and completed specific activities, being exposed to different thermal environments. In some cases, the thermal conditions were chosen, while participants were recording their thermal sensation using the 7-point ASHRAE thermal sensation scale ranging from cold (-3) to hot (3) with neutral conditions at (0).

Analyzing the parameters that compose Fanger's comfort equation, it becomes obvious that measuring temperature, humidity and air speed can be done effortlessly using sensors. On the other hand, clothing habits and activity are more subjective factors and may as well require more complex equipment for their continuous registering. Thus, clothing insulation is either measured from human subjects or mannequins [3], or an initial assumption is made using ASHRAE tables. Metabolic rate is measured either by telling human subjects to perform certain activities, or assumed from tables similarly with clothing [8]. The current work attempts to personalize the thermal comfort computation process, by predicting different clothing insulation and metabolic rate values for every subject, depending on the thermal comfort feedback that they provide. This way, the thermal comfort computation becomes more accurate, without the need of ASHRAE tables, that significantly deviate from real-life indoor thermal comfort conditions.

The remaining of the paper is structured as follows: Sect. 2 elaborates on Fanger's thermal comfort model. Section 3 describes the algorithm that was formulated for M and I_{cl} prediction. In Sect. 4, experimental results that showcase the effectiveness of the algorithm are presented while conclusions are drawn in Sect. 5.

2   Thermal Comfort Calculation

As long as the users provide no feedback, their comfort is calculated from Fanger's equation which uses the Predicted Mean Vote (PMV) index in order to quantify the degree of thermal discomfort on the 7-point ASHRAE scale. Fanger's equation is based on the general heat balance equation that describes the process of heat exchange between a man and his environment [7]:

$$M - W = C + R + E_{sk} + (C_{res} + E_{res}). \tag{1}$$

The external work W (W/m^2) in the equation is small and is generally ignored under most situations. The internal energy production M (W/m^2) is determined by metabolic activity. C (W/m^2) is the heat loss by convection. R (W/m^2) is the heat loss by thermal radiation. E_{sk} (W/m^2) is the heat loss by evaporation from the skin. C_{res} (W/m^2) and E_{res} (W/m^2) are the sensible and the evaporation heat losses due to respiration respectively.

The convection heat transfer C (W/m^2) from the human body to the environment is given by:

$$C = f_{cl} \cdot h_c \cdot (T_{cl} - T_a), \tag{2}$$

where T_{cl} (°C) is the clothing surface temperature and T_a (°C) is the ambient air temperature. The heat transfer coefficient h_c $(W/m^2 \cdot K)$ depends on the air velocity V_a (m/s) across the body and consequently also upon the position of the person and orientation to the air current while the clothing area factor f_{cl} depends on the clothing insulation.

The radiation heat transfer between the body and surrounding surfaces is given from:

$$R = \sigma \cdot \varepsilon_{cl} \cdot f_{cl} \cdot F_{vf} \cdot \left[(T_{cl} + 273.15)^4 - (T_r + 273.15)^4 \right], \tag{3}$$

where ε_{cl} is the emissivity of the clothing. F_{vf} is the view factor between the body and the surrounding surface. σ is the Stefan-Boltzmann constant, which has the numerical value of $5.67 \cdot 10^{-8}$ W/m^2K^4. T_r (°C) is the radiant temperature. The surrounding surface temperature can be taken as approximately ambient air temperature T_a (°C). The respiration heat loss is divided into evaporative heat loss (latent heat) and sensible heat loss. The rate of the heat transfer by respiration is usually at the lower level beside the other rates of the heat transfer. This rate is given by:

$$C_{res} + E_{res} = 0.014 \cdot M \cdot (34 - T_a) + 0.0173 \cdot M \cdot (5.87 - P_a), \tag{4}$$

where P_a (P_a) is the partial vapour pressure.

The rate of the heat loss by evaporation is the removal of heat from the body by evaporation of perspiration from the skin. The heat loss by evaporation is made up of two, the insensible heat loss by skin diffusion and the heat loss by regulatory sweating. This rate can be calculated by:

$$E_{sk} = 3.05 \cdot (5.73 - 0.007 \cdot M - P_a) + 0.42 \cdot (M - 58.15)). \tag{5}$$

Finally, the PMV value is determined from the following equation:

$$PMV = (0.303 \cdot e^{-0.036 \cdot M} + 0.028) \cdot L, \tag{6}$$

where L is defined as follows:

$$L = M - W - C - R - E_{sk} - (C_{res} + E_{res}). \tag{7}$$

The air speed is set to 0.1 m/s, which is a typical value used by ASHRAE standard [1]. According to [11], the difference between air temperature and mean radiant temperature is negligible for indoor environments. Sensors are utilized for the acquisition of temperature and humidity data, while metabolic rate and clothing insulation are initialized according to ASHRAE standard [1], as shown in Figs. 1 and 2. The hypothesis for the metabolic rate and clothing insulation values begins by separating the day at five intervals. Regarding the metabolic rate, mild activities such as sitting, reclining, typing, reading are considered for each user during the day, while at night the user is considered to be sleeping. The clothing insulation takes into consideration the day interval in combination with the season of the year, in order to infer a user's clothing. Day and night separation is also made here, since during sleep the bed and the sheets provide some extra insulation.

TABLE A1   Metabolic Rates for Typical Tasks

Activity	Metabolic Rate		
	Met Units	W/m^2	(Btu/h·ft^2)
Resting			
Sleeping	0.7	40	(13)
Reclining	0.8	45	(15)
Seated, quiet	1.0	60	(18)
Standing, relaxed	1.2	70	(22)
Office Activities			
Reading, seated	1.0	55	(18)
Writing	1.0	60	(18)
Typing	1.1	65	(20)
Filing, seated	1.2	70	(22)
Filing, standing	1.4	80	(26)
Walking about	1.7	100	(31)
Lifting/packing	2.1	120	(39)

Fig. 1. Metabolic rates for typical tasks [1]

3   Metabolic Rate and Clothing Insulation Prediction from User Feedback

User feedback is utilized in order to revise the initial metabolic rate and clothing insulation values and steadily converge to the user's objectives. The final goal is to create a different profile for each user, since every person may have different dressing preferences and may perform different activities in his/her house during the day. The feedback provided by the users refers to their thermal sensation in terms of the PMV index. The correction of the metabolic rate and clothing insulation values is made only for the specific interval that the feedback is given. The new M, I_{cl} are calculated according to the following methodology (Table 1):

The first step towards building this model was the formulation of a training dataset. According to ASHRAE standards [1], M and I_{cl} have upper and lower limits. Discrete values were chosen within the respective boundaries for all of the variables that compose Fanger's equation (T, H, I_{cl}, M). The step that was used for the sampling of each variable, was selected considering the variable's impact on the final PMV outcome at the $[-3, 3]$ scale of PMV (Table 2).

The next step requires solving Fanger's equation for all of the possible states that were generated. The combination of T, H, I_{cl}, M values generate a total

TABLE B2
Garment Insulation*

Garment Description[†]	I_{clu}, clo	Garment Description[b]	I_{clu}, clo
Underwear		Dress and Skirts**	
Bra	0.01	Skirt (thin)	0.14
Panties	0.03	Skirt (thick)	0.23
Men's briefs	0.04	Sleeveless, scoop neck (thin)	0.23
T-shirt	0.08	Sleeveless, scoop neck (thick), i.e., jumper	0.27
Half-slip	0.14	Short-sleeve shirtdress (thin)	0.29
Long underwear bottoms	0.15	Long-sleeve shirtdress (thin)	0.33
Full slip	0.16	Long-sleeve shirtdress (thick)	0.47
Long underwear top	0.20	Sweaters	
Footwear		Sleeveless vest (thin)	0.13
Ankle-length athletic socks	0.02	Sleeveless vest (thick)	0.22
Pantyhose/stockings	0.02	Long-sleeve (thin)	0.25
Sandals/thongs	0.02	Long-sleeve (thick)	0.36
Shoes	0.02	Suit Jackets and Vests[††]	
Slippers (quilted, pile lined)	0.03	Sleeveless vest (thin)	0.10
Calf-length socks	0.03	Sleeveless vest (thick)	0.17
Knee socks (thick)	0.06	Single-breasted (thin)	0.36
Boots	0.10	Single-breasted (thick)	0.44
Shirts and Blouses		Double-breasted (thin)	0.42
Sleeveless/scoop-neck blouse	0.12	Double-breasted (thick)	0.48
Short-sleeve knit sport shirt	0.17	Sleepwear and Robes	
Short-sleeve dress shirt	0.19	Sleeveless short gown (thin)	0.18
Long-sleeve dress shirt	0.25	Sleeveless long gown (thin)	0.20
Long-sleeve flannel shirt	0.34	Short-sleeve hospital gown	0.31
Long-sleeve sweatshirt	0.34	Short-sleeve short robe (thin)	0.34
Trousers and Coveralls		Short-sleeve pajamas (thin)	0.42
Short shorts	0.06	Long-sleeve long gown (thick)	0.46
Walking shorts	0.08	Long-sleeve short wrap robe (thick)	0.48
Straight trousers (thin)	0.15	Long-sleeve pajamas (thick)	0.57
Straight trousers (thick)	0.24	Long-sleeve long wrap robe (thick)	0.69
Sweatpants	0.28		
Overalls	0.30		
Coveralls	0.49		

Fig. 2. Garment insulation [1]

Table 1. ASHRAE thermal comfort scale

Value	Sensation
+3	Hot
+2	Warm
+1	Slightly warm
0	Neutral
−1	Slightly cool
−2	Cool
−3	Cold

of 770.400 different states. After solving the equation, a mapping table was formulated that will be utilized as the training dataset for the model (Table 3).

When a user decides to give feedback about the thermal comfort conditions in his/her house, the given value is considered to be the actual PMV value for the specific timestamp. The task of the model is to use the given feedback along with the sensor data and predict the clothing insulation and the metabolic rate. The formulated problem requires the estimation of multiple continuous variables

Table 2. Variables sampling for the training dataset

Variable	Interval	Step	Impact of variable's step change on PMV value
Temperature	[18,31]	0.3	0.1
Humidity	[20,75]	5	0.05
Metabolic rate	[44,164]	4	0.05–0.2
Clothing insulation	[0.04,1.84]	0.04	0.02–0.2

Table 3. Mapping table

Temperature	Humidity	Clothing insulation	Metabolic rate	PMV
23.4	40	0.84	52	−0.58
23.4	40	0.84	56	−0.41
23.4	40	0.84	60	−0.25
23.4	40	0.84	64	−0.17
23.4	40	0.84	68	−0.09
23.4	40	0.84	72	0.21
23.4	40	0.84	76	0.32
23.4	40	0.84	80	0.42
23.4	40	0.84	84	0.50
23.4	40	0.84	88	0.58
23.4	40	0.84	92	0.65
23.4	40	0.84	96	0.72
23.4	40	0.84	100	0.78
23.4	40	0.84	104	0.84
23.4	40	0.84	108	0.91
23.4	40	0.84	112	0.97
23.4	40	0.84	116	1.02
23.4	40	0.84	120	1.09
23.4	40	0.84	124	1.15

$y_i = (M, I_{cl})$ from a vector of k input variables $x_i = (PMV_{feedback}, T, H)$. This is a multi-target regression (MTR) problem so an appropriate regressor is selected. To this end, extremely randomized trees (extra trees), presented by Geurts et al. [5] were utilized. Extra trees is an algorithm for ensemble tree construction based on extreme randomization. It belongs to the global methods of MTR, which means that all of the target variables are predicted simultaneously using one model in contrast to the local methods that predict each target variable separately. Global methods exploit the dependencies that exist between the target variables and result in better predictive performance.

The extra trees regression algorithm builds an ensemble of unpruned regression trees according to the classical top-down procedure. It has two main differences with other ensemble tree-based methods:

- The procedure of selecting cut-points for splitting the tree nodes is performed randomly.
- The trees grow using the whole learning sample and not just a bootstrap replica.

The splitting procedure for numerical attributes includes the following parameters:

- K, which denotes the number of attributes selected at each node;
- n_{min}, which refers to the minimum sample size for splitting a node;
- L, which represents the number of trees of the ensemble.

The final prediction in regression problems is given by aggregating the predictions of all trees and then using the arithmetic average. From variance point of view, extra trees are able to reduce variance more strongly than other randomization tree methods, using explicit randomization of the cut-point and attribute, combined with ensemble averaging. Bias is also minimized by the usage of the full original learning sample, in contrast to methods that use bootstrap replicas. Assuming balanced trees, the complexity of tree growing is of order $N \cdot logN$ with respect to learning sample size. The parameters K, n_{min}, L can be adjusted manually or automatically, however it is suggested by Geurts that the default settings are used in order to maximize the computational advantages and autonomy of the method. The above claim is empirically confirmed at our case, since different settings of the algorithms were used, but finally the default settings were selected as they provided more accurate results. The default criteria for measuring the quality of a split is mean squared error.

The users are able to give their feedback through a mobile application. Then, the model uses the feedback along with temperature and humidity data and finally predicts the new values of M and I_{cl} for the current user. Those values refer to the time interval in which the feedback is given. Metabolic rate and clothing insulation values are finally stored, updating previous ones. From that point on, the user's thermal comfort will be estimated with these new M and I_{cl} values.

Summarizing, the comfort inference algorithm is executed as a whole, as described in the following steps:

1. All the necessary data are retrieved from the database. This includes: temperature, humidity, metabolic rate, clothing insulation.
2. Data are pre-processed in order to handle abnormalities such as null or duplicate values.
3. It is checked whether the user has provided feedback. If there is feedback, then the comfort feedback predictive model is loaded and new values for M and I_{cl} are calculated.
4. Thermal comfort is being calculated using Fanger's Equation (Fig. 3).

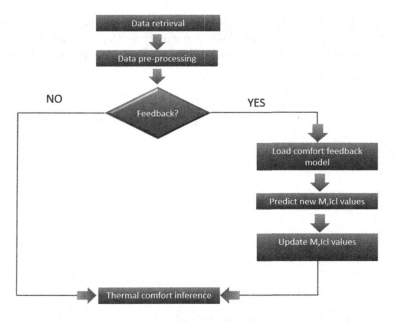

Fig. 3. Thermal comfort inference flow chart

4   Experimental Results

The algorithm is tested for the whole possible range of feedbacks, from -3 to $+3$. The feedback was set to change with a step of 0.05 at each observation, while T and H were randomly chosen to be between certain intervals. A total of 120 observations were created, that were utilized as inputs to the pre-trained predictive model, which generated a pair of M, I_{cl} predictions for each observation. The predicted M, I_{cl} values were then inserted into Fanger's equation and PMV value was calculated. Finally, the feedback was compared to the calculated PMV value for each observation in order to test the accuracy of the M, I_{cl} predictions. The comparison between thermal comfort feedback values and predicted PMV values is shown in Fig. 4.

The error that corresponds to the observations of Fig. 4, is depicted in Fig. 5. This error represents the deviation of the observed value (PMV deriving from the predicted M, I_{cl}) from the true value (actual PMV feedback).

$$error = predicted\_PMV\_value - feedback \tag{8}$$

As seen in Fig. 4, the predictions are accurate for the whole range of feedback values. This is also confirmed from the errors that are below 0.2 for the majority of the observations. Mean squared error (MSE) and mean absolute error (MAE) remain very low, at 0.0108 and 0.0739 respectively. The model's worst prediction, results to an error of 0.28 which is translated to lower than 5% error in the

[−3, 3] scale. It is deduced that the overall performance of the model is satisfying, as the predicted M, I_{cl} values approximate closely to the feedback (PMV value) from which they were derived, when inserted to Fanger's equation (Table 4).

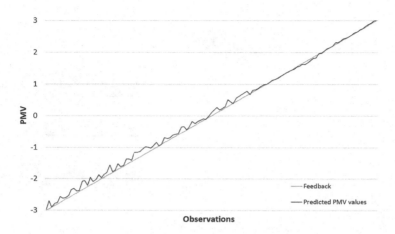

Fig. 4. Feedback and predicted PMV values comparison

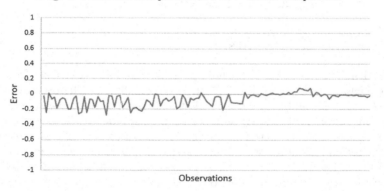

Fig. 5. Error of each observation

Table 4. Error analysis of the tested observations

Error analysis	
Mean squared error	0.0108
Mean absolute error	0.0739
Maximum error	−0.28
Minimum error	0.0009

5   Conclusions

This study elaborates on the two subjective factors that are part of the thermal comfort equation, clothing insulation and metabolic rate. The accurate prediction of these parameters, utilizing feedback provided by the study subjects, could be crucial in the field of indoor thermal comfort inference, as it allows the creation of flexible personalized models that can be more accurate as they eliminate the subjective factor enclosed in Fanger's static model. Experimental results were demonstrated that showcase the accurate prediction of M, I_{cl}. This allows the use of the proposed algorithm for the definition of M, I_{cl} values, thus enhancing the accuracy obtained by assuming initial M, I_{cl} values from ASHRAE tables. Future work may include experiments using subjects from different study groups in order to create thermal comfort models depending on age, gender etc.

Acknowledgements. This work is partially supported by the "enCOMPASS - Collaborative Recommendations and Adaptive Control for Personalised Energy Saving" project funded by the EU H2020 Programme, grant agreement no. 723059.

References

1. Thermal environmental conditions for human occupancy, vol. 55. American Society of Heating, Refrigerating and Air-Conditioning Engineers (2004)
2. Djongyang, R., Tchinda, N., Njomo, D.: Thermal comfort: a review paper. Renew. Sustain. Energy Rev. **14**, 2626–2640 (2010)
3. Fan, J., Chen, Y.: Measurement of clothing thermal insulation and moisture vapour resistance using a novel perspiring fabric thermal manikin. Meas. Sci. Technol. **13**(7), 1115 (2002)
4. Fanger, P.O., et al.: Thermal Comfort. Analysis and Applications in Environmental Engineering. Danish Technical Press, Copenhagen (1970)
5. Geurts, P., Ernst, D., Wehenkel, L.: Extremely randomized trees. Mach. Learn. **63**(1), 3–42 (2006)
6. Klepeis, N.E., et al.: The National Human Activity Pattern Survey (NHAPS): a resource for assessing exposure to environmental pollutants. J. Eposure Sci. Environ. Epidemiol. **11**(3), 231 (2001)
7. Krinidis, S., Tsolakis, A., Katsolas, I., Ioannidis, D., Tzovaras, D.: Multi-criteria HVAC control optimization. In: 2018 IEEE International Energy Conference (ENERGYCON), pp. 1–6. IEEE (2018)
8. Luo, M., Wang, Z., Ke, K., Cao, B., Zhai, Y., Zhou, X.: Human metabolic rate and thermal comfort in buildings: the problem and challenge. Build. Environ. **131**, 44–52 (2018)
9. MacPherson, R.: Studies in the preferred thermal environment. Archit. Sci. Rev. **6**(4), 183–189 (1963)
10. Mendell, M.J., Heath, G.A.: Do indoor pollutants and thermal conditions in schools influence student performance? A critical review of the literature. In: Indoor Air. Citeseer (2005)
11. Walikewitz, N., Jänicke, B., Langner, M., Meier, F., Endlicher, W.: The difference between the mean radiant temperature and the air temperature within indoor environments: a case study during summer conditions. Build. Environ. **84**, 151–161 (2015)

User-Centered Visual Analytics Approach for Interactive and Explainable Energy Demand Analysis in Prosumer Scenarios

Ana I. Grimaldo[(✉)] and Jasminko Novak[(✉)]

IACS – Institute for Applied Computer Sciences, University of Applied Sciences Stralsund,
Stralsund, Germany
{ana.grimaldo,jasminko.novak}@hochschule-stralsund.de

Abstract. As part of the energy transition, the spread of prosumers in the energy market requires utilities to look for new approaches in managing local energy demand and supply. Doing this effectively requires better understanding and managing of local energy consumption and production patterns in prosumer scenarios. This situation is particularly challenging for small municipal utilities who traditionally do not have access to sophisticated modeling and forecasting methods and solutions. To this end, we propose a user-centered and a visual analytics approach for the development of a tool for an interactive and explainable day-ahead forecasting and analysis of energy demand in local prosumer environments. We also suggest supporting this with behavioral analysis to enable the analysis of potential relationships between consumption patterns and the interaction of prosumers with energy analysis tools such as customer portals, recommendation systems, and similar. In order to achieve this, we propose a combination of explainable machine learning methods such as kNN and decision trees with interactive visualization and explorative data analysis. This should enable utility analysts to understand how different factors influence expected consumption and perform what-if analyses to better assess possible demand forecasts under uncertain conditions.

Keywords: Visual analytics · Utilities · Prosumers · User-centered design · Energy demand and supply management · Smart energy systems

1 Introduction

The spread of prosumers in smart grids has forced companies to offer sustainable energy services using bidirectional flows [1, 2]. Currently, prosumers consume their own energy, but when production is not enough, they demand it from the utilities. In contrast, when they produce a surplus, they inject it into the public grid to get an economic incentive in return [3]. According to several investigations, optimizing energy management is one of the most essential factors in the use of renewable systems [1, 4]. Efficient consumption benefits prosumers in reducing bills or even fully covering their energy needs; however, it is necessary to guide them in the transition to this process [3]. Fortunately, the interaction with technological applications (e.g. recommender systems, customer service portals)

© Springer Nature Switzerland AG 2019
D. Tzovaras et al. (Eds.): ICVS 2019, LNCS 11754, pp. 700–710, 2019.
https://doi.org/10.1007/978-3-030-34995-0_64

has demonstrated positive results by helping prosumers reduce energy costs in the long term [5–7].

On the other hand, utilities are responsible for collecting consumption data from smart meters at different intervals (e.g. hourly, daily). Analysis of this data combined with additional variables (e.g. weather and energy prices) [8, 9] can help them identify consumption patterns. Currently, there are several applications that allow the utilities to forecast energy demand and supply. Usually, these applications consider different factors such as weather conditions, socio-economic data and energy load profiles (e.g. BDEW standard load profiles).[1] Some of these applications are currently used mostly at the level of (larger) network operators or by large utilities for better short-term management of supply and demand.

Using forecasting applications involves complex machine learning processes usually based on black-box processes [10] that can be challenging to follow by users with non-technical knowledge in the area [11]. The internal functioning cannot be easily interpreted in terms of understanding how the forecasting process results obtained were actually produced. Knowing input and output data, but not being able to understand the procedure performed to calculate the forecasting can be a concerning situation, especially for small municipal utilities who are often not used to using sophisticated modeling and simulation approaches.

In order to provide an explainable forecast process (avoiding the black-box procedure), we propose to provide a method for heuristic day-ahead forecasting based on direct interaction with a visual analytics system. The idea is to model and implement analytics procedures that allow the utility analyst to inspect and compare historical data with respect to the available attributes characterizing the day ahead (e.g. weather conditions, month, day of the week and related factors) in order to identify consumption patterns that could likely correspond to the expected situation on the next day. This process could combine both manual as well as semi-automatic analysis (with the system suggesting similar days and periods based on criteria defined by the analyst). Supporting this exploratory process with appropriate interactive visualizations can aid the manual analysis to support the sense-making of the information presented and to allow the analyst to determine day-ahead forecasting in a heuristic and explainable manner.

This article presents a work in progress in the development of the main-use cases as part of the SIT4Energy project.[2] The integration of this proposal is comprised of a developed approach, method, system design and some initial mockups for the implementation of an explainable Smart Energy Dashboard to forecast energy consumption and production for the utilities. We propose the integration of decision trees and k-nearest neighbor (KNN) methods to implement an explainable forecasting process. This paper is structured as follows. Section 2 presents related work on the development of solutions to support the utilities in managing energy regarding the penetration of prosumers into the energy market. Section 3 presents the analysis of needs and requirements obtained from initial research with target end-users. Section 4 describes the approach for the development of our solution. Section 5 describes the system design for the implementation of the

---

[1] https://www.bdew.de/energie/standardlastprofile-strom/.

[2] https://sit4energy.eu/.

proposed approach. Section 6 presents the definition of guidelines for the evaluation of the proposed approach. Finally, Sect. 7 provides conclusions and outlines future work.

2   Related Work

One of the main concerns related to renewable energy is the development of technology to support the new energy market, which includes the installation of renewable systems and provides prosumers with monitoring and supporting tools [12, 13]. For example, Fiorentino et al. reported that recommendation systems have a significant impact by helping prosumers optimize energy consumption and production [5]. In addition, Yan et al. [8] proposed designing scenarios in order to define significant variables (e.g. temperature, solar radiation, humidity) to provide accurate information on energy demand. In this way, users can get a more personalized prediction based on their consumption preferences (e.g. low, medium, or high consumption) as well as storage device types and capacity.

On the other hand, due to changes in the power generation market, the operations performed by the utilities have been extended with activities such as planning energy production or purchasing it from third parties [2]. This situation leads to the necessity to put data analysis at the forefront. At the same time, real-time pricing has emerged as one of the main factors impacting the energy supply, especially during peak times [14, 15]. Therefore, the utilities have to anticipate changes in energy consumption associated with dynamic prices [16]. This situation has encouraged the development of tools to support the management of energy consumption and production [17, 18], including algorithms to forecast different conditions such as building temperature, solar radiation and gas emission, and therefore, contributing to the identification of patterns in the prediction of energy demand.

Currently, there are several machine learning methods to forecast energy consumption and production; however, it is essential to analyze the differences between explainability and accuracy (Fig. 1). As an example, neural networks are focused on providing more accurate results, but they lack interpretability because of the implementation of black-box approaches [19]. In contrast, the decision tree model is considered one of the most transparent methods based on multilayers, which have been widely applied to generate knowledge by discovery data mining [20] (Fig. 2).

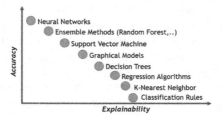

Fig. 1. Accuracy vs. explainability machine learning models (Image adapted from [22])

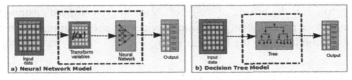

Fig. 2. (a) The Neural Network Model implements hidden and complex layers to produce results. (b) The Decision Tree Model generates an explainable tree (Image adapted from [23])

Forecasting energy consumption is a complex process that involves identifying multiple variables that together comprise energy demand, for example, weather conditions and historical consumption among others [2]. Usually, the results are integrated into charts using different periods (e.g. daily, monthly and yearly). However, this information is not always presented to the user under usability standards that facilitate understanding [21].

Another contribution proposed a decision-support tool based on predefined scenarios to allow users to analyze the environmental, economic and cultural impact of the exploitation of renewable resources available [24]. However, to validate the operation of such tools, it is important to evaluate them in different climatological regions. Climate can be classified by clustering similar conditions to facilitate learning about specific situations such as sunny days [9]. Most of these applications focus on the needs of researchers or professionals to understand and improve forecasting methods [25]. Even for such expert users, Mentler et al. [21] consider the interface design to be also of essential importance.

In addition, the use of visual analytics has shown multiple benefits in the analysis of information to support decision-making; for that reason, it is gaining the attention of researchers and developers, primarily to support decisions based on the analysis of large and complex data [26]. Several contributions have explored the design of dashboards to evaluate energy-related activities including the following features: (a) electrical power consumption monitoring using real-time systems, (b) comparing information and (c) providing educational information [27]. However, challenges in the design of such dashboards include the availability and completeness of data, its structuring and exploration to produce meaningful graphs while avoiding misinterpretation [26]. Therefore, the analysis of the information necessary to include in the dashboards is of paramount importance. Furthermore, using dashboards by building managers has demonstrated a positive impact resulting in attitudinal changes such as increased awareness, desire to learn and motivation to save energy [27]. Hence, user-centered visual analytics dashboards could also motivate utility analysts to learn from their customers' behavior regarding energy demand and production.

The majority of contributions presented above focused on facilitating information analysis by prosumers (e.g. to support their decision-making) and on improving energy forecasting models to benefit the utilities. However, most energy forecasting procedures are trained and tested with historical data [25]. According to some studies, the variation in the historical data (e.g. new appliances, new habits, a new family member) over time affects the actual performance of forecasting methods [28]. Therefore, it is paramount to consider the effects of usage of increasingly common customer applications (e.g. consumer portals, energy saving services) that can influence prosumers to change consumption habits in the long term [10].

3 Identification of Needs and Requirements

In line with the user-centered design methodology, utility employees have been involved in the design process as target users, starting with the requirements analysis and definition (e.g. interviews, feedback to mock-ups). Based on the literature review and interviews with utility representatives, initial information needs and requirements for analyzing energy demand and production were identified, and first visual mock-ups of possible solution ideas developed. The initial mock-ups were tested in a workshop with utility representatives and based on obtained feedback the requirements were consolidated and main use cases for the development of a visual analytics dashboard to support energy analysis needs for small utilities were defined.

Based on the results of this process, the main identified need was to be able to perform day-ahead forecasts of energy demand in prosumer scenarios in an easily understandable way. Another essential requirement was to enable the analysis to explore what-if scenarios by varying different input parameters that were identified as main factors influencing energy demand and production (e.g. weather conditions, season, day of the week, energy prices, etc.). This should enable the utility to better plan the energy production for the next day, a process that is currently performed manually and considering the previous experience of the people in charge, without any tools to specifically support that task.

The development of a visual tool to analyze the relationships between most important factors such as weather conditions (sun hours, temperature and wind speed), energy prices, supply, and demand, can help utilities improve their energy estimation for the next day. The simulation of different conditions, not just for the next day but also to identify specific possible situations, can support the utilities to analyze their impact on the energy demand and supply, e.g. regarding the holiday season or as a result of increasing energy prices. This could be supported through the creation of *"what-if-scenarios"* in a visual analytics process to help utilities to identify or prevent issues in the energy demand forecasting and planning process.

4 Approach

Based on the described analysis, we propose the development of a visual analytics tool to help the utilities better understand consumption and production patterns to determine a day-ahead forecast. That includes two main use cases:

(1) Using a Smart Energy Dashboard to perform dynamic analysis and interactive day-ahead forecasting of energy supply and demand,
(2) Analyzing the historical and forecasted energy consumption and production patterns to provide recommendations to prosumers for optimizing their consumption and production.

To this end, we propose the development of an interactive Smart Energy Dashboard based on a combination of visual analytics and explainable machine learning models. The implementation of explainable AI is crucial to gain users' trust regarding the inter-action with the results [20]. In this way, we expect to support the decision-making of utilities regarding energy management activities, avoiding the use of complex black-box models. Instead of the implementation of sophisticated and accurate algorithms that can be difficult to understand, we intend to implement an explainable model compatible with existing practices — e.g. heuristic assessment based on previous experience.

In order to implement the forecasting process, we selected decision tree and KNN methods (Fig. 3). We want to compare the suitability of these two methods and investigate trade-offs between the accuracy and explainability that they can provide to end users. In both cases, the overall forecasting process is modelled as follows. The analyst can set the input parameters of factors influencing consumption and production (sun hours, wind, temperature, weekday/weekend/holiday and month) to values that are expected for the next day. Based on historical data, the system then applies either kNN or decision tree methods to identify days in the past which are most similar to the conditions the analyst has defined with setting the input parameters. The consumption and production patterns corresponding to those most similar days are then displayed in a visual analytics dashboard, and accompanied with the data on corresponding context conditions (e.g. day of the week, month, weather conditions and energy prices on those days). The analyst can now inspect and explore the presented information in order to identify which of the presented historical contexts and energy demand data reflect the most comparable situation to the one expected for the next day, to derive the expected energy demand.

For the decision tree model, the input parameters are used to apply the decision rules by splitting the data in order to determine dates in the past that met the initial parameters. Explainability is applied by presenting the decision tree computed as part of the process [23]. In this way, the selected pathway is used to create the visualization in the dashboard; therefore, it should be easy to analyze the relationships between the input parameters with respect to energy consumption and production.

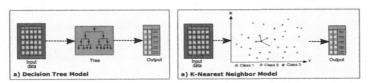

Fig. 3. Decision tree model and K-nearest neighbor model (Image adapted from [23])

On the other hand, the KNN model is a regression model used to compute the distance between data points and a query point in a multidimensional array. The model takes different input parameters from the query point to compare with the data points in the matrix by using distance equations [19]. The model is explainable because of the analysis of vicinity between data points in the matrix and the query point based on the input parameters. In this way, it should be easy for the utilities to analyze the closeness of the data in order to understand the prediction.

Finally, the described interactive forecasting process would also allow utilities to simulate *"what-if-scenarios"*: by specifying different hypothetical conditions, they could analyze the obtained forecast results for different scenarios, in order to identify or prevent issues in different possible future scenarios (e.g. peak energy demand in specific summer conditions). Combining the described forecasting features and the creation of *"what-if-scenarios"* could thus allow the utilities to understand the relationship between consumption patterns in different conditions and the factors that determine them. In this way, the utilities could also define recommendations for prosumers to help them optimize energy consumption and production.

5   System Design

The basis for the implementation of the described approach is a combination of the machine learning platform RapidMiner[3] and the visual analytics platform Qliksense[4]. The former allows the straightforward implementation of several prediction models (incl. decision trees and KNN), while the latter provides functionalities for designing and implementing customized visual analytics dashboards. The implementation of the workflow leading to the presentation and exploration of the forecasting results in the Smart Energy Dashboard is given in Fig. 4.

The first step is to collect and prepare the input data from historical datasets (energy consumption and production, weather, energy prices, etc.). Once the dataset is updated, the user enters the input parameters in the Smart Energy Dashboard and selects the forecasting model (decision tree or KNN). The dashboard is connected to the RapidMiner platform to produce the forecasting results. However, the user can analyze the model generated (e.g. the pathway obtained in the decision tree). Subsequently, the output results are visualized and can be interactively explored and analyzed in the Smart Energy Dashboard (implemented with QlikSense). The information is presented per hour to help the user analyze the progress of consumption, production and other values throughout the day (Fig. 5). The user can also interact with the explainable model generated through the machine learning methods to understand the processing of data that produced the results.

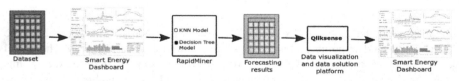

Fig. 4. The Smart Energy Dashboard system design

---

[3] https://rapidminer.com/.

[4] https://www.qlik.com/us/products/qlik-sense.

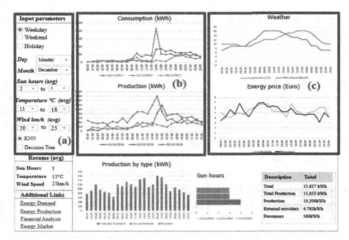

Fig. 5. The Smart Energy Dashboard prototype (a) The selection of input parameters (b) Consumption and production per hour. Each line represents a date obtained from the similarity analysis. (c) The weather conditions per hour and energy prices per hour.

The main features of the Smart Energy Dashboard have been defined in a low-fidelity prototype (Fig. 5). In the beginning, the user selects input parameters to guide the forecasting process as well as the model to be implemented (Fig. 5a). As a result, the dashboard presents production and consumption visualizations for the different dates obtained in the forecasting model (Fig. 5b). The users can also analyze the relationships between the input parameters and the information presented because of the additional visualizations such as weather conditions and energy prices (Fig. 5c). To analyze a specific day, the user can directly select the corresponding consumption line, and the dashboard will be updated only to present the corresponding information. To create *"what-if-scenarios,"* the users can simply define desired input parameters in the dashboard to describe a scenario they would like to inspect, in the same way as they define the parameters for the day-ahead forecast. The identified historical data matching the defined scenario parameters would then be presented in the dashboard in the same way as described for the day-ahead analysis.

6   Evaluation

In order to perform an initial validation of our solution design, we performed an initial "heuristic evaluation." During an interview with one utility representative, we presented the system design, method and visualizations. The participant understood the objective of the mockups and felt satisfied regarding the proposal as well as the interaction flow because this was compatible with existing practices in the company. In addition, the participant identified that the interaction with the visual tool could benefit the employees by augmenting their existing experience with the presented data-driven analysis. After completing the implementation of the functional prototype, we will perform a comprehensive evaluation of accuracy and explainability evaluation with more utility employees as follows:

(1) Accuracy Evaluation - A historical test data set and a set of test days from the past will be defined (from different seasons). The users will be asked to select input parameters to reflect the test days and to analyze the obtained system results to determine the most likely forecast for a given test day. We will then compare the difference between the forecast identified by the users and the actually measured consumption on the test day to determine the accuracy of this approach. This will be performed both for the kNN and the decision tree models, to compare their accuracy.

(2) Explainability Evaluation - A user-centered evaluation will be performed to determine perceived usefulness and explainability. When performing the test run described above, the users will also be asked about their evaluation of the perceived usefulness and performance of the system (through interviews and questionnaires). They will be given a set of questions about how well they understood the results of the system in determining similar days in the past regarding the input parameters. This will also include questions about observed relationships between specific input parameters and obtained consumption patterns, in order to verify the explainability more objectively. Finally, a control group of users will be presented with a simulation of automatic forecasting results for the same test days (common "Wizard-of-Oz" technique), without the possibility to set input parameters themselves and without the explanation of the rules that have been used to identify the forecast (as would be the case with black-box machine learning method).

7   Conclusions

We proposed an approach for interactive forecasting of day-ahead energy demand, based on historical data through a combination of explainable machine learning and visual analytics, in a way that is explainable for the users. Based on the selection of input parameters (e.g. weather conditions, energy price, day of the week), the proposed approach also allows utilities to analyze "what-if" scenarios of energy demand and their relationship to conditions that would generate them. This could help them to identify and/or prevent critical situations and issues in the future.

Furthermore, we proposed to extend this approach with a behavioral analysis of the interaction of prosumers with technological tools (e.g. customer service portal, recommendation systems) in order to identify the possible impact of such tools on consumption patterns. Understanding these relationships could be a key factor for the utilities to identify and define recommendations to optimize energy consumption. The behavioral analysis could also provide an improvement to traditional forecasting models which usually analyze historical data, but do not consider the effects of additional influences on the consumption behavior of prosumers. The evaluation approach presented in this work will be used to empirically evaluate the suitability of the proposed system design and identify possibilities for improvement. In future work, we also intend to extend the evaluation to different test years and different utilities in order to collect more substantial evidence regarding the suitability and usability of the proposed approach.

Acknowledgments. The SIT4Energy project has received funding from the German Federal Ministry of Education and Research (BMBF) and the Greek General Secretariat for Research and Technology (GSRT) in the context of the Greek-German Call for Proposals on Bilateral Research and Innovation Cooperation.

References

1. Ali, W., Razzaq, S., Shehzad, K., Zafar, R., Mahmood, A., Naeem, U.: Prosumer based energy management and sharing in smart grid. Renew. Sustain. Energy Rev. **82**, 1675–1684 (2017)
2. Stephens, J.C., Kopin, D.J., Wilson, E.J., Peterson, T.R.: Framing of customer engagement opportunities and renewable energy integration by electric utility representatives. Util. Policy **47**, 69–74 (2017)
3. Miller, W., Senadeera, M.: Social transition from energy consumers to prosumers: rethinking the purpose and functionality of eco-feedback technologies. Sustain. Cities Soc. **35**, 615–625 (2017)
4. Iria, J., Soares, F.: A cluster-based optimization approach to support the participation of an aggregator of a larger number of prosumers in the day-ahead energy market. Electr. Power Syst. Res. **168**, 324–335 (2019)
5. Fiorentino, G., Corsi, A., Fragnito, P.: A smarter grid with the internet of energy. In: Proceedings of ACM e-Energy 2015, pp. 301–306 (2015)
6. Iria, J., Soares, F., Matos, M.: Optimal bidding strategy for an aggregator of prosumers in energy and secondary reserve markets. Appl. Energy **238**, 1361–1372 (2019)
7. Starke, A., Willemsen, M., Snijders, C.: Effective user interface designs to increase energy-efficient behavior in a Rasch-based energy recommender system. In: Proceedings of ACM RecSys 2017, pp. 65–73 (2017)
8. Yan, X., Abbes, D., Francois, B.: Development of a tool for urban microgrid optimal energy planning and management. Simul. Model. Pract. Theory **89**, 64–81 (2018)
9. Ahmad, T., Chen, H.: Utility companies strategy for short-term energy demand forecasting using machine learning based models. Sustain. Cities Soc. **39**, 401–417 (2018)
10. Paulescu, M., Brabec, M., Boata, R., Badescu, V.: Structured, physically inspired (gray box) models versus black box modeling for forecasting the output power of photovoltaic plants. Energy **121**, 792–802 (2017)
11. Azaza, M., Eskilsson, A., Wallin, F.: An open-source visualization platform for energy flows mapping and enhanced decision making. Sci. Direct **158**, 3208–3214 (2019)
12. Hansen, K., Mathiesen, B.V., Skov, I.R.: Full energy system transition towards 100% renewable energy in Germany in 2050. Renew. Sustain. Energy Rev. **102**, 1–13 (2019)
13. Lund, H., Østergaard, P.A., Connolly, D., Mathiesen, B.V.: Smart energy and smart energy systems. Energy **137**, 556–565 (2017)
14. Hossein, A., Maghouli, P.: Energy management of smart homes equipped with energy storage systems considering the PAR index based on real-time pricing. Sustain. Cities Soc. **45**, 579–587 (2019)
15. Steriotis, K., Tsaousoglou, G., Efthymiopoulos, N.: A novel behavioral real time pricing scheme for the active energy consumers' participation in emerging flexibility markets. Sustain. Energy Grids Netw. **16**, 14–27 (2018)
16. Ahmad, T., Chen, H.: Nonlinear autoregressive and random forest approaches to forecasting electricity load for utility energy management systems. Sustain. Cities Soc. **45**, 460–473 (2019)
17. Zepter, J.M., Lüth, A.: Prosumer integration in wholesale electricity markets: synergies of peer-to-peer trade and residential storage. Energy Build. **184**, 163–176 (2019)

18. Taborda, M., Almeida, J., Oliveir-Lima, J.A., Martins, J.F.: Towards a web-based energy consumption forecasting platform. In: Proceedings of CPE 2015, pp. 577–580 (2015)
19. Johannesen, N.J., Kolhe, M., Goodwin, M.: Relative evaluation of regression tools for urban area electrical energy demand forecasting. J. Clean. Prod. **218**, 555–564 (2019)
20. Ding, L.: Human knowledge in constructing AI systems - neural logic networks approach towards an explainable AI. Procedia Comput. Sci. **126**, 1561–1570 (2018)
21. Mentler, T., Rasim, T., Müßiggang, M., Herczeg, M.: Ensuring usability of future smart energy control room systems. Energy Inform. **1**, 167–182 (2018)
22. Dam, H.K., Tran, T., Ghose, A.: Explainable software analytics. In: ICSE 2018 NIER, 2018, pp. 1–4 (2018)
23. Tso, G.K.F., Yau, K.K.W.: Predicting electricity energy consumption: a comparison of regression analysis, decision tree and neural networks. Energy **32**, 1761–1768 (2007)
24. Necefer, L., Wong-Parodi, G., Small, M.J., Begay-Campbell, S.: Integrating technical, economic and cultural impacts in a decision support tool for energy resource management in the Navajo Nation. Energy Strateg. Rev. **22**, 136–146 (2018)
25. Debnath, K.B., Mourshed, M.: Forecasting methods in energy planning models. Renew. Sustain. Energy Rev. **88**, 297–325 (2018)
26. Matheus, R., Janssen, M., Maheshwari, D.: Data science empowering the public: data-driven dashboards for transparent and accountable decision-making in smart cities. Gov. Inf. Q. (2018, in press). https://doi.org/10.1016/j.giq.2018.01.006
27. Timm, S.N., Deal, B.M.: Effective or ephemeral? The role of energy information dashboards in changing occupant energy behaviors. Energy Res. Soc. Sci. **19**, 11–20 (2016)
28. Gerossier, A., Girard, R., Bocquet, A., Kariniotakis, G.: Robust day-ahead forecasting of household electricity demand and operational challenges. Energies **11**, 3503–3522 (2018)

Workshop on: Vision-Enabled UAV and Counter-UAV Technologies for Surveillance and Security of Critical Infrastructures

Critical Infrastructure Security Against Drone Attacks Using Visual Analytics

Xindi Zhang$^{(\boxtimes)}$ and Krishna Chandramouli

Venaka Media Limited, 393, Roman Road, London E3 5QS, UK
zhangxindi@gmail.com, k.chandramouli@venaka.co.uk

Abstract. The recent developments in the field of unmanned aerial vehicles (UAV or drones) technology has generated a lot of interdisciplinary applications, ranging from remote surveillance of energy infrastructure, to agriculture. However, in the context of national security, low-cost drone equipment has also been viewed as an easy means to cause destructive effects against national critical infrastructures and civilian population. Addressing the challenge of real-time detection and continuous tracking, this paper proposed presents a holistic architecture consisting of both software and hardware design. The software-based video analytics component leverages upon the advancement of Region based Fully Convolutional Network model for drone detection. The hardware component includes a low-cost sensing equipment powered by Raspberry Pi for controlling the camera platform for continuously tracking the orientation of the drone by streaming the video footage captured from the long-range surveillance camera. The novelty of the proposed framework is twofold namely the detection of the drone in real-time and continuous tracking of the detected drone through controlling the camera platform. The framework relies on the capability of the long-range camera to lock into the drone and subsequently track the drone through space. The analytics processing component utilises the NVIDIA® GeForce® GTX 1080 with 8 GB GDDR5X GPU. The experimental results of the proposed framework have been validated against real-world threat scenarios simulated for the protection of the national critical infrastructure.

Keywords: Intruder drones · Surveillance camera · Critical infrastructure · Deep-learning · Drone detection and tracking · Raspberry Pi

1 Introduction

The evolution of aerial technology has seen exponential growth with the development of Unmanned Aerial Vehicles (UAV) which have found applications beyond

X. Zhang—The author is pursuing doctoral dissertation at Multimedia and Vision Research Group, School of Electronic Engineering and Computer Science, Queen Mary Univeristy of London.

military use and have become powerful business tools according to Goldman Sachs report[1]. Following the exponential increase in the commercial application of the drones, the threat associated with the misuse drones for malicious and terrorist activity has also been increased. Indeed, drones can be and are being used by terrorist groups in five primary ways: for surveillance; for strategic communications; to smuggle or transport material; to disrupt events or complement other activities; and as a weapon. This last category includes instances of a drone being piloted directly to a target, a drone delivering explosives, and a weapon being directly mounted to a drone[2].

Although the problem of detecting UAVs is considered a niche topic in the field of computer vision, there are some preliminary attempts that could be summarised. The use of morphological pre-processing and Hidden Markov Model (HMM) filters to detect and track micro-unmanned planes was reported in [10]. In addition, to the use of spatial information based methods, spatio-temporal approaches were also reported in the literature. In particular, the approach that first creates spatio-temporal cubes using sliding window method at different scales, applies motion compensation to stabilise spatio-temporal cubes, and finally utilises boosted tree and CNN based repressors for bounding box detection was reported in [13]. The reported approaches in the literature focus only on the detection of the drone in the horizon, the challenge of closely tracking the drone without the apriori knowledge of the drone trajectory has not yet been addressed. Therefore, in this paper a novel framework is proposed that is able to achieve visual lock of the drone and continuously monitor the trajectory of the drone flight path. The framework integrates a low-cost sensing equipment powered by the Raspberry Pi to control the servo motors to pan-tilt-zoom (PTZ) the camera equipment in order to track the drone. The continuous media stream captured from the real-world sensing is processed with the Region-based Fully Convolutional (RFCN) network in order to detect the presence or absence of the drone in the horizon.

The rest of the paper is structured as follows. In Sect. 2, an overview of various approaches reported in the literature for drone detection is presented. Subsequently, in Sect. 3, an overview of the overall proposed system framework is presented. The next two sections outline the training model generation for detecting drone and the implementation of tracking algorithm in Sects. 4 and 5 respectively. Section 6, provides an outline of the performance evaluation of the overall implementation followed by conclusion and future work reported in Sect. 7.

2   Literature Review

Traditionally, the task of object detection is to classify a region for any predefined objects from the training data set. Early attempts at drone detection

---

[1] http://www.goldmansachs.com/our-thinking/technology-driving-innovation/drones/.

[2] https://ctc.usma.edu/app/uploads/2016/10/Drones-Report.pdf.

also adopted a similar approach for the classification of an image region containing a drone. With formidable progress achieved in the field of deep learning algorithms within image classification tasks, similar approaches have started to be used for attacking the object detection problem. These techniques can be divided into two simple categories; region proposal based and single shot methods. The approaches in the first category differ from the traditional methods by using features learned from data with Convolutional Neural Networks (CNN) and selective search or region proposal networks to decrease the number of possible regions as reported in [4]. In the single shot approach, the objective is to compute bounding boxes of the objects in the image directly instead of dealing with regions in the image. One of the methodologies proposed in the literature considers extracting multi-scale features using CNNs and combining them to predict bounding boxes as presented in [5,9]. Another approach reported in the literature, named You Only Look Once (YOLO) divides the final feature map into a 2D grid and predicts a bounding box using each grid cell [11]. The reported techniques derive from the overall objective of object detection and are not trained specifically to address the drone classification. The large-amount of datasets that are trained on well-known objects that are repetitively encountered in real-life are used to build the detection framework. In contrast to the reported techniques in the literature, the proposed framework relies on the RFCN network for drone detection and most importantly feedback the detected outcome to the sensing equipment to enable real-time tracking of the drone flight path.

3   Conceptual Design of the Proposed Framework

The overall conceptual design of the proposed framework is presented in Fig. 1. The conceptual model is divided into two main components namely (i) the sensing unit and (ii) analytics unit. In order to address the time-critical requirement for real-time tracking of intruder drones, the proposed framework implements five stage sequence of information analysis. The first phase of implementation requires the acquisition of video data, which is achieved through the high-definition block camera. In contrast of other object detection applications, in the context of drone detection, it is critical to be able to clearly identify the objects at distance (also referred as horizon). The next stage of the framework includes the transmission of the acquired data from the sensing unit to the main analytics unit, since the sensing unit is expected to be deployed at or beyond the perimeter of the critical infrastructure. The third stage in the framework is at the heart of the application, which triggers the object detection framework (the training of the network model is presented in Sect. 4).

The outcome of the object processing algorithm is to distinguish among several objects to detect the presence of drones. The next stage of the framework implements the drone localisation component, not only in terms of the 2D position as captured from the camera but also in terms of the area of the drone occupied within the entire frame. Since the motion of low-cost drones are expected to reach between 30 km/h to 60 km/h, it is critical to ensure that tracking algorithm

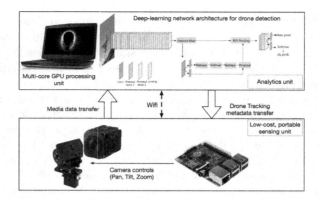

Fig. 1. Conceptual design of the proposed intruder drone detection and tracking framework

is able to swiftly adapt to the drone motion. The final stage in the framework is to trigger the overall coordination of the tracking component deployed in the sensing unit with Raspberry Pi acting as an interface to control the servo motors that provide pan-tilt orientation. It is vital to note that the communication infrastructure in the proposed framework relies on dedicated bandwidth between the sensing unit and the analytics unit due to the absence of video encoding module, which inherently introduces latency in the video transmission.

4   Training Framework for Drone Detection

The training model used in the paper extends the architecture proposed in [3]. The region based object detection can be categorised into three sub-networks. The first sub-network is a feature extractor. The second sub-network provides bounding box based on region proposal network. The third sub-network performs the final classification and bounding box regression. This sub-network is related to region, and it has to run on every single region. The feature extractor used in the RFCN is ResNet 101 as proposed in [6].

In order to move the time-consuming convolutional neural network into first two sharing sub-networks, RFCN places the 100 layers in sharing sub-network and only uses one convolutional layer to calculate prediction. The last 1000-class fully connected layer in ResNet101 is replaced by 1*1 convolutional layer with depth 1024 to achieve dimension reduction. Then use $K^2(C + 1)$ -channel convolutional layer to present position sensitive score maps which will be introduced next. C means the number of classes. Since the object detection is trained to identify drones, in contrast to other objects that could potentially appear in the sky (such as birds, aeroplanes, etc), C is 3 which stand for drones, birds and aeroplanes. To increase the accuracy, RFCN propose position-sensitive score maps. After gaining the region proposal, the Region of Interest (RoI) is divided into $K * K$ grid with $(C + 1)$ depth (1 corresponds to background). Each grid has its own score. The cumulative voting of the scores for each grid leads to the determination of the final scores of $C + 1$ classes in the respective RoI.

The loss function of RFCN is as follows, which contain classification loss and regression loss.

$$L(s, t_{x,y,w,h}) = L_{cls}(s_{c^*}) + \lambda[c^* > 0]L_{reg}(t, t^*) \tag{1}$$

Here c^* is the ground truth label of RoI ($c^* = 0$ means background). λ is balance weight which is set to 1. Classification use the cross-entropy loss $L_{cls}(s_{c^*}) = -log(s_{c^*})$. tx, y, w, h is the bounding box coordinates. $Lreg$ is doing bounding box regression. For training the models, we use 5000 annotated images of drones, aeroplanes and birds for each category. After 200,000 training steps, we export the model for prediction. In tracking progress, only when the bounding box category is drone can be shown and used to update the tracker.

5 Tracking Interface with Raspberry Pi Sensing Equipment

In order to achieve accurate and real-time visual tracking of the drone, three main challenges have to be addressed namely (i) computational time; (ii) motion blur, change in appearance of the drone and illumination changes caused due to environmental effects; (iii) drift between the object and the bounding box. Addressing these challenges, the novelty of the proposed framework relies on the integration of object detection component as described in Sect. 4 with the object tracking algorithm namely the Kernelised Correlation Filters (KCF) as presented in [7]. An overview of the implemented algorithm is presented in Algorithm 1. The integration of the KCF tracking algorithm, the tracker will update localised region of the drone based on previous frame (I_{i-1}) and previous bounding box of the drone detected (B). However, for every fifth frame (k), if the detection algorithm identifies the drone, the tracker in next frame $(k + 1)$ will be reinitialised based on this frame (k). If the detection algorithm failed to identify the drone, the tracker will continuously update based on the previous frame. The choice of updating the tracker for every 5 frames is empirically chosen in order to achieve a trade-off between the computational complexity of the detection algorithm with the need for correction of the drift caused by fast motion of the drones. If the update interval is in excess of more than 5 frames, the drift introduced by the KCF could lead to failure in tracking the drone accurately. On the other hand, if the interval is less than 5 frames, the computational complexity introduced by the detector is too high to achieve real-time visual lock on the drone.

The outcome of the detection and the tracking algorithm is fed into the sensing platform to control the motion of the pan-tilt-zoom parameters of the detection platform. The *initiate_servo_position* position(B) function calculates the relative pan-tilt-zoom parameters to trigger the servo motor controllers. The center point coordinate of detected drone B is used to determine the position according to the captured video frame. If the ratio of x coordinate is below 0.4

Input: video_frame I_i
Result: Drone detected in the horizon B
create_tracker;
if *if* $i == 1$ **then**
    $B = detection\_algorithm(I_i)$;
    $start\_tracker = false$;;
    $previous\_frame = I_i$;
else
    while *true* **do**
        read_current_frame (I_i);
        if $start\_tracker = true$ **then**
            reinitialise_tracker(I_{i-1}, B_{i-1});
            $B = calculate\_drone\_position(I_i)$;
        else
            $B = update\_drone\_position(I_i)$;
        end
        if $i\%5 == 0$ **then**
            $B = detection\_algorithm(I_i)$;
            $start\_tracker = true$;
        end
        initiate_servo_position(B)
    end
end

Algorithm 1: Pseudocode for the video analytics algorithm for drone detection

or beyond 0.6, the platform will triggered to turn left or turn right. If the ratio of y coordinates is below 0.3 or beyond 0.7, the platform will tilt up or down. The zoom parameter is calculate by frame width divided by box width. When the parameter is larger than 7 or smaller than 4, then trigger the camera to zoom in or zoom out.

6   Experimental Results

The analytics component of the proposed framework was trained the drone detection [2], with an extension of the classes to include aeroplanes. The experimental evaluation of the proposed critical infrastructure security framework against drone attacks using visual analytics is evaluated in two stages namely (i) the accuracy of the detection algorithm achieved by the use of RFCN with the inclusion of the KCF tracking algorithm and (ii) the tracking efficiency of the proposed approach to follow the trajectory of the drone flight. The experimental results were carried out on a set of video footage captured in the urban environment with DJI Phantom 3 Standard piloted from a range of 450 m from the point of attack. A total of 39 attack simulations were created with each attack ranging between 6 to 117 seconds resulting in a cumulative of 12.36 min of drone flights. The proposed RFCN Resnet101 network performance has been compared against three other deep-learning network models, namely SSD Mobilenet [8,9], SSD Inception v2 [14] and Faster RCNN Resnet101 [12]. The performance of the proposed drone detection framework integrated within the analytics component results in 81.16% as average precision. In comparison, the use of SSD Mobilenet results in 30.39%, SSD Inception v2 results in 7.78% and Faster RCNN resnet101 results in 69.49% average precision. It is worth noting that, the detection accuracy presented only refers to the classification and localisation of the drones and does not include the performance of the drone detector as deployed in the real-world. Since, the RFCN Resnet101 has yielded the best detection results, the network is further integrated within the analytics component.

Detection and Tracking. According to [1,15], the central distance curve metric for the performance assessment of object tracking is utilised to quantify the performance of the proposed detection and tracking framework. The central distance curve metric calculates the distance between the tracking bounding box centre and the ground truth box centre and summarise the proportion at different threshold. The evaluation result of the framework presented in Fig. 2 depicts the central distance curve metric for different drone flight paths. The various curves depict different scenarios under which the drone flight path took place. The scenarios include (i) the drone appears in front of the sensing platform; (ii) the drone crossing in front of sun; (iii) the drone flight takes place in the horizon; (iv) the velocity and acceleration of the drone is beyond the supported specification due to environmental facts and (v) the motion component introduced by the sensing equipment while capturing the drone footage.

A frame may be considered correctly tracked if the predicted target centre is within a distance threshold of ground truth. A higher precision at low thresholds means the tracker is more accurate, while a lost target will prevent it from achieving perfect precision for a very large threshold range. When a representative precision score is needed, the chosen threshold is 20 pixels. The overall average precision score of all footage is 95.2% is achieved for the five scenarios identified for the framework evaluation and is presented in Fig. 2.

Since the purpose of the camera platform is to keep the object at the centre of video and we do not have ground truth in practical testing, we can score the performance in every frame and calculate the average score of total frame. The evaluation methodology adopted is presented in Fig. 3. If the centre point of bounding box is inside the red region, the accuracy of the detection and tracking is assigned 100%. However, if the centre is in the yellow region, the accuracy score of 75% is assigned. Similarly, if the centre is in the blue region, 50% accuracy score is assigned to the frame. Finally, if the region contains no appearance of the drone, accuracy of 0% is assigned resulting in the loss of the drone either from the software implementation of the visual analytic module or through the latency in the camera control.

The overall functionality of pan-tilt-zoom parameters is presented in Fig. 4 mapped against the overall flight path undertaken by the drone. One of the critical challenges to be addressed in the context of real-time drone tracking is to

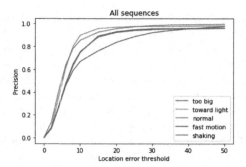

Fig. 2. Central pixel distance curve of different situations

Fig. 3. Performance calculation

ensure the size of the drone is sufficiently large enough to be suitably detected by the video analytics component. In this regard, it is vital to control and trigger the zoom parameter of the camera in order to inspect the horizon prior to the intrusion of the drone. In addition, due to the vast amount of changes in both the momentum and direction of the drone influenced by the environmental parameters such as wind speed and direction (for both head and tail wind), the accuracy of the tracking algorithm relies on the latency in controlling the

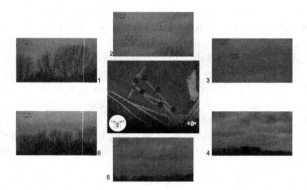

Fig. 4. Visualisation of the Pan-Tilt-Zoom functions

hardware platform. Therefore, in the proposed framework three parameters are controlled namely pan, tilt and zoom to ensure that the tracking of the drone is achieved. In order to objectively quantify the performance of the sensing platform, it is vital to consider the feedback received from the analytics component. However, the input provided to the analytics component is dependent upon the visual information acquired from the horizon.

7 Conclusion and Future Work

In this paper, a novel framework for processing real-time video footage from surveillance camera for protecting critical infrastructures against intruder drone attacks is presented. The confluence of high-precision drone detection component using deep-learning network and the low-cost sensing equipment facilitates not only the detection but also continuous tracking of the drone trajectory. The orientation of the drone flight trajectory will subsequently facilitate the deployment of drone neutralisation solution as needed by the designed countermeasure deployment. As the topic of drone detection and tracking is still in its infancy for real-time and real-world applications, there are several research directions that could be considered as a pathway for improving the proposed framework. The overall objective of the future work is to improve stability of the sensing equipment in order to achieve smooth visual lock on the intruder drone. The current configuration of the low-cost sensing equipment utilises servo motors for controlling the motion of the camera that is aligned with the drone trajectory. The control of the servo motors could be enhanced to result in a smooth aggregation of the horizon video.

Acknowledgement. This work is partially funded by the European Union Horizon 2020 research and innovation program under grant agreement No. 740898 (DEFENDER IP project) and grant agreement No. 787123 (PERSONA RIA project).

References

1. Cehovin, L., Leonardis, A., Kristan, M.: Visual object tracking performance measures revisited. CoRR abs/1502.05803 (2015). http://arxiv.org/abs/1502.05803
2. Coluccia, A., et al.: Drone-vs-bird detection challenge at IEEE AVSS2017. In: 2017 14th IEEE International Conference on Advanced Video and Signal Based Surveillance (AVSS), pp. 1–6, August 2017. https://doi.org/10.1109/AVSS.2017.8078464
3. Dai, J., Li, Y., He, K., Sun, J.: R-FCN: object detection via region-based fully convolutional networks. CoRR abs/1605.06409 (2016). http://arxiv.org/abs/1605.06409
4. Girshick, R.: Fast R-CNN. In: 2015 IEEE International Conference on Computer Vision (ICCV), Santiago, Chile, 11–18 December 2015, pp. 1440–1448 (2015). https://doi.org/10.1109/ICCV.2015.169. IEEE International Conference on Computer Vision, Amazon; Microsoft; Sansatime; Baidu; Intel; Facebook; Adobe; Panasonic; 360; Google; Omron; Blippar; iRobot; Hiscene; nVidia; Mvrec; Viscovery; AiCure

5. He, K., Zhang, X., Ren, S., Sun, J.: Spatial pyramid pooling in deep convolutional networks for visual recognition. In: Fleet, D., Pajdla, T., Schiele, B., Tuytelaars, T. (eds.) ECCV 2014. LNCS, vol. 8691, pp. 346–361. Springer, Cham (2014). https://doi.org/10.1007/978-3-319-10578-9_23

6. He, K., Zhang, X., Ren, S., Sun, J.: Deep residual learning for image recognition. CoRR abs/1512.03385 (2015). http://arxiv.org/abs/1512.03385

7. Henriques, J.F., Caseiro, R., Martins, P., Batista, J.: High-speed tracking with kernelized correlation filters. CoRR abs/1404.7584 (2014). http://arxiv.org/abs/1404.7584

8. Howard, A.G., et al.: MobileNets: efficient convolutional neural networks for mobile vision applications. CoRR abs/1704.04861 (2017). http://arxiv.org/abs/1704.04861

9. Liu, W., et al.: SSD: single shot multibox detector. CoRR abs/1512.02325 (2015). http://arxiv.org/abs/1512.02325

10. Mejias, L., McNamara, S., Lai, J., Ford, J.: Vision-based detection and tracking of aerial targets for UAV collision avoidance. In: IEEE/RSJ 2010 International Conference on Intelligent Robots and Systems (IROS 2010), Taipei, Taiwan, 18–22 October 2010, pp. 87–92 (2010). https://doi.org/10.1109/IROS.2010.5651028

11. Redmon, J., Divvala, S., Girshick, R., Farhadi, A.: You only look once: unified, real-time object detection. In: 2016 IEEE Conference on Computer Vision and Pattern Recognition (CVPR), Seattle, WA, 27–30 June 2016, pp. 779–788. IEEE Computer Society; Computer Vision Foundation (2016). https://doi.org/10.1109/CVPR.2016.91

12. Ren, S., He, K., Girshick, R., Sun, J.: Faster R-CNN: towards real-time object detection with region proposal networks. IEEE Trans. Pattern Anal. Mach. Intell. **39**(6), 1137–1149 (2017). https://doi.org/10.1109/TPAMI.2016.2577031

13. Rozantsev, A., Lepetit, V., Fua, P.: Detecting flying objects using a single moving camera. IEEE Trans. Pattern Anal. Mach. Intell. **39**(5), 879–892 (2017). https://doi.org/10.1109/TPAMI.2016.2564408

14. Szegedy, C., Vanhoucke, V., Ioffe, S., Shlens, J., Wojna, Z.: Rethinking the inception architecture for computer vision. CoRR abs/1512.00567 (2015). http://arxiv.org/abs/1512.00567

15. Wu, Y., Lim, J., Yang, M.: Online object tracking: a benchmark. In: 2013 IEEE Conference on Computer Vision and Pattern Recognition, pp. 2411–2418 (2013). https://doi.org/10.1109/CVPR.2013.312

Classification of Drones with a Surveillance Radar Signal

Marco Messina$^{(\boxtimes)}$ (iD) and Gianpaolo Pinelli$^{(\boxtimes)}$ (iD)

Ingegneria dei Sistemi S.p.A, 24, via Enrica Calabresi, 56124 Pisa, Italy
{m.messina,g.pinelli}@idscorporation.com

Abstract. This paper deals with the automatic classification of Drones using a surveillance radar signal. We show that, using state-of-the-art feature-based machine learning techniques, UAV tracks can be automatically distinguished from other object (e.g. bird, airplane, car) tracks. In fact, on a collection of real data, we measure an accuracy higher than 98%. We have also exploited the possibility of using the same features to distinguish the type of the wing of drone, between Fixed Wing and Rotary Wing, reaching an accuracy higher than 93%.

Keywords: Surveillance radar · Drone · Counter unmanned aerial vehicle · Classification · Support Vector Machines

1 Introduction

Nowadays *Unmanned Aerial Vehicles* (UAVs) make possible to imagine a multitude of previously unavailable and non-cost-effective applications, such as safeguard of human life, security and environmental monitoring. However, the exponential growth of those platforms poses new problems, making the updating of the current Aerial Traffic Management systems inevitable to maintain the same high levels of safety in presence of any aerial platform, manned or unmanned, cooperative or non-cooperative.

To this end, it is important to have an air surveillance system specifically designed to deal with UAVs. In recent years, IDS (*Ingegneria dei Sistemi*) has released a multi-sensorial counter-drone system (Black Knight) capable of detecting, tracking and, if needed, neutralizing potential UAV threats to critical infrastructures and sensitive public and private areas.

Within the ALADDIN (*Advanced hoListic Adverse Drone Detection, Identification, Neutralization*) H2020 project, IDS proposed to improve radar capability, in terms of detection range (up to 5 km for mini-UAVs@-3dBsm Radar Cross Section (RCS)), and the ability to classify drones in an automatic fashion up to the maximum distance of detection.

This paper deals with the design of an algorithm to automatically distinguish the tracks describing the Drones (i.e. UAVs) from those describing other objects (such as birds, airplanes, walking humans), and to distinguish between Fixed Wing (FW) and Rotary Wing (RW) Drone.

© Springer Nature Switzerland AG 2019
D. Tzovaras et al. (Eds.): ICVS 2019, LNCS 11754, pp. 723–733, 2019.
https://doi.org/10.1007/978-3-030-34995-0_66

In the following Sect. 2, we describe the radar signal processing chain necessary to detect and track multiple targets at the same time, the available pieces of information which can be exploited for the classification, the radar configurations and the measurement campaign performed to train the classifier. In Sect. 3 we describe the machine learning techniques adopted to design the classifier and evaluate its performance. In Sect. 4 we show the experimental results obtained on the available data.

2 Radar Signal Processing

This work aims to design a technique able to distinguish drones from other objects, and FW from RW, using the signal received by a surveillance X-band radar working with a *Linearly Frequency Modulated Continuous Wave* (LFMCW) transmitted waveform.

A *surveillance* radar operates with a rotating antenna to discover, detect and track multiple targets at the same time [1, 2]. As a surveillance radar is designed to constantly seek the space to find new targets, the *Time on Target* (ToT), i.e. the time for which the target is illuminated by the radar, is usually very small, in the order of 10 ms.

The most widely used radar architecture for classification and identification of drones is the *tracking* radar, which illuminates a single target for a fairly longer time, in the order of 1 s. The tracking radar holds the antenna in the direction of the designed target, and allows the analysis of features describing the intrinsic movements of the target through the analysis of the time variations of the Fourier spectra of received signals, which is called the micro-Doppler analysis [5, 6].

In counter-drone application, the necessity to detect and track multiple targets at the same time can only be met by a surveillance radar. For this reason, only techniques to classify the received radar signal from a surveillance radar can be applied.

2.1 Signal Processing and Available Pieces of Information

The radar processing chain to detect and track multiple targets is shown in Fig. 1. Using the same notation of [1, 2, 6], the received raw radar signal can be seen as a matrix of values defined in the *Fast Time (FT)/Slow Time (ST)* domain. FT samples identify the range sampling of the received echo (sweep) for a specific azimuth location, while ST samples identify the azimuth coordinates corresponding to the consequential transmitted radar pulses. Each sample is a complex value, identified by the I (In-phase) and Q (Quadrature) received channels.

High pass filter, 1D (along the FT direction) Fast Fourier Transform (FFT) and a calibration procedure is applied to the raw signal to obtain a Range Profile Matrix (RPM) of Radar Cross Section (RCS) values in the Range/Slow Time domain [1]. This is the first piece of information which can be used for drone classification. However, we resort to the anomalies detected by the classical radar signal processing chain of Fig. 1. The RPM are processed with 1D FFT along the Doppler direction, by taking a number of slow time samples which identifies the *Coherent Integration Time* (CIT) [1]. In our case, the CIT is of around 30 ms, and it identifies a radar azimuth cell of 4.5°. A high pass filter is also used to kill the zero-velocity components, removing the stationary clutter from the radar detections.

Finally, the Range Doppler Matrix (RDM) is obtained, which represents the RCS w.r.t. Range and Doppler (or, equivalently, radial velocity [6]) dimensions. The RDM could be directly fed to a Neural Network (NN) based algorithm for drone classification, as in [8]. In our application, RDM is used to find the local maxima points with fairly high RCS values. Those points identify the targets and are called *Detections*. The row and column of the Detection identify respectively the Range and the radial Velocity of the target.

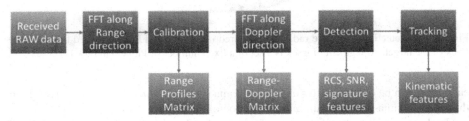

Fig. 1. Radar signal processing (blue boxes) to detect and track multiple targets, and available pieces of information (orange boxes) which can be exploited for drone classification (Color figure online)

For each Detection, the RDM can be used to define a set of signature features. For example, the Detection amplitude describes the RCS, and the ratio between the amplitude of the Detection vs the mean amplitude of pixels in the same row describes the *Signal to Noise Ratio* (SNR). Several other ratios can be defined in the RDM, describing maximum and average amplitudes between regions around the Detection and its surroundings [3, 4, 7].

Detections are then clustered using the Range/Azimuth/Radial Velocity domain. Two or more Detections very close to each other in all the three domains are grouped together in a *cluster* called *Plot* [7], and assigned to the same observation [10].

Finally, using a Kalman filter [9] designed to work in Range/Azimuth domain, the Plots are associated to one or more *Tracks* describing the trajectory and the velocity of the targets, and to predict their future positions. Tracks can be used to evaluate kinematic features of the target, and are useful for classification.

2.2 Features Definition

In this paper, we define *Detections*, *Plots* and *Tracks* (Fig. 2), according to the same notation shown in [7].

For each Detection, a set of descriptors derived from RDM has been defined, including RCS and SNR. Detections are clustered into Plots, and the number of Detections in the Plot also constitutes an useful feature which can be used for classification.

The Plots are observations for the tracking algorithm, which groups them into tracks describing the trajectory of the target. This allows the definition of the kinematic features. They can be evaluated by considering a *Segment of track*, which is a part of the track obtained after a fixed number of observations. Each observation of Kalman filter (i.e.

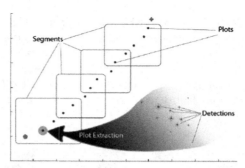

Fig. 2. The Detections from the radar are clustered into Plots, and Plots are used to define the track of the target. A Segment of track is a set of a fixed number of Plots in a track [7].

each Plot) can be obtained after one full antenna rotation, which is the time after which the radar antenna will be again in the direction of the target.

In this study, the classification performance has been analyzed w.r.t. the length of the Segment of track, in the range (4–10). We call this parameter NTREF.

In each Segment of track, a set of kinematic features [7] is defined to describe the target trajectory in the segment. For the NTREF Plots in the Segment, the mean and the standard deviation of the above mentioned signature features are considered. The total number of features for each Segment of Track is 50, of which 30 are signature-based and 20 kinematic-based.

2.3 Radar Parameters and Configurations

The radar operates in the X-band (9.35 GHz), with a Transmitted Power of 4 W and a LFMCW. Its Bandwidth can vary with the configuration, with a maximum of 100 MHz. It performs a 2D scan in Range/Azimuth domain. The central elevation angle must be set by the operator, and the antenna elevation beam is 22.8°. The Pulse Repetition Frequency (PRF) is set to 3.3 kHz.

Table 1. Radar parameters used by the three main configurations

Configuration	Operative_600 m	Operative_2 km	Operative_4 km
Parameter	Value		
Signal Start Frequency	9.3625 GHz	9.3625 GHz	9.3625 GHz
Band	75 MHz	18.867 MHz	9.75 MHz
Antenna rounds per minute	20	20	20
Max Range	624 m	2100 m	4200 m
Range Resolution	2 m	7.95 m	15.9 m
Max Target Speed	96 km/h	96 km/h	96 km/h
Pulse Repetition Frequency	3339 Hz	3339 Hz	3339 Hz
Samples in a Sweep	624	624	624

In the measurement campaigns, mainly three radar configurations have been considered, each characterized by the maximum range of the radar: 624 m, 2.1 km and 4.2 km. They are named respectively *Operative_600 m*, *Operative_2 km* and *Operative_4 km*.

Table 1 summarizes the main parameters of the three radar configurations. The classification algorithms have been designed for each configuration separately. Finally, a classification algorithm has been trained for all the data in all the configurations.

2.4 Measurement Campaign

The measurement campaign was performed by acquiring a set of UAVs:

- Commercial RW as Phantom3 Pro (DJI), Typhoon 4 K (Yuneec), Jetson (NVIDIA), Bebop2 (Parrot),
- Commercial FW as Disco (Parrot),
- IDS RW as FlySmart 2.0, Colibrì, Nik, FlyNovex,
- IDS FW as FlyFast, FlySecur.

When possible, GPS position and time of the drone flight was saved, to help the necessary labelling process of the radar signal. A semi-automatic procedure was developed to label radar data whether GPS information of the target is available or not.

A very high number of non-drone objects were recorded during the measurement campaign. They were non-cooperative targets, such as birds, airplanes, cars, helicopters, walking people. Even without GPS information, in many cases, it was possible to label them as "false alarms" (FA) using the knowledge of the position of the drone during the acquisition.

Table 2 shows the number of acquired samples for FA and Drone classes, and Table 3 for FW and RW classes, w.r.t. configuration and NTREF parameter. Of course, the higher NTREF is, the lower the number of samples to train the classifier. The number of recorded samples of the FA class is much higher for the 2 km and 4 km configurations than for the 600 m one, because the space exploited by the radar is much bigger.

Table 2. Number of acquired samples for Drone vs FA classification, for each configuration and values of the number of antenna rotations to define a Segment of track

	NTREF	4	6	8	10
Operative_600 m	FA	8914	5512	3419	2200
	Drone	2463	2007	1658	1364
Operative_2 km	FA	81824	56822	41446	31525
	Drone	3243	2819	2454	2170
Operative_4 km	FA	40833	30128	23423	18837
	Drone	1470	1258	1078	918

Table 3. Number of samples for Fixed Wing vs Rotary Wing discrimination

	NTREF	4	6	8	10
Operative_600 m	FW	689	539	419	328
	RW	1774	1468	1239	1036
Operative_2 km	FW	1496	1275	1097	961
	RW	1747	1544	1357	1209
Operative_4 km	FW	398	303	225	153
	RW	1072	955	853	765

3 Classification Algorithm

3.1 Training Process

Given an object under test, the purpose of the algorithm is to decide whether the object is a Drone or not, and if it is a Drone, to distinguish between FW and RW Drone.

Many classification algorithms have been compared to this purpose, not only in terms of performance, but also in terms of computational time for training, and overfitting avoidance. Classical algorithms from Machine Learning (ML) theory [12] have been taken into consideration, including KNN (K Nearest Neighbors), Adaboost, Gradient boost, Support Vector Machines (SVM) and Multi-Layer Perceptron.

The comparison between different ML techniques goes beyond the purpose of this paper. We choose to use SVM with radial basis kernel [11], because it obtained an acceptable trade-off between performance, training time and overfitting avoidance.

The training process of the SVM classifier has been performed trough the s-fold cross validation scheme [12]. The samples acquired during the measurement campaign have been split into s subsets. Each subset includes a number of samples such that the ratio between samples from different classes is the same as in the original dataset. The samples from the same acquisition are always included into the same subset. All experiments during the training process are thus always performed training the classifier on samples from different acquisitions w.r.t. the ones in which it is tested. This allows to design a more robust classifier, and to give a more trustful estimate of the classification performance, and thus to predict its behavior when dealing with new samples.

During the s-fold cross validation process, the hyper-parameters and the subset of features within the 50 are chosen.

The optimal subset of features would be given by the exhaustive search of all possible combinations of k features, with $1 \le k \le 50$, which is not feasible with standard hardware resources. For this reason, we adopted a suboptimal search with the *Sequential Floating Forward Feature Selection* method [12], which proved to be a very good trade-off between computational time for training and performance.

The radial basis kernel SVM needs the definition of the two hyper-parameters C and γ [11]. They have been searched choosing the best obtained by three different methods: the exhaustive search on a custom grid, the Newton-Bayes search [12] and the Automatic Model Selection method from Chapelle described in [12].

The suboptimal subset of features and the suboptimal set of SVM hyper-parameters have been searched by optimizing the classifier *accuracy*.

We choose to design two SVMs: the first to distinguish between Drone and FA, the second between FW and RW. Thus, we defined a two-stage SVM classifier. In our analyses, this approach has proven to be better in terms of performance and robustness w.r.t. the direct classification between the three classes FA/RW Drone/FW Drone.

3.2 Performance Evaluation

The performance of the classifier has been evaluated in terms of *accuracy*, per-class *recall* and *precision* indexes, which are defined as in [12].

During the s-fold cross validation process, we evaluated the mean performance (i.e. accuracy, recall, precision) among each of the s classification experiments.

A small number of acquisitions were hidden to the classifier during the training process (i.e. *holdout* [12] – 10% of the available data). This allows also to measure the performance of the classifier with a final blind test which allows to check its robustness.

After the s-fold and the blind test, all available data are re-split into new subsets, and the performance is re-evaluated. This is another precaution taken against overfitting.

Each classifier performance has been thus evaluated in three versions: the one obtained during s-fold training process, the one obtained during blind test, and the one obtained in the final tests re-partitioning the available data. The mean of the three has been called *Global Index* (GI), and can refer to all indexes (accuracy, recall, precision). In this paper, results are presented only in terms of GI.

Generally speaking, the higher the GI, the better the classifier. The comparison of the three indexes obtained during s-fold, blind test, and final re-partitioning, allows to check the robustness of the classifier. Generally speaking, the more similar the three indexes, the more robust the classifier.

The performance are presented w.r.t. the radar configuration and the number NTREF of antenna rotations to define a segment of track. We also show the performance of a classifier designed for all the radar configurations.

4   Experimental Results

In the following, we present the experimental results obtained from the SVM classifiers with the data acquired during the measurement campaign. The following tables list the mean accuracy for each configuration, and for NTREF = 4, 6, 8. For NTREF = 4, we also present the performance in terms of mean per-class precision and recall. Tables 4 and 5 list the performance for the Drone vs FA classification, and Tables 6 and 7 for FW vs RW. Figure 3 shows the trend of the accuracy for each configuration, w.r.t. NTREF in the range (4–10). All indexes are expressed in terms of GI.

Table 4 shows that all Drone vs FA classifiers have good performance, while our comparison among the three accuracies show that those classifiers are also robust. Accuracy is around 98% for the 2 km and 4 km Configurations, and for the classifier trained for all the configurations (*All_Conf*). Accuracy is lower, around 95%, for the *600 m* configuration, but this does not mean that the overall radar performance is worst in that configuration. In fact, the 2 km and 4 km configurations are characterized by a much higher number of false alarms than the *600 m* one. Those FAs are generally well classified by the algorithm, leading to a higher accuracy.

As a matter of fact, we observe that if we analyze the performance only for the samples belonging to the Drone class, the classification algorithm for the *600 m* Configuration achieves the best performance. This is shown by Table 5, which lists the mean per-class Recall and Precision indexes for NTREF = 4. The most likely error committed by the classifiers is the "missed detection", i.e. samples from Drone class erroneously assigned to FA class, and it is more likely to occur as the range increases.

The *All_Conf* classifier shows that accuracy is higher than 98%, meaning that less than 2 segments of tracks out of 100 are misclassified. Figure 3 shows that the performance of Drone/FA classifier is not afflicted by the choice of the number of antenna rotations to define a segment of track. In this case, it is preferable to use the lowest value, i.e. NTREF = 4, it is not necessary to gather more information waiting for further antenna rotations. The classifier proves also robustness for each NTREF parameter.

Table 4. Accuracy (GI) obtained for Drone vs FA classification, w.r.t. Radar Configuration and NTREF parameter. Accuracy is expressed in percentage.

Drone/FA	Accuracy %		
NTREF	4	6	8
Operative 600 m	95.46	95.40	95.62
Operative 2 km	98.82	98.79	98.74
Operative 4 km	98.32	97.69	97.99
All_Conf	98.29	98.35	98.35

Table 5. Per-class Recall and Precision (GI), obtained for Drone vs FA classification, with NTREF = 4, w.r.t Radar Configuration.

Drone/FA	Recall %		Precision %	
NTREF = 4	Drone	FA	Drone	FA
Operative 600 m	87.59	97.60	90.86	96.65
Operative 2 km	80.86	99.53	87.32	99.24
Operative 4 km	75.48	99.23	79.93	99.02
All_Conf	80.12	99.27	85.66	98.93

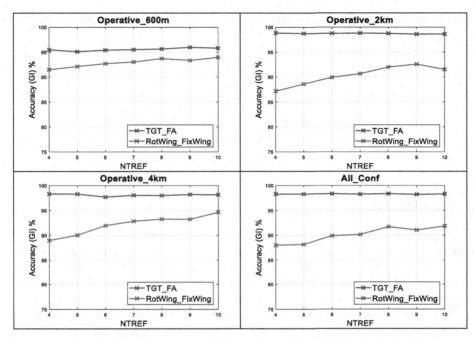

Fig. 3. Mean accuracy (GI) vs NTREF parameter, for each radar configuration

The FW vs RW classifier, instead, can take advantage of using more antenna rotations to improve both performance and robustness, as shown in Table 6 and in Fig. 3. Accuracy is around 88–90% for NTREF = 4, and improves to 92–94% for NTREF = 10.

Performance of FW/RW classifiers are good generally speaking, but not as good as the ones obtained by the Drone/FA classifiers. Due to the lower number of samples from FW class, the RW class is generally better classified, as it is shown by Table 7, which lists Recall and Precision indexes for the classifiers for NTREF = 4.

Finally, we believe that the performance of both FW/RW and Drone/FA classifiers could improve by increasing the number of samples for the Drone class in the database.

Table 6. Accuracy (GI) obtained for Fixed Wing vs Rotary wing classification.

FW/RW	Accuracy %		
NTREF	4	6	8
Operative 600 m	91.43	92.67	93.69
Operative 2 km	87.17	89.92	91.97
Operative 4 km	88.96	91.96	93.26
All_Conf	88.00	89.90	91.71

Table 7. Per-class Recall and Precision (GI), obtained for FW vs RW classification, with NTREF = 4, w.r.t Radar Configuration.

FW/RW	Recall %		Precision %	
NTREF = 4	FW	RW	FW	RW
Operative 600 m	80.71	95.34	86.21	93.19
Operative 2 km	83.20	90.58	88.37	86.24
Operative 4 km	80.49	92.09	79.01	92.73
All_Conf	82.67	90.94	83.44	90.48

5 Conclusions

In this paper, we have shown a novel radar processing algorithms designed to classify UAV versus non-UAV tracks and, within the UAV class, to discriminate among RW versus FW drone type. The multi-stage classification here proposed adheres to the SVM architecture and it is based on a proper selection of identifying signature and kinematics features.

The different stages of classification have been trained through extensive UAV measurement campaigns conducted in a controlled environment, using X-band LFMCW IDS surveillance radar with different radar parameter settings, target and scenarios. Experimental results are highly promising, showing drone/no drone average correct classification accuracy around 98% for the 2 km and 4 km radar configuration and FW/RW accuracy around 92–94% taking advantage from collection of data acquired by higher antenna rotations (NTREF = 10).

References

1. Tait P.: Introduction to Radar Target Recognition, IET (2005). https://doi.org/10.1049/pbra018e
2. Sullivan, R.J.: Radar Foundations for Imaging and Advanced Concepts, Revised edn. Electromagnetics and Radar, SciTech Publishing (2004)
3. Ghadaki, H., Dizaji, R.: Target track classification for airport surveillance radar (ASR). In: 2006 IEEE Conference on Radar (2006)
4. Dizaji, R., Ghadaki, H.: Classification System for Radar and Sonar Applications. Patent US7 567 203
5. de Wit, J.J.M., Harmanny, R.I.A., Molchanov, P.: Radar micro-Doppler feature extraction using the singular value decomposition. In: 2014 International Radar Conference, Lille, pp. 1–6 (2014)
6. Chen, V.: The Micro-Doppler Effect in Radar. Artech House Radar Library (2012)
7. Mohajerin, N., Histon, J., Dizaji, R., Waslander, S.L.: Feature extraction and radar track classification for detecting UAVs in civillian airspace. In: 2014 IEEE Radar Conference, Cincinnati, OH, pp. 0674–0679 (2014)
8. Vojtech, M.: Objects identification in signal processing of FMCW radar for Advanced Driver Assistance Systems. Diploma thesis assignment, Czech Technical University in Prague, Faculty of Electrical Engineering

9. Blackman, S., Popoli, R.: Design and Analysis of Modern Tracking Systems. Artech House, Boston (1999)

10. Klaasing, K., Wollher, D., Buss, M.: A clustering method for efficient segmentation of 3D laser data. In: 2008 IEEE International Conference on Robotics and Automation, Pasadena, CA, pp. 4043–4048, May 2008

11. Ivanciuc, O.: Applications of support vector machines in chemistry. Rev. Comput. Chem. **23**, 291 (2007)

12. Guyon, I., Gunn, S., Nikravesh, M., Zadeh, L.A.: Feature Extraction. Foundations and Applications. Springer, Heidelberg (2006). https://doi.org/10.1007/978-3-540-35488-8

Minimal-Time Trajectories for Interception of Malicious Drones in Constrained Environments

Manuel García[1]($\boxtimes$), Antidio Viguria[1], Guillermo Heredia[2], and Aníbal Ollero[2]

[1] Center for Advanced Aerospace Technologies, Calle Wilbur y Orville Wright,
19, 41300 La Rinconada, Seville, Spain
{mgarcia,aviguria}@catec.aero
[2] Grupo de Robótica, Visión y Control Escuela Superior de Ingeniería Camino de los
Descubrimientos, s/n, University of Sevilla, 41092 Seville, Spain
{guiller,aollero}@us.es

Abstract. This work is motivated by the need to improve existing systems of interception of drones by using other drones. Physical neutralization of malicious drones is recently reaching interest in the field of counter-drone technologies. The exposure time of these threats is a key factor in environments of high population densities such as cities, where the presence of obstacles can complicate the task of persecution and capture of the intruder drone. This paper is therefore focused on the development and optimization of a strategy of tracking and intercepting malicious drones in a scenario with obstacles. A simulation environment is designed in Matlab-Simulink to test and compare traditional interception methods, such as Pure Pursuit which is quite common in missile guidance field, with the proposed strategy. The results show an improvement in the interception strategy by means of a reduction in the time of exposure of the threat with the developed algorithm, even when considering obstacle environment.

Keywords: Counter-drone · Interception trajectory · Guidance

1 Introduction

Currently, there is growing concern about the malicious use of drones, and physical neutralization by other drones is considered a solution of interest by companies and security agencies. In the last years there has been a significant increase of drone usage which is making it an accessible technology for the open public, becoming highly popular among them. The proliferation of cheap and very simple drones has led to an increase in the number of incidents where the security of people and properties on ground and also other airspace users have been compromised. These drones can pose a threat to society when used with malicious intentions, which is the reason why Law Enforcement Agencies are concerned about security regarding dangerous usages of drones. Many drone countermeasures techniques have recently been developed and tested. The problem can be split in two main stages: the first one regards to detection, localization, tracking

© Springer Nature Switzerland AG 2019
D. Tzovaras et al. (Eds.): ICVS 2019, LNCS 11754, pp. 734–743, 2019.
https://doi.org/10.1007/978-3-030-34995-0_67

and classification of the intruder drone; and the second one refers to the neutralization of the intruder drone so that it cannot carry out its mission. Figure 1 shows the use case representation of the formulated problem.

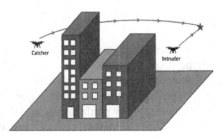

Fig. 1. Representation of the use case for physical neutralization in constrained environment

NASA has published in [1] a technical report evaluating different alternatives for the detection of intruder drones: radar, acoustic sensors, computer vision, etc. Many companies are developing their own system such as the radar of IDS [2], systems with a combination of different sensors like Ctrl+Sky of Advanced Protection Systems [3] and Boreades of CS [4].

With respect to the neutralization of drones, there are also many developments that have been emerging in recent years. The technique of jamming of communications is the one that presents greatest robustness, for example the Scrambler 1000 of MC2 [5], but they usually cannot provoke the interruption of the flight. The work shown in this paper focuses on the physical neutralization of dangerous drones, so that they are intercepted and captured to prevent them from continuing with the threat and to minimize collateral damage. There are developments made by many companies since this field arouses much interest especially among the states security agencies. Within the state of the art, it can be highlighted the ONERA drone with a net launcher in [6] to capture intruder drones and another hunter drone from Purdue University with a hanging net in [7].

The work presented here focuses on the scenario of the physical interception of the intruder drone by means of a mechanism (for example a net) driven by an approaching drone. Hence, the intruder drone must be in the range of action of the interception system of the captive drone. The main interest is improving the capabilities of interception systems by increasing the automation of the physical interception operations and improving the efficiency of the approach trajectories to the intruder drone. The promptness in this operation is an important factor due to the criticality and dangerousness of the flight of a rogue drone. The time of exposure to risk due to the presence of a malicious drone must be minimized so that the damage caused is as small as possible. Hence, the optimization of the approach trajectory of the interceptor drone to capture the malicious aircraft is considered a key factor in this paper. Only multirotors will be considered in this work since this kind of drones is the most widespread today.

Furthermore, threats with drones will be more worrying in urban environments, where there is a high population density. Given the number of buildings, towers and other obstacles that are usually present in these environments, a reactive algorithm to avoid these obstacles is required. In this way, the optimization of the approach trajectory is conditioned by the local deviations produced by the proposed obstacle avoidance algorithm to keep the catcher drone clear of collisions with obstacles.

2 Interception Capability with Obstacles

The first part of this work is focused on the generation of optimized trajectories so that the grabber drone can reach the suspicious drone in a fast and efficient manner. For this purpose, the location data of the suspicious drone which must be tracked is assumed to be an input from an external detection and localization system. Some systems such as the Boreades have integrated a multisensory platform (radar, infrared cameras, optronic cameras, etc.) which provides a localization solution in terms of local position of the intruder drone. Moreover, the trajectory of the intruder drone is unforeseeable since, as discussed above, the type of drone considered in this work is the multirotor. This adds complexity to the problem given the ability of this type of drones to execute aggressive and unpredictable maneuvers.

The problem can be formulated as an interception problem between two bodies which is quite common in physics and engineering field. Missile guidance is a widely studied field from which this study starts. Standard guidance laws like Proportional Navigation (PN) and Pure Pursuit (PP) are quite extended for interception problems and can be adapted to this scenario. Many works are available in using these guidance algorithms but mostly for fixed wing drones and for the purpose of tracking ground targets. In [8], Tan performs in his master thesis an adaptation of PN for multirotors tracking ground targets. Adaptation of PP to multirotors has also been studied in [9], where it is used for path following combining it with virtual target concept. Additionally, there are other approaches for interception problems such as the proposed in [10] with drones optimizing the time to rest after the capture.

Regarding obstacle avoidance, this is a field of great interest in aerial robotics and drones traffic management. The obstacle avoidance algorithms for drones are usually very local and reactive, executing in real time so that the drone can re-plan its trajectory to reach the objective, ensuring a certain distance from the encountered obstacles. Many approaches are available for this purpose, such as: graph search algorithms [11], methods of potential fields [12], rapidly exploring random trees [13], etc. In this work, a strategy based on the evaluation of a cost function in the future positions of the drone taking into account the approach trajectory to the intersection point based on the optimization algorithm is used.

3 Optimization Strategy

The fundamental objective is to minimize the time of arrival to the interception by the capturing drone. Successive modifications to the traditional Pure Pursuit have been made to study the improvements they produce. Several criteria have been decisive in order to minimize the interception time through the estimation of the point of interception based on the trajectory followed by the intruder drone:

– Analyze the trajectory followed by the intruder drone to predict its movements or destination
– Estimate the time to interception and optimize the approach trajectory in order to minimize the time of exposure of the threat
– Anticipate the positions to which the intruder drone is expected to move based on previous analysis

As previously stated, the position of the intruder drone is assumed to be an input, which should be given by an external system. This position is referred to the same coordinate frame as the catcher drone. The key idea behind the optimization algorithm is to take advantage of the information of the trajectory followed by the intruder drone to foresee their intentions: changes in course, speed, etc. A strategy of intelligent approach to the intruder drone is proposed. For that purpose, it is assumed a trajectory model of both drones, the intruder and the catcher, governed by the following equations:

$$\frac{dx}{dt} = v * \cos(\gamma) * \cos(\varphi)$$
$$\frac{dy}{dt} = v * \cos(\gamma) * \sin(\varphi)$$
$$\frac{dz}{dt} = v * \sin(\gamma) \tag{1}$$

Subscript "i" will refer to the intruder drone and subscript "c" will refer to the catcher drone. The positions of the drones in a world frame are defined as $[x_i y_i z_i]$ and $[x_c y_c z_c]$ respectively. The velocity of the drones is denoted as v. Finally, the climb angle and the course angle of the trajectory for each drone are defined as γ and φ respectively. All these state variables can be particularized for both drones with the specific subscript.

The control of the drones will be based on position commands. Most drones autopilots currently support commands in position. The optimization comes from an adaptation of the pure pursuit methodology for tracking a target by means of the command to the drone captor of the current position of the intruder drone with a higher cruising speed of the estimated v_i in order to reach it (Eq. 2), so that it always goes towards it.

$$[x_c, y_c, z_c]_{cmd} = [x_i, y_i, z_i] \tag{2}$$

The distance between the two drones $dist(t)$ can be calculated as a function of time based on the position of both drones. This distance is calculated for the successive positions of both drones assuming constant heading and climb angles for the two drones in the next moments (integrating Eq. 1 for both drones). The moment in which this distance is minimum is identified as candidate to be the point where the interception will take place. Apart from that, it gives an estimation of the time required for the interception t_{int} (Eq. 3).

$$dist(t) = f(x_i(t), y_i(t), z_i(t), x_c(t), y_c(t), z_c(t))$$
$$\frac{d(dist(t))}{dt} = 0 ==> t_{int} \tag{3}$$

In a similar way to the "Deviated pursuit guidance" methodology [14], it is proposed to command the catcher drone with the anticipated interception position taking into account the calculated time t_{int} and using the current velocity, heading and climb angle of the intruder drone. Then, the position command to the catcher drone would be estimated interception point based on the time calculated for interception:

$$[x_c y_c z_c]_{cmd} = [x_i(t_{int}) y_i(t_{int}) z_i(t_{int})] \qquad (4)$$

Directing the catcher drone towards the predicted interception point with the target is called Constant Bearing (CB) Guidance in missiles field, since the interceptor trajectory must be straight (see Fig. 2). This will mean a reduction in the time needed for the interception, as well as a reduction in the distance traveled by the malicious drone, and therefore its possibility of reaching its target and causing the damage. The difference between Pure Pursuit strategy (Eq. 2) and Constant Bearing (Eq. 4) can also be explained with the Line Of Sight (LOS) which joins both drones. In the Constant Bearing case, the LOS is maintained parallel during the persecution and in the nominal PP it is rotating.

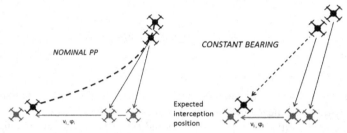

Fig. 2. Comparison between nominal approach (direct command of the target position) and the implementation of the Constant Bearing (CB)

In case of variations of course or height, the estimation of interception position is updated in real time (@ freq $= 100$ Hz which is the frequency of the controller) but there may be high deviations if these changes are abrupt. A parameter is defined to measure the abruptness of the trajectory changes of the intruder drone "var", during the recent previous instants. This parameter is based on the rate of change of climb and course angles in the previous moments. When this parameter is high, it is representative of significant changes in the intruder drone direction of movement, so there is high uncertainty in its destination. In this case, logical actions are considered in order to adapt the guidance strategy to the trajectory that the intruder drone is performing. These logical actions consist of commanding an estimation of the average position of the intruder drone in the previous moments. The time considered to compute this average position t_{av} depends on the amplitude of the orientation change and the proximity to the target.

$$var = f\left(\frac{\Delta\varphi_i}{\Delta t}, \frac{\Delta\gamma_i}{\Delta t}\right), t_{av} = f(var, dist(t)) \qquad (5)$$

Then, when the value of var is bigger than a threshold parameter k, the position command to the catcher drone is given by Eq. 6. This parameter k is defined as maximum climb and course angle rate average in the last 5 s.

$$[x_c(t)]_{cmd} = \frac{\sum_{n=0}^{t_{av}*f} x_i(t - n/f)}{t_{av}}$$

$$[y_c(t)]_{cmd} = \frac{\sum_{n=0}^{t_{av}*f} y_i(t - n/f)}{t_{av}}$$

$$[z_c(t)]_{cmd} = \frac{\sum_{n=0}^{t_{av}*f} z_i(t - n/f)}{t_{av}} \tag{6}$$

The position commanded to the catcher drone $\left[x_c(t), y_c(t), z_c(t)\right]_{cmd}$ is based on navigation NWU frame, parallel to Earth axis. In order to impose the commanded velocity to the catcher drone $v_{c,cmd}$, it is necessary to decompose the velocity vector, which joins the current position of the catcher drone to its commanded position, and impose rate of changes in each axis. In this way, using Eq. 1, it is also possible to obtain the commanded climb and course angle: $\varphi_{c,cmd}$ and $\gamma_{c,cmd}$.

4 Obstacle Avoidance

The algorithm described above has as output a position commanded to the capturing drone to approach and intercept the intruder drone. This position is calculated in real time and assuming that there are no obstacles in the environment, so that the drone executes the corresponding actions to move towards the commanded position at every moment. In the presence of obstacles, it is necessary to locally modify the trajectory in order to avoid interferences with them. For this purpose, a cost function (*cost(t)*) is defined (Eq. 7) that takes into account the distance to the target (*dist*) and the proximity of the catcher drone d_{obs} to the obstacles. The evaluation of this function has a high computational cost, so the refresh rate is 10 Hz. Through this evaluation, it will be possible to locally change the course of the catcher drone in order to avoid the present obstacles although the target position remains the same.

$$cost(t) = f(dist(t, t + \Delta t, \ldots), d_{obs}(t, t + \Delta t, \ldots)) \tag{7}$$

The strategy to avoid obstacles considers a range of commanded course $\Delta\varphi_{c,cmd}$ and climb angles $\Delta\gamma_{c,cmd}$, taking the commanded $\varphi_{c,cmd}$ and $\gamma_{c,cmd}$ as references:

$$\Delta\varphi_{c,cmd} = \varphi_{c,cmd} + \Delta\varphi$$

$$\Delta\gamma_{c,cmd} = \gamma_{c,cmd} + \Delta\gamma \tag{8}$$

The projection of the future positions of the capturing drone is calculated considering the sweep in previous angles. Hence, in each iteration, this cost function *cost(t)* is evaluated taking into account the current position of both drones and the projection of said positions in the successive instants in a certain range of orientations with respect to the current course (Eq. 8). Figure 3 shows the situation of the catcher drone (in black) in front of an obstacle that must be avoided with the objective of following the intruder drone (in red) and for this it evaluates the cost function for a range of courses. The course to follow is chosen in such a way that the cost function has the lowest value (Eq. 8).

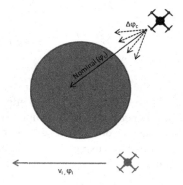

Fig. 3. Exploring different orientations to avoid obstacles (Color figure online)

$$\varphi_{c,obs} / \min cost(t) \, \forall \, \varphi \, \in \, (\varphi(t) + \Delta \varphi) \tag{8}$$

As explained in Sect. 3, where the $\varphi_{c,cmd}$ and $\gamma_{c,cmd}$ are computed, the variation of the commanded position in all its axes is limited by decomposing the maximum commanded speed according to the course and the climb angle of the aircraft and the target point. In the presence of obstacles, the computed angle $\varphi_{c,obs}$ replace $\varphi_{c,cmd}$, so the catcher drone moves according to this new angle in order to avoid the obstacles. Hence, the avoidance algorithm works by commanding changes in the course of the multirotor, but not in the position command based on the optimization strategy explained in Sect. 3. The design and development of the presented algorithms has been validated through simulations of different scenarios and trajectories. The tool used for this purpose is Matlab-Simulink. The module architecture of the developed algorithms in the Guidance block includes the approach trajectory optimization and the obstacle avoidance strategy. The modular architecture is shown in Fig. 4.

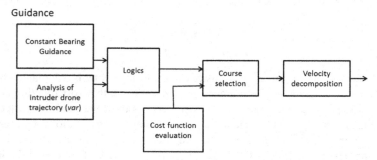

Fig. 4. Modular architecture of the algorithm

5   Results

The analysis will be centered for a random path of the intruder drone in which there are multiple changes of orientations. Then, the previous scenario will be combined with an obstacle map to evaluate the capabilities of the algorithm developed in the required environment. For the simplicity of the analysis, constant speeds for both drones will be considered and will be maintained in all simulations. The intruder drone will fly with a speed of 5 m/s and for the capturing drone to reach it, a higher speed is needed: 8 m/s. The analysis will be made according to the time required to carry out the interception. As the objective of this work is to optimize the approach path to the point of interception, this will be defined as the moment when the distance between both drones is less than a distance of 5 m.

1. Random non-straight trajectory
   A random trajectory is established in left part of Fig. 5 for the intruder drone so that it changes its course without any predefined criteria. The interception time with the nominal strategy is 235,12 s. On the other hand, if the optimization strategies and the logic for intelligent guidance are used, the required time for the interception is 195,26 s, so there is a reduction of 16,95%.

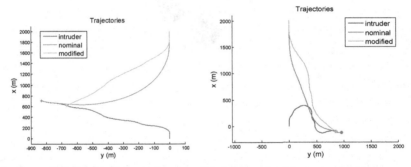

Fig. 5. Comparison between nominal and modified approach with non-straight trajectory of the intruder drone (Color figure online)

It can be seen how the trajectory with the Constant Bearing strategy directs the capturing drone towards an estimated intercept position, shortening the trip to the encounter. The estimation of the meeting point is conditioned by the changes of orientation of the intruder drone. Another result can be shown in the right case of Fig. 5 in order to remark the difference between the considered two cases with a trajectory of the intruder drone which performs a more curvilinear trajectory. In this case the improvement in performance measured as the reduction of time needed for interception is 6,7%. From the right figure it can be extracted how the red trajectory reaches the target before the nominal case in pink.

2. Obstacles environment
   In case there are obstacles in the environment where interception is desired, it will be necessary to modify the approach trajectory based on the algorithm in Sect. 4.

Below there is an example of the interception produced with the same trajectory for the intruder drone than in the previous case but with two circumferences which represent obstacles. Again, the capabilities of the two strategies analyzed in this study are compared. In the first nominal case there is a required time of interception of 225,37 s, while with the Constant Bearing the time is reduced to 200,07 s, so that the reduction of exposure time of the threat is 11,23%.

The results shown in Fig. 6 are a specific sample of certain trajectories and environments. For the validation of the work carried out, up to 100 simulations with different trajectories for the intruder drone have been carried out and considering different obstacle maps with density similar to that shown in Fig. 6. In general, the results are satisfactory with an average reduction of exposure time in an unobstructed environment around 15%; and when there are obstacles, this improvement is reduced to roughly 10%. This is due to the fact that the replanning carried out for the avoidance of obstacles makes the optimizations of the Constant Bearing algorithm less efficient.

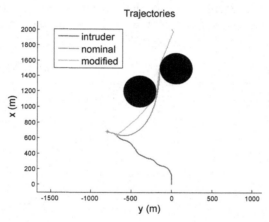

Fig. 6. Comparison between nominal and modified approach with non-straight trajectory of the intruder drone in an obstacle environment

6 Conclusions

This paper analyzes a functionality that arouses much interest in the drone community, such as designing a guiding algorithm that allows other drones to intercept malicious drones in an environment with obstacles such as urban scenario. The main objective is to reduce the exposure time of the malicious drone and minimize the threat. The location of the rogue drone is assumed as an input from an external system.

Starting from the field of missiles guidance laws, algorithms such as "Pure pursuit" have been adapted and improved to the problem of interception of multirotors taking into account the particularities of the movement of this type of aircraft. Moreover, an obstacle environment has been considered since the case of interception of drones in urban environments is of special interest, where the danger of this threat is very significant due to the high population density. The results extracted from the simulations show the improvements produced by the proposed guidance strategies compared to traditional solutions in a scenario of aerial pursuit and interception between drones.

Next work will focus on increasing the capabilities of the algorithm presented here so that with a multitude of variable trajectory inputs of the intruder drone and maps of obstacles the time of interception is minimal. Additionally, deep work is required for the final stage of the approach trajectory in which the catcher drone must capture the intruder drone, when the relative distance is small. In addition, experimental tests are intended to have a practical validation of the results obtained in simulation. One of the proposed options is to use DJI drones, that are currently the most widespread, and to command the interception trajectories through the development kit SDK by DJI.

Acknowledgements. This work was partially funded by the European Union's Horizon 2020 Research and Innovation Program within the ALADDIN project, Grant Agreement N° 740859.

References

1. Güvenç, I., Ozdemir, O., Yapici, Y., Mehrpouyan, H., Matolak, D.: Detection, localization, and tracking of unauthorized UAS and Jammers. In: 36th IEEE/AIAA Digital Avionics Systems Conference (DASC 2017) (2017)
2. NO-DRONE. https://www.idscorporation.com/pf/no-drone/
3. Ctrl+Sky. http://apsystems.tech/en/why-ctrlsky-2/
4. Boreades. http://boreades.fr/
5. Scrambler 1000. https://www.mc2-technologies.com/scrambler-1000/
6. Lefebvre, T., Dubot, T., Joulia, A.: Conceptual study of an Anti-Drone Drone through the coupling of design process and interception strategy simulations. In: 16th AIAA Aviation Technology, Integration, and Operations Conference, Washington, D.C. (2016)
7. Goppert, J.M., et al.: Hopmeier: realization of an autonomous, air-to-air counter unmanned aerial system (CUAS). In: First IEEE International Conference on Robotic Computing (IRC) (2017)
8. Tan: Tracking of Ground mobile Targets by Quadrotor Unmanned Aerial Vehicles. University of Cincinnati (2013)
9. Manjunath, A.: Path following by a quadrotor using virtual target pursuit guidance. Utah State University (2016)
10. Hehn, M., D'Andrea, R.: Real-time trajectory generation for interception maneuvers with quadcopters. In: IEEE/RSJ International Conference on Intelligent Robots and Systems (2012)
11. Garcia, M., Viguria, A., Ollero, A.: Dynamic graph-search algorithm for global path planning in presence of hazardous weather. J. Intell. Robot. Syst. **69**(1–4), 285–295 (2013)
12. Budiyanto, A., Cahyadi, A., Bharata Adji, T., Wahynggoro, O.: UAV obstacle avoidance using potential field under dynamic environment. In: International Conference on Control, Electronics, Renewable Energy and Communications (ICCEREC) (2015)
13. Ramana, M.V., Varma, S.A., Kothari, M.: Motion planning for a fixed-wing UAV in urban environments. In: IFAC Conference on Advances in Control and Optimization of Dynamical Systems, vol. 49, no. 1, pp. 419–424 (2016)
14. Ratnoo, A.: Variable deviated pursuit for rendezvous guidance. J. Guidance Control Dyn. **38**(4), 787–792 (2015)

UAV Classification with Deep Learning Using Surveillance Radar Data

Stamatios Samaras$^{(\boxtimes)}$, Vasileios Magoulianitis, Anastasios Dimou,
Dimitrios Zarpalas, and Petros Daras

Information Technologies Institute, Centre for Research and Technology Hellas,
Thessaloniki, Greece
{sstamatis,magoulianitis,dimou,zarpalas,daras}@iti.gr

Abstract. The Unmanned Aerial Vehicle (UAV) proliferation has raised
many concerns, since their potentially malicious usage renders them as a
detrimental tool for a number of illegal activities. Radar based counter-
UAV applications provide a robust solution for UAV detection and classification. Most of the existing research addresses the problem of UAV classification by extracting features from the time variations of the Fourier
spectra. Yet, these solutions require that the UAV is illuminated by the
radar for a longer time which can be only met by a tracking radar architecture. On the other hand, surveillance radar architectures don't have
such a cumbersome requirement and are generally superior in maintaining situational awareness, due their ability for constantly searching on a
360° area for targets. Nevertheless, the available automatic UAV classification methods for this type of radar sensors are relatively inefficient.
This work proposes the incorporation of the deep learning paradigm in
the classification pipeline, to provide an alternative UAV classification
method that can handle data from a surveillance radar. Therefore, a Deep
Neural Network (DNN) model is employed to discern between UAVs and
negative examples (e.g. birds, noise, etc.). The conducted experiments
demonstrate the validity of the proposed method, where the overall classification accuracy can reach up to 95.0%.

Keywords: UAV · Drones · Classification · Deep learning ·
Surveillance radar

1 Introduction

Considering their increasing popularity, UAVs have been devised to assist people
in their daily activities because of their ease of use and robustness across various
environments. They have been rendered as an efficient solution to a number of
useful applications ranging from advanced cinematography to wildlife surveillance and package delivery. On the other hand, it is not uncommon for those

This work has received funding from the European Union's Horizon 2020 Research and
Innovation Programme under grant agreement N° 740859, ALADDIN.

technological advancements to be also used for malicious purposes. In particular, UAVs have been reported in many illegal activities, such as espionage, carrying explosives for terrorist attacks or provocative messages in crowded events. Finding an effective solution to this potential threat, that can reliably detect and classify incoming UAVs, around a protected area has been an active research endeavor. Different modalities have been explored both in scientific literature and industry for UAV detection and classification, including optical, thermal, acoustic and radar data. The latter have been proved to be very effective in detecting flying vehicles and objects, and is being employed for decades in pertinent applications. Yet, typical radar sensors designed for detecting manned aircraft, with large Radar Cross Section (RCS) measurement [1] and high velocity, are not suitable for detecting very small and slow moving objects, flying at low altitude such as UAVs [2]. Moreover, UAVs share similar cues with birds and thus classifying between the two targets is another major challenge. Towards this end, specifically designed radar architectures have been devised for this demanding application that overcome such burdens.

The available radar architectures employed in literature for target detection can be divided in two broad categories: surveillance and tracking radars [3]. Surveillance radars provide 360° continuous coverage as they operate with a rotating antenna to detect and track multiple targets at the same time [4]. On the other hand, tracking radars retain their antenna to a specific direction of a designated target, illuminating it for sufficient time which enables for target classification through the analysis of time variations of the Fourier spectra (e.g. target velocity, Doppler and micro-Doppler signatures [5]). In particular, the micro-Doppler (m-D) signature of UAVs has been widely employed within literature for classification purposes [2,6–9]. The typical detection and classification pipeline is to perform radar signal processing algorithms for detecting targets (e.g. CFAR detection algorithm) [4] and extract intrinsic features from the processed signal (m-D signature) [8,10] for automatic classification.

Despite the promising classification capabilities that tracking radars have proved to withhold, the volatile nature of UAVs can make this type of solution ineffective in a real world application. Extracting the m-D signature from a non cooperative target with high maneuverability can prove to be a challenging task and most of the research has been conducted in ideal scenarios, in close flight range. On the other side, a surveillance radar seems like a more appropriate choice for the task at hand because it is designed to constantly search the protected space for new targets, hence providing 360° coverage and protection at all times. Nevertheless, the time for which a surveillance radar illuminates the target is not long enough to allow for deeper analysis of the intrinsic characteristics of the target through the Fourier spectra which creates a drawback in the classification capabilities of the radar. This has driven UAV classification methods with surveillance radars in the literature to utilize mostly target motion information by creating handcrafted kinematic features (e.g. motion tracks, velocity) [11] or building probabilistic motion estimation models [12] and less often through radar signature features derived from Range Profiles and Range Doppler matrices (e.g.

RCS, Signal to Noise Ratio (SNR), mean amplitudes, etc) [13]. However, those methods heavily rely on specific handcrafted features omitting the rich information cues available on the Range Profiles Matrix.

Deep Neural Networks (DNNs) have been proved very efficient in discovering high-level and abstract features directly from data [14]. This work proposes a novel UAV classification method based on a custom Convolutional Neural Network (CNN) [15] architecture that classifies detected targets between UAVs and negative examples (i.e. birds, noise, etc.). The proposed network learns directly from the Range Profiles Matrix signature of the detected target alongside with radar signature features derived from both Range Profiles and Range Doppler matrices, such as Radar Cross Section (RCS), Signal to Noise Ratio (SNR) and radial velocity. We train and evaluate our proposed network with real radar measurements performed from a X-band Linear Frequency Modulated Continuous Wave (LFMCW) surveillance radar operating at 9.35 GHz. To the best of authors knowledge, this is the first attempt to utilize the available radar signature information, as an input to a DNN model, so as to learn the intrinsic characteristics of the targets directly from data. The main contributions of this work can be summarized as:

- Demonstrate how the Range Profiles Matrix data can be utilized for automatic UAV classification.
- Propose a custom CNN network architecture that discriminates detected targets between UAVs and negative examples.

2   Related Work

The brief literature review is focused on UAV detection and classification methods for tracking and surveillance radars.

2.1   Tracking Radars

In recent years, the m-D signature has been the most commonly employed feature for UAV classification in the field of counter-UAV radar based applications. It is mainly utilized as a signal pre-processing step to extract features combining them with state of the art machine learning and deep learning algorithms for automatic target classification. Among the first who utilized the m-D signature for UAV classification were [7]. The authors extracted the m-D signature with the Short Time Fourier Transform (STFT) and proposed to utilize key characteristics such as rotation rate, blade tip velocity, rotor diameter and number of rotors to classify between four different rotary wing UAVs. In a similar work, Harmanny et al. [9] proposed to extract m-D features from spectrograms and cepstrograms with a low power Continuous Wave (CW) tracking radar operating at X-band for a potential two step classification process that initially classifies UAVs versus birds and subsequently classifies the type of UAVs based on the number of rotors. Molchanov et al. [2] studied the problem of UAV classification for a CW tracking radar operating in X-band at radio frequency of 9.5 GHz. Eigenpairs were extracted from the correlation matrix of the m-D signature and were employed as intrinsic features to train three classifiers, a linear and a non-linear Support Vector Machine (SVM) and a Naive Bayes Classifier (NBC) to classify between different UAV wing types and birds. De Wit et al. [6] considered a m-D feature extraction method based on singular value decomposition (SVD) to classify UAVs and birds with a X-band CW tracking radar sampled at 96 kHz. The authors proposed three main features to allow for quick classification: target velocity, spectrum periodicity, and spectrum width.

In an effort to extract the m-D signature with a non-Fourier algorithm, Oh et al. [8] proposed an empirical-mode decomposition (EMD) based method for automatic UAV classification. A nonlinear SVM was trained for target class label prediction after feature normalization and fusion. The authors validated their method on the same dataset as [2] outperforming common Fourier based m-D extraction methods.

The first work that utilized CNNs to learn directly from the m-D signature spectrograms and classify UAVs was [10]. The authors employed GoogleNet [16] and trained it with spectrograms from real UAV measurements from a FMCW radar. In addition, they proposed a method to improve the m-D signature results by merging it with its frequency domain representation, namely the cadence velocity diagram (CVD). Mendis et al. [17] utilized a deep belief network (DBN) [18] for UAV classification using a S-band CW tracking radar. The authors experimented with three different UAV types two rotary and one fixed wing. The spectral correlation function (SCF) was employed to identify unique modulations caused by the many dynamic components of the target of interest.

2.2   Surveillance Radars

UAV classification methods that utilize a surveillance radar are not as common in literature. A similar work to ours is [13] where the authors proposed a binary classification method to distinguish between UAV and bird tracks with data captured under a surveillance radar. The authors adopted a set of twenty features based on movement, velocity and target RCS extending the works of [11] and [19] that initially proposed a similar approach to classify aircraft and bird tracks. The handcrafted features are combined with a Multi Layer Perceptron (MLP) classifier demonstrating high classification accuracy. We utilize a deep learning network to learn directly from the data combining some typical radar features instead of relying on handcrafted features entirely. Furthermore, we work on each detection separately instead of utilizing tracks, which are groups of multiple detections across time.

A different approach based on motion information only is [12]. The authors proposed a probabilistic motion model estimation method based on calculating the time-domain variance of the model occurrence probability in order to classify between UAVs and birds with data originating from a surveillance radar. They validated their approach on simulated and real data showing promising results.

3   Proposed Pipeline

3.1   Radar Parameters

The radar sensor that is utilized to acquire data, constantly scans within the range and azimuth domain. Thus, this 2D radar omits altitude information from the localization of the detected target. In particular, it is a X-band LFMCW surveillance radar that operates at 9.35 GHz, with a transmitted power of 4 W and a Pulse Repetition Frequency (PRF) of 3.3 kHz. The radar antenna elevation angle is usually set to around $20°$ and the radar antenna performs 20 full revolutions per minute. The max detected range is configurable, ranging from 600 m up to 4 km, with different range resolutions. Both short range and long range configurations were picked during the data creation process.

3.2   Signal Processing

The raw data acquired by the 2D radar are a matrix of complex values, referring to In-phase and Quadrature receiver channels. The two dimensions of the matrix represent fast time and slow time signal information [4]. Each column represents a sweep, and rows represent the sampling of the LFMCW signal. The number of rows is fixed by the sampling frequency of the received signal (around 600 samples), the number of columns depends on the PRF and on the duration of the acquisition.

Raw data are processed by calibration, radar equation correction, and Fast Fourier Transform (FFT) along the direction of fast time, in order to obtain the Range Profiles Matrix of RCS values [4]. The Range Profiles Matrix is also a matrix of complex values; although, its two dimensions now represent range and slow time values. Slow time instances can be also interpreted to identify azimuth values. To detect targets, Range Profiles are divided in samples, each sample refers to a specific range cell (rows of the matrix) and the slow time or azimuth instance correspondents (columns of the matrix). These samples are processed by FFT along slow time direction, also referenced as Doppler direction in literature [4], and the result is a list of detections. This process produces the Range Doppler Matrix of RCS values in range and radial velocity dimensions. The local maximum points of the Range Doppler Matrix represent potential detected targets. These points are processed by clustering and produce a list of detections in range and radial velocity domain (rows and columns of Range Doppler Matrix). The same rows and columns correspondents can be used to extract the Range Profile Matrix signature of each detection. This is the first input to the proposed DNN model.

A set of radar signature features can be extracted from the Range Doppler Matrix and are used as the second input to our DNN model. Those would be RCS of the target, which is the detection amplitude. The Signal to Noise Ratio (SNR), which is the ratio between the detection amplitude versus the mean amplitude in the same range of the detection. And finally, the radial velocity of the detected target.

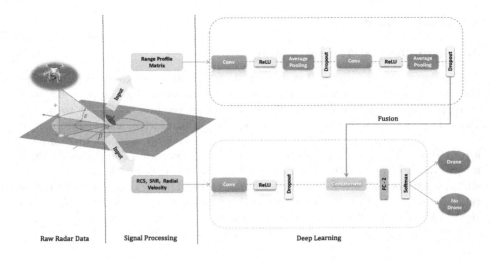

Fig. 1. UAV classification network architecture

3.3 UAV Classification Pipeline and DNN Model

UAV Classification Pipeline. The input to our proposed UAV classification method is a list of detections and radar features derived from classic radar signal processing algorithm [4]. Our DNN model yields an inference whether a detected target is a UAV or a negative example (e.g. bird, noise, etc.) based on the signature of the target, provided by the radar. The network is trained in a supervised manner, by utilizing a dataset consisting of multiple UAVs and negative example detections.

Network Architecture. The basic building block of our proposed model are the Convolutional Neural Networks (CNNs) [15]. Driven by their unprecedented success in tasks, such as image recognition [20] and object detection [21], we designed our model from scratch, following a similar architectural design, as it is depicted in Fig. 1. To the best of our knowledge, there are no recommended DNN architectures based on literature that solve the problem at hand.

We opt for a two branch architecture to handle the analysis of Range Profiles Matrix signature, separately from the radar features, and concatenate the output feature maps of both branches before the fully connected layer for joint training. The Range Profiles Matrix branch is consisted of two 1D CNN blocks with 128 and 64 output convolution filters each followed by a ReLU activation layer, an average pooling layer and a dropout layer. The radar features branch consists of a single 1D CNN layer with 64 filters followed by a ReLU activation and a dropout layer. The output of the fully connected layer is the probability of the input sample under test, to be one of the two existing classes (UAVs and negative examples) assigned by a Softmax activation function.

4   Experiments

4.1   Training and Testing Settings

The proposed DNN is trained from scratch and we use the Xavier initialization for the initial weight values. The dropout layers probability is $p = 0.5$ and is applied to prevent overfitting effects. The objective function is set to be the categorical cross entropy. We utilize the Adam optimizer with initial learning rate of 0.001. To achieve the best UAV recall possible, we keep every prediction with a probability above 0.5 for UAV class as a UAV prediction. The Range Profiles matrix input samples have size of 126×4 and we utilize the magnitude of each complex number before feeding it as input to the DNN. The three additional radar signature features are concatenated beforehand and then are being fed to the model. The network was trained for 100 epochs with a batch size of 128. All experiments were performed on a NVIDIA GeForce 1070 with 8 GB memory.

4.2   Dataset Description

To train and evaluate our proposed model, we performed real radar data acquisitions with multiple UAVs flying under different flight plans, both close to the radar (50 to 600 m), as well as far away (up to 2 km). External GPS trackers were attached on top of the deployed UAVs, to provide ground truth trajectories, thereby enabling easier annotation for the training and evaluation of the proposed method. Each radar detection has a ground truth label assigned to it, for measuring how close is the detection to the GPS ground truth trajectory. A total of 4.5 h of recorded data were collected to form a dataset with sufficient amount of data for a DNN model. By this dataset, two acquisitions were utilized for evaluating the DNN model, which include multiple UAVs (up to 5 drones) that soar in the sky with distinct patterns. The samples that were utilized for the training of the classifier were evenly selected, with respect to a balanced distribution between the available classes. The selected samples utilized for training purposes, include 2536 UAV and 2594 negative samples. Furthermore, the two test acquisitions utilized for testing the method include 2038 UAV and 1371 negative samples.

4.3   Evaluation

Table 1 illustrates the confusion matrix with detailed classification results of our DNN model, for both classes and all the detection samples under testing examination. The metrics that are utilized are the per class precision, recall and F-1 score, as well as the overall classification accuracy. Per class metrics are depicted in Table 2. It can be observed that the model yields a substantial performance in all metrics, achieving a remarkably high score, where the average precision and recall are 94.48 % and 95.55 % respectively. It turns out, the model can reliably distinguish between radar signatures, which belong to UAVs and those that can be whatever object, such as a bird or noise. Another factor that indicates the superior classification capabilities of the model, is the near perfect UAV precision score of 98.79 %. By the total number of samples, only 23 negative ones are falsely classified as UAVs. The model's overall classification accuracy (correctly classified samples to total samples) is 95.0 %.

Table 1. Confusion matrix with classification results for the testing set

	Classified UAV	Classified negative	GT samples
Label UAV	1891	147	2038
Label negative	23	1348	1371

Table 2. Per class Precision, Recall and F1 score metrics for the testing set

	Precision (%)	Recall (%)	F1 score (%)	GT samples
Label UAV	98.79	92.78	95.69	2038
Label negative	90.16	98.32	94.07	1371
AVG/total	**94.48**	**95.55**	**94.88**	**3409**

The two distinct flights employed for testing the model's performance, were carried out under close range configuration, flying in distances up to 450 m from the radar. In Fig. 2 we adduce the per distance UAV precision and recall metrics, in an attempt to examine how the range factor affects the performance of the classifier. It can be observed that relatively closer to the radar (up to 200 m) the proposed model seems to classify some of the UAV samples as negative, thus the UAV recall is below 90 %. However, at mid-range ($\geq$200 m) both UAV precision and recall are above 90 %. Finally, we present qualitative results of our model on Fig. 3, where we have included the model predictions for one of the testing acquisitions. On the left part, the flight plan GPS trajectory followed by 5 UAVs flying next to each other is depicted. On the right part, we present the predicted results of our model on the radar detections of this flight. We show UAV predictions with blue color and negative example ones with red. Our model can reliably recognize the UAVs that are flying within the area covered by the surveillance radar.

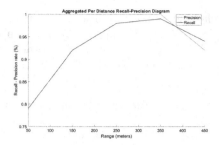

Fig. 2. Per distance precision and recall for the both recordings of the testing set.

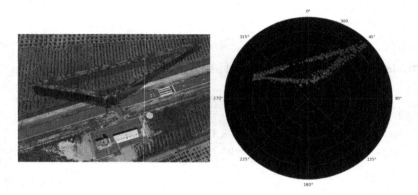

Fig. 3. Flight plan trajectories from external GPS attached on multiple UAVs flying in similar way on the left. Proposed model predictions on the right, blue color represent UAV predictions and red negative examples. (Color figure online)

5   Conclusion

This work presents a deep learning based UAV classification pipeline that utilizes data from a surveillance radar. The proposed method has been evaluated on a dataset of real radar data acquisitions with UAVs. This study provides an alternative pathway to the radar research community, for automatic target classification algorithms. As future work, we intend to extend the model's architecture, so as to recognize the wing type of UAV and also to incorporate earlier samples from the sequence, thus taking into account, besides the appearance, the relative position of the samples, as an additional cue for classification purposes.

Acknowledgments. Special thanks to IDS Ingegneria Dei Sistemi S.p.A. for providing their radar sensor, the signal processing knowledge and the assistance in the dataset creation.

References

1. Knott, E.F., Schaeffer, J.F., Tulley, M.T.: Radar Cross Section. SciTech Publishing (2004)
2. Molchanov, P., Harmanny, R.I., de Wit, J.J., Egiazarian, K., Astola, J.: Classification of small UAVs and birds by micro-doppler signatures. Int. J. Microwave Wirel. Technol. **6**(3–4), 435–444 (2014)
3. Sullivan, R.: Radar Foundations for Imaging and Advanced Concepts. The Institution of Engineering and Technology (2004)
4. Tait, P.: Introduction to Radar Target Recognition. vol. 18. IET (2005)
5. Chen, V.C., Li, F., Ho, S.S., Wechsler, H.: Micro-doppler effect in radar: phenomenon, model, and simulation study. IEEE Trans. Aerosp. Electron. Syst. **42**(1), 2–21 (2006)
6. De Wit, J., Harmanny, R., Molchanov, P.: Radar micro-doppler feature extraction using the singular value decomposition. In: International Radar Conference 2014, pp. 1–6. IEEE (2014)
7. de Wit, J.M., Harmanny, R., Premel-Cabic, G.: Micro-doppler analysis of small UAVs. In: 9th European Radar Conference 2012, pp. 210–213. IEEE (2012)
8. Oh, B.S., Guo, X., Wan, F., Toh, K.A., Lin, Z.: Micro-Doppler mini-UAV classification using empirical-mode decomposition features. IEEE Geosci. Remote Sens. Lett. **15**(2), 227–231 (2017)
9. Harmanny, R., De Wit, J., Cabic, G.P.: Radar micro-Doppler feature extraction using the spectrogram and the cepstrogram. In: 11th European Radar Conference 2014, pp. 165–168. IEEE (2014)
10. Kim, B.K., Kang, H.S., Park, S.O.: Drone classification using convolutional neural networks with merged doppler images. IEEE Geosci. Remote Sens. Lett. **14**(1), 38–42 (2016)
11. Ghadaki, H., Dizaji, R.: Target track classification for airport surveillance radar (ASR). In: 2006 IEEE Conference on Radar, 4 pp. IEEE (2006)
12. Chen, W., Liu, J., Li, J.: Classification of UAV and bird target in low-altitude airspace with surveillance radar data. Aeronaut. J. **123**(1260), 191–211 (2019)
13. Mohajerin, N., Histon, J., Dizaji, R., Waslander, S.L.: Feature extraction and radar track classification for detecting UAVs in civillian airspace. In: IEEE Radar Conference 2014, pp. 0674–0679. IEEE (2014)
14. Namatēvs, I.: Deep convolutional neural networks: structure, feature extraction and training. Inf. Technol. Manag. Sci. **20**(1), 40–47 (2017)
15. LeCun, Y., Bengio, Y., Hinton, G.: Deep learning. Nature **521**(7553), 436 (2015)
16. Szegedy, C., et al.: Going deeper with convolutions. In: Proceedings of the IEEE Conference on Computer Vision and Pattern Recognition, pp. 1–9 (2015)
17. Mendis, G.J., Randeny, T., Wei, J., Madanayake, A.: Deep learning based doppler radar for micro UAS detection and classification. In: 2016 IEEE Military Communications Conference MILCOM 2016, pp. 924–929. IEEE (2016)
18. Hinton, G.E., Osindero, S., Teh, Y.W.: A fast learning algorithm for deep belief nets. Neural Comput. **18**(7), 1527–1554 (2006)
19. Dizaji, R.M., Ghadaki, H.: Classification system for radar and sonar applications, US Patent 7,567,203, July 2009
20. Krizhevsky, A., Sutskever, I., Hinton, G.E.: ImageNet classification with deep convolutional neural networks. In: Advances in Neural Information Processing Systems, pp. 1097–1105 (2012)
21. Ren, S., He, K., Girshick, R., Sun, J.: Faster R-CNN: towards real-time object detection with region proposal networks. In: Advances in neural information processing systems, pp. 91–99 (2015)

UAV Localization Using Panoramic Thermal Cameras

Anthony Thomas[✉], Vincent Leboucher, Antoine Cotinat, Pascal Finet,
and Mathilde Gilbert

HGH Systemes Infrarouges, 91430 Igny, France
anthony.thomas@hgh.fr

Abstract. Drone detection and localization became a real challenge of many companies over the past years and for the years to come. Several technologies of different kind have given some results. Among them, thermal sensors, particularly panoramic thermal imager can provide good results. In this paper, we will address two subjects. First, we will introduce the characteristics that panoramic thermal imaging systems should reach to prove their efficiency in a C-UAV system. Then, in a second part, we will present the use of data captured from multiple 360° cameras together in order to localize targets in a 3D environment and the benefits that flow from it: distance, altitude, GPS coordinates, speed, physical dimensions can then be estimated.

Keywords: Triangulation · Stereoscopy · 3D · Thermal camera · IR · UAV detection

1 Introduction

Some governmental and European projects have been launched to identify, improve and integrate technologies with the objective to build an efficient counter drone system (commonly called C-UAV for **C**ounter – **U**nmanned **A**erial **V**ehicle). An example of a C-UAV system is BOREADES [1] (Fig. 1), which was launched in 2015 and introduced for the first time the use of panoramic thermal sensors in a C-UAV system. This project, which was launched by the national French research agency (ANR) [2], had as an objective the development and the integration of a fully operational C-UAV system, including detection, localization, decision and neutralization. In this paper we focus on the use of panoramic thermal sensors and particularly on methods for detection and localization of UAVs. Some of the key challenges for designing such a system are listed below:

Image Definition. As most the drones are both small and not really hot, panoramic thermal imaging camera applied to a C-UAV system needs to reach high performances through: (a) high image definition, and (b) high thermal sensitivity.

Image Refresh Period. UAVs are very dynamically responsive. Beyond the detection, the tracking performances are also a key indicator of the system performances. Tracking

© Springer Nature Switzerland AG 2019
D. Tzovaras et al. (Eds.): ICVS 2019, LNCS 11754, pp. 754–767, 2019.
https://doi.org/10.1007/978-3-030-34995-0_69

of UAVs requires to keep the distance between two consecutives positions as short as possible to limit the potential candidates to each track and thus keep the bad tracking as low as possible. A good compromise has been identified with 2 Hz of panoramic image refresh period.

Drone Localization Accuracy. A C-UAV system is designed to protect a defined area from unauthorized intrusions. In that context, it is very important to localize all potential intrusions in the area in terms of GPS coordinates. This information is mandatory in the decision process. In this paper we address the two topics above. The use of thermal panoramic imaging camera to achieve 3D localization and also a way to reach the expected spatial resolution.

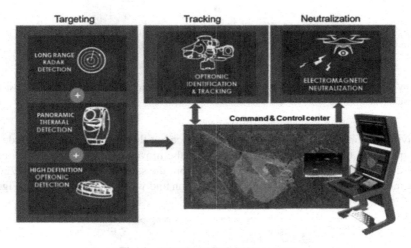

Fig. 1. BOREADES C-UAV system.

2 Panoramic Thermal Camera Concept: Application to UAV Surveillance

2.1 Fixed Versus Rotating Camera

Thermal panoramic imaging systems can be designed using two different approaches: (a) fixed head: multiple detectors and lenses and (b) rotating head: Single detector and lens. Head in rotation. In Fig. 2, an example of panoramic thermal cameras based on both fixed head and rotating head is presented. On the left part of Fig. 2 there is fixed head and multiple detector VisionView 180 from PureTech systems [3]. This camera is composed of 3 individual detectors and lenses. That approach is valid for uncooled detectors that have a much lower price and does not require any regular maintenance. The problem with uncooled detectors is that due to their limited performances, they are not applicable to UAV surveillance (thermal resolution ten times lower that cooled detector). The price and reliability of a panoramic cooled system based on that architecture would be so high that it is never retained by any manufacturer. On the right part of Fig. 2 one of the current panoramic thermal camera manufacturers is HGH and their model Spynel S2000 is presented [4]. Their camera are based on the second principle (single detector & rotating head).

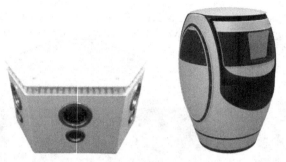

Fig. 2. VisionView 180 (Puretech Systems) on the left and Spynel S2000 (HGH Infrared Systems) on the right.

Having a rotating head seems to be an evidence but this will inevitably degrade the image quality if nothing is done to compensate the movement. An example of image quality degradation from two images captured from the same camera rotating at 720°/s at the time when utilizing movement compensation and when not is depicted in Fig. 3.

Fig. 3. How movement compensation affects image quality.

The image degradation can be roughly quantified from the reflection in Table 1. From the described data in the table, during the 3 ms of image capturing, the camera head rotation will be of **2.16°**. Considering the detector ratio (1280 px/720 px = 1.77), the resulting angular sector covered by an individual image will be (20°/1.77 = 11.25°).

As the detector horizontal resolution is of 720 px for 11.25°, the head movement will capture a slide of (2.16° * 720 px/11.25° = **138.24 px**. As a conclusion, it is mandatory to cancel this effect in order to make the camera usable for any application. Another key element in the design of such an optical system, is to be able to generate very resolved images while having the camera head in constant rotation. This requires a custom lens conception. The optical design must also consider the temporal aspect of the counter rotation system in the simulation.

Table 1. Rotation camera characteristics affecting image degradation.

Metric	Value	Unit
Average integration time	3	Ms
Camera rotation speed	720	°/sec
Detector resolution	1280 (Vertical) 720 (Horizontal)	Pixels
Expected vertical FoV	20	°
Expected horizontal FoV	360	°

2.2 Image Definition

To address C-UAV system manufacturers, panoramic thermal imaging systems must have: (a) a wide vertical field of view (to cover the maximum of the environment) and (b) an excellent image resolution (to detect and track with the longest range). In that context, it seems important to integrate the highest resolution detector into panoramic thermal cameras. SOFRADIR designed and manufactured the first MPix MWIR detector, called Daphnis HD [5]. This detector is different from the previously existing ones in the sense that the pixels are smaller (10 μm instead of 15 μm). Having smaller pixels is not without consequences because there are: (a) less photons received during the image capture and (b) the necessity for the lens to be more resolved. In Fig. 4 there are two images illustrating the difference between a 640 × 512 MWIR detector (pixel size of 15 μm) and the 1280 × 720 Daphnis HD MWIR detector manufactured by SOFRADIR (pixel size 10 μm). By examining the figure it seems very interesting to increase the image resolution to better detect and track but also to classify UAVs. Integrating a high-resolution detector also requires a high definition optical system in front of the detector. If both are not designed to meet the requirements, the produced image will not be optimized. While designing an optical system, the image resolution must not be limited by the diffraction phenomenon. This can be easily quantified by calculating the Airy disk [6] which gives the minimum point dimensions a lens or a mirror can focus a beam of light. The diameter of the Airy disk (the first in the center, depicted in Fig. 5) is given by:

$$r = 1.22 \times \lambda \times N \tag{1}$$

where r is the diameter (not radius) of the disk, λ the considered wavelength, and N the optical aperture number of the lens. Considering the average wavelength of the detector sensitivity ($\lambda = 4$ μm), and the fact that almost all lenses have an optical aperture (N = 4) the Airy disk will have a standard diameter of r = 19.52 μm. It is obvious that having a detector with a pixel size of 10 μm coupled with a lens with poor resolution will not help to generate the expected images. As a conclusion, while using 10 μm detectors in the MWIR, lenses must have an ideal optical aperture of 2.

Fig. 4. 640 × 512 MWIR detector (left image) and High resolution 1280 × 720 Daphnis HD (Sofradir) detector (right image)

Fig. 5. Illustration of Airy disk phenomenon

2.3 Existing Camera

One camera introducing the above concepts is manufactured by HGH INFRARED SYS-TEMS under the reference SP-X3500. This camera is able to produce rich and contrasted image while scanning the environment at 720°/s. Below are two pictures captured with the same field of view, same distance from a house. On the left part of Fig. 6, the sensor resolution is 640 × 512 while on the right part of the figure the latest detector with 1280 × 720px is used.

Fig. 6. Image produced by SP-S2000 (HGH) on the left and by SP-X3500 (HGH) on the right.

3 Target Localization

Range estimation methods are commonly divided into two main categories: active ones and passive ones.

3.1 Active Range Estimation

Range estimation by active methods is based on the emission of an impulsion (RF, light, microwave) and the computation of the time needed to receive the echo. The radar based example is depicted in Fig. 7. One problem with active solutions like radars concerns the authorization to use such systems in city environment. Secondly, radars can't provide an image of the environment making the target characterization more difficult. Last but not least, those systems are not furtive and can be discovered which is in certain situation is not acceptable.

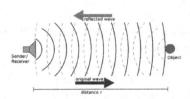

Fig. 7. Principle or range estimation used by radars courtesy of [7].

3.2 Passive Range Estimation

The use of thermal imaging devices is an efficient way to detect and track flying objects while staying fully passive. Like common radars, panoramic thermal cameras (Fig. 8) can be used as part of the detection means of an anti-drone system and can be connected to effectors (Jammer, Spoofing engine) to neutralize unauthorized UAVs. Panoramic thermal cameras usually rotate continuously and provide high resolution images of the environment. Using some image processing algorithms, they can detect objects of interest in the scene and provide their respective Azimuth and Elevation (Fig. 9). The distance will remain unknown as long as the target has no interface with the ground.

Fig. 8. Example of panoramic thermal image

Fig. 9. Example of an UAV detected and tracked by a panoramic infrared camera

3.3 3D Localization Using Thermal Cameras

In the following subsections, we introduce the methodology used to localize flying targets in a 3D space using multiple 360° cameras. As introduced by Lee and Yilmaz in Ref. [8] and by Straw in Multi-camera real-time three-dimensional tracking of multiple flying animals [9], having several cameras capturing the scene and performing object detection is not sufficient. One of the most important concerns is to be able to pair the same physical object in the different videos. This part is described in subsection 3.4. Following that, localization can be evaluated.

3.4 Object Pairing

One of the challenging questions when trying to localize targets from their position within images is to associate properly the set of targets detected by the different sensors. This is critical as most of thermal panoramic cameras will not provide a very resolved picture of the targets. Several precautions could be taken, namely: (a) ensure that the distance between the lines of sight of the same target seen from the multiple sensors is small enough (Fig. 10). And (b) ensure that the speed vector of the localized target is consistent with the predicted trajectory. Pairing must be rejected otherwise. Or (c) ensure that the attributes of the target are coherent with the camera specifications and the estimated distance.

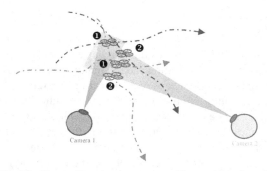

Fig. 10. Target direction impact on object pairing function

In Fig. 11, we illustrate the fact that the attributes consistency is important for the object pairing function. In the example, the UAV has physically the same size but produces a larger image for Camera 1 due to its proximity with this camera (we consider that both cameras have the same resolution). Therefore, a bird flying close to Camera 2 produces the same image size than the UAV on Camera 1. As the green UAV, the green bird and Camera 2 are all almost aligned, there is a risk of confusion if nothing is done to avoid this problem.

Fig. 11. Importance of taking the object attribute (like the size) in the object pairing function. (Color figure online)

3.5 Localization Estimations

Target localization methods use the fact that not collocated sensors will not see an object through the same angle. This basic phenomenon is used to compute several metrics like the distance. This is the same technique used by the human vision to visualize the world in 3-dimensions. In any of the below cases, the cameras location (distance between the cameras, orientation, altitude) must be well known. After a short review of 2D triangulation methods, we will quickly move to the 3D localization subject.

2D Localization. In a 2D world, we can easily introduce the mathematical concept of the triangulation using two fixed cameras.

From Fig. 12, we can extract the following mathematical relations, where d is the distance separating the two cameras:

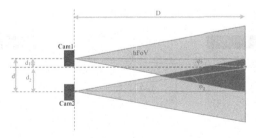

Fig. 12. A figure presenting the concept of triangulation in 2 dimensions.

$$d = d_1 + d_2 \tag{2}$$

$$tan\ \varphi_1 = \frac{d_1}{D} \tag{3}$$

$$tan\ \varphi_2 = \frac{d_2}{D} \tag{4}$$

Considering that optical axes are parallel, the distance of the object can then be calculated as:

$$D = \frac{d}{tan\ \varphi_1 + tan\ \varphi_2} \tag{5}$$

3D Localization. In three dimensions and in the general case (Fig. 13), two non-coplanar lines (D1 and D2) don't have any intersections. In that case, the distance of triangulation has to be determined using another method.

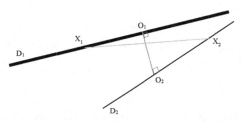

Fig. 13. General case of 3D target localization

The above considerations are only applicable to static targets. The subsections below introduce the additional metrics that needs to be considered for moving targets.

Sources of Localization Errors. Sources of localization errors can be sorted into two categories; the source of uncertainty and the error amplifiers. We also have to consider differently the case of static target from the case of moving target.

Static Targets. A poor resolution of the camera is a limitation of the localization accuracy. The impact of the camera resolution becomes larger as the target distance increases as shown in Fig. 14.

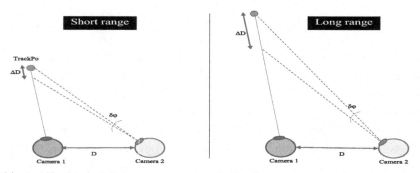

Fig. 14. A shift of 1 pixel ($\delta\varphi$) generates an error in the distance estimation of ΔD. On the right, we can easily understand that the error is larger than on the diagram on the left.

For example, considering a panoramic camera generating $360° \times 20°$ images with a resolution of W $(24 \times 512) \times$ H 640 pixels, the angular horizontal error is $0.029°$. Considering a distance between the cameras of 100 m, the impact on the localization error depending of the target distance is presented on Table 2:

Table 2. Example of distance estimation error function of the target range for a given camera resolution (iFoV $= 0.029°$)

Approximated target distance (m)	Estimated error (m)
100	0,10
250	0,37
500	1,32
1000	5,13
2000	20,29

Azimuth and Elevation Calibration Accuracy. This error is generated during the calibration process. The different cameras must be calibrated accurately in terms of Azimuth and Elevation otherwise the triangulation equations could not be resolved correctly. Considering that the error could be equivalent to 1 pixel, the impact on the distance estimation is equivalent to the one introduced by the camera resolution (see Fig. 14). When the relative positioning of the camera is not well measured and/or set in the 3D localization system, a resulting target positioning error will be introduced. This error is illustrated in Fig. 15.

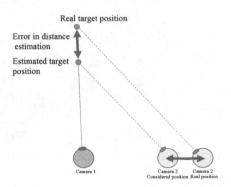

Fig. 15. Impact of camera positioning error on distance estimation

Sensors & Targets Positions. Errors on distance estimation based on triangulation are impacted by mainly three metrics:

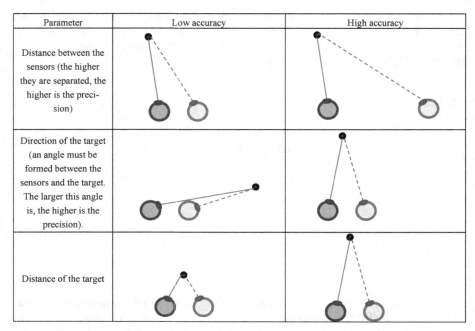

Parameter	Low accuracy	High accuracy
Distance between the sensors (the higher they are separated, the higher is the precision)		
Direction of the target (an angle must be formed between the sensors and the target. The larger this angle is, the higher is the precision).		
Distance of the target		

The diagram below (Fig. 16) shows the impact (represented in 2D to simplify the understanding) of the direction of observation on the distance estimation accuracy. The diagram represents an architecture where sensors are distant from about 140 m (100 m in both latitude and longitude). The camera iFoV used to compute the diagram is 13 mrad.

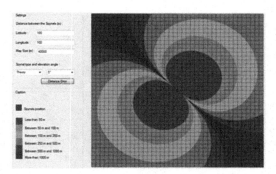

Fig. 16. Example (in 2D) of a diagram showing the impact of the target location on the positioning accuracy

Dynamic Targets. The synchronization of clocks is a key factor of good localization of moving targets. As the localization function requires to compute an estimation of the target angular position at the computing time it is mandatory to accurately synchronize the different sensors. This will be introduced below and called "relative time synchronization". As the camera turns continuously, the frame acquisition must be considered as a function of the azimuth. In addition, for a C-UAV system, it is very important to receive the position of the target in real time. This will be introduced as "absolute time synchronization".

Relative Time Synchronization. The cameras rotate continuously to acquire the panoramic images. The full panoramic image is built with stitched sectors acquired at different times. Rotation of panoramic sensors are not synchronized. The same target is imaged by the different cameras at different times (Fig. 17).

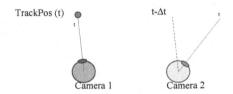

Fig. 17. Example of the same target is imaged by the different cameras at different times.

The acquisition time must then be considered as a function of the Azimuth. The image acquisition time can be represented by the below profile (Fig. 18) where each level covers the hFoV of the sensor:

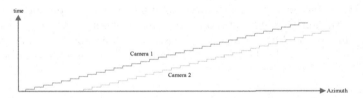

Fig. 18. Time acquisition profile of a 2D array rotating camera

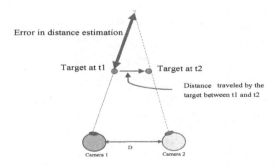

Fig. 19. Impact on the distance estimation in case of transversal moving target

To overcome the desynchronization of target imaging of the different cameras, a dynamic model of the movement of each target is computed allowing to interpolate the direction of the target at any time between actual imaged positions. If the dynamic model is accurate and the clocks synchronization effective, the errors can be perfectly compensated. We will now consider a time desynchronization of the two cameras to illustrate the impact on the target localization.

We observe that the approaching case (Fig. 20) is the most favorable situation (compared to Fig. 19). This is a good observation as approaching targets are often the interesting use case.

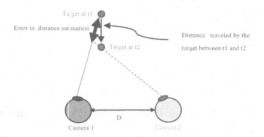

Fig. 20. Impact on the distance estimation in case of approaching target

Absolute Time Synchronization. Even if the clocks of the different cameras are well synchronized, they may be globally desynchronized with the actual time. The introduced error will be the distance traveled by the target during that period, proportionally to the target speed. Having in mind that most UAVs fly at about 60 km/h (−17 m/s) and considering that it could be possible in all cases to synchronize the cameras with an accuracy of about 0.5 s, the distance estimation could be given with an error of about 8 m.

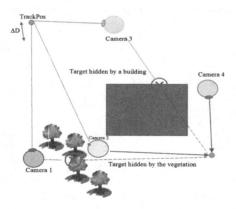

Fig. 21. In this example, four cameras are used to cover a site. Having more than two cameras is required to compensate the blind areas due to the vegetation and existing buildings.

3.6 Using More Than Two Cameras

Using more than two cameras is an efficient way to increase both the covered area and the localization accuracy. As illustrated in Fig. 21, it helps to compensate the areas of low positioning accuracy and hidden areas due to existing buildings and vegetation.

4 Conclusion

Several research projects allowed prototyping:

- The most spatially resolved high frame rate IR panoramic camera, involving a 10 μm pitch MCT megapixel detector.
- An advanced function of localization involving multiple IR panoramic cameras.

Those two major advances make IR panoramic imaging a key function of counter UAV systems. The use of 3D localization helps to communicate accurate information to a counter UAV system as it enhances the metadata that can be computed from the individual images. Combining targets distance estimation and angular data provided by the 360° cameras, the system calculates the true size and true speed of the target. Those are key parameters enabling to trigger out false alarms (make the difference between a close UAV and a far plane). Providing diverse and accurate information about targets is also one of the challenges for taking advantages of artificial intelligence algorithms, especially used for target classification. Having an accurate target classifier becomes a mandatory specification to address most of the new security markets when the false alarm rate is a critical key indicator. Even if 3D localization can be achieved in many cases by using two sensors, the integration of more than 2 sensors will allow to have optimal performances in every direction, but this also comes with new considerations, especially in the pairing function when the pairing order should not depend on the processing order.

Acknowledgements. This work was supported by the European Union's Horizon 2020 Research and Innovation Programme Advanced holistic Adverse Drone Detection, Identification and Neutralization (ALADDIN) under Grant Agreement No. 740859.

References

1. Boreades Project Website. http://boreades.fr/
2. Agence Nationale De La Recherche. https://anr.fr/Projet-ANR-15-FLDR-0001
3. PureTech Systems. https://www.puretechsystems.com/visionview.html
4. HGH Infrared Systems Website. https://www.hgh.fr/content/download/2344952/37530497/version/16/file/HGH_CaseStudy_Spynel_DroneUAVDetection.pdf
5. Lynred.com. https://www.lynred.com/produit/daphnis-hd-mw
6. University of Sheffield. https://www.sheffield.ac.uk/polopoly_fs/1.19213!/file/Lecture11.pdf
7. Wikipedia. https://en.wikipedia.org/wiki/Radar
8. Lee, Y.J., Yilmaz, A.: Real-time object detection, tracking, and 3D-positioning in a multiple camera setup. In: ISPRS (2013)
9. Straw, A.D., Branson, K., Neumann, T.R., Dickinson, M.H.: Multi-camera real-time three-dimensional tracking of multiple flying animals. J. R. Soc. Interface **8**(56), 395–409 (2011)

Multimodal Deep Learning Framework
for Enhanced Accuracy of UAV Detection

Eleni Diamantidou, Antonios Lalas, Konstantinos Votis$^{(\boxtimes)}$, and Dimitrios Tzovaras

Centre for Research and Technology – Hellas (CERTH), 6th km Harilaou - Thermi,
57001 Thessaloniki, Greece
{ediamantidou,lalas,kvotis,dimitrios.tzovaras}@iti.gr

Abstract. Counter-Unmanned Aerial Vehicle (c-UAV) systems are considered an emerging technology dedicated to address the critical issue of malicious UAV detection. Acquiring useful information from a multitude of data gathered using a topology of different sensors for UAV detection constitutes a problem with substantial importance. In this paper, we present a novel multimodal deep learning methodology to filter and combine data from a variety of unimodal approaches dedicated to UAV detection. Specifically, the aim of this work is to detect, and classify potential UAVs based on a fusion procedure of features from UAV detections provided by unimodal components. Actually, we propose a general fusion neural network framework in order to merge features extracted from unimodal modules and make deductions with increased accuracy. Our method is validated by thorough application to UAV detection and classification tasks. Our model approach achieves significant performance improvement over the unimodal detection results.

Keywords: Data fusion · UAV · Drones · Deep learning · Detection

1 Introduction

Unmanned Aerial Vehicles (UAVs) constitute a technology that has been involved in many aspects of everyday life. Indicatively, UAVs have been employed in a multitude of applications such as archaeology, cargo transport, healthcare, filmmaking, journalism, border security, law enforcement, search and rescue, surveillance, pollution monitoring, and agriculture providing useful services that were not feasible a few years ago. On the other hand, this novel technology can be used to harm people or for illegal actions in general, such as terrorist attacks by carrying explosives, trafficking of illegal objects/cargos, as well as interrupting air transports by flying drones near airports. Therefore, counter-UAV systems have been developed to address the critical issue of UAV detection when approaching an important infrastructure or event. These systems include multiple integrated sensors for detecting the threat, such as radar, electro-optical, thermal, acoustic, and radio frequency (RF) sensors. It is obvious that a critical amount of data are collected, while an accurate and robust deduction is required in very short time by combining these data to enable efficient detection of the potential threat and diminish at the same time the shortcomings of each sensor.

© Springer Nature Switzerland AG 2019
D. Tzovaras et al. (Eds.): ICVS 2019, LNCS 11754, pp. 768–777, 2019.
https://doi.org/10.1007/978-3-030-34995-0_70

Data fusion techniques have gathered significant attention in recent years mainly due to the interest of combining information from different types of sensors for a variety of applications. The target scope of data fusion is to achieve more accurate results than those derived from single sensors, while compensating for their individual weaknesses. On the other hand, artificial intelligence and deep neural networks (DNNs) have become a very attractive method for data representation. They are utilized to represent a large variety of data, because of their ability to discover high-level and abstract features that common feature extraction methods cannot. Therefore, the utilization of deep learning methods in data fusion aspects can be of significant importance in addressing the critical issue of multi-sensory data aggregation. Specifically, an indicative description of the scope of multimodal learning is presented in [1], whereas the different ways that multisensory information can be handled are described in [2]. In addition, Liu et al. [3] propose interesting deep neural network techniques for multisensory information fusion.

This work presents a novel neural network based approach for combining information from multiple sensors in order to identify the presence of a UAV or not. The main challenge of multimodal drone recognition addressed herein is to efficiently manage multisensory detections of multiple drones and fuse them into robust outcomes, taking into account the weaknesses of the various modalities. Detailed experiments are conducted in order to evaluate the performance of the proposed framework, exhibiting enhanced levels of detection capabilities in the form of prediction probabilities and classification quality metrics.

2   Related Work

A brief literature review is presented next regarding two main areas: multimodal deep learning techniques and multilayer perceptron algorithm, which constitute the base of this data fusion work.

2.1   Multimodal Deep Learning

Nowadays, there is a huge variety of information resources in the real world. In this context, multimodal learning employs techniques capable of fusing representations of different sensor type data [4]. The main ambition in these techniques is achieving accurate and optimal characterization of data that is difficult to discover with a single sensor. Deep learning is a novel area of Machine Learning that mimics the human brain behavior in processing data, detecting correlations and creating patterns for use in decision-making [5]. The concepts of Deep Learning can be associated with the fusion of multimodal data, due to the fact that deep neural networks can support multiple input streams. An essential benefit of multimodal deep learning is the ability to discover a relationship between different modalities and fuse them. However, in many circumstances multimodal learning methods are utilized to observe and complete modalities in case they are missing or corrupted. Deep Learning neural networks have the ability to learn high representations of data and relations between them. Moreover, DL neural networks have predictive capabilities and predict target variables of a given problem [6]. The predictive behavior of a DL neural network derives from multi-layered and sometimes complicated structures.

The most acknowledged neural network is considered to be the Perceptron algorithm [7].

2.2 Multilayer Perceptron

Multilayer Perceptron [8] (MLP) is a well-known neural network, connecting multiple layers in a directed graph. A characteristic representation of Multilayer Perceptron is depicted in Fig. 1. There is a large variety of Multilayer Perceptron applications. These neural networks are suitable in most of the various machine-learning problems for their ability to solve classification and regression tasks [9]. MLP neural networks consist of, at least, three layers: an input layer, a hidden layer, and output layer. The neural network parameters can vary for a given machine-learning problem. In order to initialize an MLP, it is necessary to define the number of the hidden layers and the number of nodes in input, hidden and output layers. The MLPs are fully connected neural networks, which means that each node in one layer connects to every node in the following layer [10]. Concerning this, it is easily understood that the output of any layer input or hidden layer is the input to the next hidden layers.

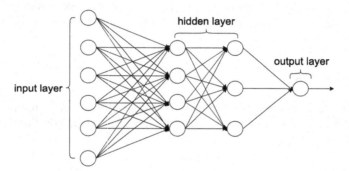

Fig. 1. Multilayer Perceptron architecture representation. Indicative interconnection between consecutive layers is depicted

3 Fusion Deep Learning Model

The purpose of the envisioned data fusion method is to learn relationships between different sensory data and combine them in order to provide detection results with enhanced accuracy, while compensating at the same time for possible weaknesses of the single sensors. In most cases, data come from different modalities, which means that they are associated with different configuration or type of information. The major challenge in data fusion is to find a viable way to combine different types of data with each other. Our approach is based on joint information from three modalities to perform UAV detection predictions. In particular, thermal image data, visual data, and 2D radar data in the form of range profile matrix data are fused within a deep neural network.

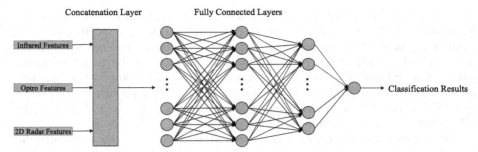

Fig. 2. Multilayer Perceptron architecture for the proposed deep learning fusion framework. The neural network consists of three input channels and one output channel

However, it is really challenging to learn to combine data from different modalities. In many cases, the information comes with noise and possibly missing or even corrupted data, especially when there is a multitude of inputs. In addition, the aforementioned unimodal data come from different sensors, thus they may have different types and dimensions, which results in different predictive power. Therefore, in order to tackle all these possible issues during the data fusion procedure, we introduce a deep neural network that efficiently combines high-level features extracted from thermal, visual and 2D radar neural networks.

3.1 Fusion Model Architecture

The deep learning fusion neural network attempts to distinguish discrete cases where a UAV is present and where it is not. Taking this into account, the Multilayer Perceptron classifier is adopted, which can support a multi-input stream and as well as a multi-class classification output by applying the softmax activation [11] as the final activation function in the neural network training procedure. Apart from the choice of the most optimal activation function, various parameters can initialize the neural network architecture. These parameters are called hyper-parameters of the neural network and they are required to be set in order to fit it. In our classification case, we present a novel approach of data fusion, using three input streams from different sensors attributed to UAV detection.

Figure 2 illustrates the concept and the design of the proposed architecture of the data fusion framework. The neural network architecture has been designed with three input streams: thermal images stream (infrared), visual images stream (optro) and 2D radar stream. The implementation of the proposed methodology involves a concatenation layer to combine the three aforementioned input streams. Each stream is actually a different input tensor. The output of the concatenation layer returns a single tensor that contains the concatenation of all inputs. Utilizing a concatenation layer facilitates to manipulate multiple input tensors due to the fact that our unimodal input features are not very closely related.

The next step in our classification problem is to employ high-level features extracted from unimodal neural networks training. Specifically, infrared and optro data are analyzed by Faster R-CNN, which is one of the state-of-the-art object detection and region proposal networks [12]. Moreover, 2D radar data are analyzed by a Convolutional Neural Network (CNN) [6]. An optimal fusion model should satisfy some specific properties to efficiently perform the classification. The form of the three neural networks outputs is large numerical arrays that represent high abstract features in the multidimensional space extracted from unimodal data. Furthermore, infrared and optro data produce features that belong to multi-dimensional space. Specifically, they produce $7 \times 7 \times 1024$ and $7 \times 7 \times 512$ (3-dimensions) features of each modality, respectively. Regarding the 2D radar data, they produce 1664 features (1-dimension).

The output of the data fusion neural network is a binary problem classification result. In more details, the output of our proposed deep neural network is a probability according to the training classes. In case that the probability result is higher than 0.5, the detection is classified as UAV. Likewise, in case the probability result of fusion procedure is less than 0.5 the detection is classified as non-UAVs.

3.2 Multilayer Perceptron Algorithm Initialization

As mentioned above, the fusion algorithm has three input streams: infrared stream, optro stream, and 2D radar stream. The extracted features of the unimodal deep neural networks are high representations of the raw input data to the neural network, whereas our implementation is based on a Multilayer Perceptron architecture. The Multilayer Perceptron algorithm consists of a concatenation layer and three fully connected layers (dense layers), as depicted in Fig. 2.

Table 1. Number of total training and testing instances for the deep neural fusion network

Modality	Classified UAV	Classified non-UAV
IR	2195 for training, 432 for test	17000 for training, 2178 for test
Optro	2195 for training, 432 for test	9875 for training, 837 for test
2D Radar	2195 for training, 432 for test	2392 for training, 460 for test

Many parameters can affect the training procedure. Some of the most important initialization parameters are the number of hidden units, the number of hidden layers, the epochs and the optimization algorithm. In essence, the number of hidden units depends on the size of the dataset. Too few hidden units will generally leave high training and generalization errors due to under-fitting. Too many hidden units will result in low training errors, but will make the training unnecessarily slow, and will result in poor generalization. Regarding the neural network initialization, the training weights are initialized randomly, which means that the neural network is training without using weights of other pre-trained models. However, all weights along the neural network training are updated according to the number of epochs. The whole purpose of training a model is to adjust its weights so it can make fine and accurate predictions. Actually, the optimization algorithm is the mechanism, with which the network updates itself [13]. Moreover, the optimizer is the algorithm that minimizes (or maximizes) a function with which the neural network is learning. Some of the most characteristic optimizers are the following: Stochastic Gradient Descent (SGD), RMSprop, Adagrad, and Adam. The Adam optimization algorithm [14] is an extension to stochastic gradient descent [15] that has recently seen broader adoption for many deep learning applications. The optimizer that is selected to train our deep neural network is the Adam, whereas it is initialized with a learning rate equal to 0.001.

3.3 Training Data

The proposed data fusion model has been successfully applied to UAV detections of multiple drones during dedicated experimental setups. The training set consists of multimodal features from positive (UAV) detections and negative (non-UAV) detections. The whole dataset is a result of 15 UAV flight recordings, which have been recorded in specific data gathering sessions in ATLAS facilities [16] in 2018 and 2019, at Spain. Unimodal deep learning techniques have been applied to complete the procedure of feature extraction. After this step, the input dataset of the fusion module is a combination of infrared, optro and 2D radar features.

An essential point in data pre-processing is the registration of the samples. The dataset contains recordings that were collected during different capturing sessions. Therefore, it contains captures for various types of UAVs under different time frames, spatial ranges, weather conditions, etc. allowing a wide diversity for the training phase. Moreover, each sensor captures data with a specific frequency and accuracy. As a result of this, the thermal camera sensor, the visual camera sensor, and the 2D radar sensor have a dissimilar number of UAV detections and non-UAV. For this reason, the application of a registration procedure to the extracted features from single modalities, such as Infrared features, Optro features, and 2D radar features, is deemed necessary. The registration is based on the detection timestamps that have been recorded and they are provided from the captures of each modality respectively. It is important to note that the non-UAV detections are very generic, without specifications about the class content. Consequently, the temporal registration is applied only in the UAV detections of each modality. The total number of training features and test features in each modality are summarized in Table 1.

According to the number of samples in each class, some balance issues between the UAV class and the non-UAV class have occurred. Briefly, this means that the two classes are not represented equally. Therefore, we under-sample the non-UAV class by randomly removing instances from non-UAVs classes. Finally, the training set results in 2195 features in UAV class and 2195 features in non-UAV class. Similarly, the test set results in 432 features in each class accordingly.

4 Experimental Results

A variety of experiments has been conducted to evaluate the proposed method. By inspecting the results, it is observed that the Multilayer Perceptron algorithm has a great potential to learn relations and dissimilarities between the input data and achieve high classification results in the UAV detection problem. The major achievement of the proposed approach is that it succeeded in increasing confidence in accuracy, while data from different modalities are combined. In order to assess the performance of the fusion model, we have also compared the fusion detection results with the unimodal detection outcomes from each modality. Figure 3[a–c] present the results of UAV detections in the form of prediction probabilities based on unimodal features. On the other hand, Fig. 3[d] presents the results of UAV detections that are derived from the fusion model. As Fig. 3[d] shows, considerably improved detection results are observed in comparison to Fig. 3[a–c]. An increased accuracy of detection is noticed, when all three modalities are combined into the fusion model.

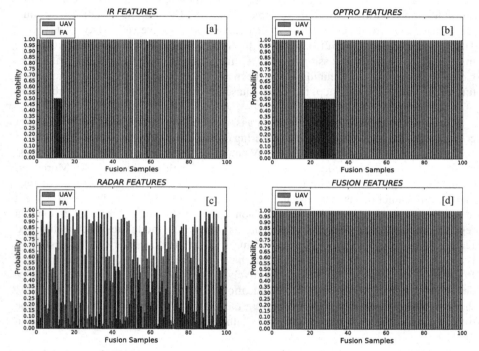

Fig. 3. Prediction probabilities as outcomes of the evaluation procedure of the unimodal features. Three modalities are assessed in terms of the different kind of features, such as [a] infrared features, [b] optro features and [c] 2D radar features. [d] Prediction results of the fusion model that involves unimodal features of infrared, optro and 2D radar modalities. Axis x: the number of testing instances. Axis y: the prediction score in probability form

For the purpose of quality estimation, we utilize two metrics that have greater importance even than the accuracy of classification. Precision and recall are both remarkably essential model evaluation metrics [17]. Precision is defined as the number of true positives (TP) over the number of true positives (TP) plus the number of false positives (FP). Moreover, recall is defined as the number of true positives (TP) over the number of true positives (TP) plus the number of false negatives (FN). The F1-Score is the harmonic mean of precision and recall. There is no doubt that our approach attempts to maximize

F1-Score. Our model has been evaluated in a real UAV flight capture. Figure 4 represents the actual plan of the UAV flight of two parrot disco drones, which has been used for the model evaluation. Table 2 describes in detail the precision and recall of our neural network evaluation on a flight performed as depicted in Fig. 4.

Fig. 4. Flight representation of two parrot disco drones in data gathering session in ATLAS facilities. UAV maximum flight distance is equal to 500 m

Table 2. Precision and Recall in a dedicated evaluation flight of two parrot disco drones

Class	Precision	Recall	F1
UAV	99%	100%	95%
FA	100%	89%	94%

Following the previous experiments with unimodal features evaluation compared with fused features evaluation, we have carried out some additional experiments regarding precision-recall metrics and the UAV flight distance. The main notion here is to validate the precision and recall of unimodal features in comparison with the fused features over the UAV flight maximum distance. As an initial assessment procedure, we decided to estimate the precision and recall from unimodal features per 100 meters of UAV flight distance. The precision and recall scores are represented in Fig. 5[a–c]. In order to enable comparison, we have measured the precision and recall scores for every 100 m of UAV flight distance for the fused feature detections also, as depicted in Fig. 5[d].

By observing the features derived only from one modality in Fig. 5[a–c], it is deduced that in the infrared features experiment, the recall score is reduced as the UAV fly in greater distances. However, in the 2D radar features experiment the recall score is

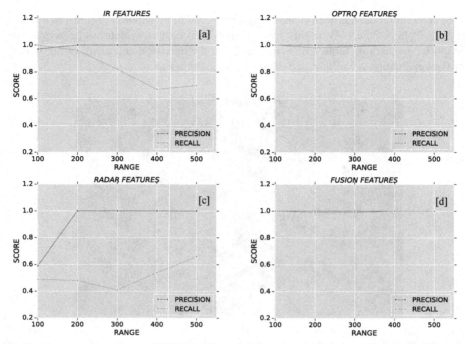

Fig. 5. Precision-Recall over UAV flight distance. Three modalities are assessed per 100 m in terms of the different kind of features, such as [a] infrared features, [b] optro features and [c] 2D radar features. [d] All three modalities as input to the data fusion model. An overall high score both in Precision and Recall metrics compared to the previous unimodal experiments has been achieved

increased as the UAV fly in greater distances. This can be explained because of the inability of the 2D radar sensor to identify objects in short distances. Finally, as illustrated in the optro features experiment the recall reaches almost the optimal, which reflects the fact that the optro sensor has a large distance zoom to recognize and determine UAVs. As presented in the previous analysis as well as in Fig. 5[d], the fusion model has reached enhanced levels of the classification metrics. It seems that our deep fusion neural network has learned the importance of each modality and reached an overall optimal precision and recall score as the UAV flight distance increases.

5   Conclusion

In this work, a multimodal neural network framework has been introduced that efficiently performs the UAV detection task based on a multitude of input features. Our proposed model architecture is based on the robust Multilayer Perceptron algorithm. We have thoroughly assessed our model on two benchmark metrics: prediction probability and evaluation quality metrics. In order to extract validated results, we have compared our fused detection results with unimodal detection outcomes. We have successfully demonstrated the effectiveness of multimodal data learning and appropriately established our efficient fusion model, which is suitable for efficient UAV detection.

Acknowledgments. This work was supported by the European Union's Horizon 2020 Research and Innovation Programme Advanced holistic Adverse Drone Detection, Identification and Neutralization (ALADDIN) under Grant Agreement No. 740859.

References

1. Ngiam, J., Khosla, A., Kim, M., Nam, J., Lee, H., Ng, A.Y.: Multimodal deep learning. In: Proceedings of the 28th International Conference on Machine Learning (ICML-11), pp. 689–696 (2011)
2. Baltrusaitis, T., Ahuja, C., Morency, L.P.: Multimodal machine learning: a survey and taxonomy. IEEE Trans. Pattern Anal. Mach. Intell. **41**(2), 423–443 (2018)
3. Liu, K., Li, Y., Xu, N., Natarajan, P.: Learn to combine modalities in multimodal deep learning. arXiv preprint. arXiv:1805.11730 (2018)
4. Scherer, S., Worsley, M., Morency, L.P.: 1st international workshop on multimodal learning analytics. In: 14th ACM International Conference on Multimodal Interaction (2012)
5. Bengio, Y., et al.: Learning deep architectures for AI. Found. Trends® Mach. Learn. **2**(1), 1–127 (2009)
6. Schmidhuber, J.: Deep learning in neural networks: an overview. Neural Netw. **61**, 85–117 (2015)
7. Rosenblatt, F.: The perceptron: a probabilistic model for information storage and organization in the brain. Psychol. Rev. **65**(6), 386 (1958)
8. Pal, S.K., Mitra, S.: Multilayer perceptron, fuzzy sets, and classification. IEEE Trans. Neural Netw. **3**(5), 683–697 (1992)
9. Gardner, M.W., Dorling, S.: Artificial neural networks (the multilayer perceptron) a review of applications in the atmospheric sciences. Atmos. Environ. **32**, 2627–2636 (1998)
10. Ramchoun, H., Idrissi, M.A.J., Ghanou, Y., Ettaouil, M.: Multilayer perceptron: architecture optimization and training. IJIMAI **4**(1), 26–30 (2016)
11. Zunino, R., Gastaldo, P.: Analog implementation of the softmax function. In: 2002 IEEE International Symposium on Circuits and Systems. Proceedings (Cat. No. 02CH37353), vol. 2, p. II. IEEE (2002)
12. Ren, S., He, K., Girshick, R., Sun, J.: Faster r-cnn: towards real-time object detection with region proposal networks. In: Advances in Neural Information Processing Systems, pp. 91–99 (2015)
13. Khalil, E., Dai, H., Zhang, Y., Dilkina, B., Song, L.: Learning combinatorial optimization algorithms over graphs. In: Advances in Neural Information Processing Systems, pp. 6348–6358 (2017)
14. Kingma, D.P., Ba, J.: Adam: a method for stochastic optimization. arXiv preprint. arXiv:1412.6980 (2014)
15. Ruder, S.: An overview of gradient descent optimization algorithms. arXiv preprint. arXiv:1609.04747 (2016)
16. ATLAS: Air Traffic Laboratory for Advanced Systems. Villacarrillo (Jaén), Spain. http://atlascenter.aero/en/
17. Buckland, M., Gey, F.: The relationship between recall and precision. J. Am. Soc. Inf. Sci. **45**(1), 12–19 (1994)

Multi-scale Feature Fused Single Shot Detector for Small Object Detection in UAV Images

Manzoor Razaak$^{(\boxtimes)}$, Hamideh Kerdegari, Vasileios Argyriou,
and Paolo Remagnino

Kingston University London, London, UK
manzoor.razaak@kingston.ac.uk

Abstract. Small object detection is a challenging computer vision problem due to their low feature representation in the images and factors such as occlusions and noise. In images captured from a camera mounted on an unmanned aerial vehicle (UAV), objects are usually acquired in small sizes depending on the UAV flight altitude. The state-of-the-art object detectors often have lower detection accuracy with small objects. New approaches of combining features at multi-levels in the network helps in improving the object detection performance. In this paper, we propose a multi-scale approach of low-level feature combinations with deconvolutional modules on a single shot multibox detection (SSD) object detector to improve the small object detection in images acquired from a UAV. The proposed SSD based architecture is evaluated on UAV datasets to compare its performance with the state-of-the-art detectors.

Keywords: Small object detection · UAV images · SSD · Deconvolution

1 Introduction

The reduction in unmanned aerial vehicles (UAV) costs and advanced modern compact imaging sensors have made aerial surveys and surveillance applications technologically and economically feasible. Computer vision methods supported by rapid advancements in deep learning are empowering various autonomous and semi-autonomous UAV applications in agriculture, security, archaeology, remote sensing, terrain modelling, crowd and disaster management domains.

Object detection algorithms are important to enable smart UAV based survey applications. You only look once (YOLO) [9], Single shot multibox detection (SSD) [7], Fast R-CNN [4] are current state-of-the-art object detection systems demonstrating high detection accuracy in real-time. Object detection algorithms

The research leading to these results have received funding from the Department for Digital, Culture, Media & Sports (DCMS), United Kingdom, under its 5G trials and testbeds program.

© Springer Nature Switzerland AG 2019
D. Tzovaras et al. (Eds.): ICVS 2019, LNCS 11754, pp. 778–786, 2019.
https://doi.org/10.1007/978-3-030-34995-0_71

thrive on prior-learning of the image features of the object and ideally require a good resolution, non-occluded representation of the object in an image to develop the capability of object prediction. However, images acquired from a sensor mounted on a UAV may often not satisfy this criteria.

UAV images essentially provide a top-down view and depending on the orientation of the camera and the UAV flight altitude, captures the object characteristics much differently from traditional ground-based images. The objects in UAV images are generally small and are acquired under varying illumination conditions and complex backgrounds. The small size of the objects in UAV images provide insufficient image features for object detection algorithms to "learn" that adversely affects the detection accuracy of the algorithms.

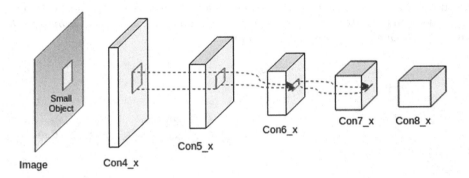

Fig. 1. Small object feature representation at deeper layers of a ConvNet. The object features are downsampled at each layer. Small objects risk losing feature representation at the prediction layer at the end.

Figure 1 illustrates the feature representations of small objects across a sequence of convolutional layers. For an image with a small object, at deeper layers the feature representations of the small object is weaker due to down-sampling and may potentially disappear at the prediction layer causing missed detection of small objects. Thus, shallow layers that hold the information of small objects are not represented at the prediction layers.

Multi-scale feature representation methods use object features at different levels in the network to improve detection of small objects [1]. The shallow layers in a neural network have better feature representations of small objects and fusing these features at the prediction layer can improve the detection performance on small objects. The SSD is one of the object detectors that makes prediction based on multi-scale features at six different levels across the neural network. Several multi-scale approach based detectors are available. For instance, Deconvolutional SSD (DSSD) [3] applies deconvolutional layers after the SSD layers to improve small object detection. MDSSD [10] combines high-level and low-level features to include the contextual information into the detector. Detectors proposed in [5,6,8] apply different approaches to concatenate features from multiple levels for small object detection.

The aim of our study is to develop deep learning models for small object detection in UAV images. In this paper, we extend the SSD detector to include a deconvolutional module at the SSD convolutional layers and combine it with a feature concatenation module from the shallow layers of the SSD to enhance the feature representations of small objects at the prediction layers and subsequently improve detection accuracy. Section 2 describes the proposed system and its components. Section 3 details the experiments conducted and the results obtained, followed by conclusion in Sect. 4.

2 System Overview

Figure 2 is the architecture of our proposed extension of the SSD network. The extended SSD architecture consists of two main modules: a *deconvolution module* and a *shallow layer feature fusion* module.

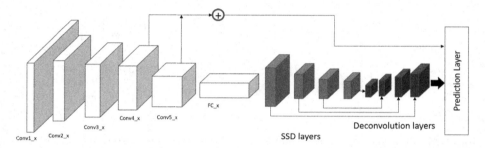

Fig. 2. Proposed SSD network with a deconvolutional module and feature concatenation module. (Color figure online)

2.1 SSD

The SSD is a state-of-the-art object detector based on the VGG16 network. The VGG network is followed by additional, progressively smaller convolutional layers, referred as SSD layers. In Fig. 2, the SSD layers are shown in blue. Unlike traditional networks that use last feature map for prediction, SSD uses a multiple layer, pyramidal feature hierarchy to predict object with different scales. The additional convolutional layers and some of the earlier layers form the prediction layers in the SSD network and enables the SSD to utilize features from multiple scales to predict scores. In the prediction layer, $3 \times 3 \times N$ channel dimensional filters for each image dimension and category box performs the predictions followed by non-maximum suppression for the final detection result. Refer [7] for more details on the SSD network.

2.2   Deconvolution Module

The feature maps at the deeper layers of the SSD network lack fine details of objects for detection. Specifically, for small objects, the fine details can be obtained at the shallower layers, however, at deeper layers the feature representation tends to be weaker and maybe completely lost at the final prediction layer. To increase the feature resolution of small objects at deeper layers, deconvolution layers can be helpful.

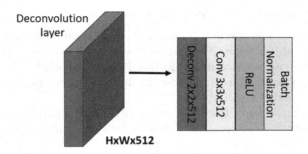

Fig. 3. A deconvolutional module unit used after the SSD layers.

To increase the feature resolution of small objects at the prediction layer, we add a deconvolution module inspired by [3] to our network. The deconvolution module includes a 2×2 deconvolutional layer with a stride of 2, followed by a convolutional layer activated by ReLU activation layer and batch normalization. Figure 3 shows the deconvolution module used in our network.

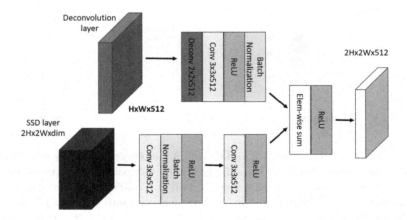

Fig. 4. A deconvolutional module unit merged with a SSD layer using element-wise sum operation.

A deconvolutional layer is added for each SSD convolutional layer, effectively up-scaling all the feature layers used by the prediction layer. The feature maps of the SSD convolutional layer and deconvolutional layer are combined through a element-wise sum operation. Each SSD layer undergoes two sets of two convolutional operation followed by batch normalization and a ReLU activation layer before combining with the deconvolution feature maps. The element-wise sum operation allows point to point combination of the feature maps at different levels into equivalent weights, as shown in Fig. 4.

2.3   Shallow Feature Concatenation Module

At deeper layers, the receptive fields for large objects are well represented, however, the small object receptive fields are weakened and may also disappear depending on the object size in the image. Due to the weak or absent feature representation of small objects, the deconvolutional layer blocks after the SSD layers may not predict small objects.

The loss of small object feature representation at the prediction layer can be avoided by fusing the feature maps from shallow layers to the prediction layer. In our proposed network, the feature maps from the $Conv4\_3$ and $Conv5\_3$ are combined and fed to the final prediction layer. This approach allows the receptive fields of small objects from shallow layers to be included in the final prediction layer.

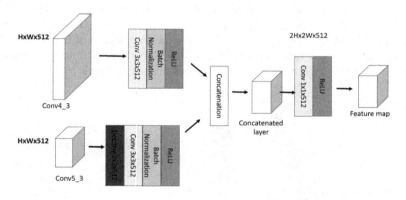

Fig. 5. Concatenation of the $Conv4\_3$ and $Conv5\_3$ feature layers.

The feature maps of $Conv4\_3$ and $Conv5\_3$ are merged using a concatenation module. The feature map of $Conv5\_3$ layer need to upsampled to the same size of $Conv4\_3$ layer. A deconvolution layer is used at $Conv5\_3$ layer for upsampling. Next, a convolutional layer is applied to both feature maps to select the best features followed by a layer of batch normalization and ReLU activation function. Finally, after concatenation, a 1×1 convolutional layer is applied for dimensionality reduction and another layer of ReLU activation. The concatenation module is illustrated in Fig. 5.

To summarise, our proposed network has six prediction layers: five prediction layers are from the deconvolutional layers combined with the original four SSD layers and one prediction layer is from the shallow feature fusion layers. This approach provides large resolution feature maps and receptive fields from small objects to the prediction layers to improve the accuracy of small object detection.

3   Experiments

3.1   Setup

The proposed SSD model was built on a Keras implementation of SSD [2] and the VGG-16 architecture. The initial tests with SSD were done on input image size of 300 × 300. Most parameters of the baseline SSD such as anchor box aspect ratios, scaling factors and others, reported in the original paper, were not modified. A batch size of 32 was used for training. The learning rate was set to 0.001 for the first 80K steps, 0.0001 for 100K and 0.00001 for 120K steps.

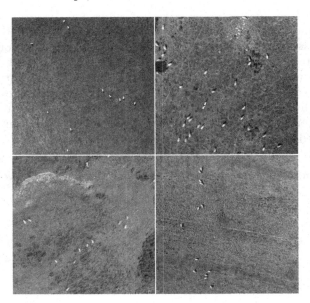

Fig. 6. Example images of the livestock dataset. In the images, sheep are small targets for the object detectors.

3.2   Dataset

The proposed network is trained on an aerial dataset captured from a UAV. The dataset consists of aerial images of livestock captured as part of the on-going UK research project, 5GRIT, that aims at integrating 5G testbed in rural areas of the United Kingdom (UK) for smart farming applications. For the initial study, the proposed network is evaluated on only one class - livestock (sheep). The custom livestock dataset, mainly consists of annotated images of sheep, captured by a UAV flown at 50 m altitude over a farm. The original dataset consists of 425 RGB images of livestock with a very high resolution of 5400 × 3600 pixels. Figure 6 shows some example images of the dataset.

The livestock dataset consists of top-down view of sheep and are small in size which makes the object detection challenging for state-of-the-art detectors. Since the original livestock dataset is of very high resolution, it is not possible to use them for training. Hence, each image was split into 300 × 300 resolution. In total, 3900 images were available for training. Data augmentation process recommended in the original SSD implementation was applied to increase the training dataset volume. The data augmentation process includes randomly sampling an entire original input image, sampling a patch such that it has a minimum jaccard overlap with the objects of around 0.1, 0.3, 0.5, 0.7, or 0.9, and randomly sample a patch.

3.3 Results

The performance of our proposed network was compared with state-of-the-art object detectors including SSD300, SSD512, and YOLO. The performance accuracy is evaluated using the mean average precision (mAP) [7] and the threshold $IoU = 0.5$ is used. Since, detection accuracy was the main criterion for our model, we do not compare the detection speed. Our network achieves a mAP of 76.2% on the livestock dataset. The network performs better than the state-of-the-art with considerable difference. Table 1 reports the detection performance of the proposed model with the other object detectors.

Table 1. The mAP results comparison of the proposed model on the livestock dataset.

Model	mAP
SSD300	74.8
SSD512	75.2
YOLO	69.4
Proposed	**76.2**

The YOLO detector shows the weakest performance in detecting livestock from our images. Both the original SSD versions, SSD300 and SSD512 perform significantly better than YOLO. Therefore, the mutli-scale feature representation helps improving the detection performance. Further, our proposed extension to the SSD architecture further improves the performance of the SSD method.

The detection results are shown in Fig. 7. It can be noted that SSD300 misses detecting some targets that were detected by our model. Our proposed detection has better detection than SSD300 and also reports better detection accuracy than SSD300.

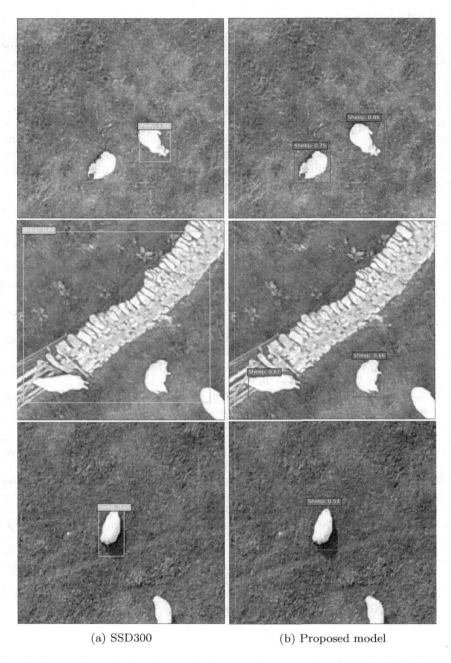

(a) SSD300   (b) Proposed model

Fig. 7. Comparison of detection performance of proposed model with SSD300. The detections of the proposed model (right-column) show that fusing features from shallow layers and deconvolutional layers improve the detection performance.

4   Conclusion

The proposed extension to SSD detector is based on applying deconvolutional module and feature fusion from shallow layers to improve the small object detection performance. The proposed model was applied on a custom dataset of livestock captured from UAV. Our approach helped in improving the feature representation of small targets and consequently improving the detection performance. A considerable improvement in detection performance was achieved in comparison to SSD and YOLO object detectors. The multi-scale feature representation provides a better representation of small targets at the prediction layer. For UAV applications, this approach is significant as targets in UAV images are often of small sizes and are difficult to detect and classify.

References

1. Cai, Z., Fan, Q., Feris, R.S., Vasconcelos, N.: A unified multi-scale deep convolutional neural network for fast object detection. In: Leibe, B., Matas, J., Sebe, N., Welling, M. (eds.) ECCV 2016. LNCS, vol. 9908, pp. 354–370. Springer, Cham (2016). https://doi.org/10.1007/978-3-319-46493-0_22
2. Ferrari, P.: A Keras port of single shot multibox detector (2018). https://github.com/pierluigiferrari/ssd_keras.git
3. Fu, C.Y., Liu, W., Ranga, A., Tyagi, A., Berg, A.C.: DSSD: deconvolutional single shot detector. arXiv preprint arXiv:1701.06659 (2017)
4. Girshick, R.: Fast R-CNN. In: Proceedings of the IEEE International Conference on Computer Vision, pp. 1440–1448 (2015)
5. Honari, S., Yosinski, J., Vincent, P., Pal, C.: Recombinator networks: learning coarse-to-fine feature aggregation. In: Proceedings of the IEEE Conference on Computer Vision and Pattern Recognition, pp. 5743–5752 (2016)
6. Kong, T., Yao, A., Chen, Y., Sun, F.: HyperNet: towards accurate region proposal generation and joint object detection. In: Proceedings of the IEEE Conference on Computer Vision and Pattern Recognition, pp. 845–853 (2016)
7. Liu, W., et al.: SSD: single shot multibox detector. In: Leibe, B., Matas, J., Sebe, N., Welling, M. (eds.) ECCV 2016. LNCS, vol. 9905, pp. 21–37. Springer, Cham (2016). https://doi.org/10.1007/978-3-319-46448-0_2
8. Liu, W., Rabinovich, A., Berg, A.C.: ParseNet: looking wider to see better. arXiv preprint arXiv:1506.04579 (2015)
9. Redmon, J., Divvala, S., Girshick, R., Farhadi, A.: You only look once: unified, real-time object detection. In: Proceedings of the IEEE Conference on Computer Vision and Pattern Recognition, pp. 779–788 (2016)
10. Xu, M., et al.: MDSSD: multi-scale deconvolutional single shot detector for small objects. arXiv preprint arXiv:1805.07009 (2018)

Autonomous Swarm of Heterogeneous Robots for Surveillance Operations

Georgios Orfanidis[1]([✉]), Savvas Apostolidis[1,2], Athanasios Kapoutsis[1],
Konstantinos Ioannidis[1]([✉]), Elias Kosmatopoulos[1,2], Stefanos Vrochidis[1],
and Ioannis Kompatsiaris[1]

[1] Centre for Research and Technology Hellas (CERTH)-Information Technologies
Institute (ITI), Thessaloniki, Greece
{g.orfanidis,kioannid}@iti.gr
[2] Democritus University of Thrace, Xanthi, Greece

Abstract. The introduction of Unmanned vehicles (UxVs) in the recent
years has created a new security field that can use them as both a potential threat as well as new technological weapons against those threats.
Dealing with these issues from the counter-threat perspective, the proposed architecture project focuses on designing and developing a complete system which utilizes the capabilities of multiple UxVs for surveillance objectives in different operational environments. Utilizing a combination of diverse UxVs equipped with various sensors, the developed
architecture involves the detection and the characterization of threats
based on both visual and thermal data. The identification of objects is
enriched with additional information extracted from other sensors such as
radars and RF sensors to secure the efficiency of the overall system. The
current prototype displays diverse interoperability concerning the multiple visual sources that feed the system with the required optical data.
Novel detection models identify the necessary threats while this information is enriched with higher-level semantic representations. Finally,
the operator is informed properly according to the visual identification
modules and the outcomes of the UxVs operations. The system can provide optimal surveillance capacities to the relevant authorities towards
an increased situational awareness.

Keywords: Unmanned vehicles (UxVs) · Visual-based operations ·
Interoperable architecture · Surveillance objectives

1 Introduction

The introduction of low-cost, yet, advanced technological tools in surveillance
applications have assisted significantly the operational effectiveness of the relevant Law enforcement agencies (LEA) and other similar authorities. The surveillance of critical infrastructures comprises an even more complex objective due
to the diverse operational scenarios and environments that the personnel has to

© Springer Nature Switzerland AG 2019
D. Tzovaras et al. (Eds.): ICVS 2019, LNCS 11754, pp. 787–796, 2019.
https://doi.org/10.1007/978-3-030-34995-0_72

monitor. Currently, most approaches involve a time and resource consuming combination of patrols with vehicles, guards and marine vessels which in most cases is proven to be insufficient. Considering also the diversity of the under surveillance territories and infrastructures that might be included, the problem becomes even more complex and challenging. Thus, the relevant personnel requires to be equipped with technologies that will be able to be adapted according to different operational and environmental needs, inter-operate with existing infrastructure, utilize multimodal sensing data and operate autonomously. Finally, due to the distinctiveness of the problem, it is essential to support effectively the personnel via decision support tools over the total situational awareness.

Towards providing a complete surveillance solution, proposed architecture's main objective is to develop a fully-functional autonomous surveillance system with various unmanned mobile robots [1] including Aerial (UAV), water Surface (USV), Underwater (UUV) and Ground (UGV) Vehicles, sufficiently adaptive to operate as a single instance or in a swarm and for either disperse or restricted areas depending on the monitored infrastructure. In particular, the developed framework relies on identifying objects and events based mostly on optical data acquired from multiple sources while the final outcomes are enriched with additional information from other sources. In order to enhance the operational capabilities, the framework includes multimodal processing of the visual data so that monitoring the required infrastructures could be accomplished accurately.

2   Detection Services and User-Interface Services of the Proposed Architecture

In order for a system to provide a complete overview of the monitoring area and thus, to support the personnel with valuable decisions, it is vital to analyze initially the tracked scene and extract information of higher-level for the detected objects. The identification of objects of interest within the operational environment is critical for the analysis of the situation, for the evaluation of the threats and for the estimation of future and imminent dangers. Furthermore, visual understanding modules in surveillance applications should combine data from different modalities, such as sensors operating in visual, infra-red and in thermal spectrum in order to augment the detection precision. Video sequences can further boost performance by combining images in a sequence acquired from multiple sources. Towards providing such capabilities in a surveillance system, framework of the proposed architecture involves the following detection modules based on visual data mostly.

2.1   Detection Services

Detection Techniques for Pollution Incidents. The framework integrates a dedicated module to provide the capacity of detecting oil spills over sea surface near relevant critical infrastructure like harbors and sea oil refineries. Pollution incidents can be identified by the UAVs of the system which are equipped with

Synthetic Aperture Radar (SAR) sensor and after transforming the acquired data from remote sensing data to visual representations. SAR sensors were integrated and utilized due to their efficiency and operations in adverse weather conditions [2]. The entire process is optimized by deploying the swarm of the UAVs in order to increase the detection accuracy and ensure a fast response from the relevant authority. The prediction model that is integrated relies on the DeepLabv2.0 architecture [3,4] and provides a semantic representation of the surveyed area based on which the criticality of the event can be evaluated properly. A simple visual result of the referred module for oil spill detection using visual semantic segmentation is provided in Fig. 1.

Fig. 1. SAR image and the corresponding mask for semantic representation of the scenery

Identification and Tracking of Suspicious Activities. The framework aims at integrating the utilization of a wide range of sensors, which typically involve visual, infrared and thermal sensors in order to identify accurately the required suspicious movements in the monitored area [5,6]. Due to the nature of the illegal activities and the diverse conditions that are involved in critical infrastructures, operational conditions play a significant role in the effectiveness of the detected targets and thus, it is important to alleviate such issues. Combined approaches of different sensors aids at the reduction of false positives in the detection system which can lead to false alarms. Figure 2 presents two examples of the object detection module on thermal and visual images, respectively. Initial results of the detector were evaluated using the Pascal Voc object detection metric [7] and are presented in Table 1. The detector that was integrated relies on the Faster R-CNN model [5] due to its high performance and robustness in identifying objects of smaller size. The identification of abnormal activities can enhance further the higher-level feedback that the system can provide the operator. A major advantage of the developed detection modules, based upon the [5] architecture, is the online adjustments of the UxV swarm routes towards increasing the detection accuracy and optimizing the surveillance objectives.

Table 1. Object detection results using Pascal Voc precision metric

Object detection results - Average precision						
Person	Car	Bus	Truck	Boat	Ship	Helicopter
0.82152	0.75726	0.57315	0.53351	0.70251	0.83586	0.71638

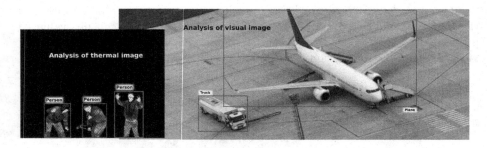

Fig. 2. Object detection applied to a thermal and a visual scene

Low-Level Fusion to Increase the Recognition Capabilities. Combining multiple information from different sources comprises a significant advantage for increasing the accuracy of the recognition modules. Apart from the late data fusion modules, the framework integrates a low-level fusion approach of visual and thermal data based on [8] in order to provide alternative advantages in comparison with the latter fusion. Typically, the fused information will derive from visual and thermal cameras with the disadvantages of each camera being complemented by the merits of the other, as depicted in Fig. 3. Visual cameras provide a clearer and easier to use representation in day-time while thermal cameras can capture scenes of the operational area in light restricted conditions (night-time operations and/or foggy environment).

Fig. 3. Image fusion of visual and thermal images

Additional Detection Module for Cyber-Physical Attacks. The corresponding module includes a framework to increase the situational awareness of agents against cyber-physical attacks [9] and secure significantly the deployed equipment. Emphasis is given on confidentiality breaches and jamming complementing the visual identification especially, in cases where they lead to incorrect actuation or unreliable sensing due to high-jacking. In many cases, the latter can lead to system failures and inability to complete the mission successfully. Thus, the additional processing capabilities of each asset may secure both their functionalities as well as their integrity.

2.2   Navigational Services

3D Virtual and Augmented Reality Interface. The system is enhanced with additional capacities in order to improve the overall experience of the operator. Towards this objective, a virtual and augmented reality [11] module was integrated and involves the use of an interaction table that allows the operators to interact with the deployed UxVs in a natural and effortless way while simultaneously, it will provide a better situation representation compared to plain view approaches.

DSL-Based Mission Specification. To effectively command the swarm of the UxVs, a dedicated command language is utilized for the description of the missions in order to command and navigate each asset. The language can cover all the aspects of a mission and includes metadata, operation commands, event oriented commands and post-operation commands (i.e. processing collected data). Based on the hierarchical architecture, the language as well as the higher-level commands are adaptable so that, the navigation of the assets could be modified according to the detection outcomes.

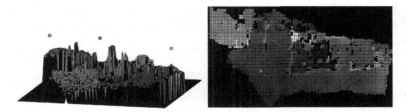

Fig. 4. Real-time surveillance of unknown/dynamically-changing environments

Resource Controller. The integrated module provides an easy-to-operate, remote-control platform, to remotely control the swarm of autonomous, heterogeneous agents (ground, underwater or aerial robots). The platform receives high level objectives and requirements from the human operator, and translates them to real-time, remote-action commands, solely provided to each member of the swarm in order to achieve the mission's objectives. For example, Fig. 4 presents a simulation instance where six UxVs were deployed having a twofold objective (which forms a trade-off). On the one hand, the part of the 3D terrain that is monitored (i.e., visible) by the robots has to be maximized and, on the other hand, for each visible point of the terrain, the closest robot has to be as close as possible to that point. Furthermore, Resource Controller supports plug-n-play additions and subtractions of members to the swarm, always with regards to the mission's requirements [12]. Another crucial factor is that the controller is highly adjustable and capable of handling changes in the missions on the fly. For the implementation of the resource controller, state-of-the-art algorithms on the field of multi-robot path planning [13], have been used and adapted to the specific framework.

CISE-Compliant Common Representation Model and Risk Models. A CISE-compliant common representation framework [14] was developed in order to cover the streaming data from different sensors and other external resources and supports the representation and mapping of data on semantic knowledge structures, valuable in enriching the extracted information. At the same time, intelligent knowledge-driven reasoning schemes are being used for the contextualized aggregation and interpretation of information, derived from different modalities, with final purpose to enrich the low-level fusion capabilities with implicit knowledge derived from hidden relations. The final purpose is the development of a "high level fusion" methodology which centrally combines data from the different geographically dispersed and heterogeneous sensors coupled with contextual, geospatial and temporal information, so as to provide a complete information mapping. In addition, the system integrates a software tool that assists in the short-term prediction of the spatial evolution of hazardous oil pollution incidents and a number of specific illegal activities. This module allows for early identification of the ongoing incidents, as for example is the case of the trajectory of an unauthorized trespasser. An adaptation of [15] strategy is selected in order to process data from diverse sensors and constantly provide better estimations to the decision support module as an extension of the visual components.

Visual Analytics and Decision Support. A decision support tool is included and takes into account data coming from higher level aggregation of sensor data and risk models in order to facilitate the operator to take the right decisions. In this aspect, a visual analytics [16] module was also deployed to visualize information from various fusion nodes and sensor streams in accordance with the necessary UI (User Interface) and HCI (Human-Computer Interaction) [17] requirements.

3   Architecture and Pilot Use Cases

3.1   System Architecture

The developed system provides a fully functional platform for the autonomous deployment of a single, or a fleet of, unmanned vehicle(s), including UAV, USV, UUV and UGV. The system is designed to be used in search-n-rescue and surveillance operations, and incorporates multimodal sensors as part of an inter-operable network, to detect, assess and respond to hazardous situations in remote-areas and border missions and tasks.

The complete network of sensors includes static networked sensors such as surveillance radars, as well as mobile sensors like cameras mounted on robotic platforms. Interoperability is supported by a common framework for the definition and interaction of agents (UxVs and/or static networked sensors) characteristics and monitored phenomena. In order to improve the detection, tracking and imaging capabilities, photonics-based technologies, as well as a passive radar, adapted for use on-board a UAV and other UxVs are used, extending the capabilities of already existing radars. In addition, a radio-frequency communication signal sensor for UAVs based on SDR (software defined radio) [10] has been customized and mounted on different unmanned mobile platforms, in order to intercept emission sources that are present in the area and enrich the overall situational awareness picture with this information, allowing the identification of unauthorized communications and an estimation of the emitter's source location.

In terms of communication, the developed platform adopts a cloudlet-based approach with open interfaces that allows the integration of a multitude of heterogeneous sensors and robots from different providers. According to the specific needs of a particular mission, the platform allows the authorities/operators to rapidly deploy the most suitable sensors and team of robots to remotely operate and collect the needed evidence. High-level, low cognitive-load tele-operation is also supported by a Domain Specific Language (DSL) that allows the operator to specify missions and invoke assets. Last, a virtual, augmented reality based multimodal interaction table allows users to interact with robots in a natural way with touch screen and other hand gestures in 3D, allowing them to understand the situation better than in plain view.

From the user point of view, the main functionalities that the platform supports are:

- Control the fleet of UAVs, UGVs, USVs and UUSVs as a team without the need for the control operators to get involved into tedious tele-operation activities.
- Monitor and control the networked border security radars.
- Monitor and interact with robots through a virtual and/or augmented reality based multimodal interaction table.
- View the results provided by each sensor on the robots positioned on the map in real time.
- View the results from the decision support module in real time for rapid assessment and intervention tailored to the situation.

- Automatic recognition of high-importance, hazardous or suspicious activities, based on multiple sensors.
- Automatic alerts when such abnormal activities are detected.
- Automatic adaptation of robot swarm patrol and radar aim according to the identified incidents.

The developed system has taken into account both maritime and remote-land-area operation scenarios (Fig. 5).

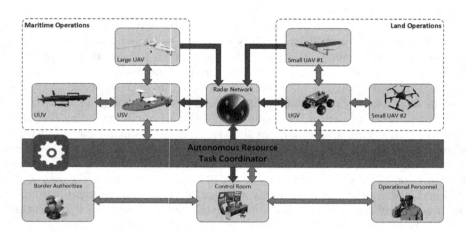

Fig. 5. Higher-level depiction of the system's architecture.

A typical operation of the system in a maritime scenario, usually includes monitoring and detection of aerial, surface or underwater incidents and threats. In such a scenario, the involved assets may be a coastal radar network, large fixed wing UAVs, USVs and UUVs. The communications of the UAV and the USV with the resource controller is direct, through RF links. Since the UUV cannot communicate directly with the resource controller, a USV will coordinate with it and act as communication relayer. Depending on the specific mission and on available assets, employing more than one unit of each type of unmanned vehicle brings proportional increase in detection capability, especially for large surveillance areas.

On the other hand, a typical operation of the system in a remote-land-area scenario, includes monitoring and detection of aerial or surface incidents and threats. Land operations may involve assets like a fixed radar network, large fixed wing UAVs, a small fixed-wing UAV, a small multi-copter UAV, UGVs and a small tethered multi-copter UAV. Depending on the use case at hand and on available assets, employing more than one unit of each type of unmanned vehicle brings proportional increase in detection capability, especially for large surveillance areas.

3.2   Use Cases

In order to assess the developed architecture, the final system will be evaluated in three real scenarios:

Identifying Pollution Incidents in Harbors and Sea Oil Refineries. The system will demonstrate the capability to track pollutants spilled at sea and to determine key environmental conditions needed for defining the response and for forecasting the fate of the pollutants within the territory of the maritime critical infrastructure. This capability will be tested using a natural phenomenon as a proxy of pollutant spill: a river plume within a port will be tracked by vehicles and environmental conditions measured.

Unauthorized Trespassing in a Maritime Infrastructure. The use case involves the monitoring of large sea territories under the responsibility of the relevant authorities. The role of the data mule is assigned to heterogeneous autonomous vehicles equipped with a plethora of sensors like optical and thermal cameras. The mobile devices interact with static infrastructure enabling the commander to determine whether an alarming situation is developing.

Unauthorized Trespassing in Land Territories. The autonomous system will allow to patrol hardly accessible territories within the area of a critical infrastructure leading to an optimized surveillance and control situation system with a maximum coverage. The exploited surveillance units will be the source for directing the patrols and tracking the illegal activities in order to mitigate personal risks and increase monitoring capabilities.

4   Conclusion

The platform presented in this paper provides the relevant monitoring authorities an efficient tool to perform surveillance operations in both remote land and maritime scenarios that includes critical infrastructures. The proposed hierarchical architecture can integrate capacities that focus on the exploitation of multi-modal cameras and identify potential threats within the operational scope. State of the art technologies have been utilized, in order to exploit legacy systems and combine them with innovative solutions, providing a user friendly operational and decision support tool, that will maximize the operational efficiency and minimize the required time to take action in critical incidents.

Acknowledgement. This work was supported by ROBORDER project funded by the European Commission under grant agreement No. 740593. The authors would like to thank the ROBORDER consortium for their valuable overall contribution.

References

1. Haddal, C.C., Gertler, J.: Homeland security: unmanned aerial vehicles and border surveillance. In: Library of Congress Washington DC Congressional Research Service (2010)
2. Fingas, M., Brown, C.: Review of oil spill remote sensing. Spill Sci. Technol. Bull. **4**(4), 199–208 (1997)
3. Chen, L.-C., Zhu, Y., Papandreou, G., Schroff, F., Adam, H.: Encoder-decoder with atrous separable convolution for semantic image segmentation. In: Ferrari, V., Hebert, M., Sminchisescu, C., Weiss, Y. (eds.) ECCV 2018. LNCS, vol. 11211, pp. 833–851. Springer, Cham (2018). https://doi.org/10.1007/978-3-030-01234-2_49
4. Badrinarayanan, V., Kendall, A., Cipolla, R.: SegNet: a deep convolutional encoder-decoder architecture for image segmentation. IEEE Trans. Pattern Anal. Mach. Intell. **39**(12), 2481–2495 (2017)
5. Ren, S., He, K., Girshick, R., Sun, J.: Faster R-CNN: towards real-time object detection with region proposal networks. In: Advances in Neural Information Processing Systems, pp. 91–99 (2015)
6. Liu, W., et al.: SSD: single shot multibox detector. In: Leibe, B., Matas, J., Sebe, N., Welling, M. (eds.) ECCV 2016. LNCS, vol. 9905, pp. 21–37. Springer, Cham (2016). https://doi.org/10.1007/978-3-319-46448-0_2
7. Everingham, M., Van Gool, L., Williams, C.K., Winn, J., Zisserman, A.: The PASCAL visual object classes (VOC) challenge. Int. J. Comput. Vision **88**(2), 303–338 (2010)
8. Bhuiyan, S.M., Adhami, R.R., Khan, J.F.: Fast and adaptive bidimensional empirical mode decomposition using order-statistics filter based envelope estimation. EURASIP J. Adv. Signal Process. **1**, 728356 (2008)
9. Pasqualetti, F., Dörfler, F., Bullo, F.: Attack detection and identification in cyber-physical systems. IEEE Trans. Autom. Control **58**(11), 2715–2729 (2013)
10. Tuttlebee, W.H.: Software Defined Radio: Enabling Technologies. Wiley, Hoboken (2003)
11. Billinghurst, M., Clark, A., Lee, G.: A survey of augmented reality. Found. Trends Hum. Comput. Interact. **8**(2–3), 73–272 (2015)
12. Kapoutsis, A.Ch., Chatzichristofis, S.A., Kosmatopoulos, E.B.: A distributed, plug-n-play algorithm for multi-robot applications with a priori non-computable objective functions. Int. J. Robot. Res. **38**(7), 813–832 (2019)
13. Kapoutsis, A.Ch., Chatzichristofis, S.A., Kosmatopoulos, E.B.: DARP: divide areas algorithm for optimal multi-robot coverage path planning. J. Intell. Rob. Syst. **86**(3), 663–680 (2017)
14. Tikanmäki, I., Ruoslahti, H.: Increasing Cooperation between the European Maritime Domain Authorities (2017)
15. Darema, F.: Dynamic data driven applications systems: a new paradigm for application simulations and measurements. In: Bubak, M., van Albada, G.D., Sloot, P.M.A., Dongarra, J. (eds.) ICCS 2004. LNCS, vol. 3038, pp. 662–669. Springer, Heidelberg (2004). https://doi.org/10.1007/978-3-540-24688-6_86
16. Keim, D., Andrienko, G., Fekete, J.-D., Görg, C., Kohlhammer, J., Melançon, G.: Visual analytics: definition, process, and challenges. In: Kerren, A., Stasko, J.T., Fekete, J.-D., North, C. (eds.) Information Visualization. LNCS, vol. 4950, pp. 154–175. Springer, Heidelberg (2008). https://doi.org/10.1007/978-3-540-70956-5_7
17. Card, S.K.: The Psychology of Human-Computer Interaction. CRC Press, Boca Raton (2018)

Author Index

Printed in the United States
By Bookmasters